3704033298

D1556609

Mobile Computing Handbook

OTHER AUERBACH PUBLICATIONS

Agent-Based Manufacturing and Control Systems: New Agile Manufacturing Solutions for Achieving Peak Performance
Massimo Paolucci and Roberto Sacile
ISBN: 1574443364

Curing the Patch Management Headache
Felicia M. Nicastro
ISBN: 0849328543

Cyber Crime Investigator's Field Guide, Second Edition
Bruce Middleton
ISBN: 0849327687

Disassembly Modeling for Assembly, Maintenance, Reuse and Recycling
A. J. D. Lambert and Surendra M. Gupta
ISBN: 1574443348

The Ethical Hack: A Framework for Business Value Penetration Testing
James S. Tiller
ISBN: 084931609X

Fundamentals of DSL Technology
Philip Golden, Herve Dedieu,
and Krista Jacobsen
ISBN: 0849319137

The HIPAA Program Reference Handbook
Ross Leo
ISBN: 0849322111

Implementing the IT Balanced Scorecard: Aligning IT with Corporate Strategy
Jessica Keyes
ISBN: 0849326214

Information Security Fundamentals
Thomas R. Peltier, Justin Peltier,
and John A. Blackley
ISBN: 0849319579

Information Security Management Handbook, Fifth Edition, Volume 2
Harold F. Tipton and Micki Krause
ISBN: 0849332109

Introduction to Management of Reverse Logistics and Closed Loop Supply Chain Processes
Donald F. Blumberg
ISBN: 1574443607

Maximizing ROI on Software Development
Vijay Sikka
ISBN: 0849323126

Mobile Computing Handbook
Imad Mahgoub and Mohammad Ilyas
ISBN: 0849319714

MPLS for Metropolitan Area Networks
Nam-Kee Tan
ISBN: 084932212X

Multimedia Security Handbook
Borko Furht and Darko Kirovski
ISBN: 0849327733

Network Design: Management and Technical Perspectives, Second Edition
Teresa C. Piliouras
ISBN: 0849316081

Network Security Technologies, Second Edition
Kwok T. Fung
ISBN: 0849330270

Outsourcing Software Development Offshore: Making It Work
Tandy Gold
ISBN: 0849319439

Quality Management Systems: A Handbook for Product Development Organizations
Vivek Nanda
ISBN: 1574443526

A Practical Guide to Security Assessments
Sudhanshu Kairab
ISBN: 0849317061

The Real-Time Enterprise
Dimitris N. Chorafas
ISBN: 0849327776

Software Testing and Continuous Quality Improvement, Second Edition
William E. Lewis
ISBN: 0849325242

Supply Chain Architecture: A Blueprint for Networking the Flow of Material, Information, and Cash
William T. Walker
ISBN: 1574443577

The Windows Serial Port Programming Handbook
Ying Bai
ISBN: 0849322138

AUERBACH PUBLICATIONS
www.auerbach-publications.com
To Order Call: 1-800-272-7737 • Fax: 1-800-374-3401
E-mail: orders@crcpress.com

Mobile Computing Handbook

Mohammad Ilyas
Imad Mahgoub
Editors

PARK LEARNING CENTRE
UNIVERSITY OF GLOUCESTERSHIRE
PO Box 220, The Park
Cheltenham GL50 2RH
Tel: 01242 714333

AUERBACH PUBLICATIONS

A CRC Press Company
Boca Raton   London   New York   Washington, D.C.

Library of Congress Cataloging-in-Publication Data

Mobile computing handbook / Mohammad Ilyas and Imad Mahgoub, editors.
   p. cm.
   Includes bibliographical references and index.
   ISBN 0-8493-1971-4 (alk. paper)
   1. Mobile computing--Handbooks, manuals, etc. I. Ilyas, Mohammad, 1953- II. Mahgoub, Imad.

QA76.59.M644 2004
004.6--dc22

2004052491

This book contains information obtained from authentic and highly regarded sources. Reprinted material is quoted with permission, and sources are indicated. A wide variety of references are listed. Reasonable efforts have been made to publish reliable data and information, but the author and the publisher cannot assume responsibility for the validity of all materials or for the consequences of their use.

Neither this book nor any part may be reproduced or transmitted in any form or by any means, electronic or mechanical, including photocopying, microfilming, and recording, or by any information storage or retrieval system, without prior permission in writing from the publisher.

All rights reserved. Authorization to photocopy items for internal or personal use, or the personal or internal use of specific clients, may be granted by CRC Press, provided that $1.50 per page photocopied is paid directly to Copyright Clearance Center, 222 Rosewood Drive, Danvers, MA 01923 USA. The fee code for users of the Transactional Reporting Service is ISBN 0-8493-1971-4/05/$0.00+$1.50. The fee is subject to change without notice. For organizations that have been granted a photocopy license by the CCC, a separate system of payment has been arranged.

The consent of CRC Press does not extend to copying for general distribution, for promotion, for creating new works, or for resale. Specific permission must be obtained in writing from CRC Press for such copying.

Direct all inquiries to CRC Press, 2000 N.W. Corporate Blvd., Boca Raton, Florida 33431.

Trademark Notice: Product or corporate names may be trademarks or registered trademarks, and are used only for identification and explanation, without intent to infringe.

Visit the Auerbach Web site at www.auerbach-publications.com

© 2005 by CRC Press
Auerbach is an imprint of CRC Press

No claim to original U.S. Government works
International Standard Book Number 0-8493-1971-4
Library of Congress Card Number 2004052491
Printed in the United States of America 1 2 3 4 5 6 7 8 9 0
Printed on acid-free paper

Contributors

AHMED ABUKMAIL *Computer and Information Science and Engineering Department, University of Florida, Gainesville, Florida*
ANDREA ACQUAVIVA *Istituto di Scienze e Tecnologie dell'Informazione, Università di Urbino, Urbino, Italy*
SYED A. AHSON *iDEN Subscriber Division, Motorola, Inc., Plantation, Florida*
JAMAL N. AL-KARAKI *Lab for Advanced Networks (LAN), Department of Electrical and Computer Engineering, Iowa State University, Ames, Iowa*
IARA AUGUSTIN *Computing and Electronic Department, Federal University of Santa Maria, Rio Grande do Sul (RS), Brazil*
RAJIVE BAGRODIA *Computer Science Department, The University of California, Los Angeles, Los Angeles, California*
SANTOSH BALAKRISHNAN *Computer and Information Science and Engineering Department, University of Florida, Gainesville, Florida*
NICHOLAS BAMBOS *Department of Electrical Engineering and Department of Management Science and Engineering, Stanford University, Stanford, California*
GURUDUTH BANAVAR *IBM TJ Watson Research Center, Hawthorne, New York*
JORGE LUIS VICTÓRIA BARBOSA *School of Computer Science, Catholic University of Pelotas, Rio Grande do Sul (RS), Brazil*
PRITHWISH BASU *BBN Technologies, Cambridge, Massachusetts*
CHRISTIAN BECKER *Institute of Parallel and Distributed Systems, University of Stuttgart, Stuttgart, Germany*
PAOLO BELLAVISTA *Dipartimento di Elettronica, Informatica e Sistemistica, Università di Bologna, Bologna, Italy*
LUCA BENINI *Dipartimento di Elettronica, Informatica e Sistemistica, Università di Bologna, Bologna, Italy*
NIKITA BORISOV *Computer Science Division, University of California at Berkeley, Berkeley, California*
DARIO BOTTAZZI *Dipartimento di Elettronica, Informatica e Sistemistica, Università di Bologna, Bologna, Italy*
SONJA BUCHEGGER *Laboratory for Computer Communications and Applications, Swiss Federal Institute of Technology EPFL-IC-LCA, Lausanne, Switzerland*

Contributors

GUOHONG CAO *Department of Computer Science and Engineering, The Pennsylvania State University, University Park, Pennsylvania*
IONUT CARDEI *Department of Computer Science and Engineering, Florida Atlantic University, Boca Raton, Florida*
GERSON GERALDO HOMRICH CAVALHEIRO *Applied Computing Interdisciplinary Post Graduation Program, University of Vale do Rio dos Sinos, Rio Grande do Sul (RS), Brazil*
EDWARD CHAN *Department of Computer Science, City University of Hong Kong, Kowloon, Hong Kong*
YIH-FARN CHEN *Research, AT&T Labs, Florham Park, New Jersey*
HARRY CHEN *Computer Science and Electrical Engineering Department, University of Maryland Baltimore County, Baltimore, Maryland*
A. CHOCKALINGAM *Department of Electrical Communication Engineering, Indian Institute of Science, Bangalore, India*
NORMAN COHEN *IBM TJ Watson Research Center, Hawthorne, New York*
ANTONIO CORRADI *Dipartimento di Elettronica, Informatica e Sistemistica, Università di Bologna, Bologna, Italy*
LUCIANO CAVALHEIRO DA SILVA *Federal University of Rio Grande do Sul, RS, Brazil*
YU DU *Department of Computer Science and Engineering, Arizona State University, Tempe, Arizona*
DING-ZHU DU *Department of Computer Science and Engineering, University of Minnesota, Minneapolis, Minnesota*
MARGARET H. DUNHAM *Department of Computer Science and Engineering, Southern Methodist University, Dallas, Texas*
HESHAM EL-REWINI *Department of Computer Science and Engineering, Southern Methodist University, Dallas, Texas*
CORRADO FEDERICI *Dipartimento di Elettronica, Informatica e Sistemistica, Università di Bologna, Bologna, Italy*
TIM FININ *Computer Science and Electrical Engineering Department, University of Maryland Baltimore County, Baltimore, Maryland*
GUSTAVO FRAINER *Federal University of Rio Grande do Sul, RS, Brazil*
CLÁUDIO FERNANDO RESIN GEYER *Federal University of Rio Grande do Sul, RS, Brazil*
SAVVAS GITZENIS *Department of Electrical Engineering, Stanford University, Stanford, California*
SANDEEP K.S. GUPTA *Department of Computer Science and Engineering, Arizona State University, Tempe, Arizona*
ZYGMUNT J. HAAS *School of Electrical and Computer Engineering, Cornell University, Ithaca, New York*
ABDELSALAM (SUMI) HELAL *Computer and Information Science and Engineering Department, University of Florida, Gainesville, Florida*
LLOYD HUANG *VSee Lab, San Jose, California*
VALÉRIE ISSARNY *INRIA-Rocquencourt, Paris, France*

Contributors

RITTWIK JANA *Department of Dependable Distributed Computing and Communication Research, AT&T Labs, Florham Park, New Jersey*
ANUPAM JOSHI *Computer Science and Electrical Engineering Department, University of Maryland Baltimore County, Baltimore, Maryland*
LALANA KAGAL *Computer Science and Electrical Engineering Department, University of Maryland Baltimore County, Baltimore, Maryland*
SRIDHAR KALUBANDI *Department of Electrical and Computer Engineering, Cleveland State University, Cleveland, Ohio*
AHMED E. KAMAL *Lab for Advanced Networks (LAN), Department of Electrical and Computer Engineering, Iowa State University, Ames, Iowa*
GEORGIA KASTIDOU *Computer Science Department, University of Ioannina, Ioannina, Greece*
WANG KE *Department of Electrical and Computer Engineering, Boston University, Boston, Massachusetts*
RAMANDEEP SINGH KHURANA *Department of Computer Science, University of Nebraska at Omaha, Omaha, Nebraska*
MYUNGCHUL KIM *School of Engineering, Information and Communications University, Daejon, Korea*
JIEJUN KONG *Computer Science Department, University of California, Los Angeles, Los Angeles, California*
UWE KUBACH *SAP Research, SAP AG, Karlsruhe, Germany*
KAM-YIU LAM *Department of Computer Science, City University of Hong Kong, Kowloon, Hong Kong*
JEAN-YVES LE BOUDEC *EPFL-IC-LCA, Laboratory for Computer Communications and Applications, Swiss Federal Institute of Technology, Lausanne, Switzerland*
BEN LEE *School of Electrical Engineering and Computer Science, Oregon State University, Corvallis, Oregon*
THOMAS D.C. LITTLE *Department of Electrical and Computer Engineering, Boston University, Boston, Massachusetts*
YOUZHONG LIU *Computer and Information Science and Engineering Department, University of Florida, Gainesville, Florida*
WEI LOU *Department of Computer Science and Engineering, Florida Atlantic University, Boca Raton, Florida*
SONGWU LU *Computer Science Department, University of California, Los Angeles, Los Angeles, California*
HAIYUN LUO *Computer Science Department, University of California, Los Angeles, Los Angeles, California*
IMAD MAHGOUB *Department of Computer Science and Engineering, Florida Atlantic University, Boca Raton, Florida*
THOMAS L. MARTIN *Department of Electrical and Computer Engineering, Virginia Tech, Blacksburg, Virginia*
SANGMAN MOH *Department of Internet Engineering, Chosun University, Gwangju, Korea*

Contributors

REBECCA MONTANARI *Dipartimento di Elettronica, Informatica e Sistemistica, Università di Bologna, Bologna, Italy*

AMY L. MURPHY *Department of Informatics, University of Lugano, Lugano, Switzerland*

PANAGIOTIS PAPADIMITRATOS *School of Electrical and Computer Engineering, Cornell University, Ithaca, New York*

JIM PARKER *Computer Science and Electrical Engineering Department, University of Maryland Baltimore County, Baltimore, Maryland*

THOMAS PHAN *Computer Science Department, The University of California, Los Angeles, Los Angeles, California*

GIAN PIETRO PICCO *Dipartimento di Elettronica e Informazione, Politecnico di Milano, Milano, Italy*

EVAGGELIA PITOURA *Computer Science Department, University of Ioannina, Ioannina, Greece*

RODRIGO ARAÚJO REAL *Federal University of Rio Grande do Sul, RS, Brazil*

GRUIA-CATALIN ROMAN *Department of Computer Science and Engineering, Washington University, St. Louis, Missouri*

NOEL RUIZ *Computer Science Department, University of California, Los Angeles, Los Angeles, California*

FRANÇOISE SAILHAN *INRIA-Rocquencourt, Paris, France*

GEORGE SAMARAS *Computer Science Department, University of Cyprus, Nicosia, Cyprus*

RAVI SANKAR *Department of Electrical Engineering, University of South Florida, Tampa, Florida*

YÜCEL SAYGIN *Faculty of Engineering and Natural Sciences, Sabanci University, Turkey*

AYŞE YASEMIN SEYDIM *Central Bank of the Republic of Turkey, Information Technology Department, Ankara, Turkey*

VINOD SHARMA *Department of Electrical Communication Engineering, Indian Institute of Science, Bangalore, India*

DANIEL P. SIEWIOREK *Institute for Complex Engineered Systems and Human Computer Interaction Institute, Carnegie Mellon University, Pittsburgh, Pennsylvania*

ASIM SMAILAGIC *Institute for Complex Engineered Systems and Human Computer Interaction Institute, Carnegie Mellon University, Pittsburgh, Pennsylvania*

DANNY SOROKER *IBM TJ Watson Research Center, Hawthorne, New York*

ILLYA STEPANOV *Institute of Parallel and Distributed Systems, University of Stuttgart, Stuttgart, Germany*

RIKY SUBRATA *Parallel Computing Research Laboratory, Department of Electrical and Electronic Engineering, University of Western Australia, Perth, Western Australia*

KIAN-LEE TAN *Department of Computer Science, National University of Singapore, Singapore*

Contributors

JING TIAN *Institute of Parallel and Distributed Systems, University of Stuttgart, Stuttgart, Germany*

DANIELA TIBALDI *Dipartimento di Elettronica, Informatica e Sistemistica, Università di Bologna, Bologna, Italy*

GEORGE VARGHESE *Department of Computer Science and Engineering, University of California, San Diego, San Diego, California*

SILVIA VECCHI *Dipartimento di Elettronica, Informatica e Sistemistica, Università di Bologna, Bologna, Italy*

JOLIN WARREN *Department of Electrical and Computer Engineering, Carnegie Mellon University, Pittsburgh, Pennsylvania*

JIE WU *Department of Computer Science and Engineering, Florida Atlantic University, Boca Raton, Florida*

CHIEN-HSING WU *Electrical Engineering Department, National Chung-Cheng University, Min-Hsiung, Taiwan*

ADENAUER CORRÊA YAMIN *Center of Computer Science, Federal University of Pelotas, Rio Grande do Sul (RS), Brazil*

HAO YANG *Computer Science Department, University of California, Los Angeles, Los Angeles, California*

FAN YE *Computer Science Department, University of California, Los Angeles, Los Angeles, California*

CHANSU YU *Department of Electrical and Computer Engineering, Cleveland State University, Cleveland, Ohio*

JOE C.H. YUEN *Department of Computer Science, City University of Hong Kong, Kowloon, Hong Kong*

PETROS ZERFOS *Computer Science Department, University of California, Los Angeles, Los Angeles, California*

LIXIA ZHANG *Computer Science Department, University of California, Los Angeles, Los Angeles, California*

ALBERT Y. ZOMAYA *Parallel Computing Research Laboratory, Department of Electrical and Electronic Engineering, University of Western Australia, Perth, Western Australia*

Contents

SECTION I INTRODUCTION AND APPLICATIONS OF MOBILE COMPUTING 1

Chapter 1
Wearable Computing 3
Asim Smailagic and Daniel P. Siewiorek
Abstract 3
  1.1 Introduction 3
  1.2 Issues in Wearable Computing 4
  1.3 Example Systems 7
      1.3.1   Procedures with Static Prestored Text/Graphics 9
      1.3.2   Master/Apprentice (Live Expert) Help Desk 10
      1.3.3   Team Collaboration 11
      1.3.4   Context-Aware Collaboration — Proactive Synthetic Assistant 12
  1.4 Evaluation 17
      1.4.1   Prestored Procedures 17
      1.4.2   Master/Apprentice Help Desk 18
      1.4.3   Team Collaboration 19
      1.4.4   Context-Aware Collaboration — Proactive Synthetic Assistant 21
  1.5 Summary and Future Challenges 21
Acknowledgments 22
References 22

Chapter 2
Developing Mobile Applications: A Lime Primer 25
Gian Pietro Picco, Amy L. Murphy, and Gruia-Catalin Roman
Abstract 25
  2.1 Introduction 25
  2.2 Linda in a Nutshell 28
  2.3 Lime: Linda in a Mobile Environment 28
      2.3.1   Model Setting and Overview 29
      2.3.2   Creating a Lime Tuple Space 30

xi

Contents

    2.3.3   Enabling Transient Sharing 31
    2.3.4   Reconciling Different Forms of Mobility 32
    2.3.5   Restricting the Scope of Operations 34
    2.3.6   Reacting to Changes 36
    2.3.7   Accessing the System Configuration 38
    2.3.8   Implementation Details 39
2.4 Application Example 39
    2.4.1   Requirements 40
    2.4.2   Design and Implementation 40
              2.4.2.1   Tuple Spaces and Tuples 40
              2.4.2.2   User Actions 41
              2.4.2.3   Display Update 43
    2.4.3   Beyond the Puzzle 44
2.5 Building Middleware Functionality on Top of Lime 44
    2.5.1   Transiently Shared Code Bases 44
    2.5.2   Service Provision 46
2.6 Related Work 47
2.7 Conclusions 50
    2.7.1   Availability 50
Acknowledgments 50
Notes 50
References 51

Chapter 3
Pervasive Application Development: Approaches and Pitfalls 53
Guruduth Banavar, Norman Cohen, and Danny Soroker

Abstract 53
3.1 What Are Pervasive Applications? 53
    3.1.1   Basic Concepts and Terms 55
3.2 Why Is It Difficult to Develop Pervasive Applications? 56
    3.2.1   Heterogeneity of Device Platforms 56
              3.2.1.1   User Interface 57
              3.2.1.2   Interaction Modalities 57
              3.2.1.3   Platform Capabilities 57
              3.2.1.4   Connectivity 57
              3.2.1.5   Development and Maintenance Complexity 58
    3.2.2   Dynamics of Application Environments 58
3.3 Approaches for Developing Pervasive Applications 59
    3.3.1   Developing Mobile Applications 59
    3.3.2   Presentation Transcoding 59
    3.3.3   Device-Independent View Component 60
              3.3.3.1   Runtime Adaptation 60
              3.3.3.2   Design-Time Adaptation 61
              3.3.3.3   Visual Tools for Constructing
                             Device-Independent Views 62

3.3.4	Platform-Independent Controller Component	63
3.3.5	Host-Independent Model Component	64
3.3.6	Developing Context-Aware Applications	66
3.3.7	Source-Independent Context Data	66

3.4 Conclusions 68
Acknowledgments 69
References 69

Chapter 4
ISAM, Joining Context-Awareness and Mobility to Building Pervasive Applications 73

Iara Augustin, Adenauer Corrêa Yamin, Jorge Luis Victória Barbosa, Luciano Cavalheiro da Silva, Rodrigo Araújo Real, Gustavo Frainer, Gerson Geraldo Homrich Cavalheiro, and Cláudio Fernando Resin Geyer

Abstract 73
4.1 Introduction 73
4.2 The ISAM Application Model 74
4.3 The ISAM Architecture 75
4.4 ISAMadapt Overview 77

4.4.1	Context	78
4.4.2	Adapters	80
4.4.3	Adaptation Commands	80
4.4.4	Adaptation Policies	81

4.5 EXEHDA Overview 81
4.6 The WalkEd Application 83

4.6.1	The GUI Being: Alternative Behaviors	86
4.6.2	The Spell Being	89
4.6.3	The Print Being	89
4.6.4	The ISAMadapt IDE	91
4.6.5	Execution Aspects	91

4.7 Conclusions 93
Acknowledgments 93
References 94

Chapter 5
Integrating Mobile Wireless Devices into the Computational Grid ... 95

Thomas Phan, Lloyd Huang, Noel Ruiz, and Rajive Bagrodia

5.1 Introduction 95
5.2 Background 98
5.3 Motivation: Mobile Devices and the Grid 99
5.4 The LEECH Architecture 104

5.4.1	Key Challenges	104
5.4.2	Overview of Architecture	105
5.4.3	Grid/Cluster and LEECH	105
5.4.4	Application Major Component and Minor Component	106

Contents

	5.4.5	Interlocutor 107
	5.4.6	Minion 108
	5.4.7	Availability Adaptation and Job Management 108
5.5	The LEECH Programming Model 110
5.6	Experiments and Analysis..................................... 111
	5.6.1	Synthetic Application for Measuring Communication Overhead.. 112
	5.6.2	RSA Decryption.................................... 113
5.7	Looking to the Future....................................... 116
References 118

Chapter 6
Multimedia Messaging Service 121
Syed A. Ahson

- 6.1 Introduction 121
- 6.2 MMS Architecture....................................... 122
  - 6.2.1 MMS Interfaces 123
  - 6.2.2 Addressing in MMS.............................. 124
  - 6.2.3 Technical Specifications 125
  - 6.2.4 Supported Formats 125
  - 6.2.5 MMS Messages 126
- 6.3 Message Submission 126
- 6.4 Message Transfer 129
- 6.5 Delivery Report 129
- 6.6 Read-Reply Reports 133
- 6.7 Message Notification 134
- 6.8 Message Retrieval 138
- 6.9 Message Forwarding 143
- 6.10 Future Directions 145
- References 146

SECTION II LOCATION MANAGEMENT 147

Chapter 7
A Scheme for Nomadic Hosts Location Management Using DNS 149
Ramandeep Singh Khurana, Hesham El-Rewini, and Imad Mahgoub

- Abstract 149
- 7.1 Introduction 149
- 7.2 Using the DNS for Location Management of Nomadic Hosts 151
  - 7.2.1 DNS Server...................................... 152
  - 7.2.2 Server Process: Web Server 154
  - 7.2.3 Client Process: Web Browser 154
  - 7.2.4 Security 155

7.3 Experiments 156
    7.3.1 Cache Time versus Time-to-Live 157
    7.3.2 Scalability Analysis 158
7.4 Concluding Remarks 160
References 161

Chapter 8
Location Management Techniques for Mobile Computing Environments 163
Riky Subrata and Albert Y. Zomaya
8.1 Introduction 163
8.2 Location Management 165
    8.2.1 Location Update 165
    8.2.2 Location Inquiry 165
        8.2.2.1 Delay Constraint 165
8.3 Location Management Cost 166
8.4 Network Topology 167
8.5 Mobility Pattern 168
    8.5.1 Memoryless (Random Walk) Movement Model 168
    8.5.2 Markovian Model 169
        8.5.2.1 Cell History 169
        8.5.2.2 Directional History 170
    8.5.3 Shortest Distance Model 171
    8.5.4 Gauss-Markov Model 171
    8.5.5 Activity-Based Model 172
    8.5.6 Mobility Trace 172
    8.5.7 Fluid-Flow Model 173
    8.5.8 Gravity Model 173
8.6 Call Arrival Pattern 174
    8.6.1 Poisson Model 174
    8.6.2 Call Arrival Trace 174
8.7 Location Update Strategies 175
    8.7.1 Always-Update Strategy 175
    8.7.2 Never-Update Strategy 175
    8.7.3 Time-Based Strategy 176
    8.7.4 Movement-Based Strategy 176
    8.7.5 Distance-Based Strategy 177
    8.7.6 Location Area 178
        8.7.6.1 Static Case 179
        8.7.6.2 Dynamic Case 181
    8.7.7 Reporting Center 182
        8.7.7.1 Static Case 182
        8.7.7.2 Dynamic Case 184
    8.7.8 Adaptive Threshold Scheme 185

Contents

    8.7.9   Profile-Based 185
    8.7.10 Compression-Based 185
    8.7.11 Hybrid Strategies 186
  8.8 Location Inquiry Strategies.. 186
    8.8.1   Simultaneous Networkwide Search 186
    8.8.2   Paging Area 187
    8.8.3   Expanding Ring Paging............................. 187
    8.8.4   Intelligent Paging 188
  8.9 Summary 190
References 190

Chapter 9
Locating Mobile Objects...................................... 197
Evaggelia Pitoura, George Samaras, and Georgia Kastidou
  9.1 Introduction 197
  9.2 Location Management 198
  9.3 Architectures of Location Directories 201
    9.3.1   Two-Tier Scheme 201
    9.3.2   Hierarchical Scheme 202
  9.4 Optimizations of the Architectures 205
    9.4.1   Call to Mobility Ratio 205
    9.4.2   Partitions.. 205
    9.4.3   Caching 206
    9.4.4   Replication 209
    9.4.5   Forwarding Pointers............................... 211
  9.5 Taxonomy and Location Management Techniques 215
  9.6 Case Studies 218
    9.6.1   Mobile IP 218
    9.6.2   Globe 220
    9.6.3   Mobile Agents Systems 222
        9.6.3.1   Ajanta...................................... 222
        9.6.3.2   Voyager 223
  9.7 Summary 223
References 224

Chapter 10
Dependable Message Delivery to Mobile Units 227
Amy L. Murphy, Gruia-Catalin Roman, and George Varghese
Abstract 227
  10.1 Introduction 227
    10.1.1 Distributed versus Mobile Computing.................. 228
    10.1.2 Algorithm Development............................. 229
  10.2 Message Delivery 229
    10.2.1 Related Work..................................... 229

10.2.2 Mobile Environment................................. 231
10.2.3 Model and Problem Definitions 231
10.3 Broadcast Search.. 232
    10.3.1 Motivation... 233
    10.3.2 From Distributed Snapshot Algorithms
            to Announcement Delivery 233
    10.3.3 Snapshot Delivery Algorithm 235
    10.3.4 Properties 238
    10.3.5 Extensions... 239
        10.3.5.1 Multiple Announcement Deliveries........... 239
        10.3.5.2 Rapidly Moving Mobile Units............... 239
        10.3.5.3 Route Discovery............................ 240
        10.3.5.4 Multicast 240
        10.3.5.5 Mobile Agents............................. 240
10.4 Tracking for Delivery.. 241
    10.4.1 From Diffusing Computations to Mobile Unit Tracking... 241
    10.4.2 Extension: Backbone-Based Message Delivery 243
10.5 Reality Check 246
    10.5.1 FIFO Channels..................................... 247
    10.5.2 Multiple RBSs per MSC 249
    10.5.3 Base Station Connectivity........................... 249
    10.5.4 Reliable Delivery on Links.......................... 249
    10.5.5 Involvement Level of MSCs......................... 250
    10.5.6 Storage Requirements 250
10.6 Conclusions 251
Acknowledgments.. 251
References 251

SECTION III LOCATION-BASED SERVICES...................253

Chapter 11
Location-Dependent Query Processing in Mobile Computing 255
Ayşe Yasemin Seydim, Margaret H. Dunham
11.1 Introduction 255
11.2 Related Work.. 256
    11.2.1 Location-Dependent Data and Queries 256
    11.2.2 Moving Object Databases Research 258
    11.2.3 Spatial Database Management 258
11.3 Location Relatedness and the Query Model 260
    11.3.1 Query Model...................................... 260
    11.3.2 Location-Aware Queries 266
    11.3.3 Location-Dependent Queries 268
    11.3.4 Moving Object Database Queries.................... 270
    11.3.5 Query Classification................................ 270

Contents

11.4 Query Translation Steps in LDQ Processing. 271
Acknowledgments 272
Note 272
References 272

Chapter 12
Simulation Models and Tool for Mobile Location-Dependent Information Access 275
Uwe Kubach, Christian Becker, Illya Stepanov, and Jing Tian
Abstract 275
12.1 Introduction 275
12.2 Spatial Model 276
    12.2.1 Location Models 277
    12.2.2 Spatial Information Models 278
12.3 Mobility 279
    12.3.1 Existing Mobility Models 279
        12.3.1.1 Random Mobility Models 280
        12.3.1.2 Advanced Models. 282
    12.3.2 Generic Mobility Model 283
12.4 Information Access Model 285
    12.4.1 Zipf Distribution 286
    12.4.2 Location-Dependent Access 287
12.5 A Tool for User Mobility Modeling. 288
    12.5.1 Objectives 288
    12.5.2 Software Architecture 288
    12.5.3 Usage 290
12.6 Conclusion. 293
References 293

Chapter 13
Context-Aware Mobile Computing 297
Rittwik Jana and Yih-Farn Chen
Abstract 297
13.1 Introduction and Motivation. 297
13.2 What Is Context?. 299
13.3 Context Acquisition 300
    13.3.1 Acquisition of Sensor Data 300
    13.3.2 Location Sensing Techniques 301
13.4 What Is Context-Awareness?. 301
    13.4.1 Technology Independent Framework and Application
             Programming Interfaces 302
        13.4.1.1 Parlay: Integration of Telecom and Internet
                   Services 302
        13.4.1.2 IETF OPES Group 303
        13.4.1.3 iMobile: a Mobile Service Platform 303

13.5 Gluing Contextware and Middleware 306
    13.5.1 Integrating Location Determination with the Service Platform 306
    13.5.2 Managing Location Information in iMobile 307
    13.5.3 Location-Based Services with iMobile 308
    13.5.4 Preserving Privacy in Environments with Location-Based Services 308
13.6 Context-Related Research Initiatives and Projects 310
13.7 The Future of Context 311
13.8 Conclusions 312
Notes 312
References 312

Chapter 14
Mobile Agent Middlewares for Context-Aware Applications 315
Paolo Bellavista, Dario Bottazzi, Antonio Corradi, Rebecca Montanari, and Silvia Vecchi

Abstract 315
14.1 Mobile Computing and Context Awareness 316
14.2 Mobile Agents and Mobile Computing 318
14.3 An Overview of MA-Based Supports for Mobile Computing 322
14.4 MA-Based Middlewares with Context Awareness: State-of-the-Art and Emerging Research Directions 326
14.5 Lessons Learned and Open Issues 330
Acknowledgments 332
References 332

SECTION IV CACHING STRATEGIES 335

Chapter 15
Cache Management in Wireless and Mobile Computing Environments 337
Yu Du and Sandeep K.S. Gupta

15.1 Introduction 337
15.2 State of the Art 343
15.3 Cache Consistency Strategies 346
    15.3.1 A Taxonomy of Cache Consistency Strategies 346
15.4 Cache Consistency Strategies in Architecture-Based Wireless Networks 347
    15.4.1 Invalidation-Based Consistency Strategy 348
        15.4.1.1 Stateful Approaches 348
        15.4.1.2 Stateless Approach 350
    15.4.2 TTL-Based Cache Consistency Strategy 355
        15.4.2.1 Handling Disconnections 355
        15.4.2.2 Achieving Energy and Bandwidth Efficiency ... 355

15.5 Open Problems 356
    15.5.1 Cache Consistency Strategy in the Ad Hoc
            Network Environment 356
    15.5.2 Cooperate Caching in Ad Hoc Network Environment 356
15.6 Summary 358
References 358

Chapter 16
Cache Invalidation Schemes in Mobile Environments 361
Edward Chan, Joe C.H. Yuen, and Kam-Yiu Lam
16.1 Introduction 361
16.2 Summary of Existing Cache Invalidation Schemes 362
16.3 Temporal Data Model for Mobile Computing Systems 364
16.4 Cache Invalidation Using AVI 366
    16.4.1 Validity Period of Data in Client Cache 366
    16.4.2 The IAVI Cache Invalidation Scheme 368
    16.4.3 Server Algorithm 369
            16.4.3.1 Invalidation Report Generation 369
            16.4.3.2 AVI Adjustment 371
    16.4.4 Client Algorithm 371
            16.4.4.1 Implicit Invalidation 371
            16.4.4.2 Explicit Invalidation 372
16.5 Performance Study 373
    16.5.1 System Model 373
    16.5.2 Performance Metrics 375
    16.5.3 Performance Evaluation 376
            16.5.3.1 Impact of Database Size 376
            16.5.4.2 Impact of Update Rate 377
16.6 Conclusion 383
References 387

Chapter 17
Hoarding in Mobile Computing Environments 389
Yücel Saygin
17.1 Introduction 389
17.2 Coda: The Pioneering System for Hoarding 391
17.3 Hoarding Based on Data Mining Techniques 392
    17.3.1 SEER Hoarding System 392
    17.3.2 Association Rule-Based Techniques 394
    17.3.3 Partitioning the History into Sessions 394
    17.3.4 Utilization of Association Rules for Hoarding 395
    17.3.5 Construction of the Candidate Sets and the Hoard Set ... 396
17.4 Hoarding Techniques Based on Program Trees 396
17.5 Hoarding in a Distributed Environment 398
17.6 A Brief Comparison of the Various Hoarding Techniques 399

17.7 Future Directions for Hoarding Techniques 400
References 401

Chapter 18
Power-Aware Cache Management in Mobile Environments 403
Guohong Cao
Abstract 403
18.1 Introduction 403
18.2 Cache Invalidation Techniques 405
    18.2.1 Cache Consistency Model............................. 406
    18.2.2 The IR-Based Cache Invalidation Model 406
        18.2.2.1 The Broadcasting Time Stamp Scheme 407
        18.2.2.2 The Bit Sequences Scheme................. 407
    18.2.3 The UIR-Based Cache Invalidation Model............. 408
    18.2.4 Using Prefetch to Improve Cache Hit Ratio and
           Bandwidth Use 409
        18.2.4.1 Remarks 410
18.3 Techniques to Optimize Performance and Power.............. 411
    18.3.1 The Basic Scheme 411
        18.3.1.1 The Basic Adaptive Prefetch Approach 412
    18.3.2 The Value-Based Prefetch Scheme.................... 413
    18.3.3 The Adaptive Value-Based Prefetch Scheme 415
        18.3.3.1 The Value of N_p......................... 415
        18.3.3.2 AVP_T: Adapting N_p to Reach a Target
                Battery Life................................ 415
        18.3.3.3 AVP_P: Adapting N_p Based on the
                Power Level 416
18.4 Conclusions and Future Work 416
Note 418
References 418

Chapter 19
Energy Efficient Selective Cache Invalidation 421
Kian-Lee Tan
19.1 Introduction 421
19.2 Preliminaries... 422
19.3 A Taxonomy of Cache Invalidation Strategies 423
    19.3.1 Content of the Invalidation Report 423
    19.3.2 Invalidation Mechanisms 425
    19.3.3 Update Log Structure................................ 426
    19.3.4 Cache Invalidation Schemes......................... 426
19.4 Selective Cache Invalidation Schemes 426
    19.4.1 Selective Dual-Report Cache Invalidation............. 427
    19.4.2 Bit-Sequences with Bit-Count 429
    19.4.3 Discussion... 431

Contents

19.5 Conclusion. 432
References 432

SECTION V   MOBILE AND AD HOC WIRELESS NETWORKS I 433

Chapter 20
Self-Policing Mobile Ad Hoc Networks 435
Sonja Buchegger and Jean-Yves Le Boudec
Abstract 435
20.1 Introduction and Overview. 435
20.2 Node Misbehavior in Mobile Ad Hoc Networks 436
    20.2.1 Reasons and Enablers for Misbehavior. 436
    20.2.2 Attacks. 438
        20.2.2.1 Traffic Diversion. 438
    20.2.3 The Effect of Misbehavior 439
20.3 Overview: Main Solution Tracks. 440
    20.3.1 Payment Systems 440
    20.3.2 Secure Routing with Cryptography 441
        20.3.2.1 Secure Routing Protocol 442
        20.3.2.2 Ariadne 442
        20.3.2.3 Secure Efficient Distance. 442
        20.3.2.4 Security-Aware Ad Hoc Routing 442
    20.3.3 Detection, Reputation, and Response Systems 442
        20.3.3.1 Watchdog and Path Rater 444
        20.3.3.2 CONFIDANT. 444
        20.3.3.3 CORE 444
        20.3.3.4 Context-Aware Inference 445
        20.3.3.5 OCEAN 445
    20.3.4 Discussion 445
20.4 Self-Policing for Mobile Ad Hoc Networks 447
    20.4.1 Enhanced CONFIDANT — a Robust Reputation
         System Approach 447
    20.4.2 Issues in Reputation Systems for Mobile
         Ad Hoc Networks 449
        20.4.2.1 Spurious Ratings 449
        20.4.2.2 Information Dissemination 450
        20.4.2.3 Type of Information 450
        20.4.2.4 Response. 451
        20.4.2.5 Redemption, Weighting of Time. 451
        20.4.2.6 Weighting of Second-Hand Information. 452
        20.4.2.7 Detection. 452
        20.4.2.8 Identity 453

20.5 Conclusions 454
References 454

Chapter 21
Securing Mobile Ad Hoc Networks 457
Panagiotis Papadimitratos and Zygmunt J. Haas
Abstract 457
21.1 Introduction 457
21.2 Security Goals... 459
21.3 Threats and Challenges...................................... 460
21.4 Trust Management... 463
21.5 Secure Routing.. 467
    21.5.1 The Secure Routing Protocol 469
        21.5.1.1 The Neighbor Lookup Protocol 469
        21.5.1.2 The Basic Secure Route Discovery
                Procedure 471
        21.5.1.3 Priority-Based Query Handling 472
        21.5.1.4 The Route Maintenance Procedure.......... 472
        21.5.1.4 The SRP Extension......................... 473
21.6 Secure Data Forwarding 473
    21.6.1 Secure Message Transmission Protocol 475
21.7 Discussion... 478
References 480

Chapter 22
Ad Hoc Network Security 483
Hao Yang, Haiyun Luo, Jiejun Kong, Fan Ye, Petros Zerfos, Songwu Lu, and Lixia Zhang
Abstract 483
22.1 Overview.. 485
22.2 Link-Layer Security 486
    22.2.1 802.11 MAC Vulnerabilities 486
    22.2.2 802.11 WEP Vulnerabilities 487
22.3 Network Layer Security...................................... 487
    22.3.1 Message Authentication Primitives 488
    22.3.2 Proactive Approach to Secure Ad Hoc Routing 489
        22.3.2.1 Source Routing............................. 489
        22.3.2.2 Distance Vector Routing 490
        22.3.2.3 Link State Routing 491
        22.3.2.4 Other Routing Protocols.................... 492
    22.3.3 Reactive Approach to Protecting Packet Forwarding.... 493
        22.3.3.1 Detection.................................. 493
        22.3.3.2 Reaction 494
    22.3.4 Sophisticated Intrusion Detection System 495

Contents

22.4 Supporting Element: Trust and Key Management 496
    22.4.1 Trusted Third Party 497
    22.4.2 Web-of-Trust 498
    22.4.3 Localized Trust 499
22.5 Future Directions 499
    22.5.1 Security in Depth 499
    22.5.2 Evaluation 500
    22.5.3 Solutions Anticipating Unknown Attacks 500
Note 500
Bibliography 500

Chapter 23
Modeling Distributed Applications for Mobile Ad Hoc Networks Using Attributed Task Graphs 503
Prithwish Basu, Wang Ke, and Thomas D.C. Little

Abstract 503
23.1 Introduction 504
23.2 Modeling Distributed Tasks with Task Graphs 507
    23.2.1 A Modeling Framework for Task Execution 507
        23.2.1.1 Preliminaries 507
        23.2.1.2 Tasks and Task Graphs 508
        23.2.1.3 A Taxonomy of Tasks 510
        23.2.1.4 A Data Flow Tuple Representation Model for Distributed Tasks 511
    23.2.2 Embedding Task Graphs onto Networks 514
    23.2.3 Metrics for Performance Evaluation 516
23.3 Algorithms and Protocols for Task Graph Instantiation 516
    23.3.1 Optimization Problem Formulation 517
    23.3.2 An Optimal Polynomial-Time Embedding Algorithm for Tree Task Graphs with Distinct Labels 517
    23.3.3 A Greedy Algorithm for Task Graph Embedding 519
    23.3.4 A Distributed Algorithm for Task Graph Instantiation ... 522
        23.3.4.1 Handling Device Mobility 525
        23.3.4.1 Impact of Disconnections on Application Layer 527
23.4 Performance Evaluation 528
    23.4.1 Dilation 528
    23.4.2 Embedding Time 531
    23.4.3 Effective Throughput 532
    23.4.4 Number and Time of Reinstantiation 535
    23.4.5 Cumulative ADU Delay Distributions 536
23.5 Related Work 537
23.6 Conclusion 539
Acknowledgment 539

Notes 539
References 540

Chapter 24
Medium Access Control Mechanisms in Mobile Ad Hoc Networks .. 543
Chansu Yu, Ben Lee, Sridhar Kalubandi, and Myungchul Kim

Abstract 543
24.1 Introduction 543
24.2 MAC Protocols 545
    24.2.1 Random Access MAC 545
    24.2.2 DCF of IEEE 802.11 MAC 546
        24.2.2.1 ACK for Collision Detection 547
        24.2.2.2 RTS/CTS and NAV for Solving Hidden Terminal Problem 547
        24.2.2.3 IFS for Prioritized Access to the Channel 547
        24.2.2.4 Backoff Algorithm with CW to Provide Fair Access with Congestion Control 549
        24.2.2.5 EIFS to Protect ACK from Collisions 549
        24.2.2.6 Performance Limit of DCF 552
24.3 Enhancing Temporal Channel Utilization 552
    24.3.1 RTS/CTS Mechanism 554
        24.3.1.1 Optimal Setting of RTSThreshold to Tradeoff between Control and Collision Overhead 554
    24.3.2 Exponential Backoff Algorithm 554
        24.3.2.1 Conservative CW Restoration to Reduce Collisions 554
        24.3.2.2 Different Treatment of New and Lost Nodes for Fairness 556
        24.3.2.3 Dynamic Tuning of CW to Minimize the Collision Probability 556
24.4 Enhancing Spatial Channel Utilization 557
    24.4.1 Busy Tone to Solve the Exposed Terminal Problem 557
    24.4.2 Transmission Power Control to Reduce Interference Range Radially 558
    24.4.3 Directional Antenna to Reduce Interference Range Angularly 560
        24.4.3.1 oRTS/oCTS-Based DMAC 562
        24.4.3.2 DRTS/oCTS-Based DMAC 562
        24.4.3.3 DRTS/DCTS-Based DMAC 562
        24.4.3.4 Other DMAC Protocols 563
24.5 Conclusions 563
Note 565
References 565

Contents

SECTION VI   MOBILE AND AD HOC WIRELESS NETWORKS II 567

Chapter 25
Quality of Service Routing in Mobile Ad Hoc Networks: Past and Future 569
Jamal N. Al-Karaki, Ahmed E. Kamal
Abstract 569
25.1 Introduction 569
25.2 Quality of Service in MANETs: The Basics 574
    25.2.1 QoS Metrics 575
    25.2.2 Challenges of QoS Routing Support in MANETs 576
25.3 QoS Routing Protocols in MANETs: Current Trends 579
    25.3.1 QoS Routing in Flat Networks 581
        25.3.1.1 Proactive QoS Routing Protocols 582
        25.3.1.2 Reactive QoS Routing Protocols 583
        25.3.1.3 Predictive QoS Routing Protocols 587
        25.3.1.4 Ticket-Based Probing Routing 589
        25.3.1.5 Bandwidth Calculation Based Routing 592
    25.3.2 Hierarchical QoS Routing Protocols 595
    25.3.3 Position-Based QoS Routing Protocol 598
    25.3.4 Power-Aware QoS Routing in MANETs 600
25.4 QoS Routing in MANETs: Future Research Directions 604
25.5 Conclusion 606
Note 607
References 607

Chapter 26
Issues in Scalable Clustered Network Architecture for Mobile Ad Hoc Networks 611
Ben Lee, Chansu Yu, and Sangman Moh
Abstract 611
26.1 Introduction 611
26.2 Classification of Cluster Architecture-Based Routing Protocols 613
    26.2.1 Flat Routing Protocols and Their Scalability 614
    26.2.2 Cluster Architectures 615
    26.2.3 Cluster Architecture-Based Routing Protocols 617
26.3 LCA for Routing Backbone 617
    26.3.1 Clustering Algorithms 617
        26.3.1.1 Master Selection Algorithms for LSG 618
        26.3.1.2 Cluster Maintenance Algorithms for LSG 620
        26.3.1.3 Master Selection Algorithms for LNG 621

26.3.2 LSG-Based Routing Protocols 622
  26.3.2.1 CGSR and HSR: Proactive Protocol with
           Conventional Master-to-Gateway Routing 623
  26.3.2.2 DSCR and LANMAR: Proactive Protocols
           with Flat Routing toward M_D. 624
  26.3.2.3 CBRP and ARC: On-Demand Protocols
           with Conventional Master-to-Gateway Routing
           (Allowing No, Single, or Joint Gateways) 625
26.3.3 LNG-Based Routing Protocols (On-Demand Protocols
       with Master-to-Master Routing) 626
26.4 Cluster Architecture for Information Infrastructures 627
  26.4.1 Clustering Algorithms 627
  26.4.2 LLog-Based Routing Protocols 629
    26.4.2.1 CEDAR Protocol 630
    26.4.2.2 Zone Routing Protocol 630
  26.4.3 LGeo-Based Routing Protocols........................ 631
    26.4.3.1 ZHLS Routing Protocol 631
    26.4.3.2 GLS Protocol 632
26.5 Summary and Conclusion 632
Note 634
References 634

Chapter 27
Routing and Mobility Management in Wireless Ad Hoc Networks .. 637
Ravi Sankar
27.1 Introduction 637
27.2 Ad Hoc Network: Definition, Characteristics, and
     Applications 638
27.3 Desired Characteristics of Routing Protocols for MANETs 640
27.4 Conventional Routing Protocols 640
  27.4.1 Problems with Conventional Routing 641
27.5 Review of Ad Hoc Routing Protocols......................... 641
  27.5.1 Table-Driven Routing Protocols 642
    27.5.1.1 Destination-Sequenced Distance Vector........ 642
    27.5.1.2 Wireless Routing Protocol 644
    27.5.1.3 Link State Routing Protocols 644
    27.5.1.4 Clusterhead Gateway Switch Routing 645
    27.5.1.5 Hierarchical Routing Protocols 646
    27.5.1.6 Summary of Table-Driven Protocols 647
  27.5.2 On-Demand Routing Protocols......................... 647
    27.5.2.1 Dynamic Source Routing..................... 647
    27.5.2.2 Ad Hoc On-Demand Distance Vector
             Routing 650

| | 27.5.2.3 | Associativity Based Routing .... 651 |
| | 27.5.2.4 | Summary of On-Demand Protocols .... 652 |

27.5.3 Hybrid Routing Protocols 652
       27.5.3.1 Zone Routing Protocol...................... 652
       27.5.3.2 Temporally Ordered Routing Algorithm 653
27.5.4 Comparison of the Routing Protocols 653
27.5.5 Other Protocols 654
       27.5.5.1 Power-Aware and QoS-Aware Routing......... 654
       27.5.5.2 Location-Based Routing 655
       27.5.5.3 Flooding and Multicasting................... 655
       27.5.5.4 Multipath Routing 656

27.6 Performance Issues and Challenges 656
27.7 Mobility Management in Ad Hoc Networks 657
27.8 Conclusion... 658
Note 658
References 658

Chapter 28
Localized Broadcasting in Mobile Ad Hoc Networks Using Neighbor Designation 663
Wei Lou and Jie Wu

28.1 Introduction 663
28.2 Classification... 665
    28.2.1 Probabilistic Algorithms 665
       28.2.1.1 Counter-Based Scheme 666
       28.2.1.2 Distance-Based Scheme...................... 666
       28.2.1.3 Location-Based Scheme 666
    28.2.2 Deterministic Algorithms 666
       28.2.2.1 Global...................................... 667
       28.2.2.2 Quasi-Global 667
       28.2.2.3 Quasi-Local 668
       28.2.2.4 Local....................................... 669
    28.2.3 Local Algorithms 669
28.3 Neighbor-Designating Broadcast Algorithms 672
    28.3.1 Forward Node Selection Process 672
    28.3.2 Multi-Point Relays.................................... 673
    28.3.3 Dominant Pruning.................................... 675
    28.3.4 Total Dominant Pruning and Partial Dominant Pruning... 675
    28.3.5 CDS-Based Broadcast Algorithm 676
28.4 Other Extensions 677
    28.4.1 Cluster-Based Broadcast Algorithm.................... 677
    28.4.2 K-hop Zone-Based Algorithm 681
    28.4.3 Reliable Broadcast Algorithm......................... 683

28.5 Summary 686
Acknowledgment 686
References 686

Chapter 29
Energy-Efficient Wireless Networks 689
Ionut Cardei and Ding-Zhu Du
Abstract 689
29.1 Introduction 689
29.2 Power-Aware Link Layer Adaptation 692
29.3 Energy Harvesting 693
29.4 Scheduling Node and Radio Activity 695
    29.4.1 Power-Aware Medium Access Control 695
    29.4.2 Energy-Efficient MAC Protocols for WSN 697
    29.4.3 Node Activity Scheduling 698
29.5 Energy Conservation in Ad Hoc Routing 700
    29.5.1 Energy-Efficient Routing Protocols 701
    29.5.2 Power-Aware Broadcast and Multicast Tree
           Construction 702
29.6 Energy-Aware Connected Network Topology 703
29.7 Conclusion 705
References 705

SECTION VII POWER MANAGEMENT 707

Chapter 30
Power Management for Mobile Computers 709
Thomas L. Martin, Daniel P. Siewiorek, Asim Smailagic, and Jolin Warren
Abstract 709
30.1 The Relationship between Power and Energy 710
30.2 Batteries 710
30.3 Power Supplies 712
30.4 Hardware Power Management States 716
30.5 Software 719
30.6 Case Study 721
    30.6.1 Memory Bottleneck and Dynamic CPU Speed-Setting .. 722
    30.6.2 Dependence of Battery Capacity on Load Power 725
30.7 General Guidelines 727
    30.7.1 An "Amdahl's Law" for Power Management 727
    30.7.2 Evaluating Power Management Options 728
30.8 Conclusions 729
References 729

Contents

Chapter 31
Power Awareness and Management Techniques 731
Ahmed Abukmail and Abdelsalam (Sumi) Helal
31.1 Introduction 731
    31.1.1 Motivation 731
    31.1.2 Taxonomy of Research and Industry Solutions 732
31.2 Hardware and Architecture Techniques 733
    31.2.1 Smart Batteries 733
        31.2.1.1 Battery Basics 733
        31.2.1.2 Intelligent Power Drainage 733
    31.2.2 Energy-Aware Processors 734
    31.2.3 Reducing Power through CMOS Circuitry Components ... 734
        31.2.3.1 The Power Consumption Equation 734
        31.2.3.2 Voltage and Frequency Scaling 735
        31.2.3.3 Capacitance Load Reduction 735
    31.2.4 Power Reduction through Architectural Design 736
31.3 Operating Systems and Communication Techniques 736
    31.3.1 Energy Management Solutions 736
    31.3.2 Memory and I/O Management 737
    31.3.3 Communication Techniques 737
    31.3.4 Scheduling 738
31.4 Software Application Techniques 738
    31.4.1 Compilation Techniques 738
        31.4.1.1 Reordering Instructions 738
        31.4.1.2 Reduction of Memory Operands 739
        31.4.1.3 Code Generation through Pattern Matching ... 739
        31.4.1.4 Remote Task Execution 739
    31.4.2 Application-Level Techniques 740
31.5 Tools and Packages for Low-Power Design and Measurement ... 741
    31.5.1 PowerScope 741
    31.5.2 Derivatives of SimpleScalar 742
        31.5.2.1 The Power Analyzer 742
        31.5.2.2 The Wattch Project 742
    31.5.3 Other Power Estimation Techniques 743
31.6 Conclusion 743
References 743

Chapter 32
Adaptive Algorithmic Power Optimization for Multimedia Workload in Mobile Environments 747
Luca Benini and Andrea Acquaviva
32.1 Introduction 747
32.2 Scalability and Energy Optimization 749
    32.2.1 Scalability in Modern Multimedia Applications 749

	32.2.1.1 Scalable Source Coding with Wavelets 750

            32.2.1.1 Scalable Source Coding with Wavelets 750
            32.2.1.2 Scalable Source Coding in MPEG-4 752
    32.2.2 Energy Scalability..................................... 753
32.3 Adaptive Algorithmic Power Optimization..................... 755
    32.3.1 Stand-Alone Power Management 756
            32.3.1.1 Adaptive Encoding Algorithms 756
            32.3.1.2 Adaptive Decoding Algorithms 760
    32.3.2 Collaborative Power Management..................... 764
            32.3.2.1 Operating System Collaborative Techniques... 765
            32.3.2.2 Server-Assisted Collaborative Techniques..... 769
32.4 Conclusion 776
References 777

Chapter 33
Energy-Aware Web Caching over Hybrid Networks 779
Françoise Sailhan and Valérie Issarny
Abstract 779
33.1 Introduction 779
33.2 Power-Aware Communication 781
    33.2.1 Energy Saving at the MAC Layer 781
            33.2.1.1 Minimizing Collisions 782
            33.2.1.2 Minimizing Channel Listening 784
    33.2.2 Energy Saving at the Routing Layer.................. 785
            33.2.2.1 Ad Hoc Routing Protocols 785
            33.2.2.2 Energy-Aware Ad Hoc Routing 787
    33.2.3 Energy Saving at the Transport Layer................. 788
33.3 Web Caching in Ad Hoc Networks 789
    33.3.1 Ad Hoc Communication for Cooperative Web Caching .. 789
    33.3.2 Ad Hoc Cooperative Caching 791
33.4 Local Caching.. 795
    33.4.1 Cache Management 795
    33.4.2 Prefetching 796
33.5 Evaluation... 798
    33.5.1 Energy Consumption of Ad Hoc Networking 799
    33.5.2 Energy Consumption of Ad Hoc Cooperative Caching ... 801
33.6 Conclusion 802
Notes... 803
References 803

Chapter 34
Transmitter Power Control in Wireless Computing 805
Savvas Gitzenis and Nicholas Bambos
34.1 Introduction 805
    34.1.1 Power Control Issues in Wireless Packet
            Communication 805

Contents

- 34.2 Packet Forwarding over Single-Mode Wireless Links 809
  - 34.2.1 Optimally Emptying the Transmitter Buffer 811
    - 34.2.1.1 The Simple Case of Independent Channel Interference — Power Phases 812
  - 34.2.2 Incorporating Packet Arrivals and Buffer Overflows..... 815
  - 34.2.3 Design of PCMA Algorithms — Responsive Channel..... 816
  - 34.2.4 The Multi-Transmitter/Multi-Receiver Case 817
- 34.3 Packet Forwarding over Multimode Transmission Links 819
  - 34.3.1 Optimally Emptying the Transmitter Buffer 820
  - 34.3.2 Structural Properties — The Independent Channel Stress Case 820
  - 34.3.3 Design of Multimode PCMA Algorithms for Responsive Channels................................. 824
- 34.4 Data Prefetching over Single-Mode Transmission Links......... 824
  - 34.4.1 System Model 825
  - 34.4.1 The Dynamic Programming Formulation 827
  - 34.4.2 The Structural Properties of the Power Decision p...... 829
  - 34.4.3 Online Look-Ahead Heuristics for Efficient Buffer Control... 830
    - 34.4.3.1 No Prefetching — Efficient Data Downloading... 831
    - 34.4.3.2 Neighbor Prefetching — Depth-1 Look-Ahead .. 833
    - 34.4.3.3 Deep Prefetching 835
- Notes 835
- References 835

SECTION VIII   PERFORMANCE AND MODELING 837

Chapter 35
A Survey on Mobile Transaction Models 839
Abdelsalam (Sumi) Helal, Santosh Balakrishnan, Margaret H. Dunham, and Youzhong Liu

- Abstract 839
- 35.1 Introduction 839
- 35.2 Reference Model................................... 841
- 35.3 Mobile Transactions: Definition, Characteristics, and Issues 842
  - 35.3.1 Characteristics 843
  - 35.3.2 Definition 844
  - 35.3.3 Issues 844
- 35.4 Applicable Transaction Models 845
  - 35.4.1 Open Nested Transactions 845
    - 35.4.1.1 Properties of Open Nested Transactions 846
  - 35.4.2 Split Transactions............................. 846
    - 35.4.2.1 Split Transaction Semantics 847
    - 35.4.2.2 Properties of Split Transactions............. 847

	35.4.3	Sagas 848
		35.4.3.1 Properties of Sagas 848
		35.4.3.2 Limitations of Sagas........................ 848
		35.4.3.3 Extensions of Saga Model 848
		35.4.3.4 Noncompensating Transactions 849
35.5	Approaches to Mobile Transaction Models 849
	35.5.1	Reporting and Cotransactions 850
		35.5.1.1 Properties of Reporting Transactions......... 851
		35.5.1.2 Properties of Cotransactions 851
	35.5.2	The Clustering Model 851
		35.5.2.1 Clusters.................................... 851
		35.5.2.2 Weak and Strict Transactions................. 851
		35.5.2.3 Transaction Migration and Proxying.......... 852
	35.5.3	The Multi-Database Transaction Processing Manager ... 852
		35.5.3.1 Architecture 852
		35.5.3.2 Transaction Model......................... 853
	35.5.4	Pro-Motion... 853
	35.5.5	Prewrite.. 855
	35.5.6	Semantic Transaction Processing 856
		35.5.6.1 Exploiting Semantics for Concurrency and Caching.................................... 856
		35.5.6.2 Fragmentable and Reorderable Objects........ 857
	35.5.7	The Kangaroo Transaction Model 857
		35.5.7.1 Reference Model 857
		35.5.7.2 Transaction Model......................... 858
		35.5.7.3 Properties................................ 858
	35.5.8	Time-Based Consistency Model 858
	35.5.9	Two-Tier Replication 859
	35.5.10	Isolation-Only Transactions 860
		35.5.10.1 The Approach Taken....................... 860
		35.5.10.2 The Coda File System 860
		35.5.10.3 What Is an IOT? 860
		35.5.10.4 IOT Execution Model...................... 860
		35.5.10.5 Why Isolation Only?....................... 861
		35.5.10.6 IOT Consistency Guarantees 861
	35.5.11	Bayou.. 862
		35.5.11.1 Two-Tier Replication and Weak Consistency... 862
		35.5.11.2 Antientropy............................... 862
		35.5.11.3 Session Guarantees 862
	35.5.12	A New Transaction Management Scheme............... 863
35.6	Comparative Analysis of Transaction Models 864
	35.6.1	Consistency and Concurrency 864
	35.6.2	Additional Infrastructure Requirements and Compatibility with Commercial Databases............. 869
	35.6.3	Communication Cost and Scalability.................. 871

Contents

35.7 Open Issues in Mobile Transactions 872
35.8 Summary 873
References 874

Chapter 36
Analytic Mobility Models of PCS Networks 877
Chien-Hsing Wu
Abstract 877
36.1 System Models 878
    36.1.2 Cellular Systems 878
    36.1.2 Markov Walk Models 879
36.2 Analysis for Location Update 881
    36.2.1 Location Tracking and Updates 882
    36.2.2 Two Renewal Processes and α_k 882
    36.2.3 Recursive Markov Analysis 883
    36.2.4 Distributions $\mu(u)$ and $\phi(c)$ 886
36.3 Paging and Cost 886
36.4 Performance Evaluation 887
References 891

Chapter 37
Battery Power Management in Portable Devices 893
Vinod Sharma and A. Chockalingam
Abstract 893
37.1 Introduction 893
    37.1.1 Relaxation Phenomenon in Batteries 896
37.2 System Model 899
    37.2.1 Battery Discharge/Recharge Model 899
37.3 Analysis 900
    37.3.1 Extensions and Generalizations 903
        37.3.1.1 Exhaustive Service with Vacations 904
        37.3.1.2 Nonexhaustive System with Vacations .. 904
        37.3.1.3 Multi-Battery System 905
37.4 Performance Results and Discussion 910
    37.4.1 Lithium Ion Battery Simulation Results 914
37.5 An Optimal Scheduling Problem 916
Notes 917
References 917

SECTION IX   SECURITY AND PRIVACY ASPECTS 921

Chapter 38
Challenges in Wireless Security: A Case Study of 802.11 ... 923
Nikita Borisov

38.1 Introduction 923
    38.1.1 Overview of 802.11 923
    38.1.2 History 924
38.2 Wireless Security Threats 926
38.3 Encryption 927
    38.3.1 Keystream Reuse 928
    38.3.2 RC4 Weaknesses 929
    38.3.3 New Standards 929
38.4 Integrity Protection 930
    38.4.1 Integrity-Based Attacks 931
    38.4.2 Replay Attacks 932
    38.4.3 New Protocols 933
38.5 Authentication and Access Control 933
    38.5.1 Authentication Extensions 934
    38.5.2 Mutual Authentication 935
    38.5.3 New Protocols 935
38.6 Key Management 936
    38.6.1 New Protocols 936
38.7 Conclusion 937
References 938

Chapter 39
Security for Mobile Agents: Issues and Challenges 941
Paolo Bellavista, Antonio Corradi, Corrado Federici, Rebecca Montanari, and Daniela Tibaldi

Abstract 941
39.1 Security: a Missing Link for MAs' Acceptance 941
39.2 Security Requirements 943
39.3 Security Countermeasures 945
    39.3.1 User–Agent Trust 945
    39.3.2 Protecting Agent Platforms 946
        39.3.2.1 Secure Agent Code 946
        39.3.2.2 Agent Authentication 947
        39.3.2.3 Agent Authorization 947
    39.3.3 Protecting Agents 948
39.4 Overview of Security Solutions in MA-Based Systems 951
39.5 Open Issues and Directions of Work in Secure MA Systems .. 953
    39.5.1 Agents and Trust 954
    39.5.2 Dynamic Configuration of Access Control 955
39.6 Conclusions 957
Acknowledgments 958
References 958

Chapter 40
Security, Trust, and Privacy in Mobile Computing Environments ... 961
Lalana Kagal, Jim Parker, Harry Chen, Anupam Joshi, and Tim Finin

- 40.1 Introduction ... 961
- 40.2 Policies and Their Role in Security in Pervasive Computing Systems ... 963
  - 40.2.1 Introduction ... 963
  - 40.2.2 Related Work ... 964
  - 40.2.3 Approach ... 966
  - 40.2.4 Discussion ... 967
- 40.3 Toward Privacy Protection in Pervasive Computing Environments ... 969
  - 40.3.1 Introduction ... 969
  - 40.3.2 Previous Work ... 969
  - 40.3.3 Design Principles for Building Privacy Systems ... 969
    - 40.3.3.1 Notice and Consent ... 970
    - 40.3.3.2 Proximity and Locality ... 970
    - 40.3.3.3 Anonymity and Pseudonymity ... 970
  - 40.3.4 Implementations of the Privacy Systems ... 970
    - 40.3.4.1 Notice and Consent ... 971
    - 40.3.4.2 Proximity and Locality ... 971
    - 40.3.4.3 Anonymity and Pseudonymity ... 971
  - 40.3.5 Context Broker Architecture ... 972
  - 40.3.6 Privacy Policy Language ... 974
  - 40.3.7 Meta-Reasoning with Policies ... 975
  - 40.3.8 Discussion ... 975
- 40.4 Intrusion Detection in Mobile Ad Hoc Networks ... 976
  - 40.4.1 Introduction ... 976
  - 40.4.2 Environments and Devices ... 976
  - 40.4.3 Intrusion Detection ... 977
  - 40.4.4 Ad Hoc Network ID ... 978
  - 40.4.5 Research ... 979
  - 40.4.6 Multiple Malicious Nodes ... 982
  - 40.4.7 Directional Antennas and Power Control ... 982
  - 40.4.8 Discussion ... 983
- 40.5 Conclusion ... 984
- References ... 984

Index ... 987

Preface

The past decade has witnessed significant advances in the technology of personal computers, wireless communication, and the Internet. Today, small, inexpensive, yet powerful portable computers are available. The Internet continues to experience exponential growth. The coming together of these trends has made it possible to use computer resources and access information anywhere and at anytime. This new computing paradigm, widely known as mobile computing, is set to drive technology over the next decade. However, there are a lot of challenges to meet. This makes mobile computing a hot research and development area. Mobile computing is being projected as the future growth area in both academia and industry.

This handbook explores the challenges in mobile computing and includes current efforts and approaches to address them. It provides technical information about various aspects of mobile computing ranging from basic concepts to research grade material, including future directions. This handbook captures the current state of mobile computing and serves as a source of comprehensive reference material on mobile computing.

This handbook is intended for researchers and practitioners in the field and for engineers and scientists involved in the design and development of mobile computing systems and their applications. This handbook can also be used as the textbook for graduate courses in the mobile computing area.

This handbook has 40 chapters written by experts from around the world. It is organized in nine sections:

1. Section I — Introduction and Applications of Mobile Computing
2. Section II — Location Management
3. Section III — Location-Based Services
4. Section IV — Caching Strategies
5. Section V — Mobile and Ad Hoc Wireless Networks I
6. Section VI — Mobile and Ad Hoc Wireless Networks II
7. Section VII — Power Management
8. Section VIII — Performance and Modeling
9. Section IX — Security and Privacy Aspects

Preface

This handbook has the following salient features:

- It serves as a comprehensive source of information and reference material on mobile computing.
- It deals with an important and timely topic of emerging computing paradigm of tomorrow.
- It presents accurate, up-to-date information on a broad range of topics related to mobile computing.
- It presents material authored by experts in the field.
- It presents the information in an organized and well-structured manner.

Many people have contributed to this handbook in their unique ways. The first and the foremost group that deserves immense gratitude is the group of highly talented and skilled researchers who have contributed chapters to this handbook. Without their expertise and effort, this handbook would never have come to fruition. It has also been a pleasure to work with Mr. Rich O'Hanley, Ms. Karen Schober, and Ms. Claire Miller of CRC Press. We are extremely grateful for their support and professionalism. Our families have extended their unconditional love and strong support throughout this project and they all deserve very special thanks.

Imad Mahgoub and Mohammad Ilyas
Boca Raton, Florida

Section I
Introduction and Applications of Mobile Computing

Chapter 1
Wearable Computing

Asim Smailagic and Daniel P. Siewiorek

Abstract

This chapter describes a taxonomy of wearable computers and their applications, focusing on their problem solving capabilities. Wearable and context-aware computers have been developed from our iterative design methodology, with a wide variety of end users, mainly mobile workers. The taxonomy is illustrated by wearable systems evolving from basic stored information retrieval through synchronous/asynchronous collaboration within a team to context-aware platforms with a proactive assistant. Example evaluation methods illustrate how user testing can quantify the effectiveness of wearable systems.

1.1 Introduction

Carnegie Mellon's Wearable Computers project is helping to define the future not only for computing technologies, but also for the use of computers in daily activities. The goal is to develop a new class of computing systems with a small footprint that can be carried or worn by a person and be able to interact with computer augmented environments [1]. Because users are an integral part of the system, techniques such as user-centered design, rapid prototyping, and in-field evaluation are used to identify and refine user interface models that are useful across a wide spectrum of applications [2, 3, 4]. Over two dozen wearable computers have been designed and built over the last 12 years, with most tested in the field. The application domains range from inspection, maintenance, manufacturing, and navigation to on-the-move collaboration, position sensing, and real-time speech recognition and language translation. At the core of these paradigms is the notion that wearable computers should seek to merge the user's information with the user's workspace. The wearable computer must blend seamlessly with existing work environments, providing as little distraction as possible. This requirement often leads to replacements for the traditional desktop paradigm, which generally requires a fixed physical relationship between the user and devices such as a keyboard and mouse. Identifying effective interaction modalities for wearable computers, as well as accurately modeling user tasks in software, are among the most significant challenges in designing wearable systems.

The goals for this chapter are:

- To present a map of wearable system functionality to application types
- To summarize examples of four user interface models

1.2 Issues in Wearable Computing

Wearable computers at Carnegie Mellon have ranged from proof of concept to customer-driven systems design based on a task specification to visionary design predicting the form and functionality of wearable computers of the future. Four wearable computers — VuMan 3, MoCCA, Digital Ink, and Promera — have been awarded prestigious international design awards.

For pervasive or ubiquitous computing to reach its potential, the average person should be able to take advantage of the information on or off the job. Even while at work, many people do not have desks or spend a large portion of their time away from a desk. Thus, mobile access is the gating technology required to make information available at any place and at any time. In addition, the computing system should be aware of the user's context, not only to be able to respond in an appropriate manner with respect to the user's cognitive and social state, but also to anticipate the needs of the user.

Today's computer systems distract a user in many explicit and implicit ways, thereby reducing their effectiveness. The systems can also overwhelm users with data leading to information overload. The challenge for human computer interaction design is to use advances in technology to preserve human attention and to avoid information saturation.

We have identified four design principles for mobile systems:

1. User interface models — what is the appropriate set of metaphors for providing mobile access to information (i.e., what is the next desktop or spreadsheet)? These metaphors typically take over a decade to develop (i.e., the desktop metaphor started in early 1970s at Xerox Palo Alto Research Center (PARC) and required over a decade before it was widely available to consumers). Extensive experimentation working with end users is required to define and refine these user interface models.
2. Input/output modalities — although several modalities mimicking the input/output capabilities of the human brain have been the subject of computer science research for decades, the accuracy and ease of use (i.e., many current modalities require extensive training periods) are not yet acceptable. Inaccuracies produce user frustrations. In addition, most of these modalities require extensive computing resources,

which will not be available in lightweight, low-power wearable computers. There is room for new, easy-to-use input devices such as the dial developed at Carnegie Mellon University (CMU) for list-oriented applications.
3. Matched capability with application requirements — many mobile systems attempt to pack as much capacity and performance in as small a package as possible. However, these capabilities are often unnecessary to complete an application. Enhancements such as full-color graphics not only require substantial resources, but also may compromise ease of use by generating information overload for the user. Interface design and evaluation should focus on the most effective means for information access and resist the temptation to provide extra capabilities simply because they are available.
4. Quick interface evaluation methodology — current approaches to evaluate a human computer interface require elaborate procedures with scores of subjects. Such an evaluation may take months and is not appropriate for reference during interface design. These evaluation techniques should focus on decreasing human errors and frustration.

Over the past 12 years, we have built wearable computers for over a dozen clients in diverse application areas. We have observed several functionalities that have proven useful across multiple applications. These functionalities form the basis for four user interface models, each with their unique user interface, input/output modality, and capability:

1. Procedures — text and graphics — maintenance and plant operation applications are characterized by a large volume of information that varies slowly over time. For example, even simple aircraft may have over a hundred thousand manual pages (like the aircraft manuals at the US Airways hangar at Pittsburgh International Airport). But due to operational changes and upgrades, half of these pages are obsolete every six months for even mature aircraft. Rather than distribute CD-ROMs for each maintenance person and running the risk of a maintenance procedure being performed on obsolete information, maintenance facilities usually maintain a centralized database to which maintenance personnel make inquiries for the relevant manual sections on demand. A typical request consists of approximately ten pages of text and schematic drawings. Changes to the centralized information base can occur on a weekly basis. Furthermore, the trend is toward more customization in manufacturing. In aircraft manufacturing, no two aircraft on an assembly line are identical. The aircraft may belong to different airlines or be configured for different missions. Customization extends to other industries. One leading manufacturer produces over 70,000 trucks per year, representing

over 20,000 different configurations. The customer can select the transmission, the engine, and even the stereo system. In the near future, trucks will be accompanied by their own documentation describing them as "built," "modified," or "repaired." When personnel carrying out manufacturing or scheduled maintenance arrive for a day's work, they receive a list of job orders that describe the tasks and include documentation such as text and schematic drawings. Thus, this information can change on a daily or even hourly basis.
2. Master/apprentice help desk — there are times, however, when an individual requires assistance from experienced personnel. Historically, an apprenticeship program, wherein a novice observes and works with an experienced worker, has provided this assistance. A simple example of this is the help desk, wherein an experienced person is contacted for audio and visual assistance in solving a problem.
3. Team maintenance/collaboration — the help desk can service many people in the field simultaneously. Today, with downsizing and productivity improvement goals, teams of people are geographically distributed, yet are expected to pool their knowledge to solve immediate problems. An extension of the help desk is a team of personnel such as field service engineers, police, and firefighters, who are joining together to resolve an emergency situation. Information can be expected to change on a minute-by-minute and sometimes even second-by-second basis.
4. Context-aware collaboration — proactive synthetic assistant — context-aware computing describes the situation where a mobile computer is aware of its user's state and surroundings, and modifies its behavior based on this information. The system can monitor a user's state and act as a proactive assistant, linking information derived from many contexts, such as location and schedule. Distractions are even more of a problem in mobile environments than desktop environments, because the user is often preoccupied with walking, driving, or other real-world interactions. A ubiquitous computing environment that minimizes distraction should be context-aware. If a human assistant were given such context, he or she would make decisions in a proactive fashion, anticipating user needs. The goal is to enable mobile computers to play an analogous role, exploiting context information to significantly reduce demands on human attention.

The remainder of the chapter is organized as follows:

- Section 1.3 describes the four user interface models and gives examples of how they address the three design principles of user interface models, input/output modalities, and functional capability requirements.
- Section 1.4 provides examples of how systems can be evaluated for each of the four user interface models.

Wearable Computing

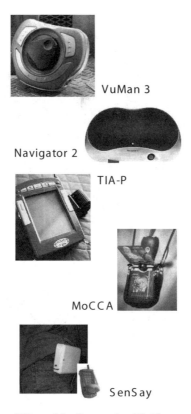

Figure 1.1A. Examples of Wearable Computer Platforms

1.3 Example Systems

Four wearable computer user interface models will be illustrated using example CMU wearable computer systems. It is interesting to note that these user interface models can also be found in systems developed by other organizations. Figure 1.1A shows examples of wearable computer platforms, corresponding to each of the four user interface models. Figure 1.1B illustrates the relationship between wearable computer platforms and their applications. This representation is based on Kiviat diagrams [5]. These examples include:

- Procedures — text and graphics
    - VuMan 3 text-based inspection of U.S. Marines heavy military vehicles, CMU [6]
    - Navigator 2 graphical inspection of Boeing aircraft, CMU [1]
    - Georgia Tech wearable computer for quality assurance inspection in food processing plants [7]

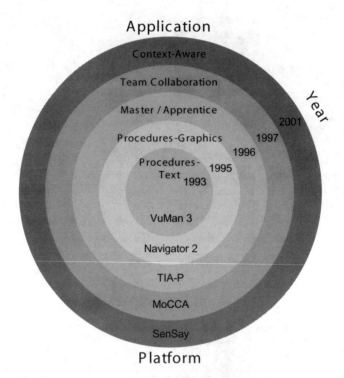

Figure 1.1B. Relationship between Wearable Platforms and Their Applications

- Master/apprentice help desk
  - TIA-P (Tactical Information Assistant — Prototype) used for C-130 help desk, CMU [8]
  - NetMan enables technicians in the field and office-based experts to collaborate in real-time using audio and video, University of Oregon [9]
- Team collaboration
  - MoCCA (Mobile Communication and Computing Architecture) to support collaboration of geographically distributed field engineers, CMU [10]
  - Land Warrior integrated infantry soldier system for close combat. Designed to avoid information overload [11]
- Synthetic collaboration — proactive synthetic assistant
  - Context-aware cell phone, which modifies its behavior based on its user's state and surroundings, CMU
  - Touring machine, which combines the overlaid 3D graphics of augmented reality with the untethered freedom of mobile computing to support users in their everyday interactions with the world, Columbia University [12]

Wearable Computing

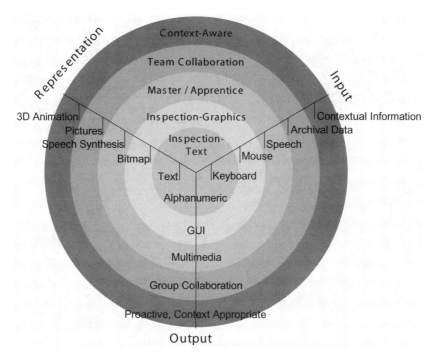

Figure 1.2. Input/Output Modalities for the User Interface Models

A synthetic assistant is a technology that allows a computer model of a human expert to interact conversationally, provide advice, read procedures, and answer questions to a human. This technology is developed at CMU [8, 13].

Figure 1.2 summarizes the four user interface models with respect to the first design principle and input/output modalities. The knowledge source and user interface models will be illustrated by system examples in Section 1.3.1 to Section 1.3.4.

The following subsections elaborate further and give examples of each problem solving capability. Figure 1.3 depicts the four problem solving capabilities in a state diagram. The system examples in the following four subsections will illustrate each capability in turn.

1.3.1 Procedures with Static Prestored Text/Graphics

The prestored capability is illustrated by the sheet metal inspection of a military aircraft. Approximately 100 defects are identified during an average 36 hour inspection. The inspection starts with selecting a region on the aircraft body and proceeds with inspecting the object, referencing information from the manuals, archival storage of observations, and recording the

MOBILE COMPUTING HANDBOOK

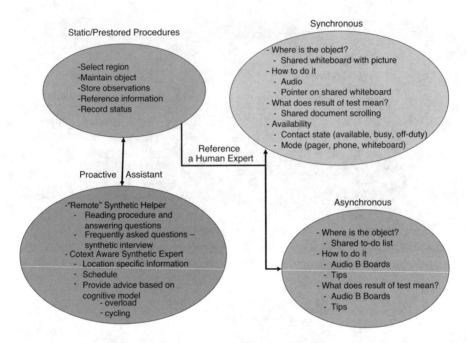

Figure 1.3. State Diagram of Problem Solving Capabilities

status. Once the user chooses to begin an aircraft inspection, the field of interest is narrowed from major features (such as left wing, right tail) to more specific details (individual panes in the cockpit window glass, aircraft body polygons), as shown in Figure 1.4. The area covered by each defect as well as the type of defect, using a "How Malfunctioned" code such as corroded, cracked, or missing, are recorded. To maximize usability, each item or control may be selected simply by speaking its name or, in the case of more complicated phonemes, a designated numeral. This two-dimensional selection method, in which defect locations are specified on a planar region, and overall user interface design have received favorable feedback from the Boeing aircraft inspectors at McClellan Air Force Base in California.

1.3.2 Master/Apprentice (Live Expert) Help Desk

The C-130 project is designed to use collaboration to facilitate training and to increase the number of trainees per trainer. Inexperienced users are being trained to perform a cockpit inspection and the trainers are remotely located. The trainee loads the inspection procedures and performs the inspection. A desktop system manages the normal job order process and is used by the instructors to observe the trainee's behavior. In collaboration, the instructor looks over the shoulder (through a small video camera attached to the top of the trainee's head-mounted display) and advises the

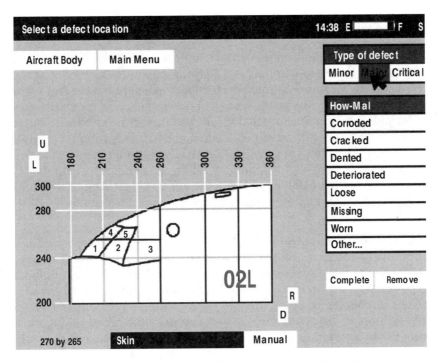

Figure 1.4. Example User Interface for Static/Prestored Information

trainee. In addition to a two-way audio channel, the instructor can provide advice via a cursor for indicating areas on a captured video image, which is being shared through a whiteboard. The instructor manages the sharing session and whiteboard. The trainee's use of the whiteboard is limited to observation.

The capability for the master/apprentice paradigm is in the synchronous communication bubble in Figure 1.3. Synchronous communications facilitate answering questions such as: Where is the object? (drawing on a captured image), How to do it? (audio guidance through prestored material), and What does the result of the test mean? (audio discussion). This model also uses the static/prestored capability.

1.3.3 Team Collaboration

MoCCA [10] is designed to support a group of geographically distributed field service engineers (FSEs). The FSEs spend up to 30 to 40 percent of their time in a car driving to customer sites. Half of what they service is third-party equipment for which they may not have written documentation. The challenge was to provide a system that allowed the FSEs to access information and advice from other FSEs while on customer sites and while commuting between sites. Synchronous and asynchronous collaboration (Figure 1.3) are supported for both voice and digitized information.

An additional challenge arose from user interviews that suggested that the FSEs desired the functionality of a laptop computer including a larger color display with an operational cycle of at least eight hours. The system had to be very light, preferably less than one pound, and required access to several legacy databases. Further discussions with the FSEs indicated that the most frequently used databases were textually oriented. Only on rare occasions is access to graphical databases required. A novel architecture combined a lightweight alphanumeric satellite computer with the high functionality of a base unit included in the FSE's tool kit. The base unit can be carried into any customer site providing instant access to the global infrastructure.

The team problem solving asynchronous capability (Figure 1.3) includes audio bulletin boards and tips for shared collaboration space between remote FSEs and their colleagues. The concept of an audio bulletin board is equivalent to a storehouse for audio clips describing the problems that the FSEs encounter while on the job. Each trouble topic contains a list of audio responses from other FSEs with the possible solution. Figure 1.5 shows the integrated user interface, which starts with the call list, list of available FSEs, and information about the incoming request for service.

1.3.4 Context-Aware Collaboration — Proactive Synthetic Assistant

A context-aware cell phone has been designed that provides the remote caller feedback on the current context of the person being called. Time (e.g., calendar), location, and audio environment sensing or interpretation are used to derive user context. We have focused on the callee being the driver of a car. The goal is for the caller to interact with the driver in a manner similar to that of a passenger in the car. For example, when there is a particularly difficult driving situation that has a high cognitive load (e.g., passing a truck on a downhill curve at night in the rain), a passenger is sensitive to the situation and suspends the conversation until the driving situation has passed. With contemporary cell phones, however, the caller is unaware of the driver's context and continues talking, perhaps causing the driver to enter a state of cognitive overload.

We have developed SenSay (*sen*sing and *say*ing) [14], a context-aware mobile phone that modifies its behavior based on its user's state and surroundings. It adapts to dynamically changing environmental and physiological states and also provides the remote caller information on the current context of the phone user. To provide context information, SenSay uses light, motion, and microphone sensors. The sensors are placed on various parts of the human body with a central hub, called the sensor box, mounted on the waist (Figure 1.1A).

Wearable Computing

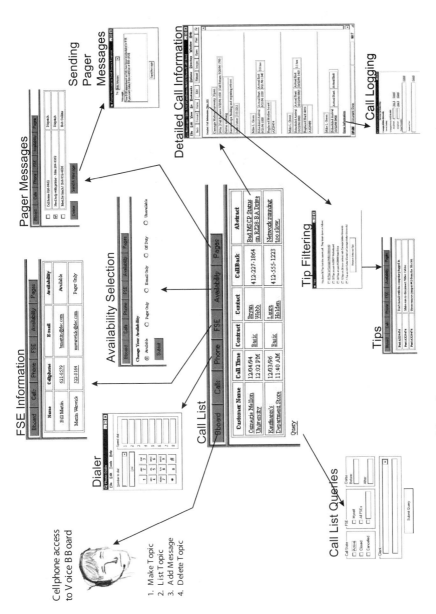

Figure 1.5. MoCCA Integrated Interface

MOBILE COMPUTING HANDBOOK

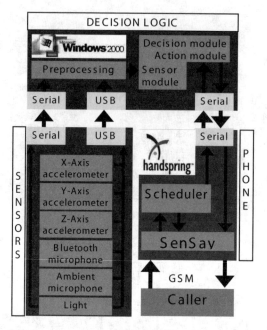

Figure 1.6. Context-Aware Cell Phone System Architecture

SenSay introduces the following four states:

1. Uninterruptible
2. Idle
3. Active
4. Normal (the default state)

A number of phone actions are associated with each state. For example, in the uninterruptible state, the ringer is turned off.

A three-tier architecture was adopted:

1. Sensor box
2. Decision module
3. Phone

The following components are shown in Figure 1.6. The sensor module, located in the bottom left, collects physical sensor data, which is then sent to the notebook computer (henceforth called the platform) through the serial port. The decision module at the top is then notified of data arrival and a series of preprocessing steps are done to the incoming data before the data is acted upon. Finally, the decision module instructs the phone to act based on the current user context. The decision module uses another serial port to communicate with the phone.

We built a custom sensor module containing two subsystems — the microcontroller and sensors. A microcontroller is used to process the queries from the sensor module and return the requested sensor data as a ten-bit word. The sensors include three accelerometers used to capture three axes of motion (x, y, z) and a light sensor. A Bluetooth® microphone is used for detecting user speech and another microphone is added to serve as an ambient noise detector.

The Bluetooth microphone communicates with a USB Bluetooth transceiver connected to the platform.

The platform software monitors sensors and phone status and makes decisions concurrently using three modules. The sensor module resides on the platform, which connects to the sensor box and the phone monitors all sensors. Similarly, the action module handles all interactions between the decision logic and the phone. The decision module determines user context and is responsible for triggering phone actions.

The decision module determines user context using sensor data and electronic calendar entries. It stores and examines the collected data and passes a phone action notification to the action model, which it may ignore. When the user is busy, Short Message Service (SMS) is used to auto respond to the caller. The action module is responsible for issuing changes and operations to the Treo™ smartphone. It accepts requests from the decision module and accesses the phone.

The decision module inspects the gathered data and determines the state that the phone should enter. Figure 1.7 shows the state diagram used by the decision module. To prevent bouncing between states too quickly, up to ten minutes of recent sensor data are stored and examined. Running averages are computed to give reasonable weight to previous data and phone state. Furthermore, four states are identified by the system, representing descending levels of uninterruptibility, as follows:

1. Uninterruptible state — the system enters this state when the user is involved in a conversation or has scheduled an important event in the electronic calendar. The ringer is disabled; vibrate is enabled only when the light level is low. The caller has an option to override this in case of emergency by calling again within three minutes.
2. Active state — high physical activity or high ambient noise level puts the system into this state. The ringer is set to high and vibrate is enabled.
3. Idle state — the system goes into this state when there is little movement and low ambient level. The system reminds the user of missed calls and provides suggestions to the user.
4. Normal state — the ringer and vibrate modes are set to the phone's default values.

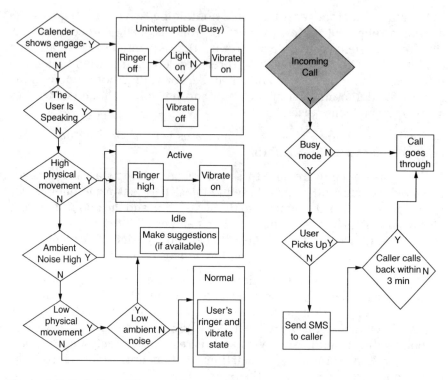

Figure 1.7. Left: State Diagram Showing What Happens When an Incoming Call Is Made in Uninterruptible Mode. Right: State Diagram Showing How the Phone Determines Which Mode to Place the Phone In

A series of threshold analysis tests were run while recording sensor values and noting trends over time. Microphones and accelerometer values were recorded from 11 subjects. In addition to the raw sensor values, average sensor values were also observed over various periods of time.

As an example of threshold experiments, consider user activity. The sensor board was taped to the user's abdomen. The physical activity test was run using the three-axis accelerometer. The maximum of the three absolute component values was used as the movement data. The data is split into three ranges.

1. Low activity includes sitting, sleeping, etc. Short, intense movements are averaged out.
2. Medium activity represents walking or other comparable activity. Medium movement indicates that the user is not idle.
3. High activity includes movements such as running.

To find generic threshold values, 11 subjects were asked to perform a test. After walking for 40 seconds, they were asked to sit down for 10 seconds.

Wearable Computing

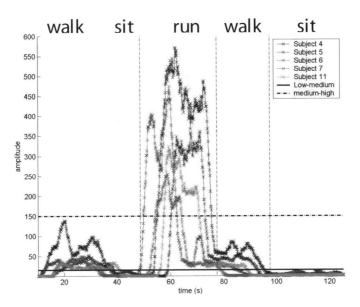

Figure 1.8. Motion State Thresholds

Then they were required to run for 30 seconds and afterward walk again. After walking for 20 more seconds, they sat down again for an additional 25 seconds. Figure 1.8 shows the resulting values. From this experiment, thresholds for differentiating between low, medium, and high activity were found and annotated on the diagram. Other tests were conducted for the microphones and light sensors.

1.4 Evaluation

Evaluation of all four applications was performed with laboratory prototypes. Metrics used in this evaluation include time on task and accuracy. These tests can be used as examples of how to evaluate wearable computer systems.

1.4.1 Prestored Procedures

Field tests were performed at the Digital Equipment Corporation facility in Forrest Hills, Pittsburgh, Pennsylvania. Five FSEs participated in tests that included performing a set of typical operations related to troubleshooting and repair operations on computing equipment. Each of the FSEs performed all of these operations. The subject systems included printers, motherboards, and networks. The use of the prototype contributed to a significant saving of time (35 to 40 percent), as shown in Figure 1.9. During these field tests, the FSEs used the system for the first time. A larger savings in time is expected with continued use. In addition, MoCCA allowed

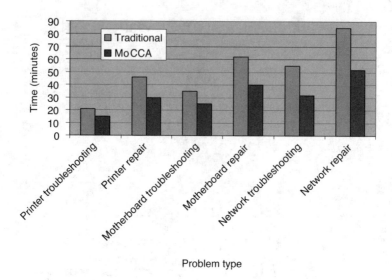

Figure 1.9. Improvement in Problem Solving with Prestored Knowledge versus Traditional Approach

the FSEs to fix some problems immediately, which otherwise would have required return trips to find and bring back manuals.

1.4.2 Master/Apprentice Help Desk

Performance on a bicycle repair task when working alone was compared to working with a helper who could provide guidance through the repair process. There was a video link between the worker and expert, as well as an audio connection. Participants consisted of 60 CMU students (69 percent male) and 2 bicycle repair experts. Workers do substantially better at performing these repair tasks with collaborative help. A repeated measures analysis of variance (ANOVA) test [15] was used to examine statistical significance of the results. In a bicycle repair experiment, average time to complete the tasks with a remote expert was half as long as in the solo condition (7.5 versus 16.5 minutes respectively; $p < .001$). This means that the probability that the two average times are the same is less than 0.1 percent. The quality of the repairs they completed was superior when they had assistance than when they worked alone (79 percent of the quality points for the collaborative condition versus 51 percent for the solo condition; $p < .001$). Although having access to an expert dramatically improved performance, having better tools for communication with the expert did not improve the number of tasks completed, the average time per completed task, or performance quality. In particular, neither video (comparison of full duplex audio/video with full duplex audio/no video) nor full duplex audio (comparison of full duplex audio/video condition with half duplex audio/video) helped workers perform more tasks, perform tasks more quickly, or perform them better [16].

1.4.3 Team Collaboration

IDEALINK provides a virtual space for groups to manipulate graphical objects related to their work task, sharing observations with each other [17]. IDEALINK is well-suited for use in wearable computers because it enables users to have ad hoc meetings and design sessions, providing them with a canvas on which they can sketch their ideas for a remote user. Asynchronous audio tags enable users to record an audio explanation or annotation of a particular object or procedure. Each session is recorded and archived for later playback, making the knowledge contained within them available for later reference.

To test IDEALINK's effectiveness, we conducted an experiment in which groups of four participants collaborated on the design of a remote control for a CD player. Each session concluded when all participants agreed upon a final design. They were encouraged, but not required, to begin by individually sketching their ideas, then proceed to discuss their sketches with the rest of the group, and conclude by collaborating on a final design based upon a combination of the sketches.

We noted occurrences of *mechanics of collaboration* as defined by Gutwin and Greenberg [18]. Explicit communication occurs when "group members intentionally provide each other with information" either "verbally" or "in combination with pointing to an artifact." Consequential communication is "information that is unintentionally given off by others as they go about their activities" including "information given off by artifacts as they are manipulated by others" and "the characteristic actions of a person's embodiment in the workspace." Coordination of action takes place when "people organize their actions in a shared workspace so that they do not conflict with others"; for example, when turns are taken. Planning occurs when people "reserve areas of the workspace for their use or consider various courses of action by simulating them." Monitoring actions are ones that gather information about others in the workspace. Assistance happens when another group member explicitly asks for help or when someone sees that assistance would be helpful. Finally, protection describes actions that a participant takes to prevent others from "inadvertently altering or destroying work" that they have done.

In addition, we also recorded instances of misunderstood illustrations, usage of body language to describe the remote control (that does not fall into any of the areas listed above), off task discussion, and equipment failure. The results of this are shown in Table 1.1.

Table 1.1. IDEALINK User Evaluation

Session Number	Whiteboard					IDEALINK				
	1	2	3	4	Total	5	6	7	8	Total
Mechanics of Collaboration Events										
Explicit communication	8	13	15	18	54	7	10	12	4	33
errors and difficulties	0	3	0	0	3	0	0	0	0	0
Implicit communication	9	8	5	18	40	2	4	1	0	7
errors and difficulties	0	2	0	0	2	0	0	0	0	0
Cooperation	7	1	0	1	9	0	0	1	0	1
errors and difficulties	0	6	0	0	6	0	1	1	0	2
Planning	2	5	4	2	13	7	1	4	1	13
errors and difficulties	0	0	0	0	0	0	1	1	0	2
Monitoring	1	0	0	0	1	1	2	4	2	9
Assistance	1	1	1	0	3	0	1	1	0	2
Protection	0	0	0	0	0	0	0	0	0	0
Other Observed Events										
Misunderstood diagram	0	1	0	0	1	1	2	5	4	12
Body language	5	0	0	0	5	0	3	7	0	10
Off task	2	0	0	1	3	3	3	1	1	8
Equipment failure	0	0	1	0	1	8	11	9	8	36

Participants tended to communicate with each other more verbally when they were using the whiteboard. In particular, they were more likely to use explicit and implicit references to positions and locations of objects. These implicit references were occasionally flawed, as occurred in session 2. Furthermore, use of the whiteboard caused more discussion of how to coordinate activities.

While using IDEALINK, participants more frequently monitored other participants' screens. They also used more body language to describe the remote control, even using the computer's shape and size as a proxy remote control.

We observed that participation was more evenly distributed among participants when they used IDEALINK than when they used the whiteboard. In the first whiteboard-based session, one participant dominated discussion. He spent a significant amount of time at the front of the room, listed his ideas on the board first, and frequently interrupted others' discussions.

While using IDEALINK, his interference with others was minimized. The other participants were able to focus their attention away from the central whiteboard at the front of the room and on the Idealink screen in front of them. They made their own sketches as he was talking and directed their attention toward these sketches. This resulted in a more effective meeting, as more time was spent focusing on the prototype sketches and less time was spent debating whether the remarks made by the dominant participant were valid.

1.4.4 Context-Aware Collaboration — Proactive Synthetic Assistant

Two experiments tested the hypothesis that a context-aware cell phone could change caller and driver behaviors. Experiment 1 tested whether remote cell phone callers would slow or stop their conversation with a driver when signaled. Experiment 2 tested whether a driver's performance while speaking on a cell phone would be improved by slowing or stopping the remote caller's conversation.

Participants (n = 24) were asked to role-play a person seeking to rent an apartment. Each participant made successive cell phone calls to three landlords, played by the experimenter. Participants were provided a list of questions to ask the landlord about each apartment (e.g., how many bedrooms the apartment had). At a prespecified point in each call, the landlord would unexpectedly pause for ten seconds.

Results show that the callers said less than half the number of sentences during the pause when they were sent a signal compared to when the driver remained silent. The spoken message: "The person you have called is busy. Please hold." was the most effective signal.

Experiment 2 used a driving simulator composed from a virtual reality authoring environment and allowed users to navigate a vehicle through a test track. Before beginning the experiment, participants (n = 20) practiced using the driving simulator until they said they felt comfortable. Participants then completed one circuit of the track on the driving simulator under each of three conditions:

1. Control (no phone call)
2. Call without pause
3. Call with pause

Results show that talking on the cell phone caused people to crash more (6.8 crashes) as compared to driving without a call (3.55 crashes). Inducing pauses during the call caused the driver to crash less (3.65 crashes) when using the cell phone.

1.5 Summary and Future Challenges

In this chapter, we have introduced and described a taxonomy of problem solving capabilities for wearable and context-aware computers. We have shown how these capabilities impact choices of input/output modalities and user interface models.

An important set of challenges must be addressed to make wearable computing effective with ubiquitous computing environments. How do we develop social and cognitive models of applications? How do we integrate input from multiple sensors and map them into user social and cognitive states? How do we anticipate user needs? How do we interact with the user?

In the future, we will focus our efforts on development of a virtual coach, which will capture a wearable augmented cognition platform and software application, as well as be able to monitor an individual's cognitive load and route tasks to less loaded individuals. Cognitive performance will be assessed online. Providing immediate suggestions to a user for cognitive augmentation and arbitration of resource redeployment will further enhance performance.

Acknowledgments

This material is based upon work supported by the National Science Foundation under Grant no. 0205266 and work supported by the Defense Advanced Research Projects Agency (DARPA) under Contract no. NBCHC030029. We would also like to acknowledge funding support received from the Pennsylvania Infrastructure Technology Alliance.

References

1. Smailagic, A. and Siewiorek, D.P., The CMU mobile computers: A new generation of computer systems, in *Proc. IEEE COMPCON 94*, P. 467, 1994.
2. Smailagic, A., Siewiorek, D.P. et al., Benchmarking an interdisciplinary concurrent design methodology for electronic/mechanical design, in *Proc. ACM/IEEE Design Automation Conference*, P. 514, 1995.
3. Siewiorek, D.P., Smailagic, A., and Lee, J.C., An interdisciplinary concurrent design methodology as applied to the Navigator wearable computer system, *J. Computer and Software Engineering*, Ablex Publishing Corporation, vol. 2, no. 3, P. 259, 1994.
4. Smailagic, A. and Siewiorek, D.P., Application design for wearable and context-aware computers, *IEEE Pervasive Computing*, vol. 1, no. 4, P. 20, 2002.
5. Ferrari, D., *Computer Systems Performance Evaluation*, Englewood Cliffs, NJ: Prentice Hall, P. 44, 1978.
6. Smailagic, A., Siewiorek, D.P., Stivoric, J., and Martin, R., Very rapid prototyping of wearable computers: A case study of custom versus off-the-shelf design methodologies, *J. Design Automation for Embedded Systems*, vol. 3, P. 217, 1998.
7. Najjar, L., Thompson, J.C., and Ockerman, J.J., A wearable computer for quality assurance in a food processing plant, in *Proc. Int. Symp. Wearable Computers*, Piscataway, NJ: IEEE Computer Society Press, P. 163, 1997.
8. Smailagic, A., An evaluation of audio-centric CMU wearable computers, *ACM Journal on Special Topics in Mobile Networking*, vol. 6, pp. 59–68, 1998.
9. Bauer, M., Heiber, T., Kortuem, G., and Segall, Z., A collaborative wearable system with remote sensing, in *Proc. Int. Symp. Wearable Computers*, Piscataway, NJ: IEEE Computer Society Press, P. 10, 1998.

10. Smailagic, A., Siewiorek, D.P. et al., MoCCA: A mobile communication and computing architecture, in *Proc. Int. Symp. Wearable Computers*, Piscataway, NJ: IEEE Computer Society Press, P. 64, 1999.
11. http://www.fas.org/man/dod-101/sys/land/land-warrior.htm.
12. Feiner, S., MacIntyre, B., and Höllerer, T., A touring machine: Prototyping 3D mobile augmented reality systems for exploring the urban environment, in *Proc. Int. Symp. Wearable Computers*, Piscataway, NJ: IEEE Computer Society Press, P. 74, 1997.
13. Marinelli, D. and Stevens, S.M., Synthetic interviews: The art of creating a 'dyad' between humans and machine-based characters, in *Proc. IEEE Workshop on Interactive Voice Technology for Telecommunications Applications*, Torino, Italy, 1998.
14. Siewiorek, D.P, Smailagic, A. et al., SenSay: A context-aware mobile phone, in *Proc. Int. Symp. Wearable Computers*, Piscataway, NJ: IEEE Computer Society Press, P. 248, 2003.
15. Girden, E.R., *ANOVA Repeated Measures,* Thousand Oaks, CA: Sage Publications, 1992.
16. Fussell, S.R., Kraut, R.E., and Siegel, J., Coordination of communication: Effects of shared visual context on collaborative work, in *Proc. ACM Conference on Computer Supported Cooperative Work*, New York: ACM Press, P. 21, 2000.
17. Garlan, D., Siewiorek, D.P., Smailagic A., and Steenkiste, P., Project Aura: Toward distraction-free pervasive computing, *IEEE Pervasive Computing*, vol. 1, P. 22, 2002.
18. Gutwin, C. and Greenberg, S., The mechanics of collaboration: Developing low cost usability evaluation metrics for shared workspaces, in *Proc. IEEE Int. Workshop on Enabling Technologies: Infrastructure for Collaborative Enterprises,* Piscataway, NJ: IEEE Computer Society Press, P. 98, 2000.
19. Billinghurst, M., Weghorst, S., and Furness III, T., Wearable computers for three-dimensional CSCW, in *Proc. Int. Symp. Wearable Computers*, Piscataway, NJ: IEEE Computer Society Press, P. 39, 1997.
20. Dey, A., Futakawa, M., Salber, D., and Abowd, G., The conference assistant: combining context-awareness with wearable computing, in *Proc. Int. Symp. Wearable Computers*, Piscataway, NJ: IEEE Computer Society Press, P. 21, 1999.
21. Healey, J. and Picard, R., StartleCam: A cybernetic wearable camera, in *Proc. Int. Symp. Wearable Computers*, Piscataway, NJ: IEEE Computer Society Press, P. 42, 1998.
22. Rekimoto, J., Transvision: A hand-held augmented reality system for collaborative design, in *Proc. Virtual Systems and Multimedia*, 1996.
23. Starner, T., Weaver, J., and Pentland, A., A wearable computer based American sign-language recognizer, in *Proc. Int. Symp. Wearable Computers*, Piscataway, NJ: IEEE Computer Society Press, P. 130, 1997.

Chapter 2
Developing Mobile Applications: A Lime Primer

Gian Pietro Picco, Amy L. Murphy, and Gruia-Catalin Roman

Abstract

Mobility poses peculiar challenges that must be addressed by novel programming constructs. Lime (Linda in a Mobile Environment) tackles the problem by adopting a coordination perspective inspired by work on the Linda model. The context for computation, represented in Linda by a single, globally accessible, persistent tuple space, is reinterpreted in Lime as the transient sharing of the tuple spaces carried by individual mobile units. Additional constructs provide increased expressiveness, by enabling programs to deal with the location of tuples and to react to specified states. The resulting model provides a minimalist set of abstractions that promise to facilitate rapid and dependable development of mobile applications. In this chapter, we illustrate the model underlying Lime, present the programming interface of the companion middleware, and discuss how applications and higher level middleware services can be built using it.

2.1 Introduction

Distributed computing has been traditionally associated with a rather static environment, where the topology of the system is largely stable and so is the configuration of the deployed application components. Today, this vision is being challenged by various forms of mobility, which are effectively reshaping the landscape of modern distributed computing. On one hand, the emergence of wireless communication and portable computing devices is fostering scenarios where the physical topology of the system is continuously modified by the free movement of the mobile hosts, whose wireless communication devices enable them to dynamically create and sever links based on proximity. In its most radical incarnation, represented by mobile ad hoc networks (MANETs), the system is entirely constituted by mobile nodes and the fixed network infrastructure is totally absent. Together with this *physical mobility* of hosts, another form of mobility has emerged, where the units of mobility are program fragments belonging to a distributed application and are relocated from one host to another. This *logical mobility* of code brings unprecedented levels of flexibility in the deployment of application components. In some domains, it enables significant improvements in the use of communication resources.

Mobility undermines several common assumptions. Disconnection is no longer an infrequent accident: in applications involving physical mobility, it is often triggered by the user in order to save battery power and hence becomes a defining characteristic of the environment. The fluidity of the physical and logical configuration of the system renders it impractical — and often impossible — to make assumptions about the availability of a specific user, host, or service. In general, the computational context is no longer fixed and predetermined, rather it becomes continuously changing in largely unpredictable ways. As a consequence, a large fraction of the body of theories, algorithms, and technology must be recast in the mobile scenario. Application development demands appropriate constructs and mechanisms to accommodate the required level of dynamicity and decoupling necessary to cope with mobility. Nevertheless, thus far the problem has been tackled only in a limited context. Pioneering work in mobile computing targeted supporting mobility at the operating system level (e.g., the work on Coda [14]) or for specific application domains (e.g., repository-based in Bayou [34]). On the other hand, many commercial applications mask mobility by relying on proxy architectures, but assume the existence of a fixed infrastructure. Clearly, these approaches are limited in that they either solve issues that are specific for a given application domain or do not address unconstrained mobile settings.

Lime [18, 26] is a model and a middleware expressly designed for supporting the development of mobile applications. In contrast with many of the existing proposals, Lime provides the application programmer with a set of general-purpose programming constructs. Moreover, by adopting a peer-to-peer architecture that does not rely on any fixed infrastructure, it addresses the needs of the most radical forms of mobility and in particular of MANETs.

The design of Lime is inspired by the realization that the problem of designing applications involving mobility can be regarded as a coordination problem [30] and that a fundamental issue to be tackled is the provision of good abstractions for dealing with, and exploiting, a dynamically changing context. Coordination is defined as a style of computing that emphasizes a high degree of decoupling among the computing components of an application. As initially proposed in Linda [9], this can be achieved by allowing independently developed components to share information stored in a globally accessible, persistent, content-addressable data structure, typically implemented as a centralized tuple space. A small set of operations enabling the insertion, removal, and copying of tuples provides a simple and uniform interface to the tuple space. Temporal decoupling is achieved by dropping the requirement that the communicating parties be present at the time the communication takes place and spatial decoupling is achieved by eliminating the need for components to be aware of each other's identity to communicate. A clean computational model, a high degree of decoupling, an abstract approach to communication, and a simple interface are the defining features of coordination technology.

Lime reinterprets Linda within the mobile scenario in an original way. Each mobile unit is permanently associated with a local tuple space, whose content is transiently shared with the content of similar tuple spaces attached to the other units within range. Hence, the tuple space used for coordination is no longer unique, global, and persistent — assumptions that macroscopically conflict with mobility. Instead, it is dynamically built out of the spaces contributed by the mobile units within range and reflects the current configuration of the system. Transiently shared tuple spaces are the key to shielding the programmer from the complexity of the system configuration, while still providing an effective abstraction for handling communication among application components. In addition, Lime defines constructs providing increased expressiveness by introducing the ability to react asynchronously to the presence of a tuple and to control the placement of a tuple and its access within the global tuple space. Finally, because no assumption is made about the nature of the mobile units, the computational model naturally encompasses both physical mobility of hosts and logical mobility of agents. The computational model is embodied in a Java™-based middleware, made available as open source [33], which has been successfully employed for developing mobile applications.

In this chapter, we present Lime by focusing on the model concepts and the programming and design techniques useful to the developer of mobile applications. Other available documents describe the design of the middleware [18] and the formal semantics of the model [19]. This chapter is organized as follows:

- Section 2.2 provides the minimal Linda background necessary to understand Lime.
- Section 2.3 contains an overview of the Lime model and of the application programming interface of the companion middleware.
- Section 2.4 walks through a case study application and shows how its requirements are satisfied through a design exploiting Lime.

- Section 2.5 presents some middleware extensions to Lime that we built entirely as an application layer on top of the original middleware.
- Section 2.6 places Lime in the context of related work.
- Section 2.7 completes our chapter with some brief concluding remarks.

2.2 Linda in a Nutshell

In Linda, processes communicate through a shared *tuple space* that acts as a repository of elementary data structures or *tuples*. A tuple space is a multi-set of tuples that can be accessed concurrently by several processes. Each tuple is a sequence of typed fields, such as <"foo", 9, 27.5>, and contains the information being communicated.

Tuples are added to a tuple space by performing an out(*t*) operation and can be removed by executing in(*p*). Tuples are anonymous, thus their selection takes place through pattern matching on the tuple content. The argument *p* is often called a *template* or *pattern* and its fields contain either actuals or formals. Actuals are values; the fields of the previous tuples are all actuals, and the last two fields of <"foo", ? integer, ? float> are formals. Formals act like wild cards, and are matched against actuals when selecting a tuple from the tuple space. For instance, the template above matches the tuple defined earlier. If multiple tuples match a template, the one returned by in is selected nondeterministically. Tuples can also be read from the tuple space using the nondestructive rd(*p*) operation. Both in and rd are blocking, i.e., if no matching tuple is available in the tuple space, the process performing the operation is suspended until a matching tuple becomes available. A typical extension to this synchronous model is the provision of a pair of asynchronous primitives inp and rdp, called probes, which allow nonblocking access to the tuple space.[1] Moreover, some variants of Linda (e.g., [32]) provide also *bulk operations,* which can be used to retrieve all matching tuples in one step. In Lime, we provide a similar functionality through the ing and rdg operations, whose execution is asynchronous like in the case of probes.[2]

2.3 Lime: Linda in a Mobile Environment

Communication in Linda is decoupled in time and space, i.e., senders and receivers do not need to be available at the same time and mutual knowledge of their identity or location is not necessary for data exchange. This form of decoupling is of paramount importance in a mobile setting, where the parties involved in communication change dynamically due to their migration or connectivity patterns. Moreover, the notion of tuple space provides a straightforward and intuitive abstraction for representing the

Developing Mobile Applications: A Lime Primer

computational context perceived by the communicating processes. On the other hand, decoupling is achieved thanks to the properties of the Linda tuple space, namely its global accessibility to all the processes and its persistence — properties that are clearly hard if not impossible to maintain in a mobile environment.

Lime [18, 26] adapts Linda to the mobile environment in an original way and provides a coordination layer that can be exploited successfully for designing applications that exhibit logical mobility, physical mobility, or both. In this section, we present the Lime model with the description of how its constructs are made available through the application programming interface of the companion middleware. The reader interested in additional programming details can find extensive documentation and examples on the Lime Web site [33].

2.3.1 Model Setting and Overview

The fundamental entities in Lime are *agents, hosts,* and *tuple spaces*. Agents are the only active components. Hosts can be mobile and are mainly roaming containers that provide connectivity and execution support for agents. Agents can be mobile as well and migrate across hosts of their own volition. The *connectivity* patterns among agents and hosts constrain coordination in Lime. Mobile hosts are connected when a communication link is available. Mobile agents are connected when they are colocated on the same host or they reside on hosts that are connected. Changes in connectivity among hosts depend only on changes in the physical communication links. Connectivity among mobile agents may depend also on arrival and departure of agents, with creation and termination of mobile agents being regarded as special cases of connection and disconnection, respectively.

Tuple spaces provide the coordination media among agents, as in Linda. Nevertheless, in Lime each agent is permanently associated with at least one *interface tuple space* (ITS). This tuple space constitutes the only access to the data context for the agent it is associated with and, at the same time, it contains the tuples the agent is willing to make available to the rest of the system. At any time, an agent can access, through its ITS, the union of the content of the ITS of all the agents currently connected. In essence, Lime shifts from the fixed context characteristic of Linda to a dynamically changing one by breaking up the single global tuple space into many and by introducing rules for transient sharing of these individual tuple spaces based on connectivity. The expressive power of the model is then increased further by the ability to execute reactive and asynchronous operations and to restrict the scope of operations based on location. We now describe in more detail the features of the Lime model and middleware.

Table 2.1. The Class LimeTupleSpace, Representing a Transiently Shared Tuple Space

```
public class LimeTupleSpace {
  public LimeTupleSpace(String name);
  public String getName();
  public boolean isOwner();
  public boolean isShared();
  public boolean setShared(boolean isShared);
  public static boolean setShared(LimeTupleSpace[] lts,
                          boolean isShared);
  public void out(ITuple tuple);
  public ITuple in(ITuple template);
  public ITuple rd(ITuple template);
  public void out(AgentLocation destination,
             ITuple tuple);
  public ITuple in(Location current, AgentLocation
             destination, ITuple template);
  public ITuple inp(Location current, AgentLocation
             destination, ITuple template);
  public ITuple[] ing(Location current, AgentLocation
             destination, ITuple template);
  public ITuple rd(Location current, AgentLocation
             destination, ITuple template);
  public ITuple rdp(Location current, AgentLocation
             destination, ITuple template);
  public ITuple[] rdg(Location current, AgentLocation
             destination, ITuple template);
  public RegisteredReaction[] addStrongReaction
    (LocalizedReaction[] reactions);
  public RegisteredReaction[] addWeakReaction
    (Reaction[] reactions);
  public void removeReaction(RegisteredReaction[]
                         reactions);
  public boolean isRegisteredReaction(RegisteredReaction
                                 reaction);
  public RegisteredReaction[] getRegisteredReactions();
}
```

2.3.2 Creating a Lime Tuple Space

Table 2.1 shows the public interface[3] of the class LimeTupleSpace, which embodies the concept of a Lime transiently shared tuple space. The association between an agent and the tuple space is established at creation time by invoking the constructor. Agents may have multiple ITSs distinguished by a name, as this is recognized [6] as a useful abstraction to separate related application data. The name of the tuple space is specified as a parameter of the constructor. In the current implementation, agents are single-threaded and only the thread of the agent that creates the tuple space is allowed to perform operations on the LimeTupleSpace object; accesses by other threads fail by returning an exception. This represents the constraint that the ITS must be permanently and exclusively attached to the corresponding mobile agent.

When the ITS is first created and bound to an agent, its sharing status is *private,* meaning that the only agent that can access the ITS's data is the one associated with it at creation time. A private ITS can be used as a stepping stone to a shared data space, allowing the agent to populate it with data prior to making it publicly accessible. Alternately, it can be useful as a primitive data structure for local data storage. Lime operations, shown in the remainder of Table 2.1 and discussed in the rest of this section, are available also on a private tuple space. For instance, the inp operation is provided through the method inp, which accepts a parameter representing the template to be matched against and returns a matching tuple, if any, or null. The only requirement for tuple objects is to implement the interface ITuple, which is defined in a separate package called LighTS [23] that provides access to a lightweight tuple space implementation. Note also that inp, and similarly the other operations, are overloaded with additional parameters representing locations, whose meaning is discussed later in this section.

One relevant difference is that blocking operations are forbidden on a private tuple space and return an exception. Because the private tuple space is exclusively associated to one agent, the execution of an in or rd when no matching tuple is present would suspend the agent forever: effectively making the agent wait for a tuple that no other agent can possibly insert.

2.3.3 Enabling Transient Sharing

When an agent is alone in the system, i.e., there are no other agents colocated on the same machine or on machines in range, the ITS of the agent is the only available data repository accessible by it. On the surface, this could seem a situation identical to an agent owning a private tuple space, but the two cases are actually rather different. Irrespective of the system configuration, a private tuple space is never shared and its existence is known only to the agent that owns it. Instead, a nonprivate tuple space is available for sharing with other units according to connectivity.

When multiple mobile units[4] are able to communicate, either directly or transitively, we say these units form a Lime *group*. We can restrict the notion of group membership beyond simple communication, but for the purposes of this document, we consider only connectivity. The semantics of Lime is such that the content of the ITSs of all group members are merged, or transiently shared, to form a single, large context that is accessed by each unit through its own ITS. The sharing itself is transparent to each mobile unit, however as the members of the group change, the content of the tuple space each member perceives through operations on the ITS changes in a transparent way. The joining of a group by a mobile unit and the subsequent merging of its local context with the group context is referred to as *engagement,* which is performed as a single, atomic operation. A mobile unit leaving a group triggers *disengagement,* that is, the atomic removal of the tuples representing its local context from the remaining group context. In general, whole groups can merge and a group can split into several groups due to changes in connectivity.

In the case of multiple tuple spaces associated to agents, the sharing rule relies on tuple space names: only identically named tuple spaces are transiently shared among the members of a group. Thus, for instance, when an agent a owning a single tuple space named X joins a group consisting of one single agent b that owns two tuple spaces named X and Y, only X becomes shared between the two agents. Tuple space Y remains accessible only to b and potentially to other agents owning Y that may join the group later on.

Because all tuple spaces are initially created private, sharing must be explicitly enabled by calling the instance method `setShared`, shown in Table 2.1. The method accepts a Boolean parameter specifying whether the transition is from private to shared (`true`) or vice versa (`false`). Calling this method effectively triggers engagement or disengagement of the corresponding tuple space. The sharing status can also be changed in a single atomic step for multiple tuple spaces owned by the same agent by using the `static` version of `setShared` (Table 2.1). Engagement or disengagement of an entire host, instead, can be triggered explicitly by the programmer by using the methods `engage` and `disengage`, provided by the `LimeServer` class (not shown here). Otherwise, they are implicitly called by the runtime support according to connectivity. The `LimeServer` class is essentially an interface to the runtime support that exports additional system-related features, e.g., loading of an agent into a local or remote runtime support, setting of properties, and so on. In particular, it also allows the programmer to limit transient sharing to the tuple spaces residing on the host, instead of spanning the whole system.

2.3.4 Reconciling Different Forms of Mobility

One of the key features of Lime is the ability to encompass both physical mobility of hosts and logical mobility of agents in a single coordination framework. The relationship between the two forms of mobility is illustrated in Figure 2.1. The transiently shared ITSs belonging to the agents colocated on a host define a *host-level tuple space.* This, in turn, can be regarded as the ITS associated to the host; hence, transient sharing determines a *federated*

Developing Mobile Applications: A Lime Primer

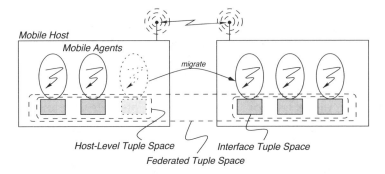

Figure 2.1. Transiently Shared Tuple Spaces Encompass Physical and Logical Mobility

tuple space. When a federated tuple space is established, a query on the shared ITS of an agent returns a tuple that may belong to the tuple space carried by that agent, to a tuple space belonging to a colocated agent, or to a tuple space associated with an agent residing on some remote, connected host.

Although Lime provides a uniform treatment of these two forms of mobility, it is worth making some observations. First, many applications do not need both forms of mobility. However, straightforward adaptations of the model are possible. For instance, applications that do not exploit mobile agents but run on a mobile host can employ one or more stationary agents (i.e., programs that do not contain migration operations). In this case, the design of the application can be modeled in terms of mobile hosts whose ITS is a fixed host-level tuple space. Instead, applications that do not exploit physical mobility — and do not need a federated tuple space spanning different hosts — can exploit only the host-level tuple space as a local communication mechanism among colocated agents.

Second, it is interesting to note how mobility is not dealt with directly in Lime (i.e., there are no constructs for triggering the movement of agents or hosts). Instead, the effect of migration is made indirectly manifest in the model and middleware only through changes observed in the connectivity among components. This choice, which sets the nature of mobility aside, keeps our model general and enables different instantiations of the model based on different notions of connectivity. This choice is retained also at the middleware level, where agent mobility is not supported directly. Instead, as with tuple spaces, agent migration is decoupled from the rest of the system by an adaptation layer that simplifies the integration of a mobile agent system. The currently available implementation relies on an adaptor built for the μCode mobile code toolkit, developed by one of the authors [24] and available as open source [22]. This adaptation layer allows a mobile agent to carry along one or more Lime tuple spaces and automatically deals with their engagement or disengagement. Upon migration, the agent tuple spaces are all toggled to private and hence disengaged. These tuple spaces are serialized as part of the agent state and migrated to the destination along with the agent, where they are deserialized and shared again before the agent code begins to execute. More details about the adaptation layer and how to integrate a mobile agent system with Lime are available in the Lime documentation and on the Lime Web site [33].

2.3.5 Restricting the Scope of Operations

Transiently shared tuple spaces foster a style of coordination that reduces the details of distribution and mobility to content changes in what is perceived as a local tuple space. This view is powerful and greatly simplifies application design in many scenarios by relieving the designer from the chore of maintaining explicitly a view of the context consistent with changes in the configuration of the system. On the other hand, this view may hide too much in situations where the designer needs more fine-grained control over the portion of context that must be accessed. For instance, the application may require control over the agent responsible for holding a given tuple, something that cannot be specified only in terms of the global context. Also, performance and efficiency considerations may come into play, as in the case where application information would enable access aimed at a specific host-level tuple space, thus avoiding the greater overhead of a query spanning the whole federated tuple space. Such fine-grained control over the context perceived by the mobile unit is provided in Lime by extending the Linda operations with tuple location parameters that operate on user-defined projections of the transiently shared tuple space. Further, all tuples are implicitly augmented with two fields, not directly accessible to the programmer, representing the tuple's *current* and *destination locations*. The current location identifies the single agent responsible for holding the tuple when all agents are disconnected; the destination location indicates the agent with which the tuple should eventually reside.

The out[λ] operation extends out with a location parameter representing the identifier of the agent responsible for holding the tuple. The semantics of out[λ](t) involve two steps. The first step is equivalent to a conventional out(t), whose effect is to insert the tuple t in the ITS of the agent calling the operation, say ω. At this point t has a current location ω and a destination location λ. If the agent λ is currently connected, the operation is completed as a single atomic step by moving the tuple t to the destination location. On the other hand, if λ is currently disconnected, the tuple remains at the current location — the tuple space of ω. This *misplaced* tuple, if not withdrawn,[5] will remain misplaced unless λ becomes connected. In this case, the tuple will migrate to the tuple space associated with λ as part of

Developing Mobile Applications: A Lime Primer

Table 2.2. Accessing Different Portions of the Federated Tuple Space Using Location Parameters

Current Location	Destination Location	Defined Projection
unspecified	unspecified	Entire federated tuple space
unspecified	λ	Tuples in the federated tuple space and destined to λ
ω	unspecified	Tuples in ω's tuple space
Ω	unspecified	Tuples in Ω's host-level tuple space, i.e., belonging to any agent at Ω
ω	λ	Tuples in ω's tuple space and destined to λ
Ω	λ	Tuples in Ω's host-level tuple space and destined to λ

ω and λ are agent identifiers.
Ω is a host identifier.

the engagement. By using out[λ], the caller can specify that the tuple is supposed to be placed within the ITS of agent λ. This way, the default policy of keeping the tuple in the caller's context until it is withdrawn can be overridden and more elaborate schemes for transient communication can be developed.

Variants of the in and rd operations using location parameters are allowed as well. These operations, of the form in[ω,λ](p) and rd[ω,λ](p), enable the programmer to refer to a projection of the current context defined by the value of the location parameters, as illustrated in Table 2.2. The current location parameter enables the restriction of scope from the entire federated tuple space (no value specified) to the tuple space associated to a given host or even a given agent. The destination location is used to identify misplaced tuples.

Location parameters are specified in the middleware by using the classes AgentLocation and HostLocation, both subclasses of Location. These classes enable the definition of globally unique location identifiers for hosts and agents and are used to specify different scopes for Lime operations. For instance, a probe inp(cur,dest,t) may be restricted to the tuple space of a single agent if cur is of type AgentLocation or it may refer the whole host-level tuple space, if cur is of type HostLocation, according to Table 2.2. The constant Location.UNSPECIFIED is used to allow any location parameter to match. For instance, in(cur,Location.UNSPECIFIED,t) returns a tuple contained in the tuple space of cur, regardless of its final destination, including also misplaced tuples. Note how typing rules allow the proper constraint of the current and destination locations according to the rules of the LIME model. For instance, the destination parameter is always an AgentLocation object, as agents are the only carriers of tuples in Lime. In the current implementation, probes are always restricted to a local subset of the federated tuple space, as defined by the location parameters. An unconstrained definition, as the one provided for in and rd, would involve a distributed transaction to preserve the semantics of the probe across the federated tuple space.

2.3.6 Reacting to Changes

Lime extends Linda not only by providing transiently shared tuple spaces, but also by introducing the notion of *reaction*. A reaction $R(s,p)$ is defined by a code fragment *s* that specifies the actions to be executed when a tuple matching the pattern *p* is found in the tuple space. Informally,[6] a reaction can *fire* if a tuple matching pattern *p* exists in the tuple space. After every regular tuple space operation, a reaction is selected nondeterministically and, if it is enabled, the statements in *s* are executed in a single, atomic step. This selection and execution continues until no reactions are enabled, at which point normal processing resumes. Blocking operations are not allowed in s, as they may prevent the execution of *s* from terminating.

Reactions provide the programmer with powerful constructs for specifying the actions that need to take place in response to a *state* change and ensure their execution in a single atomic step. In particular, it is worth noting how this model is much more powerful than many event-based ones [31], including those exploited by tuple space middleware such as TSpaces [12] and JavaSpaces™ [13], which are typically stateless and provide no guarantee about the atomicity of event reactions.

Nevertheless, this expressive power comes at a price, especially in a distributed setting. When multiple hosts are present, the content of the federated tuple space is scattered among several agents. Thus, maintaining the requirements of atomicity and serialization imposed by reactive statements requires a distributed transaction encompassing several hosts — often, an impractical solution. For specific applications and scenarios, such as those involving a limited number of hosts or those exploiting only local interactions among mobile agents, these kind of reactions, referred to as *strong reactions,* are still reasonable. For practical performance reasons, however, our implementation currently limits the use of strong reactions by restricting the current location field to be a host or agent and by enabling a reaction to fire only when the matching tuple appears on the same host as the agent that registered the reaction. As a consequence, a mobile agent can register a reaction for a host different from the one where it is residing, but such a reaction remains disabled until the agent migrates to the specified host. These constraints effectively force the *detection* of a tuple matching *p* and the corresponding *execution* of the code fragment *s* to take place (atomically) on a single host and hence do not require a distributed transaction.

To strike a compromise between the expressive power of reactions and the practical implementation concerns, we introduced another reactive construct that allows some form of reactivity spanning the whole federated tuple space but with weaker semantics. The processing of a *weak reaction* proceeds as in the case of a strong reaction, but detection and execution do not happen atomically: instead, execution is guaranteed to take place only eventually, after a matching tuple is detected. The execution of *s* takes place on the host of the agent that registered the reaction.

The use of reactions involves the operations in `LimeTupleSpace` and the classes shown in Table 2.3. Reactions can be registered on a tuple space by invoking either `addStrongReaction` or `addWeakReaction`. These methods return an object `RegisteredReaction`, which can be used to deregister a reaction with the method `removeReaction` and provide additional information about the registration process. The decoupling between the reaction used for the registration and the `RegisteredReaction` object returned allows for registration of the same reaction on different ITSs and for the same reaction to be registered with strong and, subsequently, with weak semantics.

Reactions can be annotated with location parameters, with the same meaning discussed earlier for `in` and `rd` and shown in Table 2.2. Reactions can be either of type `LocalizedReaction`, where the current and destination location restrict the scope of the operation, or `UbiquitousReaction`, which specifies the whole federated tuple space as a target for matching. In the current implementation, strong reactions are confined to a single host and hence only a `LocalizedReaction` can be passed to `addStrongReaction`. The reaction type is used to enforce the proper registration constraint through type checking. The common ancestor class `Reaction` defines a number of accessors for the properties established for the reaction at creation time. Creation of a reaction is performed by specifying the template that needs to be matched in the tuple space, a `ReactionListener` object that specifies the actions taken when the reaction fires, and a reaction *mode* that controls the extent to which a reaction is allowed to execute. A reaction registered with mode `ONCE` is allowed to fire only one time: after its execution it becomes automatically deregistered. Instead, a reaction registered with mode `ONCEPERTUPLE` is allowed to fire an arbitrary number of times, but never twice for the same tuple. The `ReactionListener` interface requires the implementation of a single method `reactsTo` that is invoked by the runtime support when the reaction actually fires. This method has access to the information about the reaction carried by the `ReactionEvent` object passed as a parameter to the method.

Table 2.3. The Classes `Reaction`, `RegisteredReaction`, `ReactionEvent` and the Interface `ReactionListener` Required for the Definition of Reactions on the Tuple Space

```
public abstract class Reaction {
  public final static short ONCE;
  public final static short ONCEPERTUPLE;
  public ITuple getTemplate();
  public ReactionListener getListener();
  public short getMode();
  public Location getCurrentLocation();
  public AgentLocation getDestinationLocation();
}
public class UbiquitousReaction extends Reaction {
  public UbiquitousReaction(ITuple template,
                            ReactionListener listener,
                            short mode);
}
public class LocalizedReaction extends Reaction {
  public LocalizedReaction(Location current, AgentLocation
                           destination, ITuple template,
                           ReactionListener listener,
                           short mode);
}
public class RegisteredReaction extends Reaction {
  public String getTupleSpaceName();
  public AgentID getSubscriber();
  public boolean isWeakReaction();
}
public class ReactionEvent extends java.util.EventObject {
  public ITuple getEventTuple();
  public RegisteredReaction getReaction();
  public AgentID getSourceAgent();
}
public interface ReactionListener extends java.util.
    EventListener {
  public void reactsTo(ReactionEvent e);
}
```

2.3.7 Accessing the System Configuration

Thus far, our extension of Linda operations with location parameters hides completely the details of the system configuration. For instance, if the probe $inp[\omega,\lambda](p)$ fails, this simply means that no tuple matching p is available in the projection of the federated tuple space defined by the location parameters $[\omega,\lambda]$. It cannot be directly inferred whether the failure is due to the fact that agent ω does not have a matching tuple or that agent ω is currently not part of the group.

Without awareness of the system configuration, only a partial context awareness can be accomplished, where applications are aware of changes only in the portion of context concerned with application data. Although this perspective is often enough for mobile applications, in many others the portion of context more closely related to the system configuration plays a key role. For instance, in some circumstances it becomes necessary to react to the departure of a mobile unit or to determine the set of units currently belonging to a Lime group. Interestingly, Lime provides this form of awareness of the system configuration by using the same abstractions discussed thus far, that is, through a transiently shared tuple space conventionally named LimeSystem to which all agents are permanently bound. The tuples in this tuple space contain information about the mobile units present in the group and their relationship, such as which host is supporting which agents or which agent is sharing which tuple spaces. Insertion and withdrawal of tuples in LimeSystem is a prerogative of the runtime support. Nevertheless, applications can read tuples and register reactions to respond to changes in the configuration of the system.

2.3.8 Implementation Details

The core Lime package is roughly 5000 noncommented source statements, resulting in an approximately 100 Kbyte jar file. The LighTS lightweight tuple space implementation and the adapter for integrating multiple tuple space engines add an additional 20 Kbyte jar file. When using mobile agents, the μCode toolkit adds approximately 30 Kbyte in a jar file. Communication is handled completely at the socket level, requiring no support for remote method invocation (RMI) or other communication mechanisms. The latest version of Lime exploits global positioning system (GPS) to automatically trigger engagement and disengagement based on physical position, along the lines of the algorithm described in [29]. Thus far, Lime has been tested successfully on personal computers running various versions of Windows® and Linux® operating systems and exploiting both wired Ethernet as well as IEEE® 802.11 wireless technology. Moreover, Lime runs successfully on personal digital assistants (PDAs) equipped with Personal-Java™ software.

2.4 Application Example

Lime has been successfully applied in the development of several applications in the mobile environment. In this section, we present a single application, a jigsaw assembly game, throughout the development process from requirements to implementation, showing the thought process applied when designing mobile applications over Lime.

2.4.1 Requirements

The goal is to build a jigsaw assembly game for multiple players in the mobile ad hoc environment. The game should reasonably emulate the physical world process of assembling a jigsaw puzzle where an individual player starts a puzzle by dumping the pieces out of the box into a common area. Other players join and the puzzle is assembled through the joint effort of the individual players.

When considering this process in a mobile environment in which each player is equipped with a palm- or laptop computer, the following requirements must be met. First, players who are currently connected should be able to see the piece assemblies of one another as soon as possible. In other words, if one player, p_1, assembles two puzzle pieces on her laptop and another player, p_2, is connected, p_2's display should be updated quickly to show that the pieces have been assembled. Second, as this is a mobile game and the players are not expected to remain connected for the duration of the puzzle assembly, it should be possible for a player to make assemblies of pieces while disconnected. This leads to the next requirement, namely that when two previously disconnected players reconnect, their displays should be updated to show the changes made by one another. Finally, the game should be able to support multiple, concurrent puzzles, and a single player should be able to participate in more than one puzzle at a time.

2.4.2 Design and Implementation

The requirements sketched above are intentionally vague, leaving many options available. The design and implementation take into consideration the programming style encouraged by Lime as well as the constraints of the wireless, mobile environment. Here we describe the puzzle application, originally assigned as a course project, outlining first the design choices for the use of the tuple spaces and tuples, followed by the implementation of the user actions, and finishing with the updating of the user interface.

2.4.2.1 Tuple Spaces and Tuples. The first choices in the design of all Lime applications are the use of the federated tuple spaces and the format of the tuples. To support multiple concurrent puzzles, an obvious choice is to use a separate tuple space for each puzzle. This effectively separates the actions and pieces of distinct puzzles and easily allows a player to participate in as many puzzles as she would like. When a player chooses to start a puzzle, she must provide a name to be used as the tuple space name. Because the names of all active puzzles are present in the `LimeSystem` tuple space, the puzzle application can query this space and display the available games to the user. Joining a puzzle is equivalent to creating a tuple space with its name.

Because each tuple space is used as the repository for the current state of a single puzzle, the next choice is how to represent this information in tuples. This is achieved with two kinds of tuples — *image* and *assembly*. An image tuple exists for every puzzle piece and contains two fields — the identifier and the bitmap of the piece. The second kind of tuple, the assembly, represents a group of connected puzzle pieces. Each such tuple contains a single field with the list of the identifiers of the connected pieces. When a new game starts, for each puzzle piece two tuples are inserted with a corresponding `out` operation: one contains the image and one contains the representation of an assembly with a single piece, i.e., a list whose single element is the piece identifier. When two pieces are assembled, the two assembly tuples of the original pieces are replaced by a single tuple representing the change.

There are two important benefits of our choice for representing the puzzle data, thanks to the fact that, when puzzle pieces are assembled, image tuples do not change. Because these tuples contain bitmaps that are likely to be large in comparison to the assembly tuples, we save the computation time necessary to remove and reinsert the large image tuple each time an assembly is made, and we also save bandwidth as the images are not repeatedly transmitted over the wireless link.

2.4.2.2 User Actions. The next decision is where the tuples should reside throughout the federated tuple space. It is clear that when all players are connected, all the tuples representing the puzzle, both images and assemblies, are present. However, when the players are disconnected, the puzzle pieces are divided among the tuple spaces of the players and therefore are not accessible to everyone. This brings us to the first of the user operations, namely the ability to select a piece and become its owner. When a player owns a piece, that piece resides in her portion of the federated tuple space and remains with the player even after disconnection. This notion of ownership is on the assembly level, but must also extend to images. In other words, when a player takes ownership of an assembly, it must also take ownership of the images associated with the pieces of the assembly.

At the user interface level, selection of a piece is achieved by right clicking on one of the pieces displayed on the screen. To show the player which pieces have been selected by which players, we associate a color with each player and outline the selected pieces with this color, as shown in Figure 2.2. In Lime, selection is accomplished by performing `inp` operations to retrieve both the image and assembly tuples, followed by `out` operations to reinsert the tuples into the tuple space. Because we do not specify a destination location for the `out` operations, the default assigns tuples a current field equal to the new player's agent. Because the `inp` operation must

MOBILE COMPUTING HANDBOOK

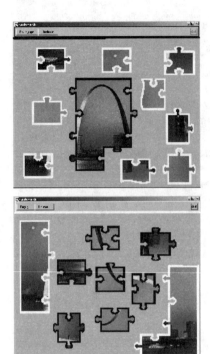 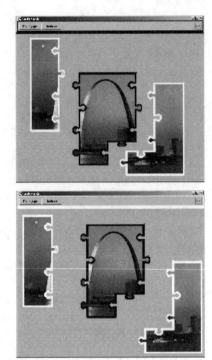

Figure 2.2. Jigsaw Assembly Game
The left two images show the puzzle trays of the black and white players while they are disconnected and able to assemble only their selected pieces. The right two images show the black and white puzzle trays after the players re-engage and see the assemblies that occurred during disconnection.

specify the (current) location from which to retrieve the tuples, this information is stored with each puzzle piece appearing on the display.

This choice to have pieces belong to players allows a player to assemble pieces while disconnected. However, it implies that a player should only be allowed to assemble pieces that she owns. It also prevents two disconnected players from using the same piece in two different assemblies. This maintains the consistency of the puzzle, despite disconnections. One can argue that because there is only one correct way to assemble a puzzle, the use of a piece in more than one assembly can easily be resolved upon reconnection of the players, thus our choice is overly restrictive. However, for the sake of the example, we have chosen to model an application where such concurrent changes are not permitted.

Piece assembly, the core of the game, is similar to selection. First, the two assembly tuples representing the pieces are removed from the player's tuple space with `inp` operations, then the new assembly tuple is written with an `out` operation. Nevertheless, concurrency issues become relevant at this point. Consider the case where player p_1 is trying to assemble pieces t_1 and t_2 while player p_2 is trying to select piece t_2 from p_1's tuple space. The following sequence of actions may take place: p_1 successfully issues the first `inp` removing t_1, p_2 successfully removes t_2 and hence becomes its owner, and finally p_1 executes its second `inp`, which at this point is bound to return `null`. In this case, p_1 cannot successfully complete the assembly and must reverse the assembly process by reinserting t_1 into the tuple space, thereby retaining consistency of the puzzle state. To the user, we indicate this failure with an audible beep.

2.4.2.3 Display Update. The main display for each player is one *puzzle tray* for each puzzle she is participating in. The requirements state that the puzzle tray must be kept up-to-date as changes are made. Unlike the previous operations, which are performed at the user's request, the updating of the puzzle tray is done in response to changes and thus is most naturally implemented with a Lime reaction. Specifically, we use a *single* ONCEPERTUPLE, ubiquitous reaction on the federated tuple space, registered for assembly tuples. When this reaction fires, the list of pieces in the assembly tuple, contained in the `ReactionEvent` object passed to the listener code, is examined. If these pieces are already in the puzzle tray, they are rearranged to reflect the connection contained in the assembly list. If they are not in the tray, the images of the pieces are retrieved with `rdp` operations and the puzzle tray is updated. Lime requires that `rdp` operations specify the current location of the tuples. In this case, the reaction contains the source of the assembly tuple; this value is used to retrieve the image tuples.

This single reaction, installed at each player's client, updates the screen in all cases including the initiation of a new puzzle, piece selection, assembly, and updating the puzzle tray upon reconnection. When a new puzzle is started, this reaction fires once for every assembly tuple, the `rdp` is executed to retrieve the image tuples, and the puzzle tray is populated with the puzzle pieces. When a player selects a piece, the `out` operation that reinserts the assembly tuple causes the reaction to fire. This time, because the graphic for the tuple has already been displayed, the image tuple is not retrieved, but the screen is updated to reflect the change in the outline color of the puzzle piece. Assembly of pieces similarly creates a new assembly tuple that is reacted to, updating the display accordingly. Finally, when two players reconnect after a disconnection, the requirements state that the puzzle trays of the players must be updated to reflect the changes made during the disconnection. Because these changes are represented in assembly tuples that are new to the previously disconnected player, the reaction fires and the display is updated. Figure 2.2 shows the appearance of the puzzle tray during disconnection and after reconnection.

It is interesting to note how a huge fraction of the application behavior is handled through a single reaction. The ability to specify asynchronous state-based reactive behavior and declaratively specify the whole system as its scope, together with the notion of transiently shared tuple spaces, greatly simplifies the programmer's task. Indeed, a look at the source code reveals that a great deal of the programming effort was devoted to proper implementation of the user interface, with only a small percentage of the code devoted to managing distribution and mobility with Lime.

2.4.3 Beyond the Puzzle

From the description, it is evident that our jigsaw assembly game embodies a pattern of interaction where the shared workspace displayed by the user interface of each player provides an accurate image of the state of all connected players, but only a weakly consistent image of the global state of the system. For instance, a user's display contains only the last known information about each puzzle piece in the tray. If a disconnected player has assembled two pieces, this change is not visible to others. However, this still allows the players to work toward achieving the global goal, i.e., the solution of the puzzle, through incremental updates of their local state.

This application is a simple game that nonetheless exhibits the characteristics of a general class of applications in which data sharing is the key element. Hence, the design strategy we exploited here may be adapted easily to handle updates in the data being shared by real applications. One example could be collaborative work applications involving mobile users, where our mechanism could be used to deal with editorial changes in sections of a document or with paper submissions and reviews evaluated by a program committee.

Other applications exhibiting diverse patterns of interaction are available, in source code form, at the Lime Web site [33].

2.5 Building Middleware Functionality on Top of Lime

In designing Lime we strived for minimality, in an attempt to identify a core of concepts and constructs general enough to be used as building blocks for higher level services and yet powerful enough to satisfy the basic needs of most mobile applications. Application development with Lime, an example of which we described in the previous section, gave us the opportunity to evaluate the expressive power of Lime constructs in building mobile applications. In this section, we report on experiences that show how Lime can be used effectively also to build high-level middleware services that, nonetheless, do not require modifications to the original middleware.

2.5.1 Transiently Shared Code Bases

In our description of Lime, we always implicitly assumed that a Lime tuple space contains data. Instead, in the work described in [25], we explored the opportunities opened by storing *code* in a Lime tuple space, while still exploiting its transient sharing and reactive features. Although the idea is simple, its implications are far reaching and hold the potential to change fundamentally the mechanisms usually exploited for supporting mobility of code.

Currently available support for mobile code is mostly limited to variations of a code on-demand approach [8] where the code is dynamically downloaded from a well-known site at name resolution time. Examples are Java applets in Web browsers and dynamic downloading of stubs in Java/RMI and Jini™ network technology. Unfortunately, in its most common incarnations this approach has at least two relevant drawbacks. First of all, the local *code base* — the set of classes locally available — is usually accessible only to the runtime support and hence it remains hidden from the applications. This prevents the development of code caching schemes with application-level policies (e.g., to intelligently cache or discard code on resource-constrained devices). Moreover, remote dynamic linking usually relies on a well-known centralized code base. This scheme evidently breaks when applied in a fluid scenario as the one defined by MANETs, but has drawbacks also in a fixed scenario, because it does not exploit the potential presence of suitable code on nearby hosts.

Using Lime tuple spaces to store code changes the situation dramatically. An agent can now manipulate its own code base using Lime primitives. Moreover, because each tuple space is permanently and exclusively associated with its agent, when the latter moves its code base migrates along with it. Finally, transient sharing effectively stretches the boundaries of an agent code base to an extent possibly covering the whole system at hand. These characteristics provide an elegant solution to the problem we mentioned earlier. A proper redefinition of the class loader, as the one described in [25], can operate on the Lime tuple space associated with the agent for which the class needs to be resolved and query it using the operations provided by Lime. Thus, the class loading mechanism can now resolve class names by leveraging off the federated code base to retrieve and dynamically link classes in a location transparent fashion (e.g., through a rd) or use location parameters to narrow the scope of searches (e.g., down to a given host or agent).

Nevertheless, the use of transiently shared tuple spaces need not be confined to the innards of the class loading mechanism, rather agents can be empowered with the ability to directly manipulate the federated code base. Hence, not only can an agent proactively query up to the whole system for a given class, but it can also insert a class tuple into the code base of another agent by using the out[λ] operation, with the semantics of engagement and misplaced tuples taking care of disconnection and subsequent reconciliation of the federated code base. This new class can then be used by the receiving agent to execute tasks in previously unknown ways or it can behave according to a new coordination protocol. Blocking operations acquire new uses, allowing agents to synchronize not only on the presence of data needed by the computation, but also on the presence of code needed to perform, or augment, the computation itself. Lime reactive operations provide additional degrees of freedom, by allowing agents to monitor the federated code base and react to changes with different atomicity guarantees. Reactions can be exploited to monitor the federated code base for new versions of relevant classes. Replication schemes can be implemented where a new class in an agent's code base is immediately replicated into the code base of all the other agents. The content of an agent's code base can be monitored to be aware of the current "skills" of the agent. The possibilities become endless.

Essentially, by exploiting the notion of transiently shared tuple space for code mobility, we defined an enhanced coordination approach that, besides accommodating reconfiguration due to mobility and providing various degrees of location transparency, enables a new form of coordination no longer limited to data exchange, but encompassing also the exchange of fragments of behavior.

2.5.2 Service Provision

Lime's flexible support for application development over ad hoc networks received renewed validation as we considered the issue of service provision, an area in which the client–server model continues to dominate. Central to supporting service provision is the notion of discovering services at runtime by relying on the service registration and discovery mechanisms. Lime made it possible to offer a solution that entails a new kind of service model built as a simple adaptation layer. The resulting veneer [10] uses Lime tuple spaces to store service advertisements and pattern matching to find services of interest and exploits the transient tuple space sharing feature of Lime to provide consistent views of the available services. The resulting system completely eliminates network awareness from the process of service discovery and utilization. The client only has to ask for the service it needs; it does not have to know how the service will be reached. Furthermore, the model provides a distributed service registry that is guaranteed to reflect the real availability of services at every moment in a mobile ad hoc environment. Consistent representation of service availability is obtained by atomically updating the view of the service repository as new connections are established or existing ones break down.

At the implementation level, a Jini-like interface [17] provides primitives for service advertisement and lookup. Every agent employs a tuple space to hold its own service registry where it advertises the services it provides. Advertisements may include proxies offering a service interface and encapsulating the communication mechanisms; the latter can be done in a manner that accommodates the mobility of both service providers and clients. As agents and hosts move, the registries of colocated agents are automatically shared. Thus, an agent requesting a service that is provided by a colocated agent can always access the service. If two hosts are within communication range, they form a community and their service registries engage, forming a federated service registry. Upon engagement, the primitives operating on the local service registry are extended automatically to the entire set of service registries present in the ad hoc network. The sharing of the service registries is completely transparent to agents as agents in the community access the federated registry via their own local registries.

The reliance on Lime concepts allowed for the fast deployment of the new service infrastructure specialized for mobile ad hoc settings with minimal programming effort. Later efforts [11] built upon this result to add secure service provision to the system by protecting tuple spaces with passwords and by using the same passwords to generate keys used to encrypt wireless traffic involving tuple spaces in general and federated registries in particular.

2.6 Related Work

A number of models and systems developed for either physical or logical mobility exhibit ideas that are somewhat similar to those put forth by Lime. Nevertheless, the concept of transiently shared tuple spaces and the semantics of reactions are unique to Lime. It is also the first system to explicitly address mobile ad hoc networks and to unite physical and logical mobility under a common coordination framework.

Distributed Linda implementations have been studied extensively, but mostly with the goal of providing fault tolerance [1, 36] and data availability [28]. These systems typically exploit replication, as opposed to transient sharing, and assume a high degree of connectivity among the nodes hosting the distributed portions of the tuple space, a property that hampers their direct use in the mobile environment.

Recent years have seen a revitalization of Linda, also from an industrial perspective. Sun and IBM have developed their tuple space implementations for client–server coordination, i.e., JavaSpaces [13] and TSpaces [12], respectively. These systems present a centralized tuple space, accessible by remote clients. It is often claimed that these systems support mobility. This is true, in that they provide the equivalent of a proxy architecture. However, as we discussed earlier, this architecture exhibits a high-degree of centralization and is inappropriate for the full-fledged mobility of ad hoc networks.

The only Linda-like system explicitly supporting physical mobility that we are aware of is Limbo [2, 35]. In this system, however, the emphasis is not on providing a general purpose programming platform for mobile computing, but on providing network-level quality of service. The information necessary to this end is stored in dedicated tuple spaces on the mobile hosts and can be made remotely accessible to agents sitting on different hosts. Interestingly, the Limbo *universal tuple space,* serving as a registry for all tuple spaces, is similar to Lime's `LimeSystem` tuple space. However, instead of describing the *current* system context, the universal tuple space remembers all tuple spaces the host has ever encountered irrespective of the current connectivity. In general, although Limbo tuple spaces may span multiple hosts, the mechanisms governing distribution (e.g., relocation of tuples) are unclear. Moreover, no form of reaction or event notification is provided.

As for logical mobility, a number of models are inspired by Linda. Nevertheless, in these models the tuple space is always exploited as a data repository explicitly accessed at a well-known location, rather than implicitly and transiently shared as in Lime. TuCSoN [21] and MARS [4] provide *programmable tuple spaces* supporting event notification and query adaptation through a notion of *reaction,* which is nonetheless rather different from that of LIME. When an agent issues a query on the tuple space, the code associated with a reaction matching the query is executed atomically, albeit asynchronously with respect to the query. Although Lime reactions form a core concept for the application programmer, MARS and TuCSoN reactions are meant to be used at the system support level to provide an intermediate adaptation layer that allows the customization of the way queries are issued and results are obtained for specific classes of agents. Moreover, a tuple space name can be fully qualified with the name of the host where it resides, hence enabling remote operations. Nevertheless, it requires connectivity and explicit agent knowledge about the tuple space location, as opposed to the Lime model that operates over the current context transparently. Furthermore, in MARS and TuCSoN, mobile agents have access only to the tuple spaces whose location they know, they do not carry tuples as they migrate, and there is no implicit data exchange among tuple spaces.

The Klaim [20] model supports a programming paradigm where code migrates during execution, using tuple spaces to provide the medium for interaction among processes. Tuple spaces have locality, but unlike in Lime, these tuple spaces are not permanently associated to a process. Instead, Klaim processes placed at a given locality implicitly interact through the colocated tuple space. There is no transient sharing among tuple spaces, but a process can explicitly interact with any tuple space by identifying its locality and a process can migrate to a new locality to interact there. Moreover, Lime leaves the details of process migration outside the model, while Klaim includes them in the formal specification making migration an integral part of the model.

As alluded to in Section 2.3, the notion of reaction put forth in Lime is profoundly different from similar event notification mechanisms such as those provided by TuCSoN, MARS, TSpaces, and JavaSpaces. In these systems, the events respond to *operations* issued by processes on the tuple spaces (e.g., out, rd, in). In Lime, instead, reactions fire based on the *state* of the tuple space itself. Further, Lime reactions execute as a single atomic step and cannot be interrupted by other operations. This makes it straightforward for a single Lime reaction to probe for a tuple, react if it is found, and register a reaction if it is not. This same operation in the other systems requires a transaction. Finally, the atomicity of strong reactions increases the power of Lime reactions. For example, with a strong, local reaction, the execution of the listener is guaranteed to fire in the same state in which the matching tuple was found. No such guarantee can be given with an event model where the events are asynchronously delivered. Nonetheless, Lime supports also this second approach through weak reactions.

As work on Lime becomes increasingly recognized, it is being used also as a basis for alternative models. At Purdue University, a group extracted the features of Lime necessary for mobile agents by removing host-level sharing and created a model referred to as CoreLime [5]. On top of this restricted model, they proposed some initial ideas for tuple space security. A group at the University of Bologna proposed a calculus-based specification [3] of a model that embeds choices different from the original Lime, including reacting to tuple space operations instead of tuple space contents and blocking agents that generate tuples destined for disconnected agents rather than creating misplaced tuples.

As we conclude this section, we note that the effort that went into developing Lime also contributed to the emergence of a more abstract and general coordination concept and methodology called *Global Virtual Data Structures* (GVDS) [27]. It is centered on the notion of constructing individual programs in terms of local actions whose effects can be interpreted at a global level. A Lime group, for instance, can be viewed as consisting of a global set of tuples and a set of agents that act on it in some constrained manner. The set has a structure that changes in accordance with a predefined set of policies. This structure governs the specific set of tuples accessible to an individual agent through its local interface at any given point in time. The analogy to the concepts of virtual memory and distributed shared memory are strong and other research projects have picked up the GVDS theme and instantiated it in their own unique ways. The XMIDDLE [15] system developed at University College of London, for instance, presents the user with a tree data structure based on Extensible Markup Language (XML) data. When connectivity becomes available, trees belonging to different users can be composed, based on the node tags. Upon disconnection, operations on replicated data are still allowed and their effect is reconciled when connectivity is restored. In addition, PeerWare [7], a project at Politecnico di Milano, exploits a tree data structure, albeit in a rather different way. In PeerWare, each host is associated with a tree of document containers. When connectivity is available, the trees are shared among hosts, meaning that the document pool available for searching under a given tree node includes the union of the documents at that node on all connected hosts.

2.7 Conclusions

In this chapter, we described Lime, a computational model and middleware specifically designed to support logical mobility of agents and physical mobility of hosts in both wired and wireless settings. Lime reinterprets the notion of tuple space introduced by Linda and adapts it in an original fashion to the mobile environment. Transparent management of tuple space sharing, contingent on connectivity, offers an effective context awareness mechanism although reactions provide an effective and uniform vehicle for responding to context changes regardless of their nature or trigger. The net result is a simple model with precise semantics and applicability in a wide range of settings, from mobile agent systems operating over wired networks, at one extreme, to mobile ad hoc networks lacking any infrastructure support, at the other. The experience to date with building applications and higher level middleware layers with Lime is encouraging and appears to confirm the value of the conceptual and technological tools put forth by Lime.

2.7.1 Availability

Lime continues to be developed as an open source project, available under GNU's Library General Public License (LGPL). Source code and development notes are available at `lime.sourceforge.net`.

Acknowledgments

This research was supported in part by the National Science Foundation under Grant no. CCR-9970939. Any opinions, findings, and conclusions or recommendations expressed in this chapter are those of the authors and do not necessarily reflect the views of the research sponsors.

Notes

1. Additionally, Linda implementations often include also an `eval` operation, which provides dynamic process creation and enables deferred evaluation of tuple fields. For the purposes of this work, however, we do not consider this operation further.
2. Hereafter, we often do not mention this pair of operations. They are useful in practice, but do not add significant complexity either to the model or to the implementation.
3. Exceptions are not shown for the sake of readability.
4. In the following, we use the term unit when we want to refer to agents and hosts, without making a distinction.

5. Specifying a destination location λ implies neither guaranteed delivery nor ownership of the tuple *t* to λ. Linda rules for nondeterministic selection of tuples are still in place; thus, it might be the case that some other agent may withdraw *t* from the tuple space before λ, even after *t* reached λ's ITS.
6. The semantics of reactions are based on the Mobile UNITY reactive statements [16]. The reader interested in formal details is redirected to [19].

References

1. D.E. Bakken and R. Schlichting. Supporting fault-tolerant parallel programming in Linda. *IEEE Trans. on Parallel and Distributed Systems*, vol. 6, no. 3, pp. 287–302, 1994.
2. G. Blair, N. Davies, A. Friday, and S. Wade. Quality of service support in a mobile environment: An approach based on tuple spaces. In *Proc. of the 5th IFIP Int'l. Workshop on Quality of Service (IWQoS'97)*, May 1997.
3. N. Busi and G. Zavattaro. Some thoughts on transiently shared dataspaces. In *Proc. of the Workshop on Software Engineering and Mobility*, colocated with the 23rd Int'l. Conf. on Software Engineering (ICSE), 2001.
4. G. Cabri, L. Leonardi, and F. Zambornelli. MARS: A programmable coordination architecture for mobile agents. *IEEE Internet Computing*, vol. 4, no. 4, pp. 26–35, 2000.
5. B. Carbunar, M.T. Valente, and J. Vitek. Lime revisited: Reverse engineering an agent communication model. In *5th Int'l. Conf. on Mobile Agents (MA2001)*, G.P. Picco, Ed., Atlanta, December 2001.
6. N. Carriero, D. Gelernter, and L. Zuck. Bauhaus-Linda. In *Object-Based Models and Languages for Concurrent Systems*, LNCS 924, Springer, 1995.
7. G. Cugola and G.P. Picco. PeerWare: Core middleware support for peer-to-peer and mobile systems. Technical report, Politecnico di Milano, Italy, 2001. Available at www.elet.polimi.it/upload/picco.
8. A. Fuggetta, G.P. Picco, and G. Vigna. Understanding code mobility. *IEEE Trans. on Software Engineering*, vol. 24, no. 5, 1998.
9. D. Gelernter. Generative communication in Linda. *ACM Computing Surveys*, vol. 7, no. 1, pp. 80–112, January 1985.
10. R. Handorean and G.-C. Roman. Service provision in ad hoc networks. In F. Arbab and C. Talcott, Eds., *Proc. of the 5th Int'l. Conf. on Coordination Models and Languages*, Lecture Notes in Computer Science, vol. 2315, pp. 207–219. New York: Springer-Verlag, 2002.
11. R. Handorean and G.-C. Roman. Secure sharing of tuple spaces in ad hoc settings. In *Proc. of the 1st Int'l. Workshop on Security Issues in Coordination Models, Languages, and Systems (SecCo 2003)*, Electronic Notes in Theoretical Computer Science (ENTCS), 2003.
12. IBM. TSpaces Web page. http://www.almaden.ibm.com/cs/TSpaces.
13. JavaSpaces. The JavaSpaces Specification Web page. http://www.sun.com/jini/specs/jini1.2htm/js-title.html.
14. J.J. Kistler and M. Satyanarayanan. Disconnected operation in the Coda file system. *ACM Trans. on Computer Systems*, vol. 10, no. 1, pp. 3–25, 1992.
15. C. Mascolo, L. Capra, S. Zachariadis, and W. Emmerich. XMIDDLE: A data-sharing middleware for mobile computing. *Kluwer Personal and Wireless Communications Journal*, vol. 21, no. 1, April 2002.
16. P.J. McCann and G.-C. Roman. Compositional programming abstractions for mobile computing. *IEEE Trans. on Software Engineering*, vol. 24, no. 2, pp. 97–110, 1998.
17. Sun Microsystems. Jini Web page. http://www.sun.com/jini.
18. A.L. Murphy, G.P. Picco, and G.-C. Roman. LIME: A middleware for physical and logical mobility. In F. Golshani, P. Dasgupta, and W. Zhao, Eds., *Proc. of the 21st Int'l. Conf. on Distributed Computing Systems (ICDCS-21)*, pp. 524–533, May 2001.

19. A.L. Murphy, G.P. Picco, and G.-C. Roman. Lime: A coordination middleware supporting mobility of hosts and agents. Technical Report WUCSE-03-21, Department of Computer Science, Washington University, St. Louis, MO, May 2003.
20. R. De Nicola, G. Ferrari, and R. Pugliese. Klaim: A kernel language for agents interaction and mobility. *IEEE Trans. on Software Engineering*, vol. 24, no. 5, pp. 315–330, 1998.
21. A. Omicini and F. Zambonelli. Tuple centres for the coordination of Internet agents. In *Proc. of the 1999 ACM Symp. on Applied Computing (SAC'00)*, February 1999.
22. G.P. Picco. µCode Web page. mucode.sourceforge.net.
23. G.P. Picco. LighTS Web page. lights.sourceforge.net.
24. G.P. Picco. µCode: A lightweight and flexible mobile code toolkit. In *Proc. of the 2nd Int'l. Workshop on Mobile Agents* (MA98), Lecture Notes in Computer Science, vol. 1477. New York: Springer-Verlag, 1998.
25. G.P. Picco and M.L. Buschini. Exploiting transiently shared tuple spaces for location transparent code mobility. In F. Arbab and C. Talcott, Eds., *Proc. of the 5th Int'l. Conf. on Coordination Models and Languages*, Lecture Notes in Computer Science, vol. 2315, pp. 258–273. New York: Springer-Verlag, 2002.
26. G.P. Picco, A.L. Murphy, and G.-C. Roman. Lime: Linda Meets Mobility. In D. Garlan, Ed., *Proc. of the 21st Int'l. Conf. on Software Engineering*, pp. 368–377, May 1999.
27. G.P. Picco, A.L. Murphy, and G.-C. Roman. On global virtual data structures. In D. Marinescu and C. Lee, Eds., *Process Coordination and Ubiquitous Computing*, pp. 11–29. Boca Raton, FL: CRC Press, 2002.
28. J. Pinakis. Using Linda as the basis of an operating system microkernel. PhD thesis, University of Western Australia, Perth, Australia, August 1993.
29. G.-C. Roman, Q. Huang, and A. Hazemi. Consistent group membership in ad hoc networks. In *Proc. of the 23rd Int'l. Conf. on Software Engineering*, pp. 381–388, Toronto, May 2001.
30. G.-C. Roman, A.L. Murphy, and G.P. Picco. Coordination and Mobility. In A. Omicini, F. Zambonelli, M. Klusch, and R. Tolksdorf, Eds., *Coordination of Internet Agents: Models, Technologies, and Applications*, pp. 254–273. New York: Springer-Verlag, 2000.
31. D.S. Rosenblum and A.L. Wolf. A design framework for Internet-scale event observation and notification. In *Proc. of the 6th European Software Engineering Conf. held jointly with the 5th ACM SIGSOFT Symp. on the Foundations of Software Engineering (ESEC/FSE97)*, Zurich, September 1997. Lecture Notes in Computer Science, vol. 1301, New York: Springer-Verlag, 1997.
32. A. Rowstron. WCL: A coordination language for geographically distributed agents. *World Wide Web Journal*, vol. 1, no. 3, pp. 167–179, 1998.
33. Lime Team. Lime Web page. lime.sourceforge.net.
34. D. Terry, M. Theimer, K. Petersen, A. Demers, M. Spreitzer, and C. Hauser. Managing update conflicts in Bayou, a weakly connected replicated storage system. *Operating Systems Review*, vol. 29, no. 5, pp. 172–183, 1995.
35. S.P. Wade. An investigation into the use of the tuple space paradigm in mobile computing environments. PhD thesis, Lancaster University, England, September 1999.
36. A. Xu and B. Liskov. A design for a fault-tolerant, distributed implementation of Linda. In *Digest of Papers of the 19th Int'l. Symp. on Fault-Tolerant Computing*, pp. 199–206, June 1989.

Chapter 3
Pervasive Application Development: Approaches and Pitfalls

Guruduth Banavar, Norman Cohen, and Danny Soroker

Abstract

In this chapter, we examine the challenges in building pervasive-computing applications — applications that support mobility and context-awareness. We summarize the key approaches being employed to address the difficulties and point out the pitfalls those approaches are trying to avoid. Using the model-view-controller (MVC) structure of applications, we identify four techniques being used to address complexity — device-independent views, platform-independent controllers, host-independent models, and source-independent context data.

3.1 What Are Pervasive Applications?

The vision of pervasive computing has been written about extensively. In a nutshell, pervasive computing is about enabling users to get access to the relevant applications and data at any location and on any device, in a manner that is customized to the user and the task at hand. This fundamentally takes computing off the desktop and into the spaces that we live in everyday. Mark Weiser [Wei91] called it "invisible" computing. This vision of pervasive computing leads to two fundamental characteristics of pervasive applications — *mobility* and *context-awareness*. Both of these characteristics are a result of the extremely dynamic nature of pervasive computing environments.

Mobility has three implications. First, applications must run on a wide variety of devices, including the devices embedded in various environments and devices carried by users. Second, because devices may be transported to locations where a high-bandwidth network connection is not available, applications must work (perhaps in a degraded mode) with low-bandwidth network connections or in the absence of any network connection. Third, applications that make use of a user's location must account for the possibility that the location will change.

The need for context-aware applications arises because pervasive computing makes applications available in contexts other than a computer workstation with a keyboard, mouse, and screen. The users of a pervasive-computing application will typically be focused upon some task other than the use of a computing device and may even be unaware that they are using a computing device. Applications must customize themselves to interact with a user in a manner appropriate to the user's current context and activities, exploiting locally available devices, without distracting the user from the task at hand.

In the simplest case, a mobile application is any standalone application that can execute on a mobile device. However, the more interesting and useful case is an application that is networked to other software components executing at different points in the network infrastructure. Ideally, an application is hosted on the network and is able to execute on any device. In this case, the application must be written in such a way as to be able to execute on multiple software platform architectures and in a manner that exploits the user interface characteristics of multiple device platforms. Although we have made great strides in network connectivity, not all devices and not all locations support continuous network connectivity. Thus, applications should support disconnected and weakly connected operation. In summary, supporting mobility implies two major technological requirements — supporting device platform heterogeneity and supporting network heterogeneity.

A context-aware application is one that is sensitive to the environment in which it is being used (e.g., the location or the particular user of the application). The application can use this information to customize itself to the particular location or the user. This implies the following technological requirements:

- Identifying and binding to data sources that provide the right information
- Composing the information from these sources to create information that is useful for an application
- Using that information in meaningful ways within the application itself

Pervasive Application Development: Approaches and Pitfalls

As a simple example, a pervasive calendar application will have the following features. First, the application will be able to run on multiple device platforms, from a networked phone (with a limited user interface and limited bandwidth, but always connected) to a smart personal digital assistant (PDA) (with a richer user interface and higher bandwidth, but not always connected) to a conference room computer (with a very rich user interface and very high bandwidth and always connected). Furthermore, I (as the user) should be able to interact with this application using multiple user interface modalities, such as a graphical user interface (GUI), a voice interface, or a combination of the two. Second, the application will be sensitive to the environment in which it is running; for example, if I bring up the calendar at home, the application might bring up my family calendar by default. If I bring up my calendar in my office when I'm almost late for my next meeting, the application might bring up my work calendar with the information about my next meeting highlighted.

In this chapter, we discuss the software engineering challenges and approaches to building pervasive applications with the characteristics mentioned above. This chapter considers the application developer's point of view (as opposed to the infrastructure developer's) and discusses the programming models and tools that can support pervasive application development. The purpose of this chapter is not to propose new techniques for addressing development issues, but rather to summarize some of the promising approaches already being developed and to point out some of the pitfalls that these approaches are trying to avoid. The software infrastructure elements for supporting the execution of such applications are discussed elsewhere [Ban00, Ban02, Coh02a, and Coh02b].

3.1.1 Basic Concepts and Terms

A *multi-device application* is one that is able to execute on devices with different capabilities. A *multimodal application* is one that supports multiple user interface modalities such as GUI, voice, and a combination of the two.

In this chapter, we consider an application model in which the application is partitioned according to the well-known MVC application structure [Kra88]. The *view* represents the presentation, and the *controller* represents the application flow, including the navigation, validation, error handling, and event handling. The view and the controller together deal with the user interaction of the application. The *model* component includes the application logic as well as the data underlying the application logic.

In this chapter, we consider only networked applications, because they represent the bulk of interesting and useful pervasive applications. In these networked applications, the application components described above are distributed across two or more physical computers with a network connection between them. A *device platform* is the distributed software platform to which a pervasive application is targeted. A *thin-client application* is a networked application in which the user interface rendering component is executing on the user's device, whereas the rest of the application is executing on a networked computer. A *thick-client application,* on the other hand, has significant application components executing on the user's device. A *disconnectable application* is one that is able to continue to execute when there are different levels of connectivity between the different components of the application.

The attributes of the environment of an application are referred to as the *context* of the application. The context of an application includes some of the user's significant attributes, such as location, destination, the identities of other people in the vicinity, and the attributes of the task being performed, such as the objective and the artifacts necessary for the task. A *context-aware application* is one that is able to sense some aspects of the environment in which it is executing and adapt its behavior to the sensed environment.

3.2 Why Is It Difficult to Develop Pervasive Applications?

There are fundamental reasons why pervasive application development is more difficult than conventional application development. One reason is the heterogeneity of environments in which a pervasive application must be able to execute. The other reason is the need for applications to adapt to dynamic environments. These reasons are discussed in more detail in this section, as are the software engineering issues that arise from them.

3.2.1 Heterogeneity of Device Platforms

End user devices, such as smart phones and PDAs, come in many varieties and have widely varying capabilities, both hardware (form factor, user interface hardware, processor, memory, and network bandwidth) and software (operating system, user interface software, services, and applications). These capabilities are so varied and broad that there are industry standards (e.g., Composite Capabilities/Preferences Profiles [CC/PP] and User Agent Profile [UAProf], by the W3C Consortium [Kly03]) being developed for describing the capabilities of individual devices. There are commercial offerings that support and maintain several hundred device profiles, with new devices being introduced at the rate of more than one every week at the current time. Furthermore, the number of applications that need to support a nontrivial number of these devices is on the rise.

The impact of device heterogeneity on application developers is that applications need to be developed (or ported) to each device and maintained separately for each device. In terms of the MVC Application Model described before, the following sections describe the specific impacts of device heterogeneity.

3.2.1.1 User Interface. The capabilities of the user interface include the output capabilities, such as the screen characteristics (e.g., size and color); the input capabilities, such as the number of hard buttons, rollers, and other controls; and the software toolkit available to manipulate these input and output capabilities. Because of differences in these capabilities from one device to another, the view component of an application will have to be rewritten for each device. In some cases, the structure of the view will also impact the structure of the controller.

3.2.1.2 Interaction Modalities. Informally, an interaction modality is a significant method of user interaction that leverages a user's natural or learned ability. Examples are keyboard or mouse, speech, pen, and tactile interfaces. (In this chapter, we consider primarily keyboard or mouse and speech.) The view and controller portions of applications may need to be significantly rewritten to enable each modality. For example, a speech-based application could have a different structure from a GUI-based application.

Furthermore, *multimodal* interfaces can use multiple modalities within a single application. For example, a single application may use GUI and speech modalities to reap the benefits of both modalities — GUI for rapid interaction and speech for eyes-free and hands-free operation. Writing such an application requires synchronizing the two modalities, so that when a particular utterance is played, the corresponding elements are displayed on the screen. This synchronization requires careful attention by the application developer.

3.2.1.3 Platform Capabilities. The *software platform* for a device is the distributed software infrastructure on which an application executes, including the device software infrastructure and the server software infrastructure. In many cases, the programming models on the device and the server are different, for example, a Java™-based Web programming model on the server and a C-based Application Processing Interface (API) on the device. Even if the programming models are the same on the device and the server, an application may need to be partitioned differently between the device and the server depending on the processor, memory, and network capabilities of a device.

3.2.1.4 Connectivity. If an application needs to execute in a dynamic environment that supports multiple levels of connectivity, the application developer needs to worry about dynamically varying the partitioning of the application between the various connectivity scenarios and resynchronizing partitioned components after reestablishing connectivity. This adds a significant amount of complexity to the application development process.

3.2.1.5 Development and Maintenance Complexity. To summarize the above discussion, there are several software engineering challenges in writing pervasive applications. Consider the development scenario for a pervasive application targeted to N device platforms. In the worst case, this requires one to build N different versions of the application. If the application is targeted to O devices that support multimodal interaction, there will be further complexity in developing versions of the application for separate modalities and for synchronizing the application across those modalities. There may be P different partitions of the application to support various platform and connectivity characteristics. Thus, the worst case development complexity for a single application is a factor of $(N + O) * P$ times the complexity of developing the application for a single platform. This results in significant increase in developer time, which is the costliest resource in a software development organization. Maintenance of the application (i.e., fixing bugs and making enhancements) has a similar complexity.

This complexity is fundamentally a scalability issue. Conventional application development methodologies do not scale for the large numbers of devices and platforms that are in existence today. To address this issue, new methods of reusing application components are being developed. These will be discussed later in this chapter.

3.2.2 Dynamics of Application Environments

In describing our vision earlier, we stated that pervasive applications should be customized to the user and task at hand — also referred to as the context of the application. The context can be highly dynamic. The data sources that provide information about the application's environment are called *context sources*. Consider the complexities of application development in the face of dynamic and heterogeneous context sources.

The context data from different context sources could have different schemas and formats. For example, location data from a cell tower is different from the location data from an IEEE® 802.11 base station. If each pervasive application that uses context data were responsible for collecting and normalizing context data from different sources, applications would indeed be quite complex.

The context information from any one source could be too low-level to be useful for an application. For example, if an application is interested in knowing whether Jane is at lunch, it is not enough to know Jane's exact latitude or longitude, but also how that lat or long corresponds to a building's map (also known as *geocoding*). If Jane's exact status is not available, it may be possible to determine whether she is at lunch from other context sources, such as the time of day, her calendar, the lights in her office, the activity on her computer, and knowledge of her normal habits. Combining these lower level forms of context into a higher level notion of "Is Jane at lunch?" should not be the responsibility of the application that requires that information.

The actual context sources themselves could be highly dynamic. For example, the location of a person can be obtained by a multitude of sources, including a cell tower, a telematics gateway, a wireless local area network (LAN) hub, and an active-badge access point. Each of these sources of location may have a different API and may be more or less applicable to different locations. Applications should not be responsible for discovering these context sources and explicitly binding to them.

In summary, the complexity of using dynamic context information boils down to the question of division of responsibility between the application and a reusable infrastructure. The reusable functions of mediation [Wie92] (including normalization, composition, and binding) should be supported by the infrastructure. The application should only be responsible for implementing the business logic, given the high-level context event.

3.3 Approaches for Developing Pervasive Applications

3.3.1 Developing Mobile Applications

Mobile applications may be standalone applications that run on mobile devices; they may be networked applications executing partly on the mobile device and partly in a networked server environment (which, by the way, does not imply that network connectivity is always available). This chapter focuses on the latter variety, because it is more relevant to realizing the pervasive computing vision. Web-based applications, whether they are browser-based or use standalone renderers that access Web services on a network, are examples of this kind of application. To understand the most common approaches to developing such mobile applications, let us keep in mind the MVC decomposition of an application.

As described earlier, the basic problem of mobile application development to multiple devices, modalities, and connectivity environments is that of complexity, because the same application may have to be rewritten multiple times. The following is a discussion of the approaches that are being used to address this problem.

3.3.2 Presentation Transcoding

An early approach to making Web applications accessible via multiple devices was transcoding. The basic idea behind transcoding is to repurpose existing content written for one device, say a desktop personal computer (PC), to different devices, via an automated runtime transcoder, typically on a server. This might involve parsing the presentation, typically represented in Hypertext Markup Language (HTML) and converting it into a markup language that is understood by a Web browser on the device, such as Website META Language (WML) [WML02] or compact HTML (cHTML) [Kam98]. In this process, images and other multimedia content may also be transformed into a format that can be handled by the target device.

This approach works to a limited extent, but has not been widely adopted in the industry. There are several reasons for the limited success of this approach:

- The input does not convey the full semantics of the content, but only the presentation, so transcoders can do no more than reformat the content in ways that are usable and pleasing to the end user on different devices.
- Content authors have little to no control on how a Web page is displayed on a device.
- Content providers are usually protective of their content and do not want runtime intermediaries to alter the carefully tailored presentation that was originally designed.

Enhancements to the basic idea of transcoding included the ability for the developer to *annotate* the content with some of the semantics behind the content. Although this may be reasonable in some cases where there is static content, this notion breaks down when there is dynamic content. Transcoding was not widely adopted because it fundamentally does not handle the deeper structure and semantics of applications.

3.3.3 Device-Independent View Component

A more widely used approach evolved, in which the view aspects of an application are conveyed in a device-independent representation. This device-independent representation describes the intent behind the user interaction within a view component (such as a page), rather than the actual physical representation of a user-interface control. For example, the fact that an application requires users to input their ages is represented by a generic INPUT element with a range constraint; an adaptation engine determines, based on the target device characteristics, usability considerations, or user preferences, whether the INPUT element should be realized as a text field, a selection list, or even voice input. Several device-independent view representations have evolved over the years, including User Interface Markup Language (UIML) [Abr99], Abstract User Interface Markup Language (AUIML, previously known as Druid) [Mer99], XForms [Dub03], and Microsoft® ASP.NET Mobile Controls [Mic03].

3.3.3.1 Runtime Adaptation. This device-independent representation is typically converted to a device-specific representation via some kind of automatic runtime adaptation. The runtime adaptation engine gets the device identifier via the request header of a Web application (specifically, the user agent field) and maps that to a database record containing detailed device information. The information in this database record guides the adaptation of the device-independent representation to device specific representations. Microsoft, Oracle, and Volantis have commercial products using some variation of runtime adaptation.

One of the pitfalls of this approach is to rely entirely on automatic runtime adaptation of the device-independent representation. Fully automatic adaptation can work in certain cases: when the content is simple or when the device variations are not too great. However, experience shows that it is extremely difficult for fully automatic adaptation to produce highly customized and usable interfaces that are comparable to hand-crafted user interfaces. This is especially true in modern, highly interactive applications. As a result, most successful systems that use this technique provide a way for developers to provide additional information to guide or augment the runtime adaptation process. The extra information can take several forms:

- Meta-information (e.g., where to split content into multiple pages)
- Style information (e.g., templates and style attributes to use for different devices or classes of devices)
- Code modules that plug into the runtime adaptation engine and alter its behavior for particular target devices

3.3.3.2 Design-Time Adaptation. Design-time adaptation is a technique that converts the device-independent representation to device-specific representations before the application is deployed to the runtime. The result of design-time adaptation is a set of target-specific artifacts that can be viewed and manipulated by the developer. At the end of this process, the developer ends up with a set of target-specific view components, similar to the components that a developer would have built by hand [Ber02]. There are two major advantages to this approach:

- The developer has full control over the adaptation process and the generated artifacts. If the developer is not satisfied with the output, the process can be rerun with different parameters, until the result is satisfactory. The generated artifacts can also be manipulated to add device-specific capabilities for particular devices.
- There is no runtime performance overhead for translating applications, because the translations have occurred at design time.

In the design-time adaptation technique, applications are converted from a higher level to a lower level representation and the generated representation can be manipulated by the developer. In this scenario, if the developer modifies the higher level representation and regenerates the application, it is critical that the changes made previously to the lower level representation be preserved. This preservation of changes to a generated artifact after the artifact is regenerated is called "round-trip" [Med99]. Failure to enable round-trip is a potential pitfall of the design-time adaptation technique. There are multiple ways to support round-trip. One way is to provide markers in the generated artifacts that indicate where the developer can make modifications. The developer modifications made within these markers are left untouched by the generation process. The other approach is to capture the history of changes to a generated artifact and to provide the capability to reapply these changes selectively to regenerated artifacts.

Design-time adaptation alone cannot be relied upon, for two reasons:

1. Design-time adaptation only supports devices that were known at design time. If there are new devices that need to be supported after an application has been deployed, it may not be reasonable to depend on the application provider to target those devices via the design-time tool.
2. For dynamic content (again, that will be unknown at design time), it is necessary to have some level of runtime adaptation.

For these reasons, some systems, such as Multi-Device Authoring Tool (MDAT) [Ban04] support a hybrid of design-time and runtime adaptation. Design-time adaptation results in one or more device-specific application versions that can be deployed to a Web application server. Additionally, devices can be classified into a hierarchy of device categories (e.g., PDAs, phones, color phones, and so on) and the application can be adapted at design time according to this classification. When a device requests the application, the runtime Web application dispatcher determines if the request can be satisfied by an existing device-specific application version or whether it falls into a category that has been defined. If not, a device-specific version of the application is generated on the fly and delivered to the device. Thus, runtime adaptation allows MDAT to service requests from devices that do not have a predefined device-specific application version.

3.3.3.3 Visual Tools for Constructing Device-Independent Views.

Regardless of the adaptation technique used, systems supporting device-independent views also provide a number of integrated development environment tools for authoring the device-independent content and for specifying the additional kinds of information described above. Consider a visual design tool for developing device-independent content. Typically, visual design tools for developing concrete device-specific content support the well-known What You See Is What You Get (WYSIWYG) paradigm. One pitfall that a visual design tool for device-independent content can fall into is

Pervasive Application Development: Approaches and Pitfalls

to attempt to support WYSIWYG capability. In a device-independent content tool, what the user sees is not what the user is going to get in general, because only a single device can be emulated in the interface. The user may be tempted to customize the design for one particular device, rather than thinking about the overall intent that is appropriate for all devices. A device-independent representation should thus be editable in an editor that displays a generic logical representation that conveys the relationships among elements, such as order, grouping, and any layout hints that may be specified. These issues are discussed in detail in [Ber01].

3.3.4 Platform-Independent Controller Component

The section above discussed adaptation of the view component of an application to multiple devices. As described earlier, the controller of an application represents the control flow, including data validation and error handling, typically via event handlers. To address the full range of applications, it is necessary to consider the role of the controller in modern interactive applications. There are several reasons why the controller of an application needs to be targeted to multiple devices:

- Different devices may have different input hardware, ranging from a keyboard, tracking device, and microphone on a PC to a pair of buttons and a scrolling wheel on a wristwatch.
- The flow of an application may be different on different devices. For example, an application that contains a secure transaction may not support this transaction on a device that does not have the appropriate level of security infrastructure. Similarly, an application that supports rich content may choose to skip those pages on devices that are not capable of presenting rich content.
- When a device-independent page is adapted and rendered on multiple devices, the page may be split into multiple device-specific pages for any device that is too small to contain the entire page.
- The controller execution framework may be different for different device platforms. Recall that a device platform is the end-to-end distributed platform that supports the execution of all components of the application. One device platform may support a Java-based Apache Struts™ framework, whereas another may support a different framework such as the base servlet framework, or a different language altogether, such as PHP or C$^\#$.

As a result, a complete solution for targeting multiple devices must include the application controller. One approach [Ban04] is to represent the controller in a declarative way using a generic graph representation, where the nodes are device-independent pages and the arcs are control flow transitions from one page to another. This representation addresses the three requirements above as follows:

1. Developers can modify the flow of the application for particular target devices. These are represented as incremental changes to the generic controller.
2. When a device-independent page is split into multiple pages, the appropriate controller elements to navigate among those pages are also automatically generated.
3. The concrete controller code for specific controller platforms (e.g., Apache Struts) is automatically generated from the declarative controller representation. The specific controller framework can be changed as necessary.

3.3.5 Host-Independent Model Component

The above sections discussed approaches for targeting the view and controller components of an application to multiple devices. In this section, we discuss how to deal with the heterogeneity of connectivity environments.

Networked mobile applications vary in the distribution of logic and data between the mobile device and the server, as illustrated in Figure 3.1.

In a thin-client application, views are generated on the server and then rendered on the client device by a component such as a Web browser. Controller logic, model logic, and model data all reside on the server, so disconnected operation is impossible. In a thick-client application, the model still resides on a server, perhaps accessed through Web services, but the rest of the application resides on the client device. Caching of data before connection and queuing of updates to be performed upon reconnection enable

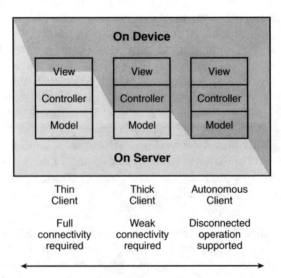

Figure 3.1. Distribution of Logic and Data between the Mobile Device and the Server

limited forms of offline operation in a weakly connected environment. The operations allowed are those that can proceed sensibly in the absence of a complete and current model. An autonomous-client application resides entirely on the client device. It maintains its own fully functional model, which may be synchronized from time to time with replicas of the model on a server. As the arrow at the bottom of the diagram suggests, thin-client, thick-client, and autonomous-client applications represent points on a continuum rather than three clearly delineated categories. For example, some nearly autonomous applications have a disconnected mode that closely resembles the connected mode, except that updates made to the model are considered tentative until the model is synchronized with a server-based replica.

Thick-client applications have been supported with varying levels of success. The main drawback of the thick-client approach is that it may not support all the needed functions to support rich interactions in disconnected mode, because the model component is missing on the mobile device. The autonomous-client approach, on the other hand, is the most general technique, because it can support varying levels of connectivity. The remainder of this section is concerned with autonomous-client applications.

A key consideration here is the programming model used for supporting disconnectable applications. We need a programming model that allows the model components of an application (like the view and controller components) to be shared by multiple versions of a disconnectable application. In this case we are concerned with connected and disconnected versions of the application.

In the ideal scenario, the logic and the data for the model component is specified once and the tools and infrastructure supporting the programming model extract the right subset of the logic and data for the disconnected mode on each supported device. In reality, this extraction process will likely need to be guided extensively by the developer. The developer will likely specify the model, view, and controller in a generic way (view and controller as described in previous sections). The tools will enable the developer to incrementally refine this generic representation to particular target environments. This is an ongoing area of work and there are significant issues that need to be resolved.

It should be noted that there is a significant level of runtime infrastructure needed for disconnectable applications:

- An application hosting and execution environment is needed on the mobile device.
- If application code is to be downloaded from the server to clients upon demand, a code-migration component is needed on the server

and device sides to coordinate the partitioning and loading of application components.
- A data synchronization component is needed for updating both the device and server instances of the application with changes to the data on the other sites and to resolve any possible conflicts.

There are difficult architectural and policy issues in the above infrastructure components. A discussion of these issues is beyond the scope of this chapter.

3.3.6 Developing Context-Aware Applications

One can think of a context-aware application as having a triggering aspect and an effecting aspect [Sow03]. The triggering aspect binds to data sources, collects data, analyzes the data, and ensures that the data is relevant to the application. If so, it notifies the effecting aspect, which takes the action corresponding to the trigger. For example, in an application that invokes a computer backup facility when a user Jane is away from her computer, the event that "Jane is at lunch" is the trigger and the act of invoking the backup utility is the effect.

Recall that context-aware applications have three sources of complexity:

1. The heterogeneous nature of data sources
2. The dynamic nature of context sources
3. The multiple sources of potentially low-level context data

Observe that these are all in the triggering component of applications. Current approaches to addressing these issues have focused on creating a reusable infrastructure (middleware or toolkit) that exposes a programming model that hides these complexities [Coh02a]. This approach is summarized below.

3.3.7 Source-Independent Context Data

An application obtaining data from heterogeneous sources with inconsistent availability and quality of service should not name a specific source of data. Rather, it should describe the kind of data that is required, so that the underlying infrastructure can discover an appropriate source for the data. This approach, known as *descriptive, data-centric,* or *intentional* naming ([Adj99], [Bow93], [Int00]), has a number of advantages. It allows the system to select the best available source of data, based on current conditions. If the selected source should fail, the infrastructure can rebind to another source satisfying the same description, thus making the application more robust. New data sources satisfying a description can be introduced, or old data sources removed, without modifying the application; likewise, the application can be ported to an environment in which there is a different set of sources for the described data.

Pervasive Application Development: Approaches and Pitfalls

The basic idea of this approach is for an application to specify the desired context data without specifying the exact location and data type of the source, or whether it is coming from multiple sources. These are considerations that will be handled transparently by the infrastructure. In some cases, the infrastructure may discover a data source, such as a device or a Web service that directly provides the described data. For example, suppose an application specifies that it is interested in a Boolean value for "Is Jane at lunch?" The infrastructure may discover a data source that directly reports whether Jane's location is the cafeteria. Alternatively, the infrastructure may discover a programmed component, called a *composer* in [Coh02a], which computes the described data from other data. In our example, some combination of Jane's calendar, office status, and computer status might be combined by a composer to determine with a degree of certainty whether she is at lunch. A composer may be reusable across multiple applications and may itself be built on top of other composers that handle lower level, more generic, data. For example, the query "Is X at lunch?" could be answered using the answer to a query of the form "Is X located at Y?" and queries of that form might themselves be answered by consulting multiple sources of location data (e.g., active badge, 802.11, or cell tower) with different resolutions, and inferring a composite location with a certain degree of confidence.

Once a composer that can answer the question "Is X at lunch?" is written and added to the infrastructure, it can be reused by all context-aware applications. A composer is itself a data source, just like a sensor, a Web service, or a database, and may be discovered by the infrastructure in response to a query for data satisfying a given description.

Some data sources, such as request-response Web services, are passive or pull-based. Other data sources, such as sensors that trigger alarms, are active or push-based. Flexible infrastructure is capable of discovering both kinds of data sources. An application can then pull the current value from a passive data source or subscribe to be notified each time an active data source generates a new value.

This application development model presents several challenges. One challenge is to define a model for the computations performed in retrieving data from pervasive sources. Another is to provide the application developer with a simple but powerful means for specifying the behavior of a composer. Still another is to devise an appropriate system for describing data-source requirements. The remainder of this section addresses each of these three challenges.

A wide variety of computation models has been proposed. Some systems, such as Tapestry [Ter92] from the Xerox Palo Alto Research Center and Cougar [Bon00] from Cornell University, view sensor data as being added to an append-only database and use Structured Query Language-like models to retrieve the data. In contrast, NiagaraCQ [Che00] defines data-retrieval compositions in terms of continuous queries over XML infosets, specified in an XQuery-like language. The Rome system [Hua99] from Stanford University and the Solar system [Che02] from Dartmouth University specify composer-like entities, called respectively *triggers* and *operators*. Both presume that all data sources are passive. The iQueue computation model [Coh02a] from IBM Research allows a composer to obtain input from lower level data sources, including both passive and active sources, and allows the composer itself to act as either a passive or an active source; this model is based on an expression that is evaluated whenever data is pulled from the composer or whenever one of the composer's input sources pushes a new value.

The means for specifying the behavior of a composer depends, of course, on the underlying computation model. For the expression-based model of [Coh02a], the appropriate specification is the expression itself. The language iQL, described in [Coh02b], is specifically tailored to the kinds of expressions that are useful in writing composers.

The description of data-source requirements poses a difficult challenge because of the wide variety of data sources. Different kinds of data sources have different interesting attributes and new kinds of data sources are continually being invented. It is untenable to adopt a fixed vocabulary of kinds of data in which current applications are interested, let alone those in which next year's applications will be interested. However, it is feasible to categorize each new data source registered with the infrastructure as belonging to a specified provider kind that can be named in a descriptive query. Some new data sources can be categorized as belonging to providers of an existing kind, although new provider kinds will have to be registered for other data sources. Provider kinds can be categorized in a superkind–subkind hierarchy, such that all attributes of a provider kind are inherited by its subkinds. A query for a provider of kind k can be satisfied a provider of any subkind of k.

3.4 Conclusions

This chapter has discussed the key difficulties in writing pervasive applications — applications that support mobility and context-awareness — and summarized the main approaches that are currently being employed to address these difficulties. The key issue is application development complexity to deal with heterogeneous devices, varying degrees of connectivity, and dynamic data sources. Reuse of application components is the fundamental means of addressing this complexity.

Four basic approaches to enhancing reuse were discussed, based on the well-known MVC application structure:

1. Device-independent views — these allow an application to capture the basic interaction structures that should be reused across multiple devices and modalities. They should be combined with the ability to fine-tune the presentation when necessary.
2. Platform-independent controllers — these allow an application to specify the overall control flow across multiple execution platforms, but still allow an application to have different control flow structures for different devices and uses.
3. Host-independent models — these allow an application to encapsulate the business logic and data in a manner that can be reused regardless of which host a component is instantiated on.
4. Source-independent context data — this allows an application to specify the intended context data to be supplied by reusable infrastructure components, which in turn are concerned with the specific data formats, locations, and combinations of physical data sources that provide the actual data.

These approaches have reached different levels of maturity (interestingly, the above order represents the highest to lowest in terms of maturity) in research projects and commercial offerings. Several challenges remain before these approaches can become widely useful.

Acknowledgments

This chapter is a compendium of many ideas that have evolved from projects and discussions with many individuals in the pervasive computing group at IBM, including Jeremy Sussman, Larry Bergman, Rich Cardone, Shinichi Hirose, Andreas Schade, and Apratim Purakayastha.

References

[Abr99] Marc Abrams, Constantinos Phanouriou, Alan L. Batongbacal, Stephen M. Williams, and Jonathan E. Shuster. UIML: An appliance-independent XML user interface language. *WWW8/Computer Networks*, vol. 31, no. 11–16, pp. 1695–1708, 1999.
[Adj99] William Adjie-Winoto, Elliot Schwartz, Hari Balakrishnan, and Jeremy Lilley. The design and implementation of an intentional naming system. Proceedings of the 17th ACM Symposium on Operating Systems Principles (SOSP'99), December 12–15, 1999, Kiawah Island Resort, SC, published as *Operating Systems Review,* vol. 33, no. 5, pp. 186–201, December 1999.
[Ban00] Guruduth Banavar, James Beck, Eugene Gluzberg, Jonathan Munson, Jeremy B. Sussman, and Deborra Zukowski. Challenges: An application model for pervasive computing. *MOBICOM 2000,* pp. 266–274.
[Ban02] Guruduth Banavar and Abraham Bernstein. Software infrastructure and design challenges for ubiquitous computing applications. *CACM*, vol. 45, no. 12, pp. 92–96, 2002.
[Ban04] Guruduth Banavar, Lawrence Bergman, Richard Cardone, Vianney Chevalier, Yves Gaeremynck, Frederique Giraud, Christine Halverson, Shin-ichi Hirose, Masahiro Hori, Fumihiko Kitayama, Goh Kondoh, Ashish Kundu, Kohichi Ono, Andreas Schade, Danny Soroker, and Kim Winz. An authoring technology for multidevice Web applications, *IEEE Pervasive Computing,* vol. 3, no. 3, July–September 2004, pp. 83–93, 2004.

[Ber01] Lawrence D. Bergman, Tatiana Kichkaylo, Guruduth Banavar, and Jeremy B. Sussman. Pervasive application development and the WYSIWYG pitfall. *EHCI 2001,* pp. 157–172.

[Ber02] Lawrence D. Bergman, Guruduth Banavar, Danny Soroker, and Jeremy Sussman. Combining handcrafting and automatic generation of user-interfaces for pervasive devices. *Proceedings of the 4th International Conference on Computer-Aided Design of User Interfaces (CADUI 2002),* Valenciennes, France, May 15–17, pp. 155–166, 2002.

[Bon00] Philippe Bonnet, Johannes Gehrke, and Praveen Seshadri. Querying the physical world. *IEEE Personal Communications,* vol. 7, no. 5, pp. 10–15, October 2000.

[Bow93] Mic Bowman, Saumya K. Debray, and Larry L. Peterson. Reasoning about naming systems. *ACM Transactions on Programming Languages and Systems,* vol. 15, no. 5, pp. 795–825, November 1993.

[Che00] Jianjun Chen, David J. DeWitt, Feng Tian, and Yuan Wang. NiagaraCQ: A scalable continuous query system for Internet databases. *Proceedings of the 2000 ACM SIGMOD International Conference on Management of Data,* Dallas, May 15–18, 2000, pp. 379–390.

[Che02] Guanling Chen and David Kotz. Context aggregation and dissemination in ubiquitous computing systems. *Proceedings of the 4th IEEE Workshop on Mobile Computing Systems and Applications (WMCSA 2002),* Callicoon, NY, June 20–21, 2002, pp. 105–114.

[Coh02a] Norman H. Cohen, Apratim Purakayastha, Luke Wong, and Danny L. Yeh. iQueue: A pervasive data-composition framework. *Proceedings of the 3rd International Conference on Mobile Data Management,* Singapore, January 8–11, 2002, pp. 146–153.

[Coh02b] Norman H. Cohen, Hui Lei, Paul Castro, John S. Davis II, and Apratim Purakayastha. Composing pervasive data using iQL. *Proceedings of the 4th IEEE Workshop on Mobile Computing Systems and Applications (WMCSA 2002),* Callicoon, NY, June 20–21, 2002, pp. 94–104.

[Dub03] Micah Dubinko, Leigh L. Klotz, Jr., Roland Merrick, and T.V. Raman, Eds. XForms 1.0. W3C recommendation, October 14, 2003. http://www.w3.org/TR/xforms/.

[Hua99] Andrew C. Huang, Benjamin C. Ling, Shankar Ponnekanti, and Armando Fox. Pervasive computing: What is it good for? *Proceedings of the International Workshop on Mobile Data Management,* Seattle, August 20, 1999, pp. 84–91.

[Int00] Chalermek Intanagonwiwat, Ramesh Govindan, and Deborah Estrin. Directed diffusion: a scalable and robust communication paradigm for sensor networks. *Proceedings of the 6th Annual International Conference on Mobile Computing and Networking (MobiCom 2000),* Boston, August 6–11, 2000, pp. 56–67.

[Kam98] Tomihisa Kamada. Compact HTML for small information appliances. W3C Note, February 9, 1998. http://www.w3.org/TR/1998/NOTE-compactHTML-19980209/.

[Kly03] Graham Klyne, Franklin Reynolds, Chris Woodrow, Hidetaka Ohto, Johan Hjelm, Mark H. Butler, and Luu Tran, Eds. Composite capability/preference profiles (CC/PP): Structure and vocabularies 1.0. W3C proposed recommendation, October 15, 2003. http://www.w3.org/TR/CCPP-struct-vocab/.

[Kra88] Glenn E. Krasner and Stephen T. Pope. A cookbook for using the model-view-controller user interface paradigm in smalltalk-80. *Journal of Object-Oriented Programming,* vol. 1, no. 3, pp. 26–49, August/September 1988.

[Med99] Nenad Medvivovic, Alexander Egyed, David S. Rosenblum. Round-trip software engineering using UML: From architecture to design and back. *Proceedings of the 2nd Workshop on Object-Oriented Reengineering (WOOR),* Toulouse, France, September 1999, pp. 1–8.

[Mer99] Roland A. Merrick. Defining user interfaces in XML. *Proceedings of the POSC Annual Meeting,* London, England, September 28–30, 1999. http://www.posc.org/notes/sep99/sep99_rm.pdf .

[Mic03] Microsoft. Mobile Web development with ASP.NET. 2003. http://msdn.microsoft.com/mobility/prodtechinfo/devtools/asp.netmc/default.aspx .

[Sow03] Daby M. Sow, David P. Olshefski, Mandis Beigi, and Guruduth Banavar. Prefetching based on Web usage mining. *Middleware 2003,* pp. 262–281.

[Ter92] Douglas Terry, David Goldberg, David Nichols, and Brian Oki. Continuous queries over append-only databases. *Proceedings of the 1992 ACM SIGMOD International Conference on Management of Data,* San Diego, June 2–5, 1992, pp. 321–330.

[Wei91] Mark Weiser. The computer for the twenty-first century. *Scientific American.* September 1991, pp. 94–104.

[Wie92] Gio Wiederhold. Mediators in the architecture of future information systems. *IEEE Computer,* vol. 25, no. 3, March 1992, pp. 38–49.

[WML02] Website META Language. October 19, 2002. http://thewml.org/.

Chapter 4
ISAM, Joining Context-Awareness and Mobility to Building Pervasive Applications

Iara Augustin, Adenauer Corrêa Yamin, Jorge Luis Victória Barbosa, Luciano Cavalheiro da Silva, Rodrigo Araújo Real, Gustavo Frainer, Gerson Geraldo Homrich Cavalheiro, and Cláudio Fernando Resin Geyer

Abstract

The essence of pervasive computing is that the user's applications are available in a suitable adapted form, wherever that user goes. Our ongoing research aims at integrating the concepts of context-aware, grid, and mobile computing toward building a pervasive computing infrastructure. The ISAM approach is the integration of programming language and the middleware that supports its execution. This integration is based on abstractions to describe context, alternative behavior, adaptation mechanisms, and policies. ISAM includes a model for writing pervasive applications and a correspondent middleware that provides a pervasive computing environment. This text exposes the most important ideas of the ISAM architecture. These ideas are illustrated by the construction of a pilot application called WalkEd (Walking Editor).

4.1 Introduction

The term mobile computing is still under definition. In our perspective, mobile computing is "the distributed computing in which the location of the involved elements may change during the computation execution." Some scenarios derive from this definition depending on which element has mobile capacity:

- Hardware — wireless computing
- User — nomadic computing
- Software — mobile code
- Computation — mobile agent

The most general scenario that joins all of them is the one provided by pervasive computing, which allows physical and logical mobility while maintaining a global network connection. This last scenario is the focus of the ISAM architecture [1, 2].

In our viewpoint about the pervasive infrastructure, the computational power is in the whole network, different from traditional networks in which each computer is a computational island. To reach this power, we believe that the pervasive applications run in an environment that manages the distributed execution and provides facilities to programming such applications. Toward supporting this view of pervasive computing, we propose a software architecture that joins many aspects focused by context-aware, grid, and mobile computing. How to integrate them in a consistent way is the challenge of the ISAM project (*Infraestrutura de Suporte às Aplicações Móveis* — Mobile Applications Support Infrastructure). The ISAM architecture aims to provide a model, language, and runtime support to build and execute pervasive applications. The initial efforts of many research projects in pervasive computing [3, 4] focus on enabling the environment and directly accessing services within that environment. None of those projects addresses application development in a general purpose way.

This chapter shortly describes a way of integrating context-aware, grid, and mobile computing to reach the management of the context-aware adaptation process and to execute the pervasive applications in the ISAM platform. This chapter is organized as follows:

- Section 4.2 introduces the ISAM Application Model.
- Section 4.3 presents the summary of ISAM architecture and its context-aware behavior.
- Section 4.4 introduces ISAMadapt, a main ISAM component.
- Section 4.5 introduces EXEHDA, a main ISAM component.
- Section 4.6 presents a pilot application, named WalkEd. This was built using the ISAM constructs to explore context adaptation strategies.
- Section 4.7 contains concluding remarks.

4.2 The ISAM Application Model

Mobile devices in nomadic computing are seen as small desktops where applications are programs running on them, accessing code and data stored locally. The ISAM Application Model is different, it considers that the computer is the whole network. The computing environment (data, device, code, service, resource) is spread in composed cells. Users can

ISAM, Joining Context-Awareness and Mobility

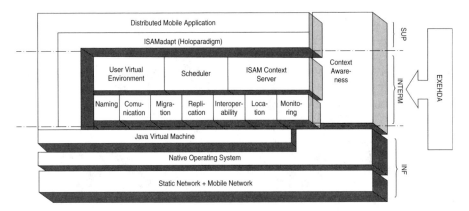

Figure 4.1. ISAM Architecture

move around, having both their applications and virtual environment following them. Mobile device is an interface device that executes a small part of the application in a collaborative way. It is an application portal, not a code or data repository. Required applications are installed on-demand and can migrate among hosts to find better resources and services. Application code explores the capabilities of the network that compose its virtual environment and adapts to it. The adaptive behavior of the application and the decisions of runtime management are defined by the current context where the application's components are inserted. Details of ISAM components, that implement this semantic, are described in the next sections.

4.3 The ISAM Architecture

Mobile software development is complex because its components change in time and space in terms of connectivity, portability, and mobility. To reduce the impact of these changes, the application must have a context-aware adaptive behavior. Context-aware adaptation is a fundamental concept for pervasive computing. An application to be run in the pervasive environment should not make undue assumptions neither about the devices upon which it will run nor the environment services it will use. However, adaptation is not of easy implementation because of its ad hoc specific nature. Traditional adaptive systems were created based on assumptions about the environment, such as permanent connection and resource availability, which are not true in the mobile environment. This fact, for example, prevents direct use of solutions from the distributed systems area.

Figure 4.1 illustrates our approach for the adaptation problem in mobile computing. We have designed the ISAM software architecture to deal with context-aware adaptation and collaborative behavior between the system

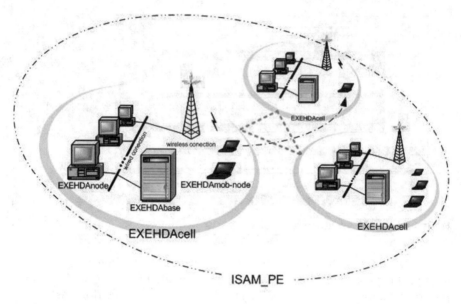

Figure 4.2. ISAM_PE

and the application. The architecture is organized in layers with three abstraction levels; it is directed to getting the maintenance of quality of service provided to the mobile user through the context-aware adaptation concept. The ISAM architecture adopts a modular organization. In Figure 4.1, the context module is virtual because it is a concept present in the design of many other modules. An important component of the architecture is the scheduling module. Adaptation decisions, from application to runtime management levels, are based on the behavior profile of three entities — mobile user, application, and the system itself — which are part of the execution context.

The applications run in the ISAM_PE (ISAM Pervasive Environment). ISAM_PE is composed of elements that have correspondents in the physical infrastructure of the mobile network, as Figure 4.2 shows. The physical organization adopted in ISAM is the cellular hierarchy. The devices belonging to the same cell communicate directly to each other (using a plane organization). In communications with resources outside the cell, a specific host (base) acts as a gateway. The hierarchical characteristic allows a cell to recursively contain other ones. This organization assists the need of context confinement, necessary to the application model adopted for ISAM.

With the motivation of exploring the concept of adaptation in pervasive environment, our system was designed to support multilevel collaborative adaptation at application and system levels. In ISAM architecture, the system adapts itself to provide quality, and the application adapts at different quality levels tolerated by mobile user. The innovation of this project is in the uniform treatment of adaptation and its suitability to general purpose mobile applications, where:

- The execution system is responsible for some adaptations related to performance or application and resource management.
- The application is responsible for decisions related to context-aware adaptations or concerning to domain specific adaptations.
- Both the applications and the execution system are responsible for adaptation decisions negotiated in many cases.

4.4 ISAMadapt Overview

ISAMadapt is the architecture component that deals with the context-aware adaptability issue through the perspective of the programming language. Distributed applications in the pervasive computing environment must be designed based on several functionalities that adapt to the current environment. These alternative functionalities are managed automatically by the application management system (ISAMadaptEngine). This new concept of applications, which modify their own behavior in function of the environmental changes, demands new language abstractions, as well as an execution system integrated with language.

The ISAM pervasive applications are modeled with Holoparadigm abstraction, which designs the application using the beings concept [5]. Beings are naturally mobile and they enable the expression of physical and logic mobility. They use a coordination model that supports implicit invocations based on reading and writing in a blackboard, call history. Each being has a history encapsulated inside itself. In the case of a composed being, the history is shared by component beings. The component beings take part in the development of shared history and they suffer the reflexes of history changes. So, the existence of several levels of encapsulated history is possible. However, the being only accesses the history in its own composition level. The history is implemented as multiple distributed tuple space and some tuples may have a reactive behavior. At the moment, the Hololanguage is mapped to the Java™ language. This strategy makes it possible to quickly explore the proposed architecture on pilot applications.

ISAMadapt extends the Hololanguage constructions with a semantic of context-awareness and dynamic adaptation to allowing your use in the design of pervasive applications. For implementing those semantics, ISAMadapt introduces new abstractions in the language — context, adapters, adaptation commands, and policies. Each abstraction is implemented in two ways — static (at programming time) and dynamic (at running time). Figure 4.3 shows the relationship among these abstractions.

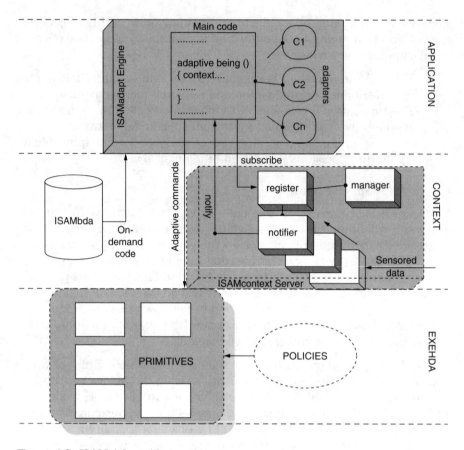

Figure 4.3. ISAM Adapt Abstractions

4.4.1 Context

In ISAM, context is defined as "all the relevant information to the application that can be obtained from the support system." The application explicitly identifies and defines the entities that characterize a situation and those that integrate its context using the ISAMadapt Development Environment software. State alterations in these entities trigger an adaptation process in the application being. So, the context definition can then be refined to "every entity for which a state alteration triggers an adaptation process in the application." This way, the context conception allows focusing on aspects relevant to a particular situation, while ignoring others.

The context-aware nature of the applications becomes explicit on three moments:

1. Description — defined at programming time and generates the application context descriptor.

ISAM, Joining Context-Awareness and Mobility

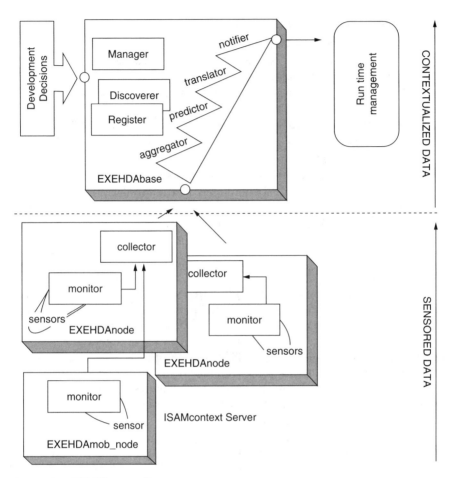

Figure 4.4. ISAMContextServer

2. Activation — executed at application load time and reconfigures and activates the ISAMcontextServer components.
3. Execution — the server will monitor the environment and notify the application when significant context changes occur.

The ISAMcontextServer component of architecture provides this service of context recognition (Figure 4.4). It runs independently of applications, being designed based on pipeline paradigm to model the sensored data treatment and publish–subscribe paradigm to model the relationship with the application. The application registers its interests within the middleware in order to receive information when elements of its context change. The server collects and interprets relevant information to the applications, generating higher level information.

The server is composed of components with the functions:
- Collector — obtains sensored data
- Aggregator — performs preliminary data treatment
- Translator — translates data to an abstract representation
- Predictor — makes a forecast analysis of data behavior
- Notifier — sends changes to the application
- Register — registers the monitoring interests of the application
- Discoverer — replies the queries about services and resources available in its scope

The information of a change in the context triggers the application reconfiguration at runtime. It is managed by the ISAMadaptEngine of the application. The reconfiguration occurs by the choice of the best alternative among the codes programmed for the current environment, which is loaded on-demand.

4.4.2 Adapters

ISAMadapt allows three forms of coding the sensibility to the context. The `onContext` command is inserted in the language, it can be associated to other commands such as `if`, `while`, `in`, `out`, and `method call`. By definition, the `onContext` command is asynchronous and schedules a task, codified in its inside block, in the ISAMadaptEngine. This task runs when the context is available. The `sync` qualifier behaves the command as synchronous. The adaptive behavior can also be expressed in the language through the definition of adapters. Adapters associate alternative code to the execution of a being's method or all the being's behavior. The adaptive being's method behaves as a generic function, whose code can be dynamically and transparently selected, among the alternatives provided, based on the current context. This process is managed by the ISAMadaptEngine, which uses EXEHDA's services that are responsible for:
- Deciding the validity of the adaptation
- Selecting one among the alternatives
- Triggering the decision execution

The ISAMadaptEngine runs in conjunction with the ISAMcontextServer, in a relation based on publish–subscribe model (registers the application's monitoring interests, asks for server reconfiguration, receives ISAMcontextServer notification, activates/deactivates the notification received). Another way of expressing context-awareness is to use the adaptation commands of the language.

4.4.3 Adaptation Commands

Some adaptation commands are made available by the system and can be used in the adapters and beings codification. The main commands with adaptive behavior are:

- Creation or replication (clone command)
- Migration (move command)
- Rescheduling (reschedule command)
- Resource and services discovering (discovery command)
- Push code (install command)
- Disconnection (disconnect command)
- Reconnection (reconnect command)

On-demand code load is a natural mechanism of ISAM architecture and it is present in the behavior of the entire application.

4.4.4 Adaptation Policies

The adaptation commands are implemented with an adaptive behavior based on collaboration between the middleware and the application. Policies express this collaboration. Policies consist of orientations given by the application to the system, contributing to the system decision, which controls the overall application behavior. Some policies are global, but others are specific to some application's beings (see Section 4.6.4).

4.5 EXEHDA Overview

EXEHDA (Execution Environment for High Distributed Applications) is the architecture component that provides a pervasive execution environment to the ISAMadapt applications. It is designed as a middleware, having as its main strategies:

- Adaptive instantiation of software — the EXEHDA loads in the nodes (mobile or fixed) a minimum set of software components, from both the middleware and the application, that guarantee the application execution based on the application's profile. The other components, if any, will be requested on-demand, characterizing a pull strategy operation. This feature is important in the presence of high heterogeneous resources.
- Prefetching of application — the instantiation process begins when the user executes its authentication in EXEHDAbase, before requesting the execution of application. It has adopted in this case a push strategy for software components and information dissemination. This instantiation can also happen with a longer anticipation, using as reference an expectation of the user's itinerary according to his mobility already consolidated. To anticipate the traffic in the structured part of the networks (with physical connection) it is an alternative proposed to increase the global acting of the mobile application and, consequently, reduce the waiting/connection time of the mobile device.
- Support to adaptive pervasive execution — EXEHDA offers to the ISAM architecture a pervasive execution environment that can be

tuned in function of both the ISAMadapt provided adaptation policies and the state of execution context. For a pervasive operation, EXEHDA integrates two mechanisms at ISAMadapt runtime:
- A manager of multiple object spaces -EXEHDA-CC-, whose control syntax is a variation of the Linda protocol [6], enabling an anonymous and asynchronous communication and coordination strategy
- An execution controller -EXEHDA-HM-, which keeps the relationships between the beings (Holotree), whose core uses a distributed naming service

The resources the application uses (databases, specific services, etc.) are also registered within the execution controller. This makes it possible to manage replicas of beings and resources, enabling a uniform handling of lookup operations for both resources and processes (beings) by the resource management (scheduling, mapping) mechanism.

The EXEHDA middleware uses a set of primitives for monitoring in both the system and application levels and also for object scheduling. These set of primitives, provided by PRIMOS (PRIMitives for Object Scheduling) [7], comprises:

- Remote object instantiation and migration
- Optimized communication primitives
- Parameterizable native host sensors, which may be extended by application provided sensors

From the ISAM physical point of view, the system is built of hosts, network segments, and computing cells. A host is the base building block and represents, as it would be expected, a machine in the system. Hosts are described by static and dynamic indexes, relatives to processor and memory capabilities, among other things. A network segment groups a set of hosts that share the same interconnection medium/technology and is described statically by its nominal latency and throughput (e.g., Ethernet bus, Token Ring, Myrinet cluster). Dynamic data is added to the static information, which reports how much of the network capabilities of a given segment are used at a given time.

Network segments are arranged in computing cells, which would be interpreted as institution boundaries in the scope of a distributed architecture (EXEHDACell). Each computing cell has an information server associated. This server keeps track of all dynamic and static sensors available in the cell. The publishing of load indexes from each host to the cell's collector is triggered by a variation of such indexes above a configured threshold (which may be changed dynamically). The computing cells are grouped into cell groups. Each cell is responsible for publishing its summarized load information to the other cells in the group. This publishing at cell group level is accomplished through a probabilistic protocol [8], which

ISAM, Joining Context-Awareness and Mobility

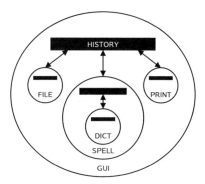

Figure 4.5. ISAM WalkEd Beings

tends, with a given probability, to keep all cells updated. To connect cell groups, one cell is configured to belong to both intended groups. Such publishing architecture has an interesting aspect to management decisions: the quality of the information about other host in the system is inversely proportional to the logical distance to that host. This property can be used to improve locality in scheduling decisions.

4.6 The WalkEd Application

To exemplify the conception and behavior of an ISAMadapt pervasive application, a well-known application was modeled using the abstractions provided. The objective of such application is to demonstrate the use of ISAMadapt constructs in order to conceive a pervasive application that is able to adapt itself to modifications in its execution context. The WalkEd application consists of a text editor with basic editing functionalities, which was modeled using the being abstraction of Holoparadigm. As Figure 4.5 shows, there are four main beings in the WalkEd application:

1. GUI being — implements the editor's graphical user interface (GUI)
2. Dict being — provides access to an external dictionary database and is used as a helper service for implementing the spell checking functionality
3. Spell being — implements the spell checking service
4. Print being — provides access to external printing services

The WalkEd application may be used from both desktops in the wired network or mobile wireless devices (e.g., laptops and personal digital assistants [PDAs]). Figure 4.6 shows a possible physical disposition of the beings that compose WalkEd, at different moments, as new beings are created.

Notice that the devices used may present differences both in connectivity and with respect to processing and visualization capabilities. In this sense, a key characteristic of the WalkEd application is being able to

MOBILE COMPUTING HANDBOOK

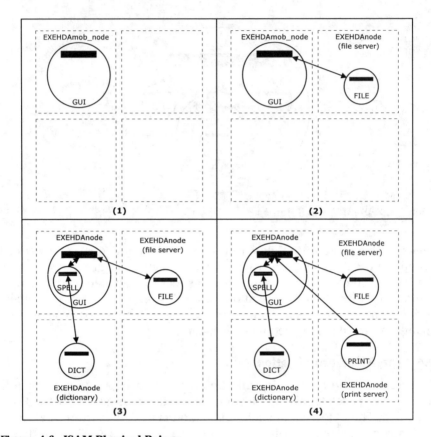

Figure 4.6. ISAM Physical Beings

adapt itself to varying execution contexts. The adaptation semantic of the application is `follow-me`: components migrate closer to the resource, following the user movements and adapting to the current state of the environment.

The WalkEd application starts its execution by the creation of the editor's graphical interface, as illustrated in the code fragment presented in Figure 4.7. In the instantiation of the being that corresponds to the editor's interface, given by the execution of the clone primitive, the first level of adaptation in the application occurs. It consists in selecting the GUI being implementation (adapter being) most appropriate to the device where the application is being launched. The GUI being is locally created, in the device where the application has been launched, keeping the edited text in its history. The interface code is loaded on-demand, using the context information — `device type` — provided by the ISAMcontextServer.

During the execution, the user may require services like spell checking and file. Such requests trigger the creation of the Spell, Dict, and Print

ISAM, Joining Context-Awareness and Mobility

```
//
// WalkEd entry point
//
being holo
{
    clone (GUI, Gui_id);
}

//
// graphic interface adapted to display device
// @context: display
//
adaptive being GUI
{
    openFile(filename)
        {
            clone( File(filename), filename );

            content = being(filename).getContents();

            history!list(filename, contents);
        }

    saveFile(filename) { ... }

    setActiveBuffer(bufferName) { ... }

    insertLine(pos,text) { ... }

    deleteLine(pos,n) { ... }

    moveCarret(line,pos) { ... }

    gotoMyHost()
        {
            move (self, "myHost");
        }

    gotoMyDesktop()
        {
            move (self, "myDesktop");
        }
}
```

Figure 4.7. ISAM WalkEd Code

beings, which may be placed on nodes distinct from those where the GUI is running. The second level of adaptation in WalkEd comes from that: the whole application adapts itself to states of connection and disconnection, because the beings that compose the application may be physically located in distinct nodes of the distributed system. The criteria used in placement of the being consider the being's dependencies to external resources. In WalkEd, for example, the Dict and Print beings are created closer to the dictionary and printer resources.

The third level of adaptation in WalkEd is related to the `follow-me` semantic. In this level of adaptation, a migration of the GUI being (activated by the `goto` interface menu) triggers a relocation of the beings Spell, Dict, and Print, aiming to reduce communication costs, although respecting its access dependencies to external resources.

4.6.1 The GUI Being: Alternative Behaviors

The WalkEd GUI adapts itself with respect to the display capabilities of the device used. The differences between the two actual implementations may be observed in Figure 4.8 and Figure 4.9.

The adapters define code alternatives for each state of the display context element — PDA and desktop — as shown in Figure 4.10. The being's methods use the adaptation commands `clone` and `move`. Communication between beings is accomplished through blackboards (history in Holoparadigm), which are accessed using the operations `in` (consume), `out` (write), and `read`. The tuples in the history are composed of strings which represent the text being edited. Because the Hololanguage currently does not provide an Application Programming Interface (API) for building GUIs, code directly written in the Java language is used for that purpose (native Java command).

The WalkEd GUI provides the more often used commands for text editing (Figure 4.8 and Figure 4.9). The `open` command creates a file system

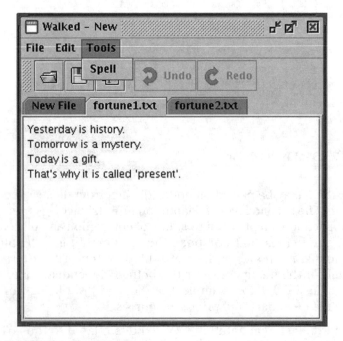

Figure 4.8. ISAM Desktop Interface

ISAM, Joining Context-Awareness and Mobility

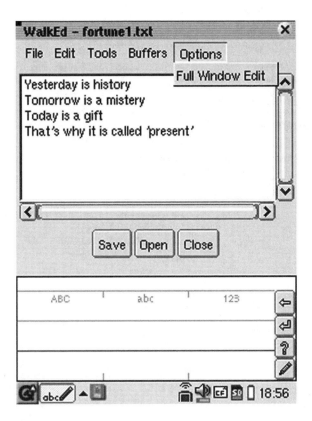

Figure 4.9. ISAM PDA Interface

being, which populates its history with records that compose the text file. The file being provides methods for other beings to acquire (getContents) and modify (setContents) the contents of the underlying file. It is also implemented using the adaptation concept of ISAM: the actual file being implementation used type of the storage (i.e., context) that serves the underlying file (e.g., database or file system). The edition commands operate on the GUI being's history.

In both target platforms, WalkEd allows the simultaneous manipulation of multiple edition buffers, although the mechanism provided to the user to access the edition buffers was conceived differently for each platform. On the PDA, the existing buffers are accessed through the Buffers GUI menu. On the other hand, in the desktop implementation, each buffer is mapped to tab in the edition pane.

At any moment, the user may request to continue the WalkEd execution at other node of the system. This is accomplished through the activation of the goto GUI submenu, after which the user is requested to select the

87

```
// file: GuiPDA.adp ////////////////////////////////////////
    //
    // WalkEd GUI implementation for PDA
    //
    // @description: implementation initialy targeted to the
    // Zaurus PDA. It uses AWT/PersonalJava APIs.
    //
    // @context:   display::PDA
    //
adapter being Gui::GuiPDA()
{
    // context state declaration
    context network::(disconnected,connected);
    // Constructor
    GuiPDA()
    {
        native Java {
            // use java.awt components to build WalkEd's GUI
            // and register menu callbacks
            ...
        }
        enableSpellChecking();
    }

    // menu call backs
    startSpellCheck(line) { ... }

    enableSpellChecking()
    {
        ...
        native Java {
            // update gui state
            ...
        }
        // schedules feature disabling to next disconnection
        onContext network::disconnected {
            disableSpellChecking()
        }
    }

    disableSpellChecking()
    {
        native Java {
            // update gui state
        }
        // schedules feature enabling to next reconnection
        onContext network::connected {
            enableSpellChecking()
        }
    }

    // other methods
    ...
}
// file: GuiDesktop.adp ////////////////////////////////////////
    //
    // WalkEd GUI implementation for desktop
    //
    // @description: It uses Swing/J2SE APIs.
```

Figure 4.10. ISAM GUI Being Adapters

```
    //
    // @context:   display::desktop, display::laptop
    //
adapter being Gui::GuiDesktop()
{
     // Constructor
    GuiDesktop()
      {
         native Java {
            // use javax.swing components to build WalkEd's GUI
            // and register menu callbacks
         }
      }

    // other methods
    ...
}
```

Figure 4.10. ISAM GUI Being Adapters (Continued)

destination node (e.g., `MyHost`, `MyDesktop`, `MyPDA`). The system evaluates the abstract targets at runtime, using the user preferences stored into the User Virtual Environment (UVE), to determine the real destination of the migration. At the destination node, the application continues its execution.

4.6.2 The Spell Being

The Spell being is created as a child of the GUI being, typically colocated, and creates a Dict being, which is placed by the system close to the dictionary resource, following the scheduling policy configured for the application (see Section 4.6.4).

The Spell being is context-aware with respect to the connectivity state of the node in which it is running. On its normal operation mode (connected), the Spell being stays in the background reading words from modified paragraphs in the GUI and sending those words to the inner Dict being. Such communication with the Dict being is done through the Spell being's history. When the node becomes disconnected, the Spell being stops sending words to the Dict being in order to reduce local memory consumption, because the produced words will not be consumed by the Dict being at that time. This way, it retains more memory to be used by the GUI being in text edition.

The source code of the Spell being is presented in Figure 4.11. Notice that the connectivity state also affects the GUI by enabling or disabling the spell checking feature.

4.6.3 The Print Being

The Print being creates a printer system being, which is colocated with the underlying print server. The location of the print server is the dynamic

```
// file: Spell.holo /////////////////////////////////////////
being Spell
    {
        context network::(disconnected,connected);

        adaptive bgSpellCheckWords();

        Spell()
            {
                // create the dictionary used to validate words
                clone (Dict, dict);

                onContext network::connected {
                    bgSpellCheckWords();
                }
            }

        spellCheckWords()
            {
                // synchronous spell checking
                ...
            }
    }

// file: bgSpellCheckWordsV1.adp ///////////////////////////////////
        //
        // @context: network::connected
        //
    adapter Spell.bgSpellCheckWords::v1()
        {
            // blocking read,
            // waits for a modified paragraph to be available
            // in the parent history
            out(history)#(modified_text);
            ...
            // send words to the Dict Being
            history!list(check,language,words);

            // schedules next verification
            onContext network::connected {
                spellCheckWords();
            }
        }


// file: bgSpellCheckWordsV2.adp ///////////////////////////////////
        //
        // @context: network::disconnected
        //
    adapter Spell.bgSpellCheckWords::v2()
        {
            // do nothing while disconnected to avoid filling up
            // the device's memory
        }
```

Figure 4.11. ISAM Spell Being Code

information provided by the resource discovery mechanism of the ISAM-contextServer. The print being provides the user with a way of sending the edited text to the system print queue.

ISAM, Joining Context-Awareness and Mobility

```
<scheduling>
   ...
   <policy composition="append">
      <being name="Spell"/>
      <max sensor="HOST_BENCH[bogomips]"/>
      <anchor being="GUI"/>
   </policy>
   <policy composition="replace">
      <anchor resource="Dictionary"/>
   </policy>
   ...
</scheduling>
```

Figure 4.12. ISAM WalkEd Policy

4.6.4 The ISAMadapt IDE

The ISAM framework provides an application development environment, the ISAMadapt IDE, for edition of beings and adapters source code. Furthermore, the ISAMadapt IDE allows the specification and specialization of the context elements the application is interested in, as well as the definition of global and per being scheduling policies and associations between adapters and adaptive beings. Such information is represented by three Extensible Markup Language (XML) documents — `context.xml`, `policy.xml`, and `adapters.xml` — which are stored into the pervasive storage base, ISAMbda. This information guides the compilation process and also the application execution.

A fragment of the scheduling policy used for the WalkEd application, related to the Spell and Dict beings is shown in Figure 4.12.

4.6.5 Execution Aspects

The `Hololanguage+ISAMadapt` code is currently translated to Java source code, which is then compiled with a standard `javac` tool to generate the executable code. The WalkEd prototype was developed with two modalities of the Java platform in mind — PersonalJava™ application for PDAs and J2SE™ platform for desktops — which have differences in the supported APIs. The application was tested using a few wired desktop personal computers and a wireless Sharp Zaurus™ 6500 PDA. The ISAM core middleware was installed on these devices prior to running the application.

The ISAM core middleware provides the user with commands for session management which, by complementing the other features provided by middleware, enable the implementation of the `follow-me` semantic for applications. In this sense, the main session management commands provided are described below:

- Login:
  - Authenticates the user using an asymmetric public key mechanism. The public key is provided in certificates, for each user, those certificates are stored in the cell base (`EXEHDAbase`).

- Activates the user default session. The activation of the default session restarts applications that were interrupted when the user last logged out.
- Provides access to the workspace with the applications already installed.
- Logout:
  - Frees the application execution context management related resources that integrate the default user session in the logout moment.
  - Notifies the ISAMadaptEngine about the end of the applications, which performs the ISAMcontextServer unsubscription.
  - Stores the `Default Session` state in the UVE for future loading when the user logs in again.
- Save/restore session — the user may have an arbitrary number of sessions additionally to its default session. Those sessions are managed through the commands — `save session` and `restore session`. The command `save session` enables the applications to move from the default session to an alternative one, stored in the UVE. The command `restore session` provides the opposite functionality, providing the user with the capability of reincorporating applications saved in sessions alternative to the default session.
- Disconnect/Reconnect — these commands take effect in the connectivity state of the user controlled device, moreover, they implement a planed disconnection protocol. Each change in the connectivity state of the device is published as a piece of information to the ISAMcontextServer, which is running on the cell management node (`EXEHDAbase`). Additionally, a broadcast to all the local ISAMadaptEngine is performed, this notification triggers the adaptations related to the local beings previously registered as sensible to this kind of context change.
- Application launching — in the conception of ISAM, the installation process of the application is consisted of the copy of the launch descriptor to the ISAM workspace. The launch descriptor of the application is a XML document; as Figure 4.13 shows, it is generated during the compilation phase of the application. The launch descriptor provides metadata that describes the application in an abstract way and independently from the ISAM_PE cell where the user is located. When the application is launched, this descriptor is expanded and the relative references are resolved for absolute references to the resources needed by the application.

After the on-demand loading of the Starter being code, the application has its context sensibility interests subscribed to the ISAMcontentServer and adaptive execution proceeds as described in Section 4.6. As long as new beings are created by the application, the application interests previously

ISAM, Joining Context-Awareness and Mobility

```xml
<?xml version="1.0" encoding="UTF-8"?>
<isamapp spec="1.0" href="WalkEd.isam">
    <info>
       <title>WalkEd</title>
       <vendor>ISAM team</vendor>
       <description>The Walking Editor</description>
       <icon href="WalkEd.png"/>
    </info>
    <code>
       <main class="isam.demo.walked.Main"/>
       <jar href="WalkEd.jar"/>
    </code>
</isamapp>
```

Figure 4.13. ISAM Launch Description

registered in the ISAMcontextServer are updated to reflect these new demands.

Our future experiments with WalkEd will lead to a better understanding of how context sensibility can be used to design adaptive behavior in pervasive applications. We want to improve the user interface and explore others adaptation strategies. The initial feedback we got from the experiments showed promising results, especially with respect to adaptation and mobility. Our next steps will be to explore more deeply the issues related to multicell, multi-institutional executions, as well as better caching strategies to improve the overall system performance.

4.7 Conclusions

The innovation of our approach is the introduction of context-aware and grid computing associated with physical mobility to provide a support infrastructure by pervasive applications. Another contribution is providing a uniform treatment of the context-aware adaptation in the whole system from the base to the programming paradigm. Besides, it is not compromised to a specific application domain. These decisions are based on our belief that there is a big potential to use mobile application in various application domains. So, it is necessary to provide an infrastructure for general purpose applications, which makes the expression of mobility and adaptability as simplified as possible. We believe that the collaborative multilevel adaptation is a real alternative to achieve a general purpose architecture, capable of guaranteeing good performance levels in different applications that are arising in the pervasive computing scenario.

Acknowledgments

ISAM project is supported in part by FAPERGS Foundation. The consortium partners are Federal University of Rio Grande do Sul, Federal University of Santa Maria, Federal University of Pelotas, and University of Vale do Rio dos Sinos, Brazil. ISAM project page: www.inf.ufrgs.br/~isam.

References

1. Yamin, A., Augustin, I., Barbosa, J., and Geyer, C., ISAM: A Pervasive View in Distributed Mobile Computing, in *Proc. Network Control and Engineering for QoS, Security and Mobility with Focus on Policy-Based Networking* (Net-Con 2002), Paris, France, 2002, Paris: IEEE/IFIP.
2. Augustin, I., Yamin, A., Barbosa, J., and Geyer, C., ISAM — A Software Architecture for Adaptive and Distributed Mobile Applications, in *Proc. 7th IEEE Symposium on Computers and Communications*, Taormina, Italy, 2002.
3. Garlan, D., Steenkiste, P., and Schmerl, B., Project Aura: Toward Distraction-Free Pervasive Computing, *IEEE Pervasive Computing*, vol. 1, no. 3, 2002.
4. Roman, M. et al., Gaia: a Middleware Infrastructure to Enable Active Spaces, *IEEE Pervasive Computing*, vol. 1, no. 4, 2002.
5. Barbosa, J. and Geyer, C., A Multiparadigm Language Oriented to Distributed Software Development, in *Proc. V Brazilian Symposium of Programming Languages (SBLP)*, 2001.
6. Gelernter, D., Generative Communication in Linda, *ACM Computing Surveys,* vol. 7, no. 1, 1985.
7. Silva, L.C. and Geyer, C.R., Primitives for Supporting Object Distribution Targeted to Pervasive Computing, Master Thesis, CPGCC — Federal University of Rio Grande do Sul, Porto Alegre, Brazil, 2003.
8. Yamin, A., Augustin, I., Barbosa, J., Silva, L., Real, R., Cavalheiro, G., and Geyer, C., A Framework for Exploiting Adaptation in High Heterogeneous Distributed Processing, in *Proc. Symposium on Computer Architecture and High Performance Computing*, Vitória, Brazil, 2002, Piscataway, NJ: IEEE Press.

Chapter 5
Integrating Mobile Wireless Devices into the Computational Grid

Thomas Phan, Lloyd Huang, Noel Ruiz, and Rajive Bagrodia

5.1 Introduction

One application domain the mobile computing community has not yet entered is that of grid and cluster computing: the aggregation of network-connected computers to form a large-scale, distributed system that can be used for resource-intensive scientific or commercial applications in a scalable and cost-effective manner [1, 3, 15]. Current grid efforts have leveraged predeployed clusters of workstations as computing nodes, while at a larger scale, grid-like distributed applications such as Seti@home and Folding@home have enlisted home personal computer (PC) enthusiasts to volunteer their desktop computers.

In this chapter, we consider grid and cluster computing from a different architectural perspective, namely from the view of using small-scale, highly heterogeneous devices that can serve as nodes within a distributed grid system. This class of small-scale devices comprises laptops, tablet PCs, personal digital assistants (PDAs), and other home consumer devices that can be connected to the Internet, potentially all through wireless links. As recent investigations have shown, using such heterogeneous devices is an interesting extension to contemporary grid computing efforts [6, 30, 40].

The integration of these devices into high-performance grid and distributed computing is not without difficulties. Small-scale devices are heterogeneous and lack the computational, storage, bandwidth, and availability characteristics [14] commonly required for high-performance distributed computing. With widespread use of these devices (including next-generation intelligent appliances whose characteristics cannot be readily foreseen), heterogeneity will only increase, making this combination ostensibly unlikely.

However, many technological, commercial, and consumer trends support the inclusion of such devices within a computational grid. First, Moore's Law of increasing transistor density will continue to drive central processing unit (CPU) performance in small-scale devices just as it has in other markets. For instance, Intel's lines of XScale® and Centrino™ processors for mobile devices continue to show improved performance with each generation. Second, availability of wide area wireless communications will be more prevalent. This increased connectivity can be seen in current 2.5G (generation) and upcoming 3G networks based on Wideband Code Division Multiple Access (W-CDMA)/CDMA2000, as well as wireless local area network (LAN) hotspots and meshes. Third, consumer use of small-scale, intelligent electronics continues to grow yearly, as can be clearly seen by increasing popularity of smart cell phones and home electronics such as networked fourth-generation video game systems and digital video recorders. Fourth, trends in ubiquitous/pervasive computing [42] suggest a future where small devices will be the predominant form-factor of choice, relegating desktop devices to the minority. We posit that these signs point toward the confluence of small, heterogeneous devices and computing in the future.

With these observations, our research project, LEECH (Leveraging Every Existing Computer out tHere), aims to identify and address the research challenge of using a wide range of heterogeneous systems as contributors to computational grids. In this chapter, we will take a look at how our LEECH system has been designed from the start to address the heterogeneity issues of small-scale devices for grid and cluster computing. Our experimental testbed used to evaluate our system adheres to our heterogeneity goal, as it includes a mix of wirelessly connected laptops and PDAs as well as a wired PlayStation® 2. We do note that the issue of power consumption in small-scale, portable devices is an ongoing, open research area that remains a significant challenge, one that we are continuing to investigate and will address in future work.

To mitigate the effects of wide heterogeneity and unpredictable availability commonly associated with small-scale devices, we designed the LEECH architecture using a hierarchical organization to abstract away the underlying devices. Specifically, we used a proxy-based clustered infrastructure to provide small-scale devices with favorable deployment, interoperability, scalability, adaptivity, and resiliency characteristics. In our design we created groups of devices clustered around a nearby proxy, as shown in Figure 5.1. Unlike contemporary peer-to-peer [40] or mobile ad

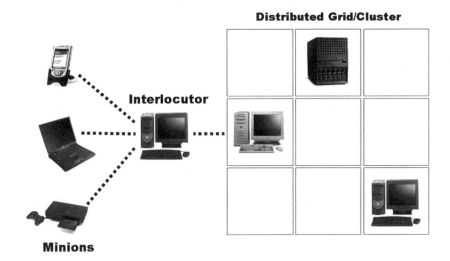

Figure 5.1. A Broad View of the LEECH Proxy-Based Clustered Architecture

hoc routing [4] approaches that also use clustering around a proxy or gateway node, in our system the proxy additionally serves the important roles of service negotiator and resource request partitioner for abstract classes of devices in its group.

Additionally, we designed and implemented a programming model, Application Programming Interface (API), and lightweight library to accompany the LEECH architecture. Distributed programming libraries, such as Message Passing Interface (MPI) [33], commonly used today for the development of grid applications, do not sufficiently support heterogeneous, small-scale, and dynamically available devices as grid nodes. The LEECH Programming Model allows a grid programmer to use whatever computational power is available through LEECH proxy machines. Although fundamentally different, the LEECH API is designed to behave, from the programmer's perspective, similarly to traditional message-passing APIs such as MPI, the de facto message-passing communication library used today.

To demonstrate the effectiveness of our architecture, programming model, and library, we use three benchmarks:

1. A RSA keybreaker using exhaustive search
2. A numerical integration program
3. A synthetic communication-intensive application

Through experimentation, we show that LEECH is lightweight with less per message overhead than MPI, an important factor for small-scale devices.

Parallel applications executed through LEECH scale well over an increasing number of added nodes. Finally, our system provides resilient operation in the face of node failure through managed replication and job control.

5.2 Background

Grid and cluster computing has roots within the field of high-performance parallel computing, which has traditionally been successful on massively parallel processor (MPP) systems designed following a nonuniform memory access (NUMA) or uniform memory access (UMA) architecture. Such MPPs have used multiple CPUs within a single chassis to produce higher performance manifest through increased throughput. However, such systems become prohibitively expensive for large CPU configurations. A different approach is taken with grid and cluster computing: by bringing together available machines, such as workstations on either a local or wide area scale, an aggregation of computational resources can be formed in a cost-effective manner.

Local area networks of workstations (NOW) [1] take advantage of clusters of uniprocessor workstations connected via a network such as Myrinet or Ethernet. Taking advantage of such commodity parts, NOWs can provide high performance at low cost. For example, Beowulf systems [3] look to leverage low-cost, high-performance Linux® PCs with LAN networking. Additionally, Condor [28] provides the capability to schedule jobs across a UNIX® NOW to achieve load balancing.

Much pioneering work in grid computing support libraries has been done with the Legion [20] and Globus [15, 18] research efforts. Globus is the *de facto* middleware standard for a number of different grid projects and provides a four-layer stack to control hardware, communications, resource sharing, and collective coordination. Our LEECH architecture is intended to integrate seamlessly with such existing systems.

To the best of our knowledge, ours is the first effort to produce a general framework for using small-scale, heterogeneous devices for grid and distributed computing. Other researchers ported the Cactus environment onto PDAs, but did not address the limitations of small-scale devices in general [6]. The University of Illinois at Urbana-Champaign recently created a cluster of 100 PlayStation 2 game consoles, outfitted with a special version of Red Hat® Linux, to take advantage of that machine's high-performance Emotion Engine graphics processor [30, 40]. However, the use of these devices is no different from that of a statically configured, homogeneous PC Beowulf cluster with no specific treatment of the issues we raise in this chapter. In our own experiments, we too use PDAs and a PlayStation 2 together as part of a heterogeneous testbed to demonstrate the architecture, programming model, and library of LEECH.

Integrating Mobile Wireless Devices into the Computational Grid

In our work we propose to use small clusters of small-scale devices to improve scalability, interoperability, and other factors we discuss later. These devices are clustered around a proxy node that interacts with the rest of grid on the devices' behalf. We note the ostensible similarity between our approach and those of other clustering techniques intended to rein in a loosely assembled group of devices. For example, many ad hoc routing schemes [4] use such clustering. In an ad hoc wireless network, particular devices are chosen as gateways or clusterheads to facilitate cluster organization and routing only, without the consideration of aggregate computation. Bluetooth®-enabled devices assemble themselves into piconets of seven or less nodes centered about a master device [22, 31]. Landmark routing [36] suggests a similar approach. Mobile Internet Protocol (IP) [38] facilitates the integration of mobile computers into the Internet by using home agents to act on behalf of mobile nodes. ALICE [21] provides a similar capability but at the application layer to support Common Object Request Broker Architecture (CORBA)-enabled applications. Our clustering approach performs request negotiation, routing, and most importantly, data partitioning and aggregation within its cluster.

Clustering is also used by the file-sharing peer-to-peer Kazaa™ program using the Fast Track infrastructure [40] to facilitate scalable searching. In this system, peer nodes are clustered around so-called supernodes. These supernodes serve as indexing repositories for search requests from peers, thereby negating the need for multicast searches in infrastructures such as Gnutella or centralized search indices as with Napster™. However, these supernodes perform only indexing (because the central objective of Kazaa is file sharing).

5.3 Motivation: Mobile Devices and the Grid

We suggest that a logical step in expanding grid and cluster computing systems to have the ability to incorporate more computational power lies with the use of heterogeneous mobile consumer devices connected through a potentially unreliable wireless or unstable wired network. In this section, we present opposing views to this assertion.

A number of problems hinder the use of contemporary small-scale devices in grid and cluster computing. For PDAs, hardware and operating system (OS) heterogeneity issues are pervasive as palmOne and Microsoft's PocketPC compete aggressively for market share. Mobile computing devices are also well-known for other inherent disadvantages, such as:

- Heterogeneous hardware
- Unreliable low-bandwidth wireless connectivity
- Unpredictable extended periods of complete disconnectivity
- Heightened power-consumption sensitivity

- Software noninteroperability
- Small secondary storage
- Incomplete security

In an ideal world, wireless networks would provide as much bandwidth and work as reliably as wired connections. Unfortunately, real-world conditions such as multipath disturbances, power-signal degradation, and intercell hand-off, among others, do not facilitate the high bandwidth, always-on characteristics expected of computation nodes. Present grid and cluster computing applications typically target idle desktop PCs that receive portions of a larger problem, perform computation, and return results within bounded time. Unreliable connectivity and prolonged periods of intended disconnectivity break this expectation. Even when connectivity is not an issue, present wireless technology cannot provide the high bandwidth typical of wired connections. Most wired LANs provide a minimum of 100 megabits per second (Mbps) commonly found with Fast Ethernet and are moving quickly to 1 gigabit per second (Gbps) Gigabit Ethernet. On the other hand, wireless bandwidth varies among different technologies, as we will discuss shortly.

Other problems are prevalent. Battery technology has matured slowly over the last decade and has failed to keep up with increased power demands from contemporary PDAs and laptops. Recent developments in lithium polymer replacements for lithium ion show promise in this field [12]. Little to no investment has been made in developing software that supports small-scale devices in grid and cluster computing, resulting in such problems as software integration, service discovery, and application-level interoperability. In terms of secondary storage, the limitation of flash memory in handhelds is a major factor against using small-scale devices. Applications need storage to place temporary and permanent data for reuse or aggregation, but contemporary PDAs typically come with only around 64 megabytes (MB) of memory or less. Grid systems also need permanent and temporary storage for system software. Perhaps the use of miniature secondary storage devices, such as IBM's 1 GB Microdrive or SanDisk®'s 4 GB CompactFlash® card, will become more prevalent in the near future. This, however, adds to the higher power requirements of the device. Finally, security is always an issue with mobile wireless devices because wireless transmission is susceptible to a wide range of attacks.

In addition to the technological issues just presented, other socioeconomic problems become evident. We raise these issues in turn and address them directly.

First, the issue of why one would even consider the use of small-scale devices with restricted resources for grid and cluster computing at this time, particularly when only a small fraction of Internet-connected desktop

Integrating Mobile Wireless Devices into the Computational Grid

Table 5.1. System Specifications for Contemporary PDAs

System	CPU	Storage	Connectivity
Casio Cassiopeia E-125	150 MHz NEC VR4122	32 MB RAM, CompactFlash	56K modem via cf.
Compaq iPAQ 3975	400 MHz Intel XScale	64 MB RAM, Secure Digital Card	Built-in Bluetooth
Compaq iPAQ 5555	400 MHz Intel XScale	128 MB RAM, Secure Digital Card	Built-in Bluetooth, 802.11b
Palm Tungsten C	400 MHz Intel XScale	64 MB RAM, Secure Digital Card	Built-in 802.11b

PCs currently contribute to the grid and grid-like distributed applications. The argument for including small-scale devices in grids and clusters is based on the sheer weight of numbers. The ubiquity of computing devices in people's pockets, briefcases, and homes has potentially become a vast new source of processing power. According to Gartner Dataquest, a market research firm, 12.1 million PDAs were shipped worldwide in 2002 [16]. Although this was a decline from the 13 million shipments in 2001, projected sales of PDAs remain promising. Of those sold annually in 2002, over 47 percent were Palm and Sony devices running the PalmOS® operating system. In Table 5.1, we list the hardware specifications of some contemporary PDAs. As can be seen from the table, the raw processing power of the handhelds is not trivial given their mobility.

The argument for laptop PCs is more intuitive. It can be informally observed that laptops are typically 0.5 generations behind desktops in terms of storage capacity and CPU performance; at the time of this writing, 2.6 gigahertz (GHz) CPUs are now available in high-end laptops. Market research showed that in 2002, sales of mobile PCs outgrew those of desktops [17]. A user who owns a small-scale device can wirelessly connect to the Internet and potentially to grids and clusters by using any of the current or emerging wireless LAN or cellular standards shown in Table 5.2. The emergence of new products using the 3G standards CDMA2000 or W-CDMA will only further strengthen the argument in favor of inclusion. An evaluation of the potential aggregate power of these machines is indeed compelling.

We add to our argument by considering five trends we believe will be prevalent in the future:

1. Moore's Law suggests increases in CPU performance for small-scale devices as has been seen for desktop PCs. Such products as Intel's XScale line of power-efficient, fast CPUs specifically for the handheld market bode well for future PDAs.

Table 5.2. Wireless LAN and Cellular Technologies

	Bandwidth	Range
Bluetooth	1 Mbps	10 meters
802.11a	54 Mbps	50 meters
802.11b	11 Mbps	100 meters
802.11g	54 Mbps	50 meters
Atheros Dynamic Turbo Mode	108 Mbps	33 meters
HomeRF	10 Mbps	50 meters
UltraWideband	200 Mbps	5 meters
Former Metricom	128 kbps	Cellular
2.5G	144 kbps	Cellular
3G	2 Mbps	Cellular

Sources: RHR, IBM, Verizon, Metricom, *EE Times*, and *The Economist*.

2. Wireless communication will grow with improved reliability for both local area (using IEEE® 802.11, Bluetooth, or Ultra Wideband) as well as wide area (using 3G technology or perhaps ad hoc meshes of wireless LANs).
3. Consumers will migrate away from tethered devices for everyday applications such as word processing, spreadsheets, and Internet browsing, as can be seen by the increased usage of desktop replacement notebooks.
4. Battery efficiency will not substantially improve.
5. Grid and cluster applications will be more widely used.

We firmly believe that careful anticipation of such future developments will lead to better preparation for later research down the road and will provide a glimpse into a future grid of completely heterogeneous machines. The time is ripe to start investigating the use of small-scale devices for grids and clusters, due largely to the expected growth of mobile processors, wireless communication, and consumer use of the first three trends. An architecture will be needed to mitigate the fourth trend of limited battery efficiency, as well as to address issues of availability, interoperability, security, and network latency. Finally, all of this is in favor of meeting the potential widespread adoption of grid and cluster technology as stated in the fifth trend.

As a second issue against small-scale devices in grid and cluster computing, it may be argued that research in this area should wait until these devices gain sufficiently powerful CPUs and other resources so that their contribution is more meaningful. Unfortunately, there will always be tiered heterogeneity, no matter what year it is. Our research addresses the problem of dealing with the lowest rung of the technological ladder, the current small-scale device, in order to address the technological issues that arise. Similar problems may be evident in the future for whatever PDA-like device may exist at that time. Research performed now helps us anticipate the long-term use of lowest-rung devices on the grid and cluster computing in the future.

Third, by their very nature, it may be doubtful that users will ever want to give up their power-limited small-scale devices for others' use. Slow improvements in battery technology only compound the problem. There are two ways to address this problem. A system architecture and programming model can be designed to assuage the problem of small-scale device overusage as perceived by the user. Our LEECH architecture can accomplish that by allowing device owners to autonomously decide whether or not to participate without adversely affecting ongoing grid or cluster activity thanks to an adaptive availability scheme, as discussed later. Additionally, the small-scale owner must be given a persuasive incentive to contribute his device. Elements of game theory suggesting commercial and monetary incentives may be needed to encourage users [5].

Fourth, users typically do not leave their small-scale devices on all the time and thus allow these machines to automatically shut off. This may substantially reduce the potential number of resource contributors. If users are motivated enough to want to contribute in the first place (as we have suggested in the previous point), they will be able to allow such devices to be always-on, a trait confluent with upcoming always-on 3G wireless technology. People who demand always-on, always-connected mobile devices can thus obtain savings by putting their machines in semi-standby mode, where, for example, the CPU clockspeed can be reduced and the energy-draining liquid crystal displays (LCDs) can be turned off while the machine continues with computations. With these techniques, battery conservation can be increased along with the amount of work that can be done in the background. Two other points are noteworthy. In contemporary society, users at their desk, either at home or work, tend to leave their small-scale device plugged into a rechargeable cradle or into the wall socket when not in use anyways. In addition, although many small-scale devices may be shut down, there will most likely always be active devices to be used due to the potentially large number of users involved.

Finally, there may not be a clear grid and cluster application domain that can leverage the use of small-scale devices. Grid and cluster computing, in general, has already established the context for its own existence: resource sharing and distributed computation in a scalable, cost-effective manner. Our research looks to preserve the grid abstraction by simply contributing small-scale device resources for contemporary and future applications. The most significant issue is that, as we have mentioned, small-scale devices are typically constrained in hardware, software, and network connectivity. Applications intended to leverage small-scale devices must be written (or be adapted retroactively) such that their problem space can be decomposed and distributed among small-scale devices accordingly to fit these limitations, as we shall show. Furthermore, as the small-scale devices evolve within the next few years, due to constant semiconductor improvements, the computational power that can be extracted from them becomes increasingly compelling.

5.4 The LEECH Architecture

Given that the use of small-scale devices in a grid is compelling and potentially useful, a system architecture must be constructed to facilitate their integration. In this section, we present an overview of our LEECH architecture designed to meet this goal.

5.4.1 Key Challenges

A naïve approach to an architecture design would be simply to run grid or cluster software on the small-scale devices, connect the devices together, and allow the devices to behave and assume the same responsibility as typical desktop PC nodes. Although this approach may work, in practice a number of significant obstacles will be encountered:

- Grid/cluster software overhead — there is a memory and CPU use overhead incurred for running grid or cluster software. Currently, Globus is available only for desktop machines. Similarly, the Seti@home distributed application runs only as a Windows® screensaver. It is unlikely that small-scale devices, particularly PDA devices with limited memory and CPU performance, will be able to operate as full-fledged nodes.
- Device heterogeneity — the variety of devices is potentially large. Workload spread across such machines cannot be generalized. For example, the workload to be performed by Seti@home nodes includes nontrivial Fast Fourier Transforms. Such an expectation for the smaller devices is not realistic.
- Scalability and management — even for existing distributed systems, scalability is a major issue. When one considers the inclusion of hundreds of thousands of small-scale devices, scalability and the management of these machines becomes an even larger issue.
- Service discovery — small-scale devices need to be able to find grid nodes in order to participate.
- Dynamic, unpredictable availability — device owners are privileged to turn off their devices at times of their own choosing.

Integrating Mobile Wireless Devices into the Computational Grid

- Power consumption — of course, the power consumption requirements of devices is a major factor.
- Multi-hop wireless network participation — if the devices participate in a wireless network, what are the ramifications with regard to routing and discovery?

5.4.2 Overview of Architecture

To address these issues, we chose a proxy-based clustered architectural approach to integrating small-scale, heterogeneous devices to computational grids and clusters. The LEECH architecture enables communication between small-scale, heterogeneous devices and a computational grid or distributed system via a proxy middleware. We term the proxy node an interlocutor and we call the small-scale, heterogeneous devices to which it is connected its minions. In our system, minions are closely associated with an interlocutor, which in turn is responsible for hiding the heterogeneity of its minions from the rest of the grid system. We suggest that interlocutors can support a large number of minion devices and that the aggregation of many minions' resources can be presented to the distributed system as an interlocutor's own resources. This hierarchical organization, similarly seen on the Internet with hierarchical routing and domain name system (DNS), improves scalability by intentionally limiting the number of devices that is globally visible.

On each interlocutor and minion, we instantiate a daemon process that facilitates communication and interactivity within the LEECH system. The interlocutor daemon provides functionality for service discovery, session management, adaptive control, and job scheduling. In a similar fashion, the minion daemon handles service discovery beaconing, application fragment management, and session control.

5.4.3 Grid/Cluster and LEECH

In Figure 5.2, we show the interaction between nodes using MPI and LEECH. In the context of MPI alone, programmers develop their parallel applications while using MPI library calls (such as `MPI_Send` and `MPI_Recv`) to send and receive buffers of data. An interconnectivity fabric, whether a communication bus within a multiprocessor, an Ethernet network within a LAN, or the Internet for distributed grids, facilitates message delivery.

We also note that distributed shared memory can be used as a programming model for distributed applications, but in this chapter we focus on message-passing applications.

Figure 5.2 also shows how the LEECH architecture fits in with existing systems. The existing grid or cluster remains unchanged, save for the execution of LEECH components on chosen nodes, which act as both a grid

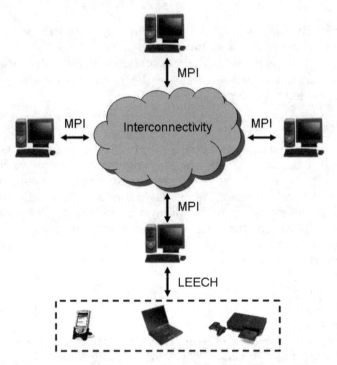

Figure 5.2. Communication Architecture of a Distributed Application Using LEECH as Part of a Grid/Cluster System

node as well as an interlocutor. Communication between the interlocutor with other grid nodes still uses MPI, although communication between the interlocutor and the minions is through our LEECH API.

5.4.4 Application Major Component and Minor Component

An application that runs atop the LEECH architecture is divided into two parts — the *major component* and the *minor component*. The major component contains the application logic and MPI communication calls as well as LEECH calls. Note that in the absence of LEECH, the major component is executed on the grid/cluster like any other MPI program. On the other hand, using the application within the LEECH architecture requires that the major component be written to contain LEECH communication calls as well as programming logic to partition tasks and data among the minions. Additionally, if the major component runs within LEECH but without minions, then the programming logic should appropriately handle normal computational activities within the cluster or grid.

The minor component runs on minion devices and should be written to contain the application logic to handle an apportioned segment of computation and data. This minor component communicates with a major component through the LEECH API. This configuration allows the interlocutor running the major component to interact with other cluster nodes (via MPI) on behalf of the minions running the minor components.

The logical division of a distributed application into our major component and minor component allows code executed on the minion nodes to be smaller than the entire distributed application and thus more easily executed on small-scale devices with limited memory.

Furthermore, the grid or cluster MPI runtime overhead may be too large for such devices, so MPI communication does not reach the minor component in our architecture. This application division also introduces a difference between LEECH applications and typical MPI applications: MPI applications are typically data parallel, but LEECH applications can be task parallel or data parallel.

5.4.5 Interlocutor

A LEECH interlocutor node is itself a grid/cluster node that has been deemed an interlocutor by its administrators based on a machine's hardware and network capabilities. Interlocutor nodes should be placed strategically such that they are geographically close to a focal point where a large number of minions often come online to reduce network latency between the minion and the interlocutor. In a wireless LAN environment, interlocutors should be colocated with wireless base stations where large numbers of wireless minions regularly connect. Current distributed/grid architectures require nodes to have fairly powerful processors, more than moderate amounts of primary and secondary storage, and reliable network connections. The small-scale devices we consider do not meet these requirements, so interlocutor nodes are typically mid-range desktop PC or workstation systems or greater. (In our experiments, we used a low-end Sun Sun Blade™ 100 workstation.) Furthermore, an interlocutor may itself be a work node and perform a part of an application's computation, but we leave that software design choice to the application programmer.

Although an interlocutor is designed to be coupled with a grid or cluster system through the distributed application itself, it is important to note that the LEECH system, in particular the daemons and the communication library, can run standalone (i.e., without MPI or Globus) outside of a grid or cluster. In this mode, a parallel message-passing application can still be written using only our APIs and daemons to run within an isolated group of minions and an interlocutor. It is also possible for a LEECH system to be organized in a hierarchical manner, such that an interlocutor can be a minion of another interlocutor. Because the focus of this chapter is on LEECH itself, we will leave a closer analysis of LEECH's interaction with other grid infrastructures like Globus for future work.

The interlocutor software is a daemon process logically divided into four main services or components. The service discovery server (SDS) allows minions to discover and register with interlocutors in the vicinity. The major component handles connections and communication with the major component running on the same machine as the interlocutor daemon. This session manager allows major components to post new work to be computed and pick up results, if any, from completed computations. The interlocutor-to-minion session manager acts as a server waiting for connection requests from minions and handles communication sessions with its minions. The interlocutor-to-minion session manager communicates directly with the minion-to-interlocutor session managers in its minions. Finally, the job manager and availability adaptation scheme handles the scheduling and assignment of jobs to minions, working closely with the interlocutor-to-minion session manager.

5.4.6 Minion

A minion executes a LEECH minion daemon process responsible for:

- Receiving a job from the interlocutor with which it has registered
- Making some computation based on the job's data
- Sending results back to the interlocutor
- Repeating the three previous steps, until going offline

The minion daemon process is the middleman between the minor component and the interlocutor. The minion, much like the interlocutor, has three main components:

1. Service discovery agent (SDA)
2. Minion component session manager
3. Minion-to-interlocutor session manager

However, unlike the interlocutor, the minion's services run serially to allow operation on small-scale devices whose operating systems may not support multi-threading. The SDA allows a minion to discover local interlocutors. The SDA advertises minion services to interlocutors it has discovered. The minion component session manager handles connection setup and teardown and communication with minor components running on the same small-scale machine as the minion daemon process. Finally, the minion-to-interlocutor session manager handles connections and communication with interlocutor daemons that have been discovered by the minion's SDA. The minion-to-interlocutor session manager works directly with the interlocutor's interlocutor-to-minion session manager.

5.4.7 Availability Adaptation and Job Management

On the small-scale class of devices we are considering, we cannot expect results returned from a node that has gone offline and will not be coming

Integrating Mobile Wireless Devices into the Computational Grid

back online. The grid community has identified this unreliability problem but has not yet addressed it [9]. Typical MPI programming idioms deal with node failures with fail-stop semantics.

Instead, resiliency can be facilitated by the parallel programming library or by the application itself. In LEECH, we provide support to handle dynamic minion availability within the library because low availability (and even periods of complete disconnectivity) is common. We have developed an availability adaptation scheme to gracefully accommodate mobile systems that come and go in and out of the network before completing the computation of some job whose results must eventually be submitted to the interlocutor. The two main features that facilitate our availability adaptation scheme are the LEECH job and replicated job assignments.

A LEECH job is a partition of the computation and data of an application. More specifically, it comprises a unique set of initial data, the computation of this data by one or more minion nodes, and its results. A job is created at the interlocutor by calling a LEECH send function and its results are collected by calling a LEECH receive function.

We chose to keep minions anonymous from the major component rather than to provide named communication. If we supplied the application programmer with a means of establishing and maintaining named communication between the interlocutor and a specific minion, providing resiliency would mean adding checkpoints for each set of communications back and forth between each interlocutor and minion pair and restoring checkpoints when failures occur. We can easily see how adapting to dynamic availability, but providing direct, named communication would cause LEECH to grow in complexity, shrink in scalability, and lose significant potential performance gains.

The LEECH job supports the communication model above by allowing interlocutors to submit and collect jobs from minion classes rather than from individual minions. Each job is individualized with a unique job ID (JID). The LEECH interlocutor generates JIDs and manages the mappings between jobs and the minions; however, this mapping is invisible to the application programmer.

A job uses one round trip communication exchange between the interlocutor and a minion. If a minion fails or goes offline unexpectedly, the interlocutor's SDS component's lease manager notifies the job manager, which simply reassigns the job to a new minion. The subsequent computation of this reassignment entails only one job's amount of additional work and only one repeated transmission.

Replicated job assignments add another level of resiliency to the LEECH architecture. Once all submitted jobs have been assigned to some set of minions, the interlocutor begins replicating job assignments across available minions. If all of the minions assigned to a replicated job eventually fail, the interlocutor begins reassigning the job to a new set of available minions. This rule guarantees that a job submitted to the LEECH system will eventually be completed, even in the case that all minion node connections are extremely unreliable. As soon as one minion returns a job's results, the interlocutor commands the minions computing the job's replications to abort and free their resources and makes the results available to the application.

5.5 The LEECH Programming Model

LEECH implements its own message-passing programming model, API, and communication library, described here. A full discussion is beyond the scope of this chapter and more detailed results can be found in our companion research papers. Our system supports the so-called embarrassingly parallel programming style also found in other grid or distributed systems. In standard grid or cluster architectures, nodes interoperate through the use of a communication library such as MPI [33]. Distributed variants of MPI include MPICH [34], a portable version commonly used in Beowulf workstation clusters, and MPICH-G2 [35], used within the Globus grid toolkit. In this chapter, we will use MPI to represent this broad family of variants.

However, unlike processor nodes in a high-bandwidth grid system, the minions in a LEECH system are small-scale, potentially mobile devices that cannot necessarily maintain high levels of availability or reliability due to limitations in network connectivity. Due to this reason, the communication semantics of message-passing libraries, such as MPI, cannot be followed. For example, a downed processor results in a MPI communication call failure; the accepted programming idiom in response to a failed MPI call is to terminate the program. With minions, intermittent connectivity or availability is the norm. Furthermore, the overhead of running the MPI library or the Globus system on small-scale devices may be prohibitively expensive. Our tiered proxy-based architecture presented in the last section suggests a means to hide minion heterogeneity, but even with this design, the communication model that we expect between the interlocutor and minions would still make MPI an inappropriate choice.

There is a need for a more responsive and adaptive programming model. Our approach uses a lightweight communication library that can be integrated in a grid/distributed application. The LEECH architecture and APIs do not require the porting of entire applications; rather, LEECH function calls are added at key points in the application code where data and computation can be decomposed and sent to minions.

The LEECH APIs are modeled after standard message passing library APIs to minimize the learning curve for application programmers. However, unlike standard message passing libraries, LEECH does not provide named communication between the interlocutor and a particular minion. For instance, the MPI runtime system assigns a unique identifier called a rank to each processing node and communication calls use this rank (or a wildcard) to identify senders and recipients. Such a model is appropriate for the homogeneous, highly available nodes common in contemporary grid or cluster systems. On the other hand, a minion contributing its resources to a LEECH system is free to come and go at any time it pleases. To gracefully facilitate this situation, our APIs provide anonymous communication between the interlocutor and classes of minions, while still giving the programmer the feel of named communication.

One of the distinguishing characteristics of our programming model is that we take into consideration the fact that available compute resources will be heterogeneous by nature. We defined an abstract virtual class ID, or VID, that allows a set of devices to be grouped into abstract categories relative to their heterogeneous characteristics. An interlocutor's minion nodes are grouped into VIDs ranging from 1 to 10, where 1 represents the lowest amount of resource power and 10 the highest. Nodes can be grouped into VIDs by a combination of 1 or more characteristics. As mentioned earlier, the use of VIDs is in contrast to direct, named communication, such as MPI's use of ranks. Although MPI allows the formation of communication groups, this capability is intended to facilitate parallel algorithm design rather than to hide heterogeneous hardware. In our current implementation, a system administrator assigns devices into VID classes, thereby allowing programmers to address devices by classes through the API. The runtime system then matches jobs with VID classes. In future work, we plan to provide an automated methodology for categorizing minions into their respective VIDs by a combination of CPU performance, network bandwidth, and storage or the specific model of a branded minion device. One particularly important metric we intend to follow in future work is that of power consumption. LEECH also gives the application programmer the option of using a built-in default algorithm to decompose or partition a distributed task and scheduling these tasks among available processors by way of the LEECH distribute and gather functions.

5.6 Experiments and Analysis

Here we analyze the performance of the LEECH library against MPI on a testbed of small-scale devices running parallel applications. Our results will show that MPI is a poor choice for communication within a small-scale device environment. In particular, we will show that LEECH has lower communication overhead (particularly relevant for communication-intensive programs), exhibits better resiliency in the face of minion failure, and provides flexible tools for more convenient data and computation partitioning.

Due to chapter space constraints, we omit results from our applications that use the interlocutor and minions within a larger grid cluster (where the interlocutor acts as a grid node). This larger scenario already encompasses the interactivity between the interlocutor and the minions, which is the critical path we are studying.

The experiments in this section were chosen to represent classes of communication-intensive or computation-intensive applications. For the latter, although the interlocutor's major component could have been written to perform application computation, we chose to have the interlocutor execute only partitioning and daemon functionality; this allowed us to focus on the performance of the minions in computationally intensive programs.

We wanted a broad range of small-scale devices that we could easily program. We thus chose a set of low-end laptops (Dell™ 3800 and 4000 models over 2 Mbps 802.11b running Red Hat Linux), PDAs (Compaq iPAQ 3650 and 3670 models over 2 Mbps 802.11b running Familiar Linux), and a PlayStation 2 (connected with Fast Ethernet) with varying CPU, storage, and communication characteristics. All of the machines have some form of Linux installed on them. The PlayStation 2 in particular used the Linux Kit from Sony, which provided a special Red Hat distribution along with a hard drive and Ethernet adapter. Except for the synthetic communication experiment, we used a Sun Blade 100 workstation running Solaris™ operating system as the interlocutor. All testbed devices used alternating current (AC) power during the course of the experiments. In future work, as our testbed is expanded, we will look at how power constraints affect the system, particularly when users are working on the devices at the same time. We will look into relevant power metrics and heuristics to react to them.

5.6.1 Synthetic Application for Measuring Communication Overhead

In our first experimental set, we wanted to show the overhead of the LEECH communication library against that of MPI. We used MPICH, a popular implementation of MPI used in current Beowulf cluster computing. We installed MPICH version 1.25 on all our machines except for the iPAQ PDAs because our cross-compiler could not compile the MPICH code. Another notable fact is that the memory footprint of MPICH's MPD messaging daemon is over 800 kilobyte (KB), whereas our LEECH minion daemon required only 400 KB. Although the PDAs could have run the MPICH daemons in theory, other more memory-constrained devices, particularly smaller embedded devices, may not.

Integrating Mobile Wireless Devices into the Computational Grid

In Figure 5.3, we show the execution of a synthetic communication benchmark, written as both LEECH and MPI versions, to reveal the round trip latencies between two laptops connected over an 802.11 network. The times measured for both versions include other 802.11 traffic. This application simply passes message buffers back and forth between one machine and another without any further computation. We varied the buffer size from 5 KB to 1000 KB, as shown on the x-axis. It can be seen in this graph that the LEECH version incurs a much lower communication overhead than does MPICH/MPD. This difference increases as the size of the buffer increases.

It has been previously noted that MPI has a high degree of communication overhead proportional to the message size [25]. Contributing factors include message headers, management of large data structures to handle unfulfilled function calls, and group organization. LEECH is much more lightweight with minimal message headers and simple internal management schemes.

5.6.2 RSA Decryption

This experimental set was chosen to show the performance of LEECH for a computationally intensive application. The application we chose, RSA keybreaking, requires a large amount of computation with little communication.

RSA is a popular public key cryptographic system [39] and decryption of its key is a computationally intensive operation that involves factoring a large integer key into its prime factors. Specifically, given a large number n that is a product of two large primes p and q, we need to find p and q. A brute force method is to check for all odd numbers in the interval between 3 and the square root of n, which can be extremely time-consuming because n is typically very large. Fortunately, this process can be parallelized by dividing the searching task to multiple machines and involves dividing the interval into nonoverlapping subintervals and letting each node work on one of them. Thus, the data transfer involves passing to each worker only the subinterval bounds and the value of n. We wrote MPI and LEECH versions of the application. They use standard C and the publicly available gmp library for handling large numbers.

Figure 5.4 shows the results of several experimental runs. We varied the key size, shown along the x-axis, and measured the application's time to completion. There are three pairs of graph lines: there is a pair from the LEECH and MPI versions for three different experiments using two, three, and four nodes. The nodes were taken from the set of laptops and the PlayStation 2.

MOBILE COMPUTING HANDBOOK

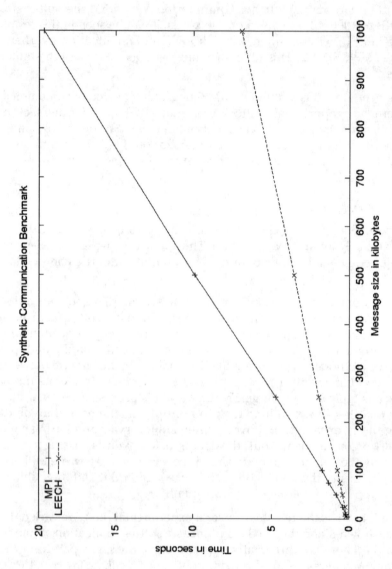

Figure 5.3. Average Round-Trip Latencies Measured from Execution of Synthetic Communication Benchmark Experiments were measured across two laptops.

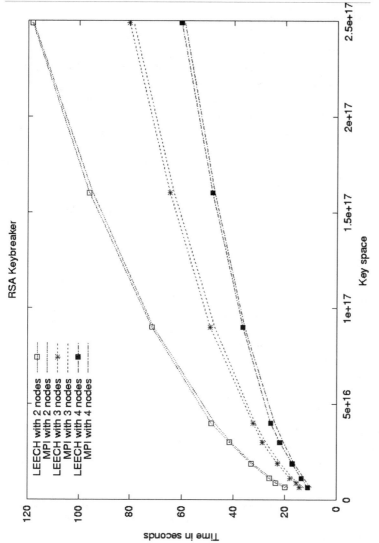

Figure 5.4. Average Execution Times of RSA Keybreaker Application
Two-node experiments were performed on two laptops. Three-node experiments were performed on three laptops. Four-node experiments were performed on three laptops and the PlayStation 2.

We note that current RSA practices suggest 512-bit keys, which is about 10^{155}. A 512-bit key supposedly takes $1 million in equipment cost to crack in about eight months [35]. In our experiments, we obviously did not want to approach these key lengths; instead, our keys ranged from 10^{15} to 10^{17}, which allowed trial runs to be completed within reasonable time bounds to facilitate experimentation.

This graph shows two important properties of the LEECH system. First, the application scales well with an increasing number of nodes, which is what one would obviously expect from a program run on parallel processing nodes. The LEECH runtime system in no way hinders this highly sought characteristic. Second, each run of the LEECH version is slightly slower than the pure MPI version on a consistent basis. We saw earlier that the communication overhead for LEECH is less than that of MPI; however, because this RSA application is computation-intensive rather than communication-intensive, this advantage is insignificant. Instead, this graph shows us that LEECH suffers from a small overhead that cannot be amortized for computationally bounded applications. This overhead comes from the job and session management subsystem. However, as can be seen in the graph, this overhead becomes less consequential with larger execution times.

5.7 Looking to the Future

Using small-scale devices for grid and cluster computing will potentially have an important role. What we have described in this chapter is LEECH, a system that supports this vision. Our use of a proxy-based architecture allows minion heterogeneity to be hidden, but our programming model and library give programmers the means to address devices as composite classes rather than extremely wide-varying individual machines. We showed through experimentation that the communication library has less overhead than MPICH and programs written to use our system demonstrate good scalability properties. Our experiments involved a heterogeneous mix of devices, including wireless laptops, PDAs, and a PlayStation 2 gaming console.

This research area is rich in technical challenges and opportunities. Potential topics include:

- Cluster topology configuration — we would like to gain insight from the field of mobile ad hoc networking to determine optimal topology strategies for our clusters. If the interlocutor is a stationary node, does it act in a similar way to a base station in an 802.11 network? If the interlocutor is mobile, is it similar to a clusterhead/gateway in a mobile ad hoc network? What if the cluster is in a multi-hop wireless network that reaches the wired grid? What if the grid system is itself in an all-wireless network? These are the types of issues that

arise when we try to find technical issues shared between our system and ad hoc networks.
- Security — we need to develop a trust model to accommodate our system architecture. Wireless security will play a major part in our attempts to provide a secure environment for small-scale device owners to participate. Network layer security protocols, such as IPsec, readily provide qualitative protection between a wireless host and a trusted LAN or a trusted host. Transport layer security protocols, such as Secure Socket Layer (SSL), Transport Layer Security (TLS), Wireless Transport Layer Security (WTLS), provide similar protection for user sessions. Nevertheless, much work remains, such as addressing the lack of security of 802.11 Wired Equivalent Privacy (WEP) and Bluetooth.
- Code deployment — code deployment of application code will need to take into account software interoperability with Globus, security, and feasibility. We are looking to leverage the research done in the field of mobile code.
- Economic model — economic modeling is a potential area of interesting research. Such a model may be needed in our plans because we are dealing with consumers and their devices. Naturally, consumers are hesitant to volunteer their machines due to concerns of power consumption, resource usage, and privacy. Will they respond to our LEECH system favorably? In one sense, this question has already been answered by the strong success of the Seti@home project: users are indeed willing to contribute their machines for the greater good of distributed computing. However, users will be less likely to allow their machines to be used than they would their resource-rich desktop PCs. As mentioned earlier, users need to be motivated on two fronts. First, a sufficiently persuasive support architecture of technical merit must be available that is proven to mitigate the issues inherent in devices. Our design will hopefully provide reassurance to the users that their limited machines are being efficiently used. However, no system architecture will completely mask these limitations. To further urge owners not fully convinced of the effectiveness of our system architecture, we suggest a second, complementary approach: users will need commercial and financial incentive to contribute what they may perceive to be their units' limited resources. We believe this project provides a great opportunity for our research group to contribute work in the field of economic modeling for distributed systems.
- Performance and scalability prediction — performance modeling and measurements will play a key role in this project. Performance modeling is traditionally used in the design of complex systems to assist in the selection of protocol/architecture alternatives; the interpretation of measured results; and the extrapolation of performance

behavior observed in a small testbed to larger, more realistic system scenarios. In our project, we will use modeling and simulation in all the above capacities. The most important (and challenging) contribution, however, will be in the extrapolation and scaling of the results. The design of a highly scalable system such as our proposed LEECH environment, where a multitude of diverse clients are executing on heterogeneous networks, requires careful *a priori* modeling to avoid serious mistakes in the implementation and deployment. These types of systems are notoriously complex to model: they are analytically intractable, abstract simulations of mobile wireless simulations are inaccurate, and detailed simulation models may be computationally impractical for any but small configurations. Further, previous performance studies of mobile and wireless systems have typically emphasized either the application or the network such that only one of the two major system components is modeled in detail and the other component is represented by an abstract model. Thus, networking-oriented simulators (e.g., GloMoSim, NS, OPNET) will tend to develop a detailed model of the network possibly including models of the protocol, together with propagation medium and radio models for wireless networks, while representing the application simply as stochastic, or possibly trace-based, traffic streams.

References

1. T. Anderson, D. Culler, D. Patterson, and the NOW Team. A Case for NOW (Networks of Workstations), *IEEE Micro*, February 1995.
2. M. Baker, R. Buyya, and D. Laforenza. The Grid: International Efforts in Global Computing, in *Proceedings of the International Conference on Advances in Infrastructure for Electronic Business, Science, and Education on the Internet,* July 31–August 6, 2000.
3. D. Becker, T. Sterling, D. Savarese, J. Dorband, U. Ranawak, and C. Packer. Beowulf: A Parallel Workstation for Scientific Computation, in *Proceedings of the International Conference on Parallel Processing,* 1995.
4. J. Broch, D. Maltz, D. Johnson, Y.-C. Hu, and J. Jetcheva. A Performance Comparison of Multi-Hop Wireless Ad-Hoc Network Routing Protocols, *Mobile Computing and Networking*, pp. 85–97, 1998.
5. R. Buyya, K. Branson, J. Giddy, and D. Abramson. The Virtual Laboratory: Enabling On-Demand Drug Design with the World Wide Grid, in *Proceedings of the IEEE International Symposium on Cluster Computing and the Grid,* May 21–24, 2002.
6. Cactus. From Supercomputers to PDAs: Cactus on an iPAQ, August 27, 2002. www.cactuscode.org/News/Ipaq.html.
7. T. Cai, P. Leach, Y. Gu, Y. Goland, and S. Albright. Simple Service Discovery Protocol, IETF Internet Draft, October 1999.
8. H. Casanova, G. Obertelli, F. Berman, and R. Wolski. The AppLeS Parameter Sweep Template: User-Level Middleware for the Grid, in *Proceedings of Supercomputing,* 2000.
9. K. Czajkowski, S. Fitzgerald, I. Foster, and C. Kesselman. Grid Information Services for Distributed Resource Sharing, in *Proceedings of the 10th IEEE International Symposium on High-Performance Distributed Computing (HPDC),* 2001.
10. H. Dail, H. Casanova, and F. Berman. A Decoupled Scheduling Approach for the GrADS Program Development Environment, in *Proceedings of Supercomputing,* 2002.

11. T. DeFanti, I. Foster, M. Papka, R. Stevens, and T. Kuhfuss. Overview of the I-WAY: Wide-Area Visual Supercomputing, *The International Journal of Supercomputing Applications and High Performance Computing,* vol. 10, no. 2, Summer–Fall 1996.
12. Hooked on Lithium, *The Economist, Technology Quarterly,* June 22, 2002.
13. J. Flinn, S. Park, and M. Satyanarayanan. Balancing Performance, Energy Conservation, and Application Quality in Pervasive Computing, in *Proceedings of ICDCS,* July 2002.
14. G. Forman and J. Zahorjan. The Challenges of Mobile Computing, *IEEE Computer,* vol. 27, no. 4, April 1994.
15. I. Foster and C. Kesselman. Globus: A Metacomputing Infrastructure Toolkit, *International Journal of Supercomputer Applications,* vol. 11, no. 2, 1997.
16. Gartner, Gartner Dataquest Says Worldwide PDA Market Suffers through a Dismal Year in 2002, *Gartner Press Release,* January 27, 2003. www3.gartner.com/ 5_about/ press_releases/pr27jan2003a.jsp.
17. Gartner, Gartner Says Worldwide PC Shipments Experienced Third Consecutive Quarter of Positive Growth, *Gartner Press Release,* April 17, 2003. www3.gartner. com/5_ about/press_releases/prapr172003a.jsp.
18. Globus, www.globus.org.
19. F. Gonzalez-Castano, J. Vales-Alonso, M. Livny, E. Costa-Montenegro, and L. Anido-Rifon. Condor Grid Computing from Mobile Handheld Devices, *ACM Mobile Computing and Communications Review,* vol. 7, no. 1, January 2003.
20. A. Grimshaw, W. Wulf, J. French, A. Weaver, and P. Reynolds, Jr. Legion: The Next Logical Step toward a Nationwide Virtual Computer, University of Virginia Technical Report no. CS-94-21, 1994.
21. M. Haahr, R. Cunningham, and V. Cahill. Supporting CORBA Applications in a Mobile Environment, in *Proceedings of the 5th International Conference on Mobile Computing and Networking,* August 1999.
22. J. Haartsen. BLUETOOTH — the Universal Radio Interface for Ad-Hoc Wireless Connectivity, *Ericsson Review,* no. 3, 1998.
23. K. Kennedy, M. Mazina, J. Mellor-Crummey, K. Cooper, L.Torczon, F. Berman, A. Chien, H. Dail, and O. Sievert. Toward a Framework for Preparing and Executing Adaptive Grid Programs, in *Proceedings of NSF Next Generation Systems Program Workshop,* April 2002.
24. T. Kimura and H. Takemiya. Local Area Metacomputing for Multidisciplinary Problems: A Case Study for Fluid/Structure Coupled Simulation, in *Proceedings of the International Conference on Supercomputing,* 1998.
25. M. Kobler, J. Kim, and D. Lilja. Communication Overhead of MPI, PVM, and Sckt Library, University of Minnesota Tech Report HPPC-98-06, 1998.
26. C. Lee, C. DeMatteis, J. Stepanek, and J. Wang. Cluster Performance and the Implications for Distributed, Heterogeneous Grid Performance, in *Proceedings of 9th Heterogeneous Computing Workshop,* May 2000.
27. C. Lee, S. Matsuoka, D. Talia, A. Sussman, M. Mueller, G. Allen, and J. Saltz. A Grid Programming Primer, Technical report, Advanced Programming Models Research Group, August 2001.
28. M. Litzkow, M. Livny, and M.W. Mutka. Condor — A Hunter of Idle Workstations, in *Proceedings of the 8th International Conference of Distributed Computing Systems,* June 1988.
29. S. Lyer, L. Luo, R. Mayo, and P. Ranganathan. Energy-Adaptive Display System Designs for Future Mobile Environments, in *Proceedings of ACM Mobisys,* May 2003.
30. J. Markoff. From PlayStation to Supercomputer, *The New York Times,* May 27, 2003.
31. G. Miklos, A. Racz, Z. Turanyi, A. Valko, and P. Johansson. Performance Aspects of Bluetooth Scatternet Formation, in *Proceedings of the 1st Annual Workshop on Mobile Ad Hoc Networking and Computing,* 2000.
32. M. Migliardi, M. Maheswarn, B. Maniymaran, P. Card, and F. Azzedin. Mobile Interfaces to Computational, Data, and Service Grid Systems, *ACM Mobile Computing and Communications Review,* vol. 6, no. 4, October 2002.

33. MPI: A Message Passing Interface Standard, June 1995. www.mpi-forum.org.
34. MPICH — A Portable MPI Implementation, www-unix.mcs.anl.gov/mpi/mpich/.
35. MPICH-G2 homepage, www3.niu.edu/mpi/.
36. M. Robshaw. Security Estimates for 512-bit RSA, Technical note, RSA Laboratories, 1995.
37. G. Pei, M. Gerla, and X. Hong. LANMAR: Landmark Routing for Large Scale Wireless Ad Hoc Networks with Group Mobility, in *Proceedings of IEEE/ACM MobiHOC*, August 2000.
38. C. Perkins and D. Johnson. Mobility Support in IPv6, in *Proceedings of the 2nd Annual International Conference on Mobile Computing and Networking,* November 1996.
39. RSA Laboratories, Cryptography Frequently Asked Questions, http://www.rsasecurity.com/rsalabs/faq/index.html.
40. K. Truelove and A. Chasin. Morpheus out of the Underworld, www.openp2p.com/pub/a/p2p/2001/07/02/morpheus.html.
41. UIUC, Scientific Computing on the Sony PlayStation 2 at the University of Illinois, Urbana-Champaign, arrakis.ncsa.uiuc.edu/ps2/index.php.
42. UPNP, Universal Plug and Play Device Architecture Reference Specification, Microsoft Corporation.
43. M. Weiser. The Computer for the Twenty-First Century, *Scientific American,* September 1991.

Chapter 6
Multimedia Messaging Service

Syed A. Ahson

6.1 Introduction

Messaging has become one of the most dominant applications in the world of mobile computing. Short Message Service (SMS) and Enhanced Messaging Service (EMS) are very successful services in second generation (2G) networks such as Global System for Mobile (GSM). Third Generation Partnership Project (3GPP) and Wireless Access Protocol (WAP) Forum have defined Multimedia Messaging Service (MMS) as the messaging service for 2.5G and 3G networks. MMS allows for the exchange of messages containing multimedia elements. Multimedia messages can also be composed as a slide show containing text, audio, video, and picture elements. MMS is designed to work well with existing multimedia messaging services such as SMS, EMS, and e-mail. The 3GPP and WAP Forum have worked together to standardize MMS. The 3GPP has been responsible for high level service requirements, architectural aspects of MMS, message structures, and content formats. The 3GPP has also produced technical realizations of selected interfaces between the network elements. WAP Forum has defined technical realizations of selected interfaces on the basis of WAP and Internet transport protocols. The 3GPP has provided four releases of service definitions and WAP Forum has completed technical realizations of two of them. MMS 1.0, the first technical realization is based on the WAP protocols. MMS 1.1, which is the current technical realization, enables devices and network elements to use Internet transport protocols.

The 3G networks are designed to provide for high bandwidth connections. High bandwidth connections will enable transfer of larger messages. MMS users will be able to compose from simple text messages, as in SMS, to complex multimedia messages as found in the Internet. MMS messages are similar to Microsoft® PowerPoint® presentations. Each slide is composed of text, audio, video, and images to be laid out on a graphical region. The time for which a slide is shown can be configured by the content

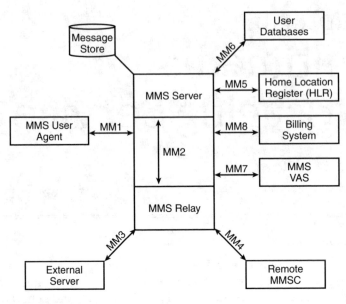

Figure 6.1. MMS Architecture

author and can be adjusted by the viewer. MMS inherits many powerful features of existing messaging systems such as group sending, delivery, and read-reply reports. MMS messages have class and priority attributes. MMS users are informed of incoming MMS messages by notifications. MMS messages can be downloaded automatically or deferred until the user requests it. MMS allows for persistent network-based storage. Messages can be stored persistently in the network and manipulated remotely by users. MMS supports content adaptation to accommodate the broad range of device capabilities. MMS messages can be manipulated for the recipient's device configuration. MMS service is now widely available across the globe and has been identified by business analysts as a major revenue generator.

6.2 MMS Architecture

Several entities work together to provide the MMS Experience. Figure 6.1 illustrates the different entities required in a typical MMS environment. MMS user agent (MMS UA) is the software application resident in the mobile handset. This software application allows for composition, viewing, sending, and retrieval of multimedia messages. MMS UA is responsible for the presentation of notifications, read-reply, and delivery reports. The MMS environment (MMSE) is a set of MMS entities under the control of a single MMS provider. MMS relays are responsible for routing MMS messages. MMS servers are responsible for storing messages. Usually the MMS server and MMS relay functionality is combined into a single element. This

Multimedia Messaging Service

Table 6.1. 3GPP to WAP Forum Terminology Mapping

3GPP Terminology	WAP Forum Terminology
MM1	MMS_M
MM2	MMS_S
MM3	E (e-mail server) and L (legacy)
MM4	MMS_R
MM5	not referred to
MM6	not referred to
MM7	not referred to
MM8	not referred to

combined element is known as the MMS center (MMSC). The MMSC also presents an interface to other existing messaging systems such as SMSC and e-mail servers. This ensures interoperability with Internet users and SMS, EMS devices. MMSC can deliver MMS messages in two modes — batch mode and streaming mode. In batch mode, entire messages are delivered to the user agent. In the optional streaming mode, messages are delivered part by part to the user agent. MMSC is responsible for content adaptation. MMSC tailor the MMS message to be delivered based on capabilities of the user agent. MMSC generate charging data records (CDR) for billing purposes. MMS may offer additional capabilities such as blocking subscribers and blacklists.

6.2.1 MMS Interfaces

In the MMS environment transactions between network entities are associated with interfaces. An MMS transaction proceeds by exchange of information elements across the interfaces. The 3GPP has termed these interfaces as MM1 through MM8. These interfaces are illustrated in Figure 6.1. MM1 is the interface between MMS UA and MMSC. MM2 is the interface between MMS relay and MMS server. MM3 is the interface between MMSC and external servers such as e-mail servers and SMSC. MM4 is the interface between two MMSCs. MM5 is the interface between MMSC and network elements such as Home Location Register (HLR). MM6 is the interface between MMSC and user databases. MM7 is the interface between MMSC and external value added service applications. MM8 is the interface between MMSC and a billing system. WAP Forum has used different terminology for the MMS interfaces. Mapping between terminology used by 3GPP and WAP Forum is given in Table 6.1. In MMS1.0 network configuration, MMSC to WAP gateway transactions use Hypertext Transfer Protocol (HTTP) as the transport protocol. WAP gateway to MMS UA transactions

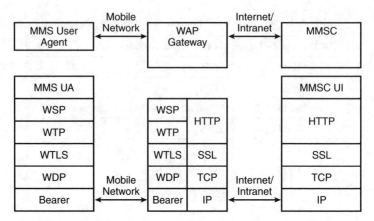

Figure 6.2. MMS 1.0 Network Configuration

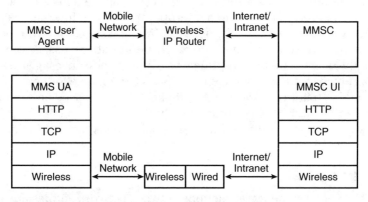

Figure 6.3. MMS 1.1 Network Configuration

are performed over the Wireless Session Protocol (WSP). The WAP gateway converts HTTP requests and responses to WSP requests and responses and vice versa. MMS1.0 network configuration and associated protocol stack are illustrated in Figure 6.2. In MMS1.1 network configuration, the MMS UA can interact directly using HTTP as the transport protocol. MMS1.1 network configuration and associated protocol stack are illustrated in Figure 6.3.

6.2.2 Addressing in MMS

MMS supports two modes of addressing — e-mail addressing and mobile station integrated services digital network (MSISDN) addressing. In e-mail addressing mode, each subscriber has an e-mail address associated with it. In MSISDN addressing, MMS subscribers are identified by their MSISDN number. An example of an e-mail address is To: Syed Ahson <syed@mms.motorola.net>/TYPE=rfc822. An example of a MSISDN address is To: + 9543709672/TYPE=PLMN. MMS also allows for address hiding. MMS messages may be delivered to recipients without providing the sender's details. MMS also supports group sending.

6.2.3 Technical Specifications

The 3GPP is responsible for the definition of high-level requirements, overall MMS architecture and MMS transaction flows. The 3GPP has also produced technical specification for several interfaces. WAP Forum has concentrated on the technical realization of MM1 interface. The 3GPP has produced three releases of technical specifications corresponding to a set of features with each set being backward compatible. The three 3GPP releases are MMS release 99, MMS release 4, and MMS release 5. WAP Forum has produced technical realizations corresponding to release 99 and release 4 that are referred to as MMS 1.0 and MMS 1.1, respectively. The technical specification [3GPP-22.140] introduced in release 99 states high-level service requirements for MMS. The technical specification [3GPP-23.140] introduced in release 99 describes MMS architecture and transaction flows. The technical specification [3GPP-26.140] introduced in release 5 describes media formats and codecs. The technical specification [3GPP-32.235] introduced in release 4 describes procedures for management of CDR. MMS Architecture Overview [WAP-205] outlines how MMS is implemented in the WAP framework. MMS Client Transaction [WAP-206] illustrates the transactions between MMS UA and MMSC over the MM1 interface. MMS encapsulation [WAP-209] details the protocol data units exchanged over the MM1 interface.

6.2.4 Supported Formats

The minimum requirement for MMS enabled devices is to support United States — American Standard Code for Information Exchange (US-ASCII) text only. MMS enabled devices may allow natural and synthetic audio content to be present in MMS messages. If the device supports natural audio content then Adaptive Multi-Rate (AMR) codec must be supported. Support for Motion Picture Experts Group (MPEG) Audio Layer-3 (MP3) is suggested. If the device supports synthetic audio content then Musical Instrument Digital Interface (MIDI) may be supported. MMS enabled devices may allow image content to be present in MMS messages. If the device supports image content then Joint Photographic Experts Group (JPEG) codec must be supported. Support for GIF87a (GIF — Graphics Interchange Format), GIF89a, and WBMP (wireless bitmap) is suggested. MMS enabled devices may allow video content to be present in MMS messages. If the device supports video content then H.263 codec must be supported. Support for MPEG4 is suggested. MMS enabled devices may allow scene description to

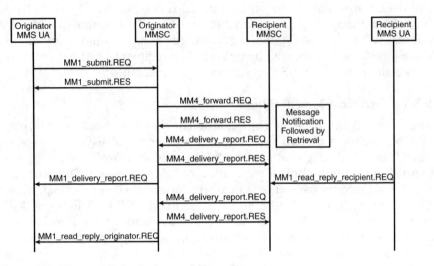

Figure 6.4. Message Submission and Transfer

be specified for MMS messages. If the device supports scene description/message presentation then Synchronized Multimedia Integration Language (SMIL) must be supported. Support for Wireless Markup Language/X Hypertext Markup Language (WML/XHTML) is suggested.

6.2.5 MMS Messages

Nontextual elements such as image, audio, and video are included in a MMS message by formatting the message as a multipart message according to Multipurpose Internet Mail Extensions (MIME). MMS messages may optionally contain graphical layout and time synchronization information for included media elements. This description is specified according to formatting languages such as SMIL, XHTML, or WML. MMS messages that have synchronized media elements are of multipart/related content type. MMS messages that have unrelated media elements are of multipart/mixed content type. The MMS protocol data unit (PDU) consists of a PDU header and a PDU data section. The MMS PDU is inserted in the data section of a WSP or HTTP request/response. The content type of a WSP/HTTP request/response containing a MMS PDU is `application/vnd.wap.mms-message`.

6.3 Message Submission

The MMS UA submits MMS messages to the MMSC over the MM1 interface. Figure 6.4 illustrates the message submission transaction. `MM1_submit.REQ` corresponds to the submission request and `MM1_submit.RES` corresponds to the response. The message originator's MMSC consults its routing tables and forwards the MMS message to the recipient's MMSC. The

Table 6.2. MM1 Submit Request

Information Elements	WAP Implementation	Type	Values
Recipient address	`To, Cc, Bcc`	String	
Sender address	`From`	String	
Date and time	`Date`	Integer	
Time of expiry	`X-MMS-Expiry`	Date	
Earliest delivery time	`X-MMS-Delivery-Time`	Date	Immediate (default)
Reply charging	`X-MMS-Reply-Charging`		Requested, Requested text only
Reply deadline	`X-MMS-Reply-Charging-Deadline`	Date	
Reply charging size	`X-MMS-Reply-Charging-Size`	Integer	
Reply charging identification	`X-MMS-Reply-Charging-ID`	String	
Delivery report	`X-MMS-Delivery-Report`		Yes, No
Read-reply	`X-MMS-Read-Reply`		Yes, No
Message class	`X-MMS-Message-Class`		Personal (default), Auto
Priority	`X-MMS-Priority`		Low, Normal (default), High
Sender visibility	`X-MMS-Sender-Visibility`		Hide, Show (default)
Subject	`Subject`	String	
Content type	`Content-type`	String	
Content	`Message body`		

recipient's MMSC stores the message and generates a notification. This notification is delivered to the recipient's MMS UA. The MMS UA can then immediately retrieve the message or defer it. WAP technical realizations have termed the `MM1_submit.REQ` as `M-Send.req` and `MM1_submit.RES` as `M-Send.conf`. Information elements present in a MM1 message submission request are shown in Table 6.2.

Table 6.3. MM1 Submit Response

Information Elements	WAP Implementation	Type	Values
Request status	`X-MMS-Response-Status`		Okay, error codes listed in Table 6.4
Request status text	`X-MMS-Response-Text`	String	
Message Identification	`Message-ID`	String	

As part of the message submission request, the message originator can request a delivery report to be generated upon delivery of the message. The message originator can also request a read-reply report to be generated once the message has been read. MMS introduces the concept of reply charging. The message's originator can indicate in the submission request that he or she will pay for the reply. The message's originator can specify certain conditions to be met such as a reply deadline and a maximum reply size. MM1 message submission request includes the mandatory recipient address (To, Cc, Bcc) information elements. The request also includes the mandatory sender address (From) information element. Contents of the message are indicated by the mandatory content type information element. Date and time of message submission by the subscriber is indicated by the optional date information element. The MMSC may overwrite the date information element. Validity duration of the MMS message is indicated by the optional `X-MMS-Expiry` information element. If the message originator requests reply charging, the reply charging (`X-MMS-Reply-Charging`) information element will be present. Reply charging deadline is specified by the `X-MMS-Reply-Charging-Deadline` information element. Reply charging size is specified by the `X-MMS-Reply-Charging-Size` information element. If a message is being submitted as reply message for which reply charging was requested in the original message, identification of the original message is indicated by the `X-MMS-Reply-Charging-ID` information element. Delivery reports may be requested by the `X-MMS-Delivery-Report` information element. Read-reply reports may be requested by the `X-MMS-Read-Reply` information element. Message class is indicated by the `X-MMS-Message-Class` information element. Message priority is indicated by the `X-MMS-Priority` information element. The message originator may request address hiding by setting the `X-MMS-Sender-Visibility` field to hide. Subject of the message is present in the subject information element. Message content type is indicated by the content-type information element. Content of the message is present in the message body.

The MMSC generates a `MM1_submit.RES` response for the `MM1_submit.REQ`. WAP Forum has termed `MM1_submit.RES` as M-Send.conf. Information elements present in a MM1 message submission request are shown in Table 6.3. Request status information element (`X-MMS-`

Response-Status) of the response indicates acceptance or rejection of the submit request. The MM1_submit.REQ may be rejected because of several permanent or transitory error conditions. Optional Request Status text information element (X-MMS-Response-Text) provides textual description of the status. Permanent and transitory error codes are listed in Table 6.4. The MMSC assigns a message identification to each accepted MMS message. This is present in the message identification (Message-ID) information element. This message identification is used to pair up delivery/read-reply reports and reply charging messages to the original message.

6.4 Message Transfer

The message originator MMSC is responsible for routing the incoming MMS message. The MMS message could be destined for another subscriber in the MMSE. Alternatively, the MMS message could be destined for a non-MMS subscriber or another MMS subscriber in a different MMSE. MMS messages destined for MMS subscribers in a different MMSE are forwarded by the MM4_forward.REQ. The message originator MMSC can request an acknowledgment from the recipient's MMSC. Figure 6.4 illustrates the message transfer transaction. In the technical realization, MM4 transactions are performed using Simple Message Transfer Protocol (SMTP). Information elements present in a MM4 message forward request are shown in Table 6.5. The recipient's MMSC acknowledges the MM4_forward.REQ with MM4_forward.RES. Information elements present in a MM4 message forward response are shown in Table 6.6. MM4_forward.RES contains the request status. MM4 Forward Response errors are listed in Table 6.7.

6.5 Delivery Report

Delivery reports may be requested by the message originator as part of the MM1_submit.REQ. The recipient's MMSC generates a MM4_delivery_report.REQ on message delivery or deletion and forwards it to the originator's MMSC. Figure 6.4 and Figure 6.5 illustrate delivery report transaction. MM4_delivery_report.REQ includes original message identification and original message time-stamping information elements that pairs the delivery report with the original message and indicates the time it was handled. The delivery status information element indicates the status of delivery such as "message retrieved," "message deleted," or "message rejected." MM4_delivery_report.REQ transaction is performed using SMTP.

The recipient's MMSC may request an acknowledgment from the originator's MMSC. Information elements present in a MM4 delivery report request are shown in Table 6.8. The originator's MMSC acknowledges the delivery report with a delivery report forward response (MM4_delivery_report.RES). Information elements present in a MM4 delivery report response are shown in Table 6.9.

Table 6.4. MM1 Submit Response Errors

Binary Value	WAP Implementation	Description
192	Error-transient-failure	Valid request, but cannot be processed due to some temporary conditions.
193	Error-transient-sending-address-unresolved	MMSC cannot resolve address due to some temporary conditions.
194	Error-transient-message-not-found	MMSC cannot retrieve message due to some temporary conditions.
195	Error-transient-network-problem	MMSC cannot process request due to some overload conditions.
224	Error-permanent-failure	Unspecified permanent error.
225	Error-permanent-service-denied	Service authorization and authentication failures.
226	Error-permanent-message-format-corrupt	Problem with message format.
227	Error-permanent-sending-address-unresolved	Unable to resolve recipient's address.
228	Error-permanent-message-not-found	Unable to retrieve the message.
229	Error-permanent-content-not-accepted	MMSC cannot process message due to content format or message size limitation.
230	Error-permanent-reply-charging-limitations-not-met	Reply charging requirements not met.
231	Error-permanent-reply-charging-request-not-accepted	Reply charging request is rejected due to service or user configuration.
232	Error-permanent-reply-charging-forward-denied	Forwarding request is denied due to reply charging requirements.
233	Error-permanent-reply-charging-not-supported	MMSC does not support reply charging.

Multimedia Messaging Service

Table 6.5. MM4 Message Forward Request

Information Elements	3GPP Header	Type	Example Values
3GPP MMS version	X-Mms-3GPP-MMS-Version	String	5.2.0
Message type	X-Mms-Message-Type	String	MM4_forward.REQ
Transaction identification	X-Mms-Transaction-ID	String	
Message Identification	X-Mms-Message-ID	String	
Recipient's address	To, Cc		
Sender's Address	From	String	
Message subject	Subject	String	
Message class	X-Mms-Message-Class	String	
Message date and time	Date	Date	
Time of expiry	X-Mms-Expiry	Date or duration	
Delivery report	X-Mms-Delivery-Report		Yes, No
Read-reply report	X-Mms-Read-Reply		Yes, No
Priority	X-Mms-Priority		Low, Normal, High
Sender visibility	X-Mms-Sender-Visibility		Hide, Show
Forward counter	X-Mms-Forward-Counter	Integer	
Previously sent by	X-Mms-Previously-Sent-By	String with index	1,syed@motorola.net 2,steve@apple.net
Previously sent date and time	X-Mms-Previously-Sent-Date-And-Time	Date with index	1,Thu Aug 07 21:00:00 2003 2,Tue Jan 07 07:00:00 2003
Request for acknowledgment	X-Mms-Ack-Request		Yes, No

Table 6.5. MM4 Message Forward Request (Continued)

Information Elements	3GPP Header	Type	Example Values
Content-type	Content-Type	String	
	X-Mms-Originator-System	String	
	Message-ID	String	
Content type	Message Body		

Table 6.6. MM4 Message Forward Response

Information Elements	3GPP Header	Type	Example Values
3GPP MMS version	X-Mms-3GPP-MMS-Version	String	5.2.0
Message type	X-Mms-Message-Type	String	MM4_forward.RES
Transaction identification	X-Mms-Transaction-ID	String	
Message identification	X-Mms-Message-ID	String	
Request status code	X-Mms-Request-Status-Code		Okay, error codes defined in Table 6.7
Status text	X-Mms-Status-Text	String	
Address of sender's MMSC	Sender	String	
Address of recipient's MMSC	To	String	
	Message-ID	String	
Message date and time	Date	Date	

Request status information element of the MM4_delivery_report.REQ indicates the status of the delivery report forwarding request. The delivery report is delivered to the message originator over the MM1 interface. The originator's MMSC generates a MM1_delivery_report.REQ. MM1_delivery_report.REQ includes information elements that indicate the original message identification, time of handling, and delivery status. Information elements present in a MM1 delivery report request are shown in Table 6.10. MM1 delivery report status codes are listed in Table 6.11.

Multimedia Messaging Service

Table 6.7. MM4 Forward Response Errors

Status Code	Description
Error-unspecified	Unspecified error.
Error-service-denied	Authentication or authorization failure.
Error-message-format-corrupt	Badly formatted message format.
Error-sending-address-unresolved	MMSC was unable to resolve one of the recipient addresses.
Error-network-problem	MMSC capacity overload.
Error-unsupported-message	MMSC does not support the request type.
Error-content-not-accepted	MMSC cannot process message due to content format or message size limitation.

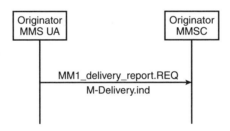

Figure 6.5. MM1 Delivery Report

6.6 Read-Reply Reports

The MMS message originator can request to be notified when the recipients have read the message. Figure 6.4 and Figure 6.6 illustrate the read-reply report transaction. The recipient's MMS UA generates a read-reply report when the recipient has read the message. The recipient's MMS UA submits a `MM1_read_reply_recipient.REQ` to the recipient's MMSC. Original message identification, original message time-stamping and read-reply status information elements pair up the read-reply report with the original message, its time of handling and status (message had been read or message has been deleted without being read). Information elements present in a `MM1_read_reply_recipient.REQ` are shown in Table 6.12. MM1 read-reply report status codes are listed in Table 6.13.

The recipient's MMSC forwards the read-reply report over the MM4 interface to the originator's MMSC. Information elements present in a MM4 read-reply report request are shown in Table 6.14. The `MM4_read_reply_report.REQ` transaction is performed using SMTP.

Table 6.8. MM4 Delivery Report Request

Information Elements	3GPP Header	Type	Example Values
3GPP MMS version	X-Mms-3GPP-MMS-Version	String	5.2.0
Message type	X-Mms-Message-Type	String	MM4_delivery_report.REQ
Transaction identification	X-Mms-Transaction-ID	String	
Message identification	X-Mms-Message-ID	String	
Recipient's address	From		
Sender's address	To, Cc	String	
Message date and time	Date	Date	
Request for acknowledgment	X-Mms-Ack-Request		Yes, No
Message status code	X-Mms-MM-Status-Code	String	Expired, Retrieved, Rejected, Deferred, Indeterminate, Forwarded, Unrecognized
Message status text	X-MM-Status-Text	String	
Address of sender's MMSC	Sender	String	
	Message-ID	String	

The recipient's MMSC may request an acknowledgment for the `MM4_read_reply_report.REQ`. The message originator's MMSC will then reply with a `MM4_read_reply_report.RES`. Information elements present in a MM4 read-reply report response are shown in Table 6.15. The message originator's MMSC delivers the read-reply report to the originator's MMS UA over the MM1 interface (`MM1_read_reply_originator.REQ`). Information elements present in a `MM1_read_reply_originator.REQ` are shown in Table 6.12.

6.7 Message Notification

The MMSC generates a notification for each incoming multimedia message. This notification indicates that a multimedia message is awaiting retrieval in the MMSC store. The MMS UA is responsible for retrieving the MMS message corresponding to the notification. The MMS UA may retrieve the MMS

Table 6.9. MM4 Delivery Report Response

Information Elements	3GPP Header	Type	Example Values
3GPP MMS version	X-Mms-3GPP-MMS-Version	String	5.2.0
Message type	X-Mms-Message-Type	String	MM4_delivery_report.RES
Transaction identification	X-Mms-Transaction-ID	String	
Message identification	X-Mms-Message-ID	String	
Request status code	X-Mms-Request-Status-Code		Okay, error codes defined in Table 6.7
Status text	X-Mms-Status-Text	String	
Address of sender's MMSC	Sender	String	
Address of recipient's MMSC	To	String	
	Message-ID	String	
Message date and time	Date	Date	

Table 6.10. MM1 Delivery Report Request

Information Elements	3GPP Header	Type	Example Values
Message identification	Message-ID	String	
Recipient address	To	String	
Event date	Date	Date	
Message status	X-MMS-Status		Expired, Retrieved, Rejected, Indeterminate, Forwarded

message without notifying the user. This is known as immediate retrieval. Alternatively, the MMS UA may present the notification to the user and the user is responsible for initiating the download of the MMS message. This is

MOBILE COMPUTING HANDBOOK

Table 6.11. MM1 Delivery Report Status Codes

Binary Value	WAP Implementation	Description
128	Expired	MMS message has expired before it could be retrieved.
129	Retrieved	MMS message has been retrieved by the recipient's MMS UA.
130	Rejected	The recipient's MMS UA is not willing to retrieve the MMS message.
133	Indeterminate	MMS message may or may not have been retrieved by the recipient's MMS UA.
134	Forwarded	The MMS message has been forwarded.

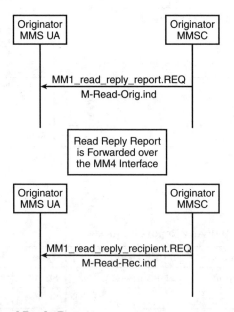

Figure 6.6. MM1 Read-Reply Report

known as deferred retrieval mode. Figure 6.7 illustrates the message notification transaction. Information elements present in a MM1 Notification Request are shown in Table 6.16.

The optional from information element of the MM1_notification.REQ indicates the message originator's address. The mandatory X-MMS-Message-Class information element indicates the message class (personal, advertisement, informational, auto) of the MMS message. The mandatory

Multimedia Messaging Service

Table 6.12. MM1 Read-Reply Report Request

Information Elements	3GPP Header	Type	Example Values
Recipient address	To	String	
Originator address	From	String	
Message identification	Message-ID	String	
Date and time	Date	Date	
Message status	X-MMS-Read-Status		Read, Deleted without being read

Table 6.13. MM1 Read-Reply Report Status Codes

Binary Value	WAP Implementation	Description
128	Read	Recipient has read the MMS message.
129	Deleted without being read	MMS message has been deleted without being read by the message recipient.

X-MMS-Expiry information element indicates the time of expiry of the MMS message. The optional X-MMS-Delivery-Report information element indicates that the message originator has requested a delivery report for that message. Reply charging attributes of the MMS message are indicated by the optional X-MMS-Reply-Charging, X-MMS-Reply-Charging-ID, X-MMS-Reply-Charging-Deadline, and X-MMS-Reply-Charging-Size information elements. Message priority and message subject are indicated by the optional X-MMS-Priority and subject information elements. Message size of the MMS message is indicated by the mandatory X-MMS-Message-Size information element. The incoming MMS message is indicated by the mandatory X-MMS-Content-Location information element.

The MMS UA acknowledges a MM1_notification.REQ with a MM1_notification.RES. Information elements present in a MM1 Notification Response are shown in Table 6.17. The mandatory X-MMS-Status information element indicates the status of the message retrieval. Possible values are Retrieved, Rejected, Deferred, Forwarded, and Unrecognized. The optional X-MMS-Report-Allowed information element indicates whether a delivery report should be generated. MM1 Notification response/status codes are listed in Table 6.18.

Table 6.14. MM4 Read-Reply Report Request

Information Elements	3GPP Header	Type	Example Values
3GPP MMS version	X-Mms-3GPP-MMS-Version	String	5.2.0
Message type	X-Mms-Message-Type	String	MM4_read_reply_report.REQ
Transaction identification	X-Mms-Transaction-ID	String	
Message identification	X-Mms-Message-ID	String	
Recipient's address	From		
Sender address	To	String	
Message date and time	Date	Date	
Request for acknowledgment	X-Mms-Ack-Request		Yes, No
Message status code	X-Mms-Read-Status-Code	String	Read, Deleted without being read
Message status text	X-MM-Status-Text	String	
	Sender	String	
	Message-ID	String	

6.8 Message Retrieval

The recipient's MMS UA requests the message by issuing a MM1_retrieve.REQ. The MMSC may alter contents of the MMS message to match the capabilities of the recipient's user agent. The MMS message is delivered to the recipient as part of the retrieve response (MM1_retrieve.RES). Figure 6.8 illustrates immediate and deferred message retrieval transaction. Information elements present in a MM1 Retrieval Response are shown in Table 6.19.

The mandatory Message-ID information element indicates the message identification of the message. The optional to, cc, and from information elements indicate address of the recipients and sender respectively. Message class (personal, advertisement, informational, auto) is indicated by X-MMS-Message-Class. Date and time of message submission or forwarding is indicated by the mandatory date information element. Request for delivery reports is indicated by the optional X-MMS-Delivery-Report information element. Reply charging options are indicated by the

Multimedia Messaging Service

Table 6.15. MM4 Read-Reply Report Response

Information Elements	3GPP Header	Type	Example Values
3GPP MMS version	`X-Mms-3GPP-MMS-Version`	String	5.2.0
Message type	`X-Mms-Message-Type`	String	`MM4_read_reply_report.RES`
Transaction identification	`X-Mms-Transaction-ID`	String	
Request status code	`X-Mms-Request-Status-Code`		Okay, error codes defined in Table 6.7
Status text	`X-Mms-Status-Text`	String	
Address of sender's MMSC	`Sender`	String	
Address of recipient's MMSC	`To`	String	
	`Message-ID`	String	
Message date and time	`Date`	Date	

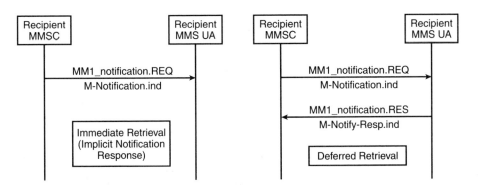

Figure 6.7. MM1 Notification

`X-MMS-Reply-Charging`, `X-MMS-Reply-Deadline`, `X-MMS-Reply-Charging-Size`, and `X-MMS-Reply-Charging-ID` information elements. Message priority is indicated by the `X-MMS-Priority` information element. Read-reply report may be requested by the optional `X-MMS-Read-Reply` information element. Message subject is indicated by the

139

Table 6.16. MM1 Notification Request

Information Elements	WAP Implementation	Type	Values
Sender address	From	String	
Message class	X-MMS-Message-Class		Personal (default), Advertisement, Informational, Auto
Time of expiry	X-MMS-Expiry	Date	
Delivery report	X-MMS-Delivery-Report		Yes, No
Reply charging	X-MMS-Reply-Charging		Requested, Requested text only
Reply deadline	X-MMS-Reply-Charging-Deadline	Date	
Reply charging size	X-MMS-Reply-Charging-Size	Integer	
Reply charging identification	X-MMS-Reply-Charging-ID	String	
Priority	X-MMS-Priority		Low, Normal (default), High
Subject	Subject	String	
Message size	X-MMS-Message-Size	Integer	
Message reference	X-MMS-Content-Location	String (URI)	

Table 6.17. MM1 Notification Response

Information Elements	WAP Implementation	Type	Values
Message status	X-MMS-Status		Retrieved, Rejected, Deferred, Forwarded, Unrecognized
Delivery report allowed	X-MMS-Report-Allowed	String	Yes (default), No

Multimedia Messaging Service

Table 6.18. MM1 Notification Response/Status Codes

Binary Value	WAP Implementation	Description
129	Retrieved	The recipient's MMS UA has already retrieved the MMS message.
130	Rejected	The recipient's MMS UA is not willing to retrieve the MMS message.
131	Deferred	The recipient's MMS UA is not willing to immediately retrieve the MMS message. The recipient's MMS UA will retrieve the MMS message at a later time.
132	Unrecognized	This status code is used for version management.

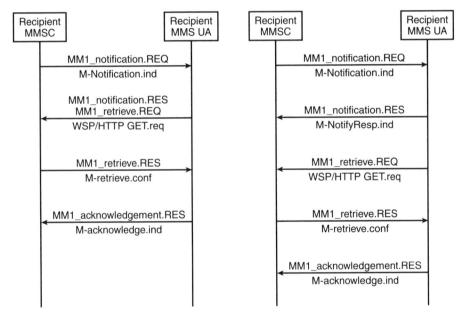

Figure 6.8. MM1 Immediate and Deferred Message Retrieval

optional subject element. Retrieve status and textual description of the retrieve status is indicated by the X-MMS-Retrieve-Status and X-MMS-Retrieve-Text fields, respectively. Retrieve status (X-MMS-Retrieve-Status) error codes are listed in Table 6.20.

Address of user agents that have submitted or forwarded the MMS message prior to the MMSC are listed in the optional X-MMS-Previously-Sent-By information element. Date and time of these user agents is specified

Table 6.19. MM1 Retrieval Response

Information Elements	WAP Implementation	Type	Values
Message identification	Message-ID	String	
Recipient address	To, Cc	String	
Sender address	From	String	
Message class	X-MMS-Message-Class		Personal (default), Advertisement, Informational, Auto
Date and time	Date	Integer	
Delivery report	X-MMS-Delivery-Report		Yes, No
Reply charging	X-MMS-Reply-Charging		Requested, Requested text only
Reply deadline	X-MMS-Reply-Charging-Deadline	Date	
Reply charging size	X-MMS-Reply-Charging-Size	Integer	
Reply charging identification	X-MMS-Reply-Charging-ID	String	
Priority	X-MMS-Priority		Low, Normal (default), High
Read-reply report	X-MMS-Read-Reply		Yes, No
Subject	Subject	String	
Retrieve status	X-MMS-Retrieve-Status		Okay, Error codes listed in Table 6.15
Status text	X-MMS-Retrieve-Text	String	
Previously sent by	X-Mms-Previously-Sent-By	String with index	1,syed@motorola.net 2,steve@apple.net
Previously sent date and time	X-Mms-Previously-Sent-Date-And-Time	Date with index	1,Thu Aug 07 21:00:00 2003 2,Tue Jan 07 07:00:00 2003
Content type	Content-type	String	
Content	Message body		

Multimedia Messaging Service

Table 6.20. MM1 Retrieval Response Errors

Binary Value	WAP Implementation	Description
192	Error-transient-failure	Valid request but cannot be processed due to some temporary conditions.
193	Error-transient-message-not-found	MMSC cannot retrieve message due to some temporary conditions.
194	Error-transient-network-problem	MMSC cannot process request due to some temporary conditions.
224	Error-permanent-failure	Unspecified permanent error.
225	Error-permanent-service-denied	Service authorization and authentication failures.
226	Error-permanent-message-not-found	Unable to retrieve the message.
227	Error-permanent-content-unsupported	MMSC cannot process message due to content format or message size limitation.

Table 6.21. MM1 Retrieval Acknowledgment

Information Elements	WAP Implementation	Type	Values
Delivery report allowed	X-MMS-Report-Allowed	String	Yes (default), No

by the optional X-MMS-Previously-Sent-Date-and-Time information element. Message content type is indicated by the mandatory message content type information element. The message itself is contained in the message body information element. The recipient's user agent acknowledges the MM1_retrieve.RES with a MM1_acknowledgment.REQ. Information elements present in a MM1 Retrieval Acknowledgment are shown in Table 6.21. X-MMS-Report-Allowed information element of the MM1_acknowledgment.REQ indicates whether the recipient allows a delivery report to be sent to the message originator.

6.9 Message Forwarding

MMS messages may be forwarded by the user agent by submitting a previously received MMS message as a MM1_submit.REQ. Additionally, MMS

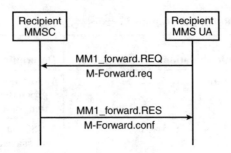

Figure 6.9. MM1 Message Forward

Table 6.22. MM1 Message Forward Request

Information Elements	WAP Implementation	Type	Values
Recipient address	To, Cc, Bcc	String	
Forwarding address	From	String	
Date and time	Date	Integer	
Time of expiry	X-MMS-Expiry	Date	
Earliest delivery time	X-MMS-Delivery-Time	Date	Immediate (default)
Delivery report	X-MMS-Delivery-Report		Yes, No
Read-reply	X-MMS-Read-Reply		Yes, No
Message reference	X-MMS-Content-Location	String (URI)	

messages that have not been retrieved may be forwarded by MM1_forward.REQ. Figure 6.9 illustrates message forwarding transaction. Information elements present in a MM1 Message Forward Request are shown in Table 6.22.

Address of recipients of the forwarded message is listed in the mandatory to, cc, bcc information elements. Address of the forwarding user agent is listed in the optional from information element. Date and time of forwarding the message is indicated by optional date information element. Time of expiry of the message is set in the optional X-MMS-Expiry information element. The earliest time of delivery may be specified by the optional X-MMS-Delivery-Time information element. Delivery reports may be requested by the optional X-MMS-Read-Reply information element.

Table 6.23. MM1 Message Forward Response

Information Elements	WAP Implementation	Type	Values
Request status	X-MMS-Response-Status		Okay, error codes listed in Table 6.4
Request status text	X-MMS-Response-Text	String	
Message identification	Message-ID	String	

The MMSC generates a `MM1_forward.RES` in response to the `MM1_forward.REQ`. Information elements present in a MM1 Message Forward Response are shown in Table 6.23. Status of the forward request is indicated by the mandatory `X-MMS-Response-Status` information element. Message identification of the message to be forwarded is indicated by the mandatory `Message-ID` information element. Textual description of the request status is indicated by optional `X-MMS-Response-Text` information element. The forwarding MMSC may route the message to the recipients MMSC over the MM4 interface to another non-MMS messaging interface.

6.10 Future Directions

MMS is a powerful messaging service. It allows for a wide range of content formats to be presented as a multimedia slideshow. MMS has been designed to be interoperable with existing messaging systems such as e-mail, SMS, and EMS. This chapter describes extensively the MMS protocol, its PDU formats, and features offered. Several features of the MMS protocol are still under investigation. MMS notifications consume precious bandwidth and an optimized method for flow control of notifications is needed. Persistent storage and manipulation of MMS messages on a remote sever are currently being looked at. The MM1 interface needs to be extended to allow a MMS UA to remotely manipulate MMS messages. The functional specification for persistent storage is termed MMBox (Multimedia Message Box) and is presented in [3GPP-23.140] release 5. The MM2 interface between the MMS server and MMS relay needs to be defined. The MM3 interface for internetworking with voice messaging systems, SMS, and e-mail systems needs to be standardized. The MM6 interface (MMSC/user databases) also needs to be defined. The MM8 interface (MMSC/billing system) also needs to be defined. Digital Rights Management (DRM) is required for providing permission to subscribers for forwarding, modification, and redistribution of digital media. Integration of Instant Messaging and Presence managing systems in a MMSE needs to be addressed. Usage of Wireless Transport Layer Security (WTLS), Secure HTTP (HTTP-S), and Secure MIME in a MMSE for providing end-to-end security is another consideration.

References

3GPP Documents

[3GPP-22.140] 3GPP TS 22.140: Multimedia Messaging Service (MMS), stage 1.
[3GPP-23.140] 3GPP TS 23.140: Multimedia Messaging Service (MMS), stage 2.
[3GPP-26.140] 3GPP TS 26.140: Multimedia Messaging Service, media formats, and codecs.
[3GPP-32.235] 3GPP TS 32.235: Charging management, charging data description for application services.

WAP Forum Documents

[WAP-205] WAP-205-MMSArchOverview-20010425-a: Multimedia Messaging Service Architecture Overview, version 25, WAP Forum, April 2001.
[WAP-206] WAP-206-MMSCTR-20020115-a: WAP MMS Client Transactions Specification, version 15, WAP Forum, January 2002.
[WAP-209] WAP-209-MMSEncapsulation-20020105-a: WAP MMS Encapsulation Protocol, version 5, WAP Forum, January 2002.

Section II
Location Management

Chapter 7
A Scheme for Nomadic Hosts Location Management Using DNS

Ramandeep Singh Khurana, Hesham El-Rewini, and Imad Mahgoub

Abstract

In this chapter, we study a simple scheme for location management of nomadic hosts on the Internet by using the existing Domain Name System (DNS) infrastructure. Most applications contact hosts on the Internet by using a Fully Qualified Host Name (FQHN) instead of the host's Internet Protocol (IP) number. The scheme presented in this chapter outlines a mechanism for dynamically updating the DNS server's name-to-IP mapping for the nomadic host thereby facilitating direct communication between the nomadic host and other hosts on the Internet. We outline the scheme and present the results of two experiments that were conducted to study the scalability and limitations of the scheme.

7.1 Introduction

The goal of mobility support is to provide the means by which computers are able to communicate even when their points of attachment to the network may have changed. Several approaches for accommodating mobile hosts on the Internet have been proposed and are described in [3–11]. Mobile IP is perhaps the most feasible approach among the various options available to support host mobility [2, 5, 10, 13, 14]. The Mobile IP working group of the Internet Engineering Task Force (IETF) has introduced this protocol to allow mobile computers equipped with wireless network interfaces to communicate with computers on the fixed network. It allows a mobile host to move around the Internet without changing its IP number.

Mobile IP involves operations such as agent discovery, location registration, and tunneling. Mobile IP is designed to support complete mobility (i.e., during its movement, a mobile host maintains all the connections, transport level sessions, and its IP number remains constant). This might add an additional requirement that a mobile node must implement Mobile IP operations, otherwise, the existence of a foreign agent becomes necessary. This will make Mobile IP dependent on operating system vendors for deployment.

It has been observed that the number of deployments of Mobile IP is still insignificant. Singh et al. believe that the lack of widespread deployment of Mobile IP is due to the lack of compelling applications, which is also a result of the lack of Mobile IP deployment [12]. They introduced the Reverse Address Translation (RAT) protocol as an attempt to help develop more applications that will in turn help achieve significant deployment of Mobile IP. Using Network Address Translation, RAT supports limited mobility of nomadic hosts that moves from one network to another, but is not constrained by the requirement of maintaining open sessions and connections during the move [12]. In this chapter, we use a similar approach to support mobility of nomadic hosts. It is not designed to be a replacement of Mobile IP; rather it is meant to provide nomadic hosts that do not have to be Mobile-IP-aware with some mobility support. It should help develop more applications that can eventually encourage more deployments of Mobile IP.

We present a simple scheme that takes advantage of the fact that most hosts on the Internet are identified by a unique, fully qualified Internet name rather than a unique IP number. Unlike Mobile IP, this scheme is not supposed to maintain transport and higher layer connections when a mobile host changes its network location. Rather, it is designed to support limited mobility of nomadic hosts. Such hosts may exhibit stop-and-go patterns, where they become part of a network for a while, then they move to another network and so on. As part of the move, a nomadic host could be assigned a different IP number. In fact, Dynamic Host Configuration Protocol (DHCP) allows a host on a network to acquire a complete IP configuration from a DHCP server. If the DHCP server is configured in synchronization with the DNS configuration, it can provide the host with its FQHN as part of its IP configuration. However, if the host moves to a foreign network, it may contact a DHCP server maintained by a different administration, which may not have access or control of the home domain for that host. Hence the DHCP server on the foreign network may not be able to assign the same FQHN to the nomadic host as part of its IP configuration. For the nomadic host to retain its original FQHN, the DNS server that controls that domain must change the name-to-address binding whenever the host receives a new IP address and must remove the binding when the lease expires. To our knowledge, there is no standard protocol for dynamic DNS

A Scheme for Nomadic Hosts Location Management Using DNS

update and until such a protocol is developed, there is no mechanism to maintain permanent hostnames while allowing DHCP to change IP addresses.

The scheme presented in this chapter uses the existing DNS infrastructure to allow a dynamic update of the name-to-IP binding for a particular host. When a nomadic host moves to a foreign network, it obtains a new IP address. It then sends its new IP configuration and an expected time for which it expects to keep that IP number, to a server in its home network. A server process, after authenticating the validity of the update request, updates the DNS name-to-IP mapping for that host. After the update, each subsequent request for the name resolution for that particular host results in the DNS server sending the new IP address. This allows all other hosts to communicate directly with the mobile host using its new IP address.

The rest of the chapter is organized as follows:

- Section 7.2 provides an overview of the scheme, which uses the existing DNS services to facilitate location management of nomadic hosts.
- Section 7.3 presents two experiments that were conducted to study the validity of the proposed scheme.
- Section 7.4 presents concluding remarks.

7.2 Using the DNS for Location Management of Nomadic Hosts

Although the IP addresses provide a unique and compact method to address hosts on the Internet, users prefer to address the hosts using pronounceable, easily remembered names. The DNS facilitates the use of names by providing a name-to-IP binding for each host. The DNS has proven that it is a robust and scalable system for providing name resolution for the Internet. However, it was not designed to support mobility of hosts. It is limited in the sense that it assumes that each host has a static IP. During server startup, each DNS server caches the binding information for all the hosts in that domain. It does not support any mechanism where this binding can be changed dynamically.

The scheme presented in this chapter uses the existing DNS infrastructure to allow a dynamic update of the name-to-IP binding for a particular host. This allows the nomadic hosts to authenticate themselves to the DNS and provide their latest IP configuration as they move. The hosts would also provide an estimate regarding how long they expect to keep that IP number.

The scheme consists of the following components:

- DNS server — authoritative for the domain to which the nomadic host belongs.

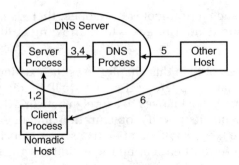

Figure 7.1. Name-to-IP Mapping Update Process

- Server process — runs on the same machine as the DNS server. This process is responsible for communicating with the clients and updating the DNS server information based on the information received from the client.
- Client process — runs on the nomadic host. This process is executed whenever the nomadic host is assigned a new IP address. It is responsible for communicating with the server process and providing the server process with the client's current IP configuration and expected time-to-live (TTL) value.

Once the nomadic host is assigned a new IP address in a foreign network, the following sequence of events typically occurs (Figure 7.1):

1. The client process on the nomadic host (client) contacts the server process using a prespecified port number and authenticates the client to the server process.
2. Once the authentication is successful, the client provides its new IP number and an estimate regarding how long it expects to keep that IP number.
3. The server process then dynamically updates the name-to-IP binding for that host in the database of the DNS server. It also updates the minimum TTL value for that binding, so that the validity of the information can follow the mobility pattern of each user.
4. After the update, the server process signals the DNS server process to reload its zone information so that it can update its cache.
5. Each subsequent request for the name resolution for that particular host results in the DNS server sending the new IP address.
6. This allows all other hosts to communicate directly with the mobile host using its new IP address.

7.2.1 DNS Server

A DNS server stores all the information regarding the domains that it is authoritative for, in multiple text files. Usually, DNS administrators maintain a

A Scheme for Nomadic Hosts Location Management Using DNS

separate text file for each domain that the DNS server is authoritative for. All such text files begin with a Start of Authority (SOA) resource record. The SOA record indicates that this name server is the best source of information for the data within the domain. The SOA record has the following information:

- Serial number — applies to all the data in the zone. It is used to synchronize the data between the primary and the secondary name servers for that domain.
- Refresh — tells the secondary how often to check that its data is up to date. All secondary servers of the domain will make one SOA query per refresh interval.
- Retry — if the secondary fails to reach the primary name server after the refresh period, it starts trying to connect every retry seconds.
- Expire — if the secondary fails to contact the primary server for expire seconds, it expires its data and stops giving out answers about the data because the data is too old to be useful.
- TTL — applies to all the resource records in the domain. The name server supplies this TTL value in query responses, if a TTL value is not explicitly specified in the resource record itself. The TTL is the amount of time any name server is allowed to cache the data. After the TTL expires, the name server must discard the cached data and get new data from the authoritative name servers.

In addition to the SOA record, the domain files contain information regarding hostname to address mapping. Such records are called Address or A records. An A record has the following structure:

```
<fully qualified internet name for the host> <TTL> IN A
<IP Address>
```

For example, an A record for the laptop1 in the domain cs.unomaha.edu with a TTL value of 1 hour and an IP address of 192.168.0.1, will be specified as:

```
laptop1.cs.unomaha.edu. 3600 IN A 192.168.0.1
```

The TTL value in the A record will override the TTL value specified in the SOA record. If the TTL value specified in the SOA record is 86,400 seconds (1 day), and the TTL value in the A record is 3600 seconds (1 hour), then all servers will cache the IP number for laptop1.cs.unomaha.edu for only 1 hour even though they will cache the other entries in the domain for one day.

The setting for the TTL value represents a trade-off between performance and consistency. If the TTL value is low, remote servers will expire their information more frequently and will be forced to query the authoritative servers more often. This will increase the load on the primary and secondary name servers for the domain. However, a low TTL value helps maintain the consistency of information among all servers. The scheme presented in this chapter manipulates the TTL value of the individual resource record of the mobile host in question to better reflect how long the user expects to maintain that particular IP address.

7.2.2 Server Process: Web Server

The server process has the following functions:
- Accept connections from clients on a prespecified port number
- Authenticate the validity of the client using some security mechanism
- Accept information regarding the new IP number and TTL value for that client
- Update the text file for that zone with the new mapping and TTL information for the client. Also, increment the serial number to denote that the information has changed
- Signal the DNS server process to reload its zone information so that it can update its cache

In the implementation of this scheme, we used a Web server with a Common Gateway Interface (CGI) as the server process. When a client submits an update request, the Web server gets the request and presents the client with the username and password screen for authentication. After it receives the correct username and password, it passes all this information to a CGI program. The CGI program then processes all the information, updates the DNS entries and passes a confirmation message back to the Web server, which passes it on to the client.

The Web server provides the information regarding the nomadic hostname, its IP address, and the TTL value in a text file that serves as the input file for the CGI program. The CGI program reads the hostname from the input file and then searches the DNS zone file to see if the hostname exists in the domain. If the hostname does not exist, it sends an error message back to the Web server. If it finds the hostname, it overwrites the TTL value and the IP number for that host in the zone file and then sends a reload signal to the DNS server, which then updates its cache with the latest information. Each subsequent request to the DNS server for the name resolution of that host results in the new IP number being sent to the requester.

7.2.3 Client Process: Web Browser

The client process has the following functions:
- Connect to the server process using a predefined address and port number.
- After connection is established, provide the security information.
- Provide the mobile hostname, current IP number, and TTL estimate for this IP.

A Scheme for Nomadic Hosts Location Management Using DNS

In the implementation of this scheme, we used a standard Web browser as the client process. The browser has a predefined URL as its address and it makes connection to the Web server on port 80 (standard for all Hypertext Transfer Protocol [HTTP] communication). After connecting to the predefined URL, the client is presented with an interface to provide the information regarding the hostname and the TTL value it expects to keep the IP address for.

When the client submits the information, the Web server authenticates the validity of the nomadic host by asking the user for a username and password. After the user has provided the correct username and password, the Web server and the CGI application process the information and then update the DNS information. It should be noted that the user does not have to explicitly enter the new IP number because the Web server gets that information automatically as part of update request. This is a step in the security process, where a mapping can only be changed to the IP number that is currently assigned to the machine that is sending the update request. After the update is completed, the Web server sends a confirmation message to the user.

7.2.4 Security

For the scheme to be accepted and used by the Internet community, it needs to address some basic security issues. It must prevent unauthorized users from performing a mapping update for a certain hostname. It should make sure that the mapping update provides the correct new IP address for that host.

Both of these security issues have been addressed in the implementation of the scheme. A username and password is created for all the hosts in the domain and each user is provided that information. Whenever the Web server receives a mapping update request, it requests the user for the correct username and password. If the user cannot provide the correct username and password, the Web server does not forward the information to the CGI program and no update is performed. The Web server is also configured to log all unsuccessful authentication requests in a log file, so that if somebody tries to hack into the system, an alarm signal can be sent to the administrator via e-mail. To address the second security issue, the CGI program is configured to use the remote IP number that the Web server received as part of the update request. The user is not provided an opportunity to specify an IP number. This prevents an unauthorized user from modifying the IP number of a host to any number other than the IP number of the machine making the update request. The implementation of this scheme is designed to use the HTTP protocol for communication thereby providing the Secure Socket Layer (SSL) level security.

Table 7.1. Hosts Used to Query Name Server

Hostname	Local DNS Server Name
cse.unl.edu	elk.unl.edu
unlinfo.unl.edu	hoss.unl.edu
cwis.unomaha.edu	dns.unomaha.edu
nrcdec.nrc.state.ne.us	nrcdec.nrc.state.ne.us
microsoft.com	dns1.microsoft.com

7.3 Experiments

The scheme was tested with an implementation of the DNS called Berkeley Internet Name Domain (BIND), on the Windows NT® platform. Although the scheme has been implemented for BIND, it will work with any DNS implementation. Several experiments were conducted to evaluate the effectiveness, scalability, and limitations of this scheme. We experimented with different TTL values to see how other name servers responded to mapping updates. To evaluate the name server responses, we used five hosts to send query messages to the name server (Table 7.1).

We used two different tools to determine the IP number for a particular hostname, namely `nslookup` and `dig`. Both of these tools are standard DNS query tools. A C program was written to send `nslookup` requests to different name servers after a specified repeat interval. This allowed us to simulate query load on the DNS server by running the C program on all of the above servers simultaneously. Querying the DNS server via `nslookup`, the user gets two responses:

1. The IP address of the hostname
2. Information on whether the response of the local name server is authoritative or nonauthoritative

If the response of the local server is nonauthoritative, this implies that the local server is providing the reply from the information that it has in its cache. If the response is authoritative, the information was obtained as a result of a resolve query that was sent to the authorized name server for that domain.

However, this information was not adequate for the requirements of this project. We are also interested in obtaining the TTL values associated with the replies. The TTL values for nonauthoritative replies would depict how long the local server would keep the data in its cache. To obtain that information, we used `dig`.

A Scheme for Nomadic Hosts Location Management Using DNS

7.3.1 Cache Time versus Time-to-Live

As mentioned earlier, the TTL value in the resource record or SOA record determines how long a name server can cache that particular information. According to the specifications of the DNS, after the TTL expires, the name server must discard the cached data and get new data from the authoritative name servers. This implies that the first resolution request for a particular host, after the TTL for that record has expired, should result in a resolve query being sent by the local host to the authoritative DNS server. To test the above implication, we performed several experiments using the following setup. We used a laptop to connect to a local Internet Service Provider (ISP) that does dynamic IP assignment. This ensured that on each connection, we would be assigned a different IP number. The laptop performed the function of the nomadic host. On each connection, we ran the Web browser to update the name-to-IP mapping for that machine and provided a different TTL value for that connection. Then, we ran the C program on all the test machines allowing them to cache the mapping. After the simulation program started on each machine, we made a new connection to the ISP and updated the DNS mapping on the DNS server. We then monitored the execution of the simulation program to determine how long it took the different machines to obtain the new IP number.

In this chapter, the term *cache time* refers to the time period, starting from the time the local server cached the name-to-IP binding for a particular host, to the time it sent a resolution request to the authoritative DNS server for the same host. Suppose that at time f, the local name server resolves and caches the name-to-IP binding for a particular host. If, at time s, the local name server sends another resolve query for the name-to-IP binding for the same host, the cache time, c, can be obtained as $c = s - f$. Our effort was to determine if the cache time was equal to the TTL specified in the resource records, at all times. If the cache time matched the TTL time for all servers, the scheme would provide a feasible method for location management of nomadic hosts on the Internet.

As we can see from Table 7.2, when the TTL value is close to five minutes and above, the cache time for all servers is identical to the TTL value. When the TTL value is below five minutes, the cache time for different servers is unpredictable. Some servers (elk.unl.edu, dns.unomaha.edu, and dns1.microsoft.com) respond as expected by keeping the cache times same as the TTL values. However, hoss.unl.edu responded by always keeping the cache time at least five minutes, irrespective of the TTL specified. Similarly, nrcdec.nrc.state.ne.us responded by keeping the cache time at least two minutes, irrespective of the TTL specified.

One possible explanation of this behavior could be the fact that most DNS servers have the negative caching implemented with a hard coded

Table 7.2. TTL Value versus Cache Time

Specified TTL (sec)	Cache Time				
	elk.unl. edu	hoss.unl. edu	nrcdec.nrc. state.ne.us	dns. unomaha. edu	dns1. microsoft. com
0	0	0	0	0	0
50	50	300	120	50	50
100	100	300	120	100	100
150	150	300	150	150	150
200	200	300	200	200	200
250	250	300	250	250	250
300	300	300	300	300	300
350	350	350	350	350	350
400	400	400	400	400	400
450	450	450	450	450	450
500	500	500	500	500	500
550	550	550	550	550	550
600	600	600	600	600	600

value of 10 minutes. In addition, keeping the TTL value very low could result in a high load on the DNS server. It appears that the designers of the different DNS servers have not followed any definite standard on determining the minimum cache time for their servers. However, when the TTL value was specified as 0, no server cached the data, irrespective of what its minimum cache time value was. A TTL value of 0 always resulted in the response from the local servers being authoritative. Hence, if the expected TTL value of any server needs to be below 30 seconds, the server should specify the TTL value as 0 so as to ensure that all remote servers would always get the latest name-to-IP mapping on all resolution requests.

7.3.2 Scalability Analysis

To perform a scalability analysis on the scheme, we increased the number of nomadic hosts in the DNS database file and tested the time it took for the client to perform a binding update on the DNS server. This time is measured from the time the Web client presses the Update button to the time the Web client displays the confirmation message. During that time, the CGI program receives the mapping update request, locks the DNS data file, parses the data file to find the entry for the nomadic host, updates the data

A Scheme for Nomadic Hosts Location Management Using DNS

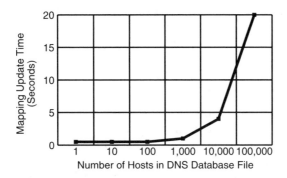

Figure 7.2. Scalability Analysis

file with the new information, and then sends the reload signal to the DNS server.

It should be noted that scalability issues arise only when large numbers of clients are trying to update their mappings, not when large numbers of servers are querying the primary DNS for mapping resolutions. The root servers maintained by InterNIC^{SM} have already proven that scalability in terms of servicing resolution requests can be resolved by increasing the compute power of the physical machines and increasing the bandwidth of their Internet connections. Statistics from InterNIC show that their root name server, ns.internic.net, receives 255,600 queries per hour, or almost 71 queries per second [1]. In addition, having multiple DNS servers and implementing some load sharing mechanism for those servers can reduce the load on the name servers.

As we can see from Figure 7.2, the mapping update time for the server is less than a second for up to a thousand hosts. It then increases by a factor of five (approximately), as the number of servers increase by a factor of ten.

With the infrastructure used in the experiments, the scalability factor introduced the following problem. When the number of hosts became very large (around 100,000), such that mapping update time was about 20 seconds, all subsequent mapping update requests within the 20 second period failed. The main reason for the failure was that the instance of the CGI program handling the first update request had locked the DNS data file and all subsequent instances of that program were unable to lock the file. This problem can be rectified by increasing the computing power of the server. This would provide faster processing of the data file and hence reduce the mapping update time. Also, implementing a queuing mechanism in the CGI program should help. Each instance of the CGI program should check if it is able to lock the data file. If the lock succeeds, the CGI program should update the mapping information. If the lock fails, the CGI program should sleep, waiting on the event that its lock will succeed and it can go ahead with its update. This way, all updates will be performed in a queued fashion.

7.4 Concluding Remarks

In this chapter, we presented a simple scheme that performs location management of nomadic hosts on the Internet. This scheme takes advantage of the existing DNS architecture to allow the nomadic hosts to update their name-to-IP binding dynamically. In addition to its simplicity, this approach has several advantages. It integrates well with the existing DNS infrastructure and can be implemented easily without requiring any changes in the Internet architecture. Because it uses the existing DNS infrastructure, which has proven to be scalable, it can be used to perform location management for a large number of hosts. This scheme allows the nomadic host to enter its own Keep Valid time, which could vary for different hosts. This allows the name-to-IP mapping validity to mirror the expected host mobility pattern. Because this scheme allows the individual hosts to specify their own TTL values, and the scheme is scalable, the name-to-IP mappings are not dependent on the frequency of movement of hosts. The hosts move and update their bindings as often as they like. It can be used for the location management of any host on the Internet and once the name-to-IP binding of the host has been updated, the changes are reflected throughout the whole Internet.

This scheme has some limitations and areas of improvements. One problem is that the queue may become very large when many requests come in for hostname-to-IP updates. One solution to this problem is to have a multi-threaded database running on a multiprocessor computer: a custom written implementation of BIND that does the standard caching of information, but instead of reading from the text files, it reads the data from the database. This would allow multiple updates and multiple requests to take place at one time.

In addition, this scheme was implemented using the BIND implementation of the DNS server. BIND supports the feature of specifying the TTL value for each resource record. If the scheme is to be ported on any DNS server that does not provide this facility, it needs to be modified to circumvent this problem. Creating a separate subdomain for the nomadic hosts and controlling the TTL value of the SOA record for the subdomain may circumvent this problem. This TTL value would then apply to all hosts in the subdomain. The TTL value of the subdomain should be the minimum of the TTL values specified by each host. In this case, the CGI program upon receiving the TTL value for the mapping update for a host would compare that value to the TTL value of the SOA record for the subdomain. If the

A Scheme for Nomadic Hosts Location Management Using DNS

specified TTL value is less than the one in the SOA record, the program would modify the TTL value in the SOA record. If the specified TTL value is the same or higher, the CGI program would just update the mapping for that host.

References

1. Paul Albitz and Cricket Liu, *DNS & Bind*, 2nd ed., Sebastopol, CA: O'Reilly & Associates, Inc., January 1997.
2. Pravin Bhagwat and Charles Perkins, A Mobile Networking System based on Internet Protocol, *IEEE Personal Communication Magazine,* February 1994.
3. Tomasz Imielinski and Henry F. Korth, *Mobile Computing*, Norwell, MA: Kluwer Academic Publishers, 1996.
4. David B. Johnson and David A. Maltz, Protocols for Adaptive Wireless and Mobile Networking, http://monarch.cs.cmu.edu.
5. David B. Johnson and Charles E. Perkins, *Route Optimization in Mobile IP*, draft-ietf-mobileip-optim-07.txt-work in progress, November 1997.
6. Kevin Lai, Mema Roussopoulos, Diane Tang, Xinhua Zhao, and Mary Baker, Experiences with a Mobile Testbed, *Proc. of the 2nd International Conference on Worldwide Computing and Its Applications (WWCA'98),* March 1998, http://mosquitonet.stanford.edu/ publications.html.
7. Ben Lancki, Abhijit Dixit, and Vipul Gupta, Mobile-IP: Supporting Transparent Host Migration on the Internet, http://anchor.cs,binghampton.edu.
8. Refik Molva, Didier Samfat, and Gene Tsudik, Authentication of Mobile Users, *IEEE Network,* vol. 8, no. 2, 1994, pp. 26–34.
9. Brian Noble, Giao Nguyen, Mahadev Satyanarayanan, and Randy Katz, *Mobile Network Tracing*, Internet RFC 2041, October 1996.
10. Charles E. Perkins, *Mobile IP — Design Principles and Practices*, Boston: Addison Wesley, 1998.
11. Mahadev Satyanarayanan, Mobile Computing, *IEEE Computer,* vol. 26, no. 9, 1993, pp. 81–82.
12. Rhandev Singh et al., RAT: A Quick (and Dirty?) Push for Mobility Support, in *Proc. of the 2nd IEEE Workshop of Mobile Computing Systems and Applications,* New Orleans, February 1999.
13. M. Spreitzer and M. Theimer, Scalable, Secure, Mobile Computing with Location Information, *Communications of the ACM,* vol. 36, no. 7, 1993.
14. Subhashini Rajgopalan and B.R. Badrinath, Adaptive Location Management for Mobile-IP, *Proc. of the 1st ACM International Conference on Mobile Computing and Networking—Mobicom'95,* November 1995.

Chapter 8
Location Management Techniques for Mobile Computing Environments

Riky Subrata and Albert Y. Zomaya

8.1 Introduction

One of the challenges facing mobile computing is the tracking of the current location of the user — the area of location management. To route incoming calls to appropriate mobile terminals, the network must from time to time keep track of the location of each mobile terminal.

Mobility tracking expends the limited resources of the wireless network. Beside the bandwidth used for registration and paging between the mobile terminal and base stations, power is also consumed from the portable devices, which usually have limited energy reserve. Furthermore, frequent signaling may result in degradation of quality of service (QoS), due to interferences. On the other hand, a miss on the location of a mobile terminal will necessitate a search operation on the network when a call comes in. Such an operation requires the expenditure of limited wireless resources. The goal of mobility tracking, or location management, is to balance the registration and search operation, so as to minimize the cost of mobile terminal location tracking.

Most, if not all, today's wireless network consists of cells. Each cell contains (or is represented by) a base station, which is wired to a fixed wire network. The base stations interact with the portable handheld devices and provide these devices the wireless link to the network. One typical cellular network plan is shown in Figure 8.1. Cells are then grouped into regions. Each region contains the whole allotted frequency spectrum, with each cell in the region using part of the frequency spectrum. The frequency

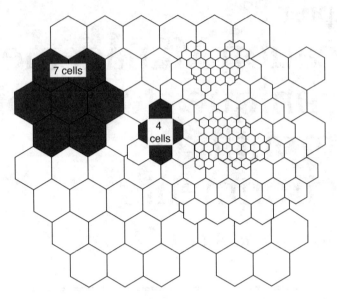

Figure 8.1. Typical Cellular Network Plan
Shown are regions having four and seven cells. The cells are smaller in the areas where more users are expected.

spectrum is then reused in the other regions, with cells in other regions reusing the same frequency band, governed by the minimum distance between cells allowed. Detailed discussions on frequency reuse and cell planning can be found in [31, 86]. Further information on wireless networks and communications can also be found in [16, 20, 70, 71, 86].

As the demand for wireless communication grows, cells sizes have continually decreased in size to achieve higher frequency reuse, especially important due to the limited frequency spectrum available. Due to the smaller cell sizes, cell crossover in mobile users' movements would undoubtedly become more frequent. As such, an efficient location management system is needed to ensure timely delivery of incoming calls to the user.

The next section is an overview of location management and its operations. The concept of location management cost, for the purpose of location management strategies evaluation and comparison is then described. This is followed by a discussion of common network topologies, as well as several common mobile users' call arrival and mobility patterns used for network simulation purposes. Finally, an overview of a number of location update and general location inquiry strategies that have been proposed in the literature over the years is given.

Location Management Techniques for Mobile Computing Environments

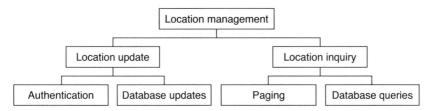

Figure 8.2. Location Management Components
(Source: I.F. Akyildiz, J. McNair, J.S.M. Ho, H. Uzunalioglu, and W. Wenye, *Proceedings of the IEEE*, vol. 87, no. 8, pp. 1347–1384, 1999. Used with permission.)

8.2 Location Management

Location management, or the tracking of mobile users' location inside the network, involves two elementary operations — location update and location inquiry (Figure 8.2). Note that there are also fixed-wire communications within the network between the different controllers for location updates as well as location inquiries — network interrogation. This is mainly for database updates during a location update and database queries during a location inquiry (Figure 8.2).

8.2.1 Location Update

Location update is initiated by a mobile terminal and is used to inform the network of its current location inside the network. This is done so as to limit the search space, should the need arise, to locate the user at a later point in time. That is, location update strategies, although not necessary, are used to reduce the amount of signaling required to locate a mobile terminal should the need arise.

8.2.2 Location Inquiry

In location inquiry, the system initiates the search for a user. The system can do this by polling cells where the user might be in. Specifically, the following procedure can be used [4]:

- Send polling signal to a target cell.
- If a reply is received before a specified time-out, the mobile terminal is in the target cell.
- If a time-out occurs and no reply is received, the mobile terminal is not in the target cell.

8.2.2.1 Delay Constraint. In many cases, in order to maintain a required QoS, it is desirable to impose a maximum allowable time delay in locating a user. Unfortunately, this added time constraint adds to the complexity of update schemes and schemes that work well with no delay constraint may need to be adjusted to work well under the delay constraint.

8.3 Location Management Cost

To be able to effectively compare the different location management techniques available, one needs to associate with each location management technique, a value or cost.

As noted above, location management involves two elementary operations of location update and location inquiry, as well as network interrogation operations. Clearly, a good location update strategy would reduce the overhead for location inquiry. At the same time, location updates should not be performed excessively, as it expends the limited wireless resources.

To determine the average cost of a location management strategy, one can associate a cost component to each location update performed, as well as to each polling/paging of a cell. The most common cost component is the wireless bandwidth used (wireless traffic load imposed on the network). That is, the wireless traffic from mobile terminals to base stations (and vice versa) during location updates and location inquiry. Although there is also fixed-wire network traffic (and database accesses and loads) between controllers within the network during location updates and location inquiry — network interrogation — this is considered much cheaper (and much more scalable) and is usually not considered.

The total cost of the above two cost components — location update and cell paging — over a period of time T, as determined by simulations (or analytically or by other means) can then be averaged to give the average cost of a location management strategy [87]. For example, the following simple equation can be used to calculate the total cost of a location management strategy:

$$\text{Total Cost} = C \cdot N_{LU} + N_P \tag{8.1}$$

where N_{LU} denotes the number of location updates performed during time T, N_P denotes the number of paging performed during time T, and C is a constant representing the cost ratio of location update and paging. The above cost formula can be used to compare the efficiency of different location management techniques. Several things, however, should be noted:

- The more complex location management strategy will almost always require more computational power at the mobile terminal, the system, or both. It may also require greater database cost (e.g., record size). These parts of the location management cost are usually ignored as they are hard to quantify.
- The cost of location update is usually much higher than the cost of paging — up to several times higher [38], mainly due to the need to setup a signaling channel. Several authors use C = 10, for example in [34, 108].

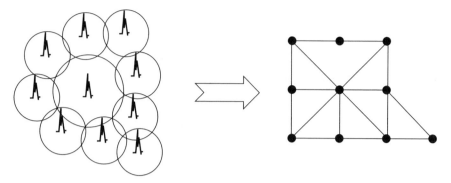

Figure 8.3. Cellular Network and Corresponding Graph Model

- If location management cost is obtained through simulations (or analytically), then depending on the location update strategy used, the total number of location updates performed over a period T may depend on the mobile users' mobility and call arrival patterns. To a certain extent, it may also depend on the network topology used in the simulation.
- As in the case for location update, if simulations are used for location management strategy evaluation, then the total cost of location inquiry over a period T would depend on the number of calls received by the users, that is, on the users' call arrival patterns. It may also be influenced by the users' mobility pattern and the network topology used in the simulation.

Noting the above issues, it is clear that any simulation results would be strongly influenced by users' mobility and call arrival patterns chosen for the simulation [59, 84]. In other words, users' mobility and call arrival patterns are especially of interest in location management.

The next few sections provide overviews of common network topologies, as well as several users' mobility and call arrival patterns commonly used in network simulations to determine the effectiveness of a location management strategy.

8.4 Network Topology

A general graph model can represent arrangement of cells in a real cellular network (Figure 8.3). In the graph model, each node represents a base station (center of cell) and neighboring cells are represented by edges connecting the nodes. Other simpler models have also been used for simulation purposes, which include one-dimensional and structured two-dimensional models.

MOBILE COMPUTING HANDBOOK

Figure 8.4. One-Dimensional Network

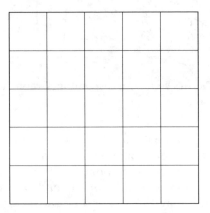

Figure 8.5. Two-Dimensional, Mesh Configuration Network

In the one-dimensional model, each user has two possible opposing moves (e.g., left or right) as shown in Figure 8.4. Common two-dimensional models include the mesh configuration (Figure 8.5), and the hexagonal configuration (Figure 8.6). In the hexagonal configuration, each cell can have a maximum of six neighboring cells. In the mesh configuration, each cell can have a maximum of either four or eight neighbors, depending on whether diagonal movements are allowed.

8.5 Mobility Pattern

In a real cellular network, one would expect the mobility level of each mobile user to be a time varying quantity. For example, users are more likely to be more mobile during rush hour and working hours, in general, than after hours. Several approaches have been proposed in the literature to model and approximate a mobile user's movement pattern. Some common approaches are described below.

8.5.1 Memoryless (Random Walk) Movement Model

In the memoryless, Random Walk Movement Model, the user's next cell location does not depend on the user's previous cell location. That is, the next cell location is selected with equal probability from the neighboring cells. Purely Random Walk Model is usually used to model pedestrian traffic,

Location Management Techniques for Mobile Computing Environments

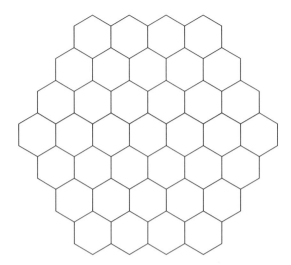

Figure 8.6. Two-Dimensional, Hexagonal Configuration Network

whose movements are usually irregular with frequent stops and directional changes.

8.5.2 Markovian Model

Unlike the memoryless movement model described above, the Markovian Movement Model incorporates memory and user's movements are influenced by the user's previous movements. Such memory can include a list of recently visited cells (cell history) or recent directions in movement (directional history).

8.5.2.1 Cell History. In this model, each of the neighboring cells has a different probability of being the user's next cell location, depending on the set of cells the user has visited.

For a discrete time, one-dimensional ring network, a first order Markov Movement Model with a geometrically distributed cell residence time can be defined as follows. Suppose a mobile user is at cell k at time t. Then at time $t + 1$, the user will stay at cell k with probability q or move to one of the two neighboring cells a or b with probability $P(a|k)$ and $P(b|k)$, respectively.

The definition above can also be extended to a higher order Markov Model. For example, for a second order Markov Model, the probabilities of moving to one of the two neighboring cells would be $P(a|jk)$ and $P(b|jk)$, where j and k denotes the last two cells the user visited. Clearly, the Random Walk Model described earlier can be thought of as a zeroth order Markov Model (that is, no memory).

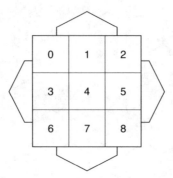

Figure 8.7. A 3×3 Network
The cells on the edge of the network wrap around and connect to the cells on the opposite side.

The definition for a continuous time Markov Movement Model is similar. For example, the user's cell residence time can be exponentially distributed with mean $1/\lambda$. In this case, only the transitional probabilities $P(a|k)$ and $P(b|k)$ — assuming a first order Markov Model, are needed. Similar situations exist for higher order Markov Model.

Finally, the definitions above can be readily extended to a two-dimensional network with arbitrary topologies. In [23, 82], a discrete time, first order Markov Movement Model is used on a two-dimensional graph model.

8.5.2.2 Directional History. The concept of memory for cells visited can also be extended to include directions of movement the user has taken. The directional information is used to model the user's movements particularly in highly structured network topology, such as the mesh configuration and hexagonal configuration network. Different ways of implementing such directional information exist.

In a mesh configuration network, one can define four possible movement directions — up, down, left, and right. To ensure that each cell has exactly four possible movement directions, the network can be made to wrap around on the edges. For example, in Figure 8.7, cell 0 connects to cell 2 and cell 6, as well as cell 1 and cell 3.

If a first order Markov Model is considered, then the movement direction the mobile user takes in the next time instant would depend on that user's last movement direction. One possible transition model, used in [112], is shown in Figure 8.8. In this case, the mobile user's residence time in each cell is modeled as an exponentially distributed random variable with mean $1/\lambda$.

Note though, the use of purely directional information for mobility modeling does not differentiate the geographical locations (e.g., attraction points such as shopping centers) of the different cells in the network.

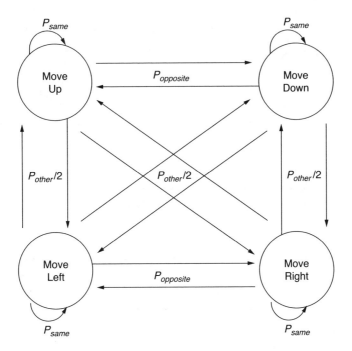

Figure 8.8. Transition Diagram for Four Possible Directional Moves

8.5.3 Shortest Distance Model

In this model [1], users are assumed to follow a shortest path from source to destination. At each intersection, a user chooses a path that maintains the shortest distance assumption. The model is particularly suited for vehicular traffic, whereby each user has a source and destination. Under such condition, the shortest distance assumption is certainly reasonable.

8.5.4 Gauss-Markov Model

The model, described in [51], captures some essential characteristics of real mobile users' behavior, including the correlation of users' velocity in time. Under a discrete time model, a mobile user's velocity v_n at time slot n can be represented as follows [51]:

$$v_n = \alpha v_{n-1} + (1-\alpha)\mu + \sqrt{1-\alpha^2}\, x_{n-1} \tag{8.2}$$

where $\alpha \in [0,1]$, μ is the asymptotic mean of v_n when n approaches infinity, and x_n is an independent, uncorrelated, and stationary Gaussian process. x_n has zero mean and standard deviation equals to the asymptotic standard deviation of v_n when n approaches infinity. In the extreme cases, the

Table 8.1. Activity Transition Probabilities for a Mobile User

Time	Current Activity	Next Activity	Probability
⋮	⋮	⋮	⋮
1	2	3	0.3
1	2	4	0.5
⋮	⋮	⋮	⋮

Gauss-Markov Model simplifies to the memoryless movement model and constant velocity Fluid-Flow Model (described below).

8.5.5 Activity-Based Model

The central concept of an activity-based model is that of activity. Each activity represents a trip purpose: that is, the activity requires the user to travel to a destination associated with the activity. New activities are then selected/generated based on such factors as the previous activities and time of day.

Certainly, several implementations of such activity-based mobility model are possible. In an implementation described in [81], each activity has with it several parameters, including time of day, duration, as well as location of the activity. New activities are then selected or generated based on the previous activity, and time of day, as shown on Table 8.1. When a new activity is selected, it is assigned a duration based on the time of day. Based on the activities information, a path taken from origin to destination, as well as the times of the cell crossings, can be determined and used for simulation. Parameter values needed for the activity-based model can be obtained from a population survey.

The activity-based mobility model captures, to an extent, movement behavior of real mobile users. The activity-based mobility model is discussed and implemented in [28, 67, 81, 84].

8.5.6 Mobility Trace

Actual mobility trace of users in a cellular network — that is, actual movement behavior of users in a real cellular network or geographical area — can also be used for simulation. Such trace is certainly more accurate and realistic than other mathematical models. However, such trace is not readily available, especially one of a large enough size to be useful for network simulation. Furthermore, movement behavior of users in one network may not be the same or valid for other network, which may depend, among other things, on the size of the network and geography. Several mobility traces have been collected and used for evaluation purposes, for example in [95, 96, 99].

In [43], a trace generator, which was corroborated using real-world data, is described. The traces include call information as well as movement information of mobile users. Output from such a generator can be obtained from [92].

8.5.7 Fluid-Flow Model

Although the above models describe an individual user's mobility, there are also models that describe systemwide (macroscopic) movement behavior. The Fluid-Flow Model is one such model. In this model, mobile users' traffic flow is modeled as fluid flow, describing the macroscopic movement pattern of the system. In this model, each mobile user is assumed to move at an average speed v and is uncorrelated with the movement of other users. Further, the direction of each mobile user's movement is uniformly distributed in the range $\{0, 2\pi\}$. For a region with length L and population density ρ, the average number of users moving out of the area per unit of time is given by:

$$N = \frac{\rho v L}{\pi} \qquad (8.3)$$

The Fluid-Flow Model is suitable for vehicular traffic, where users do not make regular stops and interruptions, as opposed to pedestrian traffic, which can be irregular with frequent stops and directional changes. Pedestrian traffic is usually modeled using a Random Walk Model. Fluid-Flow Model is discussed and used in [40, 49–51, 83, 98, 103, 110]. Because the Fluid-Flow Model describes macroscopic movement behavior, it is not suitable in cases when individual user's mobility patterns are important.

8.5.8 Gravity Model

In this model, movement traffic between two sites/regions i, j is a function of each site's gravity P_i, P_j (e.g., population) and an adjustable parameter $K(i, j)$. For example, the following simple formula can be used to model the amount of traffic from site i to site j:

$$T_{i,j} = K_{ij} P_i P_j \qquad (8.4)$$

where K_{ij} is a positive constant, and P_i and P_j can represent the population of site i and site j, respectively. As in the case of Fluid-Flow Movement Model, the gravity model describes systemwide, or macroscopic, movement behavior. As such, it cannot be used in simulations involving the individual user's mobility patterns. Gravity models have been used to model traffic in different geographical areas [29, 30, 85].

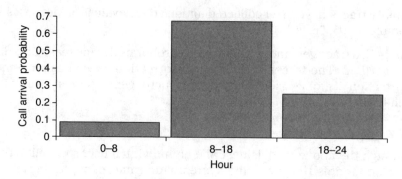

Figure 8.9. Estimated Weekday's Call Arrival Probability Distribution

8.6 Call Arrival Pattern

The call arrival rate of mobile users in a real cellular network is a time varying quantity; for example, a higher call arrival rate is expected during working hours than after hours.

8.6.1 Poisson Model

In this model, call arrivals to a particular mobile station are assumed to follow a Poisson probability distribution, resulting in exponentially distributed (continuous time) or geometrically distributed (discrete time slot) call interarrival time.

Data suggests that macroscopically (in aggregate), the Poisson call arrival rate accurately reflects call arrivals in existing cellular networks. However, an individual user may not have a Poisson call arrival rate. For an individual user, other factors come into play, such as time of day (e.g., working hours, after hours) and special days (e.g., holidays).

8.6.2 Call Arrival Trace

To overcome the problem, a time-varying call arrival model can be used, which generates call events based on the time. For example, greater call arrivals may be generated during working hours than after hours (Figure 8.9). The call arrival distribution to be used for the model can be readily obtained from users' call records.

Actual call arrival trace of an existing cellular network can also be used for simulation. Such trace certainly provides a more accurate and realistic representation of actual call arrivals than other mathematical models, such as the Poisson Model. Also, unlike mobility traces, trace data of users' call arrivals are readily available from the users' call records. However, a time-varying call arrival model described above may be enough to accurately represent call arrival patterns of actual users.

A trace generator that was corroborated using real-world data is described in [43]. The traces include call information as well as movement information of mobile users. Output from such a generator can be obtained from [92].

8.7 Location Update Strategies

Ideally, a location update strategy should efficiently use the limited resources of the network and should not require excessive computing power at the mobile terminal. This is especially true in the dynamic update strategy (discussed below), where many of the calculations need to be done at the mobile terminal. Further, it should be (easily) scalable to accommodate future expansion of the network.

Location update strategies can be classified as either static or dynamic. In the static update strategy, location update is performed independent of each user's mobility and call arrival patterns. On the other hand, dynamic strategies take into account each user's mobility and call arrival patterns. Undoubtedly, static update strategies are easier to implement in a network and require minimal processing power on the mobile terminal. However, dynamic strategies may result in lower overall signaling costs (location management cost).

Below are some of the update strategies that have been proposed over the years. Some of these strategies are static update strategies and some are dynamic update strategies. Still, some of the strategies can be implemented both statically and dynamically.

8.7.1 Always-Update Strategy

In this static location update strategy, each mobile terminal performs a location update whenever it enters a new cell. In this case, the current cell location of each user is always known. As such, no search operation would be required for incoming calls. However, the resources used (overhead) for location update would be high. Such strategy is suitable when the user is not highly mobile or the cell size is quite large and users do not move in or out of cells often.

8.7.2 Never-Update Strategy

This static location update strategy is the opposite of the always-update strategy in that no location update is ever performed. Instead, when a call comes in, a search operation is conducted to find the intended user. In this case, the overhead for the search operation would be high, but no resources would be used for the location update. This scheme may be suitable for small cell size and highly mobile users with low call arrival rates.

Although the always-update and never-update strategies represent the two extremes of location management strategies — whereby one cost is minimized and the other maximized — other location update strategies exist that use a combination of the above two strategies. These are described below.

8.7.3 Time-Based Strategy

In this dynamic location update strategy, each mobile terminal updates its location every T time units. This strategy is relatively easy to implement, as each mobile terminal needs only an internal clock to keep track of how long it has been since its last location update. More importantly, the value T can be adjusted for each mobile user, according to the individual user's mobility/movement patterns and call arrival rate/pattern.

Further, due to the nature of its periodic signaling, the network knows that the mobile terminal is powered-off or outside the coverage area if it does not perform a location update at its required times (implicit detachment). As such, this may reduce the signaling load due to unnecessary paging operations [57]. However, such a scheme would lead to unnecessary location updates for stationary or low mobility users. Furthermore, mobile users' location uncertainty is not bounded: when a call arrives, the search operation cannot be limited to a set of cells.

In [74], using a one-dimensional network model, Poisson call arrival, and assuming the user's location probability distribution as a function of time is known, it was found that the time-based method performs better than the static location area strategy (discussed later). Similar discussions can be found in [72, 73, 75, 77]. In particular, the time-based approach is extended [73, 75, 77] to a state-based approach, whereby the time since last update, as well as the user's current location, is taken into account in determining when to do a location update.

8.7.4 Movement-Based Strategy

In its simplest form, this dynamic location update strategy requires a mobile terminal to keep track (a counter) of the number of cells visited (or the number of cell-boundary crossings) since the last location update. Location update is performed when the counter exceeds a threshold value M, which can be determined on a per user basis. An example is shown in Figure 8.10, where M = 2.

This strategy is harder to implement than the time-based strategy, as each mobile terminal needs to be aware of boundary crossing. Also, cyclic user movements would trigger unnecessary location updates. One simple improvement to partially solve the cyclic movement problem is to reset the counter when the user reenters the last known cell location (e.g., the cell of the last location update performed by the mobile terminal). Such strategy is discussed in [18]. Similar strategy is discussed in [56]. One advantage of the movement-based strategy over the time-based strategy,

Location Management Techniques for Mobile Computing Environments

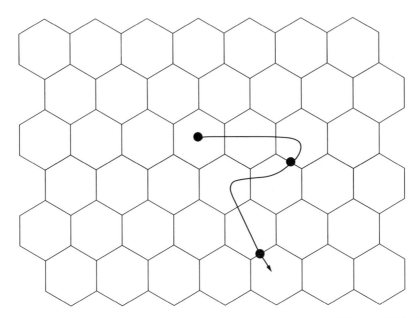

Figure 8.10. Movement-Based Location Update with Movement Threshold M = 2 Location update is triggered whenever the user crosses more than two cells. In the figure, each dot represents a location update. The first location update occurs when the mobile terminal is switched on.

however, is that a mobile terminal's location is limited to a radius of M cells. As such, when a call arrives, a search can be limited to cells within the radius M.

8.7.5 Distance-Based Strategy

In another dynamic location update strategy, the distance-based strategy (also called the distance-based location area [DBLA] strategy), each mobile terminal needs to keep track of the distance (in number of cells) it has traveled since its last location update. When the distance (in number of cells) traveled exceeds a certain threshold value D, a location update is performed (Figure 8.11). A modification of the existing IS-41 standard to incorporate the distance-based location update scheme is discussed in [112].

Clearly, the distance-based scheme limits location inquiries to the cells within the radius D. As such, in a location miss, at worst case only the cells within the radius D needs to be paged.

Undoubtedly, this strategy is harder to implement than either the time-based or the movement-based strategy above. However, the savings in the limited radio bandwidth (that may have otherwise been used for

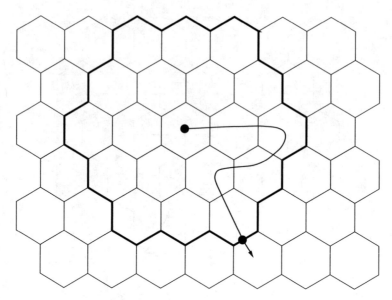

Figure 8.11. Distance-Based Location Update with Distance Threshold D = 2
Location update is triggered whenever the distance traveled is greater than two cells. In the figure, each dot represents a location update. The first location update occurs when the mobile terminal is switched on.

---

location updates and paging) may outweigh the extra complexity incurred. In [8], using Random Walk and Markovian Mobility Models, results show the distance-based strategy performs better than either the time-based strategy or the movement-based strategy.

In [107], the scheme is applied to a cellular network with arbitrary cell topologies. Using the graph model (Figure 8.3), first order Markovian Movement Model, and Poisson Call Arrival Model, optimal location update boundaries are obtained. Whenever users enter one of their update cells, location update is performed. New location update boundary is then calculated. It was shown that each location update cell might not necessarily have the same distance (in terms of the minimum number of cells to be traversed to get to the cell) to the current cell. That is, the update boundary is not circular. Distance-based strategies are discussed further in [9, 53, 66, 102].

8.7.6 Location Area

The location area (LA) method of location management is the most common and widely used location management technique used in today's existing cellular networks [82].

Location Management Techniques for Mobile Computing Environments

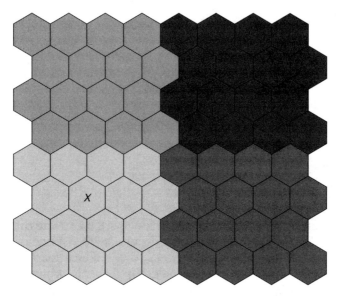

Figure 8.12. Regions Representing Location Areas and Individual Cells
Here there are 4 LAs, each consisting of 16 cells.

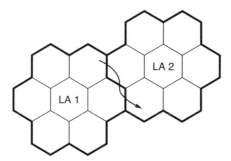

Figure 8.13. Oscillating User Location

8.7.6.1 Static Case. In this scheme, the network is partitioned into regions or LAs, with each region consisting of one or more cells (Figure 8.12). The never-update strategy can then be used within each region, with location update performed only when a user moves out to another region or LA.

One of the advantages of the LA scheme over the time-based scheme discussed earlier is that location inquiry of a mobile user is limited to within the LA.

One shortcoming of the static LA scheme is that if a mobile terminal frequently crosses the LA's boundaries, then there will be excessive, unnecessary location updates (Figure 8.13). To overcome this, the boundary of LA

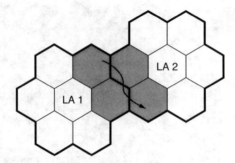

Figure 8.14. Overlapped Location Areas

1 and LA 2 can be overlapped, as shown in Figure 8.14. Note that the overlapping cells can be thicker than two cells or can be just one cell wide. If a user moves within the overlapping cells, then no location update is performed and the user is deemed to be in one LA only, which depends on where the user originally comes from.

Note however, that the use of overlapping results in the overlapping cells having to handle the paging of the LAs in which they are included, which can be quite high, especially if it involves more than two LAs.

Another problem inherent in the LA strategy is that because location update is performed at boundary crossings, more signaling traffic is generated around the boundary cells. Here are a couple of solutions for this problem:

- Provide extra bandwidth to the boundary cells to compensate for the extra traffic.
- Assign each mobile user to a group. The network would then have several groups, with each group having its own LA mapping and number of users (Figure 8.15). With appropriate parameters to each group, uniformly distributed location update traffic can be achieved. However, the use of grouping leads to increased complexity of the system, both in the planning process, and also in that the network has to be aware of each user's group.

Some of the above improvements can also be combined together to give a hybrid concept. The multilayer concept, introduced in [65] involves the overlapping (layering) improvement to prevent oscillating location update and grouping improvement to redistribute location update traffic.

Another variant is proposed in [52], called the two-location algorithm (TLA). In this strategy, the system uses two LAs, instead of one, for each mobile user. Each mobile terminal performs a location update only when it enters a new cell not within the two previously registered LAs. It was found

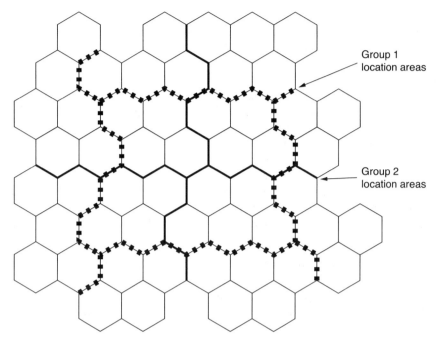

Figure 8.15. Grouping of Location Areas

---

that in the case of high mobility and low call arrival rate, the TLA might significantly outperform the standard LA strategy.

The LA planning problem can be defined as the problem of finding an optimal set of LAs, whereby the location management cost (of location update and location inquiry) is minimized. Minimizing location management cost is often a major goal for network operators. Unfortunately, most cost formulation (using different movement and call arrival patterns, among other things) for the LA planning problem results in an optimal formulation that is NP-complete (NP = nondeterministic polynomial). As such, different heuristic and approximation algorithms have been proposed for solving different versions of the LA planning problem. One popular algorithm is genetic algorithm (GA), used for example in [32, 34, 105]. In [34], a GA is used to solve a version of the LA planning problem. In this version, the location management cost is formulated using average movement between cells and average call arrivals in each cell. Near optimal solutions to the LA partitioning problem, using simulated annealing, tabu search (TS), and GA is also described in [24, 25]. Other algorithmic techniques are shown in [12, 17, 21, 22, 26, 58, 79, 97, 103].

8.7.6.2 Dynamic Case. To overcome the shortcoming of the static LA schemes above, several dynamic variants of the LA scheme have been proposed. One dynamic variant is shown in [80]. Here, each user's mobility history is used to create individualized LAs for the user. To do this, a set of counters N_{ab} (that represents the number of transitions the user has made from cell a to cell b) is maintained for each user. In the LA creation procedure, the user's current cell is automatically included in the LA. To find the next cell to be included in the LA, the following procedure is used. First, the average transition value W (from the current cell to the neighboring cells) is calculated. Neighboring cells with transition value (N_{ab}) ≥ W is then added to the LA, in descending order according to its transition value (that is, neighboring cell with the highest transition value is added first, followed by the second, and so on). Once the first ring of neighboring cells is added to the LA, the same procedure is used to calculate the second ring of neighboring cells. The procedure is repeated until the required number of cells has been included in the LA or no more neighboring cells with known transition data N_{ab} is left. Using an activity-based mobility model, results show that this strategy performs better than the static case.

Another variant is proposed in [90], whereby each user's mobility patterns are used to create individualized LAs (for each user). The proposed scheme is flexible and can be used in network with arbitrary cell topologies. Results show the proposed scheme gives better performance than the distance-based location update scheme.

There is also the method proposed in [1], whereby the design of the LA is formulated as a combinatorial optimization problem, subject to a constraint on the number of cells in the LA. Assuming independent, identically distributed cell residence time distribution, and shortest distance mobility model, it was shown that the LA design problem is NP-complete. A greedy heuristic is then proposed that gives irregular LA shapes. It was then shown that optimum rectangular LA shapes are a good approximation to the irregular LA shapes obtained from the greedy heuristic.

8.7.7 Reporting Center

Another location management scheme similar to the LA scheme described above is suggested in [7]. As in the case for the LA scheme, the reporting center scheme can be implemented statically or dynamically.

8.7.7.1 Static Case. In the static reporting center method, a subset of regions in the network is predefined as the reporting centers (Figure 8.16). Each region can represent an individual cell — in which case a reporting center corresponds to a reporting cell — or group of cells, such as a LA. Each mobile terminal performs a location update only when it enters one of these reporting centers (or reporting cells when each region consists of a single cell). When a call arrives, the search is confined to the reporting cell the user last reported and the neighboring bounded nonreporting cells. For example, in Figure 8.16, if a call arrives for user X, then search is

Location Management Techniques for Mobile Computing Environments

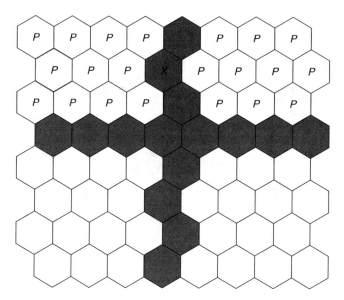

Figure 8.16. Network with Reporting Cells (Shaded Areas Represent Reporting Cells)

confined to the reporting cell the user last reported in and the nonreporting cells marked P. Obviously, certain reporting cells configuration leads to unbounded nonreporting cells, as shown in Figure 8.17, which one may want to avoid.

Following from the simple examples above, one can define the reporting center planning problem as the problem of finding an optimal set of reporting centers, such that the location management cost (of location update and location inquiry) is minimized. The cost formulation may vary depending, among other things, on the movement and call arrival patterns used. In a version of the reporting center planning problem [7], it was shown that finding an optimal set of reporting cells/centers, such that the location management cost is minimized, is a NP-complete problem. In [88, 89], GA, TS, and several variants of ant colony algorithm are used to find optimal and near optimal solutions to the reporting center problem. An evolving cellular automata system is also experimented in [91]. Other heuristics and approximation algorithms are implemented in [35, 68].

Because the subset of cells designated as the reporting cells are predefined, the scheme is easy to implement and requires minimal computing power on the mobile terminal. However, the scheme does not take into account each mobile user's mobility and call arrival patterns. For example, if a mobile terminal frequently moves in and out of a reporting cell, then there will be excessive, unnecessary location updates.

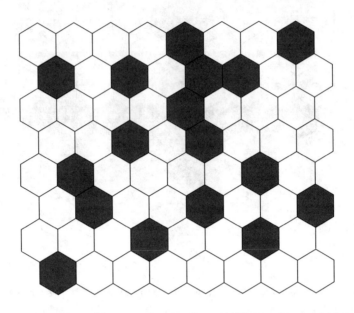

Figure 8.17. Network with Reporting Cells and Unbounded Nonreporting Cells

One study shows that this scheme at least performs better than both the always-update strategy and never-update strategy; a heuristic method to find near optimal solutions is proposed and the results are compared [35].

8.7.7.2 Dynamic Case. Each mobile user can also be assigned its own set of reporting centers. One such proposal is considered in [23, 82]. In this scheme, it is assumed the network has been partitioned into several static LAs. Using each LA as a region, the set of reporting centers is then individualized for each mobile user, whereby a mobile terminal decides, upon entering a LA, whether to perform a location update. The user's mobility and call arrival patterns are used to formulate the average location management cost and the problem is then solved using a GA.

Another similar scheme is proposed in [39], called the probabilistic location update (PLU) scheme. In this scheme, upon entering a new region (here, a region is defined as a LA), the mobile terminal will perform a location update with probability p. The optimal value for p, which varies according to the user's mobility and call arrival patterns, can then be optimized for each individual user. The authors suggest an adaptive control for the value p, based on the concept of an exponential back-off scheme described in [27].

8.7.8 Adaptive Threshold Scheme

This method was proposed in [63, 64] (there is also a similar discussion in [62]). In this adaptive threshold scheme, location update is performed not only according to each mobile user's mobility and call arrival pattern, but also according to the signaling load currently in the cell. In case of low signaling load, users transmit location updates more frequently.

To implement this scheme, each cell in the network is assigned a registration threshold level, which would depend, at any time, on the signaling load within the cell. Each mobile terminal then computes its own threshold level (according to its own mobility and call arrival pattern). The mobile terminal then sends an update message when its own threshold or priority level exceeds that of the cell. It was shown that this method reduces paging cost in comparison to the time-based method.

8.7.9 Profile-Based

In the method proposed in [69, 93], the system keeps a profile of which regions each user spends the most time in. Each region can represent an individual cell, or group of cells, such as a LA. When the user moves out of its profile's regions, a location update is performed to inform the system of its new region. However, while the user is moving around inside its profile's regions, no location update is performed.

Due to the profile kept (and maintained) for each user, the location probability of the user in each of the cells in the profile is also known. As such, in a search/paging operation (to find the user), the cells in the profile can be paged sequentially — in order of decreasing location probability — or in other ways involving the user's location probabilities (discussed later).

One condition of the scheme is a certain degree of predictability in users' movements. In [93], it was shown that introduction of the scheme results in lower location management costs, when users have medium to high predictability in their movement.

8.7.10 Compression-Based

In [14], the authors proposed the LeZi-update scheme, which uses the commonly used Ziv and Lempel compression algorithm [113]. Essentially, location update is considered as a stream of data to be transferred. The idea then, is to compress these data, resulting in fewer bits to be transferred.

In the LeZi-update scheme, it is assumed that there is a degree of predictability in users' movements. The raw location update scheme considered is the always-update coupled with periodic (time-based) location updates. Such combination of updates ensures complete knowledge of users' movements, as well as length of time spent on each cell (longer time spent results in longer character repetition in the stream). Location update is then performed only when a path is not in the dictionary/profile. When a call arrives for a user, the system pages the user based on location probabilities information in the profile. For this reason, the mobile station and the system needs to keep an identical dictionary/profile — the mobile station uses it for location updates and the system uses it for location inquiry.

Simulation done in [15] shows that there is performance improvement when users' mobility patterns remain stationary for long intervals; the improvement diminishes otherwise. It was also reported that, using the location probabilities profile, there are infrequent cases of long exhaustive searches for a mobile user.

8.7.11 Hybrid Strategies

Some of the individual location update strategies described above can also be combined to create hybrid strategies. For example, the use of LAs and time-based strategy is suggested in [44, 94]. Although a location update is performed when a user crosses a LA boundary, a location update is also performed periodically within the LAs. This allows the system to recover users' locations in case of system failure. Other hybrid strategies are discussed in [19, 33, 48, 66, 111].

8.8 Location Inquiry Strategies

In the previous section, a number of location updates, and paging/location inquiry strategies specific to particular location update strategies are discussed. In this section, general location inquiry strategies are discussed. These general strategies are applicable to most of the location update strategies discussed above.

8.8.1 Simultaneous Networkwide Search

In its simplest form, locating a mobile terminal can be done by simultaneously paging all the cells within the network. This technique will also take the least time to locate a mobile terminal. However, this technique will result in enormous signaling traffic, particularly for moderate to large networks.

It is highly desirable to be able to limit the location inquiry (search operation) to a set of cells, or region, in the network (e.g., a LA). This is because, unlike location update, in location inquiry if a cell is paged and no response is received from the mobile terminal, two possibilities exist:

1. The mobile terminal is not in the paged cell.
2. The mobile terminal is in the paged cell, but the paging signal is not received by the terminal (e.g., because of interference). That is, the paging signal is lost.

In the event that the second reason proved true (remote though it may be), cells in the network may need to be paged again. If the location possibilities of a mobile terminal can be limited to a region, however, then the repeat of the search can be confined to the region.

Besides simultaneously paging all the cells in the network (or region/LA the user is in), sequential paging (i.e., multistep paging) can be used. In sequential paging, cells (or defined regions) in the network are paged, one after another, until the required mobile terminal is found. One problem with multistep paging is the inevitable extra time delay in locating a mobile terminal, resulting in a lower QoS. Further, in a search operation, the user may move to a cell already paged previously, necessitating a simultaneous networkwide (or regionwide) search as a last stage of the search operation if the user is still not found. This implies that some cells may be paged twice. However, because such a possibility is remote, the second paging can be used to cover such a possibility and the possibility that the first paging signal is lost.

The next few sections discuss and highlight several general, sequential paging methods. Each of these strategies can be thought of as a variant of each other as they all belong to the sequential paging methods.

8.8.2 Paging Area

One simple improvement to simultaneously paging all the cells within the network is to group cells into paging areas (PAs). Cells within each PA are paged simultaneously. Each PA can then be polled sequentially, until the required user is found. The size and number of PAs can be adjusted to accommodate any delay constraint and QoS requirements of the network. Paging under delay constraints is discussed in [2, 3, 36, 76, 78, 106, 109].

In the PA technique, cells in the network need to be grouped into PAs. Further, once the cells have been grouped into PAs, the order of PAs' polling needs to be determined. Such planning, required on the network operator, can be time-consuming and expensive. One simple technique would be to arbitrarily or randomly group cells into several PAs and randomly assign the PA polling sequence. Beside arbitrary assignment, user's location probabilities can be used. The use of users' location probabilities is discussed under the intelligent paging method, discussed later below.

8.8.3 Expanding Ring Paging

In this technique, the last known cell location (or other cells deemed to have the highest location probability), also called ring 0, of a mobile terminal is paged first. On a miss, all the cells surrounding the last known cell — ring 1, are paged. This "ring of cell" paging can continue until the required mobile terminal is found. This ring of cells is illustrated for a hexagonal configuration network in Figure 8.18.

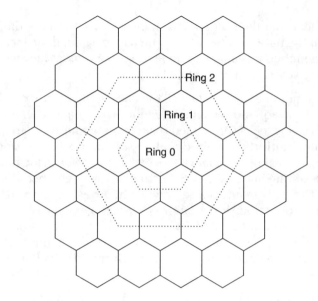

Figure 8.18. Expanding Ring Paging on Hexagonal Configuration Network

From one perspective, this technique can be thought of as an improvement to the PA scheme described above. Using this technique, time-consuming page area planning can be avoided. However, the complexity for this technique can be greater than the PA technique, as the ring must be computed or predefined for each cell in the network.

To accommodate any delay constraint, a networkwide search may be used when a user is not found within a specified maximum paging ring.

Besides paging the network ring by ring, a combination of rings can also be used at each paging step. For example, in the first paging step, ring 0 and ring 1 can both be paged simultaneously, followed by ring 2, ring 3, and ring 4 in the second paging step, and so on.

8.8.4 Intelligent Paging

Other than the simultaneous networkwide search, the other paging methods described above sequentially page groups of cells, until the required user is found. Ideally, we want to page the cells in the order of location probability (with cells that have the highest probability being paged first). In a sense, we would need to predict the current location of the user. To do this, many factors could be taken into account. This includes (but is not limited to):

- Geographical attraction points for users (e.g., shopping centers, schools), as well as road conditions (e.g., users on freeway, neighborhood streets) and layout

Location Management Techniques for Mobile Computing Environments

- User's mobility pattern
- Time, including time of day (e.g., peak hours), day of week (weekdays and weekends), as well as seasonal holidays

This extended paging strategy has been studied extensively [2, 6, 10, 11, 13, 37, 41, 45–47, 54, 60, 61, 69, 93, 100, 101, 104]. In the absence of any time delay constraint (in finding the user), minimum paging cost can be achieved by sequentially paging each cell in the order of user's location probabilities. That is, the cell with the highest probability of finding the user is paged first, followed by the cell with the second highest location probability, and so on. In real cellular networks, however, due to time delay constraints and QoS requirements, only a limited number of paging steps can be conducted.

A successful find in the first paging operation ensures limited time delay in finding a mobile terminal, thus maintaining a required QoS. As such, the probability of success in the first paging step should be sufficiently high (e.g., 90 percent [54]).

Undoubtedly, the added intelligence leads to a rather complex system with considerable computing and memory requirement. Besides the added complexity and requirements on the system, extra storage and processing is likely to be required on the mobile terminal, which usually has limited energy reserve and processing power. The ever decreasing costs of storage and processing power, however, means intelligent paging is becoming a relatively more viable strategy for location management. Very high paging cost savings have also been reported for the intelligent paging method [42], as is evident by the high success in the first paging operation.

Note that because an intelligent paging strategy takes into account each mobile user's profile (or each class of mobile users), it is clear that such strategy can be classified as an individualized and dynamic (sequential) paging strategy. It is also clear, however, that by definition, an individualized strategy requires individual information/profiles of each user, and its dynamic nature means each profile needs to be maintained and updated (on a regular basis). The kinds of data kept in the profile, as well as how such information/profile are obtained, stored, and maintained, directly affects, and at the same time is influenced by, the paging method implemented. One simple implementation is described below.

In [55], a location accuracy matrix (LAM) is used to take into account the different factors influencing the mobility of users in real mobile networks: attraction points (shopping centers, schools), road conditions, and layout (freeway, neighborhood streets). Specifically, given the last known cell location of a mobile user, the location accuracy matrix specifies the average probability of locating the mobile user in each of the cells in the network. The LAM is of size $n \times n$, where n is the number of cells in the network. Each row

$$\begin{array}{c} \quad 1 \quad\;\; 2 \quad\; \cdots \quad\; n \\ \begin{array}{c} 1 \\ 2 \\ \vdots \\ n \end{array} \left[\begin{array}{cccc} 0.3 & 0.05 & \cdots & 0.2 \\ 0.1 & 0.11 & \cdots & 0.5 \\ \vdots & \vdots & \ddots & \vdots \\ 0.02 & 0.1 & \cdots & 0.08 \end{array} \right] \end{array}$$

Figure 8.19. Location Accuracy Matrix
Each row in the matrix represents the last known cell location of the user and each column represents the probability of locating the mobile user in each cell in the network.

in the matrix represents the last known cell location of the user, with each column entry representing the probability of locating the mobile user in each cell in the network (given the last known cell location of the user) as shown in Figure 8.19. To calculate each of these probabilities, paging responses can be recorded and inputted into the LAM. For example, if a user's last known location is cell j, and a response is received from cell k (that is, the user is in cell k), then the hit counter at (j, k) of the LAM can be incremented. From these statistics, the location probabilities can be calculated.

8.9 Summary

This chapter provided an overview on location management. The concept of cost in location management, used for evaluation and comparison purposes, is discussed. Common network topologies, users' call arrival patterns, and users' mobility patterns used in network simulation for location management evaluation purposes, are also discussed.

An overview of basic location update strategies and a number of novel location update strategies that have been proposed in the literature over the years are also provided. Performances and drawbacks for several of the strategies are also discussed.

Finally, general location inquiry/paging strategies, as well as their advantages and disadvantages are discussed. The simultaneous network-wide search, PA, and expanding ring paging strategies are general in nature and should work well with most location update strategies.

References

1. A. Abutaleb and V.O.K. Li, Location update optimization in personal communication systems, *Wireless Networks*, vol. 3, no. 3, pp. 205–216, 1997.
2. A. Abutaleb and V.O.K. Li, Paging strategy optimization in personal communication systems, *Wireless Networks*, vol. 3, no. 3, pp. 195–204, 1997.

3. I.F. Akyildiz and J.S.M. Ho, A mobile user location update and paging mechanism under delay constraints, *Computer Communication Review*, vol. 25, no. 4, pp. 244–255, 1995.
4. I.F. Akyildiz, J.S.M. Ho, and B.L. Yi, Movement-based location update and selective paging for PCS networks, *IEEE/ACM Transactions on Networking*, vol. 4, no. 4, pp. 629–638, 1996.
5. I.F. Akyildiz, J. McNair, J.S.M. Ho, H. Uzunalioglu, and W. Wenye, Mobility management in next-generation wireless systems, *Proceedings of the IEEE*, vol. 87, no. 8, pp. 1347–1384, 1999.
6. D.O. Awduche, A. Ganz, and A. Gaylord, An optimal search strategy for mobile stations in wireless networks, in *Proceedings of the IEEE International Conference on Universal Personal Communications*, 1996, vol. 2, pp. 946–950.
7. A. Bar-Noy and I. Kessler, Tracking mobile users in wireless communications networks, *IEEE Transactions on Information Theory*, vol. 39, no. 6, pp. 1877–1886, 1993.
8. A. Bar-Noy, I. Kessler, and M. Sidi, Mobile users: to update or not to update? *Wireless Networks*, vol. 1, no. 2, pp. 175–185, 1995.
9. A. Bera and N. Das, Performance analysis of dynamic location updation strategies for mobile users, in *Proceedings of the 20th IEEE International Conference on Distributed Computing Systems*, 2000, pp. 428–435.
10. P.S. Bhattacharjee, D. Saha, and A. Mukherjee, An intelligent paging strategy for personal communication services network, in *Proceedings of IEEE Region 10 Conference (TENCON 99)*, 1999, vol. 2, pp. 1244–1246.
11. P.S. Bhattacharjee, D. Saha, and A. Mukherjee, Intelligent paging strategies for third generation personal communication services networks, *Journal of Interconnection Networks*, vol. 1, no. 3, pp. 153–171, September 2000.
12. P.S. Bhattacharjee, D. Saha, A. Mukherjee, and M. Maitra, Location area planning for personal communication services networks, in *Proceedings of the 2nd ACM International Workshop on Modeling, Analysis, and Simulation of Wireless and Mobile Systems*, 1999, pp. 95–98.
13. P.S. Bhattacharjee, D. Saha, and N. Mukherjee, Paging strategies for future personal communication services network, in *Proceedings 6th International Conference High Performance Computing (HiPC '99)*, 1999, pp. 322–328.
14. A. Bhattacharya and S.K. Das, LeZi-update: An information-theoretic approach to track mobile users in PCS networks, in *Proceedings of the 5th Annual ACM/IEEE International Conference on Mobile Computing and Networking*, 1999, pp. 1–12.
15. A. Bhattacharya, S.K. Das, and S. Roy, Toward a universal model for personal mobility management, in *Proceedings of the IEEE Wireless Communications and Networking Conference*, 2000, vol. 3, pp. 1578–1583.
16. U. Black, *Emerging Communications Technologies*, 2nd ed. Upper Saddle River, NJ: Prentice Hall, 1997.
17. P. Carle and G. Colombo, Sub-optimal solutions for location and paging areas dimensioning in cellular networks, in *Proceedings of the 4th IEEE International Conference on Universal Personal Communications*, 1995, pp. 672–676.
18. G.V. Casares and O.J. Mataix, On movement-based mobility tracking strategy — an enhanced version, *IEEE Communications Letters*, vol. 2, no. 2, pp. 45–47, 1998.
19. G.V. Casares and O.J. Mataix, Global versus distance-based local mobility tracking strategies: A unified approach, *IEEE Transactions on Vehicular Technology*, vol. 51, no. 3, pp. 472–485, 2002.
20. M.F. Catedra and J.P. Arriaga, *Cell Planning for Wireless Communications*. Boston: Artech House, 1999.
21. I.A. Cimet, How to assign service areas in a cellular mobile telephone system, in *Proceedings of the IEEE International Conference on Communications Serving Humanity through Communications (SUPERCOMM/ICC '94)*, 1994, vol. 1, pp. 197–200.

22. P. Curle and G. Colombo, Sub-optimal solutions for location and paging areas dimensioning in cellular networks, in *Proceedings of the 4th IEEE International Conference on Universal Personal Communications*, 1995, pp. 672–676.
23. S.K. Das and S.K. Sen, A new location update strategy for cellular networks and its implementation using a genetic algorithm, in *Proceedings of the 3rd Annual ACM/IEEE International Conference on Mobile Computing and Networking*, 1997, pp. 185–194.
24. P. Demestichas, N. Georgantas, E. Tzifa, V. Demesticha, M. Striki, M. Kilanioti, and M. Theologou, Computationally efficient algorithms for location area planning in future cellular systems, *Computer Communications*, vol. 23, no. 13, pp. 1263–1280, 2000.
25. P. Demestichas, E. Tzifa, V. Demesticha, N. Georgantas, G. Kotsakis, M. Kilanioti, M. Striki, M.E. Anagnostou, and M.E. Theologou, Control of the location update and paging signaling load in cellular systems by means of planning tools, in *Proceedings of the IEEE 50th Vehicular Technology Conference*, 1999, vol. 4, pp. 2119–2123.
26. I. Demirkol, C. Ersoy, M.U. Caglayan, and H. Delic, Location area planning in cellular networks using simulated annealing, in *Proceedings of the IEEE Conference on Computer Communications (INFOCOM 2001)*, 2001, pp. 13–20.
27. G.J. Dong and S.J. Wha, Performance of an exponential backoff scheme for slotted-ALOHA protocol in local wireless environment, *IEEE Transactions on Vehicular Technology*, vol. 44, no. 3, pp. 470–479, 1995.
28. W. Donggen and C. Tao, A spatio-temporal data model for activity-based transport demand modelling, *International Journal of Geographical Information Science*, vol. 15, no. 6, pp. 561–585, 2001.
29. J.M. Dutton, *Computer Simulation of Human Behavior*. New York: Wiley, 1971.
30. L. Fridstrom and L.H. Thune, An econometric air travel demand model for the entire conventional domestic network: The case of Norway, *Transportation Research, Part B Methodological*, vol. 3, pp. 213–223, 1989.
31. A. Gamst, Application of graph theoretical methods to GSM radio network planning, in *Proceedings of the IEEE International Symposium on Circuits and Systems*, 1991, vol. 2, pp. 942–945.
32. J.M. Gil and C.S. Hwang, A location area partitioning strategy using genetic algorithms for mobile location tracking, in *Proceedings of the 17th IASTED International Conference Applied Informatics*, 1999, pp. 349–352.
33. V.C. Giner and J.M. Oltra, Mobility tracking: Fixed location areas with hysteresis and with selective paging, in *Proceedings of Virginia Tech's 7th Symposium on Wireless Personal Communications*, 1997, pp. 1–12.
34. P.R.L. Gondim, Genetic algorithms and the location area partitioning problem in cellular networks," in *Proceedings of the IEEE 46th Vehicular Technology Conference*, 1996, vol. 3, pp. 1835–1838.
35. A. Hac and X. Zhou, Locating strategies for personal communication networks, a novel tracking strategy, *IEEE Journal on Selected Areas in Communications*, vol. 15, no. 8, pp. 1425–436, 1997.
36. J.S.M. Ho and I.F. Akyildiz, Mobile user location update and paging under delay constraints, *Wireless Networks*, vol. 1, no. 4, pp. 413–425, 1995.
37. I.L. Huey and P.L. Chien, A geography based location management scheme for wireless personal communication systems, in *Proceedings of the IEEE 51st Vehicular Technology Conference*, 2000, vol. 2, pp. 1358–1361.
38. T. Imielinski and B.R. Badrinath, Querying locations in wireless environments, in *Proceedings Wireless Communications Future Directions*, 1992, pp. 85–108.
39. D.G. Jeong and W.S. Jeon, Probabilistic location update for advanced cellular mobile networks, *IEEE Communications Letters*, vol. 2, no. 1, pp. 8–10, 1998.
40. K.W. Jyhi, S.H. Yee, H.C. Chao, and P.Y. Wei, On the traffic estimation and engineering of GSM network, in *Proceedings of the 7th IEEE International Symposium on Personal, Indoor and Mobile Radio Communications (PIMRC'96)*, 1996, vol. 3, pp. 1183–1187.

41. T.K. Kim and C. Leung, Generalized paging schemes for cellular communication systems, in *Proceedings of the IEEE Pacific Rim Conference on Communications, Computers and Signal Processing*, 1999, pp. 217–220.
42. N.E. Kruijt, D. Sparreboom, F.C. Schoute, and R. Prasad, Location management strategies for cellular mobile networks, *Electronics and Communication Engineering Journal*, vol. 10, no. 2, pp. 64–72, 1998.
43. D. Lam, D.C. Cox, and J. Widom, Teletraffic modeling for personal communications services, *IEEE Communications Magazine*, vol. 35, no. 2, pp. 79–87, 1997.
44. D.J. Lee and D.H. Cho, On optimum timer value of area and timer-based location registration scheme, *IEEE Communications Letters*, vol. 5, no. 4, pp. 148–150, 2001.
45. H.C. Lee and S. Junping, Mobile location tracking by optimal paging zone partitioning, in *Proceedings of the IEEE 6th International Conference on Universal Person Communications Record Bridging the Way to the 21st Century*, 1997, vol. 1, pp. 168–172.
46. Z. Lei and C. Rose, Probability criterion based location tracking approach for mobility management of personal communications systems, in *Proceedings of the IEEE Global Telecommunications Conference (GLOBECOM '97)*, 1997, vol. 2, pp. 977–981.
47. Z. Lei and C. Rose, Wireless subscriber mobility management using adaptive individual location areas for PCS systems, in *Proceedings of the IEEE International Conference on Communications Conference*, 1998, vol. 3, pp. 1390–1394.
48. A. Leonhardi and K. Rothermel, A comparison of protocols for updating location information, *Cluster Computing*, vol. 4, no. 4, pp. 355–367, 2001.
49. K.K. Leung, W.A. Massey, and W. Whitt, Traffic models for wireless communication networks, *IEEE Journal on Selected Areas in Communications*, vol. 12, no. 8, pp. 1353–1364, October 1994.
50. K.K. Leung, W.A. Massey, and W. Whitt, Traffic models for wireless communication networks, in *Proceedings of the IEEE Conference on Computer Communications (INFOCOM '94) Networking for Global Communications*, 1994, vol. 3, pp. 1029–1037.
51. B. Liang and Z.J. Haas, Predictive distance-based mobility management for PCS networks, in *Proceedings of the IEEE Conference on Computer Communications*, 1999, vol. 3, pp. 1377–1384.
52. Y.B. Lin, Reducing location update cost in a PCS network, *IEEE/ACM Transactions on Networking*, vol. 5, no. 1, pp. 25–33, 1997.
53. A. Lombardo, S. Palazzo, and G. Schembra, A comparison of adaptive location tracking schemes in personal communications networks, *International Journal of Wireless Information Networks*, vol. 7, no. 2, pp. 79–89, 2000.
54. G.L. Lyberopoulos, J.G. Markoulidakis, D.V. Polymeros, D.F. Tsirkas, and E.D. Sykas, Intelligent paging strategies for third generation mobile telecommunication systems, *IEEE Transactions on Vehicular Technology*, vol. 44, no. 3, pp. 543–554, 1995.
55. S. Madhavapeddy, K. Basu, and A. Roberts, Adaptive paging algorithms for cellular systems, in *Proceedings of the IEEE 45th Vehicular Technology Conference Countdown to the Wireless 21st Century*, 1995, vol. 2, pp. 976–980.
56. Z. Mao and C. Douligeris, Two location tracking strategies for PCS systems, in *Proceedings 8th International Conference on Computer Communications and Networks*, 1999, pp. 318–323.
57. J.G. Markoulidakis and M.E. Anagnostou, Periodic attachment in future mobile telecommunications, *IEEE Transactions on Vehicular Technology*, vol. 44, no. 3, pp. 555–564, 1995.
58. J.G. Markoulidakis, G.L. Lyberopoulos, D.F. Tsirkas, and E.D. Sykas, Evaluation of location area planning scenarios in future mobile telecommunication systems, *Wireless Networks*, vol. 1, no. 1, pp. 17–29, 1995.
59. J.G. Markoulidakis, G.L. Lyberopoulos, D.F. Tsirkas, and E.D. Sykas, Mobility modeling in third-generation mobile telecommunications systems, *IEEE Personal Communications*, vol. 4, no. 4, pp. 41–56, 1997.

60. S. Mishra and O.K. Tonguz, Most recent interaction area and speed-based intelligent paging in PCS, in *Proceedings of the IEEE 47th Vehicular Technology Conference Technology in Motion*, 1997, vol. 2, pp. 505–509.
61. S. Mishra and O.K. Tonguz, Analysis of intelligent paging in personal communication systems, *Electronics Letters*, vol. 34, no. 1, pp. 12–13, 1998.
62. Z. Naor, Tracking mobile users with uncertain parameters, in *Proceedings of the 6th Annual International Conference on Mobile Computing and Networking*, 2000, pp. 110–119.
63. Z. Naor and H. Levy, Minimizing the wireless cost of tracking mobile users: an adaptive threshold scheme, in *Proceedings of the IEEE Conference on Computer Communications 17th Annual Joint Conference of the IEEE Computer and Communications Societies*, 1998, vol. 2, pp. 720–727.
64. Z. Naor and H. Levy, LATS: A load-adaptive threshold scheme for tracking mobile users, *IEEE/ACM Transactions on Networking*, vol. 7, no. 6, pp. 808–817, 1999.
65. S. Okasaka, S. Onoe, S. Yasuda, and A. Maebara, A new location updating method for digital cellular systems, in *Proceedings of the 41st IEEE Vehicular Technology Conference*, 1991, pp. 345–350.
66. J.M. Oltra, V.C. Giner, and P.G. Escalle, Evaluation of tracking local strategies in wireless networks with stochastic activity networks, in *Proceedings of the IEEE International Conference on Universal Personal Communications Conference*, 1998, vol. 1, pp. 735–740.
67. A. Pal and D.S. Khati, Dynamic location management with variable size location areas, in *Proceedings of the IEEE International Conference on Computer Networks and Mobile Computing*, 2001, pp. 73–78.
68. M.C. Pinotti and L. Wilson, On the problem of tracking mobile users in wireless communications networks, in *Proceedings of the 31st Hawaii International Conference on System Sciences*, 1998, vol. 7, pp. 666–671.
69. G.P. Pollini and S. Tabbane, The intelligent network signaling and switching costs of an alternate location strategy using memory, in *Proceedings of the 43rd IEEE Vehicular Technology Conference Personal Communication Freedom through Wireless Technology*, 1993, pp. 931–934.
70. T.S. Rappaport, Cellular radio and personal communications: selected readings, Piscataway, NJ: IEEE, 1995.
71. T.S. Rappaport, *Wireless Communications: Principle and Practice,* Inglewood Cliffs, NJ: Prentice Hall, 1996.
72. C. Rose, Minimization of paging and registration costs through registration deadlines, in *Proceedings of the IEEE International Conference on Communications (ICC '95) Communications Gateway to Globalization*, 1995, vol. 2, pp. 735–739.
73. C. Rose, A greedy method of state-based registration, in *Proceedings of the IEEE International Conference on Communications (ICC '96) Converging Technologies for Tomorrow's Applications*, 1996, vol. 2, pp. 1158–1162.
74. C. Rose, Minimizing the average cost of paging and registration: a timer-based method, *Wireless Networks*, vol. 2, no. 2, pp. 109–116, 1996.
75. C. Rose, State-based paging/registration: A greedy technique, *IEEE Transactions on Vehicular Technology*, vol. 48, no. 1, pp. 166–173, 1999.
76. C. Rose and R. Yates, Minimizing the average cost of paging under delay constraints, *Wireless Networks*, vol. 1, no. 2, pp. 211–219, 1995.
77. C. Rose and R. Yates, Location uncertainty in mobile networks: A theoretical framework, *IEEE Communications Magazine*, vol. 35, no. 2, pp. 94–101, 1997.
78. C. Rose and R.D. Yates, Paging cost minimization under delay constraints, in *Proceedings of the IEEE Conference on Computer Communications 14th Annual Joint Conference of the IEEE Computer and Communications Societies*, 1995, vol. 2, pp. 490–495.

79. I. Rubin and W.C. Cheon, Impact of the location area structure on the performance of signaling channels of cellular wireless networks, in *Proceedings of the IEEE International Conference on Communications (ICC '96) Converging Technologies for Tomorrow's Applications*, 1996, vol. 3, pp. 1761–1765.
80. J. Scourias and T. Kunz, A dynamic individualized location management algorithm, in *Proceedings of the IEEE International Symposium on Personal, Indoor, and Mobile Radio Communications Technical Program*, 1997, vol. 3, pp. 1004–1008.
81. J. Scourias and T. Kunz, Activity-based mobility modeling: realistic evaluation of location management schemes for cellular networks, in *Proceedings of the IEEE Wireless Communications and Networking Conference*, 1999, vol. 1, pp. 296–300.
82. S.K. Sen, A. Bhattacharya, and S.K. Das, A selective location update strategy for PCS users, *Wireless Networks*, vol. 5, no. 5, pp. 313–326, 1999.
83. I. Seskar, S.V. Maric, J. Holtzman, and J. Wasserman, Rate of location area updates in cellular systems, in *Proceedings of the 42nd Vehicular Technology Society Conference Frontiers of Technology from Pioneers to the 21st Century*, 1992, vol. 2, pp. 694–697.
84. A.A. Siddiqi and T. Kunz, The peril of evaluating location management proposals through simulations, in *Proceedings of the 3rd International Workshop on Discrete Algorithms and Methods for Mobile Computing and Communications*, 1999, pp. 78–85.
85. P.B. Slater, International migration and air travel: global smoothing and estimation, *Applied Mathematics and Computation*, vol. 53, no. 2–3, pp. 225–234, 1993.
86. G.L. Stuber, *Principles of Mobile Communication*. Boston: Kluwer Academic, 1996.
87. R. Subrata and A.Y. Zomaya, Location management in mobile computing, in *Proceedings of the ACS/IEEE International Conference on Computer Systems and Applications*, 2001, pp. 287–289.
88. R. Subrata and A.Y. Zomaya, Artificial life techniques for reporting cell planning in mobile computing, in *Proceedings of the Workshop on Biologically Inspired Solutions to Parallel Processing Problems (BioSP3)* [published in the CD-ROM *Proceedings of the IEEE 16th International Parallel and Distributed Processing Symposium*], 2002, pp. 203–210.
89. R. Subrata and A.Y. Zomaya, A comparison of three artificial life techniques for reporting cell planning in mobile computing, *IEEE Transactions on Parallel and Distributed Systems*, vol. 14, no. 2, pp. 142–153, 2003.
90. R. Subrata and A.Y. Zomaya, Dynamic location management in mobile computing, *Telecommunication Systems*, Vol. 22, pp. 169–187, 2003.
91. R. Subrata and A.Y. Zomaya, Evolving cellular automata for location management in mobile computing networks, *IEEE Transactions on Parallel and Distributed Systems*, vol. 14, no. 1, pp. 13–26, 2003.
92. SUMATRA, http://www-db.stanford.edu/sumatra.
93. S. Tabbane, An alternative strategy for location tracking, *IEEE Journal on Selected Areas in Communications*, vol. 13, no. 5, pp. 880–892, 1995.
94. S. Tabbane, Location management methods for third generation mobile systems, *IEEE Communications Magazine*, vol. 35, no. 8, pp. 72–78, 83–84, 1997.
95. D. Tang and M. Baker, Analysis of a metropolitan-area wireless network, in *Proceedings of the 5th Annual ACM/IEEE International Conference on Mobile Computing and Networking*, 1999, pp. 13–23.
96. D. Tang and M. Baker, Analysis of a metropolitan-area wireless network, *Wireless Networks*, vol. 8, pp. 107–120, 2002.
97. D.W. Tcha, T.J. Choi, and Y.S. Myung, Location-area partition in a cellular radio network, *Journal of the Operational Research Society*, vol. 48, no. 11, pp. 1076–1081, 1997.
98. R. Thomas, H. Gilbert, and G. Mazziotto, Influence of the moving of the mobile stations on the performance of a radio mobile cellular network, in *Proceedings of the 3rd Nordic Seminar*, 1988.

99. C.K. Toh, Performance evaluation of crossover switch discovery algorithms for wireless ATM LANs, in *Proceedings of the IEEE INFOCOM '96 The Conference on Computer Communications 15th Annual Joint Conference of the IEEE Computer Societies*, 1996, vol. 3, pp. 1380–1387.
100. O.K. Tonguz, S. Mishra, and R. Josyula, Impact of random user motion on locating subscribers in mobile networks, in *Proceedings of the IEEE International Conference on Personal Wireless Communications*, 1999, pp. 491–495.
101. M. Verkama, Optimal paging — A search theory approach, in *Proceedings of the 5th IEEE International Conference on Universal Personal Communications*, 1996, vol. 2, pp. 956–960.
102. M. Verkama, A simple implementation of distance-based location updates, in *Proceedings of the IEEE 6th International Conference on Universal Personal Communications*, 1997, vol. 1, pp. 163–167.
103. M. Vudali, The location area design problem in cellular and personal communications systems, in *Proceedings of the 5th IEEE International Conference on Universal Personal Communications*, 1996, vol. 2, pp. 591–595.
104. K. Wang, J.M. Liao, and J.M. Chen, Intelligent location tracking strategy in PCS, *IEEE Proceedings Communications*, vol. 147, no. 1, pp. 63–68, 2000.
105. T.P. Wang, S.Y. Hwang, and C.C. Tseng, Registration area planning for PCS networks using genetic algorithms, *IEEE Transactions on Vehicular Technology*, vol. 47, no. 3, pp. 987–995, 1998.
106. W. Wang, I.F. Akyildiz, and G.L. Stuber, Reducing the paging costs under delay bounds for PCS networks, in *Proceedings of the IEEE Wireless Communications and Networking Conference*, 2000, vol. 1, pp. 235–257.
107. V.W.S. Wong and V.C.M. Leung, An adaptive distance-based location update algorithm for next-generation PCS networks, *IEEE Journal on Selected Areas in Communications*, vol. 19, no. 10, pp. 1942–1952, 2001.
108. H. Xie, S. Tabbane, and D.J. Goodman, Dynamic location area management and performance analysis, in *Proceedings of the 43rd IEEE Vehicular Technology Conference Personal Communication Freedom through Wireless Technology*, 1993, pp. 536–539.
109. A. Yener and C. Rose, Paging strategies for highly mobile users, in *Proceedings of the IEEE 46th Vehicular Technology Conference Mobile Technology for the Human Race*, 1996, vol. 3, pp. 1839–1842.
110. K.L. Yeung and T.S.P. Yum, A comparative study on location tracking strategies in cellular mobile radio systems, in *Proceedings of the IEEE Global Telecommunications Conference*, vol. 1, pp. 22–28, 1995.
111. W.H.A. Yuen and W.S. Wong, A contention-free mobility management scheme based on probabilistic paging, *IEEE Transactions on Vehicular Technology*, vol. 50, no. 1, pp. 48–58, 2001.
112. J.H. Zhang and J.W. Mark, A local VLR cluster approach to location management for PCS networks, in *Proceedings IEEE Wireless Communications and Networking Conference*, 1999, vol. 1, pp. 311–315.
113. J. Ziv and A. Lempel, Compression of individual sequences via variable-rate coding, *IEEE Transactions on Information Theory*, vol. IT-24, no. 5, pp. 530–536, 1978.

Chapter 9
Locating Mobile Objects

Evaggelia Pitoura, George Samaras, and Georgia Kastidou

9.1 Introduction

In distributed computing, the notion of mobility is emerging in many forms and applications. Increasingly, many users are not tied to a fixed access point but instead use mobile hardware such as cellular phones or personal digital assistants. Small devices (e.g., sensors) are mounted on moving objects such as vehicles or airplanes. Furthermore, mobile software (i.e., code or data that move among network locations) offers a new form of building distributed network-centric applications. In the presence of mobility, the cost of communicating with a mobile object or invoking it is augmented by the cost of locating it (i.e., identifying its current location).

Often, mobility is related to wireless mobile computing [9, 14, 32], because wireless communications allow the free movement of users. Future personal communication systems (PCSs) will support a huge object population and offer numerous customer services. In such systems, the signaling and directory traffic for locating mobile users is expected to increase dramatically [45]. Thus, deriving efficient strategies for *locating* mobile users (i.e., identifying their current location) is an issue central to wireless mobile computing research.

Besides mobility tied to wireless hardware, data, or code may be relocated among different network sites for reasons of performance or availability. Mobile software agents [1, 44] are a popular form of mobile software. Mobile agents are processes that may be dispatched from a source computer and be transported to remote servers for execution. Mobile agents can be launched into an unstructured network and roam around to accomplish their task [2], thus providing an efficient, asynchronous method for collecting information or attaining services in rapidly evolving networks. Other applications of moving software include the relocation of an object's personal environment to support ubiquitous computing [46] or the migration of services to support load balancing, for instance the active transfer of Web pages to replication servers in the proximity of clients [5].

The goal of this chapter is to survey and compare the various location management mechanisms proposed in the literature for storing, querying, and updating the location of both hardware and software mobile objects. The emphasis is on the underlying system architectures.

The remainder of this chapter is structured as follows:

- Section 9.2 introduces the location management problem and a taxonomy of the various approaches to its solution.
- Section 9.3 presents the basic architectures proposed.
- Section 9.4 discusses the main optimization techniques, such as caching and location forwarding that can be applied to the basic architectures to improve their efficiency.
- Section 9.5 introduces a detailed taxonomy of the proposed approaches.
- Section 9.6 presents concrete examples of location management protocols used in Mobile IP, Globe, and two mobile agents systems — Ajanta and Voyager®.
- Section 9.7 concludes the chapter.

9.2 Location Management

In mobile distributed computing, mobile objects (i.e., mobile software, data, or users using wireless hardware) may relocate from one location to another. Although, the emphasis in this chapter is on objects moving among networks sites, the approaches presented also apply to objects moving in physical space. Although in the former case, location corresponds to a network address (network point of attachment), in the latter case, location is most often specified by the spatial coordinates of the object. We shall use the term location to refer to both cases.

The exact mechanisms for locating a mobile object depend on the underlying network architecture. In cellular digital architecture as well as in wireless LAN technologies, the network configuration consists of fixed backbone networks extended with a number of mobile objects (called mobile hosts (MHs) in this context) communicating directly with stationary transceivers called mobile support stations or base stations. The area covered by an individual transceiver's signal is called a cell. Each mobile host can communicate with other hosts, mobile or fixed, only through the base station of the cell in which it resides. In this case, to locate a mobile object, the current cell must be found. In other applications, such as in the case of mobile agents, locating an agent reduces to specifying the address of the machine that is currently hosting it. Finally, mobile objects may be

Locating Mobile Objects

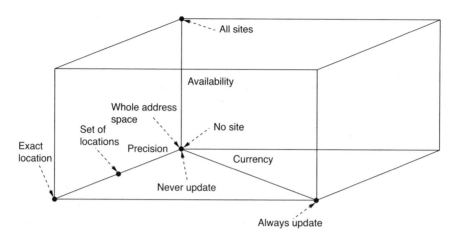

Figure 9.1. Dimensions of Quality of the Stored Location Information for a Mobile Object

equipped with location sensing devices. The most widely known location sensing system today is the global positioning system (GPS) [10, 11]. GPS enables objects equipped with a relatively cheap transceiver to deduce their latitude, longitude, and altitude with accuracy of a few meters. These spatial coordinates are computed based on time-of-flight information derived from radio signals broadcast by a constellation of satellites in earth's orbit. In this case, location refers to such information.

To efficiently locate an object, that is, identify its current position (in space or in a computer network), related information may be stored at databases at specific network sites. We call such databases that store the location of moving objects *location directories*. In abstract terms, location management involves two basic operations — lookups and updates. A *lookup* is invoked each time there is a need to locate a mobile object, in order, for example, to communicate with it, interchange data, or computation. *Updates* of the location of a mobile object stored in the location directory are initiated when the object moves.

To balance the cost of lookups against the cost of updates, the quality of the stored location information for an object varies. In particular, various approaches compromise the availability, precision, or currency of the location information stored for each mobile object (Figure 9.1).

In terms of *availability*, choices range between saving (replicating) the location of each mobile object at all network sites to not storing the location at all. In between these two approaches, location information may be maintained selectively at one or more specific network sites. There is a wide range of selection criteria for the sites that are used for saving location information for each user. For example, a choice may be to save the location of objects at the sites of their frequent callers.

Precision of location information refers to the granularity of the location information stored for each object. This may vary from maintaining the exact location to maintaining a wider region or a set of possible locations. In the case in which either a set of potential locations or a wider region is stored, there is need to search for the object either in each of the specified locations or inside the region.

Currency refers to when the stored location information is updated. For instance, for a highly mobile object, it may make sense to defer updating its stored location every time the object moves. Potential policies for initiating updates include a time-based, a movement-based, and a distance-based policy [7]. In the time-based update policy, the stored location for each mobile object is updated periodically every T time units. In the movement-based update policy, the stored location is updated after the object has performed a predefined number of moves. Finally, in the distance-based (or dead-reckoning [36]) update policy, the stored location is updated when the distance of the stored location from the actual location of the object exceeds a predefined threshold value D. Analytical performance results show that distance-based update policies outperform the other policies in most cases. However, distance-based approaches are more difficult to implement because they require knowing and computing a distance function.

When precise and current location information is stored at every network site, locating an object reduces to just querying the local location directory. On the other hand, each time the location of the object changes, a large number of associated location directories must be updated. At the other extreme, when no information about the location of the object is stored at any directory, to locate a mobile object, a global search for the object must be initiated. However, in this case, when an object moves, there is no cost associated with updating location directories. In general, when there is no current or precise information about the location of the object available locally, locating the object involves a combination of some search procedure and a number of queries posed to an appropriate set of location directories.

Besides the efficient support of location lookups and updates, a challenging issue is the management of more advanced location queries. Examples of such queries include finding the nearest service when the service or the user is mobile or queries that refer to future time, such as "find all objects that will enter a specified region within the next hour". To this end, a database management system (DBMS) with extended support for spatiotemporal data may be used as a location directory. This is an issue attracting much current research and is beyond the scope of this chapter; the interested reader is referred to [47] and other chapters of this book.

9.3 Architectures of Location Directories

In this section, we describe the basic architectures of location directories. The simplest approach is to maintain a centralized location directory in which to store the location of all mobile objects. Such an approach does not scale with the number of objects and their degree of distribution. Furthermore, it suffers from a single point of failure. Thus, the two most common approaches are a two-tier scheme, in which the current location of each mobile object is maintained at two network sites, and a hierarchically structured distributed location directory, in which the address space is hierarchically decomposed into subregions.

9.3.1 Two-Tier Scheme

In the two-tier approach, a home directory, called Home Location Register (HLR) is associated with each mobile object. The HLR is located at a site (network location, geographical region, or cell) prespecified for each object. It maintains the current location of the object. The lookup and update procedures are simple. To locate an object, its HLR is identified and queried. When an object x moves to a new location, its HLR is contacted and updated to maintain the new location.

As an enhancement to the above scheme, Visitor Location Registers (VLRs) are maintained at each site. The VLR at a site i stores the identifiers of all objects currently located at site i. When a lookup for an object x is issued at a site i, the VLR at site i is queried first and only if the object is not found there, is the HLR of x contacted. When an object x moves from site i to site j, in addition to updating x's HLR, the entry for x is deleted from the VLR at site i and a new entry for x is added to the VLR at site j.

The two prevailing existing standards for cellular technologies, the Electronics Industry Association/Telecommunications Industry Association's (EIA/TIA) Interim Standard 41 (IS-41) commonly used in North America and the Global System for Mobile Communications (GSM) used in Europe, both support carrying out location strategies using HLRs and VLRs [26].

One problem with the home location approach is that the assignment of the home register to a mobile object is permanent. Thus, long-lived objects cannot be appropriately handled, because their home location remains fixed even when the objects permanently move to a different region. Another drawback of the two-tier approach is that it does not scale well with highly distributed systems where sites are geographically widely dispersed. To contact an object, the possibly distant home location must be contacted first. Similarly, even a move to a nearby location must be registered at a potentially distant home location. Thus, locality of moves and lookups is not taken advantage of.

9.3.2 Hierarchical Scheme

Hierarchical location schemes extend two-tier schemes by maintaining a hierarchy of location directories. In this hierarchy, a location directory at a higher level contains location information for objects located at levels below it. Usually, the hierarchy is tree-structured. In this case, the location directory at a leaf serves a single site (cell) and contains entries for all objects residing in this site. A directory at an internal node maintains information about objects registered in the set of sites in its subtree. For each mobile object, this information is either a pointer to an entry at a lower level directory or the object's actual current location. In cellular architectures, the directories are usually interconnected by the links of the intelligent signaling network, e.g., a Common Channel Signaling (CCS) network. For instance, in telephony, the directories may be placed at the telephone switches. It is often the case that the only way that two sites can communicate with each other is through the hierarchy; no other physical connection exists among them.

The type of location information maintained in the location directories affects the relative cost of updates and lookups as well as the load distribution among the links and nodes of the hierarchy. We consider two cases:

1. Internal nodes that keep pointers to the appropriate lower level directory
2. Internal nodes that maintain the exact location of the objects

We use the term LCA(i, j) to denote the least common ancestor of node i and node j.

Let us discuss first the case of keeping at all internal directories pointers to lower level directories. For example, in Figure 9.2 [left] for an object x residing at site (cell) 18, there is an entry in the directory at node 0 pointing to the entry for x in the directory at node 2. The entry for x in the directory at node 2 points to the entry for x in the directory at node 6, which in turns points to the entry for x in the directory at node 18. When object x moves from site i to site j, the entries for x in the directories along the path from j to LCA(i, j) and from LCA(i, j) to i are updated. For instance, when object x moves from 18 to 20, the entries at node 20, node 7, node 2, node 6, and node 18 are updated. Specifically, the entry for x is deleted from the directories at node 18 and node 6, the entry for x at the directory at 2 is updated, and entries for x are added to the directories at node 7 and node 20. When a lookup is initiated at site i for an object x located at site j, the lookup procedure queries directories starting from node i proceeding up the tree until the first entry for x is encountered. This happens at node LCA(i, j). Then,

Locating Mobile Objects

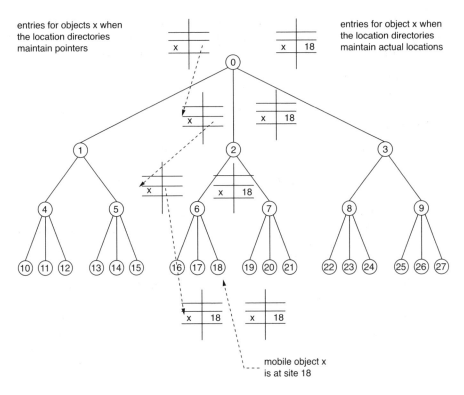

Figure 9.2. Hierarchical Location Scheme
Location directories' entries at the left are pointers at the lower level directories and location directories' entries at the right are actual locations.

the lookup procedure proceeds downward following the pointers to node j. For instance, a lookup placed at site 21 to object x located at node 18 (Figure 9.2 [left]), queries directories at node 21 and node 7 and finds the first entry for x at node 2. Then, it follows the pointers to node 6 and node 18.

Let us now consider the case of directory entries that maintain the actual location of each object. Then, for object x registered at 18 (Figure 9.2 [right]), there are entries in the directories at node 0, node 2, node 6, and node 18, each containing a pointer to location 18. In this case, a move from site i to site j causes the update of all entries along the paths from j to the root and from the root to i. For example, a relocation of object x from site 18 to site 20 involves the entries for x at 20, 7, 0, 2, 6, and 18. After the update, entries for x exist in the directories located at node 0, node 2, node 7, and node 20, each containing a pointer to 20 and the entries for x in the directories at node 6 and node 18 are deleted. On the other hand, the cost of a lookup from node i to an object located at node j is reduced, because once the LCA(i, j) is reached, there is no need to query the directories on

203

the downward path to j. For example, a lookup from node 21 to object x (Figure 9.2 [right]) queries directories at node 21, node 7, node 2, and then node 18 directly (without querying the directory at node 6).

When hierarchical location directories are used, there is no need for binding an object to a HLR. The object can be located by querying the directories in the hierarchy. In the worst case, an entry for the object will be found in the directory at the root.

A hybrid scheme using both hierarchical entries and preassigned HLRs is also possible. Assume that directory entries are maintained only at selective nodes of the hierarchy and that a HLR is used. In this case, a lookup originating from site i for an object x starts searching for the requested object from site i. It proceeds following the path from i to the LCA of i and the x's HLR and then moves downward to x's HLR, unless an entry for x is found in any directory on this path. If such an entry is encountered, it is followed instead [45].

The hierarchical scheme leads to reductions in communication cost when most calls and moves are geographically localized. In such cases, instead of contacting the HLR of the object that may be located far away from the object's current location, a small number of location directories in the object's neighborhood are accessed. However, the number of location directories that are updated and queried increases relative to the two-tier scheme.

Another problem with the hierarchical schemes is that the directories located at higher level of the hierarchy must handle a relatively large number of messages. Furthermore, they store more entries than nodes at lower levels. One solution is to partition the directories at the high-level nodes (e.g., at the root) into smaller directories at subnodes so that the entries of the original directory are shared appropriately among the directories at the subnodes [40, 41].

Table 9.1 summarizes some of the pros and cons of the hierarchical architectures when compared with the two-tier architecture.

Table 9.1. Summary of Pros (+) and Cons (-) of Hierarchical Architectures

(+)	No preassigned HLR
(+)	Support for locality
(-)	Increased number of operations in terms of both network messages and location directory accesses (lookup and update operations)
(-)	Increased processing load and storage requirements at the higher levels of the hierarchy

9.4 Optimizations of the Architectures

In this section, we describe a number of enhancements of the two basic architectures of location directories: namely partitions, caching, replication, and forwarding pointers. These approaches are orthogonal to each other in the sense that they can be combined to improve performance further. In the following, we use the term requestor object to denote the object that initiates a lookup for a mobile object and the term requested object for the object whose location is sought. We first introduce an important metric called call to mobility ratio.

9.4.1 Call to Mobility Ratio

A parameter that affects the performance of most location management schemes is the relative frequency of move and lookup (or call) operations of each object. This is captured by the call to mobility ratio (CMR). Let C_i be the expected number of searches for object P_i over a time period T and U_i the number of moves made by P_i over T, then:

$$CMR_i = C_i/U_i.$$

Another important parameter is the local call to mobility ratio, $LCMR_{i,j}$, that also involves the site of the requestor object. Let $C_{i,j}$ be the expected number of calls made from site j to an object P_i over a time period T, then the local call to mobility ratio $LCMR_{i,j}$ is defined as:

$$LCMR_{i,j} = C_{i,j}/U_i.$$

For hierarchical location schemes, the local call to mobility ratio ($LCMR_{i,j}$) for an internal node j is extended as follows:

$$LCMR_{i,j} = \sum_k LCMR_{i,k}$$

where k is a child of j. That is, the LCMR for an object P_i and an internal node j is the ratio of the number of lookups for P_i originated from any site at j's subtree to the number of moves made by P_i.

9.4.2 Partitions

To avoid maintaining location entries at all levels of the hierarchy, and at the same time reduce the search cost, partitions are deployed [4]. The partitions for each object are obtained by grouping the sites (cells) among which the object moves frequently and separating the sites between which it relocates infrequently. Thus, partitions exploit locality of movement. Partitions can be used in many ways. We describe next one such partition-based strategy.

For each partition, the information whether the object is currently in the partition is maintained at the LCA of all sites in the partition, called the representative of the partition. The representative knows that an object is in its partition but not its exact location [4]. This information is used during

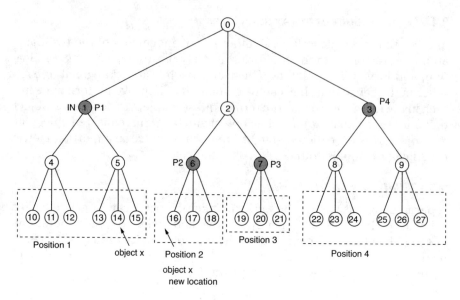

Figure 9.3. Four Partitions (P1, P2, P3, and P4)
The IN entry at node 1 indicates that object x is in partition P1.

a flat search (i.e., top-down search starting from the root) to decide which subtree in the hierarchy to search. Thus, partitions reduce the overall search cost as compared to flat searches. There is an increase however on the update cost: when an object crosses a partition, the representatives of its previous and new partitions must be informed. For example, assume that object x often moves inside four different sets of sites (i.e., partitions) and infrequently between these sets. Let the sites of each partition be {10, 12, 14, 15}, {16, 18}, {19, 20, 21} and {22, 23, 25, 26, 27} (Figure 9.3). The representative node of each partition is highlighted. When object x is at node 14 in partition 1, the representative of the associated partition, node 1, maintains the information that the user is inside its partition. When x moves to node 16, that is, outside the current partition, both node 1, the representative of the old partition, and node 6, the representative of the new partition, are updated to reflect the movement.

9.4.3 Caching

In two-tier architectures, every time a lookup for the location of an object x is initiated, x's location is cached at the VLR at the requestor's site, so that any subsequent lookups of x originated from this site can reuse this information [18]. To locate an object, the cache at the VLR of the requestor's site is queried first. If the location of the requested object is found at the cache, then a query is launched to the indicated location without contacting the object's HLR. Otherwise, the HLR is queried.

A problem associated with caching is that, when an object moves, its cached location becomes obsolete. There are two basic approaches to cache updates — eager and lazy caching.

In *eager* caching, every time an object moves to a new location, all cache entries for this object's location are updated. Thus, the cost of move operations increases for those objects whose location is cached. In this type of caching, the locations of the cache entries for an object's location must be known globally for the updates to be initiated. This leads to scalability problems as well as making the scheme susceptible to fault tolerance problems.

In *lazy* caching, a move operation signals no cache updates. Then, when at lookup, a cache entry is found there are two cases:

1. The object is still in the indicated location and there is a cache hit.
2. The object has moved out, in which case a cache miss is signaled.

In the case of a cache miss, the usual procedure is followed: the HLR is contacted and after the call is resolved, the cache entry is updated. Thus, in lazy caching, the cached location for any given object is updated only upon a miss. The basic overhead involved in lazy caching is in cases of cache misses, because the cached location must be visited first. So, for lazy caching to produce savings over the noncaching scheme, the hit ratio p for any given object at a specific site must exceed a hit ratio threshold:

$$p = C_H/C_B$$

where C_H is the cost of a lookup when there is a hit and C_B the cost of the lookup in the noncaching scheme. Among other factors, C_H and C_B depend on the relative cost of querying HLRs and VLRs. A performance study for lazy caching is presented in [12, 18] for a given signaling architecture.

The hit ratio for the cache of the object's i location at site j can also be directly related to the $LCMR_{i,j}$ of the object [18]. For instance, when the incoming calls follow a Poisson distribution with arrival rate λ and the intermove times are exponentially distributed with mean μ, then $p = \lambda/(\lambda + \mu)$ and the minimum $LCMR$, denoted $LCMR_T$, required for caching to be beneficial is found to be $LCMRT = p_T/(1 - p_T)$. So, caching can be selectively done per object i at site j, when the $LCM_{i,j}$ is larger than the $LCMR_T$ bound. In general, this threshold is lower when objects accept calls more frequently from objects located nearby.

Another approach to cache invalidation, suggested in [24], is to consider cache entries obsolete after a certain time period. To determine when a particular cache should be cleared, a threshold T is used. T is dynamically adapted to the current call and mobility patterns such that the overall network traffic is reduced.

When the cache size is limited, cache replacement policies, such as replacing the least recently used (LRU) location, may be used. Another issue is how to initialize the cache entries. Object profiles and other types of domain knowledge may be used to initially populate the cache with the locations of the objects most likely to be requested. Finally, although in the approach we have described, caching is performed on a per object basis (i.e., the cache maintains the address of the object last requested), another approach is to apply a static form of caching. For instance, one may cache the addresses of a certain group of objects or certain parts of the network where the objects' CMRs are known to be high on average.

Caching techniques can also be deployed to exploit locality of calls in hierarchical architectures. Recall that in hierarchical architectures, when a lookup is initiated from site i to object x located at site j, the search procedure traverses the tree up from i to $LCA(i, j)$ and then down to j. We also consider an acknowledgment message that returns from j to i. To support caching, during the return path, a pair of bypass pointers, called forward and reverse, is created [16]. A *forward bypass pointer* is an entry at an ancestor of i, say s that points to an ancestor of j say t; the *reverse bypass pointer* is from t to s. During the next lookup from site i to object x, the lookup message traverses the tree up until s is reached. Then, the message travels to directory t either via $LCA(i,j)$ or via a shorter route if such a route is available in the underlying network. Similarly, the acknowledgment message can bypass all intermediate pointers on the path from t to s. For example, let a lookup be placed from site 13 to object x at site 16 (Figure 9.4). A forward bypass pointer is set at node 1 pointing to node 6; the reverse bypass pointer is from 6 to 1. During the next lookup from site 13 to object x, the search message traverses the tree from node 13 up to node 1 and then at node 6, either through $LCA(1,6)$, that is node 0, or via a shorter path. In any case, no queries are posed to directories at node 0 and node 2.

The level of node s and node t where the bypass pointers are set varies. In *simple caching*, s and t are both leaf nodes, although in *level caching*, s and t are nodes belonging to any level and possibly each to a different one (as in the previous example). Placing a bypass pointer at a high-level node s makes this entry available to all lookups originated from sites at s's subtree. However, lookups must traverse a longer path to reach s. Placing the pointer to point to a high-level node t increases the cost of lookup, because to locate an object, a longer path from t to the leaf node must be followed. On the other hand, the cache entry remains valid as long as the object moves inside t's subtree. An adaptive scheme can be considered to set the levels of s and t dynamically.

Caching is orthogonal to partitions. In fact, in [40, 41] caching is used in conjunction with partitions. In particular, instead of caching the current location of the requested object, the location of its representative is

Locating Mobile Objects

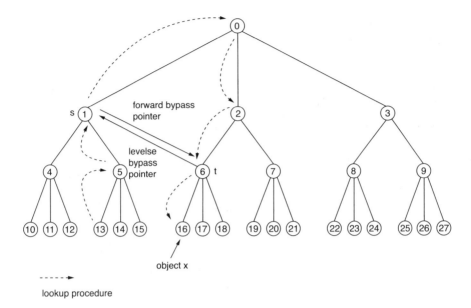

Figure 9.4. Caching in Hierarchical Location Schemes
For simplicity, the acknowledgment message is not shown; it follows the reverse route of the lookup procedure.

cached. For example, assume that partitions are defined as in Figure 9.3 and object x is at node 14. Let a lookup be placed for object x. Instead of caching location 14 (or a pointer to it), location 1 (e.g., the representative of the current partition) is cached. This significantly reduces the cost of cache updates, because a cache entry becomes obsolete only when an object moves outside the current partition.

9.4.4 Replication

To reduce the lookup cost, the location of specific objects may be replicated at selected sites. Replication reduces the lookup cost, because it increases the probability of finding the location of the requested object locally as opposed to issuing a high latency remote lookup. On the other hand, the update cost incurred increases considerably, because replicas must be consistent every time the object moves.

In general, the location of an object i should be replicated at a site j, only if the replication is judicious, that is the savings due to replication exceed the update cost incurred. As in the case of caching, the benefits depend on the *LCMR*. Intuitively, if many lookups of i originate from site j, then it makes sense to replicate i at j. However, if i moves frequently, then replica updates incur excessive costs. Let α be the cost savings when a local lookup (i.e., a query of the local VLR) succeeds as opposed to a remote

209

query and β the cost of updating a replica, then a replication of the location of object i at site j is judicious if:

$$\alpha * C_{i,j} \geq \beta * U_i, \qquad (9.1)$$

where $C_{i,j}$ is the expected number of calls made from site j to i over a time period T and U_i the number of moves made by i over T.

In addition to cost, the assignment of replicas to sites must take into account other parameters, such as the service capacity of each directory and the maximum memory available for storing replicas. The replication sites for each object may be kept at its HLR. Besides location information, other information associated with mobile objects may also be replicated [38]. Instead of the exact location of an object, more coarse location information (e.g., the object's current partition) may be replicated. The coarseness or granularity of location replicas presents location schemes with a trade-off between the update and the lookup costs. If the information replicated is coarse, then it needs to be updated less frequently in the expense of a higher lookup resolution cost.

Choosing the network sites at which to maintain replicas of the current location of a mobile object resembles the file allocation [8] and the directory allocation [29] problem. These classical problems are concerned with the selection of sites at which to maintain replicas of files or directory partitions. The selection of sites is based on the read/write pattern of each file or partition, that is the number of read and write operations issued by each site. In the case of location management, this corresponds to the lookup/update pattern of an object's locations. However, most schemes for file or directory allocation are static: they are based on the assumption that the read/write pattern does not change. The Adaptive Data Replication (ADR) algorithm [48] presents a solution to the general problem of determining an optimum in terms of the communication cost set of replication sites for an object in distributed computing when the object's read/write pattern changes dynamically.

The objective of the *per object profile approach* [39] is to minimize the total cost of moves and lookups, while maintaining constraints on the maximum number r_i of replicas per object P_i and on the maximum number p_j of replicas stored in the directory at site Z_j. Let M be the number of objects and N be the number of sites. A replication assignment of an object's profile P_i to a set of sites $R(P_i)$ is found, such that the system cost expressed as the sum:

$$\sum_{i=1}^{N} \sum_{j=1, Z_j \in R(P_i)}^{M} (\beta * U_i - \alpha * C_{i,j})$$

Locating Mobile Objects

is minimized and any given constraints on the maximum number of replicas per directory at each site and on the maximum number of replicas per object are maintained.

The *working set method* [34] relies on the observation that each object communicates frequently with a small number of other objects, called its working set, thus it makes sense to maintain copies of its location at the members of this set. The approach is similar to the per object replication except from the fact that no constraints are placed on the directory storage capacity or the number of replicas per user. Consequently, the decision to provide the information of the location of a mobile object P_i at a site Z_j can be made independently by each object P_i. Specifically, inequality (9.1) is evaluated locally at the object each time at least one of the quantities involved in the inequality changes.

In hierarchical architectures, in addition to leaf nodes, the location of a mobile object may be selectively replicated at internal nodes of the hierarchy. As in the replication schemes for two-tier architectures, the location of an object should be replicated at a node only if the cost of replication does not exceed the cost of nonreplication. However, in a hierarchical location directory scheme, if a high *LCMR* value is the determining factor for selecting replication sites, then the directories at higher levels will tend to be selected as replication sites over directories at lower levels, because they possess much higher *LCMR* values. In particular, if a directory at level j is selected, all its ancestors are selected as well. Recall that the *LCMR* for an internal node is the sum of the *LCMRs* of its children. Such a selection would result in excessive update activities at higher level directories. To compensate, replication algorithms for hierarchical directories must also set some maximum level of the hierarchy at which to replicate. To this end, Hiper proposed in [19] is a family of location management techniques with four parameters — N_{max}, S_{min}, S_{max}, and L — where N_{max} determines the maximum number of replicas per object, S_{min} and S_{max} together determine when a node may be selected as a replication site, and L determines the maximum level of the hierarchy at which replicas can be placed.

9.4.5 *Forwarding Pointers*

When the number of moves that an object makes is large relative to the number of lookups for its location, it may be too expensive to update all directory entries holding its location each time the object moves. Instead, entries may be selectively updated and lookups directed to the current location of an object through the deployment of forwarding pointers (Figure 9.5).

In particular, in the case of two-tier architectures, if the mobility of an object is high while it is located far way from its HLR, an excessive amount of messages is transmitted between the serving VLR and the HLR. Thus, to

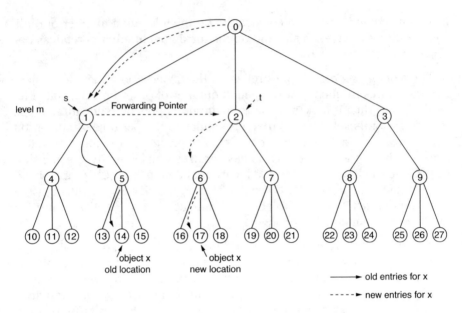

Figure 9.5. Example of Forwarding Pointers (Entries Are Pointers to Lower Level Directories)

reduce the communication overhead, as well as the query load at the HLR, the entry in x's HLR is not updated each time the mobile object x moves to a new location [17]. Instead, at the VLR at x's previous location, a forwarding pointer is set up to point to the VLR in the new location. Now, lookups for a given object will first query the object's HLR to determine the first VLR at which the object was registered and then follow a chain of forwarding pointers to the object's current VLR. To bind the time taken by the lookup procedure, the length of the chain of forwarding pointers is allowed to grow up to a maximum value of K. An implicit pointer compression also takes place when loops are formed as objects revisit the same areas. Because the approach is applied on a per object basis, the increase in the cost of call operations affects only the specific object.

The pointer forwarding strategy, as opposed to replication, is useful for those objects that receive calls infrequently relative to the rate at which they relocate. Clearly, the benefits of forwarding depend also upon the cost of setting up and traversing pointers relative to the costs of updating and contacting the HLR.

A method for dynamically determining whether to update the HLR or not is proposed in the *local anchoring scheme* [13], where a pointer chain length of, at most, one is maintained. For each mobile object, a VLR close to it is selected as its local anchor (LA). In some cases, the LA may be the same as its serving VLR. Otherwise, the LA maintains a forwarding pointer to the current VLR of the object. For each object, the HLR maintains its serving LA. To locate a mobile object, the HLR is queried first and then the associated LA is contacted. If the LA happens to be the serving VLR, no further querying is necessary, otherwise the forwarding pointer is used to locate the object. After a lookup resolution, the HLR knows the current location of the requested mobile object; therefore, the HLR is always updated after a lookup to record the current VLR. Depending on whether the HLR is updated upon a move, two schemes are proposed — static and dynamic local anchoring. In static local anchoring, the HLR is never updated at a move. In dynamic local anchoring, the serving VLR becomes the new LA if this will result in lower expected costs.

To reduce the update cost, forwarding pointer strategies may be also deployed in the case of hierarchical architectures. In a hierarchical location scheme, when a mobile object x moves from site i to site j, entries for x are created in all directories on the path from j to $LCA(j, i)$ and the entries for x on the path from $LCA(j, i)$ to i are deleted. Using forwarding pointers, instead of updating all directories on the path from j through $LCA(j, i)$ to i, only the directories up to a level m are updated. In addition, a forwarding pointer is set from node s to node t, where s is the ancestor of i at level m and t is the ancestor of j at level m (Figure 9.5). As in caching, the level of s and t can vary. In simple forwarding, s and t are leaf nodes, although in level forwarding, s and t can be nodes at any level. A subsequent lookup reaches x through a combination of directory lookups and forwarding pointer traversals.

Take for example, object x located at node 14 that moves to node 17 (Figure 9.5). Let level m = 2. A new entry for x is created in the directories at node 17, node 6, and node 2, the entries for x in the directories at node 14 and node 5 are deleted, and a pointer is set at x's entry in the directory at node 1 pointing to the entry of x in the directory at node 2. The entry for x at node 0 is not updated. When an object, say at site 23, lookups object x, the search message traverses the tree from node 23 up to the root node 0 where the first entry for x is found, then goes down to 1, follows the forwarding pointer to 2, and traverses down the path from 2 to 17. On the other hand, a lookup initiated by an object at 15 results in a shorter route: it goes up to 1, then to 2, and follows the path down to 17.

Forwarding techniques can also be deployed for hierarchical architectures in which the entries of the internal nodes are actual addresses, rather than pointers to the corresponding entries in lower level directories. The example above is repeated in Figure 9.5 for the case in Figure 9.6. Entries for x are updated up to level m = 2 and a forwarding pointer at leaf node 14 is set to redirect calls to the new location 17. Such an architecture with

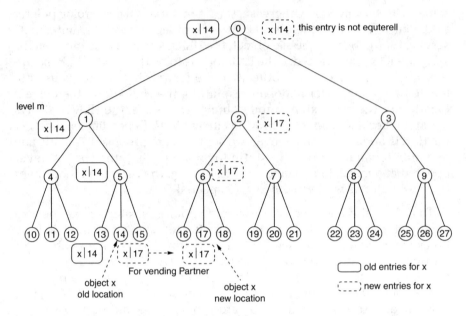

Figure 9.6. Example of Forwarding Pointers (Entries Are Actual Locations)

internal nodes storing actual addresses, rather than tree pointers, is considered in [23], where a performance analysis of forwarding is presented. Besides forwarding, the scheme in [23] also supports caching: leaf caching (i.e., caching the address of the requested object only at the site of the requestor) called jump updates and level caching (i.e., caching the address of the requested object at all nodes on the search path) called path compression.

Obsolete entries in directories at levels higher than m (e.g., the entry at node 0 in Figure 9.5 and Figure 9.6) may be updated after a successful lookup. Another possibility for updates is for each node to send a location update message to the location servers on its path to the root during off-peak hours.

To avoid the creation of long chains of forwarding pointers, some form of pointer reduction is necessary. To reduce the number of forwarding pointers, a variation of caching is proposed in [31]. After a lookup to object x, the actual location of the object is cached at the first node of the chain. Thus, any subsequent calls to x directed to the first node of the chain use this cache entry to directly access the current location of x, bypassing the forwarding pointer chain. Besides, this form of caching that reduces the number of forwarding pointers that need to be traversed to locate an object, the directory hierarchy must also be updated to avoid excessive lookup costs. Besides deleting forwarding pointers, this also involves the

Locating Mobile Objects

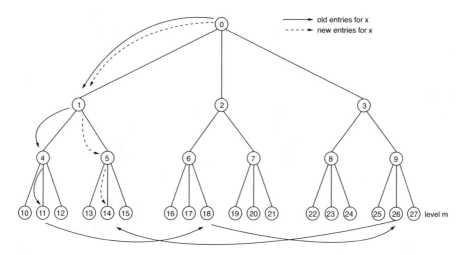

Figure 9.7. An Example of Pointer Purging

deletion of all entries in internal directories on the path from the first node, i, of the chain to the LCA of i and the current location, j, and the addition of entries in internal directories on the path from the LCA to j. Take for example, chain $11 \to 18 \to 26 \to 14$ that resulted from object x moving from node 11 to node 18, node 26, and node 14, in that order. The entries for x at node 11, node 18, and node 26 are deleted. Then, the entries in higher level directories leading to 11 are also deleted. In particular, the entry for x at 4 is deleted and entries are set at node 1 and node 5 leading to node 14, the new location (Figure 9.7). Two conditions for initiating updates are proposed and evaluated based on setting a threshold either on the number of forwarding pointers or on the maximum distance between the first node of the chain and the current location.

Forwarding pointer techniques find applications in mobile software systems to maintain references to mobile objects, such as in the Experimental Machine Example-based Reasoning and Learning Disciple (EMERALD) system and in Storage Service Provider (SSP) chains. EMERALD [21] is an object-based system in which objects can move within the system. SSP chains [37] are chains of forwarding pointers for transparently migrating object references between processes in distributed computing. The SSP chain shortcutting technique is similar to the simple update at calls method.

9.5 Taxonomy and Location Management Techniques

The techniques proposed in the previous sections are based on exploiting knowledge about the lookup and moving behavior of mobile objects. Basically, two characteristics are considered — *stability* of lookups and moves and *locality* of moves and lookups. Stability in the case of lookups means that most lookups for each object originate from the same set of locations, for example, a mobile user may receive most calls from a specific set of friends, family, and business associates. Stability of moves refers to the fact that objects tend to move inside a specific set of regions. For instance, mobile users may follow a daily routine: drive from their home to their office, visit a predetermined number of customers, return to their office, and then back to their home. This pattern can change, but remains fixed for short periods of time. Locality refers to the fact that local operations are common. In particular, in the case of lookups, an object frequently receives requests from nearby places, but in the case of moves, the object moves to neighbor locations more often than to remote ones.

Another determinant factor in designing location techniques is the relative frequency of lookups and moves expressed in the form of an appropriate CMR. In general, techniques tend to decrease the cost of either the move or lookup operation in the expense of the other. Thus, the CMR determines the efficacy of the technique. Figure 9.8 summarizes the various techniques that exploit locality, stability, and CMR. These techniques are orthogonal; they can be combined with each other.

Besides developing techniques for the efficient storage of location information, the advancement of models of movement can be used in guiding the search for the current location of a mobile object (see for example, [3, 35]), when the stored information about its location is not current or precise. For instance, potential locations may be searched in descending order of the probability of the object being there.

An important parameter of any lookup and movement model is time. The models should capture temporal changes in the movement and lookup patterns and their relative frequency as they appear during the day, the week, or even the year. For instance, the traffic volume in weekends is different than that during a workday. Thus, dynamic adaptation to the current pattern and ratio is a desirable characteristic of location techniques. Another issue is the basis on which each location technique is employed. For instance, a specific location technique may be employed on a per object basis. Alternatively, the technique may be adopted for all objects or for a group of objects based either on their geographical location (i.e., all objects in a specific region), on their mobility and lookup characteristics (i.e., all objects that receive a large number of lookups) or a combination of both.

Table 9.2 and Table 9.3 summarize, respectively, the variations of the two-tier and hierarchical location scheme and their properties.

Finally, another parameter that affects the deployment of a location strategy is the topology of network sites, how they are populated, and their

Locating Mobile Objects

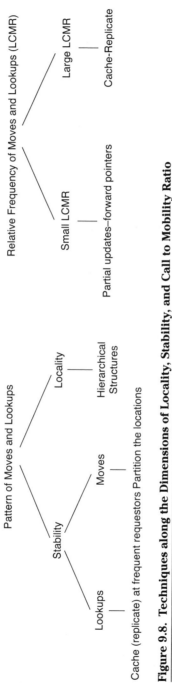

Figure 9.8. Techniques along the Dimensions of Locality, Stability, and Call to Mobility Ratio

Table 9.2. Summary of Enhancements to the Basic Two-Tier Scheme

Method	Variations		Applicable When
Caching: When x is looked up by y, cache x's location at y's site	Eager caching	Cache update overhead occurs at moves	Large *LCMR* Lookup Stability
	Lazy caching	Cache update overhead occurs at lookups	
Replication: Selectively replicate x's location at the sites from which it receives the most lookups	Per object profile replication	Additional constraints are set on the number of replicas per site and on the number of replicas per object	Large *LCMR* Lookup Stability
	Working set	Adaptive and distributed: the replication sites are computed dynamically by each mobile object locally	
Forwarding Pointers: When x moves, add a forwarding pointer from its old to its new location	Restrict the length of the chain of forwarding pointers		Small *LCMR*

geographical connectivity. How the strategy scales with the number of mobile objects, location operation, and geographical distribution is also an important consideration.

9.6 Case Studies

In this section, we present some example location management mechanisms, namely, the Mobile IP protocol that has been used for locating mobile hardware computing devices, the location mechanism of the Globe distributed system, and the location mechanisms of two mobile agent systems — Ajanta and Voyager.

9.6.1 *Mobile IP*

Mobile IP [27, 30] is a modification to wireline Internet Protocol (IP) that allows mobile devices (for simplicity, we shall call them nodes) to continuously receive messages independently of their point of attachment to the Internet. Mobile IP is designed within the Internet Engineering Task Force (IETF) and is outlined in a number of requests for comments (RFCs) [15].

Table 9.3. Summary of Proposed Enhancements to Hierarchical Location Schemes

Method	Issues/Variations	Appropriate When
Caching: When x at site i is looked up by object y at site j, cache at a node on the path from j to $LCA(i,j)$, a pointer to a node on the path from i to $LCA(i,j)$ to be used by any subsequent lookups of x from site j.	Up to which tree level to maintain cache entries When to update cache entries	Large *CMR* Lookup Stability
Replication: Selectively replicate x's location at internal and/or leaf directory.		Large *CMR* Lookup Stability
Forwarding Pointers: When x moves from site i to site j, instead of updating all directories on the path from i to $LCA(i,j)$ and from $LCA(i,j)$ to j, update all directories up to some level m and add a forwarding pointer at the level m ancestor of i to point to the level m ancestor of j.	When and how to purge the forwarding pointers Setting the level m	Small *LCMR*
Partitions: Divide the locations into sets (partitions) so that the object moves inside a partition frequently and crosses the boundary of a partition rarely. Keep information about the partition in which the object resides instead of its exact location.		Move Stability

Wireline IP assumes that the network address of a node uniquely identifies the node's point of attachment to the Internet. Thus, a node must be located on the network indicated by its IP address to receive messages destined to it. To remedy this, in Mobile IP, there are two IP addresses associated with each mobile node — the home address and the care-of address.

The *home address* is used to identify a node and is treated administratively just like a permanent IP address; the *care-of address* is associated with the object's current location and represents its current point of attachment. The purpose of the home address is to provide location transparency, in other words, to give the illusion that the mobile node can continually receive messages on its home network, while it changes its location and consequently its point of attachment. The node's home network is the network that is associated with the node's home address. The Mobile

IP protocol requires the existence of a network node, known as the home agent that receives all the packages/messages destined to the mobile node and redirects them to the mobile node's current location.

Each time a mobile object moves, it notifies its home address by registering its new point of attachment (care-of address). More specifically, the procedure begins when the mobile node sends a registration (update) request, which contains information about its new care-of address. Every time that the home agent receives a registration request, it updates its location directory (routing table) with the information included in the above request and replies with an acknowledgment.

For the case in which the mobile node cannot contact its home node, either because of a failure or a temporal disconnection, Mobile IP provides a mechanism that allows a mobile node to register with another unknown home agent located in its home network. This method is known as *automatic home agent discovery* and is based on using a broadcast IP address as a target for the registration request instead of the home agent's IP address. Registration request broadcasting aims at collecting the addresses of other home agents of the object's home network. As soon as a home agent receives the above request, it sends a rejection answer, which contains its address. Then the object selects one of the above addresses and attempts to register with a new home agent that is associated with the specific address.

The procedure for routing a packet from a node to another mobile node is quite simple. The packet is at first destined to the mobile node's home network and then the home agent redirects the packet to the node indicated by the care-of address. The redirection procedure includes the construction of a new IP header, which contains the mobile node's care-of address as the packet's destination IP address. When a packet arrives at the care-of address, the Transmission Control Protocol (TCP) or (a higher layer protocol) applies the reverse procedure and the destination IP of the packet is set back to the node's home address.

Thus, in abstract terms, the Mobile IP protocol implements a variation of the two-tier schemes, where the HLR and VLR correspond to the home address and the care-of address, respectively. Furthermore, the router optimization extensions to IETF Mobile IP protocol include pointer forwarding in conjunction with lazy caching [20].

9.6.2 *Globe*

Globe, the Global Object-Based Environment [6, 42], is an experimental worldwide distributed system. It implements a two-tier location mechanism that is organized in a hierarchical scheme based on the division of the network into regions.

Locating Mobile Objects

The location mechanism in Globe associates with each system object an identifier, called an *object handle,* which is a location–independent, universally unique reference to the object. In addition, the address of the object is described by one or more addresses called *contact addresses,* which maintain information about the location of the object and also the location of its replicas (one contact address per replica). The directory that stores the contact addresses is called *contact record.*

Each region of the system is associated with a directory node, which stores the contact record of the objects that are located in the specific region. The directories are organized in a hierarchical structure, which is in practice a distributed search tree. For each object in the system, its contact address is stored at a leaf node and all nodes in the path from the object's leaf node to the root maintain a forwarding pointer to the specific leaf node. The set of addresses that each leaf stores is determined by the domain covered by the node. In particular, each leaf node represents the smallest network in which the contact addresses can reside and an internal node represents the domain that is the union of all domains covered by its children.

Caching is used to improve the efficiency of the location management mechanism. The type of caching used is based on the observation that in the case of highly mobile objects, it is more efficient to maintain information about the node that covers the region in which the mobile object moves instead of the object's exact location. Thus, this node is cached instead of the exact location.

Object relocation (move) is a two phase procedure. In the first phase, the insertion of the object's new address takes place, while in the second phase the object's old address is deleted. The reason for distinguishing between the two phases is to allow concurrent accesses to the directory. If the old contact address about the object was deleted before its new contact address was inserted, the object would be temporarily unreachable.

During the insertion phase, the leaf node that is associated with the object's contact address receives an insertion request. A node cannot store directly the contact address of an object, but instead it must be granted permission from its parent node. In the case in which the parent node refuses to grant permission, the node will not store the contact address of the object and the procedure will be repeated with the parent node this time acting as the initial node. A parent node can refuse to grant permission to its child node, if it prefers to store the address itself (instead of just storing a forwarding pointer). When a node that managed to grant permission from its parent is reached, it stores the contact address of the object. Then, forwarding pointers are installed upward to all nodes from the path of the node where the contact address was stored to the node which already maintains another contact address of the object (or in the worst case to the root).

The deletion phase works as follows. A deletion request is sent to the leaf that is associated with the region to which the contact address of the object belongs. Then the request is forwarded until the node that stores the specific request is reached and the contact address is deleted. The contact record will be deleted too in the case in which the deleted contact address was the only record in the contact record. Finally, a recursive procedure will take place in which all the forwarding pointers that were pointing to the deleted contact address will be removed.

The lookup procedure first checks if the nearest leaf maintains information about the requested object. If not, then the lookup procedure will be repeated by asking this time the higher level node and so on until a node that stores either a forwarding pointer to the contact address of the requested object or the contact address of the object is reached. If a forwarding pointer is found, the mechanism repeats the procedure by following the forwarding pointer to discover the object's contact address.

9.6.3 Mobile Agents Systems

Mobile agents are processes that may be dispatched from a network site and be transported to another one for execution. The ability to communicate with agents in real-time as agents move from one network node to another is essential for retrieving any data or information that they have collected and for supporting coordination and cooperation among them. This subsumes the knowledge of the agent's current location.

9.6.3.1 Ajanta. Ajanta's location mechanism [22, 43] implements an HLR/VLR scheme in which a registry maintains information for all the agents located in its domain. In addition, each registry maintains the precise current location of all agents that were created in its domain.

In Ajanta, the name of each agent contains information about the registry at which the agent was created. Particularly, Ajanta implements a Uniform Resource Name (URN) scheme, which provides persistent location independent resource identifiers. An example of a URN in Ajanta is:

```
urn:ans:domain/UserName/ResourceName
```

where the domain field indicates the domain where the agent was created, the field `UserName` is a naming authority or a subdomain of the creation domain and `ResourceName` is a unique string in this subdomain.

The lookup (or name resolution) procedure in Ajanta works as follows. First a request is sent to the local registry. If the local registry cannot serve the request, thus the agent is not colocated with its requestor, the procedure

Locating Mobile Objects

Table 9.4. Summary of Location Management Mechanisms of the Presented Systems

Systems	Location Mechanism Architectures	Optimization Techniques
Mobile IP	HLR/VLR	Caching
Globe	Two-tier hierarchical scheme	Caching
Mobile agents — Ajanta	HLR/VLR	—
Mobile agents — Voyager	Centralized scheme	Forwarding pointers

continues by asking the agent's creation domain's registry, which maintains information about the exact current location.

9.6.3.2 Voyager. Voyager [28] implements a centralized schema with forwarding pointers. Every agent, that wishes to be located by other agents, registers to one or more directories called name services. Each time an agent moves, it informs all the name services that it has registered to. To locate an agent, one needs to know either one of the name services to which the agent has registered or (under some circumstances) a node that the agent has visited during its trip (these nodes will forward the request until the agent is reached).

In Voyager, one can send a request to an agent even if the agent has moved from the node where it was located. The request can be sent to the last known address of the agent. If the agent has left from the specific location, the request will search for a forwarder, which is an object that will indicate the next location of the agent. If the message locates a forwarder refering to the agent, the forwarder forwards the message to the agent's new location. The above procedure will be repeated until the request reaches the requested agent.

Table 9.4 summarizes the case studies.

9.7 Summary

Managing the location of moving objects is becoming increasingly important as mobility of users, devices, and programs becomes widespread. This chapter focuses on data management techniques for locating (i.e., identifying the current location of) mobile objects.

The efficiency of techniques for locating mobile objects is critical because the cost of communicating with a mobile object is augmented by the cost of finding its location. Location management techniques use information concerning the location of moving objects stored in location directories in combination with search procedures that exploit knowledge about the objects' previous moving behavior. The directories for storing the location of mobile objects are distributed in nature and must support high update rates because the location of objects changes as they move. Various enhancements of these techniques include partitions, caching, replication, and forwarding pointers.

References

1. Special Issue on Intelligent Agents, *Communications of the ACM*, vol. 37, no. 7, 1994.
2. Special Issue on Internet-based Agents, *IEEE Internet Computing*, vol. 1, no. 4, 1997.
3. I.F. Akyildiz and J.S.M. Ho, Dynamic Mobile User Location Update for Wireless PCS Networks, *ACM/Baltzer Wireless Networks Journal*, vol. 1, no. 2, 1995.
4. B.R. Badrinath, T. Imielinski, and A. Virmani, Locating Strategies for Personal Communications Networks, *Proceedings of the 1992 International Conference on Networks for Personal Communications,* 1992.
5. M. Baentsch, L. Baum, G. Molter, S. Rothkugel, and P. Sturm, Enhancing the Web's Infrastructure: From Caching to Replication, *IEEE Internet Computing*, vol. 1, no. 2, pp. 18–27, March 1997.
6. A. Baggio, G. Ballintijn, M. van Steen, and A.S. Tanenbaum, Efficient Tracking of Mobile Objects in Globe, *The Computer Journal*, vol. 44, no. 5, 2001.
7. A. Bar-Noy, I. Kessler, and M. Sidi, Mobile Users: To Update or Not to Update?, *ACM/Baltzer Wireless Networks Journal,* vol. 1, no. 2, 1995.
8. L.W. Dowdy and D.V. Foster, Comparative Models of the File Assignment Problem, *ACM Computing Surveys*, vol. 14, no. 2, pp. 288–313, June 1982.
9. G.H. Forman and J. Zahorjan, The Challenges of Mobile Computing, *IEEE Computer*, vol. 27, no. 6, pp. 38–47, April 1994.
10. GPS — Introduction to GPS Applications. www.redsword.com/gps/apps/index.htm.
11. GPS-USCG Navigation Center GPS Page, www.navcen.uscg.gov/gps/.
12. H. Harjono, R. Jain, and S. Mohan, Analysis and Simulation of a Cache-Based Auxiliary User Location Strategy for PCS, *Proceedings of the International Conference on Networks for Personal Communications,* March 1994.
13. J.S.M. Ho and I.F. Akyildiz, Local Anchor Scheme for Reducing Signaling Cost in Personal Communication Networks, *IEEE/ACM Transactions on Networking*, vol. 4, no. 5, 1996.
14. T. Imielinski and B.R. Badrinath, Wireless Mobile Computing: Challenges in Data Management, *Communications of the ACM*, vol. 37, no. 10, October 1994.
15. IP Routing for Wireless/Mobile Hosts Working Group, RFC Documents, http://www.ietf.org/html.charters/mobileip-charter.html.
16. R. Jain, Reducing Traffic Impacts of PCS Using Hierarchical User Location Databases, *Proceedings of the IEEE International Conference on Communications,* 1996.
17. R. Jain and Y.B. Lin, An Auxiliary User Location Strategy Employing Forwarding Pointers to Reduce Network Impacts of PCS, *Wireless Networks*, vol. 1, pp. 197–210, 1995.
18. R. Jain, Y.B. Lin, C. Lo, and S. Mohan, A Caching Strategy to Reduce Network Impacts of PCS, *IEEE Journal on Selected Areas in Communications*, vol. 12, no. 8, pp. 1434–1444, October 1994.
19. J. Jannink, D. Lam, N. Shivakumar, J. Widom, and D.C. Cox, Efficient and Flexible Location Management Techniques for Wireless Communication Systems, *ACM/Baltzer Journal of Mobile Networks and Applications,* vol. 3, no. 5, pp. 361–374, 1997.
20. D.B. Johnson and D.A. Maltz, Protocols for Adaptive Wireless and Mobile Networking, *IEEE Personal Communications,* vol. 3, no. 1, 1996.

21. E. Jul, H. Levy, N. Hutchinson, and A. Black, Fine-Grained Mobility in the EMERALD System, *ACM Transactions on Computer Systems,* vol. 8, no. 1, pp. 109–133, February 1988.
22. N.M. Karnik and A.R. Tripathi, Design Issues in Mobile Agent Programming Systems, *IEEE Concurrency,* vol. 6, no. 3, pp. 52–61, July–September 1998.
23. P. Krishna, N.H. Vaidya, and D.K. Pradhan, Static and Dynamic Location Management in Mobile Wireless Networks, *Journal of Computer Communications* [special issue on Mobile Computing], vol. 19, no. 4, March 1996.
24. Y.B. Lin, Determining the User Location for Personal Communications Service Networks, *IEEE Transactions on Vehicular Technology,* vol. 43, no. 3, August 1994.
25. Y.B. Lin and S.K. DeVries, PCS Network Signaling Using SS7, *IEEE Personal Communications,* June 1995.
26. S. Mohan and R. Jain, Two User Location Strategies for Personal Communication Services, *IEEE Personal Communications,* vol. 1, no. 1, pp. 42–50, January–February, 1994.
27. The Mobile IP tutorial, http://www.computer.org/internet/v2n1/perkins.htm.
28. ObjectSpace Voyager: Technical overview, December 1997. http://www.objectspace.com/voyager/whitepapers/VoyagerTechOview.pdf.
29. M.T. Ozsu and P. Valduriez, *Principles of Distributed Database Systems,* Upper Saddle River, NJ: Prentice Hall, 1991.
30. C.E. Perkins, *Mobile IP: Design Principles and Practices,* Boston: Addison Wesley, 1998.
31. E. Pitoura and I. Fudos, An Efficient Hierarchical Scheme for Locating Highly Mobile Users, *Proceedings of the 7th International Conference on Information and Knowledge Management (CIKM'98),* pp. 218–225, November 1998.
32. E. Pitoura and G. Samaras, *Data Management for Mobile Computing,* Norwell, MA: Kluwer Academic Publishers, 1998.
33. E. Pitoura and G. Samaras, Locating Objects in Mobile Computing, *IEEE Transactions on Knowledge and Data Engineering,* vol. 13, no. 4, pp. 571–592, July/August 2001.
34. S. Rajagopalan and B.R. Badrinath, An Adaptive Location Management Strategy for Mobile IP, *Proceedings of the 1st ACM International Conference on Mobile Computing and Networking (Mobicom'95),* October 1995.
35. C. Rose and R. Yates, Location Uncertainty in Mobile Networks: a Theoretical Framework, *IEEE Communications Magazine,* vol. 35, no. 2, 1997.
36. A.P. Sistla, O. Wolfson, S. Chamberlain, and Y. Yesha, Updating and Querying Databases that Track Mobile Units, *Distributed and Parallel Databases,* vol. 7, no. 3, 1999.
37. M. Shapiro, P. Dickman, and D. Plainfosse, SSP Chains: Robust, Distributed References Supporting Acyclic Garbage Collection, Technical Report 1799, INRIA, Rocquentcourt, France, November 1992.
38. N. Shivakumar, J. Jannink, and J. Widom, Per-User Profile Replication in Mobile Environments: Algorithms, Analysis, and Simulation Results, *ACM/Baltzer Journal of Mobile Networks and Applications,* vol. 2, no. 2, pp. 129–140, 1997.
39. N. Shivakumar and J. Widom, User Profile Replication for Faster Location Lookup in Mobile Environments, *Proceedings of the 1st ACM International Conference on Mobile Computing and Networking (Mobicom'95),* pp. 161–169, October 1995.
40. M. van Steen, F.J. Hauck, G. Ballintijin, and A.S. Tanenbaum, Algorithmic Design of the Globe Wide-Area Location Service, *The Computer Journal,* vol. 41, no. 5, pp. 297–310, 1998.
41. M. van Steen, F.J. Hauck, P. Homburg, and A.S. Tanenbaum, Locating Objects in Wide-Area Systems, *IEEE Communications Magazine,* pp. 104–109, January 1998.
42. M. van Steen, and A.S. Tanenbaum, *Distributed Systems: Principles and Paradigms,* Upper Saddle River, NJ: Prentice Hall, 2002.
43. A.R. Tripathi, N.M. Karnik, T. Ahmed, R.D. Singh, A. Prakash, V. Kakani, M.K. Vora, and M. Pathak, Design of the Ajanta System for Mobile Agent Programming, *Journal of System and Software,* May 2002.

44. J. Vitek and C. Tschudin, Eds., Mobile Object Systems: Toward the Programmable Internet, *Lecture Notes in Computer Science,* vol. 1222, New York: Springer-Verlag, 1997.
45. J.Z. Wang, A Fully Distributed Location Registration Strategy for Universal Personal Communication Systems, *IEEE Journal on Selected Areas in Communications,* vol. 11, no. 6, pp. 850–860, August 1993.
46. M. Weiser, Some Computer Science Issues in Ubiquitous Computing, *Communications of the ACM,* vol. 36, no. 7, pp. 75–84, July 1993.
47. O. Wolfson, Moving Objects Information Management: The Database Challenge, *Proceedings of the 5th Workshop on Next Generation Information Technologies and Systems (NGITS'2002),* Caesarea, Israel, June 25–26, 2002.
48. O. Wolfson, S. Jajodia, and Y. Huang, An Adaptive Data Replication Algorithm, *ACM Transactions on Database Systems,* vol. 22, no. 2, pp. 255–314, June 1997.

Chapter 10
Dependable Message Delivery to Mobile Units

Amy L. Murphy, Gruia-Catalin Roman, and George Varghese

Abstract

Mobile computing is emerging as a novel paradigm with its own characteristic problems, models, and algorithms. Much effort is being directed to integrate mobile units with fixed networks, providing bridges to connect wireless to wired. The result is a fixed core of wire-connected static nodes and a fluid fringe of wireless mobile units, a computing system similar to the cellular telephone network. The model we put forward uses the graph of fixed nodes as a foundation and models the mobile units themselves as persistent messages moving through this network graph. Such a model allows algorithms from traditional distributed computing to be directly implemented in the mobile environment; however, it has been shown that the unique properties of mobility, such as limited bandwidth and disconnection, make such direct translation impractical. This chapter presents a fundamentally different idea. Instead of recreating the functionality of distributed algorithms in the mobile domain, we show how distributed algorithms can be adapted to solve problems unique to the mobile environment. Specifically, we focus on the problem of dependably delivering a message to a moving unit. We demonstrate this technique with two new algorithms, the first based on distributed snapshots and the second on diffusing computations.

10.1 Introduction

Mobile computing reflects a prevailing societal and technological trend toward ubiquitous access to computational and communication resources. Wireless technology and the decreasing size of computer components allow users to travel within the office building, from office to home, and around the world with the computer at their side. As this new world of computing is taking form, many fundamental assumptions about the structure and the behavior of computer networks are being challenged and redefined. This results in at least two kinds of research questions:

1. What is the precise relationship between mobile computing and traditional distributed computing?
2. How are particular tasks (e.g., maintaining file consistency, point to point communication, etc.) solved in a mobile setting?

This chapter attempts to make contributions to both kinds of questions. On the modeling side, we describe a simple approach to modeling mobile units that has considerable similarity to the standard distributed computing model. This model in turn allows us to transfer results from classical distributed computing to the new mobile setting, leveraging off a large body of existing research in an emerging research area. On the computing side, we describe new algorithms for sending messages to mobile units.

10.1.1 *Distributed versus Mobile Computing*

A common model of a distributed computing system is a graph where the nodes represent computing components and the edges represent communication. With the exception of faults that can render parts of the network temporarily inoperational, the system is generally static. A mobile computing environment analogous to a cellular telephone system can be similarly modeled with two components. The first is a graph where the nodes represent base stations and the edges wired communication. The second part models the movement of mobile units among base stations as temporary, wireless connections to base stations. The resulting model is a fixed core of static components and a fluid fringe of mobile units. The similarities between the mobile computing model and the distributed computing model, as well as the ease of integrating this model with wired networks, have helped it become dominant in mobile computing research [2, 12].

Yet another model of mobility emerges from the study of code and data moving through a network of hosts [8, 9, 20]. In this case, the mobile components, commonly referred to as mobile agents, move entirely within the network, migrating by explicit message passing between hosts. We suggest a slight modification of this model to encompass both physical and logical mobility, thus moving the mobility model closer to the traditional distributed computing model. The basic idea is to treat mobile units as roving messages that preserve their identity as they travel across the network. For a cellular mobile system, this means that while a mobile unit is within a cell, it is modeled as a message residing at a node. When moving to a new cell, the handover protocol is modeled as the traversal of a channel between two nodes.

10.1.2 Algorithm Development

Our interest in this model rests with *its ability to facilitate the development of algorithms in mobile computing based on established algorithms of traditional distributed computing.* To illustrate this point, this chapter shows how snapshot algorithms can be adapted for unicast and multicast message delivery and how the idea of diffusing computations can be adapted to track and deliver messages to mobile units.

In the presentation, we bring together the two concerns of the chapter: applying techniques from distributed algorithms to mobile computing and the problem of message delivery. The rest of this chapter is organized as follows:

- Section 10.2 defines the problem we intend to solve, namely message delivery in a mobile setting and describes prior work in the area.
- Section 10.3 explores the use of snapshot algorithms as a search mechanism for message delivery, presents the motivation, algorithm properties, and possible extensions.
- Section 10.4 outlines the diffusing computation approach to tracking mobile units.
- Section 10.5 outlines adaptations that make the approach viable in a model similar to the cellular telephone system.
- Section 10.6 concludes the chapter.

10.2 Message Delivery

Although disconnected operation or working in isolation is a practical use of mobile units [13], many applications require units to communicate with one another while on the move, exchanging voice or data. Thus a fundamental problem in mobile computing is the delivery of a message from a source to a mobile unit. In this section, we discuss previous work on message passing in mobile environments, define our model of the mobile environment, and formally define the problem of message delivery.

10.2.1 Related Work

Standard solutions to message delivery to mobile units fall into two categories — tracking and searching. Fundamentally, the tracking solution involves knowing the current location of the mobile unit in either a centralized or distributed manner; the searching solution maintains no such information and instead searches for the mobile unit in order to deliver a message. Both styles apply depending on the mobility scenario. For example, tracking mechanisms are most effective in systems with low or slow mobility and high traffic levels, whereas systems with high or fast mobility and moderate traffic are more amenable to search solutions. This chapter considers solutions to both.

Most standard forms of message delivery rely on tracking. For example, in cellular systems, as a phone involved in an active session moves into an adjacent cell and detects a stronger signal from the new cellular tower, a handover is requested [10, 23]. The cellular system constantly keeps track of the association between phones and towers to forward voice packets to users. In Mobile IP [19], packet delivery is accomplished by the mobile unit registering its new location with its home agent and having the home agent forward any packets for that mobile unit to the registered location. Other approaches propose changing the routers to adapt to the movement of the mobile units, for example, intercepting packets en route to the home agent and directing them toward the mobile unit itself [18]. Such approaches involve fundamental changes to the routers and are less well accepted than Mobile IP.

One disadvantage to tracking arises if the mobile unit moves quickly from one base station to another. Each time the unit changes its point of attachment to the network, a tracking system must send update messages, even if the mobile unit is not actively receiving any messages. In fact, the transmission overhead of tracking information scales poorly with the speed of movement.

In search solutions, because the location of the mobile unit is not kept anywhere in the system, in order to deliver a message, the sender must either broadcast a search request to locate the mobile unit then forward the message to the resulting location or the sender can simply broadcast a copy of the message. The first mechanism has been suggested for mobile ad hoc environments where there is no infrastructure along which to route packets [4]. This approach takes advantage of the natural broadcasting nature of wireless radio communication to send a message to all neighboring mobile units within range. This same style of route discovery is also useful in base station environments with moderate movement of mobile units where a route to a mobile unit is viable long enough for both route discovery and message delivery. Clearly, searching the entire Internet for a mobile unit appears ludicrous; however, a search strategy can take advantage of the inherent organization of the Internet into domains and subnets to reduce the scope of the search.

Although the *unicast* problem of delivering a message to a *single* recipient is important, *multicast* has also received attention. For example, multicast support through the multicast backbone (MBONE) has become a standard part of the Internet [7] and is finding use for audio and video conferencing [11, 14] and video distribution. Additionally, the Mobile IP specification addresses the issues of enabling a mobile unit to function as either a sender or a receiver for multicast messages [19]. In this chapter, we show how our algorithms can be adapted from unicast to multicast delivery with minimal effort.

Dependable Message Delivery to Mobile Units

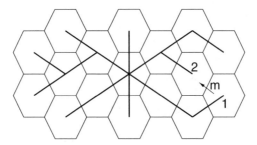

Figure 10.1. Cell-Based Broadcast Using a Spanning Tree
Each cell is a base station and adjacent cells can communicate on a wired channel.

10.2.2 Mobile Environment

We address the delivery of announcements within a network of fixed mobile support centers (MSCs) and radio base stations (RBSs). For simplicity, we assume each MSC controls only one RBS and all neighboring MSCs have a fixed communication channel between them. This channel is used by both messages and mobile units.

In Figure 10.1 each cell represents an (MSC, RBS) pair, and the MSCs of all neighboring cells are connected by a fixed network link. A mobile unit can send and receive messages from only one RBS at a time and only when it is in the cell associated with that RBS. The simplifying assumption that all neighboring base stations be physically connected can be easily removed by adding virtual channels between the physically adjacent cells. For simplicity, we also ignore the MSC/RBS distinction. We return to both of these in Section 10.5, providing details of implementing virtual channels and allowing multiple RBSs per MSC.

10.2.3 Model and Problem Definitions

As described in the introduction, the model we consider is one where the infrastructure of the mobile environment is viewed as a graph of nodes and channels; both the mobile units and the data messages are represented as messages that travel through the network. More specifically, mobile units are viewed as *persistent* messages, but data messages disappear from the system after delivery. For all temporary messages, to avoid confusion in terminology between control and data traffic, from this point forward, we use the term *announcement* to refer specifically to data messages, but a *message* can be either data or control. A mobile unit can send and receive messages and announcements only when it is present at some node in the fixed network, a situation that models the existence of an established connection between a mobile unit and a support center. When a mobile unit is on a channel, it is viewed as being temporarily disconnected from the network and unable to communicate.

Because we no longer differentiate between communication lines and physical movement, it is reasonable to question what happens when messages (both announcements and control messages) and mobile units are found on the same channel. We make the assumption that all channels preserve message ordering (i.e., they are first in, first out [FIFO]). This appears to require that mobile units travel through space and reconnect to the next support center as fast as messages can be transmitted across a network channel. The FIFO behavior, however, can be realized by integrating the handover protocol with message passing. Essentially, in the cellular model, a mobile unit moves directly between cells; however, in the graph representation, the mobile must move onto a channel before arriving at the new cell. This is a natural assumption when the details of the handover are considered, the details of which are expanded in Section 10.5. Two other assumptions we make are that a node can deliver any announcements before the mobile unit moves to a new base station and that the network is connected (i.e., there is always some path to deliver the announcement to its destination mobile unit no matter which node it is located at). Finally, we assume bidirectional channels.

The announcement delivery problem can now be formulated as follows:

> *Given a connected network with FIFO channels and guaranteed message delivery, an announcement located at one node, and a mobile unit to which the announcement is destined, develop a distributed algorithm that guarantees single delivery of the announcement and leaves no trace of the announcement, at either a node or a mobile unit, within a bounded time after delivery. Minimizing storage requirements across the network should also be considered.*

Because mobile units do not communicate directly with one another, the network must provide the mechanism to transmit the announcement. The original announcement is assumed to be in the local memory of some processing node, presumably left there by the mobile unit that is the source of the announcement. Because a mobile unit is not required to visit all nodes to gather its announcements, the announcement cannot remain isolated at the node on which it is dropped off, but instead must be distributed through the network. The specifics of this distribution mechanism are left to the algorithm and are the focus of the remainder of this chapter.

10.3 Broadcast Search

Our first approach to announcement delivery takes a broadcast search approach, meaning that a message is broadcast throughout the network in search of the mobile unit. Although this seems like a simple method, applying it directly is not trivial due to the movement of the mobile unit during the broadcast. For example, it is possible for the mobile unit to move one step ahead of the broadcast and eventually pass the announcement in the

Dependable Message Delivery to Mobile Units

opposite direction. This problem can be solved by storing the announcement at all nodes for an indefinite period; however, Internet routers have neither the storage capability nor the intention to store application announcements. Therefore, announcements must be garbage collected quickly if the scheme is to have any chance of being practical. Our solution has the attractive property of guaranteeing delivery exactly once while allowing rapid garbage collection in time proportional to one round trip delay on a single link.

10.3.1 Motivation

A straightforward broadcasting scheme designed for our model of mobility is to construct a spanning tree over the MSCs and send the announcement along this tree. In Figure 10.1, such a spanning tree is indicated by the solid lines. A disadvantage of this scheme is that a mobile unit may move and not receive the announcement. For example, consider a mobile unit located at cell 1, near the border of cell 2. Suppose the broadcast of an announcement begins at the centermost cell. Following the proposed spanning tree broadcast scheme, the MSC in the initiating cell broadcasts the announcement locally; next the announcement is forwarded on the outgoing links of the spanning tree. After successfully sending the announcement, the initiator deletes its copy of the announcement, minimizing the storage time. The MSCs downstream behave in a similar manner, broadcasting locally, forwarding the announcement to their children, and finally deleting their copy of the announcement.

If the mobile unit does not move away from cell 1, it will receive a copy of the announcement when it is broadcast by MSC_1. However, when the mobile unit is on the border between cell 1 and cell 2, it is possible for a handover to be initiated and for the mobile unit to lose contact with MSC_1 and pick up communication with MSC_2. If this handover occurs after MSC_2 deletes its copy of the announcement and before MSC_1 broadcasts its copy of the announcement, the mobile unit will not receive the announcement even though it was connected to the network during the entire broadcast lifetime of the announcement. Although in reality, messages travel through the network much faster than a mobile unit can travel through space, because a handover requires very little time to complete, and the length of the path along the spanning tree could take longer to traverse than for the handover to complete, it is reasonable for a simple broadcast mechanism such as this to fail.

10.3.2 From Distributed Snapshot Algorithms to Announcement Delivery

To guarantee delivery in any circumstance, we propose an alternative broadcast algorithm that is based on the classical notion of distributed snapshots. Before addressing announcement delivery, we first note the general properties of snapshot algorithms especially those important in announcement delivery.

The goal of a snapshot algorithm is to provide a consistent view of the state of a network of nodes and channels. The state consists of the process variables and any messages in transit among the nodes. A simple snapshot algorithm would freeze the computation until all messages are out of the channels, record the state of the processors (including outgoing message queues), and then restart the computation. Although this is an impractical solution in most distributed settings, it provides the intuition behind a snapshot algorithm, in particular that the consistent global state is constructed by combining the local snapshots from the various processors. In general, a snapshot is started by a single processor and control messages are passed to neighboring nodes informing them that a snapshot is in progress, thereby initiating local snapshots. The main property of snapshots that we exploit is that every message appears exactly once in the recorded snapshot state.

Although snapshot algorithms were developed to detect stable properties, such as termination or deadlock, by creating and analyzing a consistent view of the distributed state, minor adjustments described here adapt them to perform announcement delivery in the dynamic, mobile environment. To move from the network of nodes and channels into the mobile computing environment, we return to the mobility model for the cellular structure of mobile support centers and radio base stations. As described earlier, these components and the wires connecting them map directly to the network graph of standard distributed computing. The mobile units are simply represented as persistent messages in the distributed environment, meaning they are always somewhere in the system, either at a node (when in communication with a base station) or on a channel (during a handover).

At this point, we have a structure on which to run the snapshot. We note that because the mobile unit is a message and the snapshot records the location of messages, the global snapshot of the mobile system will show the location of the mobile unit. Therefore, one option is to simply deliver the announcement directly to this location; however, it is possible (and likely in systems with rapid movement) that the mobile unit will move between the time its position is recorded and when the announcement arrives at the recorded position. Therefore, we alter the snapshot recording to deliver the announcement by augmenting the control messages with a copy of the announcement itself and changing the recording of messages into the delivering of announcements. We further note that the global state of the system is no longer important for delivery, so no system state information is collected.

Dependable Message Delivery to Mobile Units

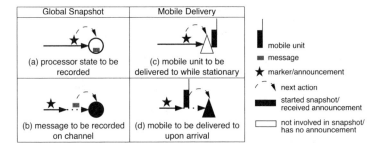

Figure 10.2. Translation of Concepts from Global Snapshots into Mobile Delivery
The curved arrow shows the processing of an element from a channel and the text describes the action triggered by such movement.

10.3.3 Snapshot Delivery Algorithm

Throughout this section, we use the Chandy–Lamport snapshot algorithm [5] and show its adaptation to announcement delivery. In making the transition to the mobile environment, we carry the restrictions of the original distributed algorithm and clarify certain characteristics of the mobile model moving from the cell structure to the graph setting.

In the Chandy–Lamport algorithm, it is possible for the snapshot to be initiated at more than one location in the graph, however, we assume that the announcement will be located initially at one point in the network; therefore the snapshot will originate from a single MSC. The Chandy–Lamport algorithm consists of two main localized actions to collect the local snapshot — the processing of the control messages (markers) and the arrival of the messages to be recorded:

1. The *marker arrival rule* states that when a marker arrives at a node not involved in a snapshot, the node begins its local snapshot by recording the processor state, then sends the marker on all outgoing channels (Figure 10.2a). In the mobile environment, this is analogous to the announcement arriving at a node. If the mobile unit is present, it will receive the announcement; otherwise the node will remain in the local snapshot state and will store the copy of the announcement until the local snapshot is complete. The local snapshot is complete when the marker (announcement) has arrived from all incoming channels.

2. The *message arrival rule* states that if the message arrives at a node from channel C before the marker arrives on channel C, and the node is in the middle of the local snapshot, the message is to be recorded as on the channel during the snapshot (Figure 10.2b). In the mobile setting, this condition is the arrival of the mobile unit at an MSC that is storing a copy of the announcement. Therefore, the

235

State	
flushed$_{A,B}$	boolean, true if announcement traversed the link from A to B; initially false everywhere
AnnAt$_A$	boolean, true if announcement stored at A; initially true only where announcement starts
MobileAt$_A$	boolean, true if mobile unit at A; initially true only where mobile located

Actions

ANNARRIVES$_A$(B)   ;arrival at A from B
  Effect:
    flushed$_{B,A}$:=TRUE
    if ¬AnnAt$_A$
      send ann. on all outgoing channels
      AnnAt$_A$:=TRUE ;save ann.
      if MobileAt$_A$
        deliver announcement
      endif
    endif

MOBILEARRIVES$_A$(B)   ;arrival at A from B
  Effect:
    MobileAt$_A$:=TRUE
    if ¬flushed$_{B,A}$ and AnnAt$_A$
      deliver announcement
    endif

MOBILELEAVES$_A$(B)   ;leaves from A to B
  Preconditions:
    MobileAt$_A$ and channel (A,B) exists
  Effect:
    MobileAt$_A$:=FALSE
    mobile unit moves onto (A,B)

CLEANUP$_A$;A finishes local snapshot
  Preconditions:
    Forall neighbors X, flushed$_{X,A}$=TRUE
  Effect:
    AnnAt$_A$:=FALSE ;delete ann.
    Forall neighbors X, flushed$_{X,A}$:=FALSE

Figure 10.3. Snapshot Delivery Code

arrival of the mobile unit triggers the transmission of the announcement to the mobile.

We capture these actions in I/O Automata-like pseudo-code shown in Figure 10.3. In addition to the announcement arrival and mobile arrival, we also include statements to terminate the local snapshot (cleaning up the state) and to allow the mobile unit to move within the network. Channels are assumed to be FIFO and hold both mobile units and all messages.

We assume the system is initialized with the location of the mobile unit (`MobileAt`) and a single announcement copy at some node (`AnnAt`). Channels are assumed to be empty. We introduce one state variable quantified over the channels (`flushed`) that is used in identifying when the local snapshot is complete. Basically, a flushed channel has received a marker and when all incoming channels have received a marker, the local snapshot is complete.

The actions of Figure 10.3 describe the local node state transitions that are sufficient for message delivery. No global information is maintained. A node will be in one of three states:

1. *Unnotified* — not yet aware of the snapshot
2. *Notified* — taking a local snapshot
3. *Finished* — finished with the local snapshot

Dependable Message Delivery to Mobile Units

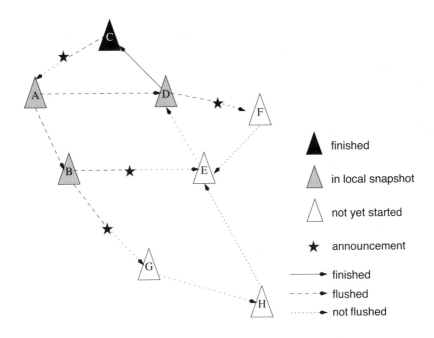

Figure 10.4. Phases of the Nodes and Channels during the Snapshot Delivery Algorithm
Each triangle represents a base station. The mobile unit could be anywhere in the network.

In Figure 10.4, these states are represented by white, grey, and black respectively. All nodes (except the node where the announcement originates) are initially unnotified. An unnotified node such as E will eventually receive an announcement (ANNARRIVES) along one of its incoming channels (such as (B, E)). This action causes it to transition to the notified state, delivering the announcement if possible, storing a copy of the announcement, marking the channel the announcement arrived on as flushed, and sending announcement copies on all outgoing channels.

Once a channel is flushed, if the mobile unit arrives on that channel, it is guaranteed to have seen the announcement at some other node (to have been recorded in some other local snapshot). Therefore, to avoid multiple delivery, if the mobile arrives on a flushed channel, delivery is not repeated (MOBILEARRIVES). If the announcement arrives at a notified node such as A (ANNARRIVES), the channel it arrives on will be marked as flushed, but because the announcement is already stored, no additional copy is made. When all incoming channels have been flushed (as in B), the node's local snapshot is complete and the local state (including the flushed status of the channels and the stored announcement) is deleted (CLEANUP).

237

The final action, MOBILELEAVES, models the movement of a mobile away from a node. The mobile is simply placed on the channel and the state variables updated to reflect this change. This models random mobile unit movement. If a particular movement pattern is desired, it can be added to this action.

10.3.4 *Properties*

Because our announcement delivery algorithm is based on a well-understood algorithm from distributed computing, we can adapt the proven properties from the distributed computing environment into the mobile environment. The three primary properties proven for the Chandy–Lamport distributed snapshot are:

1. There is no residual storage in the system at some point after the algorithm begins execution.
2. Every message is recorded once.
3. No message is recorded more than once.

We translate these three properties directly into the mobile environment stating that:

1. Eventually there is no residual storage in the system at some point after the delivery process begins.
2. The announcement is delivered to the intended recipient.
3. The announcement is delivered only once.

To more explicitly show the relationship between the snapshot algorithm and announcement delivery, we provide the outline of a reduction proof from the Chandy–Lamport distributed algorithm to the adapted snapshot delivery algorithm, showing the mapping between the actions (such as marker arrival and announcement arrival) and the system variables (such as the marker and announcement).

In the Chandy–Lamport algorithm, a processor begins its local snapshot when it receives the first marker. When this occurs, the marker is sent on all outgoing channels and the state of the processor is recorded. If there are any messages at the node, they are recorded as part of the processor state (Figure 10.2a). If the node has already started its local snapshot when a message arrives along a channel (that the node has not seen the marker on), the message is recorded as being on the channel (Figure 10.2b). Recording continues until a marker is received on all incoming links.

We translate these actions directly to the mobile environment. The announcement corresponds to the marker and the mobile unit corresponds to a message in the Chandy–Lamport algorithm. When an MSC receives the announcement for the first time, it sends copies on all outgoing channels

Dependable Message Delivery to Mobile Units

and attempts delivery to any mobile unit present. If the mobile unit is at the MSC, it will receive the announcement (Figure 10.2c).

Just as a node continues recording until it has received the marker on all links in order to record messages on channels, the delivery algorithm *will keep a copy of the announcement until it receives a copy of the announcement from all neighbors* in order to deliver to a mobile unit in transit between base stations. Intuitively, this prevents the mobile unit from hopping from node to node eluding the announcement. Thus if the mobile unit arrives prior to the announcement on a channel, the MSC delivers the data as soon as the handover is complete (Figure 10.2d). This is possible because the MSC stores a local copy that arrived on another channel.

10.3.5 Extensions

One of the strengths of our approach to algorithm development is that it rapidly produces an algorithm in the mobile environment that can be easily extended. In this section, we discuss several possible extensions including delivering multiple announcements simultaneously, delivering to rapidly moving mobile units, performing route discovery, multicasting an announcement, and working within the mobile agent environment.

10.3.5.1 Multiple Announcement Deliveries. To deliver multiple announcements simultaneously using snapshots, we can run several copies of the algorithm in parallel. This is analogous to having each MSC both index and store the incoming announcements and maintain separate channel status for each announcement in the system. This information is kept until the MSC locally determines it can be cleared. In the worst case, every node must have storage available for every potential announcement in the system, as well as maintain channel status with respect to each announcement. Although this appears excessive, we maintain that the nature of the snapshot algorithm in a real setting will not require maximum capacity. In other words, because the MSCs are able to locally determine when to delete the announcements, the nature of the network will determine how long an announcement is stored at the MSC.

10.3.5.2 Rapidly Moving Mobile Units. Another advantage of this algorithm is the ability to operate in rapidly changing environments with the same delivery guarantees. In Mobile IP, mobile units must remain in one place long enough to send a message with their new location to their home agent for forwarding purposes and remain at that foreign agent long enough for the forwarded messages to arrive. With forwarding enhancements added to the foreign agents in Mobile IP, the issue is minimized because the former location of a mobile unit becomes a kind of packet forwarder. However, even with forwarding, if the agent moves too rapidly and the system is unable to stabilize, forwarded packets will chase the mobile unit around the system without ever being delivered. Because snapshots do not maintain a notion of home or route, movements are immediately accounted for by the delivery scheme.

10.3.5.3 Route Discovery. In more moderately changing environments, the overhead of sending the announcement to every node may be excessive. In these situations, the snapshot delivery algorithm can be modified to perform route discovery. When a source mobile unit S located at MSC_S wishes to communicate with a destination mobile unit D located at MSC_D, S sends a `discovery` message using snapshot delivery. When this message is received by D, a `discoveryReply` is sent back to MSC_S, identifying D's location. Subsequent messages from S to D are sent directly to MSC_D. When a message fails to be delivered, the discovery process is repeated using snapshot delivery for the query.

10.3.5.4 Multicast. Another area of research in the mobility community is multicast, including some work on reliable multicast [1] but only under the assumption that the set of recipients is known. Our algorithm can trivially be extended to perform multicast to all mobile units in the system during the execution of the snapshot *without* knowing the list of recipients. Without changing the processing of the snapshot algorithm and by only changing the destination address from a unicast mobile unit identifier to a multicast address, it can be shown that every host accepting announcements on that address will receive the announcement. The reason for this can be found by looking back at the traditional distributed snapshots. In a snapshot, every message in the system is recorded in exactly one local snapshot. In our modified algorithm, *delivery* replaces *recording* and *mobile units* replace *messages*. Therefore, every mobile unit will be delivered to exactly one time. Although this description is concise, the importance of it should not be lost in its simplicity.

10.3.5.5 Mobile Agents. Thus far we have only considered physical movement of mobile units, but another environment that is characterized by rapid mobile movement is mobile agents where it is not a physical component that moves, but rather program code and data moving through the fixed network. Rather than connecting to a base station through a wireless mechanism, these mobile agents actually execute at a foreign host. They have the ability to move rapidly from one host to another and may not register each new location with a home. Therefore, delivering a message to a mobile code agent becomes an interesting application area in which rapid movement is not only feasible, but is the common case. The interested reader can find more details on applying snapshots in logical mobile environments in [16].

Dependable Message Delivery to Mobile Units

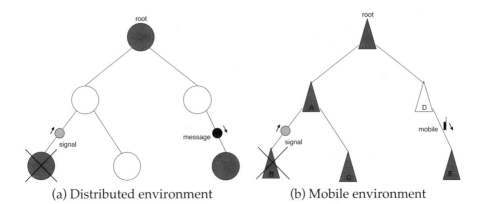

Figure 10.5. Dijkstra–Scholten Trees of Diffusing Computations
Shaded nodes are idle, white nodes are active. (a) Applied to a standard distributed environment. (b) Applied to a mobile environment, tracking a single mobile unit. Note there is only one active node and it is the node the mobile unit just left. A possible path of the mobile unit to build this tree would be: root, A, B, A, C, A, root, D, E.

10.4 Tracking for Delivery

Now we turn our focus toward an approach to message delivery based on tracking the location of the mobile unit as it moves through the network. Unlike Mobile IP tracking, our approach does not require location updates to be sent to the home node each time. This section describes a tracking and delivery approach that comes from applying our algorithm development technique to the Dijkstra–Scholten diffusing computation/termination detection algorithm. After outlining this algorithm, we present another algorithm that is not directly based on diffusing computations, but was inspired by our previous investigation. The details of this work are available in [17].

10.4.1 From Diffusing Computations to Mobile Unit Tracking

Diffusing computations have the property that the computation initiates at a single root node while all other nodes are *idle*. The computation spreads to other nodes as messages are sent from *active* nodes. Dijkstra and Scholten [6] describe an algorithm for detecting termination of such computations in which the basic idea is to maintain a spanning tree that includes all active nodes, as shown in Figure 10.5a. A message sent from an active node to an idle node (*message* in Figure 10.5a) adds the latter to the tree as a child of the former. Messages sent among tree nodes have no effect on the structure, but may activate idle nodes still in the tree. An idle leaf node can leave the tree at any time by notifying its parent (*signal* in Figure 10.5a). Termination is detected when an idle root is all that remains in the tree.

241

By applying our algorithm development technique, we adapt this tree maintenance algorithm to track the movement of a mobile unit as it travels among base stations. We define a node to be *active* when the mobile unit is present (or has started the handover process and is modeled on the channel) and therefore, when the mobile unit arrives at a node, if that node is not already part of the tree, it is added. In Figure 10.5b, this corresponds to adding E as an active node when the mobile unit arrives and changing the status of node D to idle. Because all active nodes are in the tree of the diffusing computation, the Dijkstra–Scholten algorithm guarantees that the mobile unit will always be at a node in the tree (or on a channel leaving from a node in the tree). In other words, the tree of the diffusing computation defines a subregion of the network where the mobile unit has recently traveled. As the mobile unit doubles back on its path, the node it arrives at transitions back to active and the node it departed becomes a leaf node that is cleaned up in the same way idle leaf nodes are removed in the original Dijkstra–Scholten algorithm (sending a signal message).

This tracking of a mobile unit by identification of a region containing the mobile unit is only part of our goal. Reliable message, or announcement, delivery is the other component that we achieve by designing an announcement delivery algorithm that works on top of the diffusing computation tree. Our algorithm works by placing the announcement at the root of the tree and spreading it down the tree until the mobile unit (or a leaf node) is reached. To guarantee delivery to a mobile unit that is moving during the announcement propagation, we temporarily store the announcement at the intermediate nodes and run a cleanup phase after the message is delivered to remove the extra copies.

By superimposing the delivery actions on top of the graph maintenance, the result is an algorithm that guarantees at least once delivery of an announcement while actively maintaining a graph of nodes recently visited by the mobile unit.

It is not necessary for the spanning tree to be pruned as soon as a node becomes an idle leaf. Instead this processing can be delayed until a period of low bandwidth use. An application may benefit by allowing the construction of a wide spanning tree within which the mobile units travel, similar to the graph shown in Figure 10.5b. Trade-offs include shorter paths from the root to the mobile unit versus an increase in the number of nodes involved in each announcement delivery.

By constructing the graph based on the movement of the mobile unit, the path from the root to the mobile unit may not be optimal. Therefore, a possible extension is to run an optimization protocol to reduce the length of this path. Such an optimization must take into consideration the continued movement of the mobile unit as well as any announcement deliveries in progress. The trade-off with this approach is between the benefit of a shorter route from the root to the mobile unit and the additional bandwidth and complexity required to run the optimization and simultaneously guarantee the delivery of announcements en route to the mobile unit.

Although in our algorithm only one mobile unit is tracked, the graph maintenance algorithm requires no extensions to track a group of mobile units. The resulting spanning tree can be used for unicast announcement delivery without any modifications and for multicast announcement delivery by changing only the announcement cleanup mechanism.

10.4.2 Extension: Backbone-Based Message Delivery

We now introduce a new tracking and delivery algorithm inspired by the previous investigation with diffusing computations. Our goal is to reduce the number of nodes to which the announcement propagates. To accomplish this, we note that only the path between the root and mobile unit is necessary for delivery. In the previous approach, although the parts of the tree not on the path from the root to the mobile unit can be eliminated, announcements still propagate unnecessarily down these subtrees before node deletion occurs.

To avoid this, the algorithm presented in this section maintains a graph with only one path leading away from the root and terminating at the mobile unit. This path is referred to as the *backbone*. Nodes that were once part of the backbone, but are no longer on this path between the root and the mobile unit form structures referred to as *tails*. Tails are actively removed from the graph, rather than relying on idle leaf nodes to remove themselves. Maintenance of the backbone requires additional information to be carried by the mobile unit regarding the nodes currently on the backbone, as well as the introduction of a *delete* message to remove tail nodes. The announcement delivery mechanism remains essentially the same as before, but the simpler graph reduces the number of announcement copies stored during delivery.

Intuitively, the backbone nodes are the core of the algorithm because they represent the path between the root and the mobile unit that is necessary for announcement delivery. The tail nodes are leftover pieces that were formerly part of the backbone, but the doubling back of the mobile unit to backbone nodes makes these nodes unnecessary for message delivery. If we were not concerned with leaving unnecessary state lying around in the network, we could simply ignore these tail nodes; however for completeness, we include an active mechanism to shrink tails until they disappear. The complexities of the approach lie in properly maintaining the backbone and in cleaning up only tail nodes. Because nodes only have local knowledge, all decisions about dealing with arriving messages and announcements must be based on the information held at the node and carried by the message.

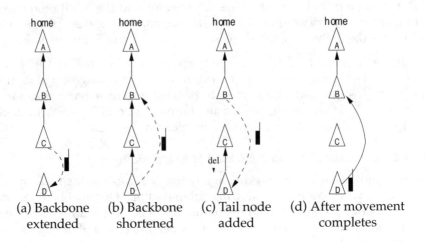

(a) Backbone extended (b) Backbone shortened (c) Tail node added (d) After movement completes

Figure 10.6. Backbone Maintenance
The parent pointers of the backbone change as the mobile moves to (a) a node not in the backbone, (b) a node higher in the backbone, and (c) a tail node. (d) Shows the state after all channels have been cleared.

To understand how the backbone is kept independent of the tails, we examine how the graph changes as the mobile unit moves. It is important to note that by the definition of the backbone, the mobile unit is always either at the last node of the backbone, or on a channel leading away from it. Figure 10.6 shows how the backbone is affected as the mobile unit moves to each of the three distinct types of nodes:

1. A node that is neither a backbone nor a tail node
2. A backbone node
3. A tail node

In Figure 10.6a, the backbone is composed of node A, node B, and node C and the dashed arrow shows the movement of the mobile unit from node C to node D, where D is not part of the graph. This is the most straightforward case in which the backbone is extended to include D by adding both the child pointer from C to D (not shown) and the parent pointer in the reverse direction (solid arrow in Figure 10.6b).

In Figure 10.6b, the mobile moves to a node B, a node already in the backbone and with a non-null parent pointer. It is clear from the figure that the backbone should be shortened to only include A and B without changing any parent pointers and that C and D should be deleted. To explicitly remove the tail composed of C and D, a *delete* message is sent to the child of B. When C receives the delete from its parent, it will nullify its parent pointer, propagate the delete to its child, and nullify its child pointer.

Dependable Message Delivery to Mobile Units

If at this point, the mobile unit moves from B onto D before the arrival of the delete (Figure 10.6c), D still has a parent pointer (C) and we cannot distinguish this case from the previous case (where B also had a non-null parent pointer). In the previous case, the parent of the node the mobile unit arrived at did not change, but in this case, we wish to have D's parent set to B (the node the mobile unit is arriving from) so that the backbone is correct. To distinguish these two cases, we require the mobile unit to carry a sequence containing the identities of the nodes in the backbone. In the first case, where the mobile unit arrives at B, B is in the list of backbone nodes maintained by the mobile unit; therefore, B keeps its parent pointer unchanged, but prunes the backbone list to remove C and D. However, when the mobile arrives at D, only A and B are in the backbone list, therefore the parent pointer of D is changed to point to B. But, what happens to the delete message moving from C to D? Because C is no longer D's parent when the delete arrives, it is simply dropped and the backbone is not affected.

The delivery algorithm is superimposed on top of the generated graph. In the algorithm of the previous section, the announcement propagated from the root down all edges of the tree. In the algorithm of this section, the announcement only propagates down the edges that are part of the backbone. It is still necessary to keep a copy of the message at every node until delivery occurs. Consider a case where the announcement is not stored and instead simply propagates down the backbone. In Figure 10.6B, if the announcement were at node C when the mobile unit moved from node D to node B, delivery would not occur because the mobile unit moved from a region below propagation to a region above propagation. Therefore, to guarantee delivery, as the announcement propagates down the backbone, a copy is stored at each node until delivery is complete. We refer to the portion of the backbone with an announcement as the *covered backbone,* see Figure 10.7b.

Delivery can occur either by the mobile unit moving to a location in the covered backbone, or the announcement catching up with the mobile unit at a node. In either case, an acknowledgment is generated and sent via the parent pointers toward the root to clean up the extra announcement copies. If the announcement is delivered when the mobile unit moves on to the covered backbone, a *delete* is generated toward the child and an *acknowledgment* is generated toward the parent. While the *acknowledgment* removes the copies of the announcement on the backbone, the *delete* removes the copies from the tails at the same time the tail nodes are removed from the graph.

Keeping the backbone sequence is a similar methodology to routing protocols passing complete paths to the destination as in Border Gateway Protocol (BGP) [21] to avoid loops. It has been argued that keeping such information in the packet greatly increases its size. However, in our case, the information is being kept by the mobile unit and we assume there is sufficient

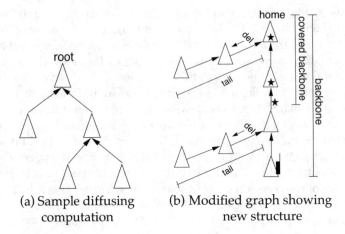

Figure 10.7. Diffusing Computations to Tracking
By adapting diffusing computations to mobility, we construct a graph reflecting the movement of the mobile. To deliver an announcement, the only part of the graph we need is the path from the root to the mobile, the backbone. Therefore we adapt the Dijkstra–Scholten algorithm to maintain only this graph segment and delete all the others.

storage on such a device for this additional information. In a mobile agent system, the path can be trivially shortened by forcing the agent to return to its home node periodically. This is not as reasonable for a physically mobile system, and in the case where the backbone sequence grows beyond a reasonable limit, a secondary, optimization algorithm can be executed to shorten its length.

A simple extension of this algorithm is to allow for multiple concurrent announcement deliveries as in sliding window protocols. The announcements and all associated acknowledgments would have to be marked by sequence numbers so that they do not interfere, but the delivery mechanism uses the same graph. Therefore, the rules governing the expansion and shrinking of the graph are not affected, but the proofs of garbage collection and acknowledgment delivery are more delicate.

10.5 Reality Check

When moving from the distributed computing environment to the mobile environment, we made several assumptions about the nature of the network and the behavior of the components in the network. In this section, we reexamine these assumptions, showing why they are reasonable or how the algorithm can be adapted to make them more reasonable. Specifically, we look at the issues of non-FIFO channels, multiple RBSs per MSC, base station connectivity, reliable delivery on links, the involvement level of MSCs, and storage requirements.

Dependable Message Delivery to Mobile Units

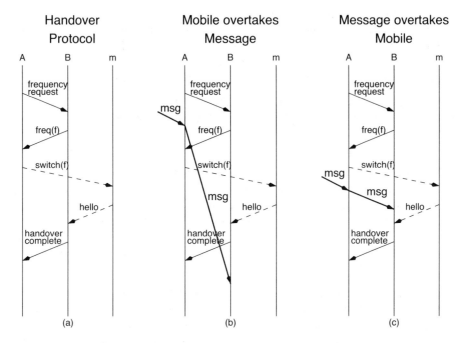

Figure 10.8. AMPS Handover Protocol
(a) For mobile unit m moving from cell A to cell B. If messages are processed (i.e., broadcast to the mobile) immediately upon receipt, it is possible (b) for the mobile to move faster than the message along the channel, or (c) for the message to move faster than the mobile, thus breaking the FIFO channel property.

10.5.1 FIFO Channels

One major issue when using the Chandy–Lamport algorithm is its reliance on FIFO channels. More specifically, in Section 10.3.2, we modeled both the mobile units and the messages as traveling on the same channel. This seems to be an unreasonable assumption given that mobile units move much more slowly through space than messages move through a fixed network. By looking in more detail at the handover protocol used when a mobile unit changes cells, we show how the FIFO assumption can be broken and propose a simple mechanism to restore it.

One of the U.S. standards for analog cellular communication is Advanced Mobile Phone Service (AMPS) [22], in which cellular telephones tune to only one frequency at a time. When the signal between the MSC and a mobile unit begins to degrade, the MSC searches for a neighboring MSC with a stronger communication signal indicating the mobile unit is moving into that particular cell. When a frequency is requested, a handover begins. Figure 10.8a shows the control messages exchanged as a mobile unit, m, moves from cell A to cell B. First the `frequency request` is exchanged

247

between the MSCs. At this point the mobile unit is made aware of the handover by receiving a new frequency from its current MSC, A. After switching to the new frequency, the mobile sends a `hello` on the new frequency, alerting B that the mobile is now listening on the new frequency. Finally, B sends a `handover complete` to A, which releases the old frequency.

With the AMPS approach, we know when a mobile unit is moving between cells and which cells it is moving between. We also note that the mobile unit is not involved in the handover until the moment it changes the frequency it is tuned to.

Our primary concern is making the channels FIFO with respect to mobile units and messages (both control messages and announcements). Even if we assume that channels between MSCs are FIFO, reordering is possible because part of the handover takes place over wireless channels that are not synchronized with the wired channels. Specifically, we address the two cases of non-FIFO behavior where the mobile overtakes a message and the message overtakes the mobile.

It is important to define the point at which the mobile logically moves onto the channel. We define this to be when communication with A is terminated by the transmission of the `switch` message. Similarly, the mobile moves off of the channel when the wireless transmission of the `hello` message is accepted at the destination cell, B in our example. As can be seen in Figure 10.8b, it is possible for a message sent on the wired channel before the `switch` message to arrive at the destination after the arrival of the mobile unit, breaking the FIFO ordering. Similarly, a message sent after the `switch` message can move quickly through the channel and arrive at the destination before the mobile (Figure 10.8c).

We propose a minor change in the protocol in order to involve both the wired and wireless channels in the handover. The only change to the source side (A in this case) is the wire transmission of a special message atomically with the wireless `switch` transmission. We call this message the `virtual mobile unit` (VMU), because it identifies the point on the wired channel at which the mobile leaves the source. All messages sent on the wired channel before the VMU were sent before the mobile unit left, and all messages after the VMU were sent after the mobile unit left. We correspondingly change the behavior of the destination (B in this case) to achieve this desired behavior; in other words, to have the VMU and the physical mobile unit arrive at the destination at the same time. Therefore, if the `hello` arrives before the VMU, all incoming messages on the wired channel are treated as if the mobile is not present even though communication is possible. Conversely, if the VMU arrives before the `hello`, all messages sent on the wired channel are buffered until the `hello` arrives. When both messages have arrived and have been processed, B continues processing all messages in the order in which they are received. By forcing the receiver to wait for both messages, the wired and wireless channels are synchronized, effectively yielding a single FIFO channel containing both mobile units and messages.

10.5.2 Multiple RBSs per MSC

Until this point, we have only allowed one RBS for each support center, however, in current cellular telephone systems, MSCs manage sets of RBSs. Because the algorithm we presented is intended to be run over the fixed network formed by the MSCs, the handover mechanisms apply only to the movement of a mobile unit from a RBS supported by one MSC to another RBS supported by a different MSC. The question remains about how to broadcast the announcement within the cells supported by a single MSC and maintain the constraints of guaranteed, single delivery. Because the MSC acts as a coordinator of the mobile units present at each of the RBSs, it is feasible to run the snapshot delivery algorithm among the RBSs, allowing it to terminate before any handovers to other MSCs are permitted. This simple solution shows how our snapshot algorithm can be used as a support layer for other algorithms.

10.5.3 Base Station Connectivity

Another possible concern with the model we presented is the necessity for physical connections between all MSCs whose cells border one another. Because of the high cost for such connectivity, it is possible that these physical wires may not exist. To allow our algorithms to function in such a setting, we propose adding virtual channels between adjacent cells and treating such channels the same as the real channels. In the implementation, however, we must be careful to ensure the FIFO nature of this virtual channel.

The same technique can be applied to support a limited form of disconnection. Suppose a mobile unit was likely to disconnect from cell A and at some time later connect to cell B. By adding a virtual channel between A and B and managing the disconnection as a long-lived handover, we can guarantee delivery even if the mobile unit disconnects during the delivery. Although this requires additional memory support at the base stations to store the announcements for the duration of this disconnection, such storage is not required at all base stations, making this a reasonable approach for guarantees in the presence of disconnection.

10.5.4 Reliable Delivery on Links

Our delivery algorithms assume that message delivery across a link is reliable. Most of the Internet uses unreliable links such as Ethernets, frame relay, and Asynchronous Transfer Mode (ATM). The probability of error on such links may be small, but packets are indeed dropped. A possible solution is to add acknowledgments for multicast messages as is done, for example, in the intelligent flooding algorithm used in Links State Routing in Open Systems Interconnection (OSI) [24] and Open Shortest Path First (OSPF) protocols [15]. Another solution is to only provide best-effort service. Because lost messages can lead to deadlock, we need to delete an announcement after a time-out even if a token is still expected along a channel.

10.5.5 Involvement Level of MSCs

For the snapshot algorithm to function, every MSC must be involved to guarantee delivery and termination. In a paper on running distributed computations in a mobile setting [3], the authors warn against requiring involvement of all mobile units in a computation, especially due to the voluntary disconnection often associated with mobile computing. Such disconnection is often done to conserve power, or in some cases, to allow disconnected operation. In either case, the mobile unit is not available for participation in the distributed algorithm. These arguments are important when designing distributed algorithms for execution over mobile units; however our goal is not to create a global snapshot containing information about the mobile units, but instead to employ the snapshot technique to a different end, namely announcement delivery. Additionally, the control messages of the snapshot are not processed by the mobile units, but rather by the fixed mobile support centers, and no resources of the mobile unit are expended, except to receive a message.

It is true that to guarantee delivery, the mobile unit must be present in the system; however, this is a reasonable assumption, because by definition there are no means to reach a disconnected mobile unit. It is worth noting that if the mobile unit is not present in the system during a delivery attempt, the algorithm will terminate normally, removing all traces of the announcement from the system, but without delivery.

10.5.6 Storage Requirements

In snapshot delivery, we assume that the MSCs hold a copy of the announcement for delivery to the mobile units for a bounded period of time limited to the duration of the local snapshot. This is more efficient than another proposal [1] in which the announcement is broadcast to all nodes, each of which stores the announcement until notified that delivery has occurred. In our approach, the time for storage is bounded by the speed of network propagation and connectivity of the network. In a system with bi-directional channels, because the local snapshot terminates when the announcement arrives on all incoming channels, the duration of a local snapshot can be as short as one round trip delay between the MSCs. One can argue that it is not the place of the MSCs to be maintaining copies of announcements when their primary purpose is routing. However, in this case, because no routing information is being kept about the mobile units, the system will be required to keep additional state to provide delivery guarantees. Therefore, keeping a copy for a short duration is a reasonable assumption.

10.6 Conclusions

This chapter makes two important contributions. First, it explores a model of mobility in which handovers are abstracted as the traversal of links among the base stations and mobile units, both physical and logical, are treated as persistent messages. The result is a model that unifies wired and cellular, wireless networking and facilitates the transfer of algorithmic knowledge between the two settings. Second, we offer a general methodology for reusing results from distributed computing in the area of mobile computing. Our main contribution is to suggest not a direct usage of the existing algorithms, a strategy shown to have limited applicability, but a way to capitalize on the intellectual investments made in the field of distributed computing. The examples presented adapt a snapshot algorithm to search for a mobile unit and deliver a message and use diffusing computations to track the movement of a mobile unit. The ease with which we built these new algorithms provides strong evidence of the efficacy of the general strategy advocated in this chapter.

Acknowledgments

This research was supported in part by the National Science Foundation Grant no. CCR-9970939. Any opinions, findings, and conclusions are those of the authors and do not necessarily reflect the views of the research sponsors.

References

1. A. Acharya and B.R. Badrinath. A framework for delivering multicast messages in networks with mobile hosts. *Journal of Special Topics in Mobile Networks and Applications (MONET)*, vol. 1, no. 2, pp. 199–219, October 1996.
2. B.R. Badrinath, A. Acharya, and T. Imielinski. Structuring distributed algorithms for mobile hosts. In *Proceedings of the 14th International Conference on Distributed Computing Systems*, Poznan, Poland, pp. 21–28, 1994.
3. B.R. Badrinath, A. Acharya, and T. Imielinski. Designing distributed algorithms for mobile computing networks. *Computer Communications*, vol. 19, no. 4, pp. 309–320, April 1996.
4. J. Broch, D.B. Johnson, and D.A. Maltz. The dynamic source routing protocol for mobile ad hoc networks. Internet draft, March 1998. IETF Mobile Ad Hoc Networking Working Group.
5. K.M. Chandy and L. Lamport. Distributed snapshots: Determining global states of distributed systems. *ACM Transactions on Computer Systems*, vol. 3, no. 1, pp. 63–75, February 1985.
6. E.W. Dijkstra and C. Scholten. Termination detection for diffusing computations. *Information Processing Letters*, vol. 11, no. 1, 1980.

7. H. Eriksson. Mbone: The multicast backbone. *Communications of the ACM*, vol. 37, no. 8, pp. 54–60, 1994.
8. A. Fuggetta, G.P. Picco, and G. Vigna. Understanding code mobility. *IEEE Transactions on Software Engineering*, vol. 24, no. 5, pp. 342–361, May 1998.
9. R. Gray, D. Kotz, S. Nog, D. Rus, and G. George. Mobile agents for mobile computing. Technical Report PCS0TR96-285, Dartmouth College, May 1996.
10. J. Ioannidis and Jr. G.Q. Maguire. The design and implementation of a mobile internetworking architecture. In *1992 Winter Usenix*, 1993.
11. V. Jacobson and S. McCanne. VAT: Visual Audio Tool, VAT manual pages, 1995.
12. D.B. Johnson. Scalable support for transparent mobile host internetworking. In H. Korth and T. Imielinski, Eds., *Mobile Computing*, Norwell, MA: Kluwer Academic Publishers, 1996, pp. 103–128.
13. J.J. Kistler and M. Satyanarayanan. Disconnected operation in the coda file system. *ACM Transactions on Computer Systems*, vol. 10, no. 1, pp. 3–25, 1992.
14. S. McCanne and V. Jacobson. vic: A flexible framework for packet video. In *ACM Multimedia'95*, pp. 511–522, San Francisco, 1995.
15. J. Moy. OSPF version 2. Internet draft, IETF, March 1994.
16. A.L. Murphy and G.P. Picco. Reliable communication for highly mobile agents. *Journal of Autonomous Agents and Multi-Agent Systems,* [Special issue on mobile agents], vol. 5, no. 1, pp. 81–100, March 2002.
17. A.L. Murphy, G.-C. Roman, and G. Varghese. Tracking mobile units for dependable message delivery. *IEEE Transactions on Software Engineering*, May 2002.
18. A. Myles and D. Skellern. Comparing four IP based mobile host protocols. *Computer Networks and ISDN Systems*, vol. 26, no. 3, pp. 349–355, 1993.
19. C.E. Perkins. IP mobility support. Technical Report RFC 2002, IETF Network Working Group, October 1996.
20. M. Ranganathan, A. Acharya, S. Sharma, and J. Saltz. Network-aware mobile programs. Technical Report CS-TR-3659, University of Maryland, College Park, 1997.
21. Y. Rekhter and T. Li. A Border Gateway Protocol 4 (BGP-4). RFC 1771, March 1995.
22. M. Steenstrup. *Routing in Communication Networks*, Ch. 10, Upper Saddle River: Prentice Hall, 1995.
23. F. Teraoka, Y. Yokore, and M. Tokoro. A network architecture providing host migration transparency. *ACM SIGCOMM Computer Communication Review (SIGCOMM'91)*, vol. 21, no. 4, pp. 209–220, September 1991.
24. H. Zimmerman. OSI reference model — The ISO model of architecture for open systems interconnection. *IEEE Transactions on Communication*, vol. 28, pp. 425–432, 1980.

Section III
Location-Based Services

Chapter 11
Location-Dependent Query Processing in Mobile Computing

Ayşe Yasemin Seydim, Margaret H. Dunham

11.1 Introduction

In a mobile environment, the need for localized data becomes inevitable with the increase in people's mobility and change of working habits. Queries asked in the mobile environment may have a slightly different structure and format from traditional database queries. For example, the query, "What are the names and addresses of the restaurants within five miles?" seeks for the restaurants within five miles of the current position of the query issuer. To provide the answer to the query, first the application has to know the location of the issuer. The query can later be bound to this location. Thus, location dependence in queries implies that the information asked is related to a location, but the location is not explicitly known when the query is asked. Therefore, there seems to be a layered approach in processing these new types of queries.

Obviously, there will be location-related and nonlocation-related attributes in any query. To answer any location-based request while the user is moving, it may become necessary to identify location-related attributes. Sometimes the statement of the query will not have any location-related attributes, but the way it is stated will have an implication that the query issuer's current position is involved in the selection criteria. If the location attribute is implied in the query, which can be called a location-dependent query (LDQ), the implied location-related attribute has to be added to the query and query is processed on location-dependent data (LDD). Once the user's location is known, the query becomes location-aware, with an explicit indication of this special location attribute. Therefore, a query including any location-related attribute in its predicates is called a location-aware query (LAQ). One can then separate the stages that a LDQ goes through prior to its processing at the data server. In this research, we do not consider implementation details of the location-dependent data, how it is stored, or the partitioning strategy used. Furthermore, we assume the query location does not change until the results are returned to the user.

In this chapter, we develop a formal query model to encompass all types of queries in a mobile environment. First, in Section 11.2, we investigate query processing in mobile computing, how the queries are classified and how they are processed in a location-dependent applications environment. We provide our formalization in Section 11.3 and we illustrate the differentiation of query types by examples. This section also examines moving object database queries and includes a summary of the query types that can be issued in this environment. The processing steps involved in the translation of the original query to the one routed to the content provider are described in Section 11.4. An earlier version of this work appeared in the MobiDE Workshop in 2001 [SDK01].

11.2 Related Work

In this section, we provide the related work in location-dependent queries and their processing. We outline the current research and the relationship of location-related applications to spatial database management. We also relate our work to moving object databases research that has been a direct outcome of the location dependence in mobile environments. These are investigated and briefly presented in the following subsections.

11.2.1 Location-Dependent Data and Queries

Querying location-dependent information in the mobile environment has been an important research area. In mobile computing, [FZ94] and [DH95] have been the first studies to view LDQs as asking values of data that change depending on a location. In [DK98], LDD has been viewed as spatial replicas depending on a data region where they are included. Query processing approaches based on physical organization of data and location binding are discussed in [KD98]. In line with these studies, a formal model has been presented in [RD00] to describe the mobility of objects. The strategies for caching in location-dependent services (LDS) applications have also been discussed in their work, where a mobility plan trajectory of the mobile user has been used.

Most of the work to date has been about data management issues of mobile objects and their location information. [IB93], [PB94], and [PS01] are some of the works that concentrate on querying location information and frequent update and inaccuracy problems that may occur mostly in wireless operator databases. Pitoura et al. [PS01] thoroughly investigates the location problem in the management of mobile users. Directory architectures that

Location-Dependent Query Processing in Mobile Computing

hold the location of moving objects are discussed along with their optimizations and variations. Data management techniques for identifying the current location of moving objects are investigated in their work, which includes efficiency of techniques, caching, replication, and partitioning of datasets. Location management techniques use both the information concerning the location of moving objects stored in location databases and the search procedures to determine the object's movement behavior.

In the classification of location-dependent or location-aware queries, many different types are defined according to their use in the applications. In addition to [SWCD97]'s instantaneous, prediction, continuous, and persistent query types, which are discussed in Section 11.3.4, there are various query classifications in the literature.

Xu and Lee [XK00] identify the queries according to their access to location information. They assume location is bound in "cell id" granularity and the data is stored in corresponding cell's server (i.e., base station computer). If the queries are local, like in the query "List the local hotels," the database search starts from the current cell to the root cell in a hierarchy of cells until the results are found. If they are nonlocal, as in "Find the weather in cell id = 8," the query is redirected to the corresponding cell to find its local replies and the result set is forwarded to the current cell. In geographically clustered queries, like in the query, "List the hospitals within five miles," spatial constraints are given in the query and should be answered by clusters of cells within a distance. Geographically dispersed queries, as in "List all hospitals with a heart surgery facility," have to be processed in every cell. The nearest query, as "Find the nearest gas station," has to be processed in the current cell and the nearest cells to the farthest until the condition is satisfied.

In a recent study, [ZL01] classifies the queries depending on the mobility of the issuer or the queried object. Here, queried objects are defined as the candidates of the results set. A mobile or a stationary client issues a query about mobile or stationary objects. They have studied mobile users querying stationary data and assumed the location of the user is obtained by a global positioning system (GPS). However, indexing of the data objects depending on the nearest neighbor relations is the main problem studied. Their approach requires change in the data management at the fixed site and does not support traditional applications.

Lee et al. [LLXZ02] classifies queries into two groups, where the first one is local versus nonlocal, similar to [XK00]. The second group contains simple versus general queries, where operators and constraints may be more complex. A general query has more than one predicate and it may be called spatial-constrained when there is a spatial constraint in it.

Samet and Aref [SA95] give a broad classification for spatial queries as *local, focal,* and *zonal* queries. A focal query seeks neighbors of locations on the same layer in a spatial database implementation. If this location is a region, then the query becomes a zonal query, which seeks the groups of locations with the same attribute on the same layer. If a query involves the intersection of layers for one location then it is a local query. This is equivalent to a service discovery application, which will find the services around one location.

11.2.2 Moving Object Databases Research

Most research to date not only has concentrated on query processing issues of location databases but also on representation of moving objects in database systems. Modeling and querying moving objects have been studied including a model to represent the spatiotemporal properties [SWCD97]. These works regard LDS applications as *moving object database* (MOD) applications. MOD applications usually access spatial objects whose position as well their extent (covered region) change with time. Because they include a temporal attribute, queries can refer to both the past and future histories of moving objects. Moving Object Spatio-Temporal (MOST) Data Model, indexing, uncertainty, and Databases for Moving Objects (DOMINO) query processing architecture are studied over time [GU00, SWCD97, WXCJ98].

In addition to the MOST Data Model and DOMINO query processing architecture, there are other modeling studies for moving objects. Guting et al. [GBE⁺00] propose a representation approach for time-dependent geometries, which provides an abstract data type extension to a database management system (DBMS) data model and query language. Formalization and the foundation for implementation of spatiotemporal DBMS extension are also given. For identifying spatial operations, [GBE⁺00] also classifies operations extensively according to nontemporal and temporal data types.

11.2.3 Spatial Database Management

The implementation of location services and applications overlap with the area of geographic information system (GIS). Any GIS-based project or application can be considered as a location-based service [EG03]. However, GIS software is more specific to geography and based on complicated processing. We do not see an efficient use of a complete GIS system for a much simpler LDQ processing.

In GIS applications, different views of the same geographical area are used for different purposes. These different layers of information can be defined in a hierarchy and the requested view would be detailed in a map. However, two areas can be defined in different layers and may be compared. In this case, a translation from one attribute to the other should be performed. The method used in this translation may utilize GIS functions or a simpler technique by using hierarchical relationships. Often, these translations are called generalization and specialization.

Samet and Aref [SA95] describe spatial data as "a term used to describe data that pertain[s] to the space occupied by the objects in a database." Spatial data is geometric and consists of points, lines, rectangles, polygons, surfaces, volumes, and data of even higher dimension. Definitely, spatial data is location related. Some attributes of spatial data, such as name and Social Security number, do not contain any embedding space [NW97] relationship. The type of queries defined in [SA95] were mentioned in the previous section.

The access methods for spatial data are also investigated by many researchers. A survey of multidimensional access methods to support search operations in spatial databases is given in [GG98]. Formal definitions of common spatial database operators are also given. We plan to cover many of these query operations including exact match, point, window, intersection (overlap), enclosure, containment, adjacency, nearest-neighbor queries, and spatial join operation in our work. Methods to process these are also studied in [GG98] as point access methods and spatial access methods.

There are combination efforts for spatial and conventional data in spatial database systems. Extending a DBMS for geographic applications has been discussed by [OSDM89]. Aref and Samet [AS91] also describe an architecture for spatial query processing with the extended operators to relational DBMS. A spatial query language has been proposed for spatial databases [Ege94]. Spatial operators are recently discussed and classified in [CDF00], including the integration issues of new-generation languages. Taxonomy of requirements to be satisfied by spatial operators and classification of them as topological, projective, and metric are also presented. Topological relationships are based on the two objects' placement in geography, however the name or type of the relationship has been defined extensively. Dunham [Dun02] mentions the types as disjoint, overlaps/intersects, equals, covered by/inside/contained in, covers/contains. Egenhofer [Ege94] defines another one, meet; whereas [DTJ01] gives more detailed set in addition to them such as interpenetrating, boundary-overlap, interior-overlap, etc. Therefore it is possible to define many relations between two spatial objects, however, the basic idea is to find a common geographic point. We have used contains/covers and its dual contained by/covered by to define the logical hierarchy.

In these previous research studies, we do not see any clear differentiation between location dependence and location awareness in queries and applications. Our research differs significantly from this earlier work. We think the main difference between a location-dependent query and a location-aware query is the location binding and the fact that this process converts the former type of query into the latter. A location query requests the location of the mobile user and we assume its processing is by the help of a location service provided by a position determining technology vendor. With this assumption in mind, our work differs in the classification of query types from the MOD queries. We also take a reasonable subset of spatial operations to serve in LDQ processing because the complicated operations are required to be processed in spatial database systems or GIS applications.

11.3 Location Relatedness and the Query Model

We examine the location and the queries to build a formal query model to encompass all types of queries in a mobile environment. Suppose a database, D, consists of a set of base relations $R_1, R_2, \ldots, R_n$, where $D = \{R_1, R_2, \ldots, R_n\} = \bigcup_{i=1}^{n} R_i$ and let each attribute set of a relation be denoted as A_{R_i}. The union of attributes of each relation will give the attribute set of the whole database D, denoted as A_D:

$$A_D = A_{R_1} \cup A_{R_2} \cup \ldots \cup A_{R_n} = \bigcup_{i=1}^{n} A_{R_i}. \tag{11.1}$$

Corresponding to each attribute a_{ij} in relation R_i, there is domain$_{ij}$, which represents the domain of the jth attribute of R_i. A domain can be an arbitrary, nonempty set, finite, or countably infinite [Mai83]. The domain of any attribute is identified by the semantics or relatedness to the meaning of the attribute in the database. For example, the domain of City attribute for Texas State contains city names as {Dallas, Houston, San Antonio, …} but not the name of any bridge or street. This relatedness for the domain is normally determined by human experts.

11.3.1 Query Model

We assume individual attributes of relations can be distinguished to be elements of location-related (LR) or Nonlocation-related (NLR) domains. For example, City is a location-related attribute name and its domain is also LR. On the other hand, LastName is a nonlocation-related attribute name and its domain has no relation to any location.

Given a database D and its attribute set A_D, if the location relatedness is distinguishable, the attribute set of the database consists of attributes from LR-domains and NLR-domains. So we can then write:

$$A_D = A_{D(NLR)} \cup A_{D(LR)} \tag{11.2}$$

$$A_{D(LR)} = A_{R_1(LR)} \cup A_{R_2(LR)} \cup \ldots \cup A_{R_n(LR)} = \bigcup_{i=1}^{n} A_{R_i(LR)} \tag{11.3}$$

Location-Dependent Query Processing in Mobile Computing

$$A_{D(NLR)} = A_{R_1(NLR)} \cup A_{R_2(NLR)} \cup ... \cup A_{R_n(NLR)} = \bigcup_{i=1}^{n} A_{R_i(NLR)} \qquad (11.4)$$

Definition 11.1

Let the relation R_i contain the attributes a_{ij}, $1 \leq j \leq k$, and be shown by $R_i = \{a_{i1}, a_{i2}, ..., a_{ik}\}$. If an attribute a_{ij} of relation R_i is distinguished as LR, we call a_{ij} a LR-Attribute and R_i a LR-Relation. Otherwise, the attribute is NLR and called a NLR-Attribute. If all attributes of R_i are NLR-Attributes, then relation R_i is called NLR-Relation. In that case, $A_{Ri(LR)} = \phi$.

Spatial data are inherently location-related and any spatial relation is a LR-Relation. Some attributes of spatial data, such as name or age, do not contain a spatial property. Thus a LR-Relation may have NLR-Attributes.

Operators applied on the attributes are implemented depending on the properties of their operands. Either unary or binary operators, which require one or two operands, respectively, can be used in both nonspatial and spatial domains [CDF00]. However, the meaning can be interpreted depending on the domain of the attributes used in the operation. For example, intersection, union, or equal operators are different operations depending on their application on nonspatial or spatial attributes. A classification for nonspatial and spatial operators is given in Table 11.1. This list is not an exhaustive list of operators. They are the operators defined on the general geometry class by the Open GIS Consortium [The98].

The spatial operators listed in Table 11.1 are defined at a general level for all basic spatial data types (points, lines, regions) [CDF00]. More complicated spatial operators can be defined based on the elementary types to perform more complex operations. It is also discussed in [CDF00] that even

Table 11.1. Nonspatial and Spatial Operators

Operator Type	Operator Group	Operators
Nonspatial	Comparison Set Boolean	$<, \leq, =, >, \geq$ union, intersection, difference ($\cup, \cap, -$) and, or, not ($\vee, \wedge, \neg$)
Spatial	Basic Topological Spatial Analysis	SpatialReference, Envelope, Export, IsEmpty, IsSimple, Boundary Equal, Disjoint, Intersect, Touch, Cross, Within, Contains, Overlap, Relate Distance, Buffer, ConvexHull, Intersection, Union, Difference, SymDifference

more theoretical research is necessary to define the complete set of operators. We examine only elementary types of operators as they are sufficient for distinguishing the location relatedness in processing of LDQs.

Definition 11.2

An LR-Operator is an operator which has LR-Attributes as its operands. Similarly, if the operands are all NLR-Attributes, then the operator is a NLR-Operator.

Spatial operators, such as `overlaps` and `contains` are LR-Operators. However, there are definitely more LR-Operators than spatial operators. This is because a LR-Operator may be stated based not only on a spatial concept, but also on the direction of movement of the query issuer. An example of such a LR-Operator is `straight ahead`. Spatial operators compare two static spatial objects, whereas a LR-Operator may compare attributes for objects that are moving. One or both of the two objects may be moving. Therefore, not only can the LR-Operator compare attributes for the spatial properties of the objects, but also the direction of them.

A LR-Operator may have filtering and movement direction arguments with it. With filtering, we mean using a restrictive area within which the desired data is selected. For a `closest` operator, one can think of a circular area around the query issuer to access the related data. Direction may not be important in this case. However, for a `straight ahead` operator, we have to define a window to select an area ahead of the direction of the user.

Different interpretation examples of a filtering window are shown in Figure 11.1. Depending on the interpretation, data related to an area of a half circle, a rectangle, or a square can be selected as the result. Figure 11.1A shows a rectangular area of selection, where Figure 11.1B and Figure 11.1C show half-circular and circular selection, respectively. In this research, we do not investigate the creation of special LR-Operators, but we give the examples for clarification. However, incorporating this window concept is a necessary step for selecting the location-related data from a traditional database.

Definition 11.3

A Simple Predicate (SP) is defined as an expression with one operator and one or two operands, which is of the form SP = op_1 p_i or SP = p_j op_2 p_k, where op_1 is a unary operator and compatible with p_i and op_2 is a binary operator and compatible with both p_j and p_k. p_j and p_k are also compatible with each other. Either one of p_j or p_k can be a constant from the same domain as the other operand. Here, operands are either attributes or constants.

Operands in a SP have to be compatible (i.e., they have to be from the same domain). For example, one can use a comparison operator and compare one `City` with constant `Dallas`, but not with constant `2001` as they

Location-Dependent Query Processing in Mobile Computing

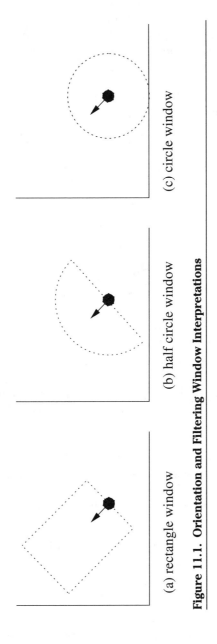

Figure 11.1. Orientation and Filtering Window Interpretations

have different domains. A Compare Predicate defined in [Ren00] is a type of SP.

Definition 11.4

A Simple LR-Predicate is a SP in which the operator is a LR-Operator and the operands are from LR-domains. A Simple NLR-Predicate is a SP that has a NLR-Operator and operands are both from NLR-domains.

Definition 11.5

A Compound Predicate (CP) is disjunction ($\vee$ predicate) of conjunctions ($\wedge$ predicates) of SPs, in which each SP_{ij} is a SP and is defined as:

$$CP = (SP_{11} \wedge SP_{12} \wedge \ldots \wedge SP_{1n}) \vee \ldots \vee (SP_{m1} \wedge SP_{m2} \wedge \ldots \wedge SP_{mr})$$

A CP can be a mixture of Simple LR-Predicates and Simple NLR-Predicates and it can be defined in BNF notation as follows:

```
SP::= <operator> <attribute>|
      <attribute> <operator> <attribute>|
      <operator> <constant>|
      <attribute> <operator> <constant>|
      <constant> <operator> <attribute>
CP1::= SP | SP ∧ CP1
CP::= CP | ¬ CP | CP ∨ CP1
```

In the standard relational operations of `selection`, `projection`, and `join`, predicates used for the production of results may be either Simple LR-Predicates or Simple NLR-Predicates or a mixture of them in a CP form. We define a query relating to these concepts as follows.

Definition 11.6

A query is defined as a set of relations, attributes, predicates, and operations used to produce a result. Therefore, a query Q is a set with elements, Q_R, Q_A, Q_P, and Q_O, which correspond to relations, attributes, and predicates in Q and operations to produce the result, respectively. We can then write:

$$Q = \{Q_R, Q_A, Q_P, Q_O\} \text{ and}$$

$$Q_S = \{issue(Q)\}$$

where Q_S is the result set after the execution of the query Q against a given database state.

Note that, Q_A designates the attributes used in the query, not all the attributes of its relation. Operations are data manipulation operations like

Location-Dependent Query Processing in Mobile Computing

select, project, join. Once we distinguish the location relatedness, we can then decompose a query to LR-related attributes, predicates, and relations and write the query as:

$$Q = \{Q_{R(LR)}, Q_{R(NLR)}, Q_{A(LR)}, Q_{A(NLR)}, Q_{P(LR)}, Q_{P(NLR)}, Q_O\}.$$

Therefore, we can define the query types accordingly. If there are no LR-Attributes in the query predicates (i.e., $Q_{A(LR)} = \phi$ and $Q_{P(LR)} = \phi$), we refer to these types of queries as Non-Location-Related Queries and denote $Q_{(NLR)}$ as:

$$Q_{(NLR)} = \{Q_{R(NLR)}, Q_{A(NLR)}, Q_{P(NLR)}, Q_O\}.$$

Definition 11.7

If all the predicates used in a query are Simple NLR-Predicates (i.e., $Q_{A(LR)} = \phi$ and $Q_{P(LR)} = \phi$), then the query is called a Non-Location-Related Query (NLR-Query). The relation may be a LR-Relation, but if only NLR-Attributes are selected, the query is still a NLR-Query.

In the selection operation, if all attributes are selected, and if the relation itself is a LR-Relation, then the result set, Q_S, will include location-related data values. Whether the relation is a LR-Relation or not, the specified and selected attributes in the query are sufficient to call the query a NLR-Query. Note that in LDD, there may be both kinds of relations. By definition, the projection operation eliminates the undesired attribute columns from the result set. If the attributes are not in the selection criteria, but they are in the desired attributes for the projection operation, then these must also be included in the whole attribute set of the query, which is represented as Q_A. Example 11.1 illustrates a NLR-Query.

Example 11.1

Suppose we have the query, "Find the lead actor's name in the movie 'Casablanca'" where the relational algebra expression is as follows:

$$\pi_{ActorName} \, \sigma_{(MovieName="Casablanca")} \, (MOVIES)$$

In this query, no location attribute is involved. We thus have:

```
QR = {MOVIES}  ⇒ QR(NLR) = {MOVIES}
QA = { MovieName, ActorName }
   ⇒ QA(NLR) = { MovieName, ActorName }
QP = { MovieName = "Casablanca"}
   ⇒ QP(NLR) = { MovieName = "Casablanca" }
QO = {select, project}
```

As can be seen, because there is no LR-Attribute or LR-Predicate, the query is a NLR-Query.

11.3.2 Location-Aware Queries

If a query includes at least one LR-Predicate or LR-Attribute, it can be referred to as location-aware. We define LAQs as follows.

Definition 11.8

If a query, Q, includes at least one Simple LR-Predicate in its predicate set, Q_P, or at least one LR-Attribute in its attribute set, Q_A, then it is called a Location-Aware Query. A LAQ produces the same result set independent of the place it is issued.

Example 11.2 illustrates a LAQ.

Example 11.2

Suppose we are given the query, "Find movie theaters in Richardson," and the Address attribute only includes the county names. The query can be stated as follows:

$$\pi_{TheaterName, Address} \sigma_{(Address="Richardson")} (THEATERS)$$

```
QR  = {THEATERS}
    ⇒ QR(LR) = {THEATERS}
QA  = {TheaterName, Address}
    ⇒ QA(LR) = {Address}, QA(NLR) = {TheaterName}
QP  = {Address = "Richardson"}
    ⇒ QP(LR) = {Address = "Richardson"}
QO  = {select, project}
```

If a query is issued to find out a location value, it is obvious that this query has a LR-Attribute in its attribute set. We differentiate these as special types of LAQs (sometimes called "Location Dips" [Les00]) and call them location queries (LQs).

Definition 11.9

If the projected attribute is the location of the selected tuple, or the result set includes only the location attribute, then we refer to these kinds of queries as a special kind of LAQ and call it a Location Query.

A query "Where is Person A?" is a LAQ and it does not depend on the query issuer's location. However, the question Where carries a meaning that we should include the location attribute from the relation USERS and find the corresponding value from the data store. Example 11.3 illustrates the set of query arguments of a LQ.

Example 11.3

"Where is Person A?" can be denoted as:

$$\pi_{ActiveLocation}\ \sigma_{(PersonId="A")}\ (USERS)$$

```
QR = {USERS}
    ⇒ QR(LR) = {USERS}
QA = {ActiveLocation, PersonId}
    ⇒ QA(LR) = {ActiveLocation}, QA(NLR) = {PersonId}
QP = {PersonId = "A"}
    ⇒ QP(NLR) = {PersonId = "A"}
QO = {select, project}
```

We assume `ActiveLocation` is a location attribute of relation USERS. If the location attribute of relation USERS were `City`, then `ActiveLocation` is to be returned as a city name for this relation.

The following examples further illustrate different LAQs:

- How is the weather in San Antonio? — all the attributes and predicates are known and there is at least one LR-Attribute. The result does not change wherever it is asked but may change by the time.
- Find hotels within five miles of (x, y) — all the attributes and predicates are known and there is at least one LR-Attribute.
- Find automatic teller machines (ATMs) within five miles of Hotel X — this is also a LAQ because it is asking for the ATMs within five miles of a known location, however, it requires a LQ for finding the location of Hotel X. We also call these kinds of queries Relative queries, which involve a LQ first.
- Find the ATMs within five miles of location X — the previous example translates into this form after the location of the hotel is found. This query is a form of Spatial Focal Query described in [SA95], which is looking at neighbors of locations on the same layer in a spatial database implementation.
- Find the services at location X — this query is a service (application) discovery query. It is defined as LQ in spatial applications [SA95]. Because the location of X is known, this becomes a LAQ and its processing involves the intersection of different context layers defined for services in spatial database or a series of queries sent to specific applications.

Note that, even though these queries may look alike, each requires a different type of processing to service.

11.3.3 Location-Dependent Queries

In LDQ processing, we see a specific type of predicate in which the location attribute is fixed, but the value to which it is compared changes depending on the query issue location. As mentioned, the predicate is actually hidden in the query statement. For example, the query, "Find hotels within five miles" has the implied meaning of "Find hotels within five miles of my current location." This current location (i.e., "Here") must be learned and bound to the query. We assume that some location service is used to provide this location. This process of binding the current location to the query is referred to as Location Binding.

Definition 11.10

Location Binding can be defined as assigning a location value to a LR-Predicate variable, which will act as a window over a set of attributes in a relation R_i.

We also define an Active Location Predicate to represent the Simple Predicate involved in Location Binding.

Definition 11.11

An Active Location Predicate is a Simple Predicate with a specific attribute name, which is of the form, `ActiveLocation = Here`, where `ActiveLocation` is the specific attribute name and `Here` is a variable that is assigned a location value after Location Binding.

An Active Location Predicate is a Simple LR-Predicate because it involves a LR-Attribute and a location value.

Definition 11.12

A query Q is called a Location-Dependent Query, if the result set of the query, Q_S, changes depending on the location of the query issuer.

A LDQ has only one Active Location Predicate, because the query issuer can not be in more than one place at a time. In a LDQ, Active Location Predicate is hidden when the query is stated and the current location of the issuer is implied. A LDQ becomes location-aware, a LAQ, when the hidden Active Location Predicate is bound to a location.

Example 11.4

Suppose the the query, "Find the closest theaters" is issued when the client is in Richardson. When this query is stated, there is no specification of Active Location Predicate. When the `closest` operator is processed, however, a second operand implied will be produced by using an Active Location Predicate, we will add this to the query and state as:

Location-Dependent Query Processing in Mobile Computing

$$\pi_{TheaterName, Address} \; \sigma_{(closest(Address, Here)} \; (THEATERS)$$

The query has the additional implicit predicate of `ActiveLocation = Here`. When location binding is done, `Here` is replaced with `Richardson`. We thus have:

```
QR = {THEATERS}
   ⇒ QR(LR) = {THEATERS}
QA = {TheaterName, Address}
   ⇒ QA(LR) = {Address, ActiveLocation, Here},
     QA(NLR) = {TheaterName}
QP = {closest(Address, Here), Address = ActiveLocation,
     ActiveLocation = Here}
   ⇒ QP(LR) = QP
QO = {select, project, closest}
```

After location binding, this LDQ becomes a LAQ, which will include {Here = "Richardson"} in its predicate set.

Notice that preprocessing is needed in these queries to add the active location predicate, which is performing the location binding. Following are other examples of LDQs:

- Where is Person A with respect to here? or How far is Person A? — the answers to these questions change depending on the value of `Here`, the location of the issuer. These queries need location binding and calculation for the distance. These are treated as LAQs in [Maa98], as opposed to our classification.
- Where is the nearest doctor? — this LDQ example given in [IB93] requires finding the location of the mobile doctor. The process is to query first the location with a LQ and then give the result accordingly.
- Where am I? — this LDQ depends on issue location, it needs location binding in which a LQ answer will be added. We treat this query as a special case. It is both LDQ and LAQ. This query resembles the Relative query example given in Section 11.3.2, but the issue location should be determined for binding. In the other example, the hotel's location is known. That is why we see this as a special case.
- Find closest restaurants and hotels — this is a Compound LDQ involving two relations, it may need fragmentation if the hotel and restaurant data are stored in different content providers. The example is from [DH95] and its process is similar to the Example 11.4. The result might be merged according to the user's needs.

- Find all cars ahead of my car in five miles — this is another Compound LDQ that needs a series of location bindings for all cars ahead and the query issuer's place.
- Find the shortest route to the hospital — this LDQ and Spatial Query mixture need location binding, which will give the initial position with a LQ. A navigation software component might be needed for directions [IB93].

11.3.4 Moving Object Database Queries

A widely studied type of queries in mobile computing is for moving objects. We have also examined the moving objects query processing from our preprocessing perspective. MOD Queries are queries that are processed on the so-called location databases that would contain frequently changing mobile objects' data. These are the queries whose answer depends not only on the database contents (location), but also on the time at which the query is asked [SWCD97]. Direction of movement and speed are considered for modeling the MODs, which may be treated as a special kind of spatiotemporal database. After a series of queries on location databases, some processing may be required to respond to the application. Some MOD query examples are as follows:

- Find the hotels that I will reach within five minutes [SWCD97] — this query also needs a location binding for a five minute period. This query is called an Instantaneous query. A projection from time to space has to be done to find the hotels within five minutes proximity.
- Find ATMs within five miles of where I will be in five minutes [SWCD97] — this query is a Prediction Query that is instantaneous with a five mile distance. It needs location binding for the present and future location.
- Find Chinese restaurants within five minutes on my path [SWCD97] — this query is a Continuous MOD query, which needs continuous location binding depending on time, direction, and speed, and the result is continuously updated. Ren and Dunham [RD00] also used the continuous query idea for querying more static data.
- Retrieve the objects whose speed in the direction of the X-axis doubles within ten minutes [SWCD97] — this Persistent query needs binding for every t seconds, and needs to keep the history in the given time period.

All the above MOD queries can be viewed as special types of LDQs or as multiple instances of LDQs. We thus view our LDQ as a building block from which the higher level MOD queries can be created.

11.3.5 Query Classification

We classify and summarize the relationships of all the query types we have discussed so far in Table 11.2. In the table, S/M denotes whether the object

Table 11.2. Summary of Relationships of Query Types

Query Type	Operators Used	Time Involved	Object Asked	Issue Location	Location Attribute	Query Result
NLR-Q	Nonspatial	No	S/M	N/A	No	Nonspatial data
LAQ	LR-Operator	Maybe	S/M	N/A	Yes	Location or LDD
LDQ	LR-Operator	Maybe	S/M	Yes (here)	No (hidden)	Location or LDD
LQ	Equal	Maybe	S/M	N/A	Yes	Location
MODQ	LR-Operator	Yes	M	N/A	Yes	Location
Spatial	Spatial Operator	Maybe	S/M	N/A	Yes	Spatial or nonspatial data
Spatio-temporal	LR-Operator	Yes	S/M	N/A	Yes	Spatial or nonspatial data

asked is Stationary or Mobile. Temporal properties are also included. LR-Operator set includes all Spatial Operators and perhaps more complex ones that would contain time and movement direction.[1]

11.4 Query Translation Steps in LDQ Processing

We envision the processing of a LDQ by translating it into an appropriate format that is suitable for processing at the content provider site. The precise translation to be performed depends on the type of query submitted at the mobile unit. Here, we assume that the content provider site stores data in a traditional relational database format.

Thus, a LDQ goes through several stages prior to being processed, such as:

```
Find the closest hotels →
Find the hotels within five miles →
Find the hotels within five mile radius of CellID = 3. →
Find the hotels with Zipcode = 75205 or Zipcode = 75206.
```

These stages may be different depending on the granularity of the bound location and the location granularity of the database. Moreover, this view of location dependency implicitly indicates a translation process and an implementation approach. The major functionality of the pre/post processing framework should include the following:

- Determine the validity (and type) of the query and request the location service to provide a location to which the LDQ is to be bound.
- Find the appropriate level of the location granularity, which will be used by the target content provider, and convert the query into the correct format.
- If needed, decompose the query for different servers and send each to the target server for processing.
- Receive the query results and perform any needed filtering to reduce the result set size.
- Combine results from multiple servers and put into a format desired by user.

Hence, a middleware can be used to perform the translation — to preprocess the location-dependent queries. We then investigate the previous architectures and propose a middleware to realize this LDQ processing model next.

Acknowledgments

This material is based upon work supported by the National Science Foundation under Grant no. IIS-9979458. This work is based on an earlier work: "Location Dependent Query Processing," in *Proceedings of the Second ACM International Workshop on Data Engineering for Wireless and Mobile Access (MobiDE)*, 2001, © ACM, http://doi.acm.org/10.1145/376868.376895.

Note

1. Some examples are straight ahead, within, closest, nearby, in, which means LR-Operators ⊂ Spatial Operators.

References

[AS91] W.G. Aref and H. Samet. Extending a dbms with spatial operations. In *Proceedings of Advances in Spatial Databases, 2nd Int. Symposium, SSD'91, Lecture Notes in Computer Science,* vol. 525, pp. 299–318. New York: Springer-Verlag, 1991.

[CDF00] E. Clementini and P. Di Felice. Spatial operators. *SIGMOD Record*, vol. 29, no. 3, pp. 31–38, September 2000.

[DH95] M.H. Dunham and A. Helal. Mobile computing and databases: Anything new? *SIGMOD Record*, vol. 24, no. 4, pp. 5–9, December 1995.

[DK98] M.H. Dunham and V. Kumar. Location dependent data and its management in mobile databases. In R. Wagner, Ed., *Proceedings of 9th International Workshop on Database and Expert System Applications, DEXA '98,* Vienna, Austria, August 1998, pp. 414–419, Piscataway, NJ: IEEE Computer Society, 1998.

[Dun02] M.H. Dunham. *Data Mining Introductory and Advanced Topics.* Upper Saddle River, NJ: Pearson Education Inc., Prentice Hall, 2002.

[EG03] ESRI and GIS.COM. About GIS: How GIS works, using geographic data. http://www.gis.com/data/usingdata/index.html, 2003.

[Ege94] M.J. Egenhofer. Spatial SQL: A query and presentation language. *IEEE Transactions on Knowledge and Data Engineering*, vol. 6, no. 1, pp. 86–95, February 1994.

[FZ94] G.H. Forman and J. Zahorjan. The challenges of mobile computing. *Computer,* pp. 38–47, April 1994.

[GBE+00] R.H. Guting, M.H. Bohlen, M. Erwig, C.S. Jensen, N.A. Lorentzos, M. Schneider, and M. Vazirgiannis. A foundation for representing and querying moving objects. *ACM Transactions on Database Systems (TODS),* vol. 25, no. 1, pp. 1–42, March 2000.

[GG98] V. Gaede and O. Gunther. Multidimensional access methods. *ACM Computing Surveys,* vol. 30, no. 2, pp. 170–231, 1998.

[GU00] G. Gok and O. Ulusoy. Transmission of continuous query results in mobile computing systems. *Information Sciences,* vol. 125, no. 1–4, pp. 37–63, 2000.

[IB93] T. Imielinski and B.R. Badrinath. Data management for mobile computing. *SIGMOD Record,* vol. 22, no. 1, pp. 34–39, March 1993.

[KD98]V. Kumar and M.H. Dunham. Defining location data dependency, transaction mobility, and commitment. Technical Report 98-CSE-01, Southern Methodist University, Dallas, 1998.

[Les00] P. Lesyk. Location dip: The location dip is a common term used within the LBS infrastructure community. It is not necessarily the official term, just the common term. Personal communication, LBS discussion list, e-mail, pleysk@wysdom.com, 2000.

[LLXZ02] D.K. Lee, W-C. Lee, J. Xu, and B. Zheng. Data management in location-dependent information services: Challenges and issues. *IEEE Transactions on Pervasive Computing,* 2002.

[Maa98] H. Maass. Location-aware mobile applications based on directory services. ACM *Baltzer Journal on Mobile Networks and Applications (MONET),* vol. 3, pp. 157–173, 1998.

[Mai83] D. Maier. *The Theory of Relational Databases.* Computer Science Press Inc., Rockville, MD, 1983.

[NW97] J. Nievergelt and P. Widmayer. Spatial data structures: Concepts and design choices. In M. van Kreveld, J. Nievergelt, T. Ross, and P. Widmayer, Eds., *Algorithmic Foundations of Geographic Information Systems, Lecture Notes in Computer Science,* vol. 1340, Ch. 6, pp. 153–197. New York: Springer-Verlag, 1997.

[OSDM89] B.C. Ooi, R. Sacks-Davis, and K.J. McDonell. Extending a dbms for geographic applications. In *Proceedings of 5th International Conference on Data Engineering (ICDE '89),* pp. 590–597, Los Angeles, February 1989. Piscataway, NJ: IEEE Computer Society, 1989.

[PB94] E. Pitoura and B. Bhargava. Building information systems for mobile environments. In *Proceedings of the 3rd International Conference on Information and Knowledge Management, CIKM'94,* Gathesburg, MD, November 1994. pp. 371–378, New York: ACM, 1994.

[PS01] E. Pitoura and G. Samaras. Locating objects in mobile computing. *IEEE Transactions on Knowledge and Data Engineering,* vol. 13, no. 4, pp. 571–591, July/August 2001.

[PTJ01] R. Price, N. Tryfona, and C.S. Jensen. Modeling topological constraints in spatial part–whole relationships. In *Conceptual Modeling — ER 2001, 20th International Conference, ER 2001,* Yokohama, Japan, November 27–30, 2001, *Lecture Notes in Computer Science,* vol. 2224, pp. 27–40, New York: Springer-Verlag, 2001.

[RD00] Q. Ren and M.H. Dunham. Using semantic caching to manage location dependent data in mobile computing. In *Proceedings of the 6th Annual International Conference on Mobile Computing and Networking, MobiCom 2000,* Boston, August 2000. pp. 210–221, ACM SIGMOBILE.

[Ren00] Qun Ren. Semantic caching in mobile computing. PhD thesis, Southern Methodist University, Computer Science and Engineering, Dallas, February 2000.

[SA95] H. Samet and W.G. Aref. Spatial data models and query processing. In W. Kim, Ed., *Modern Database Systems — The Object Model, Interoperability and Beyond.* New York: ACM Press, 1995.

[SDK01] A.Y. Seydim, M.H. Dunham, and V. Kumar. Location dependent query processing. In S. Banerjee, P.K. Chrysanthis, and E. Pitoura, Eds., *Proceedings of the 2nd ACM International Workshop on Data Engineering for Mobile and Wireless Access, MobiDE'01,* Santa Barbara, CA, pp. 47–53, May 2001.

[SWCD97] A.P. Sistla, O. Wolfson, S. Chamberlain, and S. Dao. Modeling and querying moving objects. In *Proceedings of the 13th International Conference on Data Engineering (ICDE'97),* Birmingham, UK, April 1997, pp. 422–432, Piscataway, NJ: IEEE Computer Society, 1977.

[The98] The OpenGIS Consortium. OpenGIS Simple Features Specification for SQL Revision 1.1, Technical report, OpenGIS, http://www.opengis.org/docs/99-049.pdf, May 5, 1999.

[WXCJ98] O. Wolfson, B. Xu, S. Chamberlain, and L. Jiang. Moving objects databases: Issues and solutions. In *International Conference on Scientific and Statistical Database Management, (SSDBM'98),* Capri, Italy, July 1998, pp. 111–122.

[XK00] J. Xu and D.K. Lee. Querying location-dependent data in wireless cellular environment. In *Proceedings of the W3C and WAP Workshop on Position Dependent Information Services,* France, February 15–16, 2000.

[ZL01] B. Zheng and D.L. Lee. Processing location-dependent queries in a multi-cell wireless environment. In S. Banerjee, P.K. Chrysanthis, and E. Pitoura, Eds., *Proceedings of the 2nd ACM International Workshop on Data Engineering for Mobile and Wireless Access, MobiDE'01,* Santa Barbara, CA, May 2001, pp. 54–58.

Chapter 12
Simulation Models and Tool for Mobile Location-Dependent Information Access

Uwe Kubach, Christian Becker, Illya Stepanov, and Jing Tian

Abstract

Simulating the information accesses of mobile users in location-based services or information systems is more complex than simulating those of nonmobile users in standard information systems. Additional factors, which affect the users' behavior, have to be taken into account. The two most important factors are the users' mobility and the location-dependency of the information access.

In this chapter, we will consider these two factors in detail and introduce the most common techniques to model them. We also show how these techniques can be integrated into a more generic and flexible mobility meta-model. A description of a simulation tool that we developed on the basis of this meta-model concludes the chapter.

12.1 Introduction

Location-based information systems are considered to have a huge market potential. Hence, it will become more and more important to develop new algorithms and solutions, which are especially designed for location-based systems. As with any other system, simulation is an important tool to evaluate new algorithms and solutions.

Simulating the information accesses of mobile users in location-based services or information systems is more complex than simulating those of nonmobile users in standard information systems. The reason is that additional factors, which affect the users' behavior, have to be taken into account. The two most important factors are the users' mobility and the location-dependency of the information access.

To model the users' mobility, we first have to agree on what a location is and what spatial relationships are possible between these locations; in other words, we first have to determine the spatial model that we want to build upon. Such spatial models will be discussed in Section 12.2. Based on a thorough understanding of spatial modeling, we can start looking into the mobility models. We will give an overview of existing models and show how the various aspects of these models can be integrated into a generic and flexible mobility meta-model.

Additionally, the location-dependency of the users' information access is an important aspect when modeling the access to location-based information systems. A certain information item might be of a high interest at a certain location, but be useless at another location, for example, a Web page with a specific part of a city map. Such a Web page will be preferably accessed in the area that is shown on the map, but it will be of less interest in all other areas. In Section 12.4, we show how this location-dependency can be modeled in a feasible way.

In Section 12.5, we describe a simulation tool that we developed based on our generic mobility meta-model. We conclude this chapter in Section 12.6.

12.2 Spatial Model

Location models are crucial for mobile applications. First, they provide a model of the reality where mobile objects exist. The mobility of mobile objects is constrained by the spatial structure reflected in the location model (e.g., streets or floor plans of a building). This is important for the practical application as well as for the evaluation of such systems in simulation environments. Second, the information is often assumed to have a spatial structure. This affect is typically influenced by the spatial scope of an information entity. Mobile users are typically interested in information about their current spatial vicinity. To support this locality of information, a notion for locations and their vicinity is required. Location models provide a foundation for these requirements.

We will first discuss different location models based on the underlying properties of the coordinate system before we present the extension of location models to spatial models allowing for location-dependent queries (i.e., object positions, objects in ranges, and nearest neighbors). Spatial models structure information not only with respect to an identity (e.g., primary key), but also along the spatial relation of objects. Thus, location models are an integral part of spatial models.

12.2.1 Location Models

Following the definitions in [1, 2], we define a location model to be a structure defining spatial relations on locations given by a set of one or more coordinates. Coordinates typically are related with a given positioning system, which provides information about the current position of a mobile object. Although coordinates, which do not relate to a given positioning technology, are conceivable, we will further rely on coordinates related with a positioning system.

In general, two kinds of coordinates can be distinguished. Geographic coordinates refer to a point in a metric space, which determines valid combinations, such as two-dimensional coordinates in the Universal Transverse Mercator (UTM) grid or coordinate triples in World Geodetic System 1984 (WGS 84) used by a global positioning system (GPS). The reference system of a geographic coordinate system can be local (e.g., when a room is equipped with a high precision indoor positioning system and coordinates are relative to a corner of the room) or global in the case of WGS 84. Mapping from local to global coordinate systems is possible and eases the reasoning about spatial relations [3]. Canonically, geographic coordinates allow for spatial reasoning because a distance function is present. Thus, nearest neighbors to mobile objects can be easily determined. Queries for objects within a distinct range can also be processed by including a coordinate in a given geometric object. Although geometric coordinates are quite common in the outdoor domain, due to the widespread use of GPS, the indoor domain, and distinct technologies, such as cell phone IDs, do not provide geographic coordinates, but only symbolic coordinates.

Symbolic coordinates only refer to a location without any predefined relation to other locations. A variety of positioning systems exist that can only provide support for symbolic coordinates. For an overview, see [4]. Examples are cell IDs in cellular phone networks, infrared beacons, or WLAN access point IDs in the indoor domain. To support range queries and nearest neighbor queries, an explicit location model is required. Although the distance function of geometric coordinates implicitly defines a location model, this requires some effort for symbolic coordinates. Supporting range queries leads to spatial inclusions. A location model must support a function, which determines two given locations — l1 and l2, whether l1 is contained in l2 or not. Approaches to support these spatial inclusions are typically based on hierarchical structures, such as location trees [2, 3] or lattices [1]. Location trees only support the direct inclusion of locations. For any location l1 there is at most one direct successor l2 with l1 contained in l2. For many spatial relations, as we experience them in our daily life, a location tree can be too restrictive. Consider a building, which is modeled in floors and wings where floors cross-cut wings. Clearly, any room on a floor is contained in the floor as well as in a wing of the building. A lattice is a more general concept than a tree allowing more than one inclusion of a location (e.g., a room being part of a floor as well as part of a wing).

Based on such an explicit model of the spatial inclusions, range queries can be efficiently processed. However, there is no support for the notion of "near" required for nearest neighbor queries provided by such models. A straightforward solution to this requirement is to provide a graphical structure that connects locations with weighted edges representing the distance. The distance between two locations can then be evaluated to the sum of the edge weights. Neglecting the weights, a simple heuristic could only take the number of edges into account leading to a less appropriate model of the reality. An example of such an approach can be found in [2]. In summary, both symbolic and geographic coordinates are present and have to be considered in mobile computing settings. The modeling effort of location models for symbolic coordinates is mainly determined by the required accuracy with respect to the modeled reality. Hybrid location models additionally annotate the geographic extension of locations to the location nodes in the location model. This can be used to embed a location model into a global reference system [3] or to evaluate the distance function on symbolic coordinates.

12.2.2 Spatial Information Models

Information shared between applications in mobile computing settings is typically related to a location. Users are interested in information, which is related to their physical environment. The location of users is modeled with respect to a given location model. To allow the access of information with spatial relations (i.e., information that is concerned with a location "near" to a user) requires the information to be tagged with a location where it is valid or to which it is related. The location model thus defines the structure of a spatial information model — or spatial model for short — in which mobile objects and information objects as well are stored.

Figure 12.1 provides an example for a spatial model, which contains information about objects. Queries to such spatial models can either concern information objects via their unique identifier and thus act like a common database or refer to the location of an object. Queries that relate to information retrieval and depend on a distinct location are range or nearest neighbor queries. To process queries related to the position of objects, information of a spatial model is needed. Geometric models, such as the GPS, commonly provide an implicit model via a distance function on the coordinates. Symbolic coordinates, as described above, lack such an implicit model and require a location model. Figure 12.1 sketches a location tree where the object Plant-5 is referring to Room 2.326 as its location.

To process such queries, an additional index structure for spatial models concerning the location is required. This index structure can be

Simulation Models and Tool

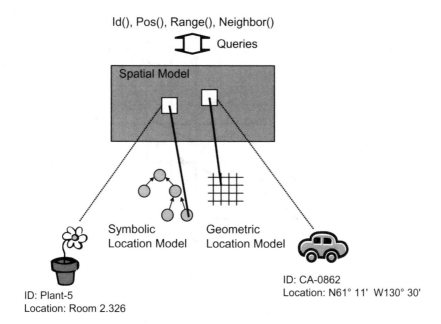

Figure 12.1. Spatial versus Location Models

obtained by using location models. Moving objects databases [5] or some context-aware platforms (e.g., Nexus [6]) or applications (e.g., GUIDE [7]) are examples of spatial models and their application.

The increasing importance of location-based services and their evaluation will lead to a variety of spatial models in use. To evaluate such services, the entire spectrum of performance related parameters has been addressed. Examples are user mobility, access patterns to information objects, and relationships between information objects. These parameters also depend on the location where the users are or the location information objects are related to and thus require a location model for the evaluation in simulation.

12.3 Mobility

Like location and spatial models, the user mobility also plays an important role for the simulation of mobile systems. A large number of approaches of mobility modeling have been proposed in the last years. In this chapter, we review commonly used mobility models and explain the drawbacks. Finally, we introduce a generic mobility meta-model that overcomes these drawbacks.

12.3.1 Existing Mobility Models

A variety of mobility models have been proposed for simulations of mobile systems. There are two types of mobility models — traces and synthetic

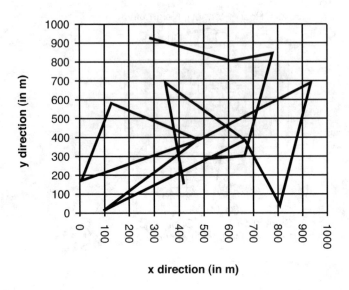

Figure 12.2. Random Walk Movement Pattern

models. Traces are generated by observing real-life systems. Thus, they provide real information about the movement of mobile users [8]. This kind of information is often stored in origin-destination matrices [9]. However, traces are difficult to get due to the high cost for data collection, as well as user privacy issues. Thus, synthetic mobility models are widely used in mobile system simulations [10–12]. In this chapter, we focus on synthetic models. Within synthetic mobility models, we further distinguish between random models and advanced models. In contrast to random mobility models, where the movement of nodes is random, advanced mobility models restrict nodes movement taking into account real-world restrictions like the infrastructure (e.g., roads, railways) the nodes can move on.

12.3.1.1 Random Mobility Models. Because of their simplicity, random mobility models are quite popular for mobility simulation. Some of the most commonly used models are:

- Random Walk Model [11] — each node initially selects a random direction θ between [θ_{min}, θ_{max}] and a random speed V between [V_{min}, V_{max}] and then moves at the speed V in the direction θ for a random period of time. At the end of this period, the node repeats this process. The Random Walk Model is simple, but does not present realistic movement behaviors due to sharp turns and sudden stops. Figure 12.2 represents a movement pattern of a mobile node using a Random Walk Model.

Simulation Models and Tool

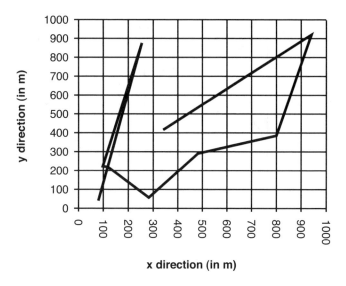

Figure 12.3. Random Waypoint Movement Pattern

- Random Waypoint Model [13] — one of the most widely used models for mobile ad hoc networks. In this model, each node selects a random point D within the simulation area as its destination point and a random speed V between $[V_{min}, V_{max}]$. The node then moves to the destination D at the speed V. When the node reaches the destination, it makes a pause for time T between $[T_{min}, T_{max}]$. After the pause, it selects a new destination and a new speed to continue its movement. However, it has been shown that the nodes tend to concentrate in the center of the simulation area, resulting in a non-uniform spatial distribution of nodes in the network. Figure 12.3 represents a movement pattern of a mobile node using the Random Waypoint Model.
- Random Direction Model [14] — similar to the Random Walk Model. The difference is that after the selection of movement direction and speed, each node continues moving until it reaches the boundary of the simulation area. It then selects a new direction in which to move. In contrast to the Random Waypoint Model, the Random Direction Model generates a constant density of nodes throughout the simulation. Figure 12.4 represents a movement pattern of a mobile node moving according to the Random Direction Model.

Random mobility models are simple and easy to implement. However, they are not always suitable to model real-world mobile application scenarios. In the real-world, people do not move randomly, but tend to select a specific destination (e.g., a room, a park, or a restaurant) and follow a

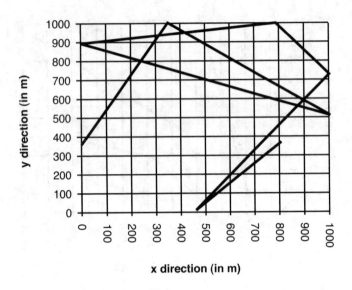

Figure 12.4. Random Direction Movement Pattern

well-defined path (like the corridor or the walking street) to reach that destination. Thus, more realistic mobility models are often needed to simulate real-world scenarios.

12.3.1.2 Advanced Models. In this section, we describe some advanced models that provide more realistic movement patterns:

- Graph-Based Mobility Model [15] — relies on a spatial model graph that describes the underlying spatial infrastructure. The vertices of the graph represent places (e.g., points of interest such as restaurants, museums, etc.) that users might visit and the edges model interconnections between the places (streets or rail connections). The users move between randomly chosen vertices of the graph on edges, thus respecting spatial constraints of the simulation area. The visit of vertices can also be determined based on a probability value that is assigned to each vertex in the graph. The Graph-Based Mobility Model can also be extended by taking into account a trip model, which describes a whole trip of a mobile node based on a sequence of vertices that are to be visited.
- Obstacle Mobility Model [16] — takes obstacles into account that might be located in the simulation area. Such obstacles can be described by polygons. The obstacles not only prevent movement of nodes, but also block wireless transmissions. The Voronoi diagram of the obstacle corners are used to generate a movement path, which is a planar graph whose edges are line segments that are equidistant from two obstacle corners. A Voronoi diagram partitions a plane with

Simulation Models and Tool

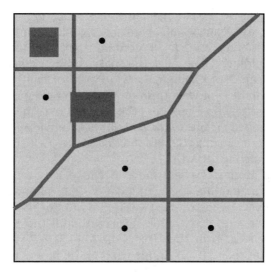

Figure 12.5. Example for Obstacle Mobility Model

n sites into n regions such that each region contains exactly one site and every point in the region around that site is closer to that site than any other sites. The node then moves between randomly selected obstacles following the shortest path in the Voronoi diagram. Figure 12.5 gives an example of the Obstacle Mobility Model with two polygons presenting buildings and a road path between them based on the Voronoi diagram.

12.3.2 Generic Mobility Model

As shown above, there exists a variety of approaches to model the mobility of users. Each of the approaches satisfies only a specific set of scenarios. For example, the Random Waypoint Mobility Model may be sufficient to simulate movement of pedestrians in an open area (e.g., rescue missions or military operations). To model movement of users in a city, it is necessary to reflect spatial area constraints (e.g., streets) using the graph-based model. To simulate movement of cars, it is necessary to include car-specific speed and direction changing dynamics to the model. The state of the art in the area of mobility modeling is the use of specific models for specific scenarios. To evaluate new scenarios, particular mobility models must be created. To minimize the model creation overhead, there is a demand to have a single meta-model, which is generic, to satisfy different scenarios and to express the existing mobility models.

A review of the existing mobility modeling approaches shows the following three major factors affecting the user movement:

1. Spatial Model — user movement takes place in a given simulation area. The area can be an open rectangular-bounded area used in random mobility models or contain spatial objects (e.g., roads, streets, buildings). If present, the spatial objects constrain the movement of users. For example, cars do not normally leave roads and mobile users do not walk through walls. To reflect this, the mobility model must rely on a model of the spatial area. The spatial area can be described using one of the standards for environment description in digital form, such as Geographic Data Files (GDF) [17] or Geometry Markup Language (GML) [18].
2. User Trip Model — users move because they want to execute a particular action in a specified location (e.g., shopping, work, sightseeing). By moving between various points of the spatial environment during the observation, users execute trip sequences. Such trip sequences form the User Trip Model. Activity-based Travel Demand Approach [9, 19, 20] is used in the model to express travel decisions. The approach defines travel as the demand to participate in activities. An activity denotes an action to be executed (e.g., shopping). The activity has an associated set of locations (e.g., shopping can be performed in supermarkets; sightseeing can be performed in museums). Each location has a relative probability that the location will be chosen as the activity execution point (attractiveness) and duration of activity execution (e.g., 30 minutes for shopping). The Trip Model interacts with the Spatial Model and matches the activities with the locations using geographic coordinates or semantic of topological elements (e.g., element class code identifier). The activities are sequences and combined into trip chains (Figure 12.6). After finishing the current activity, the next one is chosen with a certain probability and the movement continues. The movement path reflects the spatial environment constraints provided by the Spatial Model.
3. Movement Dynamics Model — different classes of mobile objects expose different movement dynamics. For example, pedestrians tend to move with low speed and frequent interruptions, but cars move with higher speeds and influence movement dynamics of neighboring vehicles. The Movement Dynamics Model defines the physics of user movement (e.g., speed and direction changes). Many approaches exist to describe movement dynamics for different classes of mobile users (e.g., [21–24]).

These three models are fundamental and are not interchangeable, because they define different aspects of user movement and have different parameters. A parameter for the Spatial Model is a description of the movement area containing topological elements (e.g., roads and buildings). A parameter for the Trip Model is a trip chain with supplementary data (e.g.,

Simulation Models and Tool

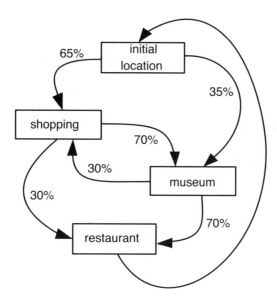

Figure 12.6. A Trip Chain

locations, their relative probabilities, and durations of activity executions). Parameters for the Movement Dynamics Model are implied by the physical model chosen to describe the user movement dynamics. The combination of the three models forms the meta-model for mobility modeling (Figure 12.7). The model is generic, because it reflects the main factors affecting user movement and can express the reviewed mobility models as its instances. For example, the Random Waypoint Mobility Model can be constructed using a rectangular-bound area in the Spatial Model, the constant-speed motion in the Movement Dynamics Model, and performing trip sequences between randomly chosen points of the area in the User Trip Model. The Graph-Based Mobility Model is similar to the Random Waypoint Mobility Model, but uses a spatial environment graph in the Spatial Model. More sophisticated mobility models (e.g., Integrated Mobility Model) rely on digital maps in the Spatial Model and appropriate trip sequences and user movement dynamics.

The described mobility meta-model is used in our simulation tool to evaluate performance of location-dependent information access (see Section 12.5).

12.4 Information Access Model

One of the major differences between a location-based information system, usually accessed by mobile users, and a standard information system is that within a location-based system the user's interest in information items might

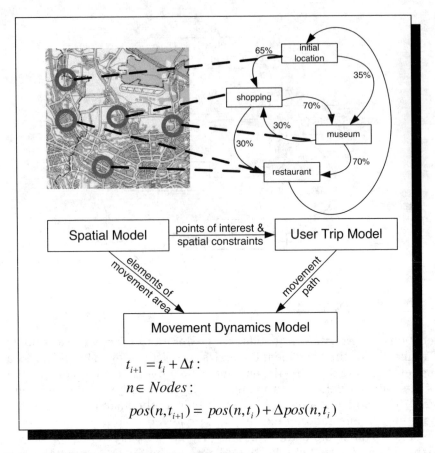

Figure 12.7. Structure of Mobility Meta-Model

change when he moves from one location to another. The aforementioned example of accessing a city map from a mobile device makes this obvious. In general, a user has a much higher interest in parts of the map which show his proximity, than in parts which reflect more distant areas.

After explaining some fundamentals about the so-called Zipf distribution in Section 12.4.1, we will show in Section 12.4.2 how such a location-dependent information access can be modeled.

12.4.1 Zipf Distribution

Thus far, a lot of work has been done on modeling information access in the World Wide Web. In many articles it is assumed [25–27] that access to a certain information space, for example the information stored on a Web server is Zipf distributed. In such a distribution, the relative probability of a request for the ith most popular Web page or more generally speaking the ith most popular information item is proportional to 1/i.

Exhaustive analyses of various Web server logs [28] have shown that the distribution of the requests over single Web pages follows a Zipf-like distribution. A Zipf-like distribution differs from the original Zipf distribution in a small correction factor α. Here, the relative probability of a request of the ith most popular page is proportional to $1/i^\alpha$. The observed value of α varies between the different considered logs, ranging from 0.64 to 0.83.

Using data collected in the GUIDE project [7], we were able to verify that this Zipf-like distribution is also valid in a mobile guide scenario. The access frequencies that have been observed at each single location have been Zipf-like distributed.

12.4.2 Location-Dependent Access

As described above, the Zipf-like distribution can be used to model the information access at a certain location. However, the Zipf-like distribution only applies to the information items that are accessed at the considered location. The Zipf-like distribution will not reflect if certain parts of the overall information space are not accessed at a specific location. However, as our analysis of the GUIDE data has shown, this is the case in real-world applications.

For example, only 10 percent of all information items within a certain information space might be accessed at a certain location. To model the access frequencies to these 10 percent of all available items, the Zipf-like distribution is the best choice. However, we additionally require a method to describe which items are accessed at least once at each location.

One simple model that meets the reality surprisingly well, at least as far as we could verify it with data available from the Guide project, is the following: For each location, for example, represented by a symbolic name or a geometric area, we assign a set of information items that are exclusively accessed there. In addition to these local information pools, we also have a global information pool, which contains information items that might be accessed from any location. A parameter controls which part of a user's information requests refers to the global information pool and which part refers to the local information pool of the location where he is currently located. This makes it possible to simulate the behavior of a widespread spectrum of location-dependent information systems beginning with location-dependent applications like map applications and ending with almost location-independent applications like wireless Web browsing. Of course more advanced models are possible, such as where the access probability of a certain information object itself is Gaussian distributed over the locations.

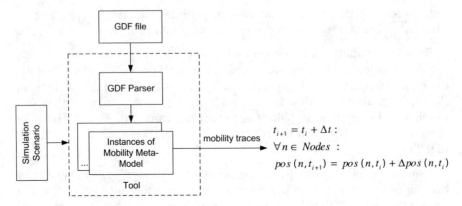

Figure 12.8. Tool Architecture

For more details on the characteristics of information spaces see [29]. There, we formally defined the most important characteristics of information spaces and measures in order to quantify them. In [30] we used the above sketched access model to evaluate a location-aware hoarding mechanism. This hoarding mechanism intelligently prefetches information that the user will probably access in the near future and can thereby significantly reduce the response time of a location-based information system.

12.5 A Tool for User Mobility Modeling

The generic approach to user mobility modeling that we described in Section 12.3 is implemented for the practical usage in a simulation tool. This chapter describes the tool design and implementation issues.

12.5.1 Objectives

The main objective of the tool is to simulate user mobility in different scenarios. The tool must be flexible enough to support a variety of scenarios from random movement in an open area to predetermined travel in a city center. The tool must provide means to easily define the scenarios and include the necessary implementations. To simplify handling of the spatial data, the tool must be capable of reading the contents of geographic data files. The tool must support existing simulation environments for mobile networks and produce mobility traces in corresponding formats.

12.5.2 Software Architecture

To handle a variety of scenarios, the tool implements the described mobility meta-model, thus integrating the spatial environment (Spatial Model), user trip sequences (User Trip Model), and user movement dynamics (Movement Dynamics Model) (Figure 12.8).

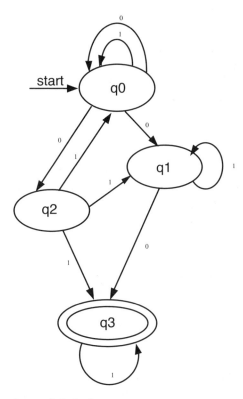

Figure 12.9. A Nondeterministic Automaton

The Spatial Model provides a map of the simulation area with its topological elements (e.g., roads, museums, restaurants, cinemas). The model uses GDF as its primary specification, because the standard explicitly denotes rules to define a variety of spatial area elements, their geometry and properties, thus easing the model interface description. To read the spatial data from a file source and to initialize the model, the tool includes a GDF parser. Converters from other geographic data formats (e.g., GML) to GDF are available [31].

The User Trip Model is an implementation of the Activity-Based Travel Demand Modeling Approach. To express trip sequences in the model, we use a nondeterministic automaton of activity sequences (Figure 12.9) [32]. In the automaton, the states denote activities to be executed. Each state contains a set of locations for activity execution, their attractiveness, and a duration of the activity execution (like pause time in the Random Waypoint Mobility Model). The automaton switches nondeterministically between the states, thus reflecting the variability factor upon selecting the next activity. The user's initial activity (associated with the user initial location) corresponds to the automaton's initial state. The final activities correspond to the automaton's final states.

The Movement Dynamics Model simulates the movement of a user (its speed and direction changes) from its current location to the target point of activity execution. The movement path between the points is calculated using the Dijkstra shortest-path algorithm taking obstructions of simulation area into account.

Using the tool, movements of every user are modeled independently with a separate instance of the meta-model. This allows mobile users with similar and dissimilar Trip or Movement Dynamics Models to coexist in a simulation. To model movements of users of the same class (e.g., belonging to the same group), it is possible to share a concrete instance of Trip or Dynamics Model by a number of instances of the meta-model, thus achieving the desired movement similarity.

To model various scenarios, the tool includes ready implementations of several User Trip Models (which differ in a degree of randomization of trip chain) and Movement Dynamics Models (for cars and pedestrians). The tool can produce mobility traces for commonly used mobile network simulation environments (e.g., Network Simulator 2 [NS-2] [33] and GloMoSim [Global Mobile Information Systems Simulation Library] [34]).

The tool is implemented on top of a discrete Java™-based simulation environment. The implementation is based on the concept of pluggable modules and thus, can be easily extended with implementations of new User Trip or Movement Dynamics Models, parsers for geographic data in other formats or simulation environments support.

12.5.3 Usage

The tool simulates user mobility and produces traces in accordance with a simulation scenario. The simulation scenario is defined in Extensible Markup Language (XML) format.

A skeleton of a simulation scenario definition for the tool is depicted in Table 12.1. The scenario description is enclosed by the `universe` element. The description includes modules used in the simulation (`extension` elements) and definitions of the meta-model components for mobile users (`node` element for a single user, `nodegroup` element for a group of users).

For example, the scenario in Table 12.2 is used to model movements of users in a rescue mission in an opened area similar to the Random Waypoint Mobility Model. The dimensions of simulation area are 1000 meters (m) by 1000 m. The mobility traces are produced in text format. Simulation time is 3600 seconds (s). The users are randomly placed in the simulation area and move between randomly chosen points of the simulation area.

Simulation Models and Tool

Table 12.1. Format of Simulation Scenario Description

```
<?xml version="1.0?">
<universe>
  <extension>extension_parameters</extension>
  <extension>extension_parameters</extension>
  <extension>extension_parameters</extension>
  ...
  <node>node_parameters</node>
  <nodegroup>nodegroup_parameters</nodegroup>
  ...
</universe>
```

Table 12.2. Simulation Scenario 1

```
<?xml version="1.0"?>
<!-- Users in a Rescue Mission -->
<universe>
  <dimx>1000.0</dimx>
  <dimy>1000.0</dimy>
  <extension class="sim.extensions.TextOutput"/>
  <extension class="sim.simulations.TimeSimulation"
        param="3600.0"/>
  <extension class="spatialmodel.core.SpatialModel"/>
  <extension name="PosGen" class="tripmodel.generators.
        RandomInitialPositionGenerator"/>
  <extension name="TripGen" class="tripmodel.generators.
        RandomTripGenerator">
    <minstay>120.0</minstay> <maxstay>600.0</maxstay>
  </extension>
  <nodegroup n="50">
    <extension class="uomm.ConstantSpeedMotion"
        initposgenerator="PosGen" tripgenerator=
        "TripGen">
      <minspeed>0.56</minspeed> <maxspeed>1.39</maxspeed>
    </extension>
  </nodegroup>
</universe>
```

The users make a pause between 120 s and 600 s, between two successive trips. Movements of 50 mobile users are simulated. Users move with constant speed movement dynamics. The speed value is chosen randomly between 0.56 m/s and 1.39 m/s.

Table 12.3. Simulation Scenario 2

```xml
<?xml version="1.0"?>
<!-- Tourists in a City Center -->
<universe>
  <extension class="canumobisim.extensions.NSOutput"/>
  <extension class="canumobisim.simulations.
        TimeSimulation" param="14400.0"/>
  <extension class="spatialmodel.core.SpatialModel"/>
  <extension class="gdfreader.GDFReader" source="Boston.
        gdf"/>
  <extension name="Gen" class="tripmodel.generators.
        ActivityBasedTripGenerator">
    <activity id="initial">
      <points>initial.txt</points>
      <minstay>0.0</minstay> <maxstay>0.0</maxstay>
    </activity>
    <activity id="shopping">
      <points>shopping.txt</points>
      <minstay>900.0</minstay> <maxstay>1800.0</maxstay>
    </activity>
    <!-More activities … -->
    <transition>
      <src>initial</src> <dest>shopping</dest>
         <p>0.65</p>
    </transition>
    <transition>
      <src>initial</src> <dest>museum</dest> <p>0.35</p>
    </transition>
    <!-More transitions … -->
  </extension>
  <nodegroup n="100">
    <extension class="uomm.ConstantSpeedMotion"
        initposgenerator="Gen" tripgenerator="Gen">
      <minspeed>0.33</minspeed> <maxspeed>0.83</maxspeed>
    </extension>
  </nodegroup>
</universe>
```

The scenario in Table 12.3 is used to model movements of tourists in a city center. The mobility traces are produced for NS-2 simulation environment. Simulation time is 14,400 s. The spatial environment is initialized from a GDF data source. An activity sequence automaton (Figure 12.6) determines trip sequences made by mobile users. Movements of 100 tourists are modeled. Users expose the constant speed movement dynamics with the speed between 0.33 m/s and 0.83 m/s.

The full tool documentation and code can be downloaded free of charge for academic use from http://canu.informatik.uni-stuttgart.de/mobisim.

12.6 Conclusion

In this chapter, we gave an overview of the aspects that need to be considered when simulating the information accesses of mobile users in location-based systems. We described the possibilities of modeling locations and spatial relationships. We also gave an overview of existing mobility models and described a more generic and flexible mobility meta-model.

We depicted an impression of how important it is to reflect the location dependency of the users' information accesses when simulating the access to a location-based information system. In addition, we described a simple method to model this location dependency.

We completed the chapter with a description of our simulation tool, which is available free of charge for academic, noncommercial use.

References

1. F. Dürr and K. Rothermel. On a Location Model for Fine-Grained Geocast. To appear in *Proceedings of the 5th International Conference on Ubiquitous Computing (UbiComp 2003)*, Seattle, WA, October 2003.
2. M. Bauer, C. Becker, and K. Rothermel. Location Models from the Perspective of Context-Aware Applications and Mobile Ad Hoc Networks, *Personal and Ubiquitous Computing*, vol. 6, pp. 322–328, New York: Springer-Verlag.
3. C. Jiang and P. Steenkiste. A Hybrid Location Model with Computable Location Identifier for Ubiquitous Computing, *Proceedings of UbiComp 2002*, Goteborg, Sweden, 2002.
4. J. Hightower and G. Borriello. *Location Systems for Ubiquitous Computing*, Piscataway, NJ: IEEE Computer Society, pp. 57–66, 2001.
5. O. Wolfson, B. Xu, S. Chamberlain, and L. Jiang. Moving Objects Databases: Issues and Solutions. In *Proceedings of the 10th International Conference on Scientific and Statistical Database Management (SSBDM '98)*, Capri, Italy, 1998.
6. F. Hohl, U. Kubach, A. Leonhardi, K. Rothermel, and M. Schwehm. Next Century Challenges: Nexus — An Open Global Infrastructure for Spatial-Aware Applications. In *Proceedings of the 5th Annual International Conference on Mobile Computing and Networking (MobiCom '99)*, pp. 249–255, Seattle, WA, August 1999.
7. K. Cheverst, N. Davies, K. Mitchell, and A. Friday. Experiences of Developing and Deploying a Context-Aware Tourist Guide: The GUIDE Project. In *Proceedings of MOBICOM'2000*, pp. 20–31, Boston: ACM Press, August 2000.
8. D. Tang and M. Baker. Analysis of a Metropolitan-Area Wireless Network. In *Proceedings of the 5th Annual International Conference on Mobile Computing and Networking (MobiCom '99)*, Seattle, WA, August 1999.
9. N. Oppenheim. Urban Travel Demand Modeling: From Individual Choices to General Equilibrium. Hoboken, NJ: Wiley-Interscience, 1995.
10. T. Camp, J. Boleng, and V. Davies. A Survey of Mobility Models for Ad Hoc Network Research. *Wireless Communication and Mobile Computing (WCMC)*, [Special issue on Mobile Ad Hoc Networking: Research, Trends and Applications], vol. 2, no. 5, pp. 483–502, 2002.

11. M. Zonoozi and P. Dassanayake. User Mobility Modeling and Characterization of Mobility Pattern, *IEEE Journal on Selected Areas in Communications*, vol. 15, no. 7, pp. 1239–1252, 1997.
12. J.G. Markoulidakis, G.L. Lyberopoulos, D.F. Tsirkas, and E.D. Sykas. Mobility Modeling in Third-Generation Mobile Telecommunications Systems, *IEEE Personal Communications*, pp. 41–56, 1997.
13. J. Broch, D.A. Maltz, D.B. Johnson, Y.C. Hu, and J. Jetcheva. A Performance Comparison of Multi-Hop Wireless Ad Hoc Network Routing Protocols. In *Proceedings of the ACM/IEEE MobiCom*, October 1998.
14. E. Royer, P.M. Melliar-Smith, and L. Moser. An Analysis of the Optimum Node Density for Ad Hoc Mobile Networks. In *Proceedings of the IEEE International Conference on Communications (ICC)*, 2001.
15. J. Tian, J. Hähner, C. Becker, I. Stepanov, and K. Rothermel. Graph-Based Mobility Model for Mobile Ad Hoc Network Simulation. In *Proceedings of the 35th Annual Simulation Symposium (ANSS)*, 2002.
16. A. Jardosh, E.M. Royer, K.C. Almeroth, and S. Suri. Toward Realistic Mobility Models for Mobile Ad Hoc Networks. In *Proceedings of the ACM Mobicom'03*, 2003.
17. Geographic Data Files (GDF) Home Page. http://www.ertico.com/links/gdf/gdfcon.htm.
18. Geography Markup Language (GML) 2.0, OpenGIS® Implementation Specification, OGC Document Number: 01-029. http://www.opengis.org/docs/01-029.pdf.
19. R. Kitamura. Applications of Models of Activity Behavior for Activity-Based Demand Forecasting. In *Proceedings of Activity-Based Travel Forecasting Conference,* June 2–5, 1996.
20. E.I. Pas. Recent Advances in Activity-Based Travel Demand Modeling, In *Proceedings of Activity-Based Travel Forecasting Conference,* June 2–5, 1996.
21. C. Bettstetter. Smooth is Better than Sharp: A Random Mobility Model for Simulation of Wireless Networks. In *Proceedings of the 4th ACM International Workshop on Modeling, Analysis, and Simulation of Wireless and Mobile Systems (MSWiM'01)*, Rome, July 2001.
22. C.W. Reynolds. Steering Behaviors for Autonomous Characters. In *Proceedings of the Game Developers Conference,* pp. 763–782, 1999.
23. I. Seskar, S.V. Marie, J. Holtzman, and J. Wasserman. Rate of Location Area Updates in Cellular Systems. In *Proceedings of IEEE VTC'92,* Denver, May 1992.
24. M. Treiber and D. Helbing. Explanation of Observed Features of Self-Organization in Traffic Flow, Preprint `cond-mat/9901239` (1999).
25. S. Glassman. A Caching Relay for the World Wide Web. In *Proceedings of the 1st International Conference on the World Wide Web (WWW '94),* Geneva, 1994
26. M.E. Crovella and A. Bestavros. Self-Similarity in World Wide Web Traffic: Evidence and Possible Causes. In *Proceedings of the 1996 ACM SIGMETRICS Conference on Measurement and Modeling of Computer Systems,* Philadelphia, pp. 160–169, 1996.
27. N. Nishikawa, T. Hosokawa, Y. Mori, K. Yoshida, and H. Tsuji. Memory-Based Architecture for Distributed WWW Caching Proxy. In *Proceedings of the 7th International World Wide Web Conference (WWW7),* Brisbane, Australia, pp. 205–214, 1998.
28. L. Breslau, P. Cao, F. Li, G. Philips, and S. Shenker. Web Caching and Zipf-like Distributions: Evidence and Implications. In *Proceedings of the IEEE INFOCOM '99,* New York, pp. 126–134, 1999.
29. U. Kubach and K. Rothermel. Estimating the Benefit of Location-Awareness for Mobile Data Management Mechanisms. In *Proceedings of International Conference on Pervasive Computing (Pervasive 2002),* Zurich, 2002.
30. U. Kubach and K. Rothermel. Exploiting Location Information for Infostation-Based Hoarding. In *Proceedings of the 7th Annual ACM SIGMOBILE International Conference on Mobile Computing and Networking (MobiCom '01),* Rome, pp. 15–27, 2001.
31. Feature Manipulation Engine (FME). http://www.safe.com/products/fme/index.htm.

32. J.E. Hopcroft and J.D. Ullman. *Introduction to Automata Theory, Languages, and Computation*, Boston: Addison-Wesley, 1979.
33. Network Simulator (NS-2). http://www.isi.edu/nsnam/ns/.
34. GloMoSim: Global Mobile Information Systems Simulation Library. http://pcl.cs.ucla.edu/projects/glomosim/.

Chapter 13
Context-Aware Mobile Computing

Rittwik Jana and Yih-Farn Chen

Abstract

This chapter provides a recent survey of context-aware mobile computing. Computers are now ubiquitously present in our everyday lives through mobile and embedded devices. In an effort to merge computing seamlessly with our surrounding environment to enrich the user experience, context-aware computing was born. We discuss the definition of context and how the inclusion of contextual information affects the computing and communication structures and broadens the appeal of an application. We continue on by describing how context can be acquired by capturing real-world situations and consequently formulating a representation of the collected information. Context can be acquired with various types of sensor systems (e.g., location, load), computer vision (i.e., gestures, cues, and activity detection), modeling of users' behaviors (i.e., thoughts, actions, and words), inferences from databases or calendars, or explicit user/device input. Once acquired, we look at how context information can be used for various applications to provide a particular service or information. Applications can use context passively to customize the application delivery or to proactively retrieve and use available context information constantly in a decision process. We then investigate the infrastructure needed to develop context-aware applications hereby named contextware [1] and discuss how it complements standard middleware. We highlight some of the recent initiatives and research projects that are context-related and provide a comprehensive overview of the various facets of context-aware applications. We conclude by predicting the future challenges of this topic and issues that warrant further investigation.

13.1 Introduction and Motivation

Humans interact with each other in an extremely efficient manner, the overriding intent being always to communicate ideas fluently and succinctly. We include information from the current situation or extrapolate from events in time (also known as context or side information) into our conversations to enrich the overall experience. Humans interacting with computers on the other hand, have difficulty including this side information particularly because of the typical modes of inputs that are currently used (keyboard and mouse). There has been a lot of emphasis on research lately to enrich the user experience with more natural inputs like voice, gestures, emotions, and facial expressions via video and situation awareness of the environment captured via a myriad of sensor enabling technologies [23]. Unfortunately, obtaining contextual knowledge is a nontrivial task. Natural voice input requires sophisticated computing engines to interpret speech using automatic speech recognition (ASR) engines and output synthesized speech using state-of-the-art text to speech (TTS) technologies [6]. Regardless of the advent of such novel technologies to enhance human computer interaction, there is still a formidable amount of work that lies ahead for computers to understand humans as accurately and efficiently as humans understand humans. Sensors can be distributed and communicate their data using a variety of protocols and formats. Raw sensor data may need to undergo complex postprocessing before being useful in any applications [7]. The overarching goal in the researchers' minds are to first facilitate computers to understand these complex input processes and second to enhance human computer interaction.

Context can be defined in a number of different ways. In [4], context is defined as any information that can be used in an entity's situation. An entity can include a person, place, or object relevant to the conversation between the end user and the application. Many researchers have also tried to define context by enumerating examples of context namely computing context, physical context, and user contexts [5]. We take a further look at the various definitions of context in Section 13.2.

So why is context so important with regard to mobile computing? Mobile computing poses difficult challenges, such as client resource constraints imposed by width and size and impoverished or fluctuating bandwidth connections. The typical process used on a desktop to specify required input parameters to request a service can be a tedious and frustrating experience for a mobile user. For example, when a cell phone user requests the current weather, the service should figure out where the mobile user is and what device limitation it has and then deliver the information appropriately (in this case, a text message less than 140 characters would be ideal).

By taking into account the relevant context (i.e., the different exposed situations that are relevant to a requested service), application delivery can be streamlined and customized to provide a more satisfying overall user experience. It is however, the responsibility of the application designer to present and use context intelligently to provide such an experience.

Acquiring context is not a trivial task. This is one of the main reasons why it is difficult to use context. Quite often this means dealing with sensor-based technologies that detect various phenomena about the environment. Acquiring context can be either explicit (i.e., provided by an end user by means of textual, voice, or video input) or automatic inference. Apart from the lack of proliferation of this kind of nonstandard technology (e.g., sensors, global positioning receivers), there is also the hesitation of ubiquitous deployment of such technologies due to privacy and other considerations. There needs to be a consensus among public utilities to standardize the rollout of such infrastructure. The recent mandatory E911 initiative by the Federal Communications Commission aims at deploying location sensing technologies in coordination with the cellular carriers to report the telephone number of a wireless 911 carrier and the precise location of the antenna to within 50 to 100 meters resolution [9].

Next we take a look at a particular methodology of modeling context information with reference to a prototype implementation of a service platform that uses this context information to provide value-added services. The notion of sharing contextual information among devices that do not have easy access to this information is interesting enough to warrant an example. This is discussed in Section 13.5.1.

Applications that use context are called context-aware. A system is context-aware if it uses contexts to provide relevant information or services to the user, where relevancy depends on the user's task [4]. Context-awareness has been defined by a number of researchers each trying to categorize its different features. For example, context-aware computing was first discussed by Schilit in [5] and [8] to be a piece of software that adapts according to its location of use, the collection of nearby people and objects, as well as changes to those objects over time. Context-awareness will be elaborated further in Section 13.4.

Section 13.5 and Section 13.6 discuss the service platforms and the more recent context-related projects, respectively, followed by highlights of the future of context-aware mobile computing.

13.2 What Is Context?

Context has been defined by a number of researchers. Mark Weiser's now legendary article defined ubiquitous computing for the first time [10]. Context is a general term and can be categorized into three categories in the realms of mobile computing — computing context, user context, and physical context.

According to [5], the three categories can be described as:

1. Computing context (what resources you have) — such as network connectivity, communication costs, and communication bandwidth and nearby resources such as printers, displays, and workstations
2. User context (whom you are with) — such as the user's profile, location, people nearby, even the current social situation
3. Physical context (where you are) — such as lighting, noise levels, traffic conditions, and temperature

Pascoe et al. [11] defines context as the user's location, environment identity, and time. Dey [12] provided examples of the user's emotional state, location and orientation, date and time, objects and people in the user's environment. Ward et al. [13] regard context as the state of applications' immediate surroundings. Ferscha [1] describes some additional interesting and useful categories and subcategories of context:

- Geographical contexts (e.g., buildings, floors, offices)
- Organizational contexts (e.g., departments, projects)
- Action contexts (e.g., tasks)
- Technological contexts (e.g., Java™ programmers)
- Time context (e.g., time of a day, week, month, season of the year)

Regardless of what contextual information a system collects, the information must be represented in a form that can be digested by the applications that depend on the context. We describe how context is acquired and represented next.

13.3 Context Acquisition

In this section, we take a brief look at how context information can be obtained using a variety of sensors and other mechanisms. The aim is to facilitate collecting this information and provide it in a reliable and secure way to middleware platforms that use it to make decisions. We also examine current standards and technologies that facilitate the use of context information during application delivery.

13.3.1 Acquisition of Sensor Data

There is a wide selection of sensors and sensing technologies that can be applied to collect contextual information. Some of the sensing technologies can be further categorized into:

- Thermal and humidity
- Vision and light
- Location, orientation, and presence
- Magnetic and electric fields
- Touch, pressure, and shock
- Audio
- Weight

Context-Aware Mobile Computing

- Smell: gas sensors
- Acceleration: motion detection

Often it is not enough to collect all the contextual information from a single sensor. In mobile computing, sensors can be in one place or distributed spatially. In an array of sensors, communication protocols need to be engineered to facilitate the fusion of such data. This can involve rather complex postprocessing. For example, in wearable computing, sensors are placed all over the body [22]. Placements of these sensors are crucial to the contextual observations. In any application, the designer has to ultimately choose the correct sensor type and its relevant positioning.

A large number of research papers, however, concentrate on location-based services using location information as context [14–16]. Location-aware computing systems respond to a user's location, either spontaneously or when activated by a user request. The next few sections summarize the state of these sensing capabilities and provide some guidance on their possible future evolution.

13.3.2 Location Sensing Techniques

The global positioning system (GPS) is the most commonly known location sensing system today, whereby a technique called triangulation is used to compute object locations. Measuring distance from an object to a particular point using time-of-flight, means measuring the time it takes to travel between the object and the point. In a system like GPS, the receiver is not synchronized with the satellite transmitters and cannot precisely measure the time it takes for the signal to reach the ground from space. Therefore, GPS satellites are perfectly synchronized with each other via accurate atomic clocks to allow for a baseline of reference that is used to calculate the difference in flight time. On the ground, all GPS receivers have an almanac programmed into their computers that tells them where in the sky each satellite is, moment by moment. Four satellites are normally required to estimate an accurate three-dimensional (3D) position. A good summary of the current state-of-the-art in location sensing techniques is provided by Hightower and Borriello in [2, 19]. There are many examples of such time-of-flight location sensing systems:

- Active Bat location systems (distance measurement from indoor mobile electronic tags, called Bats to a grid of ceiling mounted sensors) [17]
- Cricket location support system [15]
- PulsON™ Time Modulated Ultra Wideband technology [18]

13.4 What Is Context-Awareness?

Context-awareness can be classified into two groups: ones that use context passively and others that adapt themselves to context proactively. The seminal work of Schilit and Theimer explained how applications can react to context and reconfigure themselves to better solve the problem at hand [23]. The authors describe an active map service that keeps clients informed of changes in their environment. The software discovers and reacts to changes in the environment that the users are situated in. Note that this kind of active reconfiguration is similar to that advertised in software agent-based technologies, whereby agents are continually adapting to the new contextual information. Sometimes these agents are also known as context-sensitive. Context modulates behavior by affecting the actions used to achieve goals as well as the timing and manner in which those actions are carried out. By paying attention to its context, an intelligent agent can more quickly select appropriate behavior to achieve its goals, and it can more effectively focus its attention and respond to unanticipated events. Context-awareness can also be categorized into a set of features [12]:

- Proximate selection — proximate selection is an interaction technique where a list of objects (printers) or places (offices) is presented and where items relevant to the user's context are emphasized or made easier to choose.
- Automatic contextual reconfiguration — a system-level technique that creates an automatic binding to an available resource based on the current context.
- Contextual command applications — executable services made available due to the user's context or whose execution is modified based on the user's context.
- Context-triggered actions — applications that automatically understand the context they are in and execute services automatically when the right combination of context exists.

13.4.1 Technology Independent Framework and Application Programming Interfaces

The context information is best used by a middleware service platform that acts as an intermediary in orchestrating the delivery of services from the origin servers to the end clients. We describe two standardization efforts going on in this space: one closely related to the telecom industry and one from the Internet/Web community. We then describe a mobile service platform that enables the delivery of context-aware services and is agnostic to wireless carriers and devices.

13.4.1.1 Parlay: Integration of Telecom and Internet Services. In an effort to promote technology independence and link applications with the capabilities of the telecommunications world, the Parlay Group was formed in 1998 by a community of operators, information technology (IT) vendors, network equipment providers, and application developers [20]. Parlay's open Application Programming Interface (API) releases developers from having to write code for specific networks and environments. It eliminates the need for programmers to master different telecommunication protocols, at the same time supporting the evolution of second generation (2G), 2.5G, and third generation (3G) networks with the same set of APIs. For example, the WASP (Web Architecture for Service Platforms) platform supports context-aware applications based on Web services [37]. It uses Parlay-X as a Web service interface to 3G network functions. An application (such as a tourist application that depends on location-sensitive information) uses context information available from the WASP platform to provide its services to mobile users; however, contextual information is shielded from service providers by the privacy control layer, which is responsible for checking context-dependent privacy preferences based on World Wide Web Consortium's Platform for Privacy Preferences (W3C's P3P) [28] (see Section 13.5.4).

13.4.1.2 IETF OPES Group. The Internet is facilitating multiple forms of distributed applications, some of which employ application-level intermediaries. The Open Pluggable Edge Services (OPES) [21] group's primary task is to define application-level protocols enabling such intermediaries to incorporate services that operate on messages transported by Hypertext Transfer Protocol (HTTP) and Real-Time Transport Protocol/Real-Time Streaming Protocol (RTP/RTSP). At the Internet Protocol (IP) level, the participating intermediaries are endpoints that are addressed explicitly. The emergence of ideas like middleware service providers supported by a group like OPES shows an industry trend to create value-added services on the network edge. One potential use of OPES is to adapt a service based on the context of the application clients. The security model for such services involves defining the administrator roles and privileges for the application client, application server, intermediary, and auxiliary server. The data integrity model defines what operations are permitted by the content owners and what guarantees of content correctness can be made to the owners and viewers when content-related services are performed.

13.4.1.3 iMobile: a Mobile Service Platform. In this section, we briefly review the iMobile architecture, a mobile service platform. Its overall architecture (standard edition, known as iMobile) is as depicted below in Figure 13.1 [26].

iMobile implements three key abstractions — devlet, infolet, and applet. A devlet is a driver attached to the proxy that receives and sends messages through a particular protocol (America Online Instant Messenger [AIM], Short Message Service [SMS], Wireless Access Protocol [WAP], HTTP, etc.). An infolet hosted on the proxy is responsible for creating an abstract view

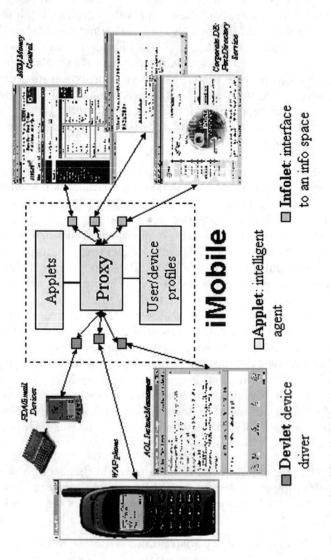

Figure 13.1. The iMobile Architecture

Context-Aware Mobile Computing

of a particular information space using appropriate protocols (e.g., HTTP for the Internet, X10 for home networks, Common Object Request Broker Architecture [CORBA] for distributed network resources). An applet implements the application logic by postprocessing information obtained by the various infolets. iMobile is a proxy-based platform designed to provide personalized mobile services. Information retrieved from infolets and applets are transcoded according to the user and device profiles.

As an example of information flow in iMobile, consider the following. The AIM devlet on iMobile starts an AIM client and listens to service requests from other AIM clients sent as instant messages. The devlet senses an incoming request via a character stream and, if validated as a correct iMobile command, forwards it to a command Dispatcher hosted on the proxy. The latter employs business logic provided by the corresponding applet to invoke specific infolets. These infolets are responsible for obtaining information from various data sources/content providers. The Dispatcher renders this information suitable for the target device by performing transcoding services. The AIM devlet then receives results from the Dispatcher formulated in a Multipurpose Internet Mail Extension [MIME] appropriate for that device, which is determined by the corresponding device profile stored at the mobile service platform.

In summary, iMobile acts as a mobile personal agent to deliver both text and application messaging to handheld devices ubiquitously regardless of the underlying communication network technologies and operating protocols. It admits personalization and has an open architecture so new devices, information sources, and protocols can be integrated. Thus, iMobile has a unique advantage to perform intelligent messaging. In Section 13.5.1, we show how to integrate location awareness into this messaging platform.

iMobile Enterprise Edition (iMobile EE)* [38] (Figure 13.2) is a redesign of the original iMobile architecture to address the security, scalability, and availability requirements of a large enterprise, such as AT&T. iMobile EE incorporates gateways that interact with corporate authentication services, replicated iMobile servers with backend connections to corporate services, a reliable message queue that connects iMobile gateways and servers, and a comprehensive service profile database that stores the user profile and device context, which governs operations of the mobile service platform. The iMobile EE architecture was also extended to provide personalized multimedia services, allowing mobile users to remotely control, record, and request video contents. iMobile EE aims to provide a scalable, secure, and modular software platform that makes enterprise services easily accessible to a growing list of mobile devices roaming among various wireless networks.

*iMobile EE has been renamed Enterprise Messaging Network (EMN).

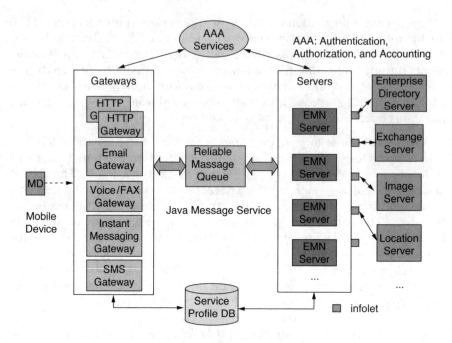

Figure 13.2. iMobile Enterprise Edition

13.5 Gluing Contextware and Middleware

The use of context information by applications in a mobile environment poses a number of challenges arising from the distributed and dynamic nature of sensors, the accuracy and resolution of sensors, and the fusion of output of multiple sensors to determine context. In addition, the mobile environment poses further challenges with regard to the dependability, predictability, and timeliness of communication. Middleware is required that provides abstractions for the fusion of sensor information to determine context, representation of context, and intelligent inference. Essential services that provide support for operation in a mobile environment, such as supporting the reliability of communication, are also required. In this section, we will review a particular middleware service platform (see iMobile in Section 13.3.3.3) developed by the authors that has been used to provide location-based services.

13.5.1 Integrating Location Determination with the Service Platform

Figure 13.3 shows one way to integrate the iMobile services platform with the location server. Another way is to integrate the service platform with the interactive voice response (IVR) in a voice application. In what follows, we will describe the concepts with the arrangement above and omit the details of the other configuration.

Context-Aware Mobile Computing

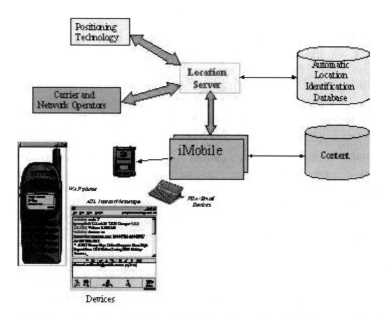

Figure 13.3. Location-Based Value-Added Services via a Middleware Service Platform

13.5.2 Managing Location Information in iMobile

Each mobile device has to be mapped to a registered iMobile user. There are two reasons for this requirement:

1. To limit access to legitimate iMobile users only
2. To personalize a service based on the user profile

A typical device-to-user map stored under iMobile maps a protocol account to a user. For example, the following map shows the iMobile platform that a service request initiated from the cell phone +19087376842 through the SMS channel or an instant message sent from the AIM screen name `webciao` should be considered a request from the user `chen`.

```
sms:+19087376842=chen
mail:dchang@research.att.com=difa
aim:webciao=chen
phone:9084321529=chen
ip:135.22.102.22=chen
```

Note that the protocol and associated context information (such as location information or HTTP header that gives device information) dictates how iMobile should transcode content to be delivered to the mobile device that initiates the request. The way iMobile authenticates a mobile user depends on the device or protocol used and is determined by its internal policy. iMobile trusts wireless networks such as Global System for Mobile (GSM) or time division multiple access (TDMA) networks to provide the correct cell phone ID when a short message is received. This is generally acceptable unless a cell phone is stolen and the user did not lock the phone with a security password. iMobile also trusts the AOL network's authentication for noncritical services. User authentication through iMobile itself is required if the user accesses iMobile through WAP or HTTP.

A typical iMobile user profile stores the username, password, and a list of the devices that the user registers with iMobile. It also stores the location information, which is updated automatically when the user invokes the appropriate location determining technology (see Section 13.3.2), probably through an explicit request from the user's GPS-equipped personal digital assistant (PDA) or via voice to update the user's crossroad or zip code. When a user later accesses iMobile through AIM using the screenname `webciao` (say from a Pocket PC device equipped with a wireless local area network [LAN] card), iMobile determines from the device-to-user map (see above) that the user is `chen` and will correctly use his user profile for subsequent location-aware service requests from this device. Note that there are other ways to keep the location information up-to-date, such as with periodic polling at the location server.

13.5.3 Location-Based Services with iMobile

The infolet abstraction in iMobile extends beyond the HTTP protocol and universal resource locator (URL) name space to provide an abstract view of various information spaces. An infolet retrieves the original content and returns it to an applet for further processing. Some of the location-aware services that have been implemented as infolets are:

- Location — reports where the user's last registered location is. It also allows one to change the location information explicitly using a zip code.
- Weather — reports the current local weather.
- Find — finds business addresses near the user. The accuracy of this result is naturally dependent on the location information. In this case we were able to deliver accuracy in metropolitan areas suitable to cell site/sector coverage.

Figure 13.4 shows the snapshot of a typical location-aware service delivered by iMobile using instant messaging.

13.5.4 Preserving Privacy in Environments with Location-Based Services

Protecting personal location information is becoming a real challenge as more value-added services emerge in the market every day. Automatic

Context-Aware Mobile Computing

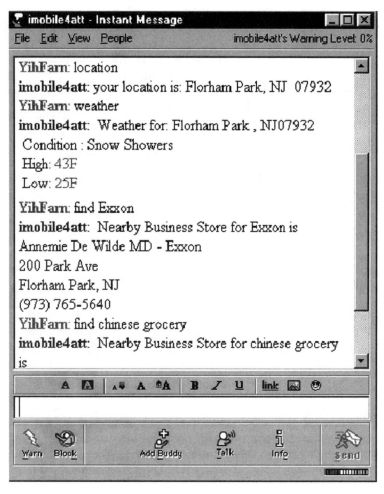

Figure 13.4. Yellow Pages Lookup from AIM via AT&T iMobile Stored Context Profile

control of their location information in a safe and controlled manner is a prerequisite to ensure that information does not get misused. An important first step in protecting users' location privacy is notifying them of requests for this information. For example, a system may ask the users for approval before releasing their location information to third parties. Third parties themselves have to be verified and guaranteed by a certificate authority that they will not tamper and misuse this information. To automate the privacy-management and decision making process, the W3C's P3P specifications should be followed closely [27, 28]. P3P is designed to support Web interactions typically involving e-commerce and business applications. It is important to note that its mechanisms for obtaining reference files and policy documents are tightly coupled with Web usage models, protocols, and deployment architectures. The policy language contains constructs for expressing information-collection and management policies appropriate for protecting information disclosed during Web browsing and user-initiated online transactions. The idea of having a user-centric privacy proxy (policy execution point) is essential to checking the various policies and constraints before disseminating such information to third-party providers. The details of one such project can be found in [29].

13.6 Context-Related Research Initiatives and Projects

This section will highlight some of the more recent projects that are related to context-aware computing. This is by no means an exhaustive list and does not show relative importance of one project over another. We found these projects interesting in their own rights and found it necessary to include them to provide the different flavors of context-aware computing:

- Massachusetts Institute of Technology (MIT)'s Oxygen project [15] — the researchers at MIT believe that computation and communication will become as pervasive and free as air. Computation will become human-centered. Anonymous devices will collect context information in the surrounding environment and systems using such information will become aware of our needs and adapt accordingly. Several key technologies in the Oxygen project are quite relevant to context-aware mobile computing:
    - Cricket — the Cricket location support system provides an indoor analog of GPS to provide information about location, orientation, and geographic spaces [15]. Cricket beacons, mounted on walls or ceilings, transmit ultrasound and radio frequency (RF) signals; compact listeners, attached to mobile or static devices, use the difference in signal arrival times to determine where they are.
    - INS — in the Intentional Naming System (INS) [30], names are *intentional;* they describe application *intent* in the form of properties and attributes of resources and data, rather than simply network addresses (such as URLs) of objects. Potentially, the use of an INS would allow us to address resources easily based on the attributes collected by the contextware.
- AT&T Labs — Cambridge:[1]
    - The Bat Ultrasonic Location System — their first experiments with context-aware systems used room-scale information generated by infrared Active Badges [31], but many applications require fine-grained 3D location and orientation information that Active Badges cannot supply. They have since developed a 3D ultrasonic location system (known as the Bat System [13]), which is low-power, wireless, and relatively inexpensive.

- Sentient Computing — the Sentient Computing [32] project uses sensors and resource status data to maintain a model of the world that is shared between users and applications. Based on the Bat location system, sentient computing can help store and retrieve context sensitive data about mobile users. Whenever information is created, the system knows who created it, where they were, and who they were with. This contextual metadata can support the retrieval of multimedia information.
- University of California — Berkeley:
  - Smart Dust — the Smart Dust [33] project is probing microfabrication technology's limitations to determine whether an autonomous sensing, computing, and communication system can be packed into a tiny device called a *mote* to form the basis of integrated, massively distributed sensor networks. Only a few cubic centimeters in size, the motes collect light, temperature, humidity, and other data about their physical environment. The data is then relayed from mote to neighboring mote until it reaches its desired destination for processing.
  - TinyOS — the networked sensor regime is an exciting new design space that is emerging as a result of innovations in RF communication technology and Micro Electrical Mechanical Systems (MEMS) technology. TinyOS [34] explores the software support that is required in that design space. TinyOS is a component-based runtime environment designed to provide support for deeply embedded systems that require concurrency-intensive operations while constrained by minimal hardware resources. For example, originally designed for the Smart Dust hardware platform, the scheduler of TinyOS fits in fewer than 200 bytes of program memory.

13.7 The Future of Context

Mobile computing devices like palmtop computers, mobile phones, and PDAs have gained widespread popularity. Even though devices and networking capabilities are becoming increasingly powerful, the design of mobile applications will continue to be constrained by the physical limitations. Mobile devices will likely continue to be battery-driven and wide-area wireless networks will have fluctuations in bandwidth depending on the physical location. Traditional middleware for fixed distributed systems cannot be used in the mobile environment particularly for the reasons outlined above. Many researchers have investigated systems that collect context information and help applications adapt to changes. Research has to continue to develop middleware that delivers better quality of services to mobile applications. Capra et al. [35] studied reflective middleware that maintains an updated representation of the context. Acquisition of context is also important and has largely focused to date in obtaining location information. There have been developments in the acquisition of other forms of context, like gesture, voice, ambience, etc. Researchers have recognized the need to create context toolkits [39, 40] or contextware that provides important abstractions and support for context-aware computing. We expect the seamless integration of contextware and middleware service platforms to emerge in the next decade to greatly enrich the mobile users' experience.

The representation of context information in a universal way is important for systems to interoperate. The W3C community is also actively involved in defining standards like Composite Capabilities/Preferences Profile (CC/PP),[2] which has the role of representing delivery context in assisting device independent presentation for the Web [36].

13.8 Conclusions

In this chapter, we conducted a survey of context-aware mobile computing. We started off by describing why context is so important and how it can help enhance human–computer interactions. Next we defined the meaning of context and how it can be acquired for use in applications that are context-aware. Tying such information to a middleware platform is an essential step in understanding how context can be modeled succinctly and how systems can take advantage of context information to customize service delivery. We motivate and describe a software architecture that provides this desired integration and extensibility of services in a context-aware application infrastructure.

To date, researchers have primarily focused on location information as context information. Context acquisition is often acquired from nontraditional devices. It is our hope that as the infrastructures of contextware and multimodal applications emerge, the integration of context with middleware service platforms will make the overall user experience more satisfying.

Notes

1. The lab ceased operation on April 24, 2002. More information can be found at http://www.xorl.org/lab/.
2. The CC/PP Work Group is now closed and the work on CC/PP will continue in the Device Independence Work Group. See http://www.w3.org/2001/di/.

References

1. A. Ferscha, Contextware: Bridging Physical and Virtual Worlds, *Proc. of the 7th International Conference on Reliable Software Technologies,* ADA-EUROPE 2002, June 17–21, 2002, Vienna.
2. J. Hightower and G. Boreiello, Location Systems for Ubiquitous Computing, *Proc. of the IEEE Computer,* August 2001, pp. 57–66.

3. A. Schmidt and M. Beigl, There is More to Context than Location: Environment Sensing Technologies for Adaptive Mobile User Interfaces, *Proc. of Interactive Applications of Mobile Computing*, Rostock, Germany, November 24, 1998.
4. G.D. Abowd and A.K. Dey, Towards a Better Understanding of Context and Context-Awareness, *Proc. of 1st International Symposium Handheld and Ubiquitous Computing*, vol. 1707, pp. 304–307, Karlsruhe, Germany, 1999.
5. N. Schilit, N. Adams, and R. Want, Context-Aware Computing Applications, *Proc. of IEEE Workshop on Mobile Computing Systems and Applications*, Santa Cruz, CA, December 1994, pp. 85–90.
6. R.V. Cox, C.A. Kamm, L.R. Rabiner, J. Schroeter, and G.J. Wilpon, Speech and Language Processing for Next-Millennium Communications Services," *Proc. of the IEEE*, vol. 88, no. 8, August 2000, pp. 1314–1337.
7. T. Urnes, A.S. Hatlen, P.S. Malm, and O. Myhre, Building Distributed Context-Aware Applications, *Proc. of Personal and Ubiquitous Computing*, vol. 5, 2001, pp. 38–41.
8. N. Schilit, System Architecture for Context-Aware Mobile Computing, PhD dissertation. Columbia University, New York, 1995.
9. Wireless Enhanced 911. http://www.fcc.gov/911/enhanced/.
10. M. Weiser, The Computer for the 21st Century, *Scientific American*, vol. 265, no. 3, pp. 66–75, 1991.
11. J. Pascoe, N. Ryan, and D. Morse, Issues in Developing Context-Aware Computing, *Proc. of 1st International Symposium on Handheld and Ubiquitous Computing*, Karlsruhe, Germany, September 1999, pp. 208–221.
12. A. K. Dey, Providing Architectural Support for Building Context-Aware Applications, Ph.D. dissertation, Georgia Institute of Technology, November 2000.
13. A. Ward, A. Jones, and A. Hopper, A New Location Technique for the Active Office, *IEEE Personal Communications,* vol. 4, no. 5, pp. 42–47, October 1997.
14. P. Bahl and V. Padmanabhan, RADAR: An In-Building RF-Based User Location and Tracking System, *Proc. of IEEE Infocom*, vol. 2, pp. 775–784, March 2000.
15. N.B. Priyantha, A. Chakraborty, and H. Balakrishnan, The Cricket Location-Support System, *Proc. of MOBICOM 2000*, Boston, August 2000, pp. 32–43.
16. R. Want and D.M. Russell, Ubiquitous Electronic Tagging, *IEEE Distributed Systems Online*, vol. 1, no. 2, September 2000.
17. A. Harter, A. Hopper, P. Steggles, A. Ward, and P. Webster, The Anatomy of a Context-Aware Application, *Proc. of IEEE Conference on Mobile Computing and Networking*, Seattle, August 1999, pp. 59–68.
18. Time Domain Corporation, PulsON Technology Time Modulated UWB Overview, Huntsville, AL, 2001.
19. J. Hightower and G. Borriello, Location Sensing Techniques, University of Washington, Computer Science and Engineering Technical Report, UW-CSE-01-07-01, July 30, 2001.
20. The Parlay Group, www.parlay.org.
21. OPES — Open Pluggable Edge Services, http://www.ietf-opes.org.
22. MIThril — The Next Generation Research Platform for Context-Aware Wearable Computing, http://www.media.mit.edu/wearables/mithril/.
23. N. Schilit and M.M. Theimer, Disseminating Active Map Information to Mobile Hosts, *IEEE Network,* vol. 8, no. 5, September–October 1994, pp. 22–32.
24. R.M. Turner, Context-Sensitive Reasoning for Autonomous Agents and Cooperative Distributed Problem Solving, *Proc. of IJCA Workshop on Using Knowledge in Its Context*, 1993, Chambéry, France.
25. J. Pascoe, Adding Generic Contextual Capabilities to Wearable Computers, *Proc. of the 2nd IEEE International Symposium on Wearable Computers (ISWC'98)*, Pittsburgh, IEEE Computer Society. October 19–20, 1998, pp. 92–99, http://www.cs.ukc.ac.uk/pubs/1998/676/index. html.

26. H. Rao, Y. Chen, D. Chang, and M. Chen, iMobile: a Proxy-Based Platform for Mobile Services, *Proc. of the 1st ACM Workshop on Wireless Mobile Internet* (WMI 2001), Rome, July 2001.
27. The Platform for Privacy Preferences 1.0 (P3P1.0) Specification, World Wide Web Consortium, September 2001, www.w3.org/TR/2001/WD-P3P-20010928.
28. A P3P Preference Exchange Language 1.0 (Appel 1.0), working draft, World Wide Web Consortium, April 2002, www.w3.org/TR/P3P-preferences.
29. G. Myles, A. Friday, and N. Davies, Preserving Privacy in Environments with Location-Based Applications, *IEEE Pervasive Computing*, January–March 2003, pp. 56–64.
30. The Intentional Naming System, http://nms.lcs.mit.edu/projects/ins/.
31. The Active Badge System, http://www.uk.research.att.com/ab.html.
32. Sentinel Computing, http://www.uk.research.att.com/spirit/.
33. B. Warneke, M. Last, B. Liebowitz, and K.S.J. Pister, Smart Dust: Communicating with a Cubic-Millimeter Computer, *Computer*, vol. 34, January 2001.
34. TinyOS, http://webs.cs.berkeley.edu/tos/.
35. L. Capra, W. Emmerich, and C. Mascolo, Reflective Middleware Solutions for Context-Aware Applications, Proc. of REFLECTION 2001. *The 3rd International Conference on Metalevel Architectures and Separation of Crosscutting Concerns*, Kyoto, Japan, September 2001, *Lecture Notes in Computer Science*, vol. 2192, New York: Springer-Verlag, pp. 126–133.
36. W3C, Composite Capability/Preference Profiles, http://www.w3.org/Mobile/CCPP/.
37. The WASP Project, http://www.w3.org/2003/p3p-ws/pp/utwente.pdf.
38. Y.F. Chen, H. Huang, R. Jana, T. Jim, M. Hiltunen, R. Muthumanickam, S. John, S. Jora, and B. Wei, iMobile EE — An Enterprise Mobile Service Platform, *ACM Journal on Wireless Networks*, vol. 9, no. 4, July 2003, pp. 283–297.
39. A. Salber, K. Dey, and G.D. Abowd, The Context Toolkit: Aiding the Development of Context-Enabled Applications, *Proceedings of CHI'99*, pp. 434–441.
40. A. Schmidt, Ubiquitous Computing — Computing in Context, PhD dissertation, Lancaster University, November 2002.

Chapter 14
Mobile Agent Middlewares for Context-Aware Applications

Paolo Bellavista, Dario Bottazzi, Antonio Corradi, Rebecca Montanari, and Silvia Vecchi

Abstract

Wireless communications and the Internet are converging toward an integrated scenario where both traditional and novel services should be ubiquitously accessible, independently of the mobility of users, terminals, resources, and service components. Mobility-enabled service provisioning introduces several challenging issues to address: from client–server location change at provision time, to wide heterogeneity of access terminals, and to unpredictable modifications in accessible resources. In this complex scenario, two main guidelines are recently emerging. The first is the need for novel middleware solutions to support service development and deployment. The second is the necessity of full visibility of the context, intended as the logical set of accessible resources depending on client location, access terminal capabilities, and system/service management policies, to adapt service provisioning to specific runtime conditions. The chapter discusses and motivates the suitability of the Mobile Agent (MA) technology to implement novel context-aware middlewares for mobile computing, mainly because of the MA properties of mobility, asynchronicity, decentralization, and location awareness. In addition, the chapter gives an extensive overview of the state-of-the-art research activities about MA-based supports for mobile computing. Especially about context-aware ones to point out, through system/prototype exemplifications, the main lessons learned and the primary directions of the on-going research work.

14.1 Mobile Computing and Context Awareness

The wide spreading of mobile computing is changing the way to develop, deploy, and expect to access Internet services. Nomadic users who disconnect from the network to reconnect to a new point of attachment after a time interval and mobile terminals that continuously roam in the network without suspending the on-going service sessions introduce new challenging issues in service design. Recent advances in wireless networking and the enlarging market of wireless-enabled portable devices further stimulate the provisioning of services to a wide set of client terminals with heterogeneous and limited resources. These services should be aware of the client location not only to yield back results to the current user/device position, but also as the basis for service customization. Service tailoring should depend on user position, on device characteristics, and also on the current state of involved resources, either within the client locality or in a wider global scope.

In other words, service providers and wired/wireless network operators have to face new challenging and state-of-the-art technical issues, toward both the deployment of novel services for mobile ad hoc networks and the full seamless integration of mobile clients with the traditional fixed Internet. The first scenario is just moving its first steps by investigating solutions mostly at the network level, e.g., for multi-hop cooperative routing [1]. However, it is attracting more and more research interest for its potential of leveraging peer-to-peer interactions between mobile clients. At the opposite, the second scenario already starts to exhibit research and commercial network-level solutions for mobile connectivity [2, 3], even if the most challenging service-level issues, such as supporting service continuity while roaming and requalifying resource bindings depending on local resource availability, still have to be addressed. We claim that, in both scenarios, mobile computing motivates rethinking traditional support solutions for distributed systems to achieve the necessary level of visibility of mobility-related properties; these properties are required to drive the runtime decisions about service adaptation to the provisioning environment.

Since the beginning of the research in mobile computing, location has been recognized as a crucial property to be aware of in order to organize effective and efficient mobile applications. Location awareness calls for mechanisms and tools to obtain the information about the physical position of all potentially mobile entities (i.e., users, access devices, resources, and service components) involved in mobile computing applications. Let us observe that support solutions in traditional distributed systems tend to hide the location information from service developers to simplify application design and implementation. On the contrary, the mobile computing scenario requires performing service management operations at provision time, such as rebinding to local resources and transcoding data depending on locally available bandwidth, which are sensibly influenced by the current client location and typically application-specific [4]. For these reasons, it is necessary that location awareness is propagated up to application-level components or, at least, to the middleware facilities that can handle the complexity of location processing/management and can provide a simplified and more usable location abstraction to mobile applications.

Location visibility is not the only crucial property that applications should be aware of in mobile computing environments. We claim that mobile computing emphasizes the need for novel methodologies to support and simplify the development of innovative classes of applications where service results and characteristics depend on the provisioning context. Context is the logical set of resources that a client can access due to runtime properties of the provisioning environment, such as client location, security permissions, access device capabilities, user preferences and trust level, resource state, and mutual relationships with currently local users, terminals, resources, and service components [5]. A notable example of context is the set of resources/services that a user can access depending on her personal preference profile and independently of her current point of attachment (Virtual Home Environment [6]). Let us note that, given the above definition of context, location is only a specific case (maybe the most important one in mobile computing) of the different kinds of information that affect the context determination.

To exemplify the relevance of having mobile applications with full context awareness, consider the case of a ubiquitously accessible stock trading service that is willing to enable its mobile users to operate on the market via laptops connected to wireless fidelity (Wi-Fi) hotspots, via personal digital assistants (PDAs) connected to Bluetooth® Local Infotainment Points [7], via Wireless Application Protocol (WAP) phones [8], and via Global System for Mobile (GSM) phones receiving simple Short Message Service (SMS)-based communication in case of abrupt changes of quotations of interest. The interface and the information content of the trading service should be adapted to the access terminal and the connectivity technology, as well as to the resource availability in the wireless access locality. In case of a local network overload, to avoid aggravating the congestion situation, the service should give priority to gold-user transactions and exclude access to bronze users. A flexible way of designing such a service is to determine the up-to-date contexts of any user currently involved in an active service session and to adapt the service behavior accordingly. For instance, if the access terminal is a GSM phone, the context should only consist of a gateway component running on a host of the fixed network infrastructure that is capable of downscaling Hypertext Markup Language (HTML) pages to text-only summaries and of delivering them as SMS messages. In the case of such a context, stock trading servers should be automatically and dynamically bounded to that gateway, with no need to hardcode this binding behavior within the service logic. As another example, in the case of congestion, the context of a stock trading bronze user should be voided to prevent any access.

In other words, high heterogeneity, dynamicity, and resource shortage/discontinuities typical of mobile computing environments stress the relevance of context-dependent services. However, the design, implementation and deployment of context-dependent mobile applications is significantly more complex than the development of traditional distributed services, thus risking to slow down this emergent service market. Therefore, we claim the need for highly flexible and innovative middleware solutions, with full context awareness, to facilitate the development and runtime support of context-dependent mobile services [5]. Several recent research efforts are paving the way to the realization of such novel middlewares. Their presentation, analysis, and comparison are the main scope of this chapter.

In the following, we analyze the research work done in the last years in the area of mobile computing middlewares with the aim of putting side by side the different solutions that have emerged and of extracting the primary lessons learned. In this overview, we focus on the innovative middleware solutions that choose the MA programming paradigm for their design and implementation:

- Section 14.2 discusses and motivates why the MA technology is to be considered particularly suitable and effective for mobile computing middlewares.
- Section 14.3 provides an overview of the existing MA-based supports for mobile computing.
- Section 14.4 specifically concentrates on the MA-based middlewares supporting differentiated forms of context awareness.
- Section 14.5 summarizes the lessons learned and sketches a possible road map for the medium-term research in the field.

14.2 Mobile Agents and Mobile Computing

It is increasingly recognized that the design and implementation of a mobility middleware can significantly take advantage of the adoption of innovative programming paradigms based on code mobility and, in particular, of the MA technology [9–11].

The appearance of the MA concept can be found in the TeleScript technology developed by General Magic in 1994 [12]. Scripting languages, such as the Tool Command Language (TCL) and its derivative SafeTCL, were gaining much interest because they enabled rapid prototyping and the generation of portable code [13]. The concept of smart messaging, based on the encapsulation of SafeTCL scripts within e-mails, made possible new kinds of distributed applications with some degrees of dynamic programmability [14]. At the same time, mobile computing, intended as the possibility of

Table 14.1. The Taxonomy of Programming Paradigms Based on Code Mobility [16]

Paradigm	Before		After	
	N_A	N_B	N_A	N_B
Client–Server	A	Know-how Resource B	A	Know-how Resource B
Remote Evaluation	Know-how A	Resource B	A	Know-how Resource B
Code on Demand	Resource A	Know-how B	Resource Know-how A	B
Mobile Agent	Know-how A	Resource	—	Know-how Resource A

supporting user/terminal mobility without suspending service provisioning, was spreading and further stimulated the research on mobile code technologies [15]. Last but not least, the same years witnessed the beginning of the Java™ programming age: portable Java bytecode is the basis for the largest part of current MA systems.

MAs are considered the most expressive among the programming paradigms based on code mobility [16]. In the traditional client–server paradigm, clients invoke functions made available by usually remote computational entities called servers. Servers execute locally the invoked services and deliver the results back to the requesting clients. Servers provide both the knowledge of how to handle client requests and the required execution resources. Code mobility can improve the traditional client–server paradigm along two different and orthogonal lines, by allowing the dynamic decision of where the service know-how is located and of which host provides the needed computational resources.

Three different paradigms based on the possibility of dynamic code migration have been identified [16] — Remote Evaluation (REV), Code on Demand (CoD), and MAs. The paradigms differ in the distribution of know-how, processor, and resources among nodes N_A and N_B (Table 14.1). In the REV paradigm [17], a component A on node N_A sends instructions specifying how a component B on node N_B should perform a service. The component B then executes the received code on its local resources. Elastic servers are a notable example of REV [18]. In the CoD paradigm, instead, component A has resources located in its execution environment, but does not know how to access and process these resources: it should obtain the needed code from component B. Java applets follow this paradigm.

The MA paradigm is an extension of REV. Whereas the latter primarily focuses on the transfer of code, the MA paradigm involves the mobility of an entire computational entity that carries its code and its reached execution state. In other words, the MA-based component A has the know-how capabilities and the processor, but it lacks the resources to operate with. It can migrate autonomously to a different computing node N_B that can offer the required resources. In addition, A is capable of resuming its execution seamlessly, because it preserves its execution state after migration. About the execution state, when a MA is capable of migrating together with its whole state (not only application-level data, but also kernel-level associated information such as program counter and register values), its mobility is called strong mobility. More often, MA platforms support only the migration of application-level execution state and simply allow MA developers to decide from which points of their code MAs have to resume their execution after migration (weak mobility). For instance, Java-based MA systems generally enable MAs to migrate together with all the serializable Java objects in their state, but do not support strong mobility, impossible to achieve without modifying the standard Java Virtual Machine (JVM™) [19].

The MA paradigm is important for network-centric systems, because it represents an alternate, or at least complementary, solution to the usual client–server model of interaction. In the last few years, several research activities have investigated the MA technology for distributed system management and encouraged its use, by showing some significant deriving benefits. The most explored and recognized advantages span the following [20]:

- Overall reduction of network traffic by exploiting resource colocality
- Flexibility of distributing software components at runtime
- Full decentralization of the monitoring, control, and management of networks, systems, and services
- Increased robustness stemming from decoupling tasks into distributed autonomous activities that can overcome temporary network/resource unavailability

By focusing specifically on mobile computing, MAs are considered a primary enabling technology to implement novel and effective middlewares. Recent research projects have proposed MA-based solutions that concentrate on different aspects related to different forms of nomadic/roaming mobility, by involving users, access terminals, and even needed resources and service components, as detailed in the following two sections. All these research projects recommend the MA technology as a crucial design and implementation choice in the field, because many MA requirements coincide with mobile computing ones [9–11].

First of all, mobility stresses the requirement of dynamicity, intended as the possibility of modifying and extending the support infrastructure, where and when needed, with new components and protocols depending

Mobile Agent Middlewares for Context-Aware Applications

on client mobility patterns and on evolving service/user requirements. Dynamic distribution/modification of code and dynamic resource binding are similar in both MAs and mobile users/terminals. Unlike other mobile code technologies, MAs benefit from the additional flexibility of moving code together with the state modified during the already performed computation.

In addition, mobile computing can greatly benefit from the possibility of asynchronicity between user/terminal requests of operations and their execution. For instance, wireless connections impose strict constraints on available bandwidth and on communication reliability and minimize the connection time of the wireless client device. The MA paradigm does not need continuous network connectivity, because connections are required only for the time needed to inject MAs from mobile terminals to the fixed network infrastructure, for instance through the current wireless point of attachment. MAs are autonomous and can carry on services even when launching users/terminals are disconnected, by delivering service results back at their reconnection.

Moreover, middleware solutions for mobile computing should give application designers full location visibility to perform application-specific optimizations and to adapt to local resource availability. For instance, a mobile service should both accommodate mobile users changing location during service provisioning and dynamically tailor its results depending on the properties of the current network location. The property of location awareness, typical of the MA programming paradigm, significantly helps in propagating allocation visibility up to the application level [10, 11]. In addition, the MA autonomy from clients simplifies dynamic personalization. For instance, MAs can act as mobile proxies working over the fixed network on the behalf of the wireless client and can follow user movements to stay colocated in the same network locality and to tailor service results depending on personal preferences and access device profiles [21].

The MA technology also has some drawbacks, often identified in the associated security and interoperability issues. However, from the beginning, these potential weaknesses have challenged MA researchers to investigate and provide rich mechanisms, tools, and strategies for security and interoperability, directly embedded in several state-of-the-art MA platforms. These MA platform-embedded solutions become building blocks to employ when dealing with the security and interoperability issues associated with mobile computing supports. For instance, as detailed in Chapter 39, many MA systems integrate with Public Key Infrastructures for securing MA communications and resource access. This security infrastructure significantly simplifies the authentication of mobile users/terminals. Analogously, the MA research has promoted interoperable and standard interfaces to interact with resources and service components available in statically unknown hosting environments (in compliance with Common Object

Request Broker Architecture [CORBA] and MA-specific standards, such as Object Management Group's [OMG] Mobile Agent System Interoperability Facility [MASIF] and Foundation for Intelligent Physical Agents [FIPA] [22, 23]); these interoperability features can help in supporting the interworking of mobile users/terminals with previously unknown local resources.

14.3 An Overview of MA-Based Supports for Mobile Computing

As already stated, MA-based middlewares are particularly suitable for supporting development deployment of mobile applications. Here, we provide an overview of the most relevant research work in the field, with the aim of exemplifying, through actual experiences of middleware design and implementation, which are the primary approaches emerging to address the different forms of mobility — user mobility, terminal mobility, and mobile access to resources.

User mobility refers to the ability of a user to seamlessly switch between different access terminals by maintaining her personal preferences and her session state independently of terminal characteristics [4]. This requires a support infrastructure capable of keeping user session information and of organizing the user working environment accordingly. MAs represent a valuable technology in implementing such a middleware support, because they can maintain user profiles and active service session data within their states, preserving that information even after migration. For instance, a personal MA can be associated to a user to manage that user's terminal switching; in particular, whenever the user changes the used access terminal, the middleware transparently reconnects that user with the associated MA (just migrated to the new access device) that is in charge of carrying on that user's service session.

The Secure and Open Mobile Agents (SOMA) middleware is specifically targeted to the support of adaptive service provisioning in pervasive environments [4, 24]. The SOMA support for user mobility provides each mobile user with a MA acting as that user's care-of entity, which encapsulates the user profile expressed according to the Extensible Markup Language (XML)-based Composite Capabilities/Preference Profiles (CC/PP) standard representation format [25]. The SOMA agent retrieves the user profile at the beginning of the user's service session and makes it available in the visited network access localities. Profile data rules service provisioning adaptation: for instance, in a museum guide service, the SOMA agent tailors the artwork description on the basis of the user's age, interests, and language. Analogous profiles, and the same CC/PP format, also express the characteristics of the currently used access device. The Adaptation Agents for Nomadic Users (Monads) project focuses on how to extend existing MA platforms with flexible and portable features to support user mobility [26, 27]. In Monads, mobile users exploit smart cards to store MAs and user profiles. When a user connects at a new access device, that user's MA jumps from the smart card to the terminal by carrying the associated user profile. Then, the Monads MA negotiates service provisioning parameters with local service agents based on the user preferences included in the profile.

MAs are also proving their effectiveness in supporting terminal mobility. A crucial advantage of MAs for this kind of mobility is that they can operate on the terminal behalf even when it is offline, by preserving its session state. Two different types of terminal mobility — nomadic mobility and roaming mobility — are identified in the literature.

Nomadic mobility assumes a scenario where typically nonlimited access terminals, such as fully equipped laptops, can disconnect from or reconnect to different network localities. The earliest MA platforms were targeted to support nomadic mobility and for this reason, most of them provide facilities for MA execution suspension and persistency. The ultimate goal of D'Agents (Dartmouth Agents) is the support of applications for distributed information retrieval in heterogeneous networks [10, 28]. The D'Agents middleware implements nomadic mobility by means of its docking system: online docking hosts, paired with mobile terminals, are permanently available in the different network localities. When a MA is unable to migrate toward a mobile terminal, it is transparently forced to wait at the associated docking node. When the nomadic terminal reconnects, the docking host is notified of the new network address of the associated terminal and forwards all waiting agents to it. Similarly, each SOMA network locality transparently freezes MAs attempting to migrate to a temporarily disconnected mobile terminal. The SOMA discovery service detects the entering and connection of a new mobile terminal to a network locality and triggers a corresponding event notification that produces the execution restart of the interested frozen MAs. The Advanced Mobile Application Support Environment (AMASE) concentrates on supporting mobile applications over a wide range of heterogeneous access terminals [9, 29]. AMASE implements a kindergarten service that hosts suspended MAs in a locally available database and registers them in a globally visible Lightweight Directory Access Protocol (LDAP) server. Suspended agents can be retrieved and awaken by exploiting this directory naming solution.

Roaming mobility identifies a service scenario where wireless access points to the fixed Internet infrastructure offer continuous connectivity to mobile terminals that roam between different wired–wireless network localities. As depicted in Figure 14.1, we can distinguish two main MA-based middleware approaches to roaming mobility, differing in positions that support MA execution: on the wireless device (on-board) or close to it over a wired node in the same network locality (on-the-dock).

Examples of systems adopting the on-board solution are JADE/LEAP and AMASE. The Java Agent Development framework (JADE) is a widely

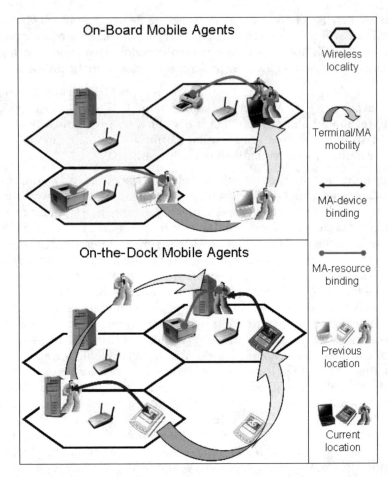

Figure 14.1. Mobile Agents at Work On-Board and On-the-Dock

adopted support for multi-agent systems and has recently been integrated with the results of the Lightweight Extensible Agent Platform (LEAP) project to obtain a FIPA-compliant agent platform with reduced footprint [30, 31]. JADE/LEAP and AMASE provide lightweight instances of their MA middleware, respectively for the Java 2 Micro Edition and PersonalJava™ software, to be installed on resource-limited access devices. These systems support the development of MA-based mobile applications, by allowing the MA execution directly on the roaming terminals.

Other MA-based middlewares, e.g., Grasshopper/Enago and SOMA, opt for the on-the-dock approach and provide a decentralized infrastructure of proxies working over the fixed network on the behalf of wireless access devices. Each proxy is implemented by one MA that follows device movements during service provisioning by maintaining the session state. The proxy can also smooth the problems due to intermittent device connections and resource limits and can exploit user/terminal profiles with personal preferences and device characteristics to customize the service provisioning session. Grasshopper is an MA platform providing interoperability with the most popular standards in the field (CORBA, MASIF, and FIPA), and is mainly targeted to network management applications, especially for third generation (3G) mobile communication systems. The Enago project supports roaming mobility by extending Grasshopper with a flexible support for a large spectrum of heterogeneous wireless devices [32, 33]. Grasshopper/Enago provides proxy components in charge of mediating between legacy distributed systems and portable devices by exploiting a multi-protocol communication service. These proxies also implement transcoding functionality to tailor service results to the specific characteristics of the client device. Analogously, adaptation is a primary feature of the SOMA shadow proxies, which provide both filtering and transcoding operations [21, 24]. Filtering permits the MA to recognize and discard parts of service responses whenever the client device cannot support their visualization and transcoding performs even complex transformations on service data flows, such as HTML-to-WML (Wireless Markup Language) conversion and multimedia format transcoding.

The support of mobile applications also calls for addressing the relevant issue of mobile access to resources: how to discover resources and service components and how to obtain, maintain, or requalify resource bindings while moving [4]. MA platforms are particularly suitable to this scope: they have traditionally treated similar problems because agent mobility forces them to provide support solutions to connect/reconnect to the needed resources after the MA migration to unknown execution environments.

A first challenging aspect of mobile access to resources rises from the impossibility to assume any *a priori* knowledge about the set of available resources and service components. This requires the adoption of articulated mobility-enabled naming solutions to maintain the information about resource availability and allocation. Several MA systems exploit either standard discovery solutions, such as Jini™ network technology, Service Location Protocol (SLP), and Salutation, or directory solutions, such as LDAP. For instance, in AMASE, local LDAP replicas provide MAs with information about locally/globally available services and about the MA location.

A second issue relates to the dynamic creation and reconfiguration of bindings between MAs and their resources during a service session. In AMASE, the MA migration is conditional to the availability of the needed resources in the destination locality. To this purpose, AMASE supports resource negotiation and reservation between MAs and the nodes where they intend to move to. In addition, once migrated, the MA is prevented from using more resources than reserved. Other recent approaches start to investigate how to support different resource binding strategies, dynamically decided depending on the runtime evaluation of the provisioning environment conditions [5]. Resource binding rearrangement after MA migration may involve voiding bindings, reestablishing new ones, or even migrating (replicas of) resources to the new network locality along with the migrating MAs. For instance, SOMA supports the dynamic reconfiguration of resource bindings according to four different strategies — resource movement, copy movement, remote reference, and rebinding. The choice of the binding strategy to apply is cleanly separated from the service application logic to increase flexibility and to favor reusability and rapid development. In particular, SOMA provides a Binder Manager middleware component that mediates between MAs and their referred resources and is also in charge of rearranging bindings after the MA migration without imposing any modification in the service logic implementation [5, 24].

14.4 MA-Based Middlewares with Context Awareness: State-of-the-Art and Emerging Research Directions

A promising and hot direction of research is the extension of MA-based mobile computing middlewares to provide full context awareness. The common idea behind all different approaches — overviewed in this section — is that developing and deploying adaptive mobile applications requires middleware solutions capable of providing mechanisms, tools, and strategies to manage dynamically the service provisioning context. These novel middlewares should be in charge of sensing, processing, controlling, and managing context-related information and of providing mobile applications with an up-to-date context view, thus significantly simplifying the design and implementation of context-dependent mobile services. In other words, the support of such middlewares is required to allow service developers to concentrate only on the application logic, on how to tailor service behavior depending on the context value at negotiation time, and on how to adapt service provisioning in response to context modifications at runtime [5].

Even if still in its infancy, the composite research area of context-aware MA-based middlewares for mobile computing tends to exhibit some common guidelines for the integration of MAs and context-specific functions. Figure 14.2 depicts the emerging middleware architecture. State-of-the-art MA platforms assume a default standard JVM (or a limited K-version of it [34]) to achieve maximum portability. However, middleware context facilities sometimes need to bypass the platform-dependence transparency of the JVM to obtain full visibility of context-related information, such as network monitoring data and operating system-dependent resource state. This requires providing several context-oriented implementation mechanisms to choose dynamically among when deploying the middleware over open and heterogeneous systems, as discussed in the following.

Mobile Agent Middlewares for Context-Aware Applications

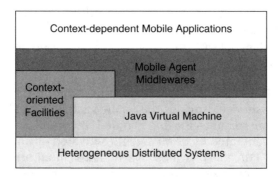

Figure 14.2. The Emerging Architecture of MA-Based Mobile Computing Middlewares with Context Awareness

We identify four main aspects that context-aware MA-based middlewares have to face:

1. How to achieve visibility and extract context-related information from heterogeneous execution environments
2. How to process and aggregate the above information to determine the context of mobile clients during their service sessions
3. How to exploit context awareness flexibly to drive decisions about service adaptation
4. How to use full context visibility to manage Quality of Service (QoS) aspects

Context-related information is intrinsically heterogeneous, spanning from data associated with the execution environment (network connectivity, communication bandwidth, current workload, and locally available resources), to user-specific information (user preference profile, location, and presence of nearby users), and to physical environment conditions (noise level and temperature). MA systems tend to achieve visibility of the above information by exploiting the portable Java environment when possible and by integrating it with different platform-specific versions of heterogeneous sensing techniques (multichannel sensing). In this way, middlewares can dynamically choose the most suitable monitoring mechanisms to exploit depending on the underlying operating system installed over the nodes of deployment [35].

For instance, by focusing on location, which is the most relevant context-related information in mobile environments, AMASE and Monads obtain terminal locations by integrating with an external global positioning system (GPS) [27, 29]. However, GPS is not well-suited to metropolitan area deployment scenarios because of shadowing problems and is often not usable with portable devices because of their strict power limitations. In addition, the GPS location precision is excessive for several application domains where a coarse-grained position estimation is enough to enable context-dependent service provisioning (e.g., location-based tourist guide assistants). For the above reasons, some middlewares dealing with wired–wireless integrated networks decide to exploit the device association with a wireless cell as the location information. For instance, the SOMA location service provides the online visibility of cell-based positioning by exploiting roaming-specific Simple Network Management Protocol (SNMP) traps generated by SNMP agents activated on any Wi-Fi/Bluetooth access point. SOMA also provides an articulated monitoring support, capable of collecting both kernel-level data (e.g., CPU usage, memory allocation, and network traffic) and application-level data (e.g., Java object instantiations and method invocations). SOMA monitoring adopts multichannel implementation of sensing mechanisms by exploiting both the Java Native Interface and the JVM Profiler Interface [35]. Another interesting approach to achieve position visibility is the Hewlett-Packard Cooltown project [36, 37], where CoolAgents adopt location sensing mechanisms based on Radio Frequency Identification (RFID) Badge Readers [38]. The Cooltown RFID Reader is a Java-based wrapper that handles the sensing heterogeneity, by encapsulating the RFID hardware device and redirecting the location information to a Web presence manager.

Due to the heterogeneity of the context-related information and to the multiplicity of needed sensing mechanisms, context-aware middlewares are also required to provide processing and aggregation functions to simplify the extraction of the interesting context data and to facilitate the determination of the service session context. These middleware functions aim at decoupling low-level monitoring mechanisms from high-level context-aware MAs, thus facilitating the MA development and potentially increasing component reusability. For instance, CoolAgents adopt the Georgia Tech Context Toolkit to simplify the handling and usage of context information [36, 39]. The Context Toolkit provides widget components to acquire heterogeneous context information from the execution environment and to make it uniformly available, independently of the specific sensing mechanism exploited. In addition, Context Toolkit aggregators can process/distill data from widgets and can provide their outputs to interpreters, responsible for performing management operations in response to context variations. A similar goal is addressed by the Solar middleware, which provides support for context data aggregation and distribution on top of the D'Agents framework [40, 41]. Solar permits context aggregation via a set of modular and reusable middleware operators: any operator processes and subscribes to one or more input event streams and produces one output event stream. Solar-based mobile applications can use a tree of possibly recursive connected operators (represented by an operator graph) to specify how to produce the needed aggregated context.

About the third aspect, which is how to decide and perform service management operations depending on the current context, recent research efforts are pointing out two main solution guidelines — policy-based and reflection-based service management.

Policy-based MA middlewares support the flexible and dynamic specification of (some aspects of) MA behavior in terms of policies: declarative rules determining the actions that subjects can or must operate upon resources when specified conditions apply. Policies are maintained completely separated from system implementation details and are generally expressed at a high-level of abstraction to simplify their specification by system administrators, service managers, and even final users. In particular, it is possible to define MA security, mobility, and binding behavior in terms of access control and obligation policies, completely separated from the MA code, thus improving the reutilization and the rapid prototyping of MA-based components. For instance, SOMA integrates with a distributed infrastructure for the enforcement of policies expressed in the Ponder language [24, 42]. SOMA policies specify which resources are accessible and belong to a service provisioning context depending on dynamically evaluated conditions (access control policies) and the actions MAs must perform when certain context-triggered events occur (obligation policies). Conditions and events, in their turn, may depend on the values/changes of context-related information. Another emerging research direction is the exploitation of policies to control the coordination and the interaction among MAs. For instance, context-related information in CoolAgents is expressed by XML-based Resource Description Framework (RDF) specifications, and is automatically translated in Prolog rules that are dynamically interpreted to customize the behavior of MA-based applications [35].

Reflection-based MA middlewares exploit reflection mechanisms for context visibility/processing and meta-objects for context-dependent service adaptation. Middlewares that follow this approach suggest a service design methodology decomposing the context-dependent service logic in two different levels:

1. A base-level including the primary MA logic
2. A meta-level in charge of transparently readjusting the MA behavior depending on the current context

For instance, Tanter and Piquer's middleware defines customizable resource binding strategies that are implemented as basic reusable meta-objects attached to any mobile application component [43]. In general, reflection represents an interesting design guideline to achieve context awareness in middleware solutions, but it is difficult to integrate with legacy systems typically implemented by nonreflective programming languages. On the contrary, policy-based approaches require the availability of monitoring and event middleware facilities to trigger the policy enforcement anytime relevant context changes occur, but can apply also to legacy services, independently of their implementation language.

Finally, context awareness is also essential to enable service management operations to monitor, control, tailor, and adapt the QoS levels provided during service sessions. This is particularly crucial when operating either over best-effort networks or with frequent client mobility (and frequent resulting modifications of the client points of attachment to the network), where it is hard to guarantee the provided QoS via resource reservation at negotiation time. In these scenarios, managing QoS means operating promptly to try to maintain the agreed upon QoS levels, independently of the changes in available resources occurring in the provisioning environment. In other words, context-aware middlewares with QoS management goals should provide the online monitoring of the context information of interest for the supported QoS-sensitive applications and should react promptly to context modifications with proper service management operations: for instance, by transcoding on-the-fly, a high-resolution MJPEG (Motion JPEG) video stream to a low-resolution MPEG-2 (Motion Picture Experts Group) version of it when the locally available bandwidth falls. For instance, the SOMA QoS support is centered on the dynamic adaptation of provided QoS over best-effort networks. SOMA includes a rich monitoring infrastructure to achieve full context visibility [35]. Context variations related to changes in network-, system-, and application-level resource states trigger corrective QoS management operations. These operations are performed by SOMA agents that compose a dynamically deploying application-specific overlay network. SOMA focuses on QoS monitoring and control, whereas Monads exemplifies an alternate, interesting approach based on context prediction [27]. Monads provides a QoS prediction facility that, on the basis of the previous context values and of its evolution history, models QoS as a function of location and time and supplies estimates about future QoS levels based on this model. Context estimates are exploited in Monads for scheduling, data prefetching, and connection management decisions.

14.5 Lessons Learned and Open Issues

As discussed above, in the last five years, several research activities have addressed different aspects of MA-based supports for mobility and have shown the suitability and the effectiveness of adopting the MA programming paradigm in this scenario. Even if these research projects are still looking for a single agreed upon MA-based killer application, they have already shown that, at least for mobile computing middlewares, where MAs work as proxies of possibly disconnected users/devices, "while none of the individual advantages of MAs is overwhelmingly strong, we believe that the aggregate advantages of MAs are overwhelmingly strong" [44].

Mobile Agent Middlewares for Context-Aware Applications

The enlarging market of wireless portable devices is further motivating this claim, because limited wireless terminals stress both the need to manage discontinuities in resource availability during service provisioning and the relevance of mobile middleware intermediaries dynamically deploying over the wired infrastructure, when and where needed.

The research work done has pointed out several lessons learned, which have become common guidelines or solutions for MA platforms in general and for MA-based middlewares in the specific case of mobile computing support. On the one hand, it is reasonable that MA platforms work on the standard JVM to achieve portability over open distributed systems. Therefore, service developers should expect weak mobility to be the typical type of MA mobility supported: it is not accidental that all Java-based MA approaches sketched in Section 14.3 and Section 14.4 provide weak mobility. On the other hand, it is crucial that MA platforms give the full visibility of location information to MA programmers, because it is fundamental, in mobile computing especially, to have location information modeled as a first class programming concept. Moreover, MA-based frameworks for mobile computing should provide a modular mobility support to accommodate differentiated forms of mobility (both nomadic and roaming), from user/terminal mobility to dynamic resource binding rearrangement, also in case of the dynamic reallocation of resources and service components.

The other two main guidelines emerged in the state-of-the-art research in the field are the adoption of proxy-based architectures and the necessity of full context awareness. On the one hand, MA-based mobile proxies can assist locally mobile clients during their whole service sessions: for instance, by tailoring and adapting service contents to the specific user preferences and access device characteristics. On the other hand, in the mobile computing scenario, the design and implementation of applications should achieve high service adaptability to fit the differentiated and heterogeneous provisioning conditions. This calls for middleware mechanisms and tools to properly manage context and its frequent runtime modifications.

The promising results obtained in MA-based middlewares for mobile computing are encouraging further activities and are likely to attract several research efforts in the next months. A relevant open issue is how to express resource binding/mobility strategies at the proper level of abstraction, by maintaining a clear separation between the currently enforced strategy and the service logic implementation to achieve the requested level of dynamicity, flexibility, and reusability of both middleware and service components. Policy-based and reflective-based approaches presented in Section 14.4 exemplify this trend. Novel middleware solutions should integrate with different types of high-level metadata to provide the needed management configurability while hiding low-level mechanisms and implementation details from service developers and system administrators [5].

Another central open issue is the support of dynamic and open service composition, crucial to fast prototype and tailor service provisioning in heterogeneous mobile environments. Dynamic service composition requires accepted models and representation formats to describe the interface, the invocation syntax, and the semantics associated with available service components. The recent standardization efforts accomplished in Web services — with their XML-based formats for registration/discovery, service description, and service access, respectively Universal Description, Discovery, and Integration (UDDI), Web Services Description Language (WSDL), and Simple Object Access Protocol (SOAP) — certainly represent an interesting first step toward open service composition, which MA supports for mobile computing are expected to integrate within their frameworks soon [45]. As a side effect, this will also extend the possibility for MA-based supports to integrate with existing systems and services, thus leveraging the diffusion of MA technologies in commercial and industrial applications.

Acknowledgments

Work supported by the Italian Ministero dell'Istruzione, dell'Università e della Ricerca (MIUR) in the framework of the FIRB WEB-MINDS Project "Wide-scale Broadband Middleware for Network Distributed Services" and by the Italian Consiglio Nazionale delle Ricerche (CNR) in the framework of the Strategic IS-MANET Project "Middleware Support for Mobile Ad hoc Networks and Their Application."

References

1. L. Blazevic, L. Buttyan, S. Capkun, S. Giordano, J.-P. Hubaux, and J.-Y. Le Boudec, Self-Organization in Mobile Ad Hoc Networks: the Approach of Terminodes, *IEEE Communications*, vol. 39, no. 6, June 2001.
2. C. Perkins (Ed.), Special Section on "Autoconfiguration," *IEEE Internet Computing*, vol. 3, no. 4, July 1999.
3. L. Bos and S. Leroy, Toward an All-IP-based UMTS System Architecture, *IEEE Network*, vol. 15, no. 1, January—February 2001.
4. P. Bellavista, A. Corradi, and C. Stefanelli, Mobile Agent Middleware for Mobile Computing, *IEEE Computer*, vol. 34, no. 3, March 2001.
5. P. Bellavista, A. Corradi, R. Montanari, and C. Stefanelli, Dynamic Binding in Mobile Applications, *IEEE Internet Computing*, vol. 7, no. 2, March–April 2003.
6. J.A. Moura, J.M. Oliveira, E. Carrapatoso, and R. Roque, Service Provision and Resource Discovery in the VESPER VHE, *IEEE Int. Conf. on Communications (ICC'02)*, Piscataway, NJ: IEEE Computer Society Press, April 2002.
7. Teleca AB — Bluetooth Local Infotainment Point (BLIP), http://www.teleca.com.
8. V. Kumar, S. Parimi, and D.P. Agrawal, WAP: Present and Future, *IEEE Pervasive Computing*, vol. 2, no. 1, January–March 2003.
9. E. Kovacs, K. Rohrle, and M. Reich, Integrating Mobile Agents into the Mobile Middleware, *Proc. of the Mobile Agents Int. Workshop* (MA'98), Berlin, 1998.
10. D. Kotz, R. Gray, S. Nog, D. Rus, S. Chawla, and G. Cybenko, Agent TCL: Targeting the Needs of Mobile Computers, *IEEE Internet Computing*, vol. 1, no. 4, July–August 1997.

11. S. Lipperts and A. Park, An Agent-Based Middleware: a Solution for Terminal and User Mobility, *Computer Networks*, vol. 31, September 1999.
12. J.E. White, Telescript Technology: the Foundation for the Electronic Marketplace, General Magic White Paper, 1994, http://www.genmagic.com.
13. N.S. Borenstein, E-Mail with a Mind of its Own: The Safe-TCL Language for Enabled Mail, *IFIP Transactions C (Communication Systems)*, vol. C-25, 1994.
14. V. Kisielius, Applying Intelligence Makes E-Commerce Pay Off, *Electronic Commerce World*, vol. 7, no. 12, December 1997.
15. D. Chess, B. Grosof, C. Harrison, D. Levine, C. Parris, and G. Tsudik, Itinerant Agents for Mobile Computing, *IEEE Personal Communications*, vol. 2, no. 5, May 1995.
16. A. Fuggetta, G.P. Picco, and G. Vigna, Understanding Code Mobility, *IEEE Transactions on Software Engineering*, vol. 24, no. 5, May 1998.
17. J.W. Stamos and D.K. Grifford, Implementing Remote Evaluation, *IEEE Transactions on Software Engineering*, vol. 16, no. 7, July 1990.
18. G. Goldszmidt, Distributed System Management via Elastic Servers, *IEEE Int. Workshop on Systems Management (SMW'93)*, Piscataway, NJ: IEEE Computer Society Press, April 1993.
19. N.M. Karnik and A.R. Tripathi, Design Issues in Mobile Agent Programming Systems, *IEEE Concurrency*, vol. 6, no. 3, July–September 1998.
20. D. Chess, C.G. Harrison, and A. Kershenbaum, Mobile Agents: Are They a Good Idea?, in G. Vigna (Ed.), *Mobile Agents and Security, Lecture Notes in Computer Science*, vol. 1419, New York: Springer-Verlag, 1998.
21. P. Bellavista, A. Corradi, and C. Stefanelli, The Ubiquitous Provisioning of Internet Services to Portable Devices, *IEEE Pervasive Computing*, vol. 1, no. 3, July–September 2002.
22. Object Management Group, *Mobile Agent System Interoperability Facility (MASIF) Specification*, http://www.fokus.fraunhofer.de/research/cc/ecco/masif.
23. Foundation for Intelligent Physical Agents (FIPA), http://www.fipa.org.
24. Department Electronics Computer Science and Systems (DEIS) — University of Bologna, *Secure and Open Mobile Agents (SOMA)*, http://lia.deis.unibo.it/research/SOMA.
25. World Wide Web Consortium, Composite Capability/Preference Profiles (CC/PP), http://w3.org/Mobile.
26. S. Campadello, H. Helin, O. Koskimies, P. Misikangas, M. Mäkelä, and K. Raatikainen, Using Mobile and Intelligent Agents to Support Nomadic Users, *Proc. of the 6th Int. Conf. Intelligence in Networks (ICIN2000)*, France, January 2000.
27. Department Computer Science — University of Helsinki, *Adaptation Agents for Nomadic Users (Monads)*, http://www.cs.helsinki.fi/research/monads.
28. Department Computer Science — Dartmouth College, *Dartmouth Agents (D'Agents)*, http://agent.cs.dartmouth.edu.
29. B. Schiemann, E. Kovacs, and K. Röhrle, Adaptive Mobile Access to Context-Aware Services, *Proc. of the 3rd Int. Workshop Mobile Agents (MA'99)*, Piscataway, NJ: IEEE Computer Society Press, October 1999.
30. F. Bellifemine, A. Poggi, and G. Rimassa, JADE — A FIPA-Compliant Agent Framework, *Proc. of the 4th Int. Conf. Practical Application of Intelligent Agents and Multi-Agent Technology (PAAM'99)*, Great Britain, 1999.
31. Telecom Italia Lab, *The Java Agent DEvelopment Framework (JADE)*, http://sharon.cselt.it/projects/jade/.
32. M. Breugst and T. Magedanz, Mobile Agents — Enabling Technology for Active Intelligent Network Implementation, *IEEE Network*, vol. 12, no. 3, August 1998.
33. IKV++, *Grasshopper 2*, http://www.grasshopper.de.
34. Sun Community Source Licensing — *K Virtual Machine (KVM)*, http://wwws.sun.com/software/communitysource/j2me/cldc/.
35. P. Bellavista, A. Corradi, and C. Stefanelli, Java for Online Distributed Monitoring of Heterogeneous Systems and Services, *The Computer Journal*, vol. 45, no. 6, November 2002.

36. H. Chen and S. Tolia, Steps Toward Creating a Context-Aware Software Agent System, HP Technical Report HPL-2001-231, September 2001.
37. Hewlett Packard, *Cooltown*, http://www.cooltown.com.
38. V. Stanford, Pervasive Computing Goes the Last Hundred Feet with RFID Systems, *IEEE Pervasive Computing*, vol. 2, no. 2, April–June 2003.
39. A.K. Dey, D. Salber, and G.D. Abowd, A Conceptual Framework and a Toolkit for Supporting the Rapid Prototyping of Context-Aware Applications, *Human-Computer Interaction (HCI) Journal*, vol. 16, 2001.
40. G. Chen and D. Kotz, Context-Sensitive Resource Discovery, *Proc. of the 1st Int. Conf. Pervasive Computing and Communications (PERCOM'03)*, Piscataway, NJ: IEEE Computer Society Press, March 2003.
41. Department of Computer Science — Dartmouth College, *Solar*, http://www.cs.dartmouth.edu/~solar.
42. R. Montanari, C. Stefanelli, and N. Dulay, Flexible Security Policies for Mobile Agent Systems, *Microprocessors and Microsystems*, Elsevier Science, vol. 25, no. 2, May 2001.
43. E. Tanter and J. Piquer, Managing References upon Object Migration: Applying Separation of Concerns, *Proc. of the 21st Int. Conf. Chilean Computer Science Society (SCCC'01)*, Chile, Piscataway, NJ: IEEE Computer Society Press, 2001.
44. R.H. Glitho, Emerging Alternatives to Today's Advanced Service Architectures for Internet Telephony: IN and Beyond, *Computer Networks*, vol. 35, no. 5, 2001.
45. F. Curbera, M. Duftler, R. Khalaf, W. Nagy, N. Mukhi, and S. Weerawarana, Unraveling the Web Services: an Introduction to SOAP, WSDL, and UDDI, *IEEE Internet Computing*, vol. 6, no. 2, March–April 2002.

Section IV
Caching Strategies

Chapter 15
Cache Management in Wireless and Mobile Computing Environments

Yu Du and Sandeep K.S. Gupta

15.1 Introduction

Caching is an important technique in the computing world. Most of us are familiar with cache used for a single processor computer, as illustrated in Figure 15.1. In a typical single processor system, a cache is a small fast memory for holding frequently used data. In fact, all the different types of memories in a computer system, such as disk, main memory, and cache, can be viewed as a part of a memory hierarchy in which the closer a memory unit is to the processor the smaller is its capacity and access time. The main problem in such a single processor computing environment with a memory hierarchy is to determine *how to reduce memory read/write latency in the presence of memory hierarchy*? To achieve this, the focus is to decide what data to cache (to copy from a farther memory unit to a closer memory unit) so that the cache (memory unit closer to the processor) can satisfy as many memory requests as possible, effectively achieving average access time close to that of the fast memory unit at the effective price of the large memory unit. Therefore, the problem for cache management is to predict what data items will most probably be used in the future and either copy them to a memory unit closer to the processor when they are accessed for the first time (i.e., on a cache miss) or to prefetch them well in advance so that they are available in the cache memory when they are needed. Many cache replacement algorithms such Least Recently Used (LRU) and prefetching algorithms have been developed to solve this problem [Silberschatz].

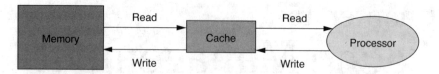

Figure 15.1. Cache Scheme for a Single-Processor Computer Showing Just Two Levels in the Memory Hierarchy

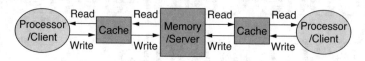

Figure 15.2. Cache Scheme in Distributed System and Network Environments

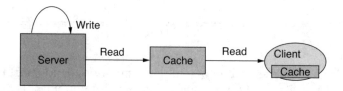

Figure 15.3. Cache Scheme in Distributed System and Network Environments

In distributed systems and network computing environments, cache management schemes need to handle two different scenarios. In the first scenario, the data in the shared memory or the server can be read/written by different processors or clients concurrently, as shown in Figure 15.2. In the second scenario the data is read-only for the clients, as shown in Figure 15.3. Further, many times the data may be replicated onto multiple servers for improving fault-tolerance and availability. An example of the former scenario is a distributed file system and the latter scenario is often seen in the World Wide Web (WWW) system. In these environments, when a client accesses a cached data item, the server or another client could have just modified that data item. Hence, in distributed and network computing environments, the most crucial problem for caching is *how to maintain data consistency among the clients and the servers?* This problem is more difficult to solve than the problem of ensuring that the copy of the data in the higher level memory is updated (through write-through or write-back schemes) when the cached copy of the data in the lower level memory has been updated. The complexity arises from the various failures that may occur — server failure, network failure, and client failure.

Cache Management in Wireless and Mobile Computing Environments

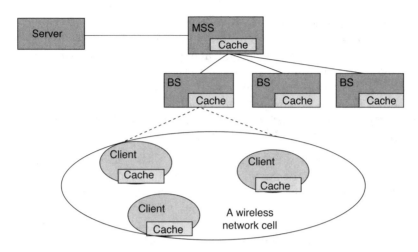

Figure 15.4. Cache Scheme in Architecture-Based Wireless Networks

To maintain cache consistency in distributed or network computing environment, different approaches have been developed. These approaches include polling-every-time, adaptive time-to-live (TTL) [Gwertz], server-invalidation [Cao], and leases-based invalidation [Cao, Yin].

With the increase in the popularity of wireless network and mobile computing environments, new problems have arisen for cache management in these environments. Broadly speaking, there are two types of mobile wireless networks — architecture-based and architectureless wireless networks. Figure 15.4 illustrates an architecture-based wireless network. In this network model, mobile switching stations (MSSs) and base stations (BSs) form a wireless cellular network through which a mobile host (MH) can communicate with remote data servers. The data servers, MSSs, and BSs are connected by static network, which has high speed and reliability compared to wireless link (last hop) of the network. Compared to wired links, the wireless links of the cellular network are of low-bandwidth and subject to frequent disconnections for various reasons, leading to weakly connected mobile clients. This feature creates a new problem for cache management: *how to ensure high data availability in mobile computing environments where frequent disconnections may occur because the clients and server may be weakly connected?*

The mobile clients can often be disconnected from the data servers either involuntarily (e.g., wireless connectivity may not be available in certain areas) or voluntarily (e.g., the user shuts off the wireless network interface to conserve battery). However, the user would like to still have access to data vital to the application he or she is working with. Further, as we will later see in the chapter, for invalidation-based cache consistency strategies, we also need to ensure the availability of the invalidation reports to the weakly connected clients so that they can maintain the consistency of the cached data. A second feature of mobile computing environment is resource constraints of the clients, because typical clients in mobile wireless network environments have limited power or processing resources. This feature leads to another problem for cache management in mobile computing environments, namely, *how to minimize energy and bandwidth overhead for cache management?* Many approaches to address these problems have been proposed. The details are discussed in Section 15.4.

The third feature of architecture-based wireless networks is asymmetric communication links, because the downstream (base station to mobiles) communication link capacity is usually much higher than the upstream (mobile to base station) capacity. Further, mobiles may have to compete with several other mobiles to get access to an upstream channel or even they may not have the capability to perform uplink communication. Competing with other mobiles for uplink channel and requesting the server for a data item is expensive in terms of battery power consumption, because a node may have to keep its network interface powered up from the time it initiates the request to the time the response arrives from the server. Some techniques for allowing the client to conserve energy during this wait period of on-demand or pull-based information access have been recently developed, for example see [Krashinsky]. Traditional client–server information systems use pull-based communication schemes for information access where clients initiate data transfers by requesting a server. Such pull-based techniques are not suitable for architecture-based wireless network because, as we pointed out above, it requires substantial upstream communications. To make use of the downstream communication capacity, push-based information system architectures have been developed, where data is pushed from the servers to clients [Acharya95, Acharya97, Imie94A, Imie94B]. The idea is that the server periodically broadcasts frequently accessed data (hot data items) on the broadcast channel. The mobile can tune in to the broadcast channel at the start of the broadcast, determine when the data items it is interested in will be available on the channel by reading the index information and then go to sleep until the time when the data item is on the channel. Such push-based architecture brings up a new problem for cache management. Traditionally, caches are used to store the most frequently used data, but in the push-based environment, the cost of obtaining the data should also be considered. Using the example from [Acharya95], data item x is accessed 1 percent of the time at a client C and is also broadcast 1 percent of the time. Data item y is accessed 0.5 percent of the time at C but is broadcast only 0.1 percent of the time. If we choose to cache x instead of y, then the client will experience a longer delay while cache miss happens for y. Therefore, new cache management algorithms have been developed for pushed-based information systems, which take into account the cost of cache miss. Not that in traditional cache systems all cache misses are assumed to have same cost. This is not so in push-based information access architecture.

In general, an important metric to evaluate these cache management algorithms is hit ratio. Hit ratio is calculated as the fraction of total data requests satisfied from the cache. Note that this metric not only depends on cache management algorithms, but also depends on the cache size and particular request pattern. However, in mobile computing environments, hit ratio should not be the only metric to evaluate cache management algorithms [Satya], because its underlying assumption is that all cache misses have equal cost. But this assumption does not necessarily hold in weakly connected environments, where cache miss cost also depends on data size and the timing. Therefore, one needs to consider new metrics representing different cache cost in mobile computing environments.

The fourth feature of mobile computing environment is that the data may be location dependent. For example, as the user moves, that user may want the yellow page, map, or traffic information specific for the user's current location. Location-dependent data imposes another problem for cache management algorithms, because the decision to cache/replace the data item not only depends on temporal locality or spatial locality in the reference pattern, but also on the location of the mobile.

The other type of mobile wireless network is called architectureless wireless network or mobile ad hoc network (MANET). This type of network is composed solely of mobile computing devices within mutual wireless communication range of each other. These types of networks are different from architecture-based wireless networks in the sense that there are no dedicated network infrastructure devices in a MANET. Because of the limited radio propagation range of wireless devices, the route from one device to another may require multiple hops. In some scenarios, to communicate with the outside world few devices that have network connections with the outside base stations (or satellite) can serve as the gateways for the ad hoc network. We would refer to such ad hoc networks as weakly connected ad hoc networks, because they do have some infrastructure support, but not as much as the mobile nodes in the infrastructure-based networks. An example of such a weakly connected ad hoc network is shown in Figure 15.5. Such weakly connected ad hoc networks may be used for emergency hospitals and for military operation in remote area. In general, the MANET environment also has the two features of wireless computing environments — weak connectivity and resource constraints. Therefore, we still need to address the problems of data availability and bandwidth/energy efficiency. However, we cannot blindly reuse the cache management schemes developed for an architecture-based wireless network to solve the cache management problems in a MANET environment, because these two environments have the following differences that we cannot overlook:

MOBILE COMPUTING HANDBOOK

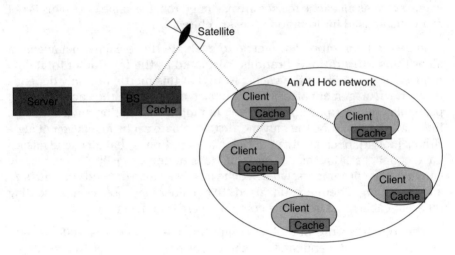

Figure 15.5. An Example of Weakly Connected Ad Hoc Networks

1. Gateways in ad hoc networks versus base stations in architecture based wireless network — gateways of ad hoc networks are unreliable mobile computing devices, yet base stations are reliable dedicated networking devices. Gateways communicate with local hosts using low-bandwidth unreliable links such as radio frequency (RF) wireless links and possibly with remote hosts through high-latency unreliable links, such as satellite channels. On the other hand, base stations communicate with remote hosts through high-speed wired networks.

2. Client–server versus peer-to-peer (P2P) architecture — MANETs are inherently P2P networks with weak connectivity. In contrast, an architecture-based wireless network follows the client–server structure. The P2P structure gives a new choice to improve cache performance in MANET environments, as we can use cooperative caching to improve cache performance. There are a few schemes for cooperative caching in MANET environments that we describe in Section 15.5.

Cooperative caching is also being studied for the Internet to provide more cache space and faster speed. For example, the National Laboratory for Applied Network Research [NLANR] caching hierarchy consists of many backbone caches and those caches can obtain data from each other using Hypertext Transfer Protocol (HTTP) and Internet Caching Protocol (ICP). However, these cooperative caching schemes do not address the special concern of MANET environments, namely, weak connectivity and severe resource constraints.

Cache Management in Wireless and Mobile Computing Environments

In summary, in mobile computing environments, the cache management schemes need to address the following problems:

- How to reduce client side latency?
- How to maintain cache consistency between various caches and the servers?
- How to ensure high data availability in the presence of frequent disconnections?
- How to achieve high energy/bandwidth efficiency?
- How to determine the cost of cache miss and how to incorporate this cost in cache management scheme?
- How to manage location-dependent data in the cache?
- How to enable cooperation between multiple peer caches?

The first two problems are not new to mobile computing environments. They have been extensively studied in distributed system and wired network computing environments. The next four problems stem from the two features of mobile computing environments. And the last problem is to take advantage of the P2P structure of ad hoc networks. In the next section, we present several cache management schemes that have been proposed to address these problems specific to mobile computing environments.

15.2 State of the Art

There is a lot of research work going on to solve the problems brought up in the previous section. The research about cache management is to study what data to cache to satisfy as many requests as possible. Consistency strategies solve the problem of how to keep cached data consistent with the original copy. Cooperative caching is to explore how to collaborate multiple caches to achieve better performance. The consideration for data availability and energy/bandwidth efficiency is embedded in the research of cache management, consistency strategies, and cooperate caching for mobile computing environments. Figure 15.6 gives an overview of the research work in this field.

Cache management is studied from three perspectives — replacement algorithms, prefetching algorithms, and cache-ability (whether or not to cache a particular data item?) determination. The focus of these algorithms is to decide what data items are worth placing in the cache. Cache replacement algorithms are used to decide what data to remove from the cache when the cache is full and a new data item needs to be placed in the cache. Some of this research work is also applicable for mobile computing environments. For example, we can use LRU as the cache replacement algorithm for mobile computing devices. However, as is shown in paper by [Acharya95], cost-based replacement policies may obtain a better performance than LRU. Another example is the Coda file system [Kistler] that is

MOBILE COMPUTING HANDBOOK

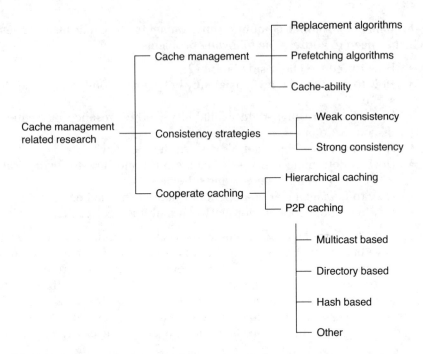

Figure 15.6. Overview of Cache Management Research

used in a weakly connected UNIX® environment. In the hoarding state of Coda, they use a prioritized algorithm to periodically reevaluate the data items and replace the data item with the one that has higher priority. Priority can be either obtained from the user preferences or from the history of access pattern. Prefetching techniques are investigated in the paper [Jiang] to improve the speed of Web access. The paper [Gitzenis] describes a modeling framework to study the problem of power-controlled data prefetching. And the paper [Grassi] proposed prefetching policies to gain a good tradeoff between latency and energy cost in a broadcasting environment. To decide whether a data item is cacheable or not, the paper [Sistla] proposed a sliding-window algorithm. The basic idea is that a data item should not be cached unless its read rate is higher than its update rate.

As it was mentioned in Section 15.2, new cache management algorithms should be developed for push-based information systems, because the cache miss cost is not simply based on the access frequency. The paper [Acharya95] proposes a cost-based page replacement algorithm using PIX. Suppose the access probability of data item d is P and the broadcast frequency is X, then the PIX value of d is (P/X). For the example in Section 15.2, the algorithm replaces item x with y, because y has a higher PIX value. Note that the broadcast pattern of the server should also be considered for the client's cache management algorithms.

For location-dependent data, cache management algorithm decides whether to cache data or not according to the location. The paper [Ren] proposed a location-dependent cache replacement strategy FAR (furthest away replacement). The basic idea of this strategy is to first replace the data that is furthest from the user and is not in the user's moving direction. The paper [Dar] exploited semantic caching. This approach maintains a semantic description for the cached data and chooses replacement victims based on the semantic dependencies of the user's queries.

From the previous discussion, we can classify cache replacement algorithms into three types [Ren]:

1. Temporal locality-based, such as LRU and CLOCK [Silberschatz]
2. Benefit-based (or cost-based), such as PIX
3. Semantic-based, such as FAR

These three types of algorithms are different in how they choose cache replacement victims. Temporal locality-based algorithms are based on the property that a recently used data item is likely to be used again in the near future and hence, access frequency is the main factor to decide the replacement. Benefit-based algorithms exploit the fact that the fetching costs are different for different data and decide the replacement by evaluating data against the fetching cost and its access probability. Semantic-based cache management is studied in [Dar] for caching database query results. It is further used in caching location-dependent data in [Ren], which decides the replacement based on the location semantic.

There exists a lot of research work to study the problem of how to maintain cache consistency in a mobile computing environment. There are two different cache consistency requirements — strong cache consistency and weak cache consistency. Strong cache consistency enforces that the cached data is always up-to-date. Polling-every-time and invalidating data upon modification are two approaches to achieve strong cache consistency. On the other hand, weak cache consistency allows some degree of data inconsistency. TTL-based consistency strategies are used when it is sufficient to guarantee weak consistency for a data item. In mobile computing environments, the main challenge for consistency strategies is caused by weak connectivity and resource constraints of mobile computing environments. We will categorize and describe different consistency strategies for mobile computing environments in Section 15.4.

Cooperative caching is an interesting problem in the mobile computing environment, especially in *ad hoc* network environment. Because an individual MH may have limited storage or power, a good cooperative caching scheme could help to achieve better performance at less storage overhead or power cost. Several cooperative caching schemes for ad hoc networks have been developed [Hara02A, Sailhan02, Sailhan03]. However, there are

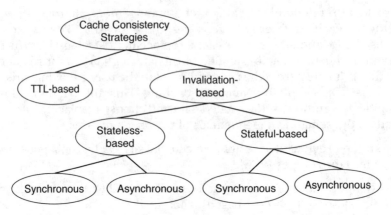

Figure 15.7. A Taxonomy of Caching Strategies

still open problems that need to be addressed in this area. Section 15.5 gives an overview of these research works and the problems that need to be addressed in the future.

15.3 Cache Consistency Strategies

In this section, we are going to discuss various cache consistency strategies. We begin this section with a taxonomy of consistency strategies. Then we describe the details of a few different types of cache consistency strategies.

15.3.1 A Taxonomy of Cache Consistency Strategies

There are different strategies for maintaining cache consistency in the literature. A taxonomy of cache management schemes is illustrated in Figure 15.7. The consistency strategies that are depicted in the figure are not restricted for mobile computing environments. According to the paper [Cao], cache consistency strategies are differentiated into three categories — polling-every-time, TTL-based, and invalidation-based. For polling-every-time and TTL-based caching strategies, the client side initiates the consistency verification: that is, the client is responsible for verifying the data consistency before using it. For the TTL-based caching strategy, every cached data item is assigned a TTL value, which can be estimated based on the data item's update history. For example, the adaptive TTL approach in [Cate] estimates TTL based on the age of a data item. When the user request arrives for a data item x, if data item x's residence time has exceeded its TTL value, the client sends a message to the server to ask if x has changed. Based on the server's response, the client may get a new copy of x from the server (if the data item x has changed since the last time the client cached x) or just use the cached copy to answer the user's request (if the data item x has not been modified since the last time the client received a copy of x). For the polling-every-time approach, every time the data is requested, the clients need to send requests to the server to verify if the cached data has changed. The polling-every-time caching strategy can be thought of as a special type of TTL-based scheme, with the TTL field equal to zero for every data item. Polling-every-time and TTL-based approaches are used in many existing Web caches.

On the other hand, for the invalidation-based strategies, the cache consistency verification is initiated by the server. Invalidation-based cache strategies are further classified into stateless and stateful approaches [Barbara0]. In a stateless approach, the server does not maintain information about the cache contents of the clients (i.e., the server does not know what data is cached or how long it has been cached by a particular client). This type of cache strategy is discussed in many papers including [Barbara0, Barbara1, Cao, Jing, Tan]. The stateless-based approach can be further categorized into synchronous and asynchronous approach. In the asynchronous approach, invalidation reports are sent out upon data modification. In the synchronous approach, the server periodically sends out invalidation reports, which may overlap with the previous invalidation. Most of the research work focuses on stateless synchronous-based approach. For a detailed comparison and evaluation of different approaches in this category, the interested reader can refer to [Tan].

In case of the stateful approach, the server keeps track of the information of the cache contents. These approaches can also be categorized into synchronous and asynchronous approaches. There are hardly any schemes in the stateful synchronous category. One example for stateful asynchronous approach is proposed in the paper [Gupta]. In that approach, a Home Location Cache (HLC) is maintained for every client and is used to record the data items cached by that client and their last modification time. Based on the information, the server can generate invalidation reports dedicated to a particular client cache.

The above taxonomy is not restricted for mobile computing environments. In the following sections, we are going to concentrate on the cache consistency strategies specified for mobile computing environments.

15.4 Cache Consistency Strategies in Architecture-Based Wireless Networks

As shown in Figure 15.8, in the Architecture-Based Network Model, MSSs and BSs form a wireless cellular network through which a MH can communicate with remote data servers. The data servers, MSSs, and BSs are connected by reliable wired networks with high speed. Whereas the wireless links of the cellular network have limited bandwidth and are subject

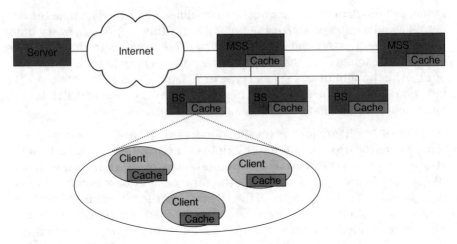

Figure 15.8. Caching in the Architecture-Based Wireless Networks

to disconnection, Coda [Kistler] studied the scenario where weakly connected clients can read/write server data, but most of the other research work has concentrated on the scenario where the server data is read-only for the clients. Because we can use similar consistency strategies for these two scenarios, the following discussion focuses on the latter scenario. We categorize and describe these consistency strategies based on the taxonomy given in Figure 15.7. Then we discuss how these strategies address the problems of weak connectivity and resource constraints.

15.4.1 *Invalidation-Based Consistency Strategy*

In invalidation-based consistency strategies, the server pushes invalidation reports to the clients to announce what data has changed. If the strategy does not require the server to maintain any information about the client's cached contents, we call it a stateless approach. If the server needs to remember what data is cached, we call it stateful approach.

15.4.1.1 Stateful Approaches. A few stateful cache consistency strategies have been proposed. One example is the asynchronous stateful (AS) algorithm [Gupta]. In this approach, the server maintains a (HLC) for every MH under its coverage. The HLC has one record for each data item cached by the corresponding MH to keep track of the item's last update time. Invalidation reports are cached in the HLCs until an explicit acknowledgment is received from the corresponding MH. At the client side, every MH maintains a cache time stamp to indicate when it received the last update report. When a MH reconnects to the network after sleeping, it first sends a probe message to the server with its cache time stamp. Then the HLC can determine which data items have changed and it can then send the updates to the MH.

The advantages of the AS approach are:

- The bandwidth is saved because it avoids unnecessary broadcasting invalidation reports.
- It supports cache consistency verification under any arbitrary sleep–wakeup pattern.

However, because a MSS needs to maintain the information about the data items cached by every MH, the storage cost at MSS has to scale up when the number of MHs or the cached data increases.

Another example is the scalable asynchronous cache consistency scheme (SACCS) from [WangZ]. In this scheme, MSS maintains a flag and a time stamp for every data item. The flag is set when the corresponding data item is cached by a MH. Otherwise, this flag is cleared. When a data item is changed, MSS will send out its invalidation report only if the flag is set. On the client side, every cached data item has an associated time stamp indicating the last update time and four possible states — valid, uncertain, uncertain with a waiting query, and ID-only. Every time the client receives an invalidation message for the data items in valid state, it changes the data item's state to ID-only. If the client is disconnected, it will set all the valid data items to uncertain state upon reconnection. When the client receives a query for an uncertain data item, it changes the data state from "uncertain" to "uncertain with a waiting query" and at the same time sends an uncertain message, including the data item ID and the time stamp, to the MSS. Then the MSS checks if the data item is still valid based on the comparison of the time stamps. If it is, then the MSS simply sends back a confirmation message. Otherwise, the MSS sends the updated data and time stamp to the client.

The advantages for the SACCS approach are:

- It saves bandwidth in the sense that it only broadcasts invalidation reports if that data is cached by some MH.
- It can deal with arbitrary sleep–wakeup pattern.
- It saves MSS storage compared to the AS approach.

However, the MH verifies the cached data items using multiple uncertain messages (one uncertain message for one data item), but the AS approach verifies the whole cache using a single probe message. Therefore, compared to AS, SACCS may cost more energy and bandwidth to verify the cache after MHs wake up. Besides, if there are a few common data cached by multiple MHs, it uses similar amount of MSS storage compared to AS approach.

Table 15.1. Example: Data Items and Their Update Time

Data	D_1	D_2	D_3	D_4	D_5	D_6	D_7	D_8	D_9	D_{10}	D_{11}	D_{12}	D_{13}	D_{14}	D_{15}	D_{16}
Time	T_{24}	T_{16}	T_{10}	T_6	T_{20}	T_{18}	T_{30}	T_{32}	T_2	T_{22}	T_{14}	T_{26}	T_8	T_4	T_{12}	T_{28}

The following example gives a simple illustration to the AS and the SACCS strategy. Suppose the data server has the data items with modification times as shown in Table 15.1.

Suppose the client initially had the cache time stamp T_0 and cached the data items D_1, D_3, D_5, D_7, D_9, D_{11}, D_{13}, D_{15}. Figure 15.9 shows the invalidation messages and the client's cache states for the AS algorithm. When a client receives an invalidation message, it will remove the data item from the cache and update its cache time stamp to the time stamp of the recent invalidation message. For example, after the client receives the first invalidation message, it removed the data item D_9 and updated its cache time stamp to T_2. While a client is disconnected, it will lose all the invalidation messages. It will ignore all the invalidation messages until it needs to send out the first query. A probe message is sent out together with the first query, which tells the client's cache time stamp. For example, after reconnection, the client sends out the cache time stamp together with the request for D_5. Then the server replies with the data items' IDs updated after the cache's time stamp, which are D_1, D_5, D_{11}, and D_{15}.

Figure 15.10 shows the invalidation messages and the client's cache state for the SACCS algorithm. When a client receives an invalidation message, it will remove the data item from the cache, but it will keep the data item's ID and set its state to ID-only, which is indicated by the "–" in the example. For example, after the client received the first invalidation message, it removed the data item D_9 and changed the data's state to ID-only. While a client is disconnected, it will lose all the invalidation messages. Upon reconnection, the client will set all the valid data items' states to uncertain, which is indicated by "*." If the client receives a query to a data item in uncertain state, it will change the data state to uncertain with a waiting query indicated by the "@" symbol and send out an uncertain message to the server, which includes the data item's time stamp indicating its last modification time. Then the server will check if there is a modification after the time stamp. If this not the case, the server will simply send back a confirmation message. Otherwise, the server will send back the data and its last modification time, which is shown in the example scenario. Accordingly, the client will change the data's state from uncertain to valid.

15.4.1.2 Stateless Approach. In stateless consistency strategies, the server does not have any information about the client's cache. Those strategies can be categorized based on the time when the server sends invalidation

Cache Management in Wireless and Mobile Computing Environments

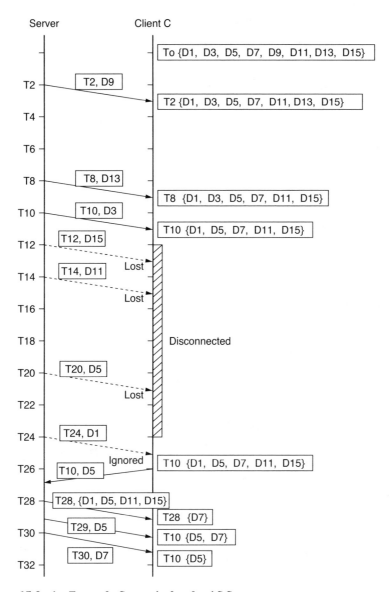

Figure 15.9. An Example Scenario for the AS Strategy

reports. For the stateless asynchronous approach, the invalidation reports are sent out only after data updates. There is no guarantee on how long a MH has to wait for the first invalidation reports after it wakes up. Because of the difficulty to verify cache consistency after a MH's sleep, most of the research work concentrates on stateless synchronous approaches. In stateless synchronous approaches, the server periodically broadcasts

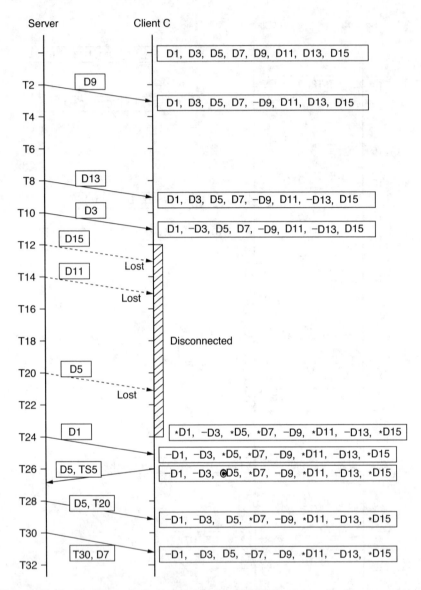

Figure 15.10. An Example Scenario for the SACCS Strategy

invalidation reports to all the MHs in its range. The major difference among those approaches is the organization of invalidation reports.

Several examples of stateless synchronous approaches (namely TS, AT, and SIG) are described in [Barbara0] [Barbara1]. In the TS (time stamps) approach, the server periodically broadcasts the invalidation reports, which consist of data item IDs and their last update times during the last w (where w is the invalidation window size) seconds. Then, based on the invalidation reports, the MH will purge the data item from its cache or update the data item's time stamp. The MH also maintains a variable to record the last time it received a report. If the difference between the current report time stamp and this variable is greater than w, the MH should drop the entire cache. The AT (amnesic terminals) approach is similar to the TS approach, but the invalidation reports are the data item IDs that have changed since the last invalidation report. Similarly in the SIG (signatures) method, the difference is that the server periodically broadcasts the signatures of the data items. In [Barbara1], they also proposed to change the TS scheme to adapt the invalidation window parameter w to different data items. The paper also compared those three strategies under different scenarios. The result is that the AT strategy is best for workaholics, but the SIG strategy is best for long sleepers. The problem is that the MH cannot decide the validity of the cache and the whole cache has to be dropped if the MH sleeps for a longer duration than what the invalidation reports covered.

In the paper [Jing], the bit-sequences approach is proposed. In this approach, the invalidation reports contain a set of binary bit sequences with a time stamp associated with each of the sequences. The bit sequences are organized in a hierarchical structure. The highest level has as many bits as the number of objects in the database and the lowest level has only two bits. A bit is set to 1 in the sequence to represent the corresponding data object has changed since the time indicated by the sequence time stamp. The nth bit in a bit sequence corresponds to the nth 1 bit in the sequence one level up. A client chooses the bit sequence with the time stamp that is immediately after the client's disconnection time and invalidates all the objects marked 1 in that sequence. The scheme was shown to be effective in reducing the number of cached data discarded; however, the size of the bit sequence broadcast may be large and may lead to poor bandwidth use.

The paper [Tan] compared several representative stateless synchronous approaches, including bit-sequence and dual-report cache invalidation (DRCI). In the DRCI approach, the data items are organized into groups and the server broadcasts a pair of invalidation reports — object invalidation report (OIR) and group invalidation report (GIR) — every L time units. The OIR report contains the object ID and the last modification time of the objects that have changed during the time [T – wL, T]. The GIR contains the group ID and the last change time of the groups that have changed during the time [T – WL, T]. Here, T stands for the current server broadcasting time, w and W are update log windows and $W > w > 0$. Therefore, the clients with short disconnection time can use the OIR to invalidate their cache and the GIR reports can help minimize invalidation of the entire cache for clients with long disconnection time.

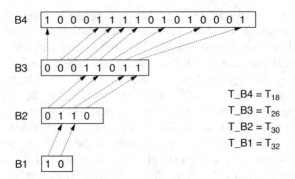

Figure 15.11. An Example for Bit-Sequence Strategy

Using the same example data items given in Table 15.1, we give a simple illustration of stateless synchronous strategies. This example is based on the example given in [Tan]. We assume that $T_{i+2} - T_i = 2$ time units for the illustration of stateless approaches.

For the TS strategy, we assume that w = 4. Then the invalidation report at T_{34} has the contents $\{D_7, D_8\}$. For the AT strategy, if the last invalidation report happened at T_{26}, then the invalidation report at T_{34} has the contents $\{D_7, D_8, D_{12}, D_{16}\}$. For the SIG strategy, the invalidation report is similar, except it uses the signature instead of the ID of the data.

For the DRCI strategy, we use the following parameters: G1 = $\{D_1, D_2, D_3, D_4\}$, G2 = $\{D_5, D_6, D_7, D_8\}$, G3 = $\{D_9, D_{10}, D_{11}, D_{12}\}$, G4 = $\{D_{13}, D_{14}, D_{15}, D_{16}\}$, L = 4, w = 2, W = 6. Then the invalidation at T_{34} will have the following OIR and GIR:

$$\text{OIR} = \{T_{34}, (D_7, T_{30}), (D_8, T_{32}), (D_{12}, T_{26}), (D_{16}, T_{28})\}$$

$$\text{GIR} = \{(G_1, T_{24}), (G_2, T_{20}), (G_3, T_{22}), (G_4, T_{12})\}$$

Because OIR is used to invalidate the data that changed during the time period [T – wL, T], that is $[T_{26}, T_{34}]$, therefore, the OIR should have the data items that changed since T_{26}, which are D_7, D_8, D_{12}, and D_{16}. The GIR should cover the time period [T – WL, T], that is $[T_{10}, T_{34}]$. Besides, the GIR should not count the data items invalidated in the OIR, therefore, the last change time for group 1 is T_{24}, group 2 is T_{20}, group 3 is T_{22}, and group 4 is T_{12}.

For the bit-sequence strategy (Figure 15.11), the bit-sequence structure is shown in this example. The first level bit sequence B4 has 16 bits with each of them representing one data item. The time stamp for B4 is T_{18} and there are 8 data items changed from that time to the invalidation broadcasting time. B3 has 8 bits with each of them representing a 1 bit in B4. Similarly, we get B2 and B1 and they show the data items that have changed since T_{30} and T_{32}, respectively.

Cache Management in Wireless and Mobile Computing Environments

Table 15.2. Summary of Characteristics of Invalidation-Based Strategies

	Stateful Consistency Strategies	Stateless Consistency Strategies
Enhance data availability	Cache invalidation information at MSS so that clients can fetch missed invalidation information from MSS after reconnection.	Repeat broadcasting history invalidation information.
Conserve bandwidth/ energy	Broadcast invalidation reports only when needed.	Coarse-grain invalidation reports.

Table 15.2 summarizes the characteristics of invalidation-based consistency strategies with regards to how they address the special problems of data availability and bandwidth/energy conservation in mobile computing environments.

15.4.2 TTL-Based Cache Consistency Strategy

In the TTL-based cache consistency strategy, the clients are responsible to poll the server to verify that the cached data is up-to-date. WebExpress [Housel] and Mobile Office Workstations using GSM Links (MOWGLI) [Lilje] are good examples of TTL-based cache consistency strategy adapted to wireless networks. Here we discuss how these approaches help to solve the problem of frequent disconnections and narrow bandwidth.

15.4.2.1 Handling Disconnections. The caching system in MOWGLI WWW supports disconnected mode in various ways:

- MOWGLI WWW chooses to validate documents only when explicitly requested by the user.
- MOWGLI WWW suffixes each hypertext link in a document with an indicator that tells the user if the referred document would have to be fetched from a server. In addition, MOWGLI WWW offers the user access to cached documents directly through a cache inventory. In this way, the user can stay safely within the bounds of the cache.
- The cache maintained by MOWGLI WWW Agent is persistent, which means that it is stored on disk and retained over multiple browsing sessions.

15.4.2.2 Achieving Energy and Bandwidth Efficiency. WebExpress uses cyclic redundancy codes (CRC) to help reduce unnecessary transmission in maintaining cache coherency. When a requested object in the client cache has exceeded the coherency interval defined by the client, the CRC code of this object is transmitted to the server to determine the difference between the fresh copy and the cached copy. A new copy will only be fetched when the difference exceeds a specified value.

WebExpress also employs the differencing technique. To update the cached data item, the whole data item is not fetched; instead WebExpress updates the cached data item based on the difference from the fresh copy. For the dynamic Web page requests, a common base object is cached on both the client and the server side. Difference is calculated and transmitted to the client to update the response to a new request.

15.4.2.2.1 Protocol Optimization. In WebExpress, each client connects to the server with a single Transmission Control Protocol/Internet Protocol (TCP/IP) connection. All requests and responses are multiplexed over the connection to avoid connection establishment overhead. Besides, when the client establishes a connection with the server, it sends its capabilities only on the first request and the server will maintain this information to avoid multiple transmission of same information. In MOWGLI, there are two levels of protocol optimization. At the transport level, TCP over the wireless part of the network is replaced by Mobile Transmission Control Protocol (MTCP), which is a lightweight protocol, has minimal packet headers, involves as few round trips over wireless link as possible, and also provides improved fault-tolerance. At the application level, HTTP is replaced by the binary encoded protocol Mobile HTTP (MHTTP), which supports the predictive upload of documents and document objects.

15.5 Open Problems

15.5.1 Cache Consistency Strategy in the Ad Hoc Network Environment

There is not much research work dealing with cache consistency maintenance in ad hoc networks. Further, as we mentioned earlier, we cannot simply copy from the strategies in architecture-based wireless networks, because ad hoc networks have distinct features compared to architecture-based wireless networks, especially the P2P nature of ad hoc networks. Therefore, some aspects of cache strategies from architecture-based wireless networks may not work well in ad hoc networks. Consequently, new cache management strategies need to be developed for ad hoc networks.

15.5.2 Cooperate Caching in Ad Hoc Network Environment

In this section, we introduce some preliminary research work that addresses the problem of how multiple MHs can cooperate with each other to deal with their energy and storage space constraints in ad hoc networks. These networks do not have network infrastructure devices like switches or routers; they are composed solely of mobile computing devices. Ad hoc

Cache Management in Wireless and Mobile Computing Environments

networks can have dynamic topology where devices may move frequently like the MANETs or they can have static topology like the sensor networks. For a static topology ad hoc network, the cooperative caching scheme can take advantage of the static feature. For example, we can use the topology information to decide where to cache the data items to minimize the overall accessing cost of all the nodes. Another factor that impacts the cooperative caching scheme is the cache consistency schemes: different cooperative cache schemes need to be developed for invalidation-based and TTL-based consistency schemes.

Under the assumption that the topology of the ad hoc network is static, the paper [Nugge] proposed the approach to determine at what nodes to place the cache. This approach considers the energy consumed by information distribution and access latency experienced at all the MHs. The goal is to minimize the weighted sum of energy cost and access delay. Under the assumption that the data does not change during the whole process (static data), the paper [Hara01] proposed three data replication methods in an ad hoc network to improve data accessibility. The system model also assumes MHs access static data items saved at other MHs and the access frequencies for each data item from each MH are known constants. The basic idea of the proposed methods is that replicas are relocated for every specific period; replica allocation is determined based on the access frequency from each MH to each data item and the network topology at the moment. The author also proposed to share the cached data item among a MH group and eliminate the duplicate copy of a data item inside the group.

The papers [Hara02A, Hara02B] discussed cooperative caching techniques in the environment where a server pushes updates to clients. The basic idea is similar to the approach in the previous paragraph.

Cooperative caching is also studied for Web browsing in wireless networks in [Sailhan02, Sailhan03]. In this approach, every MH maintains a profile for every peer that shares an interest with it. The value of the profile is characterized by the number of times the peer responds the MH's request or the peer requested the cached data items from this MH. Then the peers are weighed by the profile value divided by the number of hops from the MH. The peer that has the greatest weight is first contacted when the MH needs to fetch a data item from its peer. Besides, the MH also considers the capacity of the peers. If several peers have the same weight, the one with the most capacity is first contacted.

For reference, there are four flavors for cooperative caching in P2P network environment:

1. Hierarchically organized Web caching, such as Harvest [Chan]
2. Hash-based schemes, such as Coopnet [Pad], Squirrel [Lyer]

3. Directory-based schemes, such as Squirrel [Lyer]
4. Multicast-based schemes, such as the scheme in [Wang]

However, these works are for wired P2P network, hence they did not address the special concerns, weak connectivity, and resource constraints in mobile computing.

15.6 Summary

Mobile computing environments present several new challenges for cache management. These schemes need to be able to work efficiently in the presence of frequent disconnections while consuming a minimum amount of energy and bandwidth. This chapter has presented various cache management schemes to address these new challenges. Data caching can be instrumental in solving or mitigating problems of performance, availability, scalability, and resource paucity. Understandably, development of new cache management is an active research area in the field of mobile computing, especially in the area of mobile ad hoc networks.

References

[Acharya95] S. Acharya, R. Alonso, M. Franklin, and S. Zdonik. Broadcast disks: Data management for asymmetric communication environments. *Proceedings of the ACM SIGMOD International Conference on the Management of Data*, pp. 199–210, 1995.

[Acharya97] S. Acharya, M. Franklin, and S. Zdonik. Balancing push and pull for data broadcast. *Proceedings of the ACM SIGMOD*, pp. 183–194, May 1997.

[Barbara0] D. Barbara and T. Imielinski. Sleepers and workaholics: Caching strategies for mobile environments. *Proceedings of the ACM SIGMOD Conference on Management of Data*, pp. 1–12, 1994.

[Barbara1] D. Barbara and T. Imielinski. Sleepers and workaholics: Caching strategies for mobile environments (extended version). *The VLDB Journal — The International Journal on Very Large Data Bases*, vol. 4, Issue 4, pp. 567–602, 1995.

[Cao] P. Cao and C. Liu. Maintaining strong cache consistency in the World Wide Web. *IEEE Transcations On Computers*, vol. 47, no. 4, pp. 445–457, 1998.

[Cate] V. Cate. Alex — a global file system. *Proceedings of 1992 USENIX File System Workshop*, pp. 1–12, May 1992.

[Chan] A. Chankhunthod, P. Danzig, C. Needaels, M.F. Schwartz, and K.J. Worrell. A hierarchical Internet object cache. *Proceedings of the 1996 USENIX Technical Conference*, San Diego, January 1996.

[Chandra] R. Chandra, V. Ramasubramanian, and K.P. Birman. Anonymous gossip: Improving multicast reliability in mobile ad hoc networks. *Proceedings of the 21st International Conference on Distributed Computing Systems*, 2001.

[Dar] S. Dar, M.J. Franklin, B.T. Jonsson, D. Srivastava, and M. Tan. Semantic data caching and replacement. *Proceedings of VLDB*, pp. 330–341, Bombay, September 1996.

[Garshnek] V. Garshnek and F. Burkle. Applications of telemedicine and telecommunications to disaster medicine: Historical and future perspective. *J Am Med Inf Assoc.*, vol. 6, pp. 26–37, 1999.

[Gitzenis] S. Gitzenis and N. Bambos. Power-controlled data prefetching/caching in wireless packet networks. *Proceedings of IEEE INFOCOM*, pp. 1405–1414, 2002.

[Grassi] V. Grassi. Prefetching policies for energy saving and latency reduction in a wireless broadcast data delivery system. *Proceedings of the 3rd ACM International Workshop on Modeling, Analysis, and Simulation of Wireless and Mobile Systems*, 2000.

[Gupta] A. Kahol, S. Khurana, S. Gupta, and P. Srimani. A strategy to manage cache consistency in a disconnected distributed environment. *IEEE Transactions on Parallel and Distributed Systems*, vol. 12, no. 7, pp. 686–700, 2001.

[Gwertz] J. Gwertzman and M. Seltzer. World-Wide Web cache consistency. *Proceedings of the 1996 USENIX Technical Conference,* San Diego, January 1996.

[Hara01] T. Hara, Effective replica allocation in ad hoc networks for improving data accessibility. *Proceedings of IEEE INFOCOM,* 2001.

[Hara02A] T. Hara. Cooperative caching by mobile clients in push-based information systems. *Proceedings of the 11th International Conference on Information and Knowledge Management,* 2002.

[Hara02B] T. Hara. Replica allocation in ad hoc networks with periodic data update. *Proceedings of the 3rd International Conference on Mobile Data Management,* 2002.

[Housel] B.C. Housel, G. Samaras and D.B. Lindquist. WebExpress: A client/intercept based system for optimizing Web browsing in a wireless environment. *Mobile Networks and Applications,* vol. 3, pp. 419–431, 1998.

[i80211] IEEE Working Group for Wireless LANs. ANSI/IEEE Std 802.11, 1999 Edition. http://grouper.ieee.org/groups/802/11/.

[i80211a] IEEE Working Group for Wireless LANs. IEEE Std 802.11a-1999. http://grouper.ieee.org/groups/802/11/.

[Imie94A] T. Imielinski, S. Viswanathan, and B.R. Badrinath. Power efficient filtering of data on air. *Proceedings of EDBT Conference,* pp. 245–258, March 1994.

[Imie94B] T. Imielinski, S. Viswanathan, and B.R. Badrinath. Energy efficient indexing on air. *Proceedings of 1994 ACM SIGMOD,* pp. 25–36, May 1994.

[Jiang] Z. Jiang and L. Lkeinrock. Web prefetching in a mobile environment. *IEEE Personal Communications,* vol. 5, no. 5, pp. 25–34, 1998.

[Jing] J. Jing, A. Elmagarmid, A.S. Helal, and R. Alonso. Bit-Sequences: an Adaptive cache invalidation method in mobile client/server environments. *ACM Mobile Networks and Applications,* vol. 2, Issue 2, pp. 115–127, 1997.

[Kistler] J. Kistler and M. Satyanarayanan. Disconnected operation in the Coda file system. *ACM Transactions on Computer Systems,* vol. 10, no. 1, February 1992, pp. 3–25.

[Krashinsky] R. Krashinsky and H. Balakrishnan, Minimizing energy for wireless Web access with bounded slowdown, *Proceedings MobiCom 2002,* Atlanta, 2002.

[Lilje] M. Liljeberg, H. Helin, M. Kojo, and K. Raatikainen. Enhanced services for World-Wide Web in mobile WAN environment, University of Helsinki, Department of Computer Science, Series of Publications C, no. C-1996-28.

[Lyer] S. Lyer, A. Rowstron, and P. Druschel. Squirrel: a Decentralized peer-to-peer Web cache. *Proceedings of the 21th ACM PODC 2002 (Symposium on Principles of Distributed Computing).*

[NLANR] A distributed testbed for national information provisioning. http://ircache.nlanr.net/.

[Ns] K. Fall and K. Varadhan (Eds.). *The ns manual.* http://www.isi.edu/nsnam/ns/ns-documentation.html.

[Nugge] P. Nuggehalli, V. Srinivasan et al. Energy-efficient caching strategies in ad hoc wireless networks. *Proceedings of MobiHoc '03,* Annapolis, MD, June 2003.

[Pad] V.N. Padmanabhan and K. Sripanidkulchai. The case for cooperative networking, *Proceedings of the 1st International Workshop on Peer-to-Peer Systems (IPTPS),* Cambridge, MA, March 2002.

[Ren] Q. Ren and M.H. Dunham. Using semantic caching to manage location dependent data in mobile computing. *Proceedings of MobiCom,* 2000, Boston, 2000.

[Royer] E.M. Royer and C.E. Perkins. Multicast ad hoc on-demand distance vector (MAODV) routing. IETF Internet Draft, draft-ietf-manet-maodv-00.txt, July 2000.

[Sailhan02] F. Sailhan and V. Issarny. Energy-aware Web caching for mobile terminals. *Proceedings of the 22nd International Conference on Distributed Computing Systems Workshops,* 2002.

[Sailhan03] F. Sailhan and V. Issarny. Cooperative caching in ad hoc networks. *Proceedings of the 4th International Conference on Mobile Data Management,* 2003.

[Satya] M. Satyanarayanan. Fundamental challenges in mobile computing. *Proceedings of the 15th Annual ACM Symposium on Principles of Distributed Computing,* pp. 1–7, 1996.

[Silberschatz] A. Silberschatz and P. Galvin, *Operating System Concepts,* Boston: Addison Wesley, 1998.

[Sistla] A.P. Sistla, O. Wolfson, and Y. Huang. Minimization of communication cost through caching in mobile environments. *IEEE Transactions on Parallel and Distributed Systems,* vol. 9, no. 4, pp. 378–389, 1998.

[Tan] K. Tan, J. Cai, and B. Ooi. An evaluation of cache invalidation strategies in wireless environments. *IEEE Transactions on Parallel and Distributed Systems,* vol. 12, no. 8, pp. 789–807, 2001.

[Wang] J. Wang. A survey of Web caching schemes for the Internet. *ACM Computer Communication Review,* vol. 29, no. 5, pp. 36–46, October 1999.

[WangZ] Z. Wang, S. Das, H. Che, and M. Kumar. SACCS: Scalable asynchronous cache consistency scheme for mobile environments. *Proceedings of International Workshop on Mobile and Wireless Networks, MWN2003,* May 2003.

[Yin] J. Yin, L. Alvisi, and M. Dahlin et al. Using leases to support server-driven consistency in large-scale systems. *Proceedings of the 18th International Conference on Distributed Computing System,* May 1998.

Chapter 16
Cache Invalidation Schemes in Mobile Environments

Edward Chan, Joe C.H. Yuen, and Kam-Yiu Lam

16.1 Introduction

Rapid progress in wireless network infrastructure has led to an explosive growth in mobile applications. At the same time, increasing popularity and sophistication of portable devices such as personal digital assistants (PDAs) or intelligent mobile phones has greatly increased the ability of the mobile users to retrieve and manipulate data at a location and time of their choice. The timely dissemination of such data to the users has become the topic of much research in recent years. In a classical client–server model, the client acquires data by pulling it from the server. However, it has been argued and demonstrated by many researchers that the broadcast model (i.e., pushing data to the clients) is an efficient one in wireless networks, because a large number of users requesting the same data items can be served at no extra cost [1]. In addition, pulling data requires the transmission of an explicit request, which results in the consumption of additional battery power when compared with passive listening to the transmission channel for broadcast data. Many hybrid models incorporating both push and pull techniques have also been proposed in the literature. In these approaches, the popular items are typically disseminated using broadcast and the less popular ones pulled [2, 3].

A mobile network is typically characterized by its limited available bandwidth, as well as the possibility of frequent disconnections. With limited bandwidth, the broadcast cycle can be too long for some clients. To improve response time, frequently accessed data items can be cached by the clients. Caching also reduces the need to pull a data item when it is accessed multiple times. Although there are major advantages in adopting client side caching, the possibility of frequent disconnection can lead to serious cache inconsistency problems. Thus the design of a suitable cache invalidation scheme is a key issue in the development of a data dissemination system in a mobile environment.

A common method for cache invalidation is to broadcast invalidation reports to mobile clients. One of the most important considerations in the design of cache invalidation schemes is to minimize the size of the invalidation reports and hence, reduce the bandwidth required for broadcasting the invalidation reports. Another requirement is to reduce the response time of the client, which is the primary reason why caching is used in the first place. The final performance criterion is the need to conserve power, because the battery life of most portable devices is limited. A number of cache invalidation schemes have been proposed by researchers to satisfy one or more of these requirements and a brief discussion of their major features is given in the next section.

16.2 Summary of Existing Cache Invalidation Schemes

A large portion of the existing work on cache consistency strategies for mobile environments is based on periodic broadcast of invalidation reports. For instance, Barbara and Imielinski, in one of the earliest works in this area, proposed three different variants of this approach — broadcasting time stamp (TS), amnesic terminals (AT), and signatures (SIG) — depending on the expected duration of network disconnection [4]. However, the algorithms are only effective if the clients have not been disconnected for a period exceeding an algorithm specific parameter. Otherwise the entire cache has to be discarded, even though some of the cached data items might still be valid.

Jing et al. proposed a bit-sequence scheme (BS) in which the invalidation report consists of bit sequences associated with a set of time stamps [5]. Using the information embedded in the bit sequences, a client needs only to invalidate its entire cache if more than half of the data items have been updated in the server since its last invalidation time. This ingenious approach has the drawback of greater complexity and much larger invalidation reports than the TS or AT methods, particularly when the number of data items is larger.

Wu et al. modified the TS and AT algorithms to include a cache validity check after connection is reestablished. The proposed algorithm, called grouping with cold update set retention (GCORE) [6], however does not solve the basic problem of TS in that the entire cache will have to be dumped if the disconnection time is greater than its update history window. More recently, a family of hybrid cache invalidation algorithms is proposed [7]. The essence of these algorithms is that the type of invalidation reports to be sent (i.e., TS or BS) is determined dynamically, based on system status such as disconnection frequency and duration, as well as update and query pattern. Another recently proposed approach, called the updated invalidation report (UIR) method, focuses on reducing response time [8]. In this scheme, a small fraction of the essential information related to cache invalidation is replicated several times between successive invalidation reports. This scheme can be applied to further optimize both AT and BS. One disadvantage of this approach is that the client needs to monitor not only the invalidation reports, but the incremental updates as well. This might lead to greater consumption of energy, which is undesirable for mobile clients that rely on battery power. A modified, poweraware scheme that combines UIR with adaptive prefetching is proposed to allow the client to prefetch frequently used data items [9]. Prefetching can improve the cache hit rate, but consumes additional power; thus the client needs to determine the level of prefetching based on the popularity of the cached item and its power consumption requirements.

The above approaches deal primarily with cache invalidation for individual data items. It is also possible to design cache invalidation schemes for groups of data. The problem with this approach is that even when a group of data items has been invalidated, there may be individual data items within the group that are still valid. These issues are addressed in the dual report cache invalidation (DRCI) scheme [10] using a pair of invalidation reports called the object invalidation report (OIR) and the group invalidation report (GIR). Besides group-based invalidation, it is also possible to perform query level invalidation. Query level invalidation reduces the number of data items to be invalidated at the expense of a time stamp for every data item instead of the entire cache. An energy efficient query level cache invalidation scheme, based on DRCI, called selective dual report cache invalidation, has also been proposed [10]. The performance of these schemes, as well as a detailed taxonomy of cache invalidation strategies, has been studied by Tan et al. [11].

The cache consistency issue is also relevant for transaction processing in a mobile environment. In this case, the client needs to ensure that the set of data items (some of which are cached at the client side) used in a particular transaction are consistent. Among the approaches used is the invalidation method [12] in which an invalidation report consisting of all data items updated at the broadcast server during the previous broadcast cycle are sent before each broadcast cycle. The validity of the data items accessed by a mobile transaction is ensured by checking with the invalidation report and the transaction restarted if any of its accessed data items is found to be invalid. Another approach is to cache multiple versions of data items [13]. The multiversion broadcast method is useful for systems where the mobile clients are frequently disconnected, because the mobile clients can access consistent cached data items while it is disconnected. A major drawback is that the data items, though consistent, may not be the most current ones.

Almost all work in cache invalidation deals with data consistency caused by data updates and link disconnection. Some researchers have focused on cache inconsistency caused by location changes of a client. In effect, both temporal and spatial cache invalidation needs to be handled in such a scenario. The key issue here is that the potentially large size of the validity information of the data items (i.e., information that tells the client the set of locations or cells in a cellular network) for which the cached data will be valid. Xu et al. proposed a number of compression and implicit scoping schemes for dealing with this issue where validation information size is reduced at the expense of additional computation overhead [14].

It can be seen that the literature on cache invalidation schemes in mobile environments is extensive. However, most of these schemes fail to consider the temporal characteristics of data, which in many cases can be exploited to allow the clients to perform self-invalidation and thus improve overall system performance significantly. The next section explores an adaptive cache invalidation scheme based on this idea.

16.3 Temporal Data Model for Mobile Computing Systems

To take advantage of the temporal characteristics of data, we will introduce a temporal data model for mobile computing systems in this section. In formulating this data model, we assume that the data items in the database are used to record the instantaneous values of the objects in the external environment. Some examples are the last traded stock prices, news updates, as well as traffic and weather conditions. Because they track dynamic external events, this information may change quite rapidly. To maintain strict consistency of the data items with the actual status values of their corresponding objects in the external environment, it may be necessary to refresh the database records frequently.

The price of maintaining tight consistency between the data items in the database and the actual status of the objects in the external environment is that tracking every change in the status of the external objects will create a large number of update transactions, which may not be practical for many mobile computing systems. Actually, for many data items, such as those for weather condition, news updates, and traffic status, it is possible for updates to be performed at much longer intervals. There are two reasons for this. First, for many mobile computing systems, the value of a data item that models an object in the external environment cannot, in general, be updated continuously to perfectly track the dynamics of the real-world object as the update process itself requires time to complete. Thus, there already exists a discrepancy between the value of the data item and the real-time status of its corresponding object. An example might be the monitored traffic conditions of a particular road. In this case, by the time this information is reported to the server, the actual traffic conditions of the road will already have changed. Second, it is often unnecessary for data values to be perfectly uptodate or absolutely precise to be useful. In particular, values of a data item that are slightly different are often interchangeable and the mobile clients will consider them to be the same. For example, to many drivers, the traffic conditions at one minute past noon can be considered to be the same as the traffic condition at noon.

The dynamic properties of the data items also make the management of cached data items a difficult problem. Whenever a new version of a data item has been created in the database, all the cached versions of the data item have to be invalidated. Traditionally, invalidation is achieved by sending explicit invalidation messages to the clients that have cached the data item. However, due to the delay in invalidation as a result of the low bandwidth in mobile environments, it is difficult to maintain tight consistency between the data items at the client cache and the corresponding data items in the database. Thus the key problem is that even when the required data item is located in the client cache, it is necessary to determine whether this is the most updated version without incurring heavy overhead.

To resolve these issues, we introduce a new model to characterize the temporal properties of the data items in mobile computing systems. In our model, it is assumed that the data items are classified into different data groups based on one of their temporal properties, namely the rate at which their values change. Within each data group, the data items are assigned a timeframe, called the absolute validity interval (AVI), to characterize this property. The idea of AVI originates in research in real-time database systems where the values of the data items are highly dynamic [15]. Formally, each data item can be defined as a 3-tuple:

$$D_i = (V_i, AVI_i, UTS_i)$$

where V_i is the value for data item D_i, AVI_i is the absolute validity interval of D_i and UTS_i is the last update time stamp of D_i. Each transaction update, which captures the real-time status of an external object, is associated with a time stamp, called the update time stamp (UTS), which indicates the time at which the update transaction is created. Whenever an update transaction is completed, the UTS will also be recorded in the database along with the updated data value. A data item, D_i, is invalid if $AVI_i + UTS_i <$ current time. In this case, a new value of the data item is required.

The choice of AVI values for the data items depends on the nature of the data items; the concept of data similarity [16, 17] can be applied to determine the appropriate values. It is assumed that the application semantics allows the system designer to derive a similarity bound for each data item such that two write operations on the data item will be similar if the values come from a time interval not greater than the similarity bound. In the AVI approach, the AVIs of data items can be defined as the similarity bound of the data items. Their values can be based on the dynamic property of the data items, how the mobile clients use the data items and the importance of the data items to the mobile clients. For example, news updates are generated every 30 minutes. Thus, the AVI of news updates can be set to 30 minutes. For information whose updates are asynchronous (examples include stock prices), their AVIs can be defined based on the user requirements. For example, free delayed stock quotations might have a large AVI (on the order of minutes), whereas real-time stock information used by professional traders or paid quotation services might have a smaller AVI (probably on the order of seconds). Another type of data with a larger AVI is traffic information where changes are less drastic than stock prices.

The advantage of using AVI to define the validity of a data item is that invalidation messages are no longer needed to invalidate the data items in the client caches because the validity of the data items is explicitly defined by their AVI and their UTS. The second major advantage of the AVI approach is that it can help us to design the broadcast algorithms so that the cache hit probability can be improved. Both static and dynamic AVI-based broadcast scheduling schemes have already been proposed [18].

16.4 Cache Invalidation Using AVI

In the previous section, we introduced the concept of AVI in the context of a temporal data model for mobile computing environment. In this section, we explain how AVI can be used to facilitate cache invalidation.

16.4.1 Validity Period of Data in Client Cache

We have indicated that AVI is a useful concept for designing data broadcast strategies. However the AVI of a data item can be used also as an estimator for the validity of a data value at client caches. A data item cached at the clients has to be updated from time to time by the server to maintain consistency between the two. The update intervals can be used to define the validity periods of a cached item. A data item is updated at the beginning of each update interval. Hence, it is valid from the time it was cached until the time of the next update. In other words, as shown in Figure 15.1A, the time from the nth update on the data item to the time of the $(n + 1)^{th}$ update is the validity period of the data item after n^{th} update. The value from the nth update is stale after the arrival of the $(n + 1)^{th}$ update.

According to the above model, the validity period of a data item is only known after the arrival of the next update; in other words, the validity period of a data item after the $(n - 1)^{th}$ update is not defined until the $(n)^{th}$ update has occurred. The differences between the validity period and AVI are shown in Figure 16.1B. We define the false valid period (FVP) as the time period where AVI overestimates the validity period of the data item and the false invalid period (FIP) as the time period where AVI underestimates the

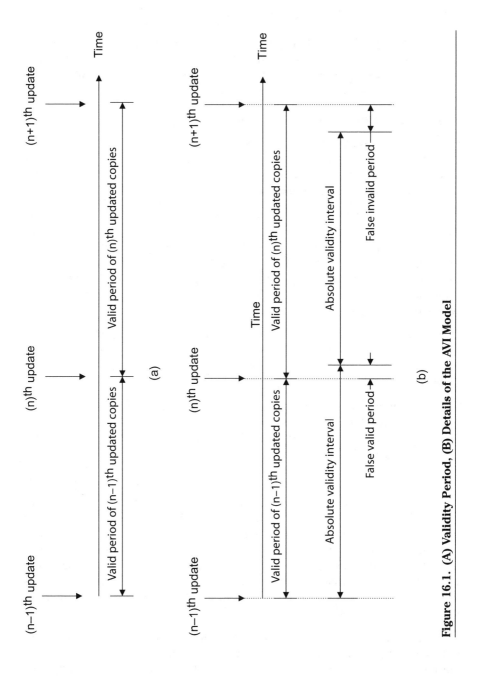

Figure 16.1. (A) Validity Period, (B) Details of the AVI Model

validity period. If FVP is greater than zero, the actual update interval of the data item is shorter than expected. The values of FVP should be kept small, because during FVP a client unwittingly accepts an invalid data item. On the other hand, if FIP is greater than zero, the actual update interval is longer than expected. A client will consider a cached item to be invalid during that period even though it is actually still valid. By suitably adjusting AVI based on the update intervals, the values of FVP and FIP can be kept rather small, as we will see in Section 16.5.

16.4.2 The IAVI Cache Invalidation Scheme

Having defined AVI and the validity period of a data item at the client cache in the previous section, we now proceed to describe how they can be used to support an efficient cache invalidation scheme, which we call invalidation by absolute validity interval (IAVI).

In our model, because a cached item is assumed to be invalid if its AVI has expired, no explicit invalidation notification is needed to invalidate the cached items in the mobile clients. In other words, a client can invalidate its cached items by calculating the items' last update times and the AVI of the data items. We call this process self-invalidation or implicit invalidation. This approach is clearly desirable, because it alleviates the need to send explicit invalidation reports.

In practice, the arrival of some data items can be aperiodic and the optimal AVI value of a data item may vary with time. This change in the AVI of a data item will either shorten its FVP or enlarge its FIP. There are two possible cases:

1. The new AVI of a data item is smaller than the previous one. In this case, the data item should be invalidated before its AVI expires. If the client uses the previous AVI, FVP will be longer than the previous estimation. The change in AVI is typically caused by an update of the data item; hence, the cached item must be invalidated.
2. The new AVI of a data item is larger than the previous one. In this case, no notification message is needed, because the cached item will be invalidated automatically when its AVI expires. FIP in this case will be increased. It is possible to send explicit notifications to mobile clients on the new AVI value. However, this must be done before the AVI of the cached item expires. Mobile clients experience frequent link disconnection and it is hard to verify the new AVI to the current cached copy of a mobile client who has been disconnected.

Thus, we conclude that an invalidation report is needed to notify the clients when the AVI value of a data item is reduced, but not when it is increased. When the AVI of a data item is reduced, the mobile client must

Cache Invalidation Schemes in Mobile Environments

be notified in order to minimize FVP. Otherwise the client might use an invalid copy of the data in its cache. However, for the case where the AVI of a data item has increased, it is better to do nothing and let the next update of the data item refresh the cache. Suppose the server sends a report to inform the clients of the new AVI (i.e., the cached item is still valid). Now if after a short time the data item is updated before the new AVI expires, an additional invalidation message will have to be sent. All these will result in a substantial increase in overhead. Moreover, all the additional effort is wasted if the cached data item is not accessed after the original AVI has expired. It is preferable to accept a larger FIP than trying to update AVI, and let the client make an explicit request when the data is actually accessed.

In the following sections, we will discuss in detail how the IAVI scheme is implemented at both the server and the client sides.

16.4.3 Server Algorithm

The server algorithm consists of two parts — invalidation report generation and AVI adjustment. The former deals with the selection of the information to be included in the invalidation report. Broadcast bandwidth is a valuable resource that is shared among a large number of mobile clients and it must be used efficiently. AVI adjustment refers to the modification of the AVI values of the data items to achieve the desired level of cache coherence.

16.4.3.1 Invalidation Report Generation. To notify the mobile clients about changes in AVI values, invalidation reports are generated and disseminated periodically. The invalidation report contains a data ID and its update time, whose update interval must satisfy the following expression:

$$T_{update(i,n)} - T_{updater(i,n-1)} < AVI_{(i)} * (1 - F_i) \qquad (16.1)$$

where
$T_{update(i,n)}$ = the time stamp of nth update on data item i
$AVI_{(i)}$ = the AVI of data item i
F_i = the AVI tolerance for data item i

According to the above expression, if the update interval of data item i is longer than its AVI plus its AVI tolerance, it will be included in the invalidation report as this implies that the life span of a data value is shorter than expected. AVI tolerance is designed to tackle the randomness in update interval. Because it is not possible to predict the occurrence of update events, the update interval and AVI will not be perfectly matched. In order words, the current AVI is valid if the difference between the update interval and the AVI of a data item is small. The tolerance limit is data dependent and different kinds of data items may have different degrees of tolerance. However, for data items with the same AVI tolerance, the one

Table 16.1. Database Snapshot at T(n)

Data Item	$T_{(n-1)}$	$T_{(n)}$	AVI	F
0	1000	1490	500	0.1
1	1000	1080	100	0.1
2	100	1000	1000	0.1
3	500	800	200	0.1
4	200	800	700	0.1

Table 16.2. Invalidation Report

Data Item	Time Stamp
1	1080
4	800

with a shorter AVI will have a smaller tolerance limit according to the above expression. This is because a shorter AVI implies that the data item is updated frequently and the value is relatively dynamic, therefore, a tighter tolerance is desirable to minimize FVP.

In general, we can divide the validity period of the cached items into three different periods. The cached data item is likely to be valid during the period between $T_{update(i,n)}$ and $T_{updater(i,n)} + AVI_{(i)} * (1 - F_i)$. Likewise, the cached data item will definitely be considered invalid after $T_{updater(i,n)} + AVI_{(i)}$, because its AVI has expired. The period in between is somewhat problematic because the validity of the cached data item cannot be determined conclusively based on its update history. Clearly, the duration of this period is a function of the AVI tolerance and the smaller the tolerance, the shorter this period of vulnerability.

We will now illustrate the idea described in this section using Table 16.1 and Table 16.2. Table 16.1 shows the snapshot of a database with five data items. $T_{(n)}$ represents the time stamp of the most recent update on the data item and the $T_{(n-1)}$ represents the previous update's time stamp. As before, F is the AVI tolerance of the data item. The invalidation report generated after $T_{(n)}$th update is shown in Table 16.2. The recent updates of data item 1 and data item 4 at time 1080 and 800, respectively, does not satisfy Equation 16.1, as their updates are predicted to occur after time 1090 and 980, respectively. Therefore, these two items are included in the invalidation report.

Cache Invalidation Schemes in Mobile Environments

16.4.3.2 AVI Adjustment. Due to the dynamic nature of a data item, continuous adjustment on its AVI is needed to minimize FVP and FIP. Because AVI is estimated by past update intervals, dramatic shifts of the mean update interval will lead to a large FVP or FIP. Furthermore, the continuous mismatch of AVI and the update intervals means that a data item will be included in the invalidation reports for an extended period of time, increasing the invalidation report size and degrading overall system performance.

The minimum, mean, and maximum values of the historical update intervals can be used to estimate the AVI of a data item depending on the desired degree of cache coherence. Using the minimum update interval as the item's AVI will result in the smallest AVI and the highest degree of cache coherence as FVP is minimized. However, the drawback is that FIP will be large. The reverse is true when the maximum historical update interval is used. In the application of minimum and maximum values for the calculation of the new AVI, the value should be taken from the set of n preceding updates. This requires more complex handling, but does allow the AVI value to adapt to changes in update intervals more effectively.

Similarly, when the mean value of update interval is used to estimate AVI, the AVI of a data item is obtained by taking its average of a number of preceding update intervals. Using the mean allows us to balance FIP and FVP and this approach is used in our simulation experiments described in Section 16.5. The server algorithm is summarized in Table 16.3.

16.4.4 Client Algorithm

Cache invalidation on the client side is divided into two parts, implicit invalidation and explicit invalidation. In implicit invalidation, a cached item is invalidated by the expiration of its AVI. This occurs when the cached item is accessed and its AVI is found to have expired. In contrast, explicit invalidation is caused by the receipt of an invalidation report from the server.

16.4.4.1 Implicit Invalidation. The validity of a cached data item is determined by the update time (the time stamp of data item) and the AVI value. When a cached item is referenced by a transaction from a mobile client, the transaction will examine the update time and the AVI value of the item. If the sum of the update date time and the AVI value is larger than the current time, the item is assumed to be invalid (although it may be actually be valid but the transaction does not know which is the case for a positive FIP).

By using implicit invalidation, traffic cost is minimized, because a client can invalidate its cached items using the method described above without waiting for additional information from the server. Moreover, when the client reconnects to the mobile network after disconnection, it can invalidate its cached copy without waiting for the next invalidation report from

Table 16.3. Server Algorithm

Notations:

IR_n:	n^{th} invalidation report (IR).
d_i:	ID of data item i.
v_i:	value of data item i.
D:	the set of data items.
$u_{(i,k)}$:	time stamp of kth update of data item i.
$avi_{(i,k)}$:	AVI for data item i after k^{th} update.
f_i:	AVI tolerance for data item i.
L:	broadcast interval of IR.
w:	window size of IR history.
R_j:	a list of data items requested by a mobile client j.
L_{dset}:	a set of data currently requested by mobile client.
P_n:	a set of data broadcast at cycle n.

Server Algorithm:

(i) Receives a request from mobile client j:

$$L_{dset} = L_{dset} \cup R_j$$

(ii) For each cycle:

$$IR_n = \{<d_i, u_{(i,k)}, avi_{(i,k)}> | (d_i \in D)$$
$$\wedge (u_{(i,k)} - u_{(i,k-1)} < avi_{(i,k-1)} * (1-f_i))$$
$$\wedge ((n-w)*L < u_{(i,k)} < n*L - avi_{(i,k)})\};$$

Broadcast IR_n, $P_n = \{<d_i, v_i, u_{(i,k)}, avi_{(i,k)}> | (d_i \in L_{dset})\};$

$$L_{dset} = \{<d_i, v_i, u_{(i,k)}, avi_{(i,k)}> | (d_i \in L_{dset}) \wedge (d_i \notin P_n)\};$$

mobile server. If AVI has expired for a cached item, it can be invalidated without extra verification. Note, however, that if the cached copy seems to be valid based on its AVI, the client needs to reference the first invalidation report after reconnection to determine the validity of its cached items.

16.4.4.2 Explicit Invalidation. Besides implicit invalidation, a cached item can also be invalidated explicitly by invalidation reports sent from the server. If the update interval of a data item violates the AVI assumption (i.e., Equation 16.1), its identifier will be included in the invalidation report and broadcast to the mobile clients. Note that even if explicit invalidation is used, the size of the invalidation report is much smaller than other techniques such as BS and TS. Unlike BS and TS, which include every update event in invalidation reports within a certain period or interval, the invalidation report of the AVI scheme only contains the entries for the data items whose AVIs have been modified and meet the condition defined in Equation

16.1. Moreover, if the AVI and the update interval of the data items are reasonably well matched, many of the update events do not generate explicit invalidation reports. This major advantage will be verified in the next section by our simulation studies. The client algorithm is summarized in Table 16.4.

16.5 Performance Study

In this section, we compare the performance of IAVI with two well-known cache invalidation schemes — the BS and TS schemes. Besides these three schemes, an idealized cache invalidation scheme called perfect server (PS) has also been developed for comparison purposes. In PS, it is assumed that the system has the full knowledge on the content of all the mobile clients' caches. As a result, the invalidation reports generated by PS will only contain the update information of the data items cached in the mobile clients' caches, thus, releasing more broadcast bandwidth for data dissemination. Such a scheme would be too costly to implement in practice, as it requires the generation of updates continuously to the mobile server regarding the content of mobile clients' caches.

16.5.1 System Model

Figure 16.2 depicts the system model of the simulation program. It consists of a database, update processes, a system monitor, a mobile server, and mobile clients.

The update processes generate transactions to update the values of data items in the database. These activities are recorded in the system monitor and retrieved by the mobile server to generate the invalidation report. This report is generated periodically and broadcast to the mobile clients via the mobile network. It is assumed that the mobile network is an unreliable low bandwidth network. Besides invalidation reports, data packets are also disseminated to fulfill the client requests. Client requests are retrieved from the system monitor and serviced by mobile servers based on a first come first service (FCFS) policy.

The mobile client generates transactions, each of which consists of a set of data requests. Once a transaction is generated, the mobile client will try to fulfill the data requirements of the transaction with its local cache. For data requests that cannot be fulfilled by the cached data, a message will be sent to the mobile server to request the data explicitly. If the validity of the cached items cannot be determined (for example after the client has just reconnected to the network after disconnection), the data request message will be deferred until the invalidation report is received. Once the data requests of a transaction are fulfilled, the mobile client will either go into a doze mode or generate another transaction after a think time. The client is disconnected from the mobile network when it enters the doze mode.

Table 16.4. Client Algorithm

Notations (note additional notations in Figure 16.2):
u_i: time stamp of the last update of data item i.
C_j: local cache of mobile client j.
t_i: transaction with transaction ID i.
T_j: a set of pending transactions generated by client j.

Client Algorithm:

(i) Received the nth invalidation report from the server
    for each data item $<d_i, v_i, u_i, avi_i>$ in the cache **do**
        if $(d_i \in IR_n) \land ((n-w)*L > u_i > n*L - avi_i)$ **then**
          invalidate d_i;
        if IR_{n-1} is missing and T_j not empty **then**
          for each t_i in T_j
            Process_Transaction(t_i);

(ii) Received data item $<d_i, v_i, u_i, avi_i>$ from air
    if $(d_i \in R_j)$ **then**
        $C_j = C_j \cup \{d_i\}$
        $R_j = \{d \mid d \in R_j \land d \neq d_i\}$
        for each t_i where $d_i \in t_i$
            Process_Transaction(t_i)
    else if $(d_i \in C_i)$ **then**
        refresh the cached data item;

(iii) Transaction generated by client-
    if just returning to connected mode after disconnection **then**
        $T_j = T_j \cup \{t_i\}$ //delay processing until next IR received
    else
        Process_Transaction(t_i);

(iv) Process_Transaction (t_i)
    $R_j = \{\}$
    for each d_x in t_i **do**
        if d_x is not found in cache or not valid **then**
            $R_j = R_j \cup \{d_x\}$
        else
            mark d_x valid
    if all d_x in t_i valid **then**
        commit the transaction;
    else
        send R_j to server;

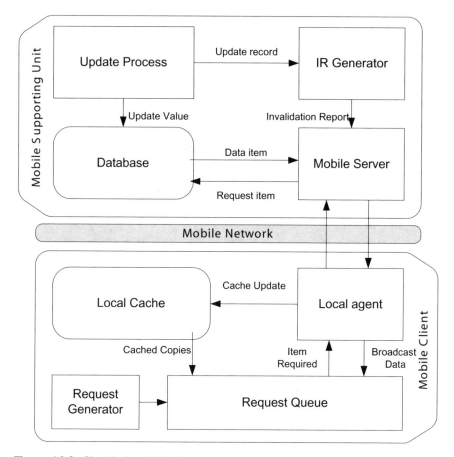

Figure 16.2. Simulation Model

16.5.2 Performance Metrics

The following measures are used to measure the system performance of the different invalidation approaches:

- Invalidation report size
- Mean response time — this is the mean elapsed time between the generation of a transaction and the time when it commits after obtaining all its data items.
- False valid rate — this metric is used to measure the probability of accessing an invalidated data item from local cache. Because the validity of cached item is affected by the cache invalidation approach used, this metric reflects the effectiveness of the approach in maintaining the coherence of the data item between the database and the client cache. It is defined as follows:

$$False\_Valid\_Rate = \frac{False\_Cache\_Hit\_Count}{Cache\_Hit\_Count}$$

- False invalid rate — this metric is used to measure the accuracy of the invalidation report in invalidating the cached item and is defined as follows:

$$False\_Invalid\_Rate = \frac{False\_Cache\_Invalid\_Count}{Total\_Cache\_Invalid\_Count}$$

- Cache hit rate — this is simply the number of data accesses satisfied by the local cache divided by the total number of data requests.

16.5.3 Performance Evaluation

The simulation experiments presented in this section compare the performance of IAVI with the other cache invalidation approaches. Owing to limited space, only the more important results are included to illustrate the idiosyncrasies of the different schemes under different server and client settings, because none of the three practical schemes outperforms all others based on all the performance metrics.

The baseline setting of the system parameters used in the simulation are listed below (Table 16.5).

16.5.3.1 Impact of Database Size. In this set of simulation experiments, we investigate the impact of database size on the performance of the different cache invalidation schemes by varying the number of data items in the database. Figure 16.3 and Figure 16.4 depict the mean response time of the transactions and invalidation report size against the database size respectively. As we can observe from Figure 16.3, the performance of PS and IAVI are much better than TS and BS (smaller mean response time). This result is not affected significantly by the changes in database size. The performance of BS depends strongly on the size of the database. When the database size is large, it is even worse than TS whose performance is consistently much worse than PS and IAVI. The poor performance of BS when database size is large is due to the large invalidation report, which can be seen in Figure 16.4. The size of invalidation report increases with the size of the database according to the formula 2N + log2N * Time_bit_size, where N is the size of the database and Time_bit_size is the number of bits needed to represent update time [5]. On the other hand, the invalidation report sizes of TS, IAVI, and PS depend on the update volume (update rate multiplied by the number of records per update).

The slightly better performance of PS when compared to IAVI is due to the higher cache hit rate as can been seen in Figure 16.5. Although the cache hit probability of TS is also higher than that of IAVI, the impact of the

Table 16.5. Simulation Parameters

Database size	100,000 items
Items size	256 bits
Broadcast bandwidth	10,000 bps
Uplink capacity	1000 bps
Broadcast cycle	30 sec. per cycle
Invalidation report	1 per cycle
Invalidation report window size	10 cycle
AVI Tolerance	0.1
Update Process Parameters	
Update interval	3600 sec.
Update interval variance	−10 percent ~ +10 percent uniform
Update processes group	4 groups
Group 1	
Relative update interval (i.e., multiple of update interval)	1.0
Group 2	
Relative update interval	2.0
Group 3	
Relative update interval	4.0
Group 4	
Relative update interval	8.0
Mobile Client Parameters	
Number of mobile clients	100
Disconnect probability	0.1
Mean disconnect interval	4000 sec.
Access hot spot	100 items
Hot item access probability	0.8
Mean think time	100 sec.
Mean query length	10 data items
Cache size	50 items

much larger invalidation report offsets the benefits from its higher cache hit rate.

16.5.4.2 Impact of Update Rate. Figure 16.6 and Figure 16.7 show the mean response time and invalidation report size at different update intervals. As shown in Figure 16.6, the mean response time of TS decreases dramatically with an increase in update interval whereas other schemes, especially BS, are much less sensitive to the changes in update interval. The drop in response time of TS is due to the decrease in the invalidation report size when the update interval is longer as can be seen in Figure 16.7. Decreasing the update interval, thereby increasing the update rate, will increase the size of the invalidation reports. As the invalidation report of TS includes all update records within the invalidation time window, its size will increase when the update interval is decreased. On the other hand, IAVI only includes the update records that violate its AVI value. Therefore,

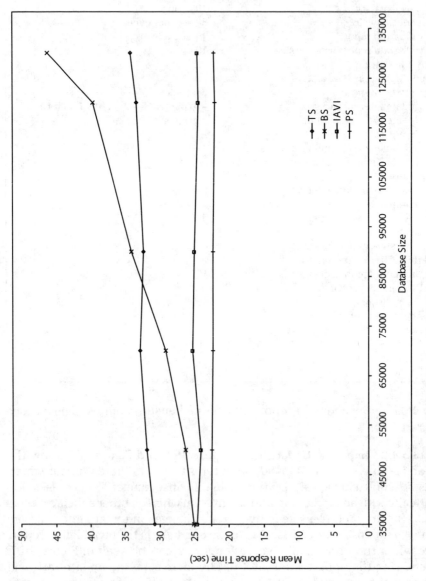

Figure 16.3. Mean Response Time versus Database Size

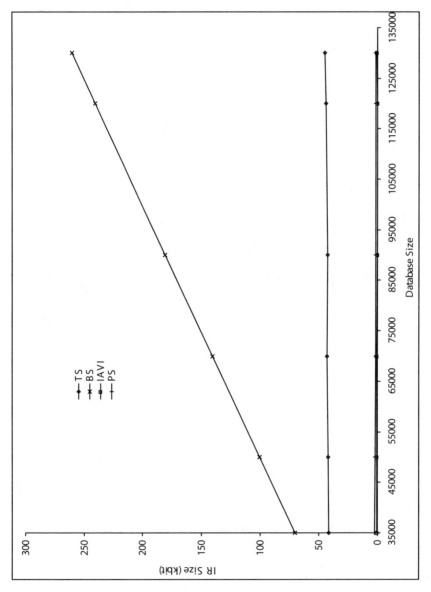

Figure 16.4. Invalidation Report Size versus Database Size

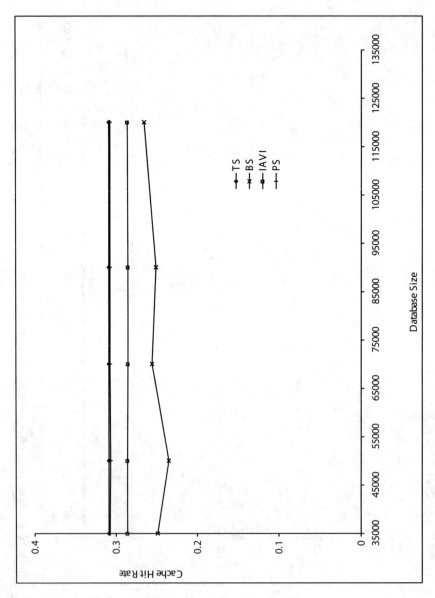

Figure 16.5. Cache Hit Rate versus Database Size

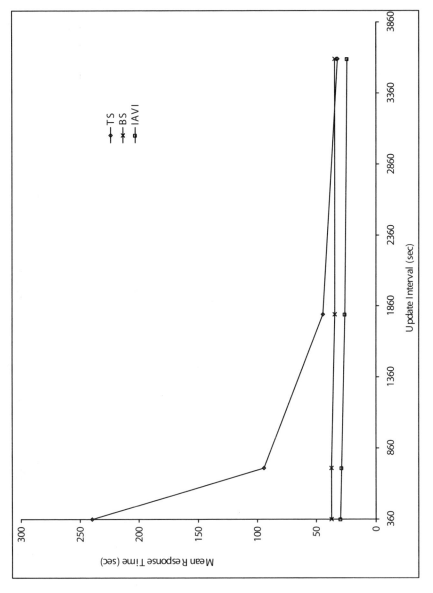

Figure 16.6. Mean Response Time versus Update Interval

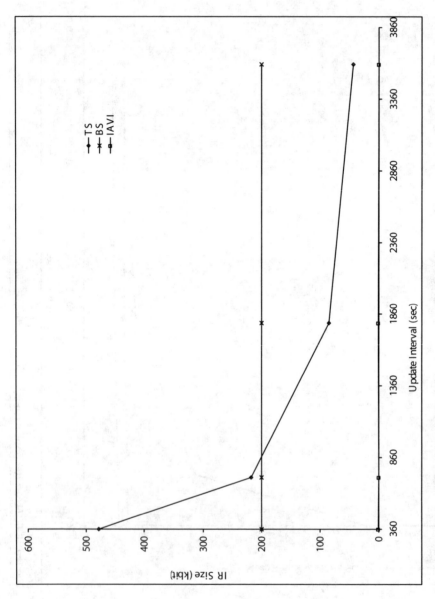

Figure 16.7. Invalidation Report Size versus Update Interval

it is relatively immune to changes in update rate. Note that the mean response time of IAVI decreases with update interval due to the decrease in the invalidation report size. In addition, the increase in cache hit rate also helps to decrease the mean response time, as shown in Figure 16.8. When the update rate is lower, cached items will remain valid for a long time. Thus, the cache hit rate will be higher.

The false valid rate and false invalid rate are shown in Figure 16.9 and Figure 16.10, respectively, for different database sizes. The higher false valid rate of TS as shown in Figure 16.9 is due to low bandwidth in a mobile network: after a data item is updated at the server, it takes a long time before a client receives the new value. However, for IAVI the cached items expire according to their AVI values and hence IAVI is comparatively less affected by the high network latency. Therefore, IAVI enjoys a low false valid rate and a low false invalid rate. On the other hand, BS has the highest false invalid rate, but at the same time also has the lowest false valid rate. Because the invalidation report of BS only has `log(Database_Size)` time slots for all update records in the report, most of the update time stamps of the items are marked early. Consequently, the false invalid rate is the highest among all the cache invalidation methods. Note also that the false valid rates of all three schemes decrease with an increase in update interval. This is because a lower update rate results in a smaller invalidation report and in turn decreases the latency in transmitting the report. Furthermore, a lower update rate decreases the probability of an invalid access of the cached item in spite of the presence of an invalidation report.

16.6 Conclusion

In a mobile computing environment, mobile clients are usually equipped with a local cache for reducing latency in data accesses. However, frequent disconnection of mobile clients from the network and updates occurring at the mobile server introduce the problem of cache incoherence. A large number of cache invalidation schemes have been proposed to handle this problem. However, most of them do not take into consideration the temporal properties of data.

In this chapter, we described the IAVI scheme for cache invalidation based on the real-time properties of the data items. We defined an AVI for each data item and used this property to self-invalidate items in the client cache. When a mobile client accesses a cached item, the update time stamp and AVI of the data item can verify the validity of the item. The cached item is invalidated if the access time is greater than the last update time by its AVI. With this self-invalidation mechanism, IAVI uses the invalidation report to inform the mobile client about changes in AVI rather than relying on individual update events of the data item. As a result, the size of the typical invalidation report can be reduced significantly.

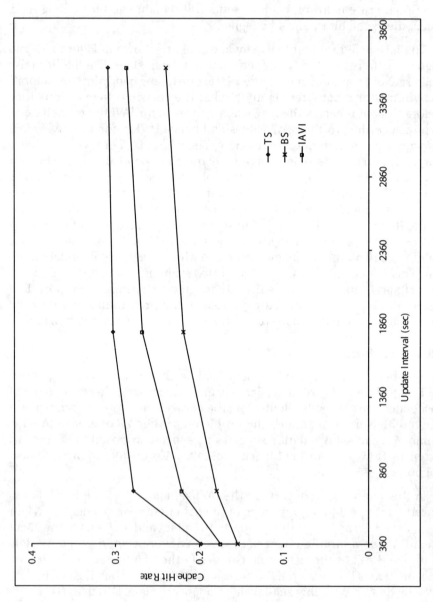

Figure 16.8. Cache Hit Rate versus Update Interval

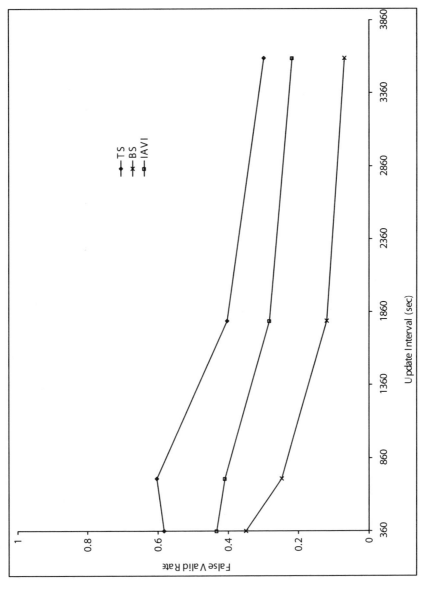

Figure 16.9. False Valid Rate versus Update Interval

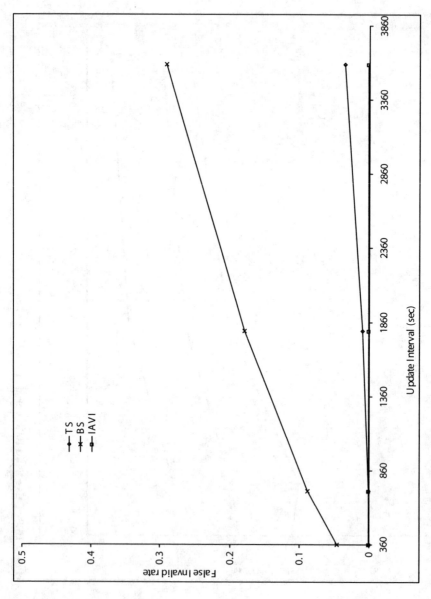

Figure 16.10. False Invalid Rate versus Update Interval

Performance studies based on extensive simulation show that the IAVI scheme can significantly reduce the mean response time and invalidation report size under various system parameters. Furthermore, the simulation results also show that the highest cache hit rate alone is not enough to get the best system performance. In practice it is important to also minimize invalidation report size so as to allocate as much broadcast bandwidth to data broadcast. By balancing these two factors, IAVI produces good results even when system parameters, such as database size and update rate, are varied.

It is possible to improve the performance of IAVI at the expense of additional complexity. It has been pointed out that a user query cannot be answered until the next invalidation report interval, which leads to an unnecessary long delay [8]. A possible enhancement is to combine IAVI with the UIR scheme [8], which will reduce response time and improve power efficiency.

References

1. Imielinski, T., Viswanathan, S., and Badrinath, B., Data on Air: Organization and Access, *IEEE Trans. Knowledge and Data Engineering*, 9, 353, 1997.
2. Acharya, S., Alonso, R., Franklin, M., and Zdonik, S., Broadcast Disk: Data Management for Asymmetric Communication Environments, in *Proc. ACM SIGMOD*, pp. 199–210, May 1995.
3. Fernanadez, J. and Ramamritham, K., Adaptive Dissemination of Data in Real-Time Asymmetric Communication Environments, in *Proc. Euromicro Conference on Real-Time*, pp. 195–203, 1998.
4. Barbara, D. and Imielinski, T., Sleepers and Workaholics: Caching Strategies in Mobile Environments, in *Proc. ACM SIGMOD*, pp. 1–12, 1994.
5. Jing, J., Elmargarmid, A., Helal, A., and Alonso, F., Bit-Sequences: an Adaptive Cache Invalidation Method in Mobile Client/Server Environments, *Mobile Networks and Applications,* 2, 1997.
6. Wu, K.L., Yu, P.S., and Chen, M.S., Energy-Efficient Caching for Wireless Mobile Computing, in *Proc. 20th International Conference on Data Engineering*, pp. 336–343, February 1996.
7. Hu, Q. and Lee, D.L., Adaptive Cache Invalidation Methods in Mobile Environments, in *Proc. 6th IEEE Int'l. Symp. High Performance Distributed Computing Environments*, pp. 264–273, 1997.
8. Cao, G., A Scalable Low-Latency Cache Invalidation Strategy for Mobile Environments, *IEEE Trans. Knowledge and Data Engineering*, 15, 1, 2003.
9. Cao, G., Proactive Power-Aware Cache Management for Mobile Computing Systems, *IEEE Trans. on Computers*, 51, 608, 2002.
10. Cai, J. and Tan, K.L., Energy Efficient Selective Cache Invalidation, *Wireless Networks*, 5, 489, 1999.
11. Tan, K.L, Cai, J., and Ooi, B.C., An Evaluation of Cache Invalidation Strategies in Wireless Environments, *IEEE Trans. Parallel and Distributed Systems*, 12, 789, 2001.
12. Pitoura, E. and Chrysanthis, P.K., Scalable Processing of Read-Only Transactions in Broadcast Push, in *Proc. International Conference on Distributed Computing Systems*, pp. 432–439, 1999.
13. Pitoura, E. and Chrysanthis, P.K., Exploiting Versions for Handling Updates in Broadcast Disks, in *Proc. VLDB Conference*, pp. 114–125, 1999.

14. Xu, J., Tang, X., and Lee, D.L., Performance Analysis of Location-Dependent Cache Invalidation Schemes for Mobile Environments, *IEEE Trans. on Knowledge and Data Engineering*, 15, 474, 2003.
15. Ramamrithan, K., Real-Time Databases, *Parallel and Distributed Databases*, 2, 199, 1993.
16. Kuo, T.W. and Mok, A.K., Application Semantics and Concurrency Control of Real-Time Data-Intensive Applications, in *Proc. IEEE 13th Real-Time Systems Symposium*, pp. 35–45, 1992.
17. Kuo T.W. and Mok, A.K., SSP: a Semantics-Based Protocol for RealTime Data Access, in *Proc. IEEE 14th RealTime Systems Symposium*, pp. 76–83, 1993.
18. Lam, K.Y., Chan, E., and Yuen, C.H., Approaches for Broadcasting Temporal Data in Mobile Computing Systems, *Journal of Systems and Software,* 51, 175, 2000.

Chapter 17
Hoarding in Mobile Computing Environments

Yücel Saygin

17.1 Introduction

Portable computers that are equipped with wireless communication devices enabled users to access global data services from any location. Mobile devices are now supporting applications such as multimedia and World Wide Web, which make it possible for the mobile computer users to surf the multimedia content on the Internet and read their e-mails while traveling. However, mobile computers with wireless communication are frequently disconnected from the network due to the cost of wireless communication or the unavailability of the wireless network. Consider a businessperson who frequently travels around by plane. Before getting on the plane that person would have to disconnect and then continue to work in the air without wireless network support. The same businessperson may travel by car and knows that after a certain point, no wireless network in a certain region exists. Thus this person's operation must go to disconnected mode. Another person may not be willing to pay for a continuous wireless connection, but would prefer to connect intermittently to access files of interest. Such scenarios suggest that the disconnected mode of operation will be in high demand and the mobile computer systems should provide support for it.

Disconnection can be voluntary or nonvoluntary. For example, a mobile user unplugging the network cable and packing for a trip or a user deciding to disconnect from the wireless network due to energy concerns or connection costs means voluntary disconnection. Involuntary disconnection occurs when the user goes out of the coverage area of the wireless network while traveling. A mobile computer should allow its users to work in disconnected mode seamlessly for both short-term and long-term disconnections, as well as voluntary and involuntary disconnections. Depending on the nature of disconnection (i.e., if it is voluntary/involuntary or short-term/long-term), seamless disconnection can be achieved by loading the files that a user will access in the future from the network to the local storage. This preparation process for disconnected operation is called hoarding.

Some important factors that complicate the hoarding process are listed in [3] as:

- The difficulty of prediction of the future access behavior of users
- Unpredictable disconnections
- The difficulty in cost evaluation of a cache miss in case of disconnection, which depends on the time to reconnect as well as the network latency
- Version control (which is also a problem in any distributed system that allows replication)
- How to avoid filling the limited local cache with noncritical objects

Hoarding may seem similar to prefetching in cache systems. Therefore we should make a clear differentiation of these two concepts. First of all, the purpose of prefetching is different from hoarding. With prefetching, the idea is to use the idle central processing unit (CPU) and spare bandwidth to load the data that will possibly be needed in the future with the aim of reducing the network latency. However, with hoarding, the latency will be practically infinite when the required file is not in local storage during disconnected operation. Therefore the cost of a hoard miss is not comparable to a cache miss considered in prefetching. In addition to that, CPU use is not the main concern during hoarding. The mobile users may sacrifice CPU time to make sure that files that will be needed in the future are loaded into the mobile cache, keeping in mind that a hoard miss may be fatal. A final point that differentiates hoarding from prefetching is that hoarding is done in a bulk manner, but prefetching is done continuously whenever the CPU is idle and the bandwidth is available.

In this chapter, we will try to give the reader a broad view of how hoarding is done in different systems and different environments together with a historical perspective. The early work on disconnected operation, done by the Coda project group at Carnegie Mellon University, is discussed in Section 17.2 as the pioneer of all hoarding systems. More recent approaches are provided in Section 17.3, Section 17.4, and Section 17.5. Hoarding methods based on data mining techniques are discussed in Section 17.3 and an approach based on the notion of program execution trees is explained in Section 17.4. In Section 17.5, we discuss a hoarding technique that assumes a distributed information system as the underlying architecture and that uses spatial locality to hoard the data. We provide a brief comparison of the various hoarding methods in Section 17.6 and finally in Section 17.7, we list some future research directions to improve the existing hoarding systems.

17.2 Coda: The Pioneering System for Hoarding

Coda is a distributed file system based on client–server architecture, where there are many clients and a comparatively smaller number of servers [3, 4]. It is the first system that enabled users to work in disconnected mode. The concept of hoarding was introduced by the Coda group as a means of enabling disconnected operation. Disconnections in Coda are assumed to occur involuntarily due to network failures or voluntarily due to the detachment of a mobile client from the network. Voluntary and involuntary disconnections are handled the same way. The cache manager of Coda, called Venus, is designed to work in disconnected mode by serving client requests from the cache when the mobile client is detached from the network. Requests to the files that are not in the cache during disconnection are reflected to the client as failures. The hoarding system of Coda lets users select the files that they will hopefully need in the future. This information is used to decide what to load to the local storage. For disconnected operation, files are loaded to the client local storage, because the master copies are kept at stationary servers, there is the notion of replication and how to manage locks on the local copies. When the disconnection is voluntary, Coda handles this case by obtaining exclusive locks to files. However in case of involuntary disconnection, the system should defer the conflicting lock requests for an object to the reconnection time, which may not be predictable.

The cache management system of Coda, called Venus, differs from the previous ones in that it incorporates user profiles in addition to the recent reference history. Each workstation maintains a list of pathnames, called the hoard database. These pathnames specify objects of interest to the user at the workstation that maintains the hoard database. Users can modify the hoard database via scripts, which are called hoard profiles. Multiple hoard profiles can be defined by the same user and a combination of these profiles can be used to modify the hoard database. Venus provides the user with an option to specify two time points during which all file references will be recorded. Due to the limitations of the mobile cache space, users can also specify priorities to provide the hoarding system with hints about the importance of file objects. Precedence is given to high priority objects during hoarding where the priority of an object is a combination of the user specified priority and a parameter indicating how recently it was accessed. Venus performs a hierarchical cache management, which means that a directory is not purged unless all the subdirectories are already purged.

In summary, the Coda hoarding mechanism is based on a least recently used (LRU) policy plus the user specified profiles to update the hoard database, which is used for cache management. It relies on user intervention to determine what to hoard in addition to the objects already maintained by the cache management system. In that respect, it can be classified as semi-automated. Researchers developed more advanced techniques with the aim of minimizing the user intervention in determining the set of objects to be hoarded. These techniques will be discussed in the following sections.

17.3 Hoarding Based on Data Mining Techniques

Data mining techniques aim to find interesting patterns from a large collection of data or try to build a descriptive model over the data [10]. In the context of hoarding, clustering and association rule mining techniques were adopted from data mining domain. Association rules describe the associations between a set of items with certain significance measures [12]. Clustering tries to group the data items in such a way that the data items in the same group are similar to each other or close to each other in space [11]. Section 17.3.1 explains the hoarding mechanism in SEER, which is based on grouping files into projects using clustering techniques. Section 17.3.2 describes a hoarding technique based on association rules.

17.3.1 SEER Hoarding System

To automate the hoarding process, Kuenning et al. developed a hoarding system called SEER that can make hoarding decisions without user intervention [5, 6]. The basic idea in SEER is to organize users' activities as projects in order to provide more accurate hoarding decisions.

A distance measure needs to be defined in order to apply clustering algorithms to group related files. SEER uses the notion of semantic distance based on the file reference behavior of the files for which semantic distance needs to be calculated. Once the semantic distance between pairs of files are calculated, a standard clustering algorithm is used to partition the files into clusters. The developers of SEER also employ some filters based on the file type and other conventions introduced by the specific file system they assumed.

The basic architecture of the SEER predictive hoarding system is provided in Figure 17.1. The observer monitors user behavior (i.e., which files are accessed at what time) and feeds the cleaned and formatted access paths to the correlator, which then generates the distances among files in terms of user access behavior. The distances are called the semantic distance and they are fed to the cluster generator that groups the objects with respect to their distances. The aim of clustering is, given a set of objects and a similarity or distance matrix that describes the pairwise distances or similarities among a set of objects, to group the objects that are close to each other or similar to each other. Calculation of the distances between files is done by looking at the high-level file references, such as open or status inquiry, as opposed to individual reads and writes, which are claimed to obscure the process of distance calculation. The semantic distance

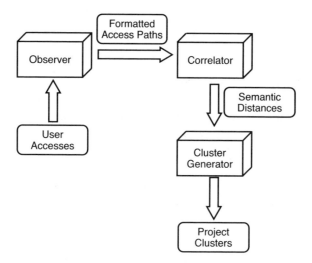

Figure 17.1. Architecture of the SEER Predictive Hoarding System

between two file references is based on the number of intervening references to other files in between these two file references. This definition is further enhanced by the notion of lifetime semantic distance. Lifetime semantic distance between an open file A and an open file B is the number of intervening file opens (including the open of B). If the file A is closed before B is opened, then the distance is defined to be zero. The lifetime semantic distance relates two references to different files; however it needs to be somehow converted to a distance measure between two files instead of file references. Geometric mean of the file references is calculated to obtain the distance between the two files. Keeping all pairwise distances takes a lot of space. Therefore, only the distances among the closest files are represented (closest is determined by a parameter K, K closest pairs for each file are considered).

The developers of SEER used a variation of an agglomerative (i.e., bottom up) clustering algorithm called k nearest neighbor, which has a low time and space complexity. An agglomerative clustering algorithm first considers individual objects as clusters and tries to combine them to form larger clusters until all the objects are grouped into one single cluster. The algorithm they used is based on merging subclusters into larger clusters if they share at least k_n neighbors. If the two files share less than k_n close files but more than k_f, then the files in the clusters are replicated to form overlapping clusters instead of being merged.

SEER works on top of a user level replication system such as Coda and leaves the hoarding process to the underlying file system after providing the hoard database. The files that are in the same project as the file that is currently in use are included to the set of files to be hoarded. During disconnected operation, hoard misses are calculated to give a feedback to the system.

17.3.2 Association Rule-Based Techniques

Association rule mining is a well researched area in the data mining domain [12]. The basic idea in association rule-based hoarding is to find patterns of user data access behavior expressed in the form of association rules by mining the collective request history of mobile clients.

An association rule is an implication of the form $X \Rightarrow Y$ where X and Y are sets of items. In our case, X and Y are sets of data items or files. An association rule captures two important forms of information about the items involved:

1. That they are referenced together frequently
2. That the items on the left hand side mimic the items on the right hand side

The second form of information helps us with the prediction process.

The client request history is preprocessed so that state-of-the-art efficient and incremental association rule mining algorithms can be used. The extracted association rules, which represent client access patterns, can be used to predict future client requests. The predicted request set is what should be loaded prior to disconnection so that the future client requests are satisfied locally without requiring a connection to the server. Association rules provide guides, such as support, confidence, and size of the rules, which are crucial in limiting the size of the data set to be hoarded.

The notion of a session is used, which consists of a group of continuous client requests and represents a period of user interest for a particular topic. In theory such sessions are independent of each other. They assume that client requests consist of sessions and that client request history could be partitioned to obtain those sessions. Each session contains some patterns of client requests. Data mining techniques find these patterns and produce rules that can be used to build a rule base of associations for prediction purposes. Before the disconnection occurs, rules are used to infer user's future requests to complete the current user session automatically. Hoarding is limited to the context of a session.

17.3.3 Partitioning the History into Sessions

There can be two basic approaches for mining the client request history depending on how the history of user requests is partitioned:

1. Flat approach
2. User-based partitioning approach

Hoarding in Mobile Computing Environments

The flat approach tries to extract data item request patterns regardless of who requested them. The user-based partitioning approach, on the other hand, divides the client request history into subsets with respect to the user who requested them. The analysis is done for each subset of the client request history corresponding to a user independent of the other requests.

To make use of the existing data mining algorithms, sessions need to be constructed out of the existing history. When session boundaries are not determined, a sliding window approach is used to obtain associations. A gap-based approach was also proposed to determine the session boundaries. In a gap-based approach, a new session starts when the time delay between two consecutive requests is greater than a prespecified threshold called the gap threshold.

In both flat- and user-based partitioning approaches, the request history is divided into sessions and association rules are extracted from the set of windows. Associations are extracted incrementally by first finding the frequently occurring requests and then finding the pairs using the frequently occurring singleton requests and counting their frequencies. After that, triples are found similarly. The process ends when the maximal sets of associated requests are found. From these sets of associations, the association rules are derived. For example, if requests for file A, file B, and file C frequently occur together, then A, B, C is an associated set of requests. From the this set of associated requests, rules of the form A => B C, A B => C, A C => B are derived, and those rules with a significant confidence (i.e., probability of the occurrence of the right hand side, given that the left hand side requests occur) are selected for the prediction process.

17.3.4 Utilization of Association Rules for Hoarding

The association rules obtained after mining the request history are used for determining the candidate set and the hoard set of the client upon disconnection. The candidate set is defined as the set of all candidates for hoarding for a specific client. Hoard set is the set of all data items actually loaded to the client prior to disconnection. A candidate set is constructed using inferencing on association rules as explained later on. Some other heuristics are used to prune the candidate set to the hoard set so that it fits to the cache of the mobile client.

The process of automated hoarding via association rules can be summarized as follows:

- Requests of the client in the current session are used through an inferencing mechanism to construct the candidate set prior to disconnection.
- Candidate set is pruned to form the hoard set.
- Hoard set is loaded to the client cache.

The need to have separate steps for constructing the candidate set and the hoard set arises from the fact that users also move from one machine to another that may have lower resources. The construction of the hoard set must adapt to such potential changes. Details about how the hoarding process constructs candidate and hoard sets are provided in the rest of this section.

17.3.5 Construction of the Candidate Sets and the Hoard Set

An inferencing mechanism is used to construct the candidate set of data items that are of interest to the clients to be disconnected. The candidate set of the client is constructed in two steps:

1. The inferencing mechanism finds the association rules whose heads (i.e., left hand side) match with the client's requests in the current session.
2. The tails (i.e., the right hand side) of the matching rules are collected into the candidate set.

The inferencing mechanism examines the current requests and predicts future ones. Priorities need to be assigned for the items obtained as a result of inferencing. The priority metric is based on the rule confidence and support values (i.e., the items inferred by a rule with a high confidence).

The client that issued the hoard request has limited resources. The storage resource is of particular importance for hoarding, because we have limited space to load a candidate set. Therefore, the candidate set obtained in the first phase of the hoarding set should shrink to the hoard set so that it fits the client cache. Each data item in the candidate set is associated with a priority. These priorities together with various heuristics must be incorporated for determining the hoard set. The data items are used to sort the rules in descending order of priorities. The hoard set is constructed out of the data items with the highest priority in the candidate set just enough to fill the cache.

For an effective hoarding, the cache misses during disconnection should be recorded and reflected to the history upon reconnection. In this manner, the whole picture of client requests, both connected and disconnected mode can be captured.

17.4 Hoarding Techniques Based on Program Trees

A hoarding tool based on program execution trees was developed by Tait et al. running under OS/2® operating system. Their method is based on analyzing program executions to construct a profile for each program depending on the files the program accesses. They proposed a solution to the hoarding problem in case of informed disconnections: the user tells the

Hoarding in Mobile Computing Environments

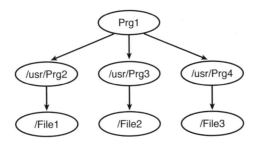

Figure 17.2. Sample Program Tree

mobile computer that there is an imminent disconnection to fill the cache intelligently so that the files that will be used in the future are already there in the cache when needed.

The proposed hoarding mechanism lets the user make the hoarding decision. They present the hoarding options to the user through a graphical user interface and working sets of applications are captured automatically. The working sets are detected by logging the user file accesses at the background. During hoarding, this log is analyzed and trees that represent the program executions are constructed. A node denotes a file and a link from a parent to one of its child nodes tells us that either the child is opened by the parent or it is executed by the parent. Roots of the trees are the initial processes. Program trees are constructed for each execution of a program, which captures multiple contexts of executions of the same program. This has the advantage that the whole context is captured from different execution times of the program. Finally, hoarding is performed by taking the union of all the execution trees of a running program. A sample program tree is provided in Figure 17.2.

Due to the storage limitations of mobile computers, the number of trees that can be stored for a program is limited to 15 LRU program trees. Hoarding through program trees can be thought of as a generalization of a program execution by looking at the past behavior. The hoarding mechanism is enhanced by letting the user rule out the data files. Data files are automatically detected using three complementary heuristics:

1. Looking at the filename extensions and observing the filename conventions in OS/2, files can be distinguished as executable, batch files, or data files.
2. Directory inferencing is used as a spatial locality heuristic. The files that differ in the top level directory in their pathnames from the running program are assumed to be data files, but the programs in the same top level directory are assumed to be part of the same program.

3. Modification times of the files are used as the final heuristic to determine the type of a file. Data files are assumed to be modified more recently and frequently than the executables. They devised a parametric model for evaluation, which is based on recency and frequency.

17.5 Hoarding in a Distributed Environment

Another hoarding mechanism, which was presented in [2], assumes a specific architecture, such as infostations where mobile users are connected to the network via wireless local area networks (LANs) that offer a high bandwidth, which is a cheaper option compared to wireless wide area networks (WANs). The hoarding process is handed over to the infostations in that model and it is assumed that what the user wants to access is location-dependent. Hoarding is proposed to fill the gap between the capacity and cost trade-off between wireless WANS and wireless LANs. The infostations do the hoarding and when a request is not found in the infostation, then WAN will be used to get the data item. The hoarding decision is based on the user access patterns coupled with that user's location information. Items frequently accessed by mobile users are recorded together with spatial information (i.e., where they were accessed). A region is divided into hoarding areas and each infostation is responsible with one hoarding area. The previous systems investigated and developed hoarding techniques from an operating system perspective. However in this one, a generic mobile unit is assumed and hoarding is done without regarding the file system details.

The hoarding process is divided into three steps:

1. Download — the mobile user connects to a proxy server via the wireless LAN for that infostation. The information items that are requested with high probability by the users are loaded to the users mobile device. Only those items that are usually requested from the current infostation to the next infostation are loaded, which provides the location information.
2. Disconnected operation — the users are now out of the range of the infostation and are on their own until reaching the next infostation. All the requests will now be answered through the hoarded data set. Whenever there is a miss, the data item can be loaded through the wireless WAN if the user accepts the associated costs. All the request history of the user is logged for updating the popularity of the item, which is calculated by the frequency of requests for that item.
3. Last phase — the user enters the range of the next infostation. The user logs are transferred to the proxy at the infostation, which distributes the logs to the infostations where a related hoard miss occurred in the hoarding area of that infostation. After that process, the log of the mobile client is flushed.

Access probabilities are maintained at each infostation with the arriving logs and the hoarding decision is made by simply looking at the frequently accessed items and loading as much as possible until the capacity of the mobile user is reached. The link capacity determines the time period until the user leaves and the bandwidth of the LAN determines the maximum number of items that can be loaded together with the space capacity of the user.

Some external information is also used for the user. If the destination of the user can be predicted, then the data items can be downloaded depending on the destination, which will be even more useful. By looking at the past behavior of the mobile users, the probabilities of a user to go from one location to the other is calculated and the hoarding decision is made based on the linear combination of the access probabilities at each location of hoard.

17.6 A Brief Comparison of the Various Hoarding Techniques

The hoarding techniques discussed above vary depending on the target system and it is difficult to make an objective comparative evaluation of their effectiveness. We can classify the hoarding techniques as being automated or not. In that respect, being the initial hoarding system, Coda is semiautomated and it needs human intervention for the hoarding decision. The rest of the hoarding techniques discussed are fully automated; however, user supervision is always desirable to give a final touch to the files to be hoarded.

Among the automated hoarding techniques, SEER and program tree-based ones assume a specific operating system and use semantic information about the files, such as the naming conventions, or file reference types and so on to construct the hoard set. However, the ones based on association rule mining and infostation environment do not make any operating system specific assumptions. Therefore, they can be used in generic systems.

Coda handles both voluntary and involuntary disconnections well. The infostation-based hoarding approach is also inherently designed for involuntary disconnections, because hoarding is done during the user passing in the range of the infostation area. However, the time of disconnection can be predicted with a certain error bound by considering the direction and the speed of the moving client predicting when the user will go out of range. The program tree-based methods are specifically designed for previously informed disconnections.

The scenario assumed in the case of infostations is a distributed wireless infrastructure, which makes it unique among the hoarding mechanisms. This case is especially important in today's world where peer-to-peer systems are becoming more and more popular.

17.7 Future Directions for Hoarding Techniques

The trend in hoarding started with semi-automated hoard set construction in Coda and went on with automated hoarding techniques. However, the mobile computing systems also became more complicated, supporting new types of applications and environments. Data mining techniques should be incorporated more into the hoarding mechanisms to achieve more accurate predictions. Below we listed some future suggestions for improvement of the current systems with data mining techniques.

Hoarding in the case of infostations used spatial locality, which assumed a different environment than the rest of the techniques. Infostation-based hoarding can be extended to capture temporality of the access patterns as well. In addition to associating requests with space, a time component can also be included. This requirement comes from the fact that some data items can be of high demand at a certain time at a certain place. For example, financial news is interesting particularly in the morning mostly in a business district and households are interested more in tabloid type news in residential areas during the daytime. Infostation hoarding assumes a model for the visit probabilities of mobile clients. Another extension to infostation hoarding would be to include path information as opposed to just having transition probabilities. Spatial and temporal data mining techniques are the approaches that can be used to achieve that.

We can extend the tree-based hoarding techniques as well with data mining methods. Frequent pattern trees can be constructed out of the program trees. Mining subgraphs in a graph structure is a growing area of research in data mining [13]. It can be used to extract frequent subgraphs from program executions graphs or trees.

With the advances in ad hoc networking technology, users can form communities of mobile clients with wireless ad hoc network connections. Data is shared in that case and available in the network, however a mobile client can be disconnected from the network and the rest of the clients should consider the fact that the data at the disconnecting client may be needed by them. Although they are not going to be disconnected, they may need to hoard the data at the disconnecting client, which can be viewed as reverse hoarding as the data is flowing from the disconnecting client to another client. This is an important case for wireless peer-to-peer information sharing systems.

A hoarding and reintegration mechanism involves the transfer of all contents. To perform efficient hoarding and reintegration, an incremental approach was proposed in [11] to do data transfers, which is also beneficial in the weakly connected mode of operation. This is also an important issue that still needs further investigation.

References

1. Carl Tait, Hui Lei, Swarup Acharya, and Henry Chang. Intelligent File Hoarding for Mobile Computers. In *Proceedings of the 1st Annual International Conference on Mobile Computing and Networking (MOBICOM'95)*, Berkeley, CA, 1995.
2. Uwe Kubach and Kurt Rothermel. Exploiting Location Information for Infostation-Based Hoarding. In *Proceedings of the 7th Annual International Conference on Mobile Computing and Networking (MOBICOM'01)*, Rome, 2001.
3. James J. Kistler and Mahadev Satyanarayanan. Disconnected Operation in the Coda File System. *ACM Transactions on Computer Systems,* vol. 10, no. 1, pp. 3–25, 1992.
4. Mahadev Satyanarayanan. The Evolution of Coda. *ACM Transactions on Computer Systems*, vol. 20, no. 2, pp. 85–124, 2002
5. Geoffrey H. Kuenning and Gerald J. Popek. Automated Hoarding for Mobile Computers. In *Proceedings of the 16th ACM Symposium on Operating System Principles (SOSP 1997)*, October 5–8, St. Malo, France, pp. 264–275, 1997.
6. Geoffrey H. Kuenning, Peter L. Reiher, and Gerald J. Popek. Experience with an Automated Hoarding System. *Personal and Ubiquitous Computing,* vol. 1, no. 3, 1997.
7. Yücel Saygin, Ozgur Ulusoy, and Ahmed K. Elmagarmid. Association Rules for Supporting Hoarding in Mobile Computing Environments. In *Proceedings of the 10th IEEE Workshop on Research Issues in Data Engineering (RIDE 2000),* February 28–29, San Diego, pp. 71–78, 2000.
8. Abdelsalam Helal, Abhinav Khushraj, and Jinsuo Zhang. Incremental Hoarding and Reintegration in Mobile Environments. In *Proceedings of Symposium on Applications and the Internet (SAINT 2002)*, January 28–February 1, Nara City, Japan, *IEEE Computer Society,* 2002.
9. Abdelsalam Helal, Joachim Hammer, Jinsuo Zhang, and Abhinav Khushraj. A Three-Tier Architecture for Ubiquitous Data Access, *ACS/IEEE International Conference on Computer Systems and Applications*, 2001, June 25–29, pp. 177–180, 2001.
10. Ming-Syan Chen, Jiawei Han, and Philip S. Yu. Data Mining: An Overview from a Database Perspective, *IEEE Transactions on Knowledge and Data Engineering,* vol. 8, no. 6, 1996.
11. Raymond A. Jarvis and Edward A. Patrick. Clustering Using a Similarity Measure Based on Shared Near Neighbors. *IEEE Transactions on Computers,*vol. 22, pp. 1025–1034, 1973.
12. Rakesh Agrawal and Ramakrishna Srikant, Fast Algorithms for Mining Association Rules. In *Proceedings of the 20th International Conference on Very Large Databases*, Santiago, Chile, pp. 487–499, 1994.
13. Michihiro Kuramochi and George Karypis. Frequent Subgraph Discovery. In *Proceedings of the International Conference on Data Mining*, San Jose, CA, 2001.
14. Jin Jing, Abdelsalam Helal, and Ahmed K. Elmagarmid. Client–Server Computing in Mobile Environments. *ACM Computing Surveys*, vol. 31, no. 2, 1999.

Chapter 18
Power-Aware Cache Management in Mobile Environments

Guohong Cao

Abstract

Recent work has shown that invalidation report (IR)-based cache management is an attractive approach for mobile environments. To improve the cache hit ratio of the IR-based approach, clients should proactively prefetch the data that is most likely to be used in the future. Although prefetching can make use of the broadcast channel and improve cache hit ratio, clients still need to consume power to receive and process the data. In this chapter, we first present a basic scheme to dynamically optimize performance and power based on a novel prefetch-access ratio (PAR) concept. Then, we extend the scheme to achieve a balance between performance and power considering various factors such as access rate, update rate, and data size.

18.1 Introduction

With the advent of the third generation wireless infrastructure and the rapid growth of wireless communication technology such as Bluetooth® and IEEE® 802.11, wireless Internet becomes possible: people with battery powered mobile devices (personal digital assistants [PDAs], hand-held computers, cellular phones, etc.) can access various kinds of services at any time any place. However, the goal of achieving ubiquitous connectivity with small-size and low-cost mobile devices (clients) is challenged by the power constraints. Most mobile clients are powered by battery, but the rate at which battery performance improves is fairly slow [19]. Aside from major breakthrough in battery technology, it is doubtful that significant improvement can be expected in the foreseeable future. Instead of trying to improve the amount of energy that can be packed into a power source, we can design power-aware protocols so that the mobile clients can perform the same functions and provide the same services while minimizing their overall power consumption.

Understanding the power characteristics of the wireless network interface (WNI) used in mobile clients is important for the efficient design of communication protocols. A typical WNI may operate in four modes — transmit, receive, idle, and sleep. Many studies [25, 28] show that the power consumed in the receive or idle mode is similar, but are significantly higher than the power consumed in the sleep mode. As a result, most of the work on power management concentrates on putting the WNI into sleep, when it is in the idle mode. This principle has been applied to different layers of the network hierarchy [8, 20, 28] and can also be applied to data dissemination techniques such as broadcasting. With broadcasting, mobile clients access data by simply monitoring the channel until the required data appears on the broadcast channel. To reduce the client power consumption, techniques such as indexing [12] were proposed to reduce the client tune-in time. The general idea is to interleave index (directory) information with data on the broadcast channels such that the clients, by first retrieving the index information, are able to obtain the arrival time of the desired data. As a result, a client can enter sleep most of the time and only wake up just before the desired data arrives.

Although broadcasting has good scalability and low bandwidth requirement, it has some drawbacks. For example, because a data item may contain a large volume of data (especially in the multimedia era), the data broadcast cycle may be long. Hence, the clients have to wait for a long time before getting the required data. Caching frequently accessed data at the client side is an effective technique to improve performance in mobile computing systems. With caching, the data access latency is reduced, because some data access requests can be satisfied from the local cache, thereby obviating the need for data transmission over the scarce wireless links. When caching is used, cache consistency must be addressed. Although caching techniques used in file systems such as Coda [23], Ficus [21] can be applied to mobile environments, these file systems are primarily designed for point-to-point communication environment and may not be applicable to the broadcasting environment.

Recently, many works [3–6, 14, 26, 29, 30] have shown that IR-based cache management is an attractive approach for mobile environments. In this approach, the server periodically broadcasts an invalidation report in which the changed data items are indicated. Rather than querying the server directly regarding the validation of cached copies, the clients can listen to these IRs over the wireless channel and use them to validate their local cache. Because IRs arrive periodically, clients can go to sleep most of time and only wake up when the IR comes. The IR-based solution is attractive, because it can scale to any number of clients who listen to the IR.

However, the IR-based solution has some drawbacks, such as long query latency and low cache hit ratio. In our previous work [6], we addressed the long latency problem with a UIR-based approach, where a small fraction of the essential information (called updated invalidation report [UIR]) related to cache invalidation is replicated several times within an IR interval and hence the client can answer a query without waiting until the next IR. However, if there is a cache miss, the client still needs to wait for the data to be delivered. To improve the cache hit ratio, we proposed a proactive cache management scheme [5], where clients intelligently prefetch the data that is most likely to be used in the future. Prefetching has many advantages in mobile environments because wireless networks such as wireless LANs or cellular networks support broadcasting. When the server broadcasts data on the broadcast channel, clients can prefetch interested data to increase the cache hit ratio without increasing the bandwidth consumption. Although prefetching can make use of the broadcast channel and improve cache hit ratio, clients still need to consume power to receive and process the data. Further, they cannot power off the wireless network interface, which consumes a large amount of power even when it is in the idle mode [24]. Because most mobile clients are powered by battery, it is important to prefetch the right data. Unfortunately, most of the prefetch techniques used in the current cache management schemes [6, 7] do not consider power constraints of the mobile clients and other factors such as the data size, the data access rate, and the data update rate.

To address the power consumption issue, we first present a basic adaptive scheme to save power during prefetch. Based on a novel PAR concept, the proposed scheme can dynamically optimize performance or power based on the available resources and the performance requirements. Then, we extend the basic scheme and present a value-based prefetch (VP) scheme, which makes prefetch decisions based on the value of each data item considering various factors such as access rate, update rate, and data size. Finally, we extend the VP scheme and present two adaptive value-based prefetch (AVP) schemes, which can achieve a balance between performance and power based on different user requirements.

The rest of the chapter is organized as follows:

- Section 18.2 develops the necessary background.
- Section 18.3 proposes power-aware cache management techniques to balance the trade off between performance and power.
- Section 18.4 concludes the chapter and points out future research directions.

18.2 Cache Invalidation Techniques

In this section, we define our Cache Consistency Model and describe techniques to improve the performance of the IR-Based Cache Invalidation Model.

18.2.1 Cache Consistency Model

When cache techniques are used, data consistency issues must be addressed. The notion of data consistency is, of course, application dependent. In database systems, data consistency is traditionally tied to the notion of transaction serializability. In practice, however, few applications demand or even want full serializability and more efforts have gone into defining weaker forms of correctness. In this chapter, we use the latest value consistency model [2, 6, 16], which is widely used in dissemination-based information systems. In the latest value consistency model, clients must always access the most recent value of a data item. This level of consistency is what would arise naturally if the clients do not perform caching and the server broadcasts only the most recent values of items. Note that the Coda file system [23] does not follow the latest value consistency model. It supports a much weaker consistency model to improve performance. However, some conflicts may require manu-configuration and some updated work may be discarded.

When client caching is allowed, techniques should be applied to maintain the latest value consistency. Depending on whether or not the server maintains the state of the client's cache, two invalidation strategies are used — the stateful server approach and the stateless server approach. In the stateful server approach, the server maintains the information about which data is cached by which client. Once a data item is changed, the server sends invalidation messages to the clients with copies of that particular data. The Andrew File System [17] is an example of this approach. However, in mobile environments, the server may not be able to contact the disconnected clients. Thus, a disconnection by a client automatically means that its cache is no longer valid. Moreover, if the client moves to another cell, it has to notify the server. This implies some restrictions on the freedom of the clients. In the stateless server approach, the server is not aware of the state of the client's cache. The clients need to query the server to verify the validity of their caches before each use. The Network File System (NFS) [22] is an example of this approach. Obviously, in this option, the clients generate a large amount of traffic on the wireless channel, which not only wastes the scarce wireless bandwidth, but also consumes a lot of battery energy. Next, we present the IR-Based Cache Invalidation Model, which has been widely used in mobile environments.

18.2.2 The IR-Based Cache Invalidation Model

In the IR-based cache invalidation strategy, the server periodically broadcasts IRs, which indicate the updated data items. Note that only the server can update the data. To ensure cache consistency, every client, if active, listens to the IRs and uses these IRs to invalidate its cache accordingly. To answer a query, the client listens to the next IR and uses it to decide whether its cache is still valid or not. If there is a valid cached copy of the

Power-Aware Cache Management in Mobile Environments

requested data, the client returns the data immediately. Otherwise, it sends a query request through the uplink (from the client to the server). The server keeps track of the recently updated information and broadcasts an IR every L seconds. In general, a large IR can provide more information and is more effective for cache invalidation, but a large IR occupies a large amount of broadcast bandwidth and the clients may need to spend more power listening to the IR, because they cannot switch to power save mode while listening. In the following, we look at two IR-based algorithms.

18.2.2.1 The Broadcasting Time Stamp Scheme. The time stamp (TS) scheme was proposed by Barbara and Imielinski [3]. In this scheme, the server broadcasts an IR every L seconds. The IR consists of the current time stamp T_i and a list of tuples (d_x, t_x), such that $t_x > (T_i - w * L)$, where d_x is the data item id, t_x is the most recent update time stamp of d_x, and w is the invalidation broadcast window size. In other words, IR contains the update history of the past w broadcast intervals.

To save energy, a MT may power off most of the time and only turn on during the IR broadcast time. Moreover, a MT may be in the power off mode for a long time to save energy and hence, the client running the MT may miss some IRs. Because the IR includes the history of the past w broadcast intervals, the client can still validate its cache as long as its disconnection time is shorter than $w * L$. However, if the client disconnects longer than $w * L$, it has to discard the entire cached data items, because it has no way to tell which parts of the cache are valid. Because the client may need to access some items in its cache, discarding the entire cache may consume a large amount of wireless bandwidth in future queries.

18.2.2.2 The Bit Sequences Scheme. In the bit-sequence (BS) scheme [14], the IR consists of a sequence of bits. Each bit represents a data item in the database. Setting the bit to 1 means that the data item has been updated. The update time of each data item is also included in the IR. To reduce the length of the IR, some grouping methods are used to make one bit coarsely represent several data items. Instead of including one update time stamp for each data item, the BS scheme uses one time stamp to represent a group of data items in a hierarchical manner. Let IR be $\{[B_0, TS(B_0)], \ldots, [B_k, TS(B_k)]\}$ where $B_i = 1$ means that half of the data items from 0 to 2^i at time $TS(B_i)$ have been updated. The clients use the bit sequences and the time stamps to decide what data items in their local cache should be invalidated. The scheme is flexible (no invalidation window size is needed) and it can be used to deal with the long disconnection problem by carefully arranging the bit sequence. However, because the IR represents the data of the entire database (half of the recently updated data items in the database, if more than half data items have been updated since the initial time), broadcasting the IR may consume a large amount of downlink bandwidth.

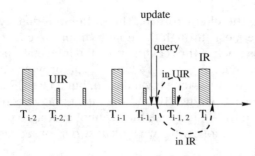

Figure 18.1. Reducing the Query Latency by Replication UIRs

Many solutions [11, 14, 27] are proposed to address the long disconnection problem, and Hu et al. [11] has a good survey of these schemes. Although different approaches [3, 14] apply different techniques to construct the IR to address the long disconnection problem, these schemes maintain cache consistency by periodically broadcasting the IR. The IR-based solution is attractive because it can scale to any number of MTs who listen to the IR. However, this solution has long query latency, because the client can only answer the query after it receives the next IR to ensure cache consistency. Hence, the average latency of answering a query is the sum of the actual query processing time and half of the IR interval.

18.2.3 The UIR-Based Cache Invalidation Model

To reduce the query latency, we [6] proposed to replicate the IRs m times; that is, the IR is repeated every $(1/m)^{th}$ of the IR interval. As a result, a client only needs to wait at most $(1/m)^{th}$ of the IR interval before answering a query. Hence, latency can be reduced to $(1/m)^{th}$ of the latency in the previous schemes (when query processing time is not considered).

Because the IR contains a large amount of update history information, replicating the complete IR m times may consume a large amount of broadcast bandwidth. To save the broadcast bandwidth, after one IR, $m-1$ UIRs are inserted within an IR interval. Each UIR only contains the data items that have been updated after the last IR was broadcast. In this way, the size of the UIR becomes much smaller compared to that of the IR. As long as the client downloads the most recent IR, it can use the UIR to verify its own cache. The idea of the proposed technique can be further explained by Figure 18.1. In Figure 18.1, $T_{i,k}$ represents the time of the kth UIR after the ith IR. When a client receives a query between $T_{i-1,1}$ and $T_{i-1,2}$, it cannot answer the query until T_i in the IR-based approach, but it can answer the query at $T_{i-1,2}$ in the UIR-based approach. Hence, the UIR-based approach can reduce the query latency in case of a cache hit. However, if there is a cache miss, the client still needs to fetch data from the server, which increases the query latency. Next, we present a cache management algorithm to improve the cache hit ratio and the bandwidth use.

18.2.4 Using Prefetch to Improve Cache Hit Ratio and Bandwidth Use

In most previous IR-based schemes, even though many clients cache the same updated data item, all of them have to query the server and get the data from the server separately. Although the approach works fine for some cold data items, which are not cached by many clients, it is not effective for hot data items. For example, suppose a data item is frequently accessed (cached) by 100 clients, updating the data item once may generate 100 uplink (from the client to the server) requests and 100 downlink (from the server to the client) broadcasts. Obviously, it wastes a large amount of wireless bandwidth and battery energy.

We address the problem by asking the clients to prefetch data that may be used in the near future. For example, if a client observes that the server is broadcasting a data item, which is an invalid entry[1] of its local cache, it is better to download the data; otherwise, the client may have to send another request to the server and the server will have to broadcast the data again in the future. To save power, clients may only wake up during the IR broadcasting period and then how to prefetch data becomes an issue. As a solution, after broadcasting the IR, the server first broadcasts the id list of the data items whose data values will be broadcast next and then broadcasts the data values of the data items in the id list. Each client should listen to the IR if it is not disconnected. At the end of the IR, a client downloads the id list and finds out when the interested data will come and wakes up at that time to download the data. With this approach, power can be saved because clients stay in the sleep mode most of the time; bandwidth can be saved because the server may only need to broadcast the updated data once.

Prefetching also consumes power, so it is important to identify which data should be included in the id list. Based on whether the server maintains information about the client or not, two cache invalidation strategies are used — the stateful server approach and the stateless server approach. In [4, 7], we studied the stateful server approach. In the proposed solution, a counter is maintained for each data item. The counter associated with a data item is increased by 1 when a new request for the data item arrives. Based on the counter, the server can identify which data should be included in the id list. Novel techniques are designed to maintain the accuracy of the counter in case of server failures, client failures, and disconnections. However, the stateful approach may not be scalable due to the high state maintenance overhead, especially when handoffs are frequent. Thus, we adopt the stateless approach in this chapter. Because the server does not maintain any information about the clients, it is difficult, if not impossible, for the server to identify which data is hot. To save broadcast bandwidth, the server does not answer the client requests immediately; instead, it waits for the next IR interval. After broadcasting the IR, the server broadcasts the id list (L_{bcast}) of the data items that have been requested during the last IR interval. In addition, the server broadcasts the values of the data items in the id list. At the end of the IR, the client downloads L_{bcast}. For each item id in L_{bcast}, the client checks whether it has requested the server for the item or the item becomes an invalid cache entry due to server update. If any of the two conditions are satisfied, it is better for the client to download the current version as the data will be broadcast.

One important reason for the server not to serve requests until the next IR interval is due to energy consumption. In our scheme, a client can go to sleep most of the time and only wake up during the IR and L_{blist} broadcast time. Based on L_{blist}, it checks whether there is interested data that will be broadcast. If not, it can go to sleep and only wake up at the next IR. If so, it can go to sleep and only wakes up at that particular data broadcast time. For most of the server initiated cache invalidation schemes, the server needs to send the updated data to the clients immediately after the update and the clients must keep awake to get the updated data. Here we trade-off some delay for more battery energy. Due to the use of UIR, the delay tradeoff is not that significant; most of the time (cache hit), the delay can be reduced by a factor of m, where $(m-1)$ is the number of replicated UIRs within one IR interval. Even in the worst case (for cache miss), our scheme has the same query delay as the previous IR-based schemes, where the clients cannot serve the query until the next IR. To satisfy time constraint applications, we may apply priority requests as follows: when the server receives a priority request, it serves the request immediately instead of waiting until the next IR interval.

18.2.4.1 Remarks. Prefetching has been widely used to reduce the response time in the Web environment. Most of these techniques [13, 15, 18] concentrate on estimating the probability of each file being accessed in the near future. They are designed for the point-to-point communication environment, which is different from our broadcasting environment. Although the objective of prefetching is the same — to improve cache hit ratio and reduce the response time — there are many differences between our prefetch technique and the existing work. First, prefetch in the Web environment will increase the Web traffic, but not in our UIR-Based Model due to the broadcast environment. Second, the prefetch technique used in UIR is not a simple prefetch. We consider power consumption issues and carefully design L_{blist} so that clients can still stay sleep most of time. With our careful design, most clients stay sleep when not prefetching, but the clients that need to prefetch still consume power to download and process the data. Next, we present techniques to further reduce this part of power consumption.

18.3 Techniques to Optimize Performance and Power

In this section, we first present a basic scheme to optimize performance and power, then extend it to consider various factors such as access rate, update rate, and data size.

18.3.1 The Basic Scheme

The advantage of the prefetch depends on how hot the requested data item is. Let us assume that a data item is frequently accessed (cached) by n clients. If the server broadcasts the data after it receives a request from one of these clients, the saved uplink and downlink bandwidth can be up to a factor of n when the data item is updated. Because prefetching also consumes power, we investigate the trade off between performance and power and propose an adaptive scheme to efficiently use the power in this subsection.

Each client may have different available resources and performance requirements and these resources, such as power, may change with time. For example, suppose the battery of a laptop lasts three hours. If the user is able to recharge the battery within three hours, power consumption may not be an issue and the user may be more concerned about the performance aspects such as the query latency. However, if the user cannot recharge the battery within three hours and wants to use it a little bit longer, power consumption becomes a serious concern. As a design option, the user should be able to choose whether to prefetch data based on the resource availability and the performance requirement. This can be done manually or automatically. In the manual option, the user can choose whether the query latency or the power consumption is the primary concern. In the automatic approach, the system monitors the power level. When the power level drops below a threshold, power consumption becomes the primary concern. If query latency is more important than power consumption, the client should always prefetch the interested data. However, when the power drops to a threshold, the client should be cautious about prefetching.

There are two solutions to reduce the power consumption. As a simple solution, the client can reduce its cache size. With a smaller cache, the number of invalid cache entries reduces and the number of prefetches drops. Although small cache size reduces prefetch power consumption, it may also increase the cache miss ratio, thereby degrading performance. In a more elegant approach, the client marks some invalid cache entries as nonprefetch and it will not prefetch these items. Intuitively, the client should mark those cache entries that need more power to prefetch, but are not accessed too often.

18.3.1.1 The Basic Adaptive Prefetch Approach.
To implement the idea, for each cached item, the client records how many times it accessed the item and how many times it prefetched the item during a period of time. The PAR is the number of prefetches divided by the number of accesses. If the PAR is less than 1, prefetching the data is useful because the prefetched data may be accessed multiple times. When power consumption becomes an issue, the client marks those cache items that have PAR > β as nonprefetch, where $\beta > 1$ is a system tuning factor. The value of β can be dynamically changed based on the power consumption requirements. For example, with a small β, more energy can be saved, but the cache hit ratio may be reduced. On the other hand, with a large β, the cache hit ratio can be improved, but at a cost of more energy consumption. Note that when choosing the value of β, the uplink data request cost should also be considered.

When the data update rate is high, the PAR may always be larger than β and clients cannot prefetch any data. Without prefetch, the cache hit ratio may be dramatically reduced and this results in poor performance. Because clients may have a large probability to access a small amount of data, marking these data items as prefetch may improve the cache hit ratio and does not consume too much power. Based on this idea, when PAR > β, the client marks N_p number of cache entries, which have high access rate as prefetch.

Because the query pattern and the data update distribution may change over time, clients should measure their access rate and PAR periodically and refresh some of their history information. Assume N^x_{acc} is the number of access times for a cache entry d_x. Assume $N^x_{c\_acc}$ is the number of access times for a cache entry d_x in the current evaluation cycle. The number of access times is calculated by:

$$N^x_{acc} = (1-\alpha) * N^x_{acc} + \alpha * N_{c\_acc}$$

where $\alpha < 1$ is a factor that reduces the impact of the old access frequency with time. A similar formula can be used to calculate PAR.

Although the basic adaptive prefetch scheme can achieve a better trade off between performance and power, it does not consider varying data size and the data update rate. Also, there is no clear methodology as to how and when N_p should be changed. In the next two subsections, we will address this problem by an AVP scheme, which consists of two parts. The first part is the VP scheme, which identifies valuable data for prefetching. The second part is the AVP scheme, which determines how many data items should be prefetched.

18.3.2 The Value-Based Prefetch Scheme

To consider the effects of data size, we need to introduce a new performance metric. One widely used performance metric is the response time: the time between sending a request and receiving the reply. It is a suitable metric for homogeneous settings where different data requests have the same size. However, the data requirements of users and applications are inherently diverse and to encapsulate all responses into a single-size broadcast would be unreasonably wasteful. Therefore, unlike some previous work [5, 10], we do not assume that the data items have the same size. When data requests are heterogeneous, response time alone is not a fair measure given that the individual requests significantly differ from each other in their service time, which is defined as the time to complete the request if it was the only job in the system. We adopt an alternate performance measure, namely the stretch [1] of a request, defined as the ratio of the response time of a request to its service time. The rationale behind this choice is based on our intuition; for example, clients with larger jobs should be expected to be in the system longer than those with smaller requests. The drawback of minimizing response time for heterogeneous workloads is that it tends to improve the system performance of large jobs, because they contribute the most to the response time. Minimizing stretch, on the other hand, is more fair to all job sizes. Note that in broadcast systems, the service time for a request is the requested data size divided by the bandwidth. For simplicity, we remove the constant bandwidth factor and use the data size to represent the service time.

Next, we present a value-based function that allows us to gauge the worth of a data item when making a prefetch decision. The following notations are used in the presentation:

p_{a_i} — the access probability of data item i
p_{u_i} — the probability of invalidating cached data item i before next access
f_i — the delay of retrieving data item i from the server
s_i — the size of data item i
v — the cache validation delay

The value function is used to identify the data to be prefetched. Intuitively, the ideal data item for prefetching should have a high access probability, a low update rate, a small data size, and a high retrieval delay. Equation 18.1 incorporates these factors to calculate the value of a data item i:

$$value(i) = \frac{p_{a_i}}{s_i}(f_i - v - p_{u_i} \cdot f_i) \qquad (18.1)$$

This value function can be further explained by the data access cost model shown in Figure 18.2. If item i is not in the cache, in terms of the

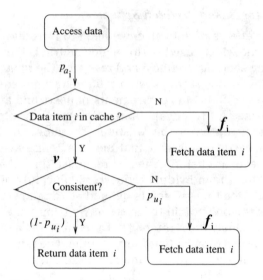

Figure 18.2. The Data Access Cost Model

stretch value, it takes f_i/s_i to fetch item i into the cache. In other words, if i is prefetched to the cache, the access cost can be reduced by f_i/s_i. However, it also takes $((v + P_{u_i} f_i))/s_i$ to validate the cached item i and update it if necessary. Thus, prefetching the data can reduce the cost by $((f_i - v - p_{u_i} \cdot f_i))/s_i$ for each access. Because the access probability is p_{a_i}, the value of prefetching item i is $(P_{a_i}/s_i)(f_i - v - p_{u_i} \cdot f_i)$.

The VP scheme decides which data item should be prefetched based on the value function. The VP scheme is defined as follows. Suppose a client can prefetch N_p data items, the VP scheme prefetches the N_p items that have the highest value based on the value function. Note that VP is not responsible for determining how many items (N_p) should be prefetched. N_p is determined by the adaptive scheme, which will be discussed in Section 18.3.3.

Theorem 18.1

Prefetching items with high value can achieve lower stretch than any other prefetch schemes given that the number of prefetches is limited.

Details of the proof and how to estimate the parameters can be found in [29]. The proposed value-based function is calculated in terms of stretch because the performance metric is stretch. Actually, this value-based function can be easily extended for other performance metrics. For example, if the performance metric is query delay, the value function will be changed to $value(i) = p_{a_i}(f_i - v - p_{u_i} \cdot f_i)$. Similar techniques can be used to prove that this value function can minimize the query delay.

18.3.3 The Adaptive Value-Based Prefetch Scheme

Due to limitations of battery technology, the energy available for a mobile client is limited and must be used prudently. If the prefetched data item is not accessed or is invalidated before it is accessed, the energy spent on downloading this item will be wasted. To avoid wasting power, it is important that clients only download the data with high value, but such a strict policy may adversely affect the performance of the system and increase the query delay.

Each client may have different available resources and performance requirements and these resources, such as power, may change over time. Because N_p controls the number of data to be prefetched and then affects the trade-off between performance and power, we propose adaptive schemes to adjust N_p to satisfy different client requirements.

18.3.3.1 The Value of N_p. When N_p reduces to 0, there will be no prefetch. As N_p increases, the number of prefetches increases and the power consumption also increases. Because the maximum number of data items to be prefetched is limited by the cache size, N_p is also limited by this number. Intuitively, the query delay decreases as the number of prefetches increases. However, this is not always true considering the overhead to maintain cache consistency. In our Cache Invalidation Model, a client needs to wait for the next IR to verify the cache consistency. This waiting time may increase the query delay compared to the approaches without prefetch. The cost has been quantified in Equation 18.1, where v is the cache invalidation delay. Due to the cost of v, the value of a data item may be negative. If value(i) is negative, prefetching item i not only wastes power, but also increases the average stretch. Therefore, N_p should be bounded by N_p^{max}, which is limited by the client cache size and the data value: a client will not prefetch items with negative values.

The trade-off between performance and power can be achieved by adjusting N_p. In the following subsections, we present two adaptive schemes:

1. The AVP_T (T for time) scheme, which dynamically adjusts N_p to reach a target battery life time
2. The AVP_P (P for power) scheme, which dynamically adjusts N_p based on the remaining power level

18.3.3.2 AVP_T: Adapting N_p to Reach a Target Battery Life. A commuter normally knows the amount of battery energy and the length of the trip between home and office. With these resource limitations, the commuter wants to achieve the lowest query delay. This is equivalent to the problem of adapting N_p to reach a target battery life and minimize the average stretch. Suppose a battery with E joule lasts T_1 seconds when $N_p = N_p^{max}$

and T_2 seconds when $N_p = 0$. It is possible to adjust N_p to reach a target battery life time $T \in [T_1, T_2]$. In AVP_T, the client monitors the power consumed in the past. If it consumed too much power in the past and cannot last T seconds, N_p is reduced. On the other hand, it increases N_p when it found that it had too much power left. Certainly, N_p should be bounded by N_p^{max}.

18.3.3.3 AVP_P: Adapting N_p Based on the Power Level.
When the energy level is high, power consumption is not a major concern and then trading off energy for performance may be a good option if the user can recharge the battery soon. On the other hand, when the energy level is low, the system should be power-aware to prolong the system running time to reach the next battery recharge time. Based on this intuition, the AVP_P scheme dynamically changes N_p based on the power level. Let a_k be the percentage of energy left in the client. When a_k drops to a threshold, the number of prefetches should be reduced to some percentage, say $f(a_k)$, of the original value. Some simple discrete function can be as follows:

$$f(a_k) = \begin{cases} 100\% & 0.5 < a_k \leq 1.0 \\ 70\% & 0.3 < a_k \leq 0.5 \\ 50\% & 0.2 < a_k \leq 0.3 \\ 30\% & 0.1 < a_k \leq 0.2 \\ 10\% & a_k \leq 0.1 \end{cases} \quad (18.2)$$

At regular intervals, the client reevaluates the energy level a_k. If a_k drops to a threshold value, $N_p = N_p \cdot f(a_k)$. The client only marks the first N_p items in the cache, which have the maximum value, as prefetchable. In this way, the number of prefetches can be reduced to prolong the system running time. Because this is a discrete function, N_p does not need to be frequently updated and the computation overhead is low.

Note that a simple policy that is neither too aggressive nor conservative might result in similar average stretch and lifetime as the VAP_P scheme if the battery runs out before recharge. However, if the user recharges the battery frequently, this simple policy may not be a good option because it saves power at the cost of delay, but power consumption is not a concern at this time. In contrast, our adaptive scheme tries to trade off power for performance at the beginning and become power-aware when the client cannot recharge in time.

18.4 Conclusions and Future Work

Prefetching can be used to improve the cache hit ratio and reduce the bandwidth consumption in our UIR-Based Cache Invalidation Model. However, prefetching consumes power. In mobile environments where power is limited, it is essential to correctly identify the data to be prefetched in

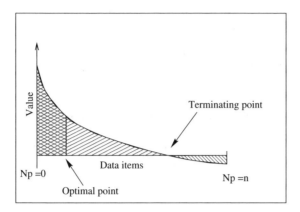

Figure 18.3. Prefetch Value of the Data Items

order to provide better performance and reduce the power consumption. In this chapter, we first presented a basic scheme to dynamically optimize performance or power based on a novel PAR concept. Then, we extended it to achieve a balance between performance and power considering various factors such as access rate, update rate, and data size. Although we explored various adaptive approaches, many issues still need further investigation. For example, the channel state information [9] can be used when making prefetch decisions (i.e., N_p can be dynamically adjusted based on the channel state). If the target battery life time is not known or only known with some probability, AVP_T needs to be extended to factor into these uncertainties. If the battery recharge cycle or the user profile is known or known with a high probability, how to enhance AVP_T and AVP_P needs further investigation.

Currently, we are investigating the optimal value of N_p. Figure 18.3 shows the effects of N_p (n is the database size) when data items are sorted according to their values from high to low. Suppose the value of the data can be correctly identified, then increasing N_p means more data items with high value will be prefetched. If the uplink power consumption is much higher than the downlink power, prefetching the data that will be accessed in the future can actually reduce the power consumption because the uplink power consumption can be saved. Thus, as N_p increases from 0, the system performance increases while the power consumption is not increased or even decreased. Improving performance without sacrificing power will continue until N_p reaches the optimal point shown in Figure 18.3.

After N_p passes through the optimal point, the system performance can still be improved. However, the power consumption will be increased, because some less valuable data are prefetched. Although prefetching these data can improve the performance, they are not frequently accessed and are more likely to be invalidated; hence, prefetching them costs more power than getting them when being requested.

Intuitively, the system performance should be improved when more data is prefetched if the power consumption is not an issue. However, this may not be true if the IR-Based Cache Invalidation Model is used to ensure the latest value consistency model. In our stretch value function (Equation 18.1), v is the cache invalidation delay. This v can make the value of a data item be negative; hence, prefetching the data will increase the average stretch compared to not prefetching it. Therefore, N_p should not be increased beyond a certain point (terminating point) even if the goal is to achieve the best performance.

In summary, before reaching the optimal point, prefetching can save power and improve performance. Between the optimal point and the terminating point, there is a trade-off between performance and power. After the terminating point, the system should not prefetch. However, how to determine the optimal point and the terminating point still needs further investigation. In some cases (e.g., when priority requests are not allowed in the UIR-Based Model or when using Difference Cache Consistency Model), the terminating point may not exist.

Note

1. We assume cache locality exists. When cache locality does not exist, other techniques such as profile-based techniques can be used to improve the effectiveness of the prefetch.

References

1. S. Acharya and S. Muthukrishnan, Scheduling On-Demand Broadcasts: New Metrics and Algorithms, *ACM MobiCom'98*, pp. 43–54, October 1998.
2. S. Acharya, M. Franklin, and S. Zdonik, Disseminating Updates On Broadcast Disks, *Proc. 22nd VLDB Conf.*, September 1996.
3. D. Barbara and T. Imielinski, Sleepers and Workaholics: Caching Strategies for Mobile Environments, *ACM SIGMOD*, pp. 1–12, 1994.
4. G. Cao, On Improving the Performance of Cache Invalidation in Mobile Environments, *ACM/Baltzer Mobile Networks and Application (MONET)*, vol. 7, no. 4, pp. 291–303, August 2002.
5. G. Cao, Proactive Power-Aware Cache Management for Mobile Computing Systems, *IEEE Transactions on Computer*, vol. 51, no. 6, pp. 608–621, June 2002.
6. G. Cao, A Scalable Low-Latency Cache Invalidation Strategy for Mobile Environments, *IEEE Transactions on Knowledge and Data Engineering*, vol. 15, no. 5, September/October 2003 (A preliminary version appeared in *ACM MobiCom'00*).
7. G. Cao and C. Das, On the Effectiveness of a Counter-Based Cache Invalidation Scheme and its Resiliency to Failures in Mobile Environments, *The 20th IEEE Symposium on Reliable Distributed Systems (SRDS)*, pp. 247–256, October 2001.
8. J.C. Chen, K.M. Sivalingam, P. Agrawal, and R. Acharya, Scheduling Multimedia Services in A Low-Power MAC for Wireless and Mobile ATM Networks, *IEEE/ACM Transactions on Multimedia*, vol. 1, no. 2, June 1999.

9. S. Gitzenis and N. Bambos, Power-Controlled Data Prefetching/Caching in Wireless Packet Networks, *INFOCOM*, 2002.
10. V. Grassi, Prefetching Policies for Energy Saving and Latency Reduction in a Wireless Broadcast Data Delivery System, *ACM MSWIM*, 2000.
11. Q. Hu and D. Lee, Cache Algorithms Based on Adaptive Invalidation Reports for Mobile Environments, *Cluster Computing*, pp. 39–48, February 1998.
12. T. Imielinski, S. Viswanathan, and B. Badrinath, Data on Air: Organization and Access, *IEEE Transactions on Knowledge and Data Engineering*, vol. 9, no. 3, pp. 353–372, May/June 1997.
13. Z. Jiang and L. Kleinrock, An Adaptive Network Prefetch Scheme, *IEEE Journal on Selected Areas in Communications*, vol. 16, no. 3, pp. 1–11, April 1998.
14. J. Jing, A. Elmagarmid, A. Helal, and R. Alonso, Bit-Sequences: An Adaptive Cache Invalidation Method in Mobile Client/Server Environments, *Mobile Networks and Applications*, pp. 115–127, 1997.
15. A. Joshi, On Proxy Agents, Mobility, and Web Access, *Mobile Networks and Applications*, vol. 5, no. 4, pp. 233–241, December 2000.
16. A. Kahol, S. Khurana, S. Gupta, and P. Srimani, An Efficient Cache Management Scheme for Mobile Environment, *The 20th Int'l. Conf. on Distributed Computing Systems*, pp. 530–537, April 2000.
17. M. Kazar, Synchronization and Caching Issues in the Andrew File System, *USENIX Conf.*, pp. 27–36, 1988.
18. V. Padmanabhan and J. Mogul, Using Predictive Prefetching to Improve World Wide Web Latency, *Computer Communication Review*, pp. 22–36, July 1996.
19. R. Powers, Batteries for Low Power Electronics, *Proc. IEEE*, vol. 83, no. 4, pp. 687–693, April 1995.
20. B. Prabhakar, E. Uysal-Biyikoglu, and A. El Gamal, Energy-Efficient Transmission over a Wireless Link via Lazy Packet Scheduling, *IEEE INFOCOM'01*, March 2001.
21. P. Reiher, J. Heidemann, D. Ratner, G. Skinner, and G.J. Popek, Resolving File Conflicts in the Ficus File System, *Proc. of the USENIX Summer 1994 Technical Conference*, pp. 183–195, 1994.
22. S. Sandberg, D. Goldberg, S. Kleiman, D. Walsh, and B. Lyon, Design and Implementation of the Sun Network File System, *Proc. USENIX Summer Conf.*, pp. 119–130, June 1985.
23. M. Satyanarayanan, J. Kistler, P. Kumar, M. Okasaki, E. Siegel, and D. Steere, Coda: A Highly Available File System for a Distributed Workstation Environment, *IEEE Transactions on Computers*, vol. 39, no. 4, April 1990.
24. M. Stemm and R. Katz, Measuring and Reducing Energy Consumption of Network Interfaces in Hand-Held Devices, *IEICE Trans. on Communications*, vol. 80, no. 8, pp. 1125–1131, August 1997.
25. M. Stemm and R.H. Katz, Measuring and Reducing Energy Consumption of Network Interfaces in Handheld Devices, *IEICE Transactions on Communications*, vol. E80-B, no. 8, August 1997.
26. K. Tan, J. Cai, and B. Ooi, Evaluation of Cache Invalidation Strategies in Wireless Environments, *IEEE Transactions on Parallel and Distributed Systems*, vol. 12, no. 8, pp. 789–807, 2001.
27. K. Wu, P. Yu, and M. Chen, Energy-Efficient Caching for Wireless Mobile Computing, *The 20th Int'l Conf. on Data Engineering*, pp. 336–345, February 1996.
28. Y. Xu, J. Heidemann, and D. Estrin, Geography-Informed Energy Conservation for Ad Hoc Routing, *MobiCom'01*, July 2001.
29. L. Yin, G. Cao, C. Das, and A. Ashraf, Power-Aware Prefetch in Mobile Environments, *IEEE International Conference on Distributed Computing Systems (ICDCS)*, pp. 571–578, 2002.
30. J. Yuen, E. Chan, K. Lam, and H. Leung, Cache Invalidation Scheme for Mobile Computing Systems with Real-Time Data, *ACM SIGMOD Record*, December 2000.

Chapter 19
Energy Efficient Selective Cache Invalidation

Kian-Lee Tan

19.1 Introduction

Today's users, equipped with portable and handheld devices, can access data stored at information servers located at the static portion of the network without space and time restriction [1, 2]. However, there are two obstacles to the widespread adoption of this technology:

1. The limited bandwidth of wireless communication channels
2. The short battery lifespan of portable computers

Caching of frequently accessed data at the portable devices is a promising mechanism to reduce wireless bandwidth requirement as well as energy consumption. The challenge is to keep the cache content consistent with those stored at the server. This is, unfortunately, difficult to enforce due to the frequent disconnection and mobility of clients.

Existing cache invalidation strategies can be classified into stateless and synchronous [3–9] and stateful and asynchronous [10, 11]. For the former, the server does not maintain the states of the clients' cache. Instead, the server periodically broadcasts invalidation reports containing information on objects that are most recently updated. Based on the reports, clients can invalidate objects that have been updated and salvage their cache contents that are still valid. The latter category, on the other hand, requires the server to maintain state information about clients resulting in a more heavyweight server architecture. Because of the asynchronous dissemination of invalidation information, clients must always monitor the channel for updates. As such, it is difficult, if not impossible, to support selective tuning.

In this chapter, we provide an overview of cache invalidation strategies in wireless environments. The focus, however, is on stateless and synchronous cache invalidation strategies that organize invalidation reports to facilitate selective tuning. We look at the issues to be considered in designing selective cache invalidation strategies and present some representative strategies in the literature.

The rest of this chapter is organized as follows:

- Section 19.2 provides some preliminaries.
- Section 19.3 examines the issues and solutions related to cache invalidation schemes.
- Section 19.4 presents two representative algorithms obtained from the taxonomy that supports selective tuning.
- Section 19.5 provides the conclusion and discusses some promising future work.

19.2 Preliminaries

The wireless environment consists of two distinct sets of entities: a larger number of mobile clients (MC) and relatively fewer, but more powerful, fixed hosts (or database servers) (DS). The fixed hosts are connected through a wired network and may also be serving local terminals. Some of the fixed hosts, called mobile support stations (MSS), are equipped with wireless communication capability. A MC can connect to a server through a wireless communication channel. It can operate in full operational active model or disconnect from the server by operating in a doze mode or a power-off mode.

Each MSS can communicate with MCs that are within its radio coverage area called a wireless cell. A wireless cell can either be a cellular connection or a wireless local area network. At any time, a MC can be associated with only one MSS and is considered to be local to that MSS. A MC can directly communicate with a MSS if the mobile client is physically located within the cell serviced by the MSS. A MC can move from one cell to another. The servers manage and service on-demand requests from MCs. Based on the requests, the objects are retrieved and sent via the wireless channel to the MCs. The wireless channel is logically separated into two subchannels:

1. An uplink channel, which is used by clients to submit queries to the server via MSS
2. A downlink channel, which is used by MSS to pass the answers from server to the intended clients

We assume that updates only occur at the server and MCs only read the data.

Energy Efficient Selective Cache Invalidation

To conserve energy and minimize channel contention, each MC caches its frequently accessed objects in its nonvolatile memory such as a hard disk. Thus, after a long disconnection, the content of the cache can still be retrieved. To ensure cache coherency, each server periodically broadcasts invalidation reports. All active MCs listen to the reports and invalidate their cache content accordingly. We assume that all queries are batched in a query list and are not processed until the MC has invalidated its cache with the most recent invalidation report. We also assume that each server stores a copy of the database and broadcasts the same invalidation reports. In this way, clients moving from one cell to another will not be affected. Thus, it suffices for us to restrict our discussion to just one server and one cell.

There are two metrics that are used to characterize information retrieval in the wireless computing environment. The first is the access time, which is the time elapsed from the moment the client submits a request to the point when all the resultant objects are downloaded by the client. The second deals with the energy efficiency of the retrieval mechanisms. There are two measures for this:

1. Tuning time — the amount of time the client spent on listening to the channel
2. Number of uplink bits transmitted — reflects the amount of energy consumed in transmitting data

Traditionally, once a client submits a query, it listens to the channel until all the resultant objects are received. This leads to the tuning time being equal to the access time. Because listening/transmitting operations require the central processing unit (CPU) to be in full operation, they should be minimized to keep the power consumption low.

19.3 A Taxonomy of Cache Invalidation Strategies

As we examined existing cache invalidation strategies, we identified three issues that were considered in designing them:

1. The content of the invalidation reports
2. How invalidation is performed
3. The support the server provides

Here, we shall briefly look at some solutions to these issues.

19.3.1 Content of the Invalidation Report

The invalidation report contains update information that clients can use to determine the validity of its cache content. Ideally, the report should reflect all updates. But, this is costly and impractical in view of the limited bandwidth and short battery life of mobile clients. Instead, the report typically reflects only a short (or reasonable length) history of updates on the most recent changes. Each entry in the report can be of the form:

- (id,TS) pair [3] that indicates that object with identifier id is updated at time stamp TS
- ($\{id_1, ..., id_n\}$, TS) pair [8] that reflects that the set of objects with identifiers $id_1, ..., id_n$ have been updated since TS
- (group-id, TS) pair [4–6, 9] that indicates the most recent time stamp TS at which an object in the group identified by group-id (the dataset is assumed to have been partitioned into groups) has been updated

We note that instead of object and group identifiers, the report could have contained the (updated) objects themselves. In this case, clients can immediately update the invalid objects. But, because the object size is typically much larger than the identifier, the report can take up a significant portion of the downlink channel capacity. Thus, most of the existing works broadcast invalidation reports.

There is a clear trade-off among the three types of invalidation reports. The first method is precise but consumes more bandwidth. The third (group-based) scheme minimizes bandwidth consumption but may result in unnecessary invalidation as valid objects within a group may have to be invalidated. The second approach is a compromise between these two extremes: it reduces the size of the report by using fewer time stamps at the expense of some unnecessary invalidation.

The report can be of fixed [7, 8] or variable length [3–6, 9]. It can be fixed by the amount of bandwidth allocated for the report. It can also be fixed by the number of entries to be included in the report. Obviously, under these cases, the update history cannot be predetermined (i.e., it has to be variable) because the number of updates varies over a fixed period of time. On the other hand, the report size can vary from broadcast to broadcast by fixing the update history to be reflected. As an example, an update report broadcast at time T may reflect all updates during the interval [T − wL, T] for some w > 0 and L; w is referred to as the broadcast window and L is the fixed interval at which the report is periodically broadcast.

Finally, the report may be organized to facilitate selective tuning to conserve energy [4–6, 7]. The basic idea is to interleave the content of the report with metadata (or indexes) that can provide direct access to targeted portions of the report. Thus, only the desired portions of the report need to be examined. More importantly, because the client knows exactly when the desired portion of the report will arrive, it can potentially switch to operate in the doze mode to minimize energy consumption while waiting for the arrival of the targeted data. When it is time to receive the targeted portion of the report, it will switch back to the active mode. Given that the invalidation report may be large, and clients only need to validate a relatively smaller number of objects, such a mechanism can keep the energy consumption low (compared to listening to the entirety of the report). Although there is some work being done on indexing broadcast data [2], these schemes cannot be readily applied to facilitate selective tuning of invalidation reports. This is because broadcast data are typically static — the values of the objects may change but the objects do not — but invalidation information cannot be predetermined and vary from report to report. Moreover, many of these schemes are complex (e.g., hash-based, B⁺-tree) and costly to construct.

There is yet another trade-off in deploying selective tuning mechanisms. Selective tuning allows clients to operate in a power-saving mode to conserve energy. However, the metadata inadvertently lengthen the size of the invalidation report. This translates to more bandwidth consumption and longer access time.

19.3.2 Invalidation Mechanisms

When a client receives an invalidation report, it can invalidate its cache content in two ways. First, it can perform cache-level invalidation [3, 5, 8, 9], in which cache validity is performed for all objects cached. This requires scanning a large portion of the invalidation report, if not the entire report. As a result, it is not particularly suited for selective tuning (unless the number of cached objects is very small). Under this approach, the cache content is associated with one time stamp only: the time stamp of the most recent invalidation report read.

On the other hand, the client can perform query-level invalidation [4, 6, 7], where validation is performed only on the objects queried. This reduces the number of objects to be invalidated and hence the report can be organized for selective tuning. However, each cached object has to be associated with a time stamp (compared to a single time stamp for all cached object in cache-level invalidation). The time stamp of an object represents the time stamp at which the object is last known to be valid. This is usually the time stamp of the invalidation report that was last used to validate the object. Thus, different cached objects will have different time stamps. The consequence of this is that each queried object may use a different list of objects for invalidation and query processing becomes more involved.

Invalidating the cache content can be performed by the client alone [3–8]. This requires that the client based its invalidation purely on the invalidation reports. Thus, the effectiveness of such approaches is dependent on the content of the report. On the other extreme, we can allow the server to perform the invalidation alone. This, however, will require the client to inform the server about its cache content, which can be costly because transmitting this information consumes energy and bandwidth. Finally, the client and server can collaborate to identify the cache content that should be invalidated: the client uses the invalidation report to invalidate its cache content; for those that remain uncertain, the client submits their information to the server for invalidation [9]. The latter has the advantage that the invalidation report can be small (containing only very recent updates), whereas the server maintains a longer history of updates.

19.3.3 Update Log Structure

Another important issue in the design of a cache invalidation scheme concerns the information (update logs) maintained at the server to reflect the updates on the database. Clearly, this information will influence the choice of the invalidation report. The update logs can contain:

- (id, TS) pairs to reflect that object with identifier id has been updated at time stamp TS
- (group-id, TS) pair where TS is the most recent time stamp that an object in the group identified by group-id has been updated

Like the invalidation report, the size of the log and its history (duration that the update logs should be maintained) are interrelated. The size can be fixed by restricting updates to be maintained for a fixed number of objects [7, 8]. In this case, the log history changes depending on the updates. On the other hand, variable sized logs can be maintained by fixing the log history to a fixed interval, say [T − WL, T] for some W > 0; W is the update log window, L is the fixed interval of broadcasting the invalidation report, and T is the time stamp at which the report is broadcast [3–9]. In other words, only updates in the interval [T − WL, T] are maintained. Clearly, we only need to keep the updates sufficient for the invalidation report; keeping all updates is ideal but can be costly in terms of storage, whereas keeping only very recent updates may not be effective as it may lead to false invalidation.

19.3.4 Cache Invalidation Schemes

Based on the above discussion, we can derive a large number of cache invalidation schemes from the products of the various alternatives in addressing each issue. It should be pointed out that some of the options may not be practical or feasible. For example, it is not possible to have a fixed-size report that reflects all updates in a fixed update history [T − wL, T]. On the other hand, we can design techniques that combine multiple solutions. For example, techniques exist that reflect both recent updates for individual items, as well as group updates (e.g., Selective Cache Invalidation scheme in [4]).

19.4 Selective Cache Invalidation Schemes

In this section, we describe two representative selective cache invalidation schemes. Both schemes built on existing schemes by providing selective

Energy Efficient Selective Cache Invalidation

tuning capabilities. For each scheme, we shall first review the base scheme and then discuss the extension.

19.4.1 Selective Dual-Report Cache Invalidation

The Selective Dual-Report Cache Invalidation Scheme (SDCI) [7] extends the Dual-Report Cache Invalidation Scheme (DRCI) [7]. DRCI is a broadcast-based version of grouping with cold update set retention (GCORE) [9] that eliminates the need for the server to participate in the invalidation process.

In DRCI, the invalidation report comprises two components — an object invalidation report (OIR) and a group invalidation report (GIR). The former is a list of (id, TS) pairs and the latter a list of (group-id, TS) pairs. The size and the update history vary. The invalidation is at the cache level and is performed by the client only. The server maintains logs for objects. The size of the log varies though the log history is fixed at [T − WL, T] for objects, where T is the current time, w and W are update log windows, and L is the fixed interval at which the reports are broadcast.

Under DRCI, the server broadcasts the two invalidation reports every L time units. Let the current time stamp be T. To generate the reports, the server keeps track of all the objects updated in the interval [T − WL, T], where W is the update log window and W > 0. The most recent OIR broadcast contains the update history of the past w broadcast intervals, w < W. The contents of OIR are the current time stamp T and a list of (o_{id}, t_{id}) pairs where o_{id} is an object identifier of an object updated during the interval [T − wL, T] and t_{id} is the corresponding most recent update time stamp and t_{ld} > (T − wL). The GIR, on the other hand, is a fixed-size report that contains the update history of the past W broadcast intervals. GIR contains for each group a time stamp that reflects the most recent time stamp in which the group is valid. In other words, the contents of the GIR is a list of (G_{id}, T_{id}) pairs where G_{id} is a group identifier and T_{id} is the most recent time stamp in which the group G_{id} is valid. The time stamps of the groups in GIR are determined only at the time when the invalidation reports are to be broadcast and are assigned as follows:

1. Based on the observation that objects that are updated during [T − wL, T] would be reflected in the OIR, these objects are ignored when determining the time stamp of a group.
2. For each group, the following is performed. Among the group's remaining objects in the update history, the one with the largest time stamp that is less than T − wL is determined. Let this time stamp be t. If no such objects exist (i.e., no updates during [T − WL, T] or the updates are already included in OIR), then t = 0. The time stamp for the group is max(T − WL, t).

The DRCI scheme facilitates invalidation in two ways. First, for those clients with a small disconnection time, a direct cache checking is performed

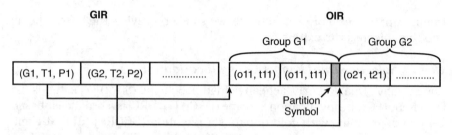

Figure 19.1. Structure of Invalidation Reports in SDCI

using the OIR. Second, for those disconnected before the time T − wL, OIR and GIR can work together so that the clients need not discard the entirety of its cache: the OIR is used to invalidate individual objects in the cache first, the GIR is then used to invalidate the remaining objects whose groups have been updated. Because the GIR excludes objects in OIR, the OIR effectively helps to minimize false invalidation. The GIR can further minimize invalidation of the entire cache for long disconnection time. Note that the entire cache content will be discarded if the client disconnection time is longer than T − WL.

Although the basic DRCI scheme is independent of the way the database objects are grouped, its performance can be influenced by the grouping scheme. In particular, the whole database can be split into four data sets — hot update-hot demand (HH), hot update-cold demand (HC), cold update-hot demand (CH), and cold update-cold demand (CC) — and groups are formed from objects in the same category.

To facilitate selective tuning, SDCI organizes the pair of invalidation reports (the OIR and GIR) as follows. First, the GIR is broadcast before the OIR. Second, the entries in the OIR are ordered and broadcast based on the groups (i.e., updates in the same group will appear together). A partition symbol will separate continuous groups. Third, an additional pointer is added to each element of the GIR; this pointer reflects the starting position of the objects within this group in the OIR. Thus, the GIR consists of triplets of the form (group-id, TS, ptr) where group-id represents the group identifier, TS is the time stamp of the most recent update (excluding those in the OIR) of the group, and ptr is an offset to the starting position of the objects in OIR corresponding to the group identified by group-id. Figure 19.1 shows the structure of the invalidation reports.

SCDI is also distinguished from DRCI in that the invalidation process is query-level-based. That means for the objects queried, the client first selectively tunes to the GIR and keeps the pointers of the associated groups in memory; once all the desired groups are determined, it selectively tunes to the respective positions in the OIR using the pointers. Thus, only the desired groups and objects are examined.

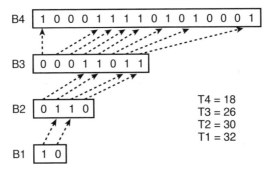

Figure 19.2. Bit Sequence Structure

19.4.2 Bit-Sequences with Bit-Count

The bit-sequences with bit-count (BB) scheme [6] extends the bit sequences (BS) algorithm [8]. In the BS scheme, the report consists of a list of (list of ids, TS) pairs in a compact form. This allows the report size to be fixed, though the update history varies. The invalidation is performed by the client at cache level. The logs maintained at server keep track of individual object update information (using [id, TS] pairs) for up to half the database size (i.e., the size of the log is fixed but the update history is variable).

Figure 19.2 shows a BS structure. Let the number of database objects be $N = 2^n$. In the BS algorithm, the invalidation report reflects updates for n different times: $T_n, T_{n-1}, \ldots, T_1$, where $T_i < T_{i-1}$ for $1 < i \le n$. The report comprises n binary BS, each of which is associated with a time stamp. Each bit represents a data object in the database. A 1 bit in a sequence means that the data object represented by the bit has been updated since the time specified by the time stamp of the sequence. A 0 bit means that the object has not been updated since that time. The n BSs are organized as a hierarchical structure with the highest level (i.e., BS B_n) having as many bits as the number of objects in the database, and the lowest level (i.e., BS B_1) having only two bits. For the sequence B_n, as many as half of the N bits (i.e., N/2) can be set to 1 to indicate the N/2 objects that have been updated (initially, the number of 1 bits may be less than N/2). The time stamp of the sequence B_n is T_n. The next sequence in the structure, B_{n-1}, has N/2 bits. The kth bit in B_{n-1} corresponds to the kth 1 bit in B_n and $N/2^2$ bits can be set to 1 to indicate that $N/2^2$ objects have been updated since T_{n-1}. In general, for sequence B_{n-i}, $0 \le i \le n\ 1$, there are $N/2^i$ bits, and the sequence will reflect that $N/2^{i+1}$ objects have been updated after the time stamp T_{n-i}. The kth bit in sequence B_{n-i} corresponds to the kth 1 bit in the preceding sequence (i.e., B_{n-i+1}). An additional dummy sequence B_0 with time stamp T_0 is used to indicate that no object has been updated after T_0. In general, N does not need to be a power of 2 and the number of lists can also be any value other than n. Furthermore, the list associated with time stamp T_i

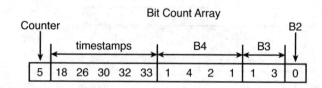
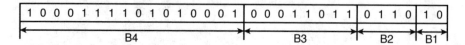

Figure 19.3. Bit Sequences with Bit Count

does not need to reflect the updates for half the number of objects in the list associated with T_{i+1}, $1 \leq i \leq n-1$.

The BB method extends the BS scheme by providing some additional metadata (a bit count array) so that only the relevant bits need to be examined. Figure 19.3 shows the organization of the BB method for the BSs in Figure 19.2. Let N be the number of objects in the database.

As in the BS scheme, the BB structure comprises a set of n BSs: sequence B_n has a time stamp T_n that indicates that updates after T_n are reflected and comprises N bits, half of which are set to 1; sequence B_{n-1} has timestamp T_{n1} and N/2 bits, of which $N/2^2$ bits are set to 1; and so on. In fact, the content of the BSs are exactly the same as those of the BS scheme. Like the BS scheme, if the BS B_{n-i} is to be used to invalidate the cache, then the sequences B_{n-i}, B_{n-i+1}, ..., B_n may have to be examined. However, the BB strategy adopts a top-down examination of the sequences (i.e., from B_n to B_{n-i}), rather than the bottom-up approach (i.e., B_{n-i} to B_n) of BS scheme. Moreover, for some valid objects, it may not be necessary to examine all the sequences from B_n to B_{n-i}, as it may be possible to determine their validity and terminate the search before sequence B_{n-i}. In addition, it only examines the relevant bits in each sequence. As the kth 1 bit in B_{n-i} corresponds to the kth bit in B_{n-i-1}, a mechanism that can count the number of 1 bits in a sequence, say B_{n-i}, without examining the entire sequence is needed. Before we discuss this mechanism, let us illustrate how selective tuning can be facilitated with such a mechanism. To validate an object O, the client first identifies the BS that should be used. This is accomplished by examining the set of time stamps. Suppose this is sequence B_{n-i}. This means that we need to examine sequence B_n, followed by B_{n-1}, and so on until B_{n-i}. From object O, the client can selectively tune to the corresponding bit in B_n (without scanning the entire B_n). If the bit is set to 0, then the object is valid (because object O will not be found in any subsequent sequences B_{n-1}, B_{n-2}, ...); otherwise, the client determines (we shall discuss shortly how this can be done) the number of 1 bits from the beginning of B_n to the bit corre-

Energy Efficient Selective Cache Invalidation

sponding to O. From this number, it can again selectively tune to B_{n-1} and examine the corresponding bit of O in B_{n-1}. Again, if the bit is 0, then the object O is valid and the search terminates; otherwise, its position in the B_{n-2} is determined and this process is repeated until sequence B_{n-i}. We can terminate when we encounter 0 bits at any of the sequences from B_n to B_{n-i}. If the relevant bit at B_{n-i} is 1, then the object is invalid; otherwise, it is valid.

Now, the mechanism to facilitate selective tuning is simple. We associate with each BS a bit count array, all of which have entries that are j bits. For BS B_{n-i}, $0 \leq i \leq n-1$, the sequence is partitioned into packets of 2^j bits. In other words, there are $\lceil (N/2^i)/2^j \rceil$ packets. In general, for sequence B_{n-i}, the number of array entries is $\lceil (N/2^i)/2^j \rceil$. (We note that sequences with fewer than j bits do not need to be associated with a bit count array as all bits have to be examined anyway.) Essentially, the kth entry in the bit count array of sequence B_{n-i} represents the number of 1 bits that have been set for the kth packet in the sequence. Selective tuning is achieved as follows: We know which bit/position in a sequence we should be looking for. Let packet i contain the bit that we are interested in. From the bit count array, we can determine the number of 1 bits that have been set for packets 1 to i − 1. The client can then tune into the ith packet and scan the ith packet until the relevant bit. In this way, we will be able to compute the number of 1 bits.

The invalidation report is organized as follows. The counter is broadcast first, the time stamps are broadcast next, followed by the bit count arrays for sequences B_n, B_{n-1}, ..., and finally the BSs B_n, B_{n-1},..., B_1.

In BB, a client validates objects using a query level approach. It first downloads the metadata — the counter, the time stamps, and the bit count arrays. Based on the time stamps, it can determine the BSs that need to be examined to validate each object (recall that because the validation is query level, each object may have different time stamps in the cache). Objects that need to be validated by a certain BS are done so together (to minimize repeated scanning). As described above, a client only needs to tune to the targeted portion of the data. As such, the queried objects must also be sorted in the same order as their positions in the BS structure.

19.4.3 Discussion

In [7], a simulation study was reported on the relative performance of SCDI and BB. It was shown that SCDI outperforms BB in most cases in terms of both access time and tuning time. BB performs poorly in access time because of the bandwidth overhead of the BS structure is significant. It also consumes more energy because more (complex) work has to be done to validate the data objects. However, if the update rate is high, SCDI becomes inferior in the access time. This is expected because given a fixed update log window, as the number of updates increases, the size of the report increases accordingly (compare to the fixed and compact BS structure of BB).

19.5 Conclusion

In this chapter, we have looked at the design of energy efficient cache invalidation schemes in a wireless computing environment. We have identified the issues that have to be considered in designing cache invalidation strategies. From the existing solutions to these issues, a large set of cache invalidation schemes can be constructed. We reviewed two representative algorithms that employ selective tuning to conserve energy. There are several directions in which existing schemes can be improved. First, most of the existing schemes require the client to tune to the invalidation report before processing the query. This means that the access time is unnecessarily lengthened as the client must first wait for the report. Developing mechanisms to minimize this waiting time is important. Second, wireless channels are error prone. As such, the reports received may be corrupted. Mechanisms to handle errors in such an environment are necessary. Finally, it may be interesting to see how a recently stateful-based mechanism can benefit from selective tuning.

References

1. J. Jing, A. Helal, and A. Elmagarmid, Client–Server Computing in Mobile Environments, *ACM Computing Surveys,* vol. 31, no. 2, pp. 117–157, June 1999.
2. K.L. Tan and B.C. Ooi, *Data Dissemination in Wireless Computing Environments,* Norwell, MA: Kluwer Academic Publishers, 2000.
3. D. Barbara and T. Imielinski, Sleepers and Workaholics: Caching in Mobile Distributed Environments, In *Proceedings of the 1994 ACM-SIGMOD International Conference on Management of Data,* June 1994, pp. 1–12.
4. J. Cai and K.L. Tan, Energy Efficient Selective Cache Invalidation, *Wireless Networks,* vol. 5, no. 6, pp. 489–502, 1999.
5. K.L. Tan and J. Cai, Broadcast-Based Group Invalidation: An Energy Efficient Cache Invalidation Scheme, *Information Sciences,* vol. 100, no. 1–4, August 1997, pp. 229–254.
6. K.L. Tan, Organization of Invalidation Reports for Energy Efficient Cache Invalidation in Mobile Environments, *Mobile Networks and Applications,* vol. 6, pp. 279–290, 2001.
7. K.L. Tan, J.Cai, and B.C. Ooi, An Evaluation of Cache Invalidation Strategies in Wireless Environments, *IEEE Transactions on Parallel and Distributed Systems,* vol. 12, no. 8, pp. 789–807, 2001.
8. J. Jing, A. Elmagarmid, A. Helal, and R. Alonso, Bit-Sequences: An Adaptive Cache Invalidation Method in Mobile Client/Server Environments, *Mobile Networks and Applications,* vol. 2, no. 2, pp. 115–127, 1997.
9. K.L. Wu, P.S. Yu, and M.S. Chen, Energy-Efficient Caching for Wireless Mobile Computing, In *Proceedings of the 12th International Conference on Data Engineering,* pp. 336–343, February 1996.
10. A. Kahol, S. Khurana, S.K.S. Gupta, and P.K. Srimani, A Strategy to Manage Cache Consistency in a Distributed Mobile Wireless Environment, *IEEE Transaction on Parallel and Distributed System,* vol. 12, no. 7, pp. 686–700, 2001.
11. Z. Wang, S. Das, H. Che, and M. Kumar, SACCS: Scalable Asynchronous Cache Consistency Scheme for Mobile Environments, *International Workshop on Mobile and Wireless Networks,* pp. 797–802, 2003.

Section V
Mobile and Ad Hoc Wireless Networks I

Chapter 20
Self-Policing Mobile Ad Hoc Networks

Sonja Buchegger and Jean-Yves Le Boudec

Abstract

Misbehavior in mobile ad hoc networks occurs for several reasons. Selfish nodes misbehave to save power or to improve their access to service relative to others. Malicious intentions result in misbehavior as exemplified by denial of service attacks. Faulty nodes simply misbehave accidentally. Regardless of the motivation for misbehavior, its impact on the mobile ad hoc network proves to be detrimental, decreasing the performance and the fairness of the network, and in the extreme case, resulting in a nonfunctional network. Countermeasures to prevent or to combat misbehavior have been proposed, such as payment schemes for network services, secure routing protocols, intrusion detection and reputation systems to detect and isolate misbehaved nodes. We discuss the trade-offs and issues of self-policing mobile ad hoc networks and give an overview of the state of the art, discussing and contrasting several solution proposals.

20.1 Introduction and Overview

In mobile ad hoc networks, nodes act as both routers and terminals. For the lack of routing infrastructure, they have to cooperate to communicate. Cooperation at the network layer takes place at the level of routing (i.e., finding a path for a packet) and forwarding (i.e., relaying packets for other nodes).

Misbehavior means aberration from regular routing and forwarding behavior resulting in detrimental effects on the network performance. Misbehavior arises for several reasons. When a node is faulty, its erratic behavior can deviate from the protocol and thus produce nonintentional misbehavior. Intentional misbehavior aims at providing an advantage for the misbehaved node. An example for an advantage gained by misbehavior is power saved when a selfish node does not forward packets for other nodes. An advantage for a malicious node arises when misbehavior enables it to mount an attack.

The detrimental effects of misbehavior result in unfairness and degraded performance and they can endanger the functioning of the entire network. To avoid these adverse effects, mobile ad hoc network routing protocols have to be able to cope with misbehavior attempts. Section 20.2 discusses types of misbehavior, their payoff for the attacker, and their effect on the network. It thus gives a motivation for the need of enhancements of mobile ad hoc networks to counteract misbehavior.

In the literature, three main approaches to strengthen mobile ad hoc networks have been proposed, namely payment systems, secure routing using cryptography, and detection and reputation systems for self-policing. Payment schemes serve as an incentive to forward packets for other nodes. Secure routing aims at the prevention of attacks by using cryptography to secure the routing messages themselves. Detection and reputation systems identify misbehaved nodes and isolate them from the network by monitoring and keeping records of past behavior. Section 20.3 gives an overview and comparison of these three solution tracks.

In Section 20.4, we discuss detection and reputation systems in more detail. They address the cases that have not been prevented. Not all types of misbehavior can be prevented, however, for misbehavior reputation systems it suffices that misbehavior can be detected. We present our approach to a self-policing mobile ad hoc network and use it as a basis to compare with other proposed solutions.

There are several challenges to the design of misbehavior reputation systems, a fundamental example being that the system should not add vulnerabilities to the mobile ad hoc routing protocol it is built to protect. There are trade-offs to take into account when considering which type of reputation to use, whether and how to use second-hand information, and so forth. We discuss the main challenges and issues in Section 20.4.2 and conclude in Section 20.5.

20.2 Node Misbehavior in Mobile Ad Hoc Networks

This section gives a more detailed problem description and reasons why it is a problem worth investigating. Mobile ad hoc networks have properties that render them more vulnerable to attacks and misuse, as we show in Section 20.2.1. Several attacks on routing and forwarding in mobile ad hoc networks are described in Section 20.2.2. Finally, in Section 20.2.3, we illustrate the effects of misbehavior on the mobile ad hoc network as well as the potential effects of countermeasures such as incentives for cooperation.

20.2.1 Reasons and Enablers for Misbehavior

The lack of infrastructure and organizational environment of mobile ad hoc networks offer special opportunities to intentionally misbehaved nodes.

Without proper countermeasures, it is possible to gain various advantages by malicious behavior: better service than cooperating nodes, monetary benefits by exploiting incentive measures or trading confidential information, saving power by selfish behavior, preventing someone else from getting proper service, extracting data to get confidential information, and so on. Even if the misbehavior is not intentional, as in the case of a faulty node, the effects can be detrimental to the performance of a network.

Mobile ad hoc networks have the following properties that can be exploited:

- Lack of infrastructure — nodes have to cooperate in the routing and forwarding of packets.
- No organizational authorities — any node can join an unmanaged mobile ad hoc network, there is no access control and no specific entry point.
- No central authorities — no permanent access to central services such as certification authorities can be assumed.
- Wireless network — nodes can promiscuously eavesdrop on communications by others. Collisions can be intentional or accidental.
- Mobility — with high mobility, routes are not valid over extended periods of time. Link errors can be ambiguous, communications can fail due to a node having moved out of range or due to an intentional interruption.
- Routing protocols lack security — most of the proposed routing protocols, such as DSR [12] and AODV [21], do not provide any security. Routing messages can be modified or fabricated, sent at inappropriate times or be omitted when needed. We discuss some proposals to add security in Section 20.3.2 and more detailed misbehavior descriptions in Section 20.2.2.
- Potentially low battery power — truly mobile and not merely portable devices have to be reasonably small and lightweight and therefore are often assumed to have limited battery power. This results in communications and computations being relatively expensive in power, opening the door to attacks aiming at excessive resource consumption of the target node, selfish behavior of resource conscious nodes, and limited ability to perform cryptographic computations.

In addition to authentication, integrity, confidentiality, availability, access control, and nonrepudiation (see [28] for details), which have to be addressed differently in a mobile, wireless, battery-powered, and distributed environment, mobile ad hoc networks raise the issue of cooperation and fairness.

There is a trade-off between good citizenship (i.e., cooperation) and resource consumption, so nodes have to economize on their resources.

Assuming rational behavior with a node maximizing its utility, the best strategy is not forward for other nodes. If several nodes, however, follow this strategy, the performance of the network deteriorates. In the extreme case of all nodes choosing this strategy, no communications can take place. This outcome is clearly unfavorable for the nodes. In game theoretic terms, this is a dilemma. Incentives are required to stimulate the cooperation among nodes.

20.2.2 Attacks

Ning and Sun [19] classify attacks on routing protocols as atomic attacks (modifying or forging one message) and compound attacks (combining or repeating several atomic attacks).

They argue that preventive security may not be enough to cope with insider attacks, where nodes can be compromised despite tamper-proof hardware.

They give a list of goals for an attacker, then look at the atomic attacks to see whether they can achieve them and also pick some compound attacks and investigate their effectiveness in reaching the four goals:

1. Route disruption
2. Route invasion
3. Node isolation
4. Resource consumption

Simulation results confirm the theoretical success of even atomic attacks, at least temporarily. For sustained success (e.g., to circumvent local route repair mechanisms), atomic attacks can be repeated.

Total noncooperation with other nodes and only exploiting their readiness to cooperate is one of several boycotting behavior patterns. Therefore, there has to be an incentive for a node to forward messages that are not destined to itself. Attacks include incentive mechanism exploitation by message interception, copying, or forging; incorrect forwarding; and bogus routing advertisement.

20.2.2.1 Traffic Diversion. Routes should be advertised and set up adhering to the chosen routing protocol and should truthfully reflect the knowledge of the topology of the network. By diverting the traffic in the following ways, nodes can work against that requirement.

To get information necessary for successful malicious behavior, nodes can attract traffic to themselves or their colluding nodes by means of false routing advertisements. Although only suitable for devices that have enough power, a lot of information can be gathered this way by malicious nodes for later use to enable more sophisticated attacks.

Denial-of-service attacks can be achieved by bogus routing information (injecting of incorrect routing information or replay of old routing information or black hole routes) or by distorting routing information to partition the network or to load the network excessively, thus causing retransmissions.

Nodes can decide to forward messages to partners in collusion for analysis, disclosure, or monetary benefits; or may decide not to forward messages at all, thus boycotting communications.

In general, the following types of misbehavior can be indicated:

- No forwarding (of control messages or data)
- Unusual traffic attraction (advertises many very good routes or advertises routes very fast, so they are deemed good routes)
- Deflecting traffic in order not to be used on a route
- Route salvaging (i.e., rerouting to avoid a broken link), although no error has been observed
- Lack of error messages, although an error has been observed
- Fabricating error messages, although no error has been observed
- Unusually frequent route updates
- Silent route change (tampering with the message header of either control or data packets)

Several more attacks have been proposed in the literature, such as the following:

- Black hole — reroute a path so that it ends up or passes a nonexisting node.
- Grey hole — like the black hole, but only performed sporadically.
- Sleep deprivation — make a node send messages excessively to decrease its resources.

20.2.3 The Effect of Misbehavior

Without appropriate countermeasures, the effects of misbehavior have been shown by several simulations [4, 17, 19] to dramatically decrease network performance. Depending on the proportion of misbehaved nodes and their specific strategies, network throughput can be severely degraded, packet loss increases, nodes can be denied service, and the network can be partitioned. Quantitative measures make more sense in comparison to routing protocols that have been enhanced with measures against misbehavior. We discuss these in Section 20.3.

In a theoretical analysis of how much cooperation mechanisms can help by increasing the probability of a successful forward, Lamparter, Plaggemeier, and Westhoff find that increased cooperation super-proportionally increases the performance for small networks (i.e., fairly short routes). Cooperation increases more if the initial probability e (the probability to cooperate by forwarding) is fairly acceptable (above 0.6). Even small increases in e, as given by δi, the change of the probability to cooperate in the presence of an incentive mechanism such as a reputation system, can have a dramatic improvement.

Zhang and Lee [30] argue that prevention measures, such as encryption and authentication, can be used in ad hoc networks to reduce intrusions, but cannot eliminate them. For example, encryption and authentication cannot defend against compromised mobile nodes, which carry the private keys. No matter how many intrusion prevention measures are inserted in a network, there are always some weak links that one could exploit to break in. Intrusion detection presents a second wall of defense and it is a necessity in any high-survivability network.

20.3 Overview: Main Solution Tracks

The main solution tracks addressing the problem of misbehavior in mobile ad hoc networks are secure routing, economic incentives, and detection and reputation systems. Economic incentives such as payment or counter schemes specifically address forwarding of packets for other nodes. Secure routing aims at securing the establishment and maintenance of routes.

Self-policing schemes aim at reactively detecting misbehavior and proactively isolating misbehaved nodes to prevent further damage. They are not restricted to any particular kind of misbehavior. The only requirement is that the misbehavior be detectable (i.e., observable) and classifiable as such with a high probability.

In the following sections, we describe the main features of some proposals within the respective solution tracks, briefly describe how they work, what they protect, and what the open problems are.

20.3.1 Payment Systems

Several approaches to provide economic incentives for cooperation have been proposed. They thus target the problem of selfish misbehavior. The main assumption is that nodes are economically rational.

Buttyán and Hubaux proposed incentives to cooperate by means of so-called nuglets [6] that serve as a per hop payment in every packet in a secure module in each node to encourage forwarding. The secure module is required to ensure the correct number of nuglets is withdrawn or deposited. They propose two models for the payment of packet forwarding, the Packet Purse Model and the Packet Trade Model. In the Packet Purse Model the sender pays and thus loads the packet with a number of nuglets. Each intermediate node takes one nuglet when it forwards the packet. If there are no nuglets left at an intermediate node, the packet is dropped. If there are nuglets left in the packet once it reaches the destination, the nuglets are lost. In the Packet Trade Model, the destination pays for the packet. Each intermediate node buys a packet from the previous hop and sells it to the next for more nuglets. Because charging the destination and not the sender can lead to an overload of the network and the destination receiving packets it does not want, mainly the Packet Purse Model is considered. This model, however, can lead to the loss of nuglets that have to be reintroduced into the network by a central authority.

To address this problem, the authors introduced another approach based on credit counters [7], also implemented in tamper-proof hardware. In this approach, each node keeps track of its remaining battery power and credit. One of their findings of a simulation study of four different rules is that increased cooperation is beneficial not only for the entire network but also for individual nodes.

Zhong, Chen, and Yang proposed Sprite [31]. As opposed to nuglets or counters they do not require tamper-proof hardware to prevent the fabrication of payment units, but their payment scheme requires a central credit clearance service (CCS) to be available eventually. Nodes keep a receipt of a message when they receive it. The receipt contains a hash of the message itself so it can be verified which message the receipt belongs to. To claim their payment, nodes have to send this receipt to the CCS. The CCS charges the sender based on the number of receipts, the number of intermediate nodes left to reach the destination, if any, and whether the destination has sent a receipt. The specific calculation of the fee is designed to make misbehavior in Sprite itself economically undesirable, even in the case of collusion. The sender then pays the nodes that sent a receipt to the CCS. For the nodes that were on the route but did not send a receipt, the sender has to pay a small fee to the CCS. In addition to the availability of a central authority, Sprite assumes source routing and a public key infrastructure. They do not explain how the payment from the sender to nodes is done, whether nodes have accounts with the CCS that transfer the payment, or whether nodes remunerate one another directly. In the latter case, the money has to be unforgeable and payment has to be ensured.

Raghavan and Snoeren propose priority forwarding as incentives against selfish misbehavior. In their approach, potential dangers for ad hoc networks are distinguished as misbehaving and greedy, where misbehavior constitutes a deviation from the protocol and should be taken care of by secure routing mechanisms. For greedy behavior, which is located at a higher layer in this approach, incentives to get priority forwarding are proposed to be given by payment.

20.3.2 Secure Routing with Cryptography

Secure routing proposals have been proposed mainly as modifications to existing routing protocols such as Dynamic Source Routing (DSR) [12] and Ad Hoc On-Demand Distance Vector (AODV) [21]. They aim at securing the routing messages by cryptographic means to prevent misbehavior by malicious nodes.

20.3.2.1 Secure Routing Protocol. The Secure Routing Protocol (SRP) by Papadimitratos and Haas [24], guarantees correct route discovery, so that fabricated, compromised, or replayed route replies are rejected or never reach the route requester. SRP assumes a security association between end points of a path only, so intermediate nodes do not have to be trusted for the route discovery. This is achieved by requiring that the request along with a unique random query indentifier reach the destination, where a route reply is constructed and a message authentication code is computed over the path and returned to the source. The correctness of the protocol is proven analytically.

20.3.2.2 Ariadne. Ariadne is a secure on-demand routing protocol by Hu, Perrig, and Johnson [11] that prevents attackers from tampering with uncompromised routes consisting of uncompromised nodes. It is based on DSR and relies on symmetric cryptography only. It uses a key management protocol called Timed Efficient Stream Loss-Tolerant Authentication (TESLA) that relies on synchronized clocks. Simulations have shown that the performance is close to DSR without optimizations.

20.3.2.3 Secure Efficient Distance. Secure Efficient Distance (SEAD) vector routing for mobile ad hoc networks by Hu, Johnson, and Perrig [10] is based on the design of destination-sequenced distance-vector routing (DSDV) and uses one-way hash functions to prevent uncoordinated attackers from creating incorrect routing state in another node. Performance evaluation has shown that SEAD outperforms DSDV-SQ (sequence number) in terms of packet delivery ratio, but SEAD adds overhead and latency to the network.

20.3.2.4 Security-Aware Ad Hoc Routing. The Security-Aware Ad Hoc Routing (SAR) protocol by Yi, Naldburg, and Kravets [29] modifies AODV to include security metrics for path computation and selection. They define trust levels according to organizational hierarchies with a shared key for each level, so that nodes can state their security requirements when requesting a route and only nodes that meet these requirements (trust level, metrics), participate in the routing. Questions not yet addressed by this protocol include the mechanism for key distribution, knowledge of the keys of the other nodes, what happens when a node leaves the group with the shared trust level, and how trust hierarchies are defined in the first place, especially in civilian applications. SAR relies on tamper-proof hardware.

20.3.3 Detection, Reputation, and Response Systems

A method for thwarting attacks is prevention. According to Schneier [27], a prevention-only strategy only works if the prevention mechanisms are perfect; otherwise, someone will find out how to get around them. Most of the attacks and vulnerabilities have been the result of bypassing prevention mechanisms. Given this reality, detection and response are essential.

Combining misbehavior detection with a reputation system and appropriate response leads to what we call here a self-policing mobile ad hoc network. Self-policing means that there are no authorities higher than the nodes themselves. Each node can make their own decisions on how to react to the behavior of other nodes. As opposed to the Byzantine Generals problem, the nodes in a self-policing system for mobile ad hoc networks do not have to reach a consensus on which nodes misbehave. Each node can keep its own rating of the network denoted by the reputation system entries and it can choose to consider the ratings of other nodes or to rely solely on its own observations. One node can have varying reputation records with other nodes across the network. The subjective view of each node determines its actions. Byzantine robustness [22] in the sense of being able to tolerate a number of erratically behaving servers or, in this case, nodes is the goal of a self-policing system in mobile ad hoc networks. Here, the detection of malicious nodes by means of observation has to be followed by a response to render these nodes harmless.

Because mobile ad hoc networks have properties that differ from wired networks, such as the lack of infrastructure, misbehavior detection has to be adapted. Every node is its own authority. Nodes can cooperate to compare notes, but contrary to a wired organized network, one cannot assume that the nodes are under the control of the same organization.

Reputation systems are used to keep track of the quality of behavior of others. In mobile ad hoc networks, we are interested in the routing and forwarding behavior of nodes. To keep track of behavior and to classify it according to whether it is regular or misbehavior for instance, nodes have to be able to observe other nodes. The main goal of reputation systems in mobile ad hoc networks is to differentiate between regular and misbehaved nodes in order to react accordingly, e.g., by isolating misbehaved nodes from the network.

Only good behavior should pay off in terms of service and reasonable power consumption. Detection of misbehavior has to trigger a response: a reaction of other nodes that results in a disadvantage for the misbehaved node.

The terms reputation and trust are being used for various concepts in the literature, also synonymously. We define the term reputation here to mean the performance of a principal in participating in the base protocol as seen by others. For mobile ad hoc networking, this means participation in the routing protocol and forwarding. By the term trust we denote the performance of a principal in the policing protocol that aims at protecting the base protocol. For reputation systems this means the reliability as a witness to provide honest reports, in a game-theoretic sense, it entails the willingness for retribution and in payment systems, the participation in the payment itself.

Self-policing provides a disincentive for cheating by excluding nodes from the network. This isolation also protects the regular nodes. Misbehaved nodes are shunned in two ways. First, nodes route around suspected misbehaved nodes and thus select more reliable routes which increases their throughput. Second, nodes do not provide service to suspected misbehaved nodes, hence their misbehavior ceases to have an impact. The first prevents the misbehaved nodes from being used, the second prevents them from using other nodes.

Reputation systems are not restricted to any one type of misbehaved node, such as selfish, malicious, or faulty.

We now briefly describe some of the protocols proposed in the literature.

20.3.3.1 Watchdog and Path Rater. Marti, Giuli, Lai, and Baker [15] have proposed Watchdog and Path Rater components to mitigate routing misbehavior. They observed increased throughput in mobile ad hoc networks by complementing DSR with a Watchdog for detection of denied packet forwarding and a Path Rater for trust management and routing policy rating every path used, which enable nodes to avoid malicious nodes in their routes as a reaction. Ratings are kept about every node in the network and the rating of actively used nodes is updated periodically. Their approach does not punish malicious nodes that do not cooperate, but rather relieves them of the burden of forwarding for others, whereas their messages are forwarded without complaint. This way, the malicious nodes are rewarded and reinforced in their behavior.

20.3.3.2 CONFIDANT. CONFIDANT (see our papers [2–4]) stands for Cooperation Of Nodes, Fairness In Dynamic Ad-hoc NeTworks and it detects malicious nodes by means of observation or reports about several types of attacks and thus allows nodes to route around misbehaved nodes and to isolate them from the network. Nodes have a monitor for observations, reputation records for first-hand and trusted second-hand observations, trust records to control trust given to received warnings, and a path manager for nodes to adapt their behavior according to reputation. Simulations for no forwarding have shown that CONFIDANT can cope well even with half of the network population misbehaving.

20.3.3.3 CORE. CORE, a collaborative reputation mechanism proposed by Michiardi and Molva [16], also has a Watchdog component; however, it is complemented by a reputation mechanism that differentiates between subjective reputation (observations), indirect reputation (positive reports by others), and functional reputation (task-specific behavior), which are weighted for a combined reputation value that is used to make decisions about cooperation or gradual isolation of a node. Reputation values are obtained by regarding nodes as requesters and providers, and comparing the expected result to the actually obtained result of a request. Nodes only exchange positive reputation information.

20.3.3.4 Context-Aware Inference. A context-aware inference mechanism has been proposed by Paul and Westhoff [20], where accusations are related to the context of a unique route discovery process and a stipulated time period. A combination is used that consists of unkeyed hash verification of routing messages and the detection of misbehavior by comparing a cached routing packet to overheard packets. The decision of how to treat nodes in the future is based on accusations of others, whereby a number of accusations pointing to a single attack, the approximate knowledge of the topology, and context-aware inference are claimed to enable a node to rate an accused node without doubt. An accusation has to come from several nodes, otherwise a single node making the accusation is itself accused of misbehavior.

20.3.3.5 OCEAN. OCEAN [1] by Bansal and Baker relies exclusively on first-hand observations. Directly observed positive behavior increases the rating, directly observed negative behavior decreases it by an amount larger than that used for positive increments. If the rating is below the faulty threshold, the node is added to the faulty list. This faulty list is appended to the route request by each node broadcasting it to be used as an avoid list. A route is rated good or bad depending on whether the next hop is on the faulty list. As a response to misbehavior, nodes reject all traffic coming from a suspected misleading node, even if it is not the source of the traffic. The second chance mechanism for redemption employs a time-out after an idle period. Then a node is removed from the faulty list, its rating remaining unchanged. In addition to the rating, nodes keep track of the forwarding balance with their neighbors by maintaining a chip count for each node, which increases when requesting a node to forward a packet and decreases with an incoming request from that node.

20.3.4 Discussion

Payment systems serve as an incentive to provide a well-defined service, such as packet forwarding, to others for remuneration. The payment has to be unforgeable. To ensure this, tamper-proof hardware and trusted third parties have been suggested. With payment systems, the issue of pricing and other economic questions, such as how to deal with lost payment, arise. They can prevent selfish forwarding misbehavior, however, they do not address malicious or faulty misbehavior.

Secure protocols prevent preconceived deviations from specific protocol functions. They do, however, not aim at serving as incentives for cooperation or dealing with novel types of misbehavior that occur by going around the protected functions.

Reputation systems apply to a broader range of desired behavior as long as it is observable and classifiable. They can, if they use second-hand information and have means to cope with false accusations or false praise, partially prevent misbehavior by excluding misbehaved nodes. This way, nodes can protect themselves before encountering the misbehaved node. If the reputation systems rely exclusively on first-hand experience to build reputation ratings, they can only prevent more of the misbehavior experienced by a node after it occured.

Preventive schemes can only protect what they set out to protect from the start. There can, however, be unanticipated attacks that circumvent the prevention. It is vital that this misbehavior be detected and prevented from happening again in the future. Self-policing schemes are only as limited as their intrusion detection component regarding detected attacks. The schemes themselves are flexible and can accommodate an evolving intrusion detection component. If the detection of a new attack is conceived of, the detection component can be changed to reflect this added knowledge. This does not in any way change the protocol. If a preventive scheme needs to be extended to accommodate the advent of a new attack, a new version of the routing protocol is required.

As opposed to payment systems, reputation systems do not assume that nodes have to forward for others at least as many packets as they generate themselves. A self-policing system in the sense of an intrusion detection component with a reputation system merely penalizes a node if it does not do what it is supposed to do according to its own promises. This difference offers an advantage in situations where a node is simply not in the position to cooperate (e.g., when it is at the edge of the network and does not get many requests). In any of the payment systems described here, the node would run out of means to afford having its own packets forwarded by others. This problem is prevented in a self-policing system.

Economic systems assume a rational node that aims at maximizing its utility expressed in power or payment units. The node misbehavior targeted by payment systems is thus selfish concerning utility, but it is not malicious.

A malicious node is not necessarily aiming at economizing on its resources. Its interest lies in mounting attacks on others. Secure routing protocols aim at preventing malicious nodes from mounting attacks.

Although some reactive systems focus on selfish (Watchdog) or malicious misbehavior (intrusion detection), this is not an intrinsic limitation. Self-policing networks can cope with both selfish and malicious, and, in addition, with nonintentional fauly misbehavior, the only requirement being that such misbehavior be detectable (i.e., observable and classifiable).

We deem the consideration of nonintentional misbehavior such as bugs of high importance. We think it is vital to protect the network against misbehaved nodes regardless the nature of their intentions. Nonintentional misbehavior can result from a node being unable to perform correctly due to a lack of resources, due to its particular location in the network, or simply because of the node being faulty. Self-policing misbehavior detection, reputation, and response systems can be applied irrespective of the actual cause of the misbehavior, be it intentional or not. When a node is classified as misbehaved it simply means that the node performs badly at routing or forwarding. No moral judgment is implied.

The question of a tamper-proof security module remains controversial [23], but might prove inevitable. As opposed to nuglets and counters, the self-policing reputation systems do not need tamper-proof hardware for themselves, because a malicious node neither knows the entries of its reputation in other nodes nor does it have access to all other nodes for potential modification. The secure module might still be necessary for complementary protection such as authentication.

20.4 Self-Policing for Mobile Ad Hoc Networks

In this section, we explore the properties and trade-offs of self-policing misbehavior detection, reputation, and response systems in more detail. We first describe our own approach to use it as a basis for comparison. We then discuss several issues and contrast the way the proposed approaches address them.

20.4.1 Enhanced CONFIDANT — a Robust Reputation System Approach

The main properties of a reputation system are the representation of reputation, how the reputation is built and updated, and for the latter, how the ratings of others are considered and integrated. The reputation of a given node is the collection of ratings maintained by others about this node. In our approach, a node i maintains two ratings about every other node j that it cares about. The reputation rating represents the opinion formed by node i about node j's behavior as an actor in the base system (for example, whether node j correctly participates in the routing protocol). The trust rating represents node i's opinion about how honest node j is as an actor in the reputation system (i.e., whether the reported first-hand information summaries published by node j are likely to be true).

We represent the ratings that node i has about node j as data structures $R_{i,j}$ for reputation and $T_{i,j}$ for trust. In addition, node i maintains a summary record of first-hand information about node j in a data structure called $F_{i,j}$.

To take advantage of disseminated reputation information (i.e., to learn from observations made by others before having to learn by our own experience), we need a means of incorporating the reputation ratings into the views of others. We do this as follows. First, whenever node i makes a first hand observation of node j's behavior, the first hand information $F_{i,j}$ and the reputation rating $R_{i,j}$ are updated. Second, from time-to-time, nodes publish their first-hand information to their neighbors. Say that node i receives from k some first hand information $F_{k,j}$ about node j. If k is classified as trustworthy by i, or if $F_{k,j}$ is close to $R_{i,j}$ then $F_{k,j}$ is accepted by i and is used to slightly modify the rating $R_{i,j}$. Else, the reputation rating is not updated. In all cases, the trust rating $T_{i,k}$ is updated; if $F_{k,j}$ is close to $R_{i,j}$, the trust rating $T_{i,k}$ slightly improves, else it slightly worsens. The updates are based on a modified Bayesian approach and a linear model merging heuristic.

Note that, with our method, only first-hand information $F_{i,j}$ is published; the reputation and trust ratings $R_{i,j}$ and $T_{i,j}$ are never disseminated.

The ratings are used to make decisions about other nodes, which is the ultimate goal of the entire self-policing system. For example, in a mobile ad hoc network, decisions are about whether to forward for another node, which path to choose, whether to avoid another node and delete it from the path cache, and whether to warn others about another node. In our framework, this is done as follows. Every node uses its rating to periodically classify other nodes, according to two criteria:

1. Regular or misbehaved
2. Trustworthy or not trustworthy

Both classifications are performed using a Bayesian approach, based on reputation ratings for the former, trust ratings for the latter.

Because we apply our reputation system approach to the CONFIDANT [4] protocol, we briefly describe its main features here. The approach we use in CONFIDANT is to find the selfish or misbehaved nodes and to isolate them, so that misbehavior will not pay off, but rather result in isolation and thus cannot continue. CONFIDANT detects misbehaved nodes by means of observation or reports about several types of attacks, thus allowing nodes to route around misbehaved nodes and to isolate them.

Nodes have a monitor for observations, reputation records for first-hand and trusted second-hand observations about routing and forwarding behavior of other nodes, trust records to control trust given to received warnings, and a path manager to adapt their behavior according to reputation and to take action against misbehaved nodes.

The dynamic behavior of CONFIDANT is as follows. Nodes monitor their neighbors and change the reputation accordingly. If they have reason to believe that a node misbehaves (i.e., when the reputation rating is bad),

Self-Policing Mobile Ad Hoc Networks

they take action in terms of their own routing and forwarding. They thus route around suspected misbehaved nodes. Depending on the rating and the availability of paths to the destination, the routes containing the misbehaved node are either reranked or deleted from the path cache. Future requests by the badly rated node are ignored. Simulations for no forwarding have shown that CONFIDANT can cope well, even if half of the network population misbehaves.

Note that simply not forwarding is just one of the possible types of misbehavior in mobile ad hoc networks. Several others, mostly concerned with routing rather than forwarding have been suggested, such as black hole routing, gray hole routing, worm hole routing. Other kinds of misbehavior aim at draining energy, such as the sleep deprivation attack. CONFIDANT is not restricted to handling any particular kind of misbehavior, but can handle any attack that is observable. Even if the observation cannot precisely be attributed to an attack but is the result of another circumstance in the network such as a collision, CONFIDANT can make use of it. If it is a rare accident, it will not influence the reputation rating significantly, and if it happens more often, it means the observed node has difficulties performing its tasks.

20.4.2 Issues in Reputation Systems for Mobile Ad Hoc Networks

The self-policing systems proposed in the literature differ in several aspects, which we explain in the following.

20.4.2.1 Spurious Ratings. If second-hand information is used to influence reputation, nodes could lie and give spurious rating information. The benefits of false accusations for an adversary are that they can lead to a denial of service of another node by being excluded, false praise can benefit a colluding node. False accusations are not an issue in positive reputation systems, because no negative information is kept [8, 14], however, the disseminated information could still be false praise and result in a good reputation for misbehaved nodes. Moreover, even if the disseminated information is correct, one cannot distinguish between a misbehaved node and a new node that just joined the network. Many reputation systems build on positive reputation only [26], some couple privileges to accumulated good reputation, e.g., for exchange of gaming items or auctioning [25]. Positive reputation systems are thus used where one has a choice of transaction partners and wishes to find the best one. In mobile ad hoc networks, the requirements are different; the focus is on the isolation of misbehaved nodes.

When allowing second-hand information, the question arises whether liars should be punished just as misbehaved nodes are isolated. If we punish nodes for their seemingly inaccurate testimonials, we might end up punishing the messenger and thus discourage honest reporting of observed misbehavior. Note that we evaluate testimonial accuracy according to affinity to the belief of the requesting node along with the overall belief of the network as gathered over time. The accuracy is not measured as compared to the actual true behavior of a node, because the latter is unknown and cannot be proved beyond doubt. Even if it were possible to test a node and obtain a truthful verdict on its nature, a contradicting previous testimonial could still be accurate. Thus, instead of punishing deviating views, we restrict our system to merely reduce their impact on public opinion. Some node is bound to be the first witness of a node misbehaving, thus starting to deviate from public opinion. Punishing this discovery would be counterproductive, as the goal is precisely to learn about misbehaved nodes even before having had to make a bad experience in direct encounter. Therefore, in our design, we do not punish a node when it is classified as not trustworthy.

20.4.2.2 Information Dissemination. There is a trade-off between the speed of detection of misbehaved nodes by use of second-hand information and the classification vulnerability introduced by it. CONFIDANT makes use of second-hand information to proactively isolate misbehaved nodes before actual encounter. This would make it vulnerable to spurious ratings, notably false accusations, also referred to as blackmailing, and false praise in the case that trusted nodes lie. To prevent that but still retain the advantage of earlier detection, only compatible second-hand information is used and then only slightly influences the reputation rating.

Collaborative reputation (CORE) [16] permits only positive second-hand information, which makes it vulnerable to spurious positive ratings and misbehaved nodes increasing each other's reputation. Observation-based cooperation enforcement in ad hoc networks (OCEAN) [1] relies exclusively on first-hand information for its ratings, trading off detection speed for robustness against spurious ratings. However, it disseminates information about suspected misbehaved nodes by adding them to the avoid list in the route request. Context-aware detection [20] accepts negative second-hand information on the condition that at least four separate sources make such a claim, otherwise the node spreading the information is considered misbehaved. Although this distributes the trust given into accusations over several nodes and thus spreads the risk, it inadvertently serves as a disincentive to share ratings and warn others by accusation. Depending on the network density it is also not guaranteed to have at least four witnesses of any event present, let alone four that report it.

20.4.2.3 Type of Information. The original CONFIDANT used only negative information for the consideration for the reputation system. In the enhanced version, as described in [5], positive information is also used to discriminate between active nodes that misbehave sometimes and rarely active nodes that misbehave most of the time. We are thus interested in the relative rate of misbehavior, not the absolute number of misbehavior incidents. OCEAN also uses both positive and negative information for ratings and chip counts. Path Rater and the context-aware detection only consider negative information.

20.4.2.4 Response. Except for Watchdog and Path Rater [15], all other schemes here have a punishment component in their way of isolating nodes, thus the isolation is twofold: misbehaved nodes are avoided in routes and are denied cooperation when they request it. Not using misbehaved nodes, but allowing to be still used by them only increases the incentive for misbehavior, because it results in power saving due to the decrease in number of packets they have to forward for others.

20.4.2.5 Redemption, Weighting of Time. CORE gives more weight to the past behavior of a node and less to its current behavior. The rationale behind this is that wrong observations or rare behavior changes should not have too much influence on the reputation rating. This holds true only under the assumption that the behavior of a node is constant over time. CONFIDANT takes the opposite approach of discounting the past behavior. This is to ensure that a node cannot leverage on its past good performance with its misbehavior gone unpunished. It also ensures that the system is able to react more quickly to changes of behavior. The other reputation systems do not weight ratings according to time.

Ratings are not only weighted to shift emphasis to the past or the present, but also to add importance to certain kinds of observation. CONFIDANT gives the most weight to first-hand observations and less to reported second-hand information. CORE also uses weights to distinguish between types of observations.

Path Rater, context-aware detection, and OCEAN do not weight ratings according to time.

Redemption has the purpose of mitigating misclassification of a node as misbehaved, either by deceptive observation, spurious ratings, or a fault in the reputation system. Another case that requires redemption is when a node that has been correctly isolated as misbehaved should be allowed back into the network because the root of its misbehavior has been removed (e.g., a faulty node has been repaired, a compromised node has been recaptured by its rightful user).

CONFIDANT allows for redemption of misbehaved or misclassified nodes by reputation fading (i.e., discounting the past behavior even in the absence of testimonials and observations) and periodic reevaluation (i.e., checking from time to time whether the rating of a node is above or below the acceptable threshold). Hence, even if a node has been isolated by all nodes, it can get back into the network eventually. Whether it then remains in the network depends on its behavior. Because the ratings do not get erased, but only discounted, the rating of the formerly misbehaved node is still close to the threshold value and thus the reaction to renewed misbehavior is swift, resulting in earlier isolation than the misbehavior of a new node. It is thus possible for a node to redeem itself, given that nodes have each their own reputation belief, which is not necessarily shared by all the others. Because their opinions can differ, a node is most probably not excluded by all other nodes and can thus partially participate in the network with the potential of showing its good behavior. Even if this is not the case and the suspect is excluded by everyone it can redeem itself by means of the reputation fading.

In CORE, an isolated node should get redemption if it behaves well again, but because it cannot prove itself when isolated, it remains isolated unless there is a sufficient number of new nodes arriving in the network that have no past experience with the isolated node.

OCEAN, like the initial version of CONFIDANT, relies on a time-out of reputation. The sudden lapse back into the network can pose a problem if several nodes set the timer at roughly the same time.

Path Rater and context-aware detection have no notion of redemption.

20.4.2.6 Weighting of Second-Hand Information. The schemes that use second-hand information have to administer trust of the witnesses (i.e., the sources of second-hand information) to prevent blackmailing attacks. The initial CONFIDANT weighted second-hand information according to the trustworthiness of the source and by setting a threshold that had to be exceeded before taking second-hand information into consideration. Second-hand information had to come from more than one trusted source, several partially trusted sources, or any combination thereof provided that the trust times number exceeds the trust threshold. This adds a vulnerability of trusting untrustworthy nodes. The notion of trust has been more specifically defined in the enhanced version of CONFIDANT, where it means a consistent good performance as a witness, measured as the compatibility between first- and second-hand information. This dynamic assessment allows it to keep track of trustworthiness and to react accordingly. If the second-hand information is accepted, it still only has a small influence on the reputation rating. More weight is given to its own direct observation.

The other schemes have no trust management component.

20.4.2.7 Detection. Reputation systems require a tangible object of observation that can be categorized as good or bad. In online auction or trading systems, this is the sale transaction with established and measurable criteria such as delivery or payment delay. For reputation systems on misbehavior in mobile ad hoc networks, the analogy to a transaction is not straightforward due to the limited observability and detectability in a mobile and, even more importantly, wireless environment. To detect misbehavior, which translates into being able to classify the behavior node as regular (i.e., according to the protocol) or misbehaving (i.e., deviating from the protocol), nodes promiscuously overhear the communications of their neighbors. The component used for this kind of observation is called Watchdog [15], Monitor [4], or Neighbor Watch [1].

The function most used to implement the detection component in the proposed reputation systems is passive acknowledgment [13], where nodes register whether their next-hop neighbor on a given route has attempted to forward a packet. Assuming bidirectional links, a node can listen to the transmissions of a node that is within its own radio range. If within a given time window a node hears a retransmission of a packet by the next-hop neighbor it has sent the packet to previously, the behavior is judged to be good. Note that this does not necessarily mean that the packet has been transmitted successfully, because the observing node cannot know what goes on outside of its radio range (e.g., there could still be a collision on the far side of the next-hop neighbor).

Several problems with Watchdogs have been identified in [15], such as the difficulty of unambiguously detecting that a node does not forward packets in the presence of collisions or in the case of limited transmission power.

In addition to a Watchdog-like observation, CORE nodes do not only rely on promiscuous mode, but they can also judge the outcome of a request by rating end-to-end connections [16].

CONFIDANT uses passive acknowledgment not only to verify whether a node forwards packets, but also as a means to detect if a packet, e.g., a routing control message, has been illegitimately modified before forwarding.

20.4.2.8 Identity. The question of identity is central to reputation systems. They ideally can assume three properties of identity — persistent, unique, and distinct. The requirement to be persistent means that a node cannot easily change its identity. One way of achieving this is by expensive pseudonyms, another is to have a security module. Identity persistence is desirable for reputation systems to enable them to gather the behavior history of a node. An identity is unique if no other node can use it and thus impersonate another node. One way to ensure this is the use of cryptographically generated unique identifiers, as proposed by Montenegro and Castelluccia [18]. This property is needed to ensure that behavior observed was indeed that of the node observed. The requirement of distinct identities is the target of the so-called Sybil attack analyzed by Douceur [9], where nodes generate several identities for themselves to be used at the same time. This property does not so much concern the reputation system itself, because those identities that exhibit misbehavior will be excluded, but other identities stemming from the same node will remain in the network as long as they behave well. The Sybil attack can, however, influence public opinion by having its rating considered more than once. In the scenario where the mobile ad hoc network is not completely cut off the Internet, we can make use of certification authorities. Examples for such a scenario are publicly accessible wireless LANs with Internet connection. The detection and isolation of misbehaved nodes as achieved by a distributed reputation system for mobile ad hoc networks are still necessary, even in the presence of network operators.

20.5 Conclusions

Mobile ad hoc routing and forwarding are vulnerable to misbehavior, which can occur due to selfish, malicious, or faulty nodes. Solutions to the problem of misbehavior have so far been classifiable into three main categories — payment systems, secure routing, and detection and reputation systems. Payment systems target selfish misbehavior by providing economic incentives for cooperation. Secure routing proposals aim at the prevention of malicious misbehavior. Self-policing systems that consist of detection, reputation, and response components target at the isolation of misbehaved nodes regardless of the reason for misbehavior. None of these solution approaches alone can do prevention, detection, and response. A combination, however, for example of self-policing systems with secure routing can be beneficial to obtain a prevention mechanism along with the advantage of detecting selfish and faulty misbehavior and providing an adequate response.

References

1. Sorav Bansal and Mary Baker. Observation-based cooperation enforcement in ad hoc networks. Technical Report, 2003.
2. Sonja Buchegger and Jean-Yves Le Boudec. IBM Research Report: The selfish node: Increasing routing security in mobile ad hoc networks. RR 3354, 2001.
3. Sonja Buchegger and Jean-Yves Le Boudec. Nodes bearing grudges: Towards routing security, fairness, and robustness in mobile ad hoc networks. In *Proceedings of the 10th Euromicro Workshop on Parallel, Distributed, and Network-Based Processing,* pp. 403–410, Canary Islands, Spain. January 2002. Piscataway, NJ: IEEE Computer Society, 2002.
4. Sonja Buchegger and Jean-Yves Le Boudec. Performance analysis of the CONFIDANT protocol: Cooperation of nodes — Fairness in dynamic ad hoc networks. In *Proceedings of IEEE/ACM Symposium on Mobile Ad Hoc Networking and Computing (MobiHOC),* Lausanne, Switzerland, June 2002. Piscataway, NJ: IEEE Computer Society, 2002.
5. Sonja Buchegger and Jean-Yves Le Boudec. A robust reputation system for mobile ad-hoc networks. EPFL Technical Report no. IC/2003/50, July 2003.
6. Levente Buttyán and Jean-Pierre Hubaux. Enforcing service availability in mobile ad-hoc WANs. In *Proceedings of IEEE/ACM Workshop on Mobile Ad Hoc Networking and Computing (MobiHOC),* Boston, August, 2000.

7. Levente Buttyán and Jean-Pierre Hubaux. Stimulating cooperation in self-organizing mobile ad hoc networks. Technical Report DSC/2001/046, EPFL-DI-ICA, August 2001.
8. Chrysanthos Dellarocas. Immunizing online reputation reporting systems against unfair ratings and discriminatory behavior. In *Proceedings of the ACM Conference on Electronic Commerce,* pp. 150–157, 2000.
9. John R. Douceur. The Sybil attack. In *Proceedings of the IPTPSO2 Workshop,* Cambridge, MA, March 2002.
10. Yih-Chun Hu, David B. Johnson, and Adrian Perrig. SEAD: Secure efficient distance vector routing for mobile wireless ad hoc networks. In *Proceedings of the 4th IEEE Workshop on Mobile Computing Systems and Applications (WMCSA 2002),* Calicoon, NY, June 2002.
11. Yih-Chun Hu, Adrian Perrig, and David B. Johnson. Ariadne: A secure on-demand routing protocol for ad hoc networks. Technical Report Technical Report TRO1-383, Department of Computer Science, Rice University, December 2001.
12. Dave B. Johnson and David A. Maltz. The dynamic source routing protocol for mobile ad hoc networks. Internet Draft, Mobile Ad Hoc Network (MANET) Working Group, IETF, Version 9, April 2003.
13. John Jubin and Janet D. Tomow. The DARPA packet radio network protocols. In *Proceedings of the IEEE,* vol. 75, no. 1, pp. 21–32, January 1987.
14. Peter Kollock. The production of trust in online markets. In *Advances in Group Processes,* E.J. Lawler, M. Macy, S. Thyne, and H.A. Walker, Eds., Greenwich, CT, 1999.
15. Sergio Marti, T.J. Giuli, Kevin Lai, and Mary Baker. Mitigating routing misbehavior in mobile ad hoc networks. In *Proceedings of MobiCom 2000,* pp. 255–265, 2000.
16. Pietro Michiardi and Refik Molva. CORE: A collaborative reputation mechanism to enforce node cooperation in mobile ad hoc networks. In *Proceeding of the 6th ISP Conference on Security Communications, and Multimedia (CMS 2002),* Portoroz, Slovenia, 2002.
17. Pietro Michiardi and Refik Molva. Simulation-based analysis of security exposures in mobile ad hoc networks. In *Proceeding of the European Wireless Conference,* 2002.
18. G. Montenegro and C. Castelluccia. Statistically unique and cryptographically verifiable (sucv) identifiers and addresses. In *Proceeding of NDSS'02,* February 2002.
19. Peng Ning and Kun Sun. How to misuse AODV: A case study of insider attacks against mobile ad-hoc routing protocols. In *Proceedings of the 4th Annual IEEE Information Assurance Workshop,* West Point, June 2003.
20. Krishna Paul and Dirk Westhoff. Context aware inferencing to rate a selfish node in DSR based ad-hoc networks. In *Proceedings of the IEEE Globecom Conference,* Taipeh, Taiwan, 2002.
21. Charles E. Perkins, Elizabeth M. Royer, and Santanu Das. Ad hoc on demand distance vector (AODV) routing. RFC 3561, IETF, July 2003.
22. Radia Perlman, Network layer protocols with byzantine robustness. Ph.D. Thesis Massachussetts Institute of Technology, 1988.
23. Andreas Pfitzmann, Birgit Pfitzmann, Matthias Schunter, and Michael Waidner. Trusting mobile user devices and security modules. IEEE *Computer,* pp. 61–68, February 1997.
24. P. Papadimitratos and Z.J. Haas. Secure routing for mobile ad hoc networks. In *Proceedings of SCS Communication Networks and Distributed Systems Modeling and Simulation Conference (CNDS 2002),* San Antonio, TX, January 27–31, 2002.
25. Paul Resnick and Richard Zeckhauser. Trust among strangers in Internet transactions: Empirical analysis of eBay's reputation system. Working Paper for the NBER Workshop on Empirical Studies of Electronic Commerce, 2001.
26. Paul Resnick, Richard Zeckhauser, Eric Friedman, and Ko Kuwabara. Reputation systems. *Communications of the ACM,* vol. 43, no. 12, pp. 45–48, 2000.
27. Bruce Schneier. *Secrets and Lies. Digital Security in a Networked World.* 1st ed., Hoboken, NJ: John Wiley & Sons, 2000.

28. William Stallings. *Network and Internetwork Security.* 2nd ed., Piscataway, NJ: IEEE Press, 1995.
29. Seung Yi, Prasad Naldurg, and Robin Kravets. Security-aware ad-hoc routing for wireless networks. In *Proceedings of MobiHOC Poster Session,* 2001.
30. Yongguang Zhang and Wenke Lee. Intrusion detection in wireless ad-hoc networks. In *Proceedings of MobiCom 2000,* pp. 275–283, 2000.
31. S. Zhong, Y. Yang, and J. Chen. Sprite: A simple, cheat-proof, credit-based system for mobile ad hoc networks. In *Proceedings of Infocom,* 2003.

Chapter 21
Securing Mobile Ad Hoc Networks

Panagiotis Papadimitratos and Zygmunt J. Haas

Abstract

The vision of nomadic computing with its ubiquitous access has stimulated much interest in the mobile ad hoc networking (MANET) technology. These infrastructureless, self-organized networks, which either operate autonomously or as an extension to the wired networking infrastructure, are expected to support new MANET-based applications. However, the proliferation of this networking paradigm strongly depends on the availability of security provisions, among other factors. The absence of infrastructure, the nature of the envisioned applications, and the resource-constrained environment pose some new challenges in securing the protocols in ad hoc networking environments. The security requirements can differ significantly from those for infrastructure-based networks and the provision of security enhancements may take completely different directions as well. In this chapter, we study the schemes proposed to secure mobile ad hoc networks. We explain the primary goals of security enhancements, shed light on the commensurate challenges, survey the current literature on this topic, and finally introduce our approach to this multifaceted and intriguing topic.

21.1 Introduction

Mobile ad hoc networks comprise freely roaming wireless nodes that cooperatively make up for the absence of fixed infrastructure; that is, the nodes themselves support the network functionality. Nodes transiently associate with peers that are within the radio connectivity range of their transceivers and implicitly agree to assist in provision of the basic network services. These associations are dynamically created and torn down, often without prior notice or the consent of the communicating parties. MANET technology targets networks that can be rapidly deployed or formed in an arbitrary environment to enable communications or to serve a common objective dictated by the supported application. Such networks can be highly heterogeneous, with various types of equipment, usage, transmission, and mobility patterns.

Secure communication, an important aspect of any networking environment, becomes an especially significant challenge in ad hoc networks. This is due to the particular characteristics of this new networking paradigm and due to the fact that traditional security mechanisms may be inapplicable.

The absence of a central authority deprives the network of the administrative and management services that would otherwise greatly facilitate its operation. MANET has to rely on continuous self-configuration, especially because of the highly dynamic nature of the network. Problems such as scheduling, address assignment, provision of naming services, and formation of network hierarchy cannot be solved by traditional centralized protocols. Instead, distributed operation is necessary in all aspects of network control, including basic security-related operations, such as the validation of node credentials. In the fully distributed and open environment of ad hoc networking, the provision of such services may not only incur a high overhead, but also provide additional opportunities for misbehaving nodes to harm the network operation.

In general, nodes participate in a protocol execution as peers, which implies that potentially any network node can abuse the protocol operation. As a result, it is fairly difficult to identify trustworthy and supportive nodes based on the network interaction. Additionally, determining the protocol or network components that have to be safeguarded is far from straightforward, something that makes the design of adequate security countermeasures even more difficult.

Meanwhile, the practically invisible (or nonexistent) administrative or domain boundaries make the enforcement of any security measures an even more complex problem. Migrating nodes may face varying rules even when they run the same application, as they move through different network areas and become associated with different groups of nodes. Or, they may lack the ground for the establishment of trust associations, that is, the establishment of some type of a secret, so that cryptographic mechanisms can be employed.

Below, we discuss in further detail the vulnerability of mobile ad hoc networks, clarify how security goals may have to be modified, and explain which types of solutions are plausible for different network instances. Although the discussion throughout a great part of the chapter applies to all types of ad hoc networks, it is important to realize that strictly not all solutions can be applied in all ad hoc networking environments. Moreover, it is necessary to emphasize the relative importance of addressing certain security issues, which can be considered, to some extent, as prerequisites for solutions to other security problems. In the following sections, we present the challenges posed by the MANET environment, survey the relevant literature, identify the limitations of the proposed approaches, and suggest directions for future solutions.

21.2 Security Goals

The overall problem of securing a distributed system comprises the security of the networked environment and the security of each individual network node. The latter issue is important due to the pervasive nature of MANET, which does not allow us to assume that networked devices will always be under the continuous control of their owners. As a result, the physical security of the node becomes an important issue, leading to the requirement of tamper-resistant nodes [24], if comprehensive security is to be provided. However, security problems manifest themselves in a more emphatic manner in a networked environment and especially in mobile ad hoc networks. This is why in this work we focus on the network-related security issues.

Security encompasses a number of attributes that have to be addressed — availability, integrity, authentication, confidentiality, nonrepudiation, and authorization. These goals, which are not MANET-specific only, call for approaches that have to be adapted to the particular features of MANETs.

Availability ensures the survivability of network services despite misbehavior of network nodes; for instance, when nodes exhibit selfish behavior or when denial-of-service (DoS) attacks are mounted. DoS attacks can be launched at any layer of an ad hoc network. For example, an adversary could use jamming to interfere with communication at the physical layer or, at the network layer, it could disable the routing protocol operation, by disrupting the route discovery procedure. Moreover, an adversary could bring down high-level services. One such target is the key management service, an essential service for implementation of any security framework.

Integrity guarantees that an in-transit message is not altered. A message could be altered because of benign failures, such as radio propagation impairments, or because of malicious attacks on the network. Integrity viewed in the context of a specific connection, that is, the communication of two or more nodes, can provide the assurance that no messages are removed, replayed, reordered (if reordering would cause loss of information), or unlawfully inserted.

Authentication enables a node to ensure the identity of the peer node that it is communicating with. Without authentication, an adversary could masquerade a node, possibly gain unauthorized access to resources and sensitive information, and interfere with the operation of other nodes.

Confidentiality ensures that certain information is never disclosed to unauthorized entities. Confidentiality is required for the protection of sensitive information, such as strategic or tactical military information. However, confidentiality is not restricted to user information only; routing information may also need to remain confidential in certain cases. For example, routing information might be valuable for an enemy to identify and to locate targets on a battlefield.

Nonrepudiation ensures that the originator of a message cannot deny having sent the message. Nonrepudiation is useful for detection and isolation of compromised nodes. When node A receives an erroneous message from node B, A can use this message to accuse B and convince other nodes that B is compromised.

Finally, authorization establishes rules that define what each network node is or is not allowed to do. In many cases, it is required to determine which resources or information across the network a node can access. This requirement can be the result of the network organization or the supported application when, for instance, a group of nodes or a service provider wishes to regulate the interaction with the rest of the network. Another example could be when specific roles are attributed to nodes to facilitate network operation.

The security of mobile ad hoc networks has additional dimensions, such as privacy, correctness, reliability, and fault tolerance. In particular, the resilience to failures, which in our context can be the result of malicious acts, and the protection of the correct operation of the employed protocols are of critical importance and should be considered in conjunction with the security of the mobile ad hoc network.

21.3 Threats and Challenges

Mobile ad hoc networks are vulnerable to a wide range of active and passive attacks that can be launched relatively easily, because all communications take place over the wireless medium. In particular, wireless communication facilitates eavesdropping, especially because continuous monitoring of the shared medium, referred to as promiscuous mode, is required by many MANET protocols. Impersonation is another attack that becomes more feasible in the wireless environment. Physical access to the network is gained simply by transmitting with adequate power to reach one or more nodes in proximity, which may have no means to distinguish the transmission of an adversary from that of a legitimate source. Finally, wireless transmissions can be intercepted. An adversary with sufficient transmission power and knowledge of the physical and medium access control layer mechanisms can obstruct its neighbors from gaining access to the wireless medium.

Assisted by these opportunities that wireless communication offers, malicious nodes can meaningfully alter, discard, forge, inject, and replay control and data traffic; generate floods of spurious messages; and, in general, avoid complying with the employed protocols. The impact of such malicious behavior can be severe, especially because the cooperation of all network nodes provides for the functionality of the absent fixed infrastructure. In particular, as part of the normal operation of the network, nodes are transiently associated with a dynamically changing, over time, subset of their peers; that is, the nodes within the range of their transceivers or the ones that provide routing information and implicitly agree to relay their data packets. As a result, a malicious node can obstruct the communications of potentially any node in the network, exactly because it is entitled or even expected to assist in the network operation.

In addition, freely roaming nodes join and leave MANET subdomains independently, possibly frequently, and without notice, making it difficult in most cases to have a clear picture of the ad hoc network membership. In other words, there may be no ground for an *a priori* classification of a subset of nodes as trusted to support the network functionality. Trust may only be developed over time, although trust relationships among nodes may also change, when, for example, nodes in an ad hoc network dynamically become affiliated with administrative domains. This is in contrast to other mobile networking paradigms, such as Mobile IP or cellular telephony, where nodes continue to belong to their administrative domain in spite of mobility. Consequently, security solutions with static configuration would not suffice, and the assumption that all nodes can be bootstrapped with the credentials of all other nodes would be unrealistic for a wide range of MANET instances.

From a slightly different point of view, it becomes apparent that nodes cannot be easily classified as internal or external (i.e., nodes that belong to the network or not): nodes that are expected to participate and be dedicated to supporting a certain network operation and those that are not. In other words, the absence of an infrastructure impedes the usual practice of establishing a line of defense, separating nodes into trusted and not trusted. As a result, attacks cannot be classified as internal or external either, especially at the network layer. Of course, such a distinction could be made at the application layer, where access to a service or participation in its collaborative support may be allowed only to authorized nodes. In the latter example, an attack from a compromised node within the group (i.e., a group node under the control of an adversary) would be considered an internal one.

The absence of a central entity makes the detection of attacks a difficult problem, because highly dynamic large networks cannot be easily monitored. Benign failures, such as transmission impairments, path breakages, and dropped packets, are a fairly common occurrence in mobile ad hoc networks and, consequently, malicious failures will be more difficult to distinguish. This will be especially true for adversaries that vary their attack pattern and misbehave intermittently against a set of their peers that also changes over time. As a result, short-lived observations will not allow detection of adversaries. Moreover, abnormal situations may occur frequently because nodes behave in a selfish manner and do not always assist the network functionality. It is noteworthy that such behavior may not be malicious, but only necessary when, for example, a node shuts its transceiver down to preserve its battery.

Most of the currently considered MANET protocols [30] were not originally designed to deal with malicious behavior or other security threats. Thus, they are easy to abuse. Incorrect routing information can be injected by malicious nodes that respond with or advertise nonexistent or stale routes and links. In addition, compromised routes (i.e., routes that are not free of malicious nodes), may be repeatedly chosen with the encouragement provided by the malicious nodes themselves. The result is that pairs of communicating end nodes will experience DoS and may have to rely on cycles of time-out and new route discovery to find operational routes, with successive query broadcasts imposing additional overhead. Or even worse, the end nodes may be easily deceived for some period of time that the data flow is undisrupted, although no actual communication takes place. For example, an adversary may drop a route error message, hiding a route breakage, or forge network and transport layer acknowledgments.

Finally, mobile or nomadic hosts have limited computational capabilities, due to constraints stemming from the nature of the envisioned MANET applications. Expensive cryptographic operations, especially if they have to be performed for each packet and over each link of the traversed path, make such schemes implausible for the vast majority of mobile devices. Cryptographic algorithms may require significant computation delays, which in some cases would range from one to several seconds for low-end devices [5, 11]. These delays, imposed, for example, by the generation or verification of a single digital signature, affect the data rate of secure communication. More important, mobile devices could become targets of DoS attacks due to their limited computational resources. An adversary could generate bogus packets, forcing the device to consume a substantial portion of its resources. Even worse, a malicious node with valid credentials could frequently generate control traffic, such as route queries, at a high rate not only to consume bandwidth, but also to impose cumbersome cryptographic operations on a sizable portion of the network nodes.

21.4 Trust Management

The use of cryptographic techniques is necessary for the provision of any type of security services; mobile ad hoc networks are not an exception to this rule. The definition and the mechanisms for security policies, credentials, and trust relationships (i.e., the components of what is collectively identified as trust management) are a prerequisite for any security scheme. A large number of solutions have been presented in the literature for distributed systems, but they cannot be readily transplanted into the MANET context, because they rely on the existence of a network hierarchy and a central entity. Envisioned applications for the ad hoc networking environment may require a completely different notion of establishing a trust relationship, although the network operation may impose additional obstacles to the effective implementation of such solutions.

For small-scale networks, of the size of a personal or home network, trust can be established in a truly ad hoc manner, because relationships can be static and sporadically reconfigured manually. In such an environment, the owner of a number of devices or appliances can imprint them, that is, distribute their credentials along with a set of rules that determine the allowed interaction with and between devices [24]. The proposed security policy follows a master–slave model, with the master device being responsible for reconfiguring slave devices, issuing commands, or retrieving data. The return to the initial state can be performed only by the master device or by some trusted key escrow service.

This model naturally lends itself to represent personal area networking, in particular network instances such as Bluetooth® wireless technology [4], in the sense that within a piconet, the interactions between nodes can be determined by the security policy. The model can be extended by allowing partial control or access rights to be delegated, so that the secure interaction of devices becomes more flexible [25]. However, if the control over a node can be delegated, the new master should be prevented from eradicating prior associations and assuming full control of the node.

A more flexible configuration, independent of initial bindings, can be useful when a group of people wish to form a collaborative computing environment [9]. In such a scenario, the problem of establishing a trust relationship can be solved by a secure key agreement, so that any two or more devices are able to communicate securely. The mutual trust among users allows them to share or establish a password using an offline secure channel or perform a preauthentication step through a localized channel [1]. Then, they can execute a password-based authenticated key exchange over the nonsecure wireless medium. Schemes that derive a shared symmetric key could use a two- or a multiparty version of the password authenticated Diffie–Hellman key-exchange algorithm [3].

Human judgment and intervention can greatly facilitate the establishment of spontaneous connectivity among devices. Users can select a shared password or manually configure the security bindings between devices, as seen above. Furthermore, they could assess subjectively the security of their physical and networking environment and then proceed accordingly. However, human assistance may be impossible for the envisioned MANET environment with nodes acting as mobile routers, even though the distinction between an end device and a router may be only logical, with nodes assuming both roles. Frequently, the sole requirement for two transiently associated devices will be to mutually assist each other in the provision of basic networking services, such as route discovery and data forwarding. This could be so because mobile nodes do not necessarily pursue collectively a common goal. As a result, the users of the devices may have no means to establish a trust relationship in the absence of a prior context.

However, there is no reason to believe that a more general trust model would not be required in the MANET context. For instance, a node joining a domain may have to present its credentials to access an available service and, at the same time, authenticate the service itself. Similarly, two network nodes may wish to employ a secure mode of multi-hop communication and verify each other's identity. Clearly, support for such types of secure interaction, either at the network or at the application layer, will be needed.

A public key cryptosystem can be a solution, with each node bound to a pair of keys, one publicly known and one private. However, the deployment of a public key infrastructure (PKI) requires the existence of a certification authority (CA), a trusted third party responsible for certifying the binding between nodes and public keys. The use of a single point of service for key management can be a problem in the MANET context, especially because such a service should always remain available. It is possible that network partitions or congested links close to the CA server, although they may be transient, could cause significant delays in getting a response. Moreover, in the presence of adversaries, access to the CA may be obstructed or the resources of the CA node may be exhausted by a DoS attack. One approach is not to rely on a CA and thus abolish all the advantages of such a facility. Another approach is to institute the CA in a way that answers the particular challenges of the MANET environment.

The former approach can be based on the bootstrapping of all network nodes with the credentials of every other node. However, such an assumption would dramatically narrow the scope of ad hoc networking, because it can be applied only to short-lived mission-oriented, and thus, closed networks. An additional limitation may stem from the need to ensure a sufficient level of security, which implies that certificates should be refreshed from time to time, requiring, again, the presence of a CA.

Securing Mobile Ad Hoc Networks

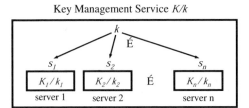

Figure 21.1. Configuration of a Key Management Service Comprising n Servers
The service, as a whole, has a public/private key pair K/k. The public key K is known to all nodes in the network, whereas the private key k is divided into n shares $s_1, s_2, \ldots, s_n$, with one share for each server. Moreover, each server has a public/private key pair K_i/k_i and knows the public keys of all nodes.
(Source: L. Zhou and Z.J. Haas, Securing Ad Hoc Networks, *IEEE Network Magazine*, November/December 1999. © 1999 IEEE) Used with permission.)

Alternatively, it has been suggested that users certify the public keys of other users. One such scheme proposes that any group of K nodes may provide a certificate to a requesting node. Such a node broadcasts the request to its one-hop neighborhood, each neighbor provides a partial certificate, and if sufficient K such certificates are collected, the node acquires the complete certificate [14, 29]. Another scheme proposes that each node select a number of certificates to store, so that, when a node wants the public key of one of its peers, the two certificate repositories are merged, and if a chain of certificates is discovered, the public key is obtained [13].

The solution of a key management facility that meets the requirement of the MANET environment has been proposed in [29]. To do so, the proposed instantiation of the public key infrastructure provides increased availability and fault tolerance. The distributed instantiation of the CA is equipped with a private/public key pair. All network nodes know the public key of the CA and trust all certificates signed by the CA's private key. Nodes that wish to establish secure communication with a destination query the CA and retrieve the required certificate, thus being able to authenticate the other end and establish a secret shared key for improved efficiency. Similarly, nodes can request an update from the CA (i.e., change their own public key and acquire a certificate for the new key).

The distributed CA is instantiated by a set of nodes (servers), as shown in Figure 21.1, for enhanced availability. However, this is not done through naïve replication, which would increase the vulnerability of the system, because the compromise of a single replica would be sufficient for the adversary to control the CA. Instead, the trust is distributed among a set of nodes that share the key management responsibility. In particular, each of the n servers has its own pair of public/private keys and they collectively share the ability to sign certificates. This is achieved with the use of threshold cryptography,

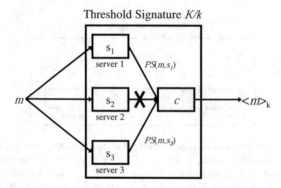

Figure 21.2. Calculation of a Threshold Signature
As an example, the service consists of three servers and uses a (3, 2) threshold cryptography scheme. K/k is the public/private key pair of the service and each server has a share s_i of the private key. To calculate the threshold signature on a message m, each server generates a partial signature $PS(m, s_i)$ and correct server 1 and server 3 forward their partial signatures to a combiner c. Even though server 2 fails to submit a partial signature, c is able to generate the signature $<m>_k$ (m signed by the service private key k).
(Source: L. Zhou and Z.J. Haas, Securing Ad Hoc Networks, *IEEE Network Magazine,* November/December 1999. © 1999 IEEE) Used with permission.)

which allows any $t + 1$ out of n parties to perform a cryptographic operation, although t parties cannot do so. To accomplish this, the private key of the service, as a whole, is divided into n shares, with each of the servers holding one share. When a signature has to be computed, each server uses its share and generates a partial signature. All partial signatures are submitted to a combiner, a server with the special role to generate the certificate signature out of the collected partial signatures, as shown in Figure 21.2. This is possible only with at least $t + 1$ valid partial signatures.

The application of threshold cryptography provides protection from compromised servers, because more than t servers have to be compromised before it assumes control of the service. If fewer than $t + 1$ servers are under the control of an adversary, the operation of the CA can continue, because purposefully invalid partial signatures, contributed by rogue servers, will be detected. Moreover, the service provides the assurance that the adversary will not be able to compromise enough servers over a long period of time. This is done with the help of share refreshing, a technique that allows the servers to calculate new shares from the old ones without disclosing the private key of the service. The new shares are independent from the older ones and cannot be combined with the old shares in an attempt to recover the private key of the CA. As a result, to compromise the system, all $t + 1$ shares have to be compromised within one refresh period, which can be chosen appropriately short in order to decrease vulnerability. The vulnerability can be decreased even further, when a quorum of correct servers detects compromised or unavailable servers and reconfigures the service. In other words, the quorum generates and distributes a new set of n shares, $t + 1$ of which need be combined now to calculate a valid signature. It is noteworthy that the public/private key pair of the service is not affected by share refreshing and reconfiguration operations, which are transparent to all clients.

The threshold cryptography key management scheme can be adapted further by selecting different configurations of the key management service for different network instances. For example, the numbers of servers can be selected according to the size or the rate of membership changes of the network; for a large number of nodes within a large coverage area, the number of servers should also be large, so that the responsiveness of the service can be high. Nodes will tend to interact with the closest server, which can be only a few hops away, or with the server that responds with the least delay. Another possibility is to alternate among the servers within easy reach of the client, something that can happen naturally in a dynamically changing topology. This way, the load from queries and updates will be balanced among different servers and the chances of congestion near one of the servers will be reduced. At the same time, the storage requirements can be traded off for interserver communication, by storing at each server a fraction of the entire database.

Additionally, the efficient operation of the CA can be enhanced when it is combined with secure route discovery and data forwarding protocols. Such protocols could, in fact, approximate the assumption of reliable links between servers in [29] even in the presence of adversaries. In particular, two of the protocols that will be discussed below — Secure Routing Protocol (SRP) and Secure Message Transmission (SMT) — lend themselves naturally to this model. Any two servers can discover and maintain routes to each other and forward service-related traffic, regardless of whether or not intermediate nodes are trusted.

21.5 Secure Routing

The secure operation of the MANET routing protocol is of central importance because of the absence of a fixed infrastructure. Instead, nodes are transiently associated and will cooperate with virtually any node, including those that could potentially disrupt the route discovery and data forwarding operations. In particular, the disruption of the route discovery may be an effective means to systematically obstruct the flow of data. Adversaries can respond with stale or corrupted route replies or broadcast forged control packets to obstruct the propagation of legitimate queries and routing updates.

However, the usual practice for securing Internet routing protocols [19] cannot be applied in the MANET context. The schemes proposed to secure Internet routing rely mainly on the existence of a line of defense, separating the fixed routing infrastructure from all other network entities. This is achieved by distributing a set of public keys/certificates, which signify the authority of the router to act within the limits of the employed protocol (e.g., advertise certain routes) and allow all routing data exchanges to be authenticated, not repudiated, and protected from tampering. However, such approaches cannot combat a malicious router disseminating incorrect topological information. More importantly, they are not applicable in the MANET context because of impediments such as the absence of a fixed infrastructure and a central entity.

Although the appropriate design could provide increased assurances of the availability of an online CA, the requirement to authenticate all nodes assisting the route discovery and data communication may not be practical. The interaction with the CA could become a limiting factor, because a node would have to acquire and validate the credentials of all intermediate network nodes along the discovered routes. Clearly, at least one route to the server has to be discovered before the node can contact the node instituting the CA server. But the problem is that, in the presence of adversaries, forged replies would still require the server's response to be validated. Another important limitation arises from the frequently changing topology and network membership, which would incur frequent queries addressed to the CA. In addition, congested links close to the server, although they may be transient or intermittent, could result in significant delays or even total failure to provide the certification services.

The protection of the route discovery process has been regarded as an additional Quality-of-Service (QoS) issue [28], by choosing routes that satisfy certain quantifiable security criteria. In particular, nodes in a MANET subnet are classified into different trust and privilege levels. A node initiating a route discovery sets the sought security for the route, that is, the required minimum trust level for nodes participating in the query/reply propagation. Nodes at each trust level share symmetric encryption and decryption keys. Intermediate nodes of different levels that cannot determine whether the required QoS parameter can be satisfied or decrypt in-transit routing packets drop them. This scheme provides protection (e.g., integrity) of the routing protocol traffic against adversaries outside a specific trust level.

An extension of the Ad hoc On-Demand Distance Vector (AODV) [20] routing protocol has been proposed [10] to protect the routing protocol messages. The Secure-AODV scheme assumes that each node has the certified public keys of all network nodes, because intermediate nodes validate all in-transit routing packets. The basic idea is that the originator of a control message appends a RSA signature [23] and the last element of a hash chain [15] (i.e., the result of n consecutive hash calculations of a random number). As the message traverses the network, intermediate nodes cryptographically validate the signature and the hash value, generate the k-th element of the hash chain, with k being the number of traversed hops, and place it in the packet. Route replies are provided either by the destination or by intermediate nodes that have an active route to the sought destination.

A second proposal to secure AODV makes use of public key cryptography as well and operates in two stages — an end-to-end authentication and an optional secure shortest path discovery [7]. First, a signed route request propagates to the sought destination, which returns a signed response to the querying node. At each hop, for either direction, the receiving node validates the received control packet and forwards it after signing it. At the second stage, a shortest path confirmation packet is sent toward the destination, while now intermediate nodes sign the message in an onion-like manner to disallow changes of the path length.

21.5.1 The Secure Routing Protocol

The SRP [17] for mobile ad hoc networks provides correct end-to-end routing information over an unknown, frequently changing network, in the presence of malicious nodes. It is assumed that any two nodes that wish to employ SRP have a security association (SA), such as a symmetric shared secret key. Communication takes place over a broadcast medium and it is assumed that malicious nodes, which may concurrently corrupt the route discovery, cannot collude during a single route discovery. Moreover, we assume that nodes have a single data link interface, with a one-to-one correspondence between data link and IP addresses. Under these assumptions, the protocol is proven robust.

SRP provides one or more route replies, the correctness of which is verified by the route geometry itself, and compromised and invalid routing information is discarded. The route request packets verifiably propagate to the destination, and route replies are returned to the querying node strictly over the reversed route, as accumulated in the route request packet. To guarantee this crucially important functionality, SRP employs an explicit interaction with the network layer (i.e., the IP-related functionality). Moreover, a number of novel features allow SRP to safeguard the route discovery operation, as explained below.

21.5.1.1 The Neighbor Lookup Protocol. The Neighbor Lookup Protocol (NLP) is an integral part of SRP responsible for the following tasks:

1. It maintains a mapping of Media Access Control and IP layer addresses of the node's neighbors.
2. It identifies potential discrepancies, such as the use of multiple IP addresses by a single data-link interface.
3. It measures the rates at which control packets are received from each neighbor, by differentiating the traffic primarily based on Media Access Control addresses.

The measured rates of incoming control packets are provided to the routing protocol as well. This way control traffic originating from nodes that selfishly or maliciously attempt to overload the network can be discarded.

Basically, NLP extracts and retains the 48-bit hardware source address for each received (overheard) frame along with the encapsulated IP address. This requires a simple modification of the device driver [27], so that the data link address is passed up to the routing protocol with each packet. With nodes operating in promiscuous mode, the extraction of such pairs of addresses from all overheard packets leads to a reduction in the use of the neighbor discovery and query/reply mechanisms for medium access control address resolution. Each node updates its neighbor table by retaining both addresses. The mappings between data-link and network interface addresses are retained in a table as long as transmissions from the corresponding neighboring nodes are overheard; a time-out period is associated with each entry removed from the table upon expiration.

NLP issues a notification to SRP in the event that according to the content of a received packet:

1. A neighbor used an IP address different from the address currently recorded in the neighbor table.
2. Two neighbors used the same IP address (i.e., a packet appears to originate from a node that may have spoofed an IP address).
3. A node uses the same medium access control address as the detecting node (in that case, the data link address may be spoofed).

Upon receiving the notification, the routing protocol discards the packet bearing the address that violated the aforementioned policies.

Even without cryptographic validation, NLP thwarts adversaries from presenting themselves at the routing layer as more than one node. This would have been possible if different IP addresses were inserted in or used as the source address of the control traffic the adversary relays or originates. However, the effectiveness of NLP relies on the fact that medium access control addresses are either hardwired or may be changed only with substantial latency. In the former case, NLP can provide strong assurances; in the latter one, it will be a significant line of defense, deterring, for example a malicious node from flooding the network with spurious traffic. Moreover, NLP can use cryptography and establish pair-wise security associations (e.g., symmetric secret keys) among neighbors, so that nodes can authenticate traffic received from their neighbors. In any case, we should note that it is not of interest for SRP whether a relay node indeed presented itself with its actual IP address, but whether the node participated in the discovery of the route.

21.5.1.2 The Basic Secure Route Discovery Procedure. The querying node maintains a query sequence number, *Qseq,* for each destination it securely communicates with. The monotonically increasing sequence number allows the destination to detect outdated route requests. At the same time, route requests are assigned a pseudorandom query identifier, Q_{ID}, which is used by intermediate nodes. Q_{ID} is statistically indistinguishable from a random number and thus unpredictable by an adversary with limited computational power. As a result, broadcasted fabricated requests will fail to cause subsequent legitimate queries to be dropped as previously seen, if, for example, the forged packets carry a higher sequence number.

Both Q_{ID} and Q_{seq} are placed in the SRP header, along with a message authentication code (MAC) that covers the shared key, $K_{S,T}$, and the protocol header. Fields that are updated as the packet propagates toward the destination, such as the accumulated addresses, are excluded from the MAC calculation.

Nodes compare the last entry in the accumulated route to the IP datagram source address, which belongs to the neighboring node that relayed the request. If there is a mismatch, or NLP provides a notification that the relaying neighbor violated one of the enforced policies, the query is dropped. Otherwise the Q_{ID} and the source and destination addresses are placed in the query table, so that previously seen queries are discarded. Fresh route requests are rebroadcast, with intermediate nodes inserting their IP address in the request packet.

The destination validates the integrity and freshness of queries originating from nodes it is securely associated with. It generates a number of replies that does not exceed the number of its neighbors, so that a malicious neighbor does not control more than one route. The reversed accumulated route serves as the source route of the reply packet, which is identified by Q_{ID} and *Qseq*. The appended MAC covers the SRP header, including the source route. This way the source can be provided with evidence that the request had reached the destination and, in conjunction with the source route, that the reply was indeed returned along the reverse of the discovered route.

As the reply propagates along the reverse route, each intermediate node simply checks whether the source address of the route reply datagram is the same as the one of its downstream node, as determined by the route reply; if not, the reply is discarded. Ultimately, the source validates the reply by first checking whether it corresponds to a pending query. Then, it is sufficient to validate the MAC, because the IP source-route already provides the (reversed) route itself.

21.5.1.3 Priority-Based Query Handling. To guarantee the responsiveness of the routing protocol, nodes maintain a priority ranking of their neighbors according to the rate of queries observed by NLP. The highest priority is assigned to the nodes generating (or relaying) requests with the lowest rate and vice versa. Quanta are allocated proportionally to the priorities and low-priority queries that are not serviced are eventually discarded. Within each class, queries are serviced in a round robin manner. Selfish or malicious nodes that broadcast requests at a very high rate are throttled back, first by their immediate neighbors and then by nodes farther from the source of potential misbehavior. Nonmalicious queries (i.e., queries originating from benign nodes that regulate in a nonselfish manner the rate of query generation) will be affected only for a period equal to the time it takes to update the priority (weight) assigned to a misbehaving neighbor. In the mean time, the round robin servicing of requests provides the assurance that benign requests will be relayed even amid a storm of malicious or extraneous requests.

21.5.1.4 The Route Maintenance Procedure. The route-error packets are source-routed to the end node along the prefix of the route that is being reported as broken. The intermediate upstream nodes, with respect to the point of breakage, check whether the source address of the route error datagram is the same as the one of their downstream node as reported in the broken route. Then, if there is no notification from NLP that the relaying neighbor violated one of the enforced policies, they relay the packet toward the source. In this case, NLP prevents an adversary that does not belong to the route, but lies at a one-hop distance from it, from generating an error message, because an inconsistency with the addresses already used (during the route discovery) by the actual downstream neighbor will be detected.

The notified source compares the source-route of the error message to the prefix of the corresponding active route. This way, it verifies that the provided route error message refers to the actual route and that it is not generated by a node that is not part of the route. The correctness of the feedback (i.e., whether it reports an actual failure to forward a packet) cannot be verified, though. As a result, a malicious node lying on a route can mislead the source by corrupting error messages generated by another node or by masking a dropped packet as a link failure. However, it can harm only the route it belongs to, something that was possible in the first place if it simply dropped or corrupted the data packets. To ensure that a nonoperational route is detected or to verify that a route remains operational, the source can utilize feedback from the trusted destination. This functionality can be provided by secure data communication protocols, such as the Secure Message Transmission (SMT) protocol. We discuss this in Section 21.6.1, with route maintenance providing complementary information.

21.5.1.4 The SRP Extension. The basic operation of SRP can be extended to allow for nodes other than the destination to provide route replies. This would be possible only under additional trust assumptions, when, for example, nodes sharing a common objective belong to the same group and mutually trust all the group members. In particular, this could be the case when all group members share a secret key.

Under this assumption, a querying node appends to each query an additional MAC calculated with the group key, which we call Intermediate Node Reply Token (INRT). The functionality of SRP remains as described above, with the following addition: each group member maintains the latest Q_{ID} seen from each of its peers and can thus validate both the freshness and origin authenticity of queries generated from other group nodes.

If a node other than the sought destination receives such a valid query, it can respond to the request if it has knowledge of a route to the destination in question. However, the correctness of such a route is conditional upon the correctness of the second portion of the route, which is provided by the intermediate node.

This functionality can be provided independently from and in parallel with the one relying solely on the end-to-end security associations. For example, it could be useful for frequent intragroup communication; any two members can benefit from the assistance of their trusted peers, which may already have useful routes.

21.6 Secure Data Forwarding

The frequent interaction with a CA and the frequent use of computationally expensive cryptographic tools are restrictive assumptions, especially for secure data-forwarding schemes. Such protocols must also take into account the inherent limitations of the MANET paradigm, exploit its features, and incorporate widely accepted and evaluated techniques to be efficient and effective. The above SRP for mobile ad hoc networks satisfies the above-stated goals.

However, SRP or any other underlying routing protocol cannot guarantee that the nodes along a correctly discovered route will, indeed, relay the data as expected. An adversary may misbehave in an intermittent manner (i.e., provide correct routing information during the route discovery stage and later forge or corrupt data packets during the data-forwarding stage). This is exactly the function that is required by any secure data-forwarding protocol: to secure the flow of data traffic in the presence of malicious nodes, after the routes between the source and the destination have been discovered.

One of the solutions targeting the MANET environment proposes two mechanisms that detect misbehaving nodes and report such events, by maintaining a set of metrics reflecting the past behavior of other nodes [16].

To alleviate the detrimental effects of packet dropping, nodes choose the best route, comprised of relatively well-behaved nodes (i.e., nodes that do not have a history of avoiding forwarding packets). Among the assumptions in the above-mentioned work are a shared medium, bidirectional links, use of source routing (i.e., packets carry the entire route that becomes known to all intermediate nodes), and no colluding malicious nodes. Nodes operating in promiscuous mode overhear the transmissions of their successors and may verify whether the packet was forwarded intact to the downstream node. Upon detection of a misbehaving node, a report is generated, and nodes update the rating of the reported misbehaving node. The ratings of nodes along a well-behaved route are periodically incremented, although reception of a misbehavior alert dramatically decreases the nodes rating. When a new route is required, the source node calculates a path metric equal to the average of the ratings of the nodes in each of the route replies and selects the route with the highest metric.

A different approach is to provide incentive to nodes so that they comply with protocol rules, i.e., properly relay user data. The concept of fictitious currency is introduced in [6] to endogenize the behavior of the assumed greedy nodes, which would forward packets in exchange for currency. Each intermediate node purchases from its predecessor the received data packet and sells it to its successor along the path to the destination. Eventually, the destination pays for the received packet. This scheme assumes the existence of an overlaid geographic routing infrastructure and a PKI. All nodes are preloaded with an amount of currency, have unique identifiers, and are associated with a pair of private/public keys. Finally, the cryptographic operations related to the currency transfers are performed by a physically tamper-resistant module.

Another approach appropriate for MANET, which departs significantly from the two above-mentioned schemes, is presented below. Low-cost cryptography is used to protect the integrity and origin authenticity of exchanged data, without placing any overhead at intermediate nodes. Moreover, the feedback that determines the security of the chosen paths originates only from trusted destinations, thus allowing safe inferences on the quality of the paths. Finally, the reliability and fault tolerance of data transmissions is enhanced significantly.

Securing Mobile Ad Hoc Networks

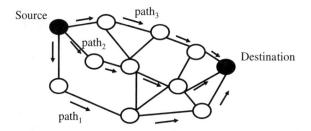

Figure 21.3. SMT Protocol Uses Multiple Diverse Paths Connecting the Source and the Destination
In particular, the APS contains paths that have not been detected as failed, either due to path breakage or because of the presence of an adversary on the path.

21.6.1 Secure Message Transmission Protocol

The SMT protocol [18, 31] is a network-layer secure and fault-tolerant data-forwarding scheme, tailored to the MANET characteristics. In short, SMT is provided with routing information by a protocol such as SRP. This allows SMT to determine a set of diverse paths connecting the source and the destination, as shown in the example of Figure 21.3. Then, it introduces limited transmission redundancy across the paths, by dispersing a message into N pieces, so that successful reception of any M-out-of-N pieces allows the reconstruction of the original message at the destination. Each piece, transmitted over one path, is equipped with a cryptographic header that provides origin authentication, integrity, and replay protection. Upon reception of a number of pieces, the destination informs the source of which pieces, and thus routes, were intact. To enhance the robustness of the feedback mechanism, the small-sized acknowledgments, also protected by a cryptographic header, are maximally dispersed, so that successful reception of one piece is sufficient. If less than M pieces of the message were received, the source retransmits the remaining pieces over the intact routes or in general the ones deemed as more secure. If too few pieces were acknowledged or too many messages remain outstanding, the protocol adapts its operation by determining a different path set, reencoding undelivered messages, and reallocating pieces over the path set. Otherwise, it proceeds with subsequent message transmissions.

SMT exploits MANET features such as the topological redundancies, interoperates widely with accepted techniques such as on-demand route discovery and source routing, relies on a security association only between the source and the destination, and makes use of highly efficient symmetric-key cryptography. Moreover, the routing decisions are made by the querying node, based on the feedback that the destination and the underlying SRP provide. At the same time, no additional processing overhead is imposed on intermediate nodes, which do not perform any cryptographic operation but simply relay the message pieces. However, the use of multiple paths and the resultant greater number of nodes involved in the forwarding of a single message can be admittedly considered as the price to pay to achieve the sought robustness.

On the one hand, SMT's robustness can be enhanced by the adaptation of parameters such as the number of paths and the ratio of the numbers of transmitted to required pieces, termed as the redundancy or dispersion factor. On the other hand, in a low-risk environment with limited malicious failures, the same parameters can be adjusted so that the imposed transmission overhead is reduced to a level close to that of a single-path scheme. An additional element that contributes to the flexibility of SMT is that different algorithms can be implemented for the selection of the path set, based on different metrics and interpretations of the network feedback. SMT can yield 100 percent successful message reception even in a highly adverse environment, when, for example, 20 percent of the network nodes are malicious, while keeping the message and computation overhead low.

The two communicating end nodes make use of the Active Path Set (APS), comprising diverse paths that are not deemed failed. The sender invokes the underlying route discovery protocol, updates its network topology view, and then determines the APS for a specific destination. This model can be extended to multiple destinations, with one APS per destination. At the receiver's side, the APS is used for the transmission of the feedback, but if links are not bidirectional, the destination will have to determine its own reverse APS.

The dispersion of messages, which is performed by the information dispersal algorithm (IDA) [22], is coupled to the APS characteristics through an appropriate selection of the dispersion algorithm parameters. For example, in low connectivity conditions (small number of disjoint paths), the sender may increase the redundancy factor to provide increased assurance and possibly low transmission delay. The adaptation of the protocol is the result of the interplay among the following parameters:

- K, the (sought) cardinality of APS
- k, the S,T-connectivity, i.e., the maximum number of S,T node-disjoint paths from the source (S) to the destination (T)
- r, the redundancy factor of the IDA encoding
- x, the maximum number of malicious nodes

Clearly, the condition for successful reception is $x \leq [K \cdot (1 - r^{-1})]$ which demonstrates the coupling among choices of parameters.

In particular, K can be determined as a function of r, so that the probability of successful transmission is maximized. (Note that K is equal to N

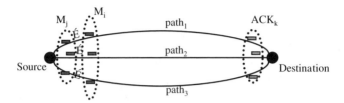

Figure 21.4. APS with Three Paths
For an APS with three paths, the source can disperse each message into three pieces and transmit them across APS. The destination responds to each message M_k with an acknowledgment ACK_k, notifying explicitly which pieces were received. This feedback allows the source to quickly update the rating of the APS paths and retransmit lost pieces across the operational paths if the message cannot be reconstructed at the destination.

when one message piece is allocated per path.) To do so, the source starts by determining an APS of the k shortest, in number of hops, node disjoint paths. Then, let P_{GOAL} be the sought probability of successful reconstruction of a dispersed message. P_{GOAL} can be provided from the application layer and may correspond to the features of the supported application, for example. Given P_{GOAL} and k, the node calculates the required redundancy factor, r_{GOAL}, for a given or estimated fraction of present adversaries. The source disperses outgoing messages with the redundancy value closest to r_{GOAL}, with M and N selected to minimize the transmission overhead.

Once dispersed, the message pieces are transmitted across APS. If $N < k$, the node selects the N paths of the APS with the highest rating. If the receiver cannot reconstruct the message, the source retransmits the pieces that were not received, according to the feedback provided by the destination. Message pieces are retransmitted by SMT a maximum number of times, $Retry_{MAX}$, which is a protocol selectable parameter. If all retransmissions fail, the message is discarded. This way, limited retransmissions enhance the efficiency of SMT by alleviating the overhead from retransmitting the entire amount of data while SMT promptly detects failures and adapts its operation to remain effective and efficient (Figure 21.4).

The transmission of data is continuous over the APS, with retransmissions placed at the head of the queue, upon reception of the feedback. The continuous usage of the APS allows SMT to update quickly its assessment on the quality of the paths. For each successful or failed piece, the rating of the corresponding path is increased or decreased, respectively. When the rating drops below a threshold, the path is discarded. The path rating is also decreased slowly as time goes by to reduce the chance of using a stale path. Moreover, the simultaneous routing over a number of paths, if not the entire APS, provides the opportunity for low-cost probing of the paths. In particular, the source can easily tolerate the loss of a piece that was transmitted over a low-rated path.

21.7 Discussion

The fast development of mobile ad hoc networking technology over the last few years, with satisfactory solutions to a number of technical problems, supports the vision of widely deployed mobile ad hoc networks with self-organizing features and without the necessity of a preexisting infrastructure. In this context, the secure operation of such infrastructureless networks becomes a primary concern. Nevertheless, the provision of security services is dependent on the characteristics of the supported application and the networked environment, which may vary significantly. At one extreme, we can think of a library or an Internet café, which provides short-range wireless connectivity to patrons, without any access constraint other than the location of the mobile device. At the other extreme, a military or public safety unit can make use of powerful mobile devices capable of performing expensive cryptographic operations. Such devices would communicate only with other trusted devices.

Between these two ends of the spectrum, a multitude of MANET instances will provide different services, assume different modes of interaction and trust models, and admit solutions such as the ones surveyed above. Moreover, it is probable that instead of a clear-cut distinction among network instances, devices and users with various security requirements will coexist in a large, open, frequently changing ubiquitous network.

In this context, an important related issue is the IP addressing scheme employed in the MANET environment. The common assumption that node credentials (e.g., certificates) are bound to IP addresses may need to be revisited, because one can imagine that roaming nodes will join MANET subdomains and IP addresses will be assigned dynamically (e.g., DHCP [8] or IPv6 auto-configuration [26]) or even randomly (e.g., Zero-Configuration [12]).

A type of ad hoc network with particular requirements is a sensor network, which requires multi-hop communication throughout a network of hundreds or even thousands of nodes, with relatively infrequent topological changes. It is expected that a single organization will undertake the deployment and administration of these networks. Moreover, sensing devices have limited computational capabilities, network transmission rates are relatively low, and communications are mostly data driven. These requirements may affect, in different ways, the design of security measures for sensor networks, as demonstrated by the schemes proposed in the literature.

One of the proposals to secure sensor networks provides a protocol for data authentication, integrity, and freshness and a lightweight implementation of an authenticated broadcast protocol [21]. The scheme targets a restricted, infrastructure-oriented environment, with a trusted central entity instituted by a set of base stations. Sensor nodes communicate only with a base station, which broadcasts messages toward the sensors. The base station and all nodes initially possess a symmetric encryption and authentication key, which secures the exchanged traffic, and later, the base station periodically broadcasts the key that was used to authenticate transmissions during the last period.

An approach that has similarities but targets a more general setting proposes a key management scheme for sensor networks [2]. The focus is on resource-constrained large sensor networks, comprising nodes that are assumed tamper-resistant and equipped with a secret group key. Similarly to the previous scheme, the use of symmetric key cryptography is proposed as the only feasible, low-cost solution. However, frequent rekeying (i.e., periodic regeneration of the single key that is used to encrypt all data transmitted by sensors) is proposed to protect it from possible compromise. To make this reconfiguration operation efficient, the sensors are organized into clusters with a two-hop diameter and clusterheads are elected and form a backbone. Then, from a subset of the backbone, a randomly elected node generates the new key.

The simplified trust models of sensor networks, which, nonetheless, lead to efficient solutions, may not necessarily be usable in other ad hoc networking instances. The circumstantial coexistence of disparate nodes or the requirement of fine-grained trust relationships call for solutions that can adapt to specific contexts and support the corresponding application. However, although the requirements of the application are expected to dictate the characteristics of the required security mechanisms, some aspects of security, such as confidentiality, may not be different at all in the MANET context. Instead, the greatest challenge is to safeguard the basic network operation.

In particular, the securing of the network topology discovery and data forwarding is a prerequisite for the secure operation of mobile ad hoc networks in any adverse environment. Furthermore, the protection of the functionality of the networking protocols will be in many cases orthogonal to the security requirements and the security services provided at the application layer. For example, a transaction can be secured when the two communicating end nodes execute a cryptographic protocol based on established mutual trust, with the adversary being practically unable to attack the protocol. But this does not imply that the nodes are secure against denial of service attacks; the adversary can still abuse the network protocols, and in fact, do it with little effort compared to the effort needed to compromise the cryptographic protocol.

The self-organizing networking infrastructure has to be protected against misbehaving nodes, with the use of low-cost cryptographic tools, under the least restrictive trust assumptions. Moreover, the overhead stemming from such security measures should be imposed mostly, if not entirely, on nodes that communicate in a secure manner and that directly benefit from these security measures. Furthermore, we believe that the salient MANET features and the unique operational requirements of these networks call for security mechanisms that are primarily present at, and closely interwoven with, the network-layer operation, to realize the full potential of this promising new technology.

References

1. D. Balfanz, D.K. Smetters, P. Stuart, and H.C. Wang, Talking to Strangers: Authentication in Ad Hoc Networks, *Proceedings of Network and Distributed System Security Symposium,* San Diego, CA, February 2002.
2. S. Basagni, K. Herrin, E. Rosti, and D. Bruschi, Secure Pebblenets, *Proceedings of 2nd MobiHoc,* Long Beach, CA, October 2001.
3. S.M. Bellovin and M. Merritt, Encrypted Key Exchange: Password-Based Protocols Secure Against Dictionary Attacks, *Proceedings of the IEEE Symposium on Security and Privacy,* Oakland, CA, May 1992.
4. Bluetooth Special Interest Group, Specifications of the Bluetooth System, http://www.bluetooth.com.
5. M. Brown, D. Cheung, D. Hankerson, J.L. Hernadez, M. Kirkup, and A. Menezes, PGP in Constrained Wireless Devices, *Proceedings of 9th USENIX Symposium,* Denver, August 2000.
6. L. Buttyan and J.P. Hubaux, Enforcing Service Availability in Mobile Ad Hoc WANs, *Proceedings of 1st MobiHoc,* Boston, August 2000.
7. B. Dahill, B.N. Levine, E. Royer, and C. Shields, A Secure Routing Protocol for Ad Hoc Networks, Technical Report UM-CS-2001-037, EE&CS, University of Michigan, Ann Arbor, August 2001.
8. R. Droms, Dynamic Host Configuration Protocol, IETF RFC 2131, March 1997.
9. L.M. Feeney, B. Ahlgren, and A. Westerlund, Spontaneous Networking: An Application-Oriented Approach to Ad Hoc Networking, *IEEE Communications Magazine,* June 2001, pp. 176–181.
10. M. Guerrero Zapata and N. Asokar, Securing Ad Hoc Routing Protocols, *Proceedings of the ACM Wireless Security Workshop (WiSe 2002),* Atlanta, GA, 2002.
11. V. Gupta and S. Gupta, Securing the Wireless Internet, *IEEE Communications Magazine,* December 2001, pp. 68–74.
12. M. Hattig, Ed., Zero-Conf IP Host Requirements, Draft-ietf-zeroconf-reqts-09.txt, IETF MANET Working Group, August 2001.
13. J.P. Hubaux, L. Buttyan, and S. Capkun, The Quest for Security in Mobile Ad Hoc Networks, *Proceedings of 2nd MobiHoc,* Long Beach, CA, October 2001.
14. J. Kong, P. Zerfos, H. Luo, S. Lu, and L. Zhang, Providing Robust and Ubiquitous Security Support for Mobile Ad-Hoc Networks, *Proceedings of IEEE ICNP (International Conference on Network Protocols) 2001,* Riverside, CA, November 2001.
15. L. Lamport, Password Authentication with Insecure Communication, *Comm. of ACM,* vol. 24, pp. 770–772, 1981.
16. S. Marti, T.J. Giuli, K. Lai, and M. Baker, Mitigating Routing Misbehavior in Mobile Ad Hoc Networks, *Proceedings of 6th MobiCom,* Boston, August 2000.

17. P. Papadimitratos and Z.J. Haas, Secure Routing for Mobile Ad Hoc Networks, *Proceedings of SCS Communication Networks and Distributed Systems Modeling and Simulation Conference (CNDS 2002),* San Antonio, TX, January 27–31, 2002.
18. P. Papadimitratos and Z.J. Haas, Secure Data Transmission in Mobile Ad Hoc Networks, *Proceedings of the ACM Wireless Security Workshop (WiSe 2003),* San Diego, CA, 2003.
19. P. Papadimitratos and Z.J. Haas, Securing the Internet Routing Infrastructure, *IEEE Communications Magazine,* vol. 40, no. 10, October 2002.
20. C.E. Perkins, E.M. Royer, and S.R. Das, Ad hoc On-Demand Distance Vector Routing, Draft-ietf-manet-aodv-08.txt, IETF MANET Working Group, June, 2001.
21. A. Perrig, R. Szewczyk, V. Wen, D. Culler, and J.D. Tygar, SPINS: Security Protocols for Sensor Networks, *Proceedings of 7th Ann. Int'l. Conf. Mobile Computing and Networks (MobiCom 2001),* Rome, 2001, pp. 189–199.
22. M.O. Rabin, Efficient Dispersal of Information for Security, Load Balancing, and Fault Tolerance, *Journal of ACM,* vol. 36, pp. 335–348, 1989.
23. R. Rivest, A. Shamir, and L. Adleman, A Method for Obtaining Digital Signatures and Public Key Cryptosystems, *Comm. of ACM,* vol. 21, pp. 120–126, 1978.
24. F. Stajano and R. Anderson, The Resurrecting Duckling: Security Issues for Ad Hoc Wireless Networks, *Proceedings of Security Protocols, 7th Int'l. Workshop, Lecture Notes in Computer Science,* vol. 1796, Berlin: Springer-Verlag, 2000.
25. F. Stajano, The Resurrecting Duckling — What Next? *Proceedings of Security Protocols, 8th Int'l. Workshop, Lecture Notes in Computer Science,* vol. 2133, Berlin: Springer-Verlag, 2001.
26. S. Thomson and T. Narten, IPv6 Stateless Address Autoconfiguration, IETF RFC 1971, www.ietf.org.
27. G.R. Wright and W. Stevens, *TCP/IP Illustrated, Volume 2: The Implementation,* Boston: Addison Wesley, 1997.
28. S. Yi, P. Naldurg, and R. Kravets, Security-Aware Ad-Hoc Routing for Wireless Networks, UIUCDCS-R-2001-2241 Technical Report, August 2001.
29. L. Zhou and Z.J. Haas, Securing Ad Hoc Networks, *IEEE Network Magazine,* November/December 1999.
30. IETF Mobil Ad-hoc Networks (MANET) Working Group, http://www.ietf.org/html.charters/manet-charter.html.
31. P. Papadimitratos and Z.J. Haas, Secure Message Transmission in Mobile Ad hoc Networks, *Elsevier Ad Hoc Networks Journal,* vol. 1, no. 1, July 2003.

Chapter 22
Ad Hoc Network Security

Hao Yang, Haiyun Luo, Jiejun Kong, Fan Ye,
Petros Zerfos, Songwu Lu, and Lixia Zhang

Abstract

Security is a basic requirement for mobile ad hoc networks (MANETs) in order for users to perform protected peer-to-peer communication over multi-hop wireless channel. Depending on the application context, a user may desire various security services such as confidentiality, authentication, integrity, nonrepudiation, and access control. As a basis to support these services, the functionality of a MANET to deliver data bits from one node to another must be protected at the first place. Unlike wired networks that have dedicated routers, MANET has no infrastructure support; each mobile node may function as a router and forward packets for other nodes. The wireless medium is open and incurs far more dynamics than in wired networks. These characteristics present a set of unique challenges to secure a MANET:

- No clear line of defense — MANET does not offer a clear line of defense. There is no well-defined place/infrastructure where we may deploy a single security solution. Moreover, the wireless channel is accessible to both legitimate users and malicious attackers. The boundary that separates the inside network from the outside world becomes blurred.
- Resource constraints — the wireless channel is bandwidth constrained and shared among multiple networking entities. The computation and energy resources of a mobile node are also constrained. For example, although some devices, such as notebook computers, may be capable of performing computation-intensive tasks, other low-end devices, such as personal digital assistants (PDAs), may have limited computation capability and energy supply.

- Network dynamics — the topology of MANETs is highly dynamic as nodes freely roam in the network, join or leave the network on their own will, and fail occasionally. The wireless channel is also subject to interferences and errors, exhibiting volatile characteristics in terms of bandwidth and delay. Despite such dynamics, mobile users roaming in the network may request for anytime, anywhere security services.
- Device compromise or theft — portable devices, as well as the system security information they store, are vulnerable to compromises, especially for those low-end devices with weak protection. These subverted nodes may further pose as the weakest link in the system and incur the domino effect for security breaches.

Security solutions for MANET have to meet the following design goals while addressing the above mentioned challenges:

- Comprehensive protection — the solution should thwart threats from both outsiders, which launch attacks on the wireless channel and network topology, and insiders, which sneak into the system through compromised or stolen devices and gain access to certain system knowledge.
- Efficiency — the solution should be efficient in terms of communication overhead and energy consumption and computationally affordable by a portable networking device.
- Scalability — the design should scale well to a large number of nodes in terms of state maintenance complexity, packet exchange amount, etc.
- Robustness — the design should adapt well to channel errors and network dynamics due to node mobility, arrival, departure, failure, etc.
- Availability — the security service should be highly available to network nodes at any time and at any place.

In this chapter, we survey the state-the-art solutions for securing a MANET. We describe the main designs of these solutions, critique their strength and limitations, and identify a few new directions for future research. The rest of the chapter is organized as follows:

- Section 22.1 provides an overview.
- Section 22.2 discusses the link-layer security solutions that ensure one-hop connectivity.
- Section 22.3 describes the proactive and reactive approaches to securing routing protocols and packet forwarding operations.
- Section 22.4 further describes solutions for key management, a critical supporting subsystem.
- Section 22.5 identifies possible future directions.

22.1 Overview

To fulfill the main functionality of delivering bits between two communicating nodes, a MANET must provide network connectivity over potentially multi-hop wireless channels through the following two steps:

1. Ensure one-hop connectivity between a node and its neighbors, typically through link-layer protocols (e.g., wireless Media Access Control [MAC])
2. Extend connectivity to multiple hops through network-layer routing and data forwarding-protocols (e.g., ad hoc routing)

Accordingly, current MANET security proposals can be roughly categorized as link-layer solutions and network-layer solutions. Both types are required for the ultimate solution to protect the whole system.

The MANET link layer, specifically the *de facto* standard IEEE® 802.11, assumes cooperation among neighboring nodes for channel contention and reservation, which is not true in a malicious environment. An attacker that does not follow the rules can launch DoS attacks and deny wireless channel access to its neighbors or selectively jam the channel. Implementing and enforcing a fair MAC protocol is a necessary, but not sufficient, condition to solve this problem. The built-in security mechanism Wired Equivalent Privacy (WEP) protocol in 802.11 is also vulnerable to a number of sophisticated attacks due to its misuse of cryptographic algorithms. Recent research efforts such as 802.11i/WPA and robust security network/AES counter mode with CBC-MAC protocol (RSN/AES-CCMP) [17] have mended most such loopholes.

The network-layer security solutions protect the network-layer functionality of MANET (i.e., routing and packet forwarding). They seek to ensure that the routing information exchanged between nodes is consistent with the protocol specification and that the data packet forwarding behavior of each node is consistent with its routing states. Based on their perspectives in tackling the problem, the existing solutions can be classified into proactive and reactive approach. The proactive approach attempts to thwart security threats at the first place, typically using various cryptographic techniques. However, the reactive approach seeks to detect threats *a posteriori* and react accordingly. Each approach has its own merits and is suitable for different subproblems in this domain. For example, most secure routing protocols adopt the proactive approach to secure routing messages and the reactive approach is widely used to protect the packet forwarding operations.

Due to the absence of a clear line of defense, a complete security solution for MANETs should integrate both proactive and reactive approaches and encompass all three components of prevention, detection, and reaction.

The prevention component deters the attacker by significantly increasing the difficulty to penetrate the system. However, the history of security has clearly shown that a completely intrusion-free system is infeasible, no matter how carefully the prevention mechanisms are designed. This is especially true in MANET, because mobile devices are prone to compromise or physical capture. Therefore, the detection and reaction components, which discover the occasional intrusions and take reactions to avoid persistent adverse effects, are indispensable for the security solutions to operate in the presence of limited intrusions.

In the MANET context, the prevention component is mainly achieved by secure ad hoc routing protocols that prevent the attacker from installing incorrect routing states at other nodes. These protocols enhance current ad hoc routing protocols such as Dynamic Source Routing (DSR), Ad Hoc On-Demand Distance Vector (AODV), and Destination-Sequenced Distance-Vector (DSDV), and employ cryptographic primitives (e.g., HMAC, digital signature, hash chain) to authenticate the routing messages. The detection component discovers ongoing attacks through identification of abnormal behavior exhibited by the attacker. Such misbehaviors can be detected by the neighboring nodes through overhearing the channel and reaching collaborative consensus. Once the attacker has been detected, the reaction component makes adjustments in routing and forwarding operations, ranging from avoiding the node in route selection to collectively excluding the node from the network.

22.2 Link-Layer Security

The link-layer security solutions protect one-hop connectivity between neighboring nodes. Here we focus on the security concerns of 802.11, the *de facto* standard MAC protocol for MANETs.

22.2.1 802.11 MAC Vulnerabilities

The 802.11 DCF mode uses Carrier Sense Multiple Access (CSMA) with collision avoidance (CSMA/CA) to arbitrate wireless medium access among multiple local transmitters. Before data transmission, a sender must reserve the channel by broadcasting a request to send (RTS) frame, and only the intended unicast receiver replies with a clear to send (CTS) frame. Any cooperative node overhearing either frame defers transmission based on the channel reserve duration information (NAV) contained in RTS/CTS frames.

One basic assumption of such operations is that *all* neighbors are cooperative. This is unfortunately not true in a hostile environment. The adversarial nodes can exploit this loophole to deny channel access to their neighbors. In the denial-of-service (DoS) attack described in [25], a malicious neighbor of either the sender or the receiver can intentionally introduce 1-bit error into a victim's link-layer frame by wireless interference. The corrupted

Ad Hoc Network Security

frame has to be discarded by the receiver upon error detection. Because the adversary knows the transmission duration based on the NAV field, it only needs to transmit a few bits to disrupt a legitimate frame of thousands or even tens of thousands of bits. Thus the power consumption battle favors the adversary side rather than the legitimate side. Continuous attacks on a link can cause the channel goodput degrades to zero and disconnect all multi-hop connections going through this link. Even worse, by deploying a limited number of nodes at the right locations, the adversary can disable critical links and partition the network.

The attacker can also exploit the binary exponential back-off scheme of 802.11 to launch DoS attacks [11, 21]. Because the last winner is always favored among the local contending node set, a continuously transmitting node can always capture the channel and cause other nodes to back off again and again. Moreover, the back-offs at the link layer can incur a chain reaction in upper layer protocols using back-off schemes (e.g., Transmission Control Protocol [TCP]'s window management). Gupta et al. [11] use simulations to show that implementing a fair MAC protocol is a necessary, but not sufficient means to solving the problem. Ensuring fairness only at link layer is not the final answer to the problem. A more robust MAC protocol with fairness guarantees is demanded to secure the link-layer operations in MANET.

22.2.2 802.11 WEP Vulnerabilities

It is well-known that 802.11 WEP [16] is vulnerable to attacks of two categories:

1. Message privacy and message integrity attacks [4]. These attacks are based on various mechanisms like short IV, linear CRC-32 checksum, and key stream recovery by known plaintext attacks.
2. Probabilistic cipher key recovery attacks such as Fluhrer–Mantin–Shamir attack [31]. These attacks are based on the fact that the initial output in the RC4 key stream is disproportionately affected by a small number of key bits, especially the prefix and postfix parts of the key [9].

Fortunately, the recently proposed 802.11i/WPA [17] has mended all obvious loopholes in WEP. Future countermeasures like RSN/AES-CCMP [17] are also being developed to improve the strength of wireless security. We do not provide more details here because these cryptographic problems are not unique to ad hoc networks and have been extensively studied in the context of wireless LANs.

22.3 Network Layer Security

The network-layer security designs for MANET are concerned with protecting both routing and packet forwarding operations. The secure ad hoc routing protocols take the proactive approach and prevent the attacker from poisoning the routing states at other nodes, typically through authentication of routing messages exchanged between nodes. On the other hand, the packet forwarding operations are protected by the reactive approach that detects and reacts to ongoing attacks. Before we introduce these security solutions, we first describe the trade-off behind different cryptographic primitives for message authentication, the essential component in any security design.

22.3.1 Message Authentication Primitives

There are three cryptographic primitives widely used for authenticating the content of messages exchanged among nodes.

1. HMAC[1] — message authentication codes. If two nodes share a secret symmetric key K, they can efficiently generate and verify a message authenticator $h_K(\cdot)$ using a cryptographic one-way hash function h. The computation is efficient, even affordable by low-end devices such as small sensor nodes [32]. However, an HMAC can be verified by only the intended receiver, making it unappealing for broadcast message authentication. Besides, establishing the secret key between any two nodes is a nontrivial problem. If a pairwise shared key is used, a total number of $(n \cdot (n-1))/2$ keys have to be maintained in a network with n nodes. Secure Routing Protocol (SRP) for DSR [27] takes this approach with pairwise shared keys.
2. Digital signature — digital signature is based on asymmetric key cryptography (e.g., RSA), which involves much more computation overhead in signing/decrypting and verifying/encrypting operations. It is less DoS resilient because an attacker may feed a victim node with a large number of bogus signatures to exhaust the victim's computation resource for verifying them. Each node also needs to keep a certificate revocation list (CRL) of the revoked certificates. However, one signature can be verified by any node given that it knows the public key of the signing node. This makes digital signature scalable to large numbers of receivers. Only a total number of n public/private key pairs need to be maintained in a network of n nodes. Secure AODV (SAODV) [35] and Authenticated Routing for Ad hoc Networks (ARAN) [7] take the digital signature approach.
3. One-way HMAC key chain — many cryptographic one-way functions exist such that given the output f(x), it is computationally infeasible to find the input x. By applying $f(\cdot)$ repeatedly on an initial input x, one can obtain a chain of outputs $f^i(x)$. These outputs can be used in the reverse order of generation to authenticate messages: a message with an HMAC using $f^i(x)$ as the key is proven to be authentic when the sender reveals $f^{i-1}(x)$. Commonly used are one-way hash chains [30]. Secure Efficient Distance (SEAD) protocol for DSDV [12],

Ariadne for DSR [13], and packet leashes [14] for wormhole attacks all take this approach.

The computation involved in one-way key chain based authentication is lightweight and one authenticator can be verified by large numbers of receivers. However, these benefits come with certain costs. First, hash-chain-based authentication requires clock synchronization at granularities that may need special hardware support. Second, receivers need to buffer a message to verify them when the key is revealed. The delay in the verification of routing messages may greatly decrease the responsiveness of the routing protocol. If immediate authentication is desired, tight clock synchronization and large storage are necessary (e.g., TESLA with instant key disclosure (TIK) [14]). Third, the release of the key involves a second round of communication. The timer has to be carefully gauged according to the specific context. Finally, the storage of the hash chain is non-trivial for long chains, as required in the scenarios with large rekeying intervals.

22.3.2 Proactive Approach to Secure Ad Hoc Routing

Using the cryptographic authentication primitives, mobile nodes can proactively sign their routing and data messages so that other collaborative nodes are able to efficiently differentiate legitimate traffic from external outsiders' unauthenticated packets. However, an authenticated node may be compromised and controlled by an attacker. We have to further ensure proper compliance with the routing protocols, even for an authenticated node. In the following, we describe how different types of routing protocols are secured.

22.3.2.1 Source Routing. For MANET source routing protocols such as DSR, the main challenge is to ensure that each intermediate node cannot remove existing nodes from or add extra nodes to the route. The basic technique is to attach a per hop authenticator for the source routing forwarder list so that any altering of the list can be immediately detected (or after the key is disclosed for HMAC key chain-based authentication).

A secure extension of DSR is Ariadne [13]. It uses one-way HMAC key chain (i.e., Timed Efficient Stream Loss-tolerant Authentication [TESLA]) for the purpose of message authentication. By key management and distribution, a receiver is assumed to possess the last released key of the sender's TESLA key chain. Take the following example as an illustration. Source node S using source routing to connect to the destination D through three intermediate node A, node B, and node C. The protocol will establish a hash chain at the destination

$$H(C, H(B, H(A, HMAC_{K_{SD}}(S, D))))$$

where $HMAC_{KSD}(M)$ denotes message M's HMAC code generated by a key shared between S and D. The well-known one-way hash function H authenticates the contents in the chain and $HMAC_{KSD}(S,D)$ authenticates the source–destination relation. The propagation of the route request (RREQ) and route reply (RREP) messages is described below, where $*$ denotes a local broadcast and $HMAC_{K_X}(\cdot)$ denotes HMAC code generated on node X.

$S : p_S = (RREQ,S,D) , m_S = HMAC_{K_{SD}}(p_S)$

$S \to * : (p_S, m_S)$

$A : h_A = H(A,m_S) , p_A = (RREQ,S,D,[A],h_A,[]) , m_A = HMAC_{K_A}(p_A)$

$A \to * : (p_A, m_A)$

$B : h_B = H(B,h_A) , p_B = (RREQ,S,D,[A,B],h_B,[m_A]) , m_B = HMAC_{K_B}(p_B)$

$B \to * : (p_B, m_B)$

$C : h_C = H(C,h_B) , p_C = (RREQ,S,D,[A,B,C],h_C,[m_A,m_B]) , m_C = HMAC_{K_C}(p_C)$

$C \to * : (p_C, m_C)$

$D : p_D = (RREP,D,S,[A,B,C],[m_A,m_B,m_C]) , m_D = HMAC_{K_{DS}}(p_D)$

$D \to C : (p_D, m_D, [])$

$C \to B : (p_D, m_D, [K_C])$

$B \to A : (p_D, m_D, [K_C, K_B])$

$A \to S : (p_D, m_D, [K_C, K_B, K_A])$

At the destination, D can compute m_S because information of p_S is contained in p_C. D dynamically computes h_C's value according to the explicit node list embedded in p_C, then compares this h_C to the one embedded in p_C for forgery detection. At RREP phase, there is no need to generate separated authentication code for every RREP packet. By trapdoor commitment, any forwarder X already committed the one-way function output $m_X = HMAC_{K_X}(\cdot)$ at RREQ phase, then at RREP phase the commitment $m_X \to K_X$ is fulfilled by revealing the key K_X.

22.3.2.2 Distance Vector Routing. For distance vector based routing protocols such as DSDV and AODV, the main challenge is that each intermediate node has to advertise the routing metric correctly. For example, when hop count is used as the routing metric, each node has to increase the hop count by one exactly. Hop count hash chain [12, 35] is devised so that an intermediate node cannot *decrease* the hop count in a routing update. Note that hash chain for this purpose does not need time synchronization, which is different from one-way HMAC key chain for authentication.

Ad Hoc Network Security

Assuming the maximum hop count of a valid route is n, a node generates a hash chain of length n every time it initiates a RREP message:

$$h_0, h_1, h_2, \cdots, h_n$$

where $h_i = H(h_{i-1})$ and $H(\cdot)$ is a well-known one-way hash function. The node then adds $h_x = h_0$ and h_n into the routing message, with Hop_Count set to 0. Note that h_n and Hop_Count are authenticated with an authenticator, according to the adopted authentication strategy as we discussed at the beginning of this section.

When a node receives a RREQ or RREP packet, it first checks if:

$$h_n = H^{n-Hop\_Count}(h_x)$$

where $H^m(h_0)$ denotes the result of applying $H(\cdot)$ m times on h_x. Then the node sets:

$$h_x = H(h_x).$$

Finally, the node increments the Hop_Count by 1, updates the authenticator, and forwards the route discovery packet.

This approach provides authentication for the lower bound of a hop count, but it does not prevent a forwarder from advertising the same hop count as the one from another forwarder. In [14], a more-complicated mechanism called hash tree chain is proposed to ensure a monotonically increasing hop count as the routing update traverses the network. One general limitation of the above approaches is that they can be only used to protect discrete metrics. For continual metrics that take noninteger values, the one-way chain is ineffective.

22.3.2.3 Link State Routing. Secure Link State Routing (SLSP) [28] is a link state routing protocol for ad hoc networks. Its operations are similar to Internet link state routing protocols (e.g., Open Shortest Path First [OSPF]): each node seeks to understand its neighborhood by Neighbor Lookup Protocol (NLP) and periodically floods link state update (LSU) packets to propagate link state information. NLP is responsible for:

- Maintaining a mapping of MAC and IP addresses of a node's neighbors
- Identifying potential discrepancies, such as the use of multiple IP addresses on a single link
- Measuring the control packet rates from each neighbor

Neighbors use one-hop hello message to discover each other and connectivity is assumed to be lost if hello message is not received within a time-out.

A node collects LSUs from all over the network to construct the global topology and calculate the route to any destination. Based on NLP, one LSU packet is constructed for each neighbor. Each LSU packet contains a sequence number and a hop count. Like DSR and AODV, duplicate LSU packets with previously seen sequence numbers are suppressed. The hop count determines the packet's time-to-live, so that an LSU packet only travels within a zone, as in hybrid routing protocols like ZRP. An LSU-receiving node adds a link to its global topology only if two valid LSUs from both nodes of the link are received. Thus one malicious node alone cannot inject false link information successfully.

SLSP adopts a digital signature approach in authentication. NLP's `hello` messages and LSU packets are signed with sender's private key. Any verifier can use the public key vouched by the sender's valid certificate to verify message's veracity. A certificate can be delivered to verifiers either by attachment to LSU packet or by dedicated public key distribution (PKD) packets. Unlike ARAN, SLSP's NLP/LSU/PKD components employ various rate control mechanisms like time-to-live and rate throttle. Thus SLSP is less vulnerable to DoS attack when attackers present large amount of bogus messages to consume the victim's resource.

22.3.2.4 Other Routing Protocols.
ARAN [7] ensures each node knows the correct next hop on a route to the destination by public key cryptography. Each message is signed and the sender's certificate is attached to prove the authenticity of its public keys. A source S floods the network with a signed RREQ packet. Upon receiving the first copy of RREQ, a node sets up state of reverse path, pointing to the node from which it receives the RREQ. It then signs and broadcasts the packet. Upon receiving the RREQ, the destination D signs a RREP and unicasts it back on the reverse path. Each node along the reverse path signs the RREP and sends it to the next hop, which verifies the signature of the previous hop, until S receives the RREP. Thus the discovered path is the one along which the first copy of RREQ reaches D from S; each node on this path knows the correct next hop, but not the whole path. It does not use any metric such as hop count, so the discovered path may not be optimal.

Awerbuch et al. [2] propose to flood both RREQ and RREP to defend against byzantine failures. When a source S needs a route to a destination D, it signs and floods a RREQ throughout the network. When D receives the first copy of the request, it signs and floods a RREP, which carries a route list so that each intermediate hop can appends its identifier. When a node receives the reply, it computes the total cost of the path as contained in the route list of RREP. If the cost is smaller than that of any previously received RREP, it verifies the packet, appends its own identifier to the route list, signs the packet and broadcasts it. Finally when S receives a reply, it can

Ad Hoc Network Security

verify that it is from D and each hop in the route list is signed properly. Different from ARAN where only one, possibly nonoptimal path, is discovered, here S may receive multiple replies for different routes. Each route contains the full list of intermediate nodes and has a total cost. S can choose the one with minimum cost or hop count for actual data delivery.

22.3.3 Reactive Approach to Protecting Packet Forwarding

The protection of a routing message exchange is only part of the network-layer security solution for MANET. A malicious node may correctly participate in the route discovery phase, but fail to correctly forward data packets. The security solution should ensure that each node indeed forwards packets according to its routing table. This is typically achieved by reactive approach, because attacks on packet forwarding cannot be prevented: an attacker may simply drop all packets passing through it, even though the packets are carefully signed. At the heart of the reactive solutions are a detection technique and a reaction scheme, which are described as follows.

22.3.3.1 Detection. Because the wireless channel is open, each node can perform localized detection by overhearing the ongoing transmissions and evaluating the behavior of its neighbors. However, its accuracy is limited by a number of factors such as channel error, interference, mobility, etc. A malicious node may also abuse the security solution and intentionally accuse legitimate ones. In order to address such issues, the detection results at individual nodes can be integrated and refined in a distributed manner to achieve consensus among a group of nodes. An alternative detection approach relies on explicit acknowledgment from the destination or intermediate nodes to the source, so that the source can figure out where the packet was dropped.

- Localized detection — Marti et al. [24] propose Watchdog to monitor packet forwarding on top of source routing protocols like DSR. It assumes symmetric bidirectional connectivity (i.e., if A can hear B, then B can also hear A). Because the whole path is specified, when node A forwards a packet to the next hop B, it knows B's next hop C. It then overhears the channel for B's transmission to C. If it does not hear the transmission after a time-out, a failure tally associated with B is increased. If the tally exceeds a threshold bandwidth, A sends a report packet to the source notifying B's misbehavior.

    Yang et al. [33] follows the same concept but works with distance-vector protocols such as AODV. It adds a next_hop field in AODV packets so that a node can be aware of the correct next hop of its neighbors. It also considers more types of attacks, such as packet modification, packet duplication, and packet jamming DoS

attacks. Each independent detection result is signed and flooded; multiple such results from different nodes can collectively revoke a malicious node of its certificate, thus excluding it from the network.
- ACK-based detection — the fault detection mechanism proposed in [2] is based on explicit acknowledgments (ACKs). The destination sends back ACKs to the source for each successfully received packet. The source can initiate a fault detection process on a suspicious path that has recently dropped more packets than an acceptable threshold. It performs a binary search between itself and the destination and sends out data packets piggybacked with a list of intermediate nodes, also called probes, which should send back ACKs. The source shares a key with each probe and the probe list is onion encrypted. Upon receiving the packet, each probe sends back an ACK, which is encrypted with the key shared with the source. The source in turn verifies the encrypted ACKs and attributes the fault to the node closest to the destination that sends back an ACK.

22.3.3.2 Reaction. Once a malicious node is detected, certain actions are triggered to protect the network from future attacks launched by this node. The reaction component typically is related to the prevention component in the overall security system. For example, the malicious node may be revoked of its certificate or have a lower chance to be chosen in future forwarding paths. Based on their scope, the reaction schemes can be categorized as global reaction and end-host reaction. In the former scheme, all nodes in the network react to a malicious node as a whole. In other words, the malicious node is excluded from the network. On the contrary, in the end-host reaction scheme, each node may make its own decisions about how to react to a malicious node (e.g., putting this node in its own blacklist or adjusting the confidentiality weight of this node).

- Global reaction — the reaction scheme in [33] falls into the global reaction category. It is based on the ubiquitous and robust access control (URSA) framework [18]. Once multiple nodes in a local neighborhood have reached the consensus that one of their neighbors is malicious, they collectively revoke the certificate of the malicious node. Consequently, the malicious node is isolated in the network as it cannot participate in the routing or packet forwarding operations in the future.
- End-host reaction — the Pathrater in [24] allows each node to maintain its own rating for every other node that it knows about. A node slowly increases the rating of well-behaving nodes over time, but dramatically decreases the rating of a malicious node that is detected by its Watchdog. Based on the rating, the source always picks up a path with highest average rating. Clearly each node may have different opinion about whether another node is malicious and

make its independent reaction accordingly. Awerbuch et al. [2] extend their idea with security protection of the routing messages, as we discussed earlier.

22.3.4 Sophisticated Intrusion Detection System

Intrusion detection systems (IDSs) can be used to detect a wide range of security violations ranging from attempted break-ins by outsiders to system penetrations and abuses by insiders. The packet forwarding monitoring techniques described earlier are simple prototypes of IDS, but there are also more sophisticated designs that are not limited by packet forwarding misbehaviors.

In general, an IDS can be classified as anomaly detection and misuse detection. An anomaly detection IDS records the system usage patterns and evaluates the current system behavior with a statistical model, which is built over time based on previously recorded patterns. However, a misuse detection IDS proactively establishes an *a priori* misuse profile database based on the knowledge about the well-known attacks or loopholes. Zhang and Lee [36] argue to mainly use reactive anomaly detection techniques for MANET intrusion detection, because it is generally hard to obtain an *a priori* misuse profile that characterizes the wireless communication pattern due to the significant dynamics in the network.

Zhang and Lee [36] provide an architectural overview of MANET intrusion detection, which is based on localized detection by each node and collaborative decision making by multiple nodes. In this architecture, each node implements an intrusion detection agent that performs two tasks:

1. Local data collection, detection, and response modules
2. Global cooperative communication, detection, and response modules

An anomaly detection model is used in the local detection engine to find anomalies in the agent-residing node's locality. If an agent detects a known intrusion or anomaly with strong evidence, then the local response module launches a corresponding countermeasure. However, it is likely the evidence is not strong, hence multiple nodes collaborate to decide whether the intrusions are indeed real threats or false alarms. They use their communication and global detection modules to evaluate the suspicious event following a rule-based distributed consensus algorithm, which computes a probabilistic estimation value for a reported intrusion event. Local and global responses (e.g., channel reinitialization and node excommunication) are triggered if the intrusion is confirmed.

We have described in Section 22.3.3 the Watchdog technique [24] in the DSR context and its variants [33] in the AODV context. These overhearing-based detection schemes typically require each node to buffer the overheard packets for certain amount of time. Despite the memory requirement, they are easy to implement and lightweight in terms of communication and computation overhead. However, due to factors such as channel contention, link congestion, interference, mobility, etc., the information gathered from channel overhearing is not sufficient to clearly differentiate malicious nodes from legitimate ones. The detection accuracy can be improved by distributed consensus algorithms, but how to adapt the threshold used by these algorithms remains an open problem.

The packet leash [14] is concerned with wormhole attacks in which an attacker tunnels packets from one point to another in the network. It embeds a time–space virtual bound in a packet flow. When sending a packet, the sender includes in the packet its own location stamp, p_s, and its sending time stamp, t_s. When receiving a packet, a receiver compares these values to its own location, p_r, and its receiving time, t_r. For packet leash to function, the sender's and the receiver's clocks must be synchronized to within $\pm\Delta$. Then the distance traveled by a packet is bounded by the maximum mobility as well as the computable physical distance between the sender and the receiver.

Kong et al. [19] proposed an adaptive design to accommodate both centralized and fully distributed decision making. Every legitimate node is capable of public key crypto-processing and holds an authority-signed certificate before it joins the network. A local report is a record of the suspicious node's behavior so that the agent-residing node will not serve the suspicious node in the future. A global report is a signed packet against the suspicious node. In both centralized and fully distributed scenarios, the signed reports function as signed votes in decision making. In the centralized case, the signed votes are sent to centralized authority. If the number of signed votes collected from legitimate members exceeds a threshold, the authority responds by signing a counter-certificate against the convicted suspect, who is excommunicated from the network because its certificate is revoked. In the fully distributed case, each legitimate member is capable of being a partial authority using a localized threshold cryptosystem. Therefore, in a locality, multiple nodes can locally exchange their reports and sign a counter-certificate right on site to excommunicate an intruder.

22.4 Supporting Element: Trust and Key Management

Trust and key management is a critical supporting element in any security system. Its basic operations include establishing trust and secret connections, as well as key exchange and update. Keys are the basic blocks of symmetric and asymmetric cryptographic functions, which in turn furnish authentication, confidentiality, integrity, and nonrepudiation security services.

Ad Hoc Network Security

The main body of trust and key management in MANET is concerned with a hybrid of asymmetric and symmetric cryptosystems, where trust is established via credential verification, and shared secrets are exchanged for latter use in efficient symmetric cryptosystems. An inherent issue in trust management is the trust graph, where the nodes correspond to the network entities and edges to the verifiable credentials. A directed edge is added from node X to node Y if X's credentials are verifiable by Y. Although such trust can be established by noncryptographic means (e.g., by checking physical IDs), digitally signed credentials (e.g., certificates or tickets) are widely used in computer networks to implement verifiable credentials. The concept of a trust graph provides a means for classifying the various key management systems proposed in the literature. In general, there are three popular trust graphs:

1. Trusted third party (TTP) [29] — a centralized authority (e.g., a key distribution center [KDC] or certification authority [CA]) is trusted by every entity and an entity A is trusted by another if the authority claims A is trustworthy. This scheme is centrally managed, thus the neighborhood of the central point is potentially the bottleneck of a scalable network and subject to DoS attacks.
2. Web-of-trust [38] — no particular structure exists in such trust graphs. Each entity manages its own trust based on direct recommendation from others. The scheme is fully distributed, making it resilient to attacks, but also difficult to achieve consensus among various entities.
3. Localized trust [23] — this model is the middle ground of the previous two. A node is trusted if any k trusted entities among the node's one-hop neighbors claim so, within a bounded time period. As trust management and maintenance are fully distributed in both space and time domains, the model fits in large dynamic ad hoc networks with mobility and on-demand authentication requirements.

Based on the above trust models, different solutions have been proposed.

22.4.1 Trusted Third Party

Zhou and Haas [37] propose a certification service distributed among n special nodes designated as the server infrastructure. It does not use a centralized server as the CA or a hierarchical scheme as is the current practice on the Internet [1]. It employs threshold cryptography [8] to distribute n shares of the private key of the key management service among the server nodes in a secure way, even though the respective public key is assumed to be known to all members of the network. Consequently, the ability to sign certificates is also shared and performed by generating and combining partial signatures from $t + 1 < n$ servers. The system is able to tolerate up to t server compromises and in order to protect against mobile adversaries [26] and adapt its configuration to changes in the network, proactive share refreshing periodically updates the shares without disclosing the service private key to any subset of servers of the size $< t + 1$.

Similar to [37], the MOCA framework in [34] also establishes trust through a TTP entity (the CA service) and employs threshold cryptography among n MOCA servers to collectively provide the functionality of a CA. But different from [37], MOCA employs on-demand service discovery to contact $t + 1$ MOCA servers without assuming a server infrastructure. Certification service is rendered by selecting $t + 1$ best routes to MOCA servers via dynamic on-demand route discovery, which is implemented by certification request (CREQ) and certification reply (CREP) packet flows, similar to route request and reply flows used in on-demand routing discovery.

22.4.2 Web-of-Trust

A fully distributed, self-organized public-key management system for mobile ad hoc networks that shares conceptual similarities with the Pretty Good Privacy (PGP) [38] web-of-trust framework is proposed in [5, 15]. In the scheme, there is no concept of a trusted authority: every node acts as its own authority. The main difference between PGP and the scheme is that there is no longer a well-known certificate directory where all certificates are stored. Instead, each node maintains a local certificate repository. When two nodes want to verify the public keys of each other, in PGP they must find a trust chain between them based on the well-known certificate directory. In self-organized web-of-trust, this problem is translated into using a shortcut search algorithm to find an intersecting point between the partial trust chains carried on the two nodes. Through simulation results, they show that after a short convergence time, a node's repository contains almost the whole certificate trust graph. Although the web-of-trust approach suits for totally self-organized networks relying on informal recommendations, there is no definite trust anchor like the CA in other CA-based PKI approaches.

In [6], web-of-trust is built on top of direct communication through secure side channels. The direct point-to-point contact works only when two nodes are within a secure range of each other, thus being able to directly exchange information and credentials (e.g., using infrared interfaces). This idea essentially mimics human behavior and is similar to that proposed in [3], where a preauthentication phase between two mobile devices makes possible the secure exchange of public keys of the nodes over location-limited channels. A distinction of [6] is that friend nodes that have established security associations via direct contact also trust other nodes with whom either of them has already directly encountered. This expedites the process of setting up trust and security associations. But unlike [5], the friend relationship is assumed to be nontransitive.

22.4.3 Localized Trust

According to recent analysis [10, 22], per node capacity of ad hoc wireless networks rapidly approaches zero as network size increases linearly, particularly when network traffic is not localized. To achieve balance among global consensus on trust, system scalability, and availability of security service, [20] proposes a framework that employs threshold secret sharing to distribute the functionality of a CA into any node that carries a share of the private key of the key management service. Now, any $t + 1$ nodes in the local neighborhood of the requesting node can collectively provide certification services. The localized traffic pattern of security service provisioning minimizes its impact on network scalability and performance concerns and meanwhile maximizes security service's availability.

In addition, a localized secret share refresh mechanism protects the private key of the service by updating all shares periodically. Contrary to nonlocal approaches, it is sufficient to refresh only the first $t + 1$ nodes in a locality to hold appropriate shares, then nodes that do not yet have a refreshed share (including nodes just newly joined the network) can locally obtain one from any refreshed group of $t + 1$ nodes. For bootstrapping of the entire network, the first $t + 1$ shares are allocated by an out-of-band CA, which is not needed afterward. To complete the set of key management operations provided by this scheme, two certificate revocation mechanisms are also described: the explicit revocation in which an explicit counter-certificate carries out the task of invalidating previously issued certificates and the implicit, which is based on the expiration time each certificate carries.

22.5 Future Directions

The research on MANET security is still in its early stage. There are many possible new directions, the following provides a sample set.

22.5.1 Security in Depth

From the systems perspective, the security solution should provide multiple fences of defense. It is hard to believe that a single fence is sufficient to thwart all threats. Moreover, even though it may exist, the resource constraint at each device may prohibit the deployment of such a solution. In the envisioned multi-fence security system for wireless networks, security is built into possibly every component, resulting in an in-depth protection solution that offers multiple lines of defense against many possible security threats. The individual fence adopted by a device may vary in security strength depending on the available resources, deployment cost, and complexity concerns of the device. The system does not stipulate the minimum requirement that a component must have. Instead, it expects a best-effort approach from each component. The more strength a component has, the higher degree of security it has. The system security strength relies on the collective behaviors of individual fences.

22.5.2 Evaluation

Though many solutions provide preliminary security and network performance analysis, it is recognized that the current evaluation for the state-of-the-art wireless security solutions is quite ad hoc. The research community still lacks effective analytical tools, particularly in a large-scale wireless network setting. The multidimensional trade-offs among security strength, communication overhead, computation complexity, energy consumption, and scalability, still remain largely unexplored.

22.5.3 Solutions Anticipating Unknown Attacks

The state-of-the-art solutions are typically designed with some attack model in mind. Therefore, they work well in the presence of designated attacks but may collapse with unanticipated attacks. On the other hand, it may be feasible to design security solutions that can handle unexpected attacks. The solution may incorporate additional information and perform additional operations for security purpose. At each step of the protocol operation, the design makes sure what it has done is completely along the right track. Anything deviating from the valid operations is treated with caution. This way, the protocol tells right from wrong because it knows right with higher confidence, not necessarily knowing what is exactly wrong. The design strengthens the correct operations and may handle even unanticipated threats at the runtime operations.

Note

1. In network literatures, MAC normally refers to Media Access Control protocol at link layer. To avoid ambiguity we use MAC to refer to link layer medium access control, and use HMAC to refer to keyed hashing for message authentication.

Bibliography

1. Aresenault, A. and Turner, S. Internet x.509 public key infrastructure. Internet Draft, draft-ietf-pkix-roadmap-06.txt (Work in Progress), 2000.
2. Awerbuch, B., Holmer, D., Nita-Rotaru, C., and Rubens, H. An On-Demand Secure Routing Protocol Resilient to Byzantine Failures. In *Proc. ACM Workshop on Wireless Security (WiSe)*, Atlanta, 2002.
3. Balfanz, D., Smetters, D., Stewart, P., and Wong, H. Talking to Strangers: Authentication in Ad-Hoc Wireless Networks. In *Proc. 9th Annual Network and Distributed System Security Symposium (NDSS)*, San Diego, 2002
4. Borisov, N., Goldberg, I., and Wagner, D. Intercepting Mobile Communications: The Insecurity of 802.11. In *ACM International Conference on Mobile Computing and Networking (MobiCom)*, Rome, 2001.

5. Capkun, S., Buttyan, L., and Hubaux, J. Self-Organized Public-Key Management for Mobile Ad Hoc Networks. *IEEE Transactions on Mobile Computing*, vol. 2, no. 1, pp. 52–64, 2003.
6. Capkun, S., Hubaux, J., and Buttyan, L. Mobility Helps Security in Ad Hoc Networks. In *Proc. ACM International Symposium on Mobile Ad Hoc Computing and Networking (MobiCom)*, Rome, 2001.
7. Dahill, B., Levine, B., Royer, E., and Shields, C. A Secure Protocol for Ad Hoc Networks. In *10th IEEE International Conference on Network Protocols (ICNP)*, Paris, 2002.
8. Desmedt, Y. Threshold Cryptography. *European Transactions on Telecommunications*, vol. 5, no. 4, pp. 449–457, 1994.
9. Fluhrer, S., Mantin, I., and Shamir, A. Weakness in the Key Scheduling Algorithm of RC4. In *Proc. Annual Workshop on Selected Areas in Cryptography (SAC)*, Toronto, 2001.
10. Gupta, Piyush and Kumar, P.R. The Capacity of Wireless Networks. *IEEE Transactions on Information Theory*, IT-46, no. 2, pp. 388–404, 2000.
11. Gupta, V., Krishnamurthy, S., and Faloutsos, M. Denial of Service Attacks at the MAC Layer in Wireless Ad Hoc Networks. In *Proc. IEEE Military Communication Conference (Milcom)*, Anaheim, CA, 2002.
12. Hu, Y., Johnson, D., and Perrig, A. SEAD: Secure Efficient Distance Vector Routing for Mobile Wireless Ad Hoc Networks. In *Proc. IEEE Workshop on Mobile Computing Systems and Applications (WMCSA)*, Callicoon, NY, 2002.
13. Hu, Y., Perrig, A., and Johnson, D. Ariadne: A Secure On-Demand Routing Protocol for Ad Hoc Networks. In *Proc. ACM International Conference on Mobile Computing and Networking (MobiCom)*, Atlanta, 2002.
14. Hu, Y., Perrig, A., and Johnson, D. Packet Leashes: A Defense against Wormhole Attacks in Wireless Networks. In *Proc. Annual Joint Conf. of the IEEE Computer and Communications Societies (InfoCom)*, San Francisco, 2003.
15. Hubaux, J., Buttyan, L., and Capkun, S. The Quest for Security in Mobile Ad Hoc Networks. In *Proc. ACM International Symposium on Mobile Ad Hoc Networking and Computing (MobiCom)*, Atlanta, 2001.
16. IEEE Computer Society. Wireless LAN Medium Access Control (MAC) and Physical Layer (PHY) Specifications. IEEE Standard 802.11, 1997.
17. IEEE Computer Society. Wireless Medium Access Control (MAC) and Physical Layer (PHY) Specifications: Specification for Enhanced Security. IEEE Standard 802.11i/D30, 2002.
18. Kong, J., Zerfos, P., Luo, H., Lu, S., and Zhang, L. Providing Robust and Ubiquitous Security Support for MANET. In *Proc. IEEE ICNP*, 2001a.
19. Kong, J., Luo, H., Xu, K., Gu, D., Gerla, M., and Lu, S. Adaptive Security for Multi-Layer Ad-Hoc Networks. *Special Issue of Wireless Communications and Mobile Computing*, vol. 2, no. 5, pp. 533–547, Hoboken, NJ: Wiley Interscience Press, 2002.
20. Kong, J., Zerfos, P., Luo, H., Lu, S., and Zhang, L. Providing Robust and Ubiquitous Security Support for Mobile Ad-Hoc Networks. In *9th IEEE International Conference on Network Protocols (ICNP)*, 2001b.
21. Kyasanur, P. and Vaidya, N. Detection and Handling of MAC Layer Misbehavior in Wireless Networks. In *Proc. IEEE Dependable Computing and Communication Symposium (DCC)*, San Francisco, 2003.
22. Li, Jinyang, Blake, Charles, Couto, Douglas De, Lee, Hu Imm, and Morris, Robert. Capacity of Ad Hoc Wireless Networks. In *Proc. ACM International Conference on Mobile Computing and Communication Symposium (MobiCom)*, Rome, 2001.
23. Luo, H., Zerfos, P., Kong, J., Lu, S., and Zhang, L. Self-Securing Ad Hoc Wireless Networks. In *Proc. IEEE International Symposium on Computers and Communications (ISCC)*, Taormina, Italy, 2002.
24. Marti, S., Giuli, T., Lai, K., and Baker, M. Mitigating Routing Misbehavior in Mobile Ad Hoc Networks. In *Proc. ACM International Conference on Mobile Computing and Networking (MobiCom)*, Boston, 2000.

25. Noubir, G. and Lin, G. Low-Power DoS Attacks in Data Wireless LANs and Countermeasures. In *4th ACM International Symposium on Mobile Ad Hoc Networking and Computing (MobiHoc), Poster Session,* Annapolis, MD, 2003.
26. Ostrovsky, R. and Yung, M. How to Withstand Mobile Virus Attacks. In *Proc. ACM Symposium on Principles of Distributed Computing,* Montreal, 1991.
27. Papadimitratos, P. and Haas, Z. Secure Routing for Mobile Ad Hoc Networks. In *Proc. Communication Networks and Distributed Systems Modeling and Simulation Conference (CNDS),* San Antonio, TX, 2002.
28. Papadimitratos, P. and Haas, Z. Secure Link State Routing for Mobile Ad Hoc Networks. In *Proc. IEEE Workshop on Security and Assurance in Ad Hoc Networks (SAINT),* Orlando, FL, 2003.
29. Perlman, R. An Overview of PKI Trust Models. *IEEE Network,* vol. 13, no. 6, pp. 38–43, 1999.
30. Perrig, A., Canetti, R., Tygar, D., and Song, D. The TESLA Broadcast Authentication Protocol. *RSA CryptoBytes,* vol. 5, no. 2, pp. 2–13. 2002.
31. Stubblefield, A., Ioannidis, J., and Rubin, A. Using the Fluhrer, Mantin, and Shamir Attack to Break WEP. In *Proc. Network and Distributed System Security Symposium (NDSS'02),* San Diego, 2002.
32. Wen, Victor, Perrig, Adrian, and Szewczyk, Robert. SPINS: Security Suite for Sensor Networks. In *Proc. ACM International Conference on Mobile Computing and Networking, (MobiCom),* Rome, 2001.
33. Yang, H., Meng, X., and Lu, S. Self-Organized Network Layer Security in Mobile Ad Hoc Networks. In *Proc. ACM Workshop on Wireless Security (WiSe),* Atlanta, 2002.
34. Yi, S. and Kravets, R. MOCA: Mobile Certificate Authority for Wireless Ad Hoc Networks. In *Proc. Annual PKI Research Workshop,* Gaithersburg, MD., 2003.
35. Zapata, M. and Asokan, N. Securing Ad Hoc Routing Protocols. In *Proc. ACM Workshop on Wireless Security (WiSe),* Atlanta, 2002.
36. Zhang, Y. and Lee, W. Intrusion Detection in Wireless Ad-Hoc Networks. In *Proc. ACM International Conference on Mobile Computing and Networking (MobiCom),* Boston, 2000.
37. Zhou, L. and Haas, Z. Securing Ad Hoc Networks. *IEEE Network,* vol. 13, no. 6, pp. 24–30, 1999.
38. Zimmermann, P. *The Official PGP User's Guide.* Cambridge, MA: MIT Press, 1995.

Chapter 23
Modeling Distributed Applications for Mobile Ad Hoc Networks Using Attributed Task Graphs

Prithwish Basu, Wang Ke, and Thomas D.C. Little

Abstract

Mobile ad hoc networks (MANETs) have received significant attention from the research community recently owing to the growth in popularity of portable computing and wireless networking. Although researchers have primarily focused on developing lower layer mechanisms such as channel access and routing for making MANETs operational, higher layer issues such as application modeling have largely remained ignored. In this chapter, we present a novel distributed application framework based on attributed task graphs that enables a large class of resource discovery-based applications on mobile, failure-prone environments such as MANETs. A distributed application is represented as a task comprising smaller subtasks that need to be performed on different classes of computing devices with specialized roles. Execution of a particular task on a MANET requires several logical patterns of data flow between nodes representing such device classes. As a result, dependencies are induced between the different classes of devices that need to cooperate to execute the application. Such dependencies yield a task graph, representation of the distributed application.

We consider the problem of executing distributed tasks on a MANET by means of dynamic selection of specific devices that are needed to execute the subtasks. We present simple and efficient algorithms for dynamic discovery and selection of suitable devices in a MANET from among a number of them providing the same functionality. This is carried out with respect to the proposed task graph representation of the application. We call this process task embedding or anycasting. Because MANETs are prone to disconnections, we advocate periodic monitoring of the selected devices by one another. In the event of an application disruption owing to node mobility or failures, our algorithms adapt to the situation and dynamically rediscover the affected parts of the task graph, if possible. We propose metrics for evaluating the performance of these algorithms and report simulation results for a variety of application scenarios differing in complexity, traffic, and device mobility patterns. We demonstrate by simulation that our protocol can instantiate and reinstantiate task group nodes effectively in mobile scenarios; also the delivered effective throughput is near perfect at low to medium degrees of mobility and moderately high for high mobility scenarios.

23.1 Introduction

The shrinking size of tetherless computing devices and increasing diversity of their capabilities has dramatically increased the value of pervasive computing. Exploiting the full potential of a large network of such devices while not frustrating the end user with interminable configuration tasks poses several interesting challenges for a developer of distributed applications. Wireless networking technologies such as IEEE® 802.11b (or wireless fidelity [Wi-Fi]) [12], Bluetooth® [9], and ZigBee™ [19] have begun to enable several distributed applications on truly tetherless computing environments for end users. These technologies are capable of enabling connectivity between possibly mobile users through infrastructureless networking, also known as mobile ad hoc networking. Formally, a mobile ad hoc network (MANET) is a rapidly deployable, autonomous system of mobile devices that are connected by wireless links to form an arbitrary graph at any instant of time.

With the ubiquity of portable devices and wireless network connectivity, MANETs are likely to gain popularity in the near future, especially in settings where a networking infrastructure is impossible, cumbersome, or expensive to establish. We can conceive scenarios in which the environment surrounding us consists of a large number of specialized as well as multipurpose devices, many of them portable, and linked through wireless connections, albeit with fluctuating link availability. When a large number of computing devices become equipped with wireless connectivity and they form an ad hoc network, they can offer services to other devices for performing several tasks. Ideally, such pervasive networks can enable a broad range of distributed applications that need exchange of information between multiple devices. In such scenarios, because the service providing devices may themselves be mobile, a user cannot rely on one particular device for a certain service as its reachability or availability is not guaranteed. Instead, a user must be prepared to access the required service from

Modeling Distributed Applications for Mobile Ad Hoc Networks

any of several devices in the MANET providing similar services. Besides, the user may not have a preference for a specific device as long as her task is accomplished in a seamless manner.

To realize the above features, we advocate the decoupling of the logical structure of a distributed application or task (consisting of subtasks) from the actual physical devices that execute the application. We propose the use of the task graph abstraction for representing the structure of the user applications in terms of smaller subtasks. It is a graph composed of nodes and edges, where the nodes represent the classes[1] of devices or services needed for processing data related to the task and the edges represent necessary associations between different nodes for performing the task.

Thus when a task is to be executed, specific devices are selected (instantiated) at runtime and are made to communicate with one another according to the specifications of the task graph. More specifically, for each class of device in task graph, one suitable instance needs to be chosen to take part in task execution. We call this process dynamic task-based anycasting or embedding [6].

When a participating device becomes unavailable, a new substitute device with similar capabilities is selected to continue the task. Therefore, a basic proposition in our model is that as long as there is one accessible device in the entire network capable of performing a particular subtask as requested by the user-level application, the latter can proceed. Obviously, the application should be elastic enough to adapt to the changing conditions of the mobile multi-hop network.

The task graph abstraction of a distributed task is advantageous in many ways. It is inherently distributed, as most pervasive applications and services of the future are likely to be, because more and more specialized devices will need to communicate with one another to offer more and more powerful services. It also offers hierarchical composability, as collections of devices can be logically grouped together to constitute a single node in a task graph [5].

The rest of the chapter is organized as follows:

- Section 23.2 introduces the basic modeling framework with the necessary terminology.
- Section 23.3 presents task graph instantiation algorithms for mapping applications onto MANETs.
- Section 23.4 presents simulation results under various degrees of mobility.
- Section 23.5 presents related work on the topic.
- Section 23.6 concludes the chapter.

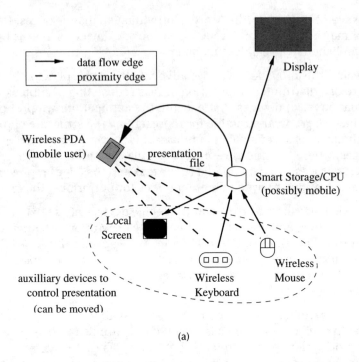

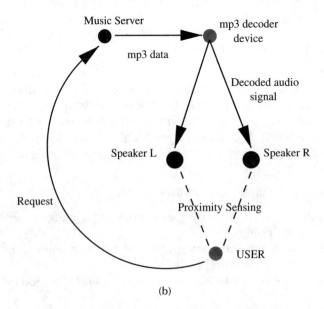

Figure 23.1. Smart Office and Home Applications
(A) Smart presentation task, (B) Stereo music service.

23.2 Modeling Distributed Tasks with Task Graphs

In the past few decades, a variety of distributed applications have been enabled by many advances in computer networking. A distributed networked application or task is composed of several components or subtasks. These components often execute on different hardware devices and communicate among each other in order to yield a desired result. Traditional parallel and distributed computing platforms are composed of high performance nodes internetworked with static high capacity links. However, as mentioned in Section 23.1, the computation and communication substrate offered by a MANET is potentially mobile and hence, prone to link failures. Therefore, it is necessary to develop a model for a distributed application that decouples the bindings between its logical components and the actual hardware devices that they are executed on until application runtime. Additionally, the model should use the component-level structure of an application to dynamically discover and select appropriate devices in the network with desired capabilities for hosting and executing the aforementioned application components.

23.2.1 A Modeling Framework for Task Execution

In this section, we propose the modeling framework which advocates the decoupling of the needs and structure of a distributed task from the physical network. We begin with an introduction of the necessary terminology.

23.2.1.1 Preliminaries. A device in our context is a physical entity that performs at least one particular function such as interaction with its physical surroundings, computation, and communication with other devices. It may be equipped with an embedded processing element, sensors and actuators for interacting with the physical environment, a wireless communication port, or a user interface.

If a device primarily performs one specific function, it is called a specialized device, otherwise, it is referred to as a multipurpose device. Examples of the former type include digital cameras, speakers, printers, keyboards, display devices, etc.; examples of the latter include personal digital assistants (PDAs) and portable notebook computers.

The capabilities of each device can be summarized in their attributes. Attributes can be static (i.e., time-invariant) or dynamic (i.e., time-variant). For example, a network digital camera can have a static attribute resolution that can take values like 320×240, 640×480, etc. Examples of dynamic attributes include location (absolute or relative, depending on the availability of GPS), available computational power, and current load. In this dissertation, we only consider devices with their principal attribute (i.e., their primary function). Multi-attribute extensions are possible and are considered elsewhere [1].

A service is a functionality provided by a device or a collection of cooperating devices. A service provided by a single device is referred to as a simple service, whereas one provided cooperatively by a collection of devices is referred to as a composite service.

Multiple devices can exist in the MANET for providing the same service. For example, there can be multiple wireless cameras in the network from which a user can choose for taking a picture. We refer to this situation as multiple instances of wireless camera services.

Service composition is the process of constructing an instance of a composite distributed service from other simple or composite service instances available in the current networked physical space. In this chapter, we concentrate on the composition of composite services from simple services only. However, service composition can be carried out in a hierarchical manner: complex services can be constructed from composite services using hierarchical task graphs [5].

A node is an abstract representation of a device or a collection of devices characterized by a minimal set of attributes that can offer a particular service.

A node is simple when it represents a single physical device. It is complex when it represents multiple simple nodes. We refer to the principal attribute of a node or a device as its class or category or type. Examples of classes include printer, speaker, joystick, etc.

An edge is a necessary association between two nodes with attributes that must be satisfied for the completion of a task. Examples of edge attributes include causal ordering, relative importance in the overall task, required data rate between nodes, allowable bit error rate, and physical proximity.

23.2.1.2 Tasks and Task Graphs. A task can be described as work executed by a node with a certain expected outcome. The work done by a component of a complex node is considered a subtask of the larger task. An atomic task is an indivisible unit of work that is executed by a simple node. Atomicity is related to the core capability of a device, described through its attributes, and is partially constrained by subjective design choices.

A task graph is a graph $TG = (V_T, E_T)$ where V_T is the set of nodes that need to participate in the task T and E_T is the set of edges denoting data flow between participating nodes.

Instantiation or embedding of a task graph TG on a MANET represented by a graph G is the process of mapping all nodes of TG to nodes in G such that their attributes match. The process also maps edges in TG to paths (single-hop or multi-hop) in G.

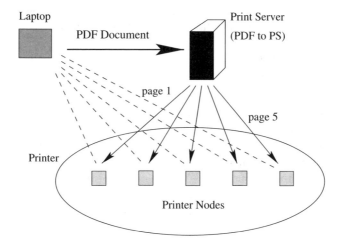

Figure 23.2. A Smart Printing Service

We explain the abstractions developed so far by means of a simple example. Consider a scenario in which there is a PostScript® (PS) printer connected to a computer (print server) running conversion software that can convert portable document format (PDF) files to printable PS format. The printer node and the computer node each represent devices that offer particular services. The printer is considered a specialized device offering the service of converting PS files into printed pages, but the computer is a multipurpose device that has among its many offered services the one service of converting PDF files into PS format. This example is illustrated in Figure 23.2 where the task of printing a PDF document to a single or multiple printers has been logically represented as a task graph.

The printer is a physical device representation of a simple node with certain attributes (e.g., print resolution, color capabilities) and it offers the service of converting PS files into printed pages. Analogously, the print server computer plus its conversion software can be viewed as a representation of a PDF → PS converter node. By taking these two nodes together we can form a complex node that offers a PDF printing service. Let a task be the printing of one PDF document. In this specific case, based on subjective criteria, we define an atomic task to be the printing of one page of the document.[2] The entire document can be then printed on a set of available printers as shown in Figure 23.2. The mechanisms of how appropriate physical devices are discovered and selected to perform a subtask are discussed later in this section.

Note that in the above scenario, we formed a new composite service — PDF printing — by composing simpler existing service instances. Although this example is simplistic, we believe that research that enables such capability in today's MANETs for arbitrary device types and quantities is essential for exploiting the network's full potential.

23.2.1.3 A Taxonomy of Tasks.
We broadly classify tasks into the following distinct categories.

23.2.1.3.1 Preassigned Tasks. In this category of tasks, specific devices need to participate; nodes in the task graph already have devices mapped to them and hence discovery is not required. These nodes are referred to as bound nodes. Therefore, the problem of embedding a task is equivalent to finding suitable (not necessarily the shortest) routes between pairs of devices that are directly connected by an edge in the task graph. If the optimization variable is load on intermediate forwarding devices instead of delay, algorithms for load balancing should be executed instead of a shortest path algorithm.

23.2.1.3.2 Nonpreassigned Tasks. This category of tasks entails a number of homogeneous or heterogeneous computing devices in the network providing specific services. Unlike the preassigned case, nodes in the task graph are logical entities and do not signify devices with specified physical addresses. In fact, any device that can satisfy the requirements specified in a task graph node's attribute set is a candidate for participating in the task. We, therefore, refer to such tasks as anycastable. Communication between selected devices needs to satisfy the edge attributes as well. Because all nodes in a task graph corresponding to such a task are free to be chosen, we refer to them as free nodes, as opposed to those in a preassigned task, which are referred to as bound nodes. Optimization of certain performance measures is desirable during the process of instantiation of task graphs. This is described in more detail in Section 23.2.3.

Partially preassigned tasks have a subset of task graph nodes that are bound. These bound devices have to be selected in the physical network, whereas the remaining free nodes can be chosen smartly. As in anycastable tasks, the choice of free nodes is governed by certain optimization criteria.

Most existing networked distributed applications fall into the preassigned category, as there is no freedom in the choice of devices and the user decides beforehand which devices participate in the application. We believe that with the advent of pervasive computing, a whole class of anycastable tasks will emerge by exploiting the philosophy of loose coupling between services and the devices offering them.

In the context of the smart presentation application, a pocket PDA containing the presentation slides and a particular overhead display can be bound devices, but the keyboard, the mouse, and the smart storage are free devices, instances of which can be smartly chosen from the available network.

Modeling Distributed Applications for Mobile Ad Hoc Networks

23.2.1.4 A Data Flow Tuple Representation Model for Distributed Tasks. In this section, we propose a simple data flow tuple-based model for the high-level representation of the logical relationships between different components of a distributed application. The entire application is modeled by a set of tuples, each corresponding to a particular data flow in the application. In other words, each tuple corresponds to a logical unit of data processing that is needed between the distributed components of an application. Every application component is characterized by a tuple node with the same semantics as that of a node described in Section 23.2.1. Each unit of data flow is originated at a certain tuple node and is consumed at one or more terminal tuple nodes (called sinks) after being processed and relayed by a set of intermediate tuple nodes. Consider the smart presentation application described in Section 23.1. The following data flows can characterize a sample presentation:

1. Presenter's PDA (U) sends presentation data (e.g., a PowerPoint® slide) to Smart Storage (SS) device, which hosts appropriate presentation software.
2. Keystrokes are originated at a wireless keyboard (K) by the presenter.
3. Mouse commands are originated at a wireless mouse (M) by the presenter.
4. SS receives presentation data, keystrokes, and mouse clicks; processes the data; and displays it on a projected display (D) and a local screen (LS). SS also extracts and sends the ASCII part of the presentation and some corresponding notes to the user on that user's PDA screen (U).

To represent such application data flow between nodes, we employ a generalized tuple architecture. If a node of type X receives data from nodes of types A, B, and C and sends the processed data to nodes of types D and E for a certain application flow (e.g., mouse commands or keystrokes or something more application specific), we can represent this data flow schematically using the following tuple:

$$X : [A, B, C; \{processing\}; D, E]_{tag}$$

Each data flow can be uniquely identified at any node by its tag attribute. We denote by $\{processing\}$, the transformation of the incoming data units from source nodes before they are transmitted to the destination nodes.

23.2.1.4.1 Generating Task Graphs from Tuples. The user node is expected to specify the data flows in the distributed application as a set of tuples using a standardized language. A task graph representation can be easily generated from a tuple representation: each task graph node is derived directly from the corresponding tuple node as it bears one-to-one correspondence with the

Table 23.1. Data Flow Tuples for the Smart Presentation Task

ID	Node	Data-Flow Tuples
1	U	$[-;SS]_{ppt}[SS;-]_{notes}$
2	SS	$[U;LS,D]_{ppt}[K;LS,D]_{keys}[M;LS,D]_{clicks}[U;\{ppt \rightarrow notes\};U]_{notes}$
3	K	$[-;SS]_{keys}$
4	M	$[-;SS]_{clicks}$
5	LS	$[SS;-]_{ppt,keys,clicks}$
6	D	$[SS;-]_{ppt,keys,clicks}$

latter. A task graph edge is created between task graph nodes X_i and X_j if a data flow exists between the tuple nodes corresponding to X_i and X_j, respectively.

The application data flows for the smart presentation application can be depicted as tuples as shown in Table 23.1 and they translate to the task graph shown in Figure 23.1A.

23.2.1.4.2 Advantages of the Tuple Representation. Having a data-flow tuple representation for a task serves two purposes:

1. It is a natural and structured specification of the data flows in a task from which a task graph can be derived easily.
2. After the logical resources specified in the task graph are mapped to physical devices in the MANET, tuples govern the flow of actual application data at each participating device.

Examples of data-flow tuples presented in this section contain only the essential information for data exchange, namely the data source and the data destination, and whether the incoming data needs any processing before it is relayed to another device. In general, the edges in a task graph can have attributes such as upper bounds on channel error rates, bandwidth, etc., which reflect the quality-of-service (QoS) needs of a distributed application. These, and requirements such as proximity (devices like keyboard, mouse, etc., should be located as near the user as possible), can also be integrated in the task graph via the tuple architecture. A direct way of incorporating such requirements and task constraints is by specification of edge attributes in the tuple. For example, consider a scenario where a node of type X needs to communicate with another node of type D such that the separation between them is no more than three MANET hops and that the average delay over that path does not exceed ten milliseconds. These two requirements are specified as attributes of the edge e = (X, D) in the corresponding task graph: $e.separation \leq 3$ and $e.delay \leq 0.01s$. Implementation details of most of these edge attributes are beyond the scope of this research and are not considered further.

Now we give another example of an application — location based wireless polling — that can be enabled by the proposed attributed task graph framework. Imagine a full capacity Fenway Park (approximately 34,000) hosting a Red Sox game. As Nomar Garciaparra hits a home run, the stadium authorities decide to poll the people in the stadium with a question: "Was Ted Williams a better hitter than Nomar?" Polling can be achieved over the wireless ad hoc network in the stadium formed by the PDAs owned by the fans.

Suppose that one wants to conduct a poll in a scientific or controlled fashion. Instead of broadcasting the query to all PDAs in the stadium and processing all responses, one wants only a fraction of people in the audience to reply as long as people from most profiles are represented proportionally in the poll results. The advantages of doing this are twofold:

1. Less wireless bandwidth will be consumed in the polling process.
2. The poll results are likely to represent samples from different sections of the population in a fair and controlled fashion.

The extent of fairness and control in the polling process can be defined by the poller quantitatively by means of a task graph.

A sample task graph depicting a structured poll is shown in Figure 23.3. The poller wants a specified proportion of votes (specified by parameters $k, l, m, \ldots$) from spectators in particular age groups sitting in specific sections of Fenway Park as shown in the figure. The simplest way to perform the poll would be as mentioned before: flood the query throughout the MANET and collect the responses. In addition, only $k, l, m, \ldots$ responses need to be processed by the poller. Because this wastes wireless bandwidth, expanding ring search (also known as TTL scoping) can be used until the requisite amount of responses have been gathered. However, even this suffers from one problem that virtually all pollees will respond to the single poller node, which will be swamped with incoming traffic. In fact, the nodes within a few wireless hops of the poller will be busy routing the incoming packets toward it.

A task–graph-based solution can mitigate the above problems by delegating the task of polling to an intermediate layer of nodes that have enough computing resources and are less power constrained in their operation. We call these nodes poll managers. They conduct the polls based on the set of profiles that they are responsible for and act as aggregators of poll results, which are processed and then returned back to the poller. If the poll managers are spatially spread out uniformly across the network (they are selected based on their location attributes), it will result in less channel contention and hence reduce hot spots in the network. Another advantage of using intermediate poll managers is that they can localize the detection of mobility of a device in the middle of a poll transaction.

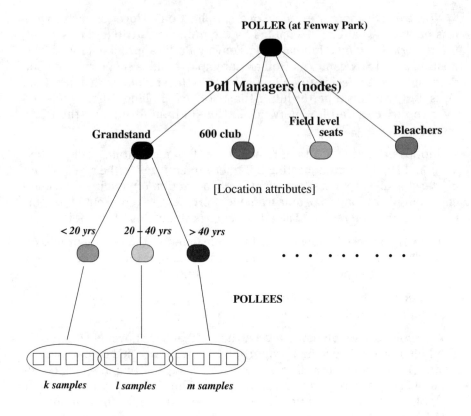

Figure 23.3. A Task Graph for Location-Based Wireless Polling

23.2.2 Embedding Task Graphs onto Networks

The first step in executing a distributed application on a set of specialized devices is to discover appropriate devices in the network and to select from the ones who responded, the devices that are suitable for the execution of the more complex application. Mathematically speaking, embedding a task graph $TG = (V_T, E_T)$ onto a MANET graph $G = (V_G, E_G)$ involves finding a pair of mappings (φ, ψ) such that $\varphi : V_T \rightarrow V_G$ and $\psi : E_T \rightarrow P_G$, where the type or *class* of $v \in V_T$ is the same as that of $\varphi(v)$ and P_G is the set of all source–destination paths in G. Figure 23.4A depicts a hypothetical task graph. Figure 23.4B and Figure 23.4C show a sample network topology with two possible embeddings of TG on it.

The complete process of device discovery, selection of a device from multiple instances of devices in the same category, and the assignment of a physical device to a logical node in the task graph is referred to as instantiation. We also refer to the collective process of instantiating all task graph nodes as task embedding or task-based anycasting [6].

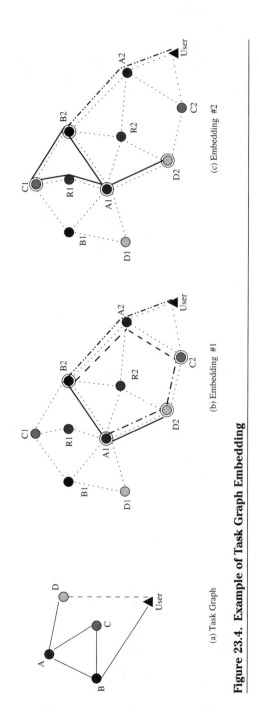

Figure 23.4. Example of Task Graph Embedding

23.2.3 Metrics for Performance Evaluation

The embedding function (φ,ψ) maps nodes and edges in $TG = (V_T, E_T)$ to devices and paths in G. Average (Maximum) Dilation of an embedding is the average (maximum) length of such paths taken over all edges in TG. Mathematically, if $\|a, b\|_G$ denotes the length of a shortest path between node a and node b in G, average and maximum dilation are respectively given by:

$$D_{avg} = \frac{1}{|E_T|} \sum_{e \in E_T} \|\psi(e)\|_G = \frac{1}{|E_T|} \sum_{(x,y) \in E_T} \|\varphi(x), \varphi(y)\|_G \qquad (23.1)$$

$$D_{max} = \max_{e \in E_T} \|\psi(e)\|_G = \max_{(x,y) \in E_T} \|\varphi(x), \varphi(y)\|_G \qquad (23.2)$$

Average dilation is a significant metric because it impacts the throughput between instantiated devices. An embedding with large dilation signifies long paths between directly communicating devices, which is undesirable in MANETs because Transmission Control Protcol (TCP) throughput drops significantly with increase in hop distance [16]. In contrast, an embedding with low dilation results in better task throughput. We consider the weighted version of the metric in Section 23.3.1 where we formally describe the optimal embedding problem.

Instantiation time is a metric that measures the time taken to embed or instantiate all nodes in TG onto G. Reinstantiation time measures the time taken to find a replacement device after an embedding is disrupted owing to node, link, or route failures.

Average Effective Throughput, (AvgEffT), is the average number of application data units (ADUs) actually received at instantiated data sinks divided by the number of ADUs that were supposed to be received at the intended targets in an ideal situation.[3] Therefore, $0 \leq AvgEffT \leq 1$. It is a useful metric for measuring the resilience of the protocols to failures.

Source-to-sink delay is the latency suffered by an ADU as it funnels itself through various intermediate relay nodes in the instantiated task graph. This metric is useful for measuring application performance during transmission of task data.

The above metrics are useful in the performance evaluation of our embedding algorithms (see Section 23.4). Additional metrics that have not been investigated in this research have been listed in [4].

23.3 Algorithms and Protocols for Task Graph Instantiation

In this section, we describe how task graphs can be mapped onto MANETs with respect to certain optimization criteria. First, we formulate an optimization problem and show how it can be efficiently solved exactly for tree

Modeling Distributed Applications for Mobile Ad Hoc Networks

task graphs. We then give a greedy heuristic than is amenable to a simple distributed implementation.

23.3.1 Optimization Problem Formulation

We formulated the constrained task graph embedding problem (CC-EMBED) as the following optimization problem:

If C be a set of principal attributes (or classes) of specialized devices; $G = (V_G, E_G)$ represents the MANET topology, with the class of each device in V_G belonging to C; $TG = (V_T, E_T)$ is a task graph such that the class of each node in V_T belongs to some $S \subseteq C$; and function $w : E_T \to \mathbb{R}^+$ defines edge weights which could signify application data-flow requirements, find mappings $\varphi : V_T \to V_G$ and $\psi : E_T \to P_G$, where the class of $v \in V_T$ is same as that of $\varphi(v)$ and P_G is the set of all paths in the network G, such that the weighted average dilation given by:

$$D_{avg}^{(wt)} = \frac{1}{\sum_{e \in E_T} w(e)} \sum_{e=(x,y) \in E_T} w(e) \| \varphi(x), \varphi(y) \|_G \qquad (23.3)$$

is minimized, where $\|a,b\|_G$ denotes the shortest path between devices a and b in G.

The computational complexity of the general version of the problem where a task graph can have multiple nodes belonging to the same class and then that of a more specialized version of the problem where all nodes in a task graph belong to distinct classes have been investigated in [4]. The above problem has been shown to be NP-complete in both these situations. However, the problem becomes tractable when the task graph is a tree with nodes belonging to distinct classes; we give an exact polynomial time algorithm for this scenario in Section 23.3.2. The solution approach in Section 23.3.2 assumes that the user node possesses the knowledge of the entire network topology as well as that about the capabilities of the devices in the network. In Section 23.3.4, we propose distributed algorithms for embedding, which albeit suboptimal, operate locally and are efficient.

23.3.2 An Optimal Polynomial-Time Embedding Algorithm for Tree Task Graphs with Distinct Labels

Although the CC-EMBED problem is NP-complete with respect to the average dilation metric for the general graphs, there is an interesting special case of a tree which lends itself to an optimal polynomial time solution.

We present below TREEEMBED, an optimal algorithm (with respect to D_{avg}) for embedding a tree task graph TG onto a host network G. The running time is polynomial in $|G|$ as well as $|TG|$. The algorithm minimizes searching in

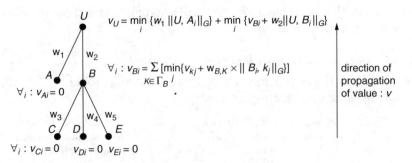

Figure 23.5. Outline of the Exact Optimal Polynomial–Time Algorithm

the solution space by exploiting the tree structure of TG and is based on the principle of optimality.[4] The algorithm requires that the node executing the algorithm have complete knowledge of the snapshot of the network topology at the given instant of time.

For each node X in TG, Algorithm 23.1 seeks to discover the best embedding for each child node Z at every instance (x) of X in G. After the best child candidates are known for all instances, the optimal cost embedding φ is selected starting at root node U.

The algorithm proceeds by the propagation of a certain value function v(.) from the leaf nodes of TG toward the root node U. The crux of the idea is that the principle of optimality holds because of the tree structure of TG: if a device instance x of node X is selected by its parent and is optimal, then the choice of instance z (of X's child Z) is optimal too. This greatly reduces the search space for an exact optimal embedding. Moreover, embedding of children nodes can proceed independently of each other because they possess distinct attributes. After carrying out this step for all children of X for each instance x, assign the sum of the calculated minimum values to v(x). Figure 23.5 illustrates the procedure for a task graph of six nodes. Γ_B = child(B) is the set of children of B in TG. k_j is an instance in G of child K of B in TG.

The running time of the TreeEmbed algorithm can be calculated as follows: assigning levels to TG nodes takes $O(|V_T|)$ time. In the worst case, the maximum level of a TG, $L_{max} = |V_T| = O(|V_T|)$; although in more balanced trees, $L_{max} = O(\log|V_T|)$. Suppose there are $|C|$ classes of devices in G with $|V|/|C|$ instances of each class on average. For every parent instance $x \in V$, each child instance $z \in V$ is considered by the embedding algorithm: the shortest path between x and z is computed (in $O(|V|^2)$ time); the minimization step in line 11 of Algorithm 23.1 is performed (in $O(|V|/|C|)$ time). Because this process is performed for all edges in TG, the time complexity of Algorithm 23.1 (line 7 through line 17) is

$$O\left(|E_T| \times \frac{|V|}{|C|}\left(\frac{|V|}{|C|} \times |V|^2 + \frac{|V|}{|C|}\right)\right) = O\left(|E_T|\frac{|V|^4}{|C|^2}\right) = O\left(|V_T|\frac{|V|^4}{|C|^2}\right).$$

Note that the `for` loops in Algorithm 23.1 (line 7 and line 8) are subsumed in this calculation and because |V| is the dominant term, the time complexity is given by the above expression itself.

If Warshall–Floyd's all-pairs shortest path algorithm is used (running time is $O(|V|^3)$) and extraction of shortest path cost is $O(1)$ assuming random access storage), then the running time of `TreeEmbed` is

$$O\left(|E_T|\frac{|V|^2}{|C|^2} + |V|^3\right) = O\left(|V|^3\right).$$

23.3.3 A Greedy Algorithm for Task Graph Embedding

If TG is a general graph (and not a tree), then the task embedding problem is much harder because the principle of optimality may not hold in that case. This is because the optimal embedding of every pair of nodes and the edge connecting them in TG cannot be done independently of other edges and nodes in TG, as can be done if TG were a tree. In the case of a tree TG, as we propagate the values from the leaves to the root, the optimal embeddings of each subtree are retained and used later while embedding a node closer to the root. This is not possible for any general task graph with greater connectivity than a tree.

Algorithm 23.1 suffers from large time complexity even though it is polynomial, the main reason for this being that all devices in the network G are considered as candidates for embedding and the dynamic programming algorithm chooses the best subset among them systematically. Moreover, the algorithm may often fail to run in polynomial time if a few nodes of the same class occur more than once in TG. Due to these reasons, we developed a simple greedy algorithm `GreedyEmbed`, which is suboptimal (even for trees) but has lower time complexity and works for the case where all node types in TG are not distinct. We briefly describe it below.

The greedy algorithm begins the search for candidate devices from the user node U itself and conducts it in a breadth-first manner. At every step of the breadth first search (BFS) process, an unvisited task graph node is instantiated greedily by the nearest candidate device in G, which matches the requested attributes. Ties are broken arbitrarily and there is no lookahead. Because only nearby devices in G (from the current location) are considered as candidates, searching for the nearest suitable instance of a

Algorithm 23.1. TREEEMBED(TG, G, w, c_1, c_2)

1: Given: Tree Task Graph, $TG = (V_T, E_T); w : E_T \rightarrow \mathbb{R}^+; c_1 : V_T \rightarrow C$,

   Host Network Graph $G = (V, E); c_2 : V \rightarrow C$;

   /* C: attribute universe; c1, c2: attribute fns.; c1 is injective; */

2: $\forall X \in V_T : X$ is a leaf in $TG, L[X] \leftarrow 0$; /* assign levels to each leaf node */

3: $\forall X \in V_T : X$ is not a leaf in $TG, L[X] \leftarrow 1 + \max_{Z \in child(X)} L[Z]$; /* and the rest */

4: for all $(X : L[X] == 0)$ do

5:     $\forall x : (c_2(x) == c_1(X)), v(x) \leftarrow 0$; /* assign value to matching instances */

6: end for

7: for $(\ell \leftarrow 1; \ell \leq L_{max}; \ell \leftarrow \ell + 1)$ do

8:     for all $(X \in V_T : L[X] == \ell)$ do

9:         for all $(x \in V : (c_2(x) == c_1(X))$ do

10:            for all $(Z : Z \in child(X))$ do

11:                $z^* \leftarrow \arg\min_{z \in V \wedge c_2(z) == c_1(Z)} \{v(z) + w_{(X,Z)} \| x, z \|_G \} ; \ldots [v(z) + w_{(x,z)} \| x, z \|_G$

12:                $\varphi_x(Z) \leftarrow z^*$; /* best instance of child node Z for x */

```
13:         v(x) ← v(x) + v(z*) + w_{(x,z)} ||x,z*||_G ;  /* update value of x */
14:       end for
15:     end for
16:   end for
17: end for
18: for (ℓ ← L_{max}; ℓ ≥ 0; ℓ ← ℓ-1) do
19:   S ← {X | X ∈ V_T ∧ L[X] == ℓ};
20:   while (X ∈ S ∧ child(X) ≠ φ) do
21:     x ← φ*(X);  /* note that φ*(U) = U */
22:     for all (Z : Z ∈ child(X)) do
23:       φ*(Z) ← φ_x(Z); ψ*(X,Z) ← (φ*(X),φ*(Z))_G ;  /* optimal embedding */
24:     end for
25:   end while
26: end for
```

task graph node may not require a complete traversal of G. Hence the algorithm trades off optimality for time efficiency.

`GreedyEmbed` also possesses a few clear advantages over `TreeEmbed` in its functionality and implementation. Unlike the latter, `GreedyEmbed` can handle the case in which multiple nodes in TG possess the same attributes. Moreover, distributed implementations of `GreedyEmbed` are facilitated easily due to the nature of BFS. We describe a distributed approach based on these principles in the next section.

23.3.4 A Distributed Algorithm for Task Graph Instantiation

In this section, we present a distributed approach for solving the task graph embedding problem in a MANET with an objective of minimizing D_{avg}. We assume here that each heterogeneous device can provide a single type of service and that all nodes in the network are simple. The additional nuances of the homogeneous case have been elaborated upon in [4]. We assume the presence of a MANET routing protocol — Dynamic Source Routing (DSR) — and a reliable transport protocol — TCP — for control and application data packet transmission.

All devices in the network execute copies of the same algorithm except the user node U, which executes a different algorithm because it acts as a state synchronizer or coordinator in the initial phases of the embedding process. In our opinion, the user devices are best suited for acting as coordinators because they usually originate the application data flows, and even under mobility, always remain near the user.

The embedding process begins at U with a distributed search that proceeds through the MANET G hand-in-hand with a BFS through TG. Figure 23.10 depicts a task graph with its BFS and non-BFS edges. We call the spanning tree on TG induced by BFS and rooted at U, a BFS tree ($BFST_{TG}$) of TG. We propose a greedy solution much like the `GreedyEmbed` algorithm described in Section 23.3.3 to keep the dilation of the embedding low: the algorithm begins from U by progressively mapping the nodes of $BFST_{TG}$ to nearest devices and the edges to shortest paths in G. Instantiation of any pair of nodes $x, y \in V_T$ cannot affect each other if x is not a parent of y in $BFST_{TG}$ or vice versa. Hence, the search can proceed in a distributed manner along the branches of $BFST_{TG}$.

A sample path of the instantiation protocol helps illustrate the salient steps of the algorithm. These have been shown in Figure 23.6 as a message exchange diagram and are also summarized below. Details of the protocol including finite state machine descriptions can be found in [4, 6].

1. U broadcasts search queries for each neighbor category in TG[5] (A and B).
2. Available instances of each queried node reply to U. Candidate devices that reply first (A_i, B_j) become the chosen instances at U.

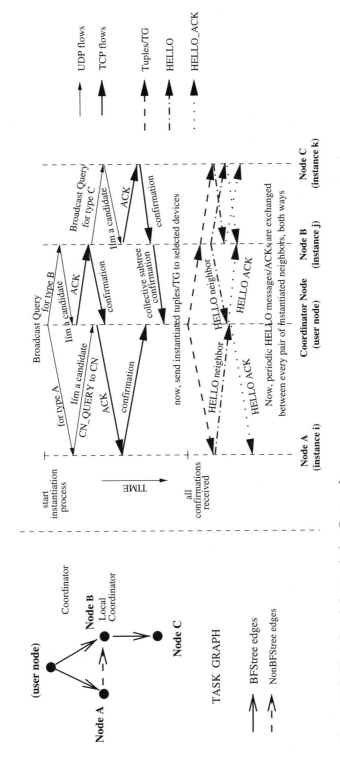

Figure 23.6. Task Graph Instantiation Protocol

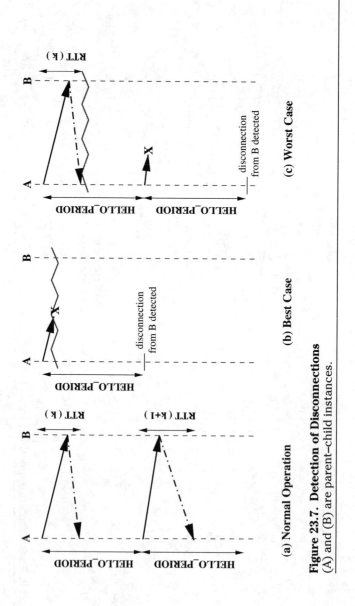

Figure 23.7. Detection of Disconnections
(A) and (B) are parent–child instances.

Modeling Distributed Applications for Mobile Ad Hoc Networks

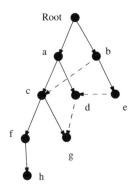

Node	Parents	Children	Grandchildren	Siblings
Root	—	a,b	c,d,e	—
a	Root	c,d	f,g	b
b	Root	{c},e	d,f,g	a
c	a,{b}	f,g	h	d
d	a,{e}	{g}	—	c
f	c	h	—	—
g	c,{d}	—	—	—
h	f	—	—	—

Figure 23.8. Logical Neighbor Table Information at Instantiated Nodes
Letters enclosed in { } represent devices that are non-BFS parents and non-BFS children.

3. U sends an ACK to these selected devices which send back confirmations.
4. If there are any uninstantiated nodes rooted at any instance in TG (such as C below B_j), then it broadcasts search query packets for all those node categories and the instantiation proceeds further.[6]
5. When confirmations from all nodes reach U, the data transmission can begin.[7]

The task graph itself is sent as control data during the instantiation process. After the selection of a device, control packets and application data are transmitted using TCP because packet losses due to route errors are common in MANETs.

23.3.4.1 Handling Device Mobility. When mobility causes network partitions or disconnections, the instantiated devices may no longer be able to communicate if the partition breaks all paths between them. In such situations, new instances need to be selected. The necessary first step in this direction is the detection of disconnections. We propose a lightweight, soft-state exchange protocol for detecting disconnections in an instantiated task graph. The protocol requires each instantiated device to send periodic HELLO messages (with period T) to its logical neighbor instances in TG, which reply with a HELLO-ACK. This has been demonstrated in Figure 23.9.

Specifically, each instantiated device keeps track of its BFS parent and BFS children and some additional information. Figure 23.8 shows the information that each instantiated device stores for detecting disconnections and performing recovery. We refer to this as the two-hop logical neighborhood information [4]. The instance of node C (denoted by c) keeps track of the instance of node A (its BFS parent) as well as of the instances of node

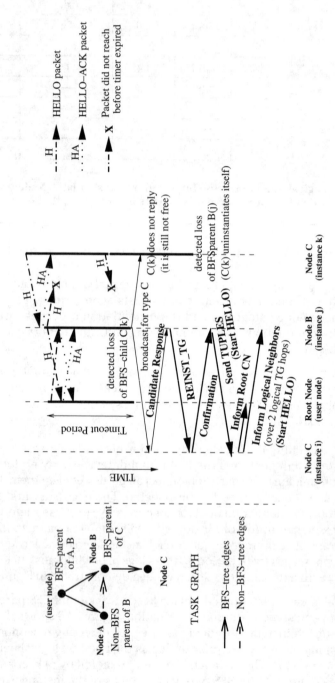

Figure 23.9. Handling Task Disruptions by Reinstantiation of Devices

F and node G (BFS children). If a BFS parent device stops hearing from one of its BFS children,[8] it uninstantiates its child and starts looking for a replacement of the same type. The child meanwhile would stop hearing HELLO-ACKs from the parent (assuming bidirectional links) and will uninstantiate itself. This has been illustrated in Figure 23.9. On average, if the HELLO timer at every instantiated device is set to time period T, disconnections will be detected after time $(3/2)T$.

Although one-hop logical neighborhood information is sufficient for detecting disconnections in the instantiated task graph, two-hop logical neighborhood information as illustrated in Figure 23.8 is necessary and sufficient to recover from single persistent disconnections without any extra unnecessary reinstantiations of downstream nodes in TG [4]. The process of reinstantiation involves:

1. Discovery of a suitable replacement
2. Transfer of instantiated logical neighborhood information from parents and children
3. Resumption of application data transfer

We do not discuss these schemes in detail in this chapter and direct the reader to references [4, 6].

23.3.4.1 Impact of Disconnections on Application Layer. The application layer of every participating device keeps up-to-date (in–out) tuple information for parent and children devices. If disconnection of some participating devices disrupts a running task, then it is the responsibility of the BFS parent node to transfer the application state to the newly instantiated replacement device and then resume the application data flow. Meanwhile, data packets reaching old node instances are dropped by those devices. The average effective throughput (AvgEffT) metric tries to capture the effectiveness of our disruption handling algorithm by measuring the fraction of the data that actually reached the current data sinks from the source. An application layer buffer management scheme at the BFS parent node instance can increase the reliability of task completion. We plan to investigate these issues in future.

Mobility of devices may also result in lengthening or shortening of routes between device instances, and ideally, if there is no disconnection/partition, the application should proceed without disruption. But such ideal conditions may not hold in reality where route failures can trigger route discovery, which along with TCP retransmissions after time-outs, may sometimes take several seconds to complete. Hence, this can result in HELLO-ACKs not coming back in T seconds, which results in the conclusion that a disconnection has happened, even when the nodes are reachable from one another.

Researchers have proposed solutions to the above problem based on explicit notification of route errors to TCP [11]. However, in this work, we do not attempt to alter TCP or DSR (including their default timer settings); we simply develop our protocol on top of these protocols. Hence, if a device does not receive a HELLO-ACK from its neighbor in T seconds, we deem the neighbor to be disconnected. A reasonable value of T is one which is not low enough to cause significant control overhead,[9] and not high enough such that disconnections are not detected fast enough. For our simulations, we chose T = 7 seconds (> 6s, the default TCP retransmission timer).

23.4 Performance Evaluation

We simulated the greedy instantiation/reinstantiation algorithms proposed in Section 23.3.4 using the network simulator ns-2 [2]. 100 mobile devices with specialized roles (represented by their principal attributes) were simulated in a 1500 × 600 area. Node motion was governed by the Random Waypoint Mobility Model; the velocity was randomly chosen from $[0, v_{max}]$ where $v_{max} = \{1, 5, 10, 15, 20\}$ m/s. We assume that the devices are constantly moving between waypoints. The simulated transmission range for each node was 250 m. We show simulation results for task graphs in Figure 23.10. We refer to them as Tree TG, Non-Tree TG-1, and Non-Tree TG-2, respectively. The principal attribute of each device belonged to 1 of 12 different classes. The attributes were uniformly distributed across the MANET.

The total simulation time was 400 s — the instantiation process began at 200 s, and at 600 s, the user/root node started sending data to the data sinks. The data flow consisted of a constant bit rate (CBR) source, with a burst of S bytes of data every T seconds. We report results for (S, T) = (2500, 1). Devices that are not part of the instantiated task graph do not forward packets and such packets are not buffered; in other words, if a device that was part of an instantiated task graph becomes disconnected while there is a packet in transit, the packet is lost.

23.4.1 Dilation

First we analyze the constant mobility scenarios for different simulation parameters. We first evaluate the quality of embedding using the average dilation metric. For every mobility scenario, dilation is measured initially after completion of instantiation and subsequently after every reinstantiation event. These values are then averaged over the simulation time period to yield one number. We observe from Figure 23.11 that average dilation for the embedding scheme does not vary greatly with speed; in fact, d_{avg} lies between 1.25 and 2 for all three task graphs at all different values of Max-Speed. This means that the average number of physical hops between two

Modeling Distributed Applications for Mobile Ad Hoc Networks

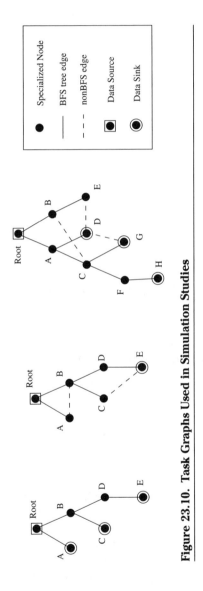

Figure 23.10. Task Graphs Used in Simulation Studies

529

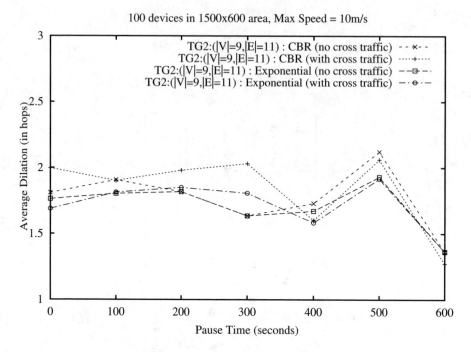

Figure 23.11. Average Dilation versus Maximum Speed

instantiated nodes in TG is low and remains approximately constant under mobility. This is because of the approximately uniform spatial distribution of device categories and the reasonable abundance of devices of each category in the network (5 to 13 of each type).

However, we do observe that d_{avg} increases when the maximum speed is increased above 1 m/s. The principal reason for this is the following: at 1 m/s speeds, reinstantiations are rare and the d_{avg} does not deviate too much from its value after initial instantiation. However, at greater speeds, reinstantiation events occur more frequently because of logical neighbor instances either having moved far away from each other or having been disconnected by a network partition. Either of these events disrupts the usual smooth exchange of HELLO messages resulting in reinstantiations. Owing to the uniform distribution of device categories in space, the reinstantiation process will find another device with similar attributes within its vicinity. Although that keeps the contribution of the new path length toward d_{avg} low, the hop distances between existing instances along other task graph edges are likely to have increased over time (although not high enough to cause reinstantiations along those edges). This causes d_{avg} to increase at higher speeds on the whole.

Modeling Distributed Applications for Mobile Ad Hoc Networks

Table 23.2. Task Embedding Time

TG/Scenario	Minimum (s)	Maximum (s)	Median (s)
Tree (1 m/s)	0.795719	6.561610	1.435320
TG-1 (1 m/s)	0.810867	6.819640	1.399530
TG-2 (1 m/s)	2.170060	7.957830	6.674960
Tree (5 m/s)	0.670853	6.111210	1.728970
TG-1 (5 m/s)	0.536686	7.708620	6.278840
TG-2 (5 m/s)	1.742180	9.537000	7.827160
Tree (10 m/s)	0.643709	1.438240	1.216280
TG-1 (10 m/s)	0.842213	6.694860	1.530080
TG-2 (10 m/s)	3.337190	9.168950	7.275040
Tree (15 m/s)	0.749414	4.039460	1.062140
TG-1 (15 m/s)	0.446600	6.511620	0.909011
TG-2 (15 m/s)	1.520370	4.090240	3.241610
Tree (20 m/s)	0.651414	2.062220	1.088190
TG-1 (20 m/s)	0.717359	4.022630	1.484370
TG-2 (20 m/s)	1.361380	7.674460	5.262870

Another observation from Figure 23.11 is that at lower speeds, d_{avg} is lower for TG-1 (a tree) than TG-2. This is obvious because, our heuristic algorithm attempts to minimize the hop count only along the BFS tree edges of a task graph both during instantiation as well as reinstantiation; because TG-2 has extra edges, the minimization does not occur along those edges, thus yielding a higher dilation, in general. The above reasoning does not hold at high rates of mobility as all instantiated paths break more often and device category distribution is spatiotemporally more uniform in the neighborhood of a device. Hence, non-BFS tree edges are likely to be mapped onto paths with similar lengths as BFS tree edges quite often and that causes d_{avg} to be similar for both TG-1 and TG-2.

23.4.2 Embedding Time

Table 23.2 compares the times taken for embedding each task graph on the network. We depict the minimum, maximum, and median times for each task graph for three different maximum speeds. We show the median instantiation time instead of the average instantiation time because the time samples are skewed. Generally, the times for TG-2 exceed those for TG-1 and Tree, because the former is a larger task graph and it needs exchange of packets between a larger number of devices during instantiation.

Some samples are much greater than the rest owing to the role of TCP (over DSR routing protocol) in the instantiation process. After the candidate's response reaches a coordinator node, it sends ACKs encapsulated in TCP packets, because they can be lost if sent using an unreliable transport protocol. TCP is also used in all subsequent communication (except broadcast and candidate response packets).

Now, if for some reason a route error occurs while a TCP transmission has not completed, TCP attempts redelivery only after waiting for a period of time even if a new route is rediscovered immediately by DSR. This period can be as large as 6 seconds (default retransmission timer of TCP) if no prior communication has happened between the two communicating devices. If a route error occurs shortly after two devices have communicated using TCP but before another TCP transmission is completed, the retransmission timer is set based on the round trip time estimate between those two devices and hence it can be lower than 6 seconds. Hence, we see instantiation time samples greater than 6 seconds on several occasions. If mechanisms such as explicit feedback [11] are added to TCP, then these times can be reduced significantly. Also, no monotonic pattern is observed as a result of the increasing mobility of devices. This can be attributed to the uniform spatial distribution of device classes in all random mobility patterns as well as the large variability in TCP timers during the multiple steps of the instantiation process.

23.4.3 Effective Throughput

After the completion of the instantiation process, we begin data transmission from the user node (source) to the various sinks shown in Figure 23.10 according to particular tuple specifications. In Tree TG, instances of A, C, and E receive one flow each. In TG-1, the instance of E receives four flows through instances of various relay nodes. In TG-2, instances of D and H receive one flow each and the instance of G receives four flows. We plot normalized AvgEffT for all three task graphs in Figure 23.12. We generate task data traffic using two different patterns: periodic CBR bursts and bursts with exponentially distributed sizes after exponentially distributed interarrival times (resulting in Poisson distributed bursts). The mean burst sizes and interarrival times are kept constant for both cases. A maximum aggregate throughput of 300 kbps can be reached for the TG-2 scenario assuming simultaneous transmission at all instantiated devices in accordance with the underlying tuple architecture.

In Figure 23.12 we can see that at low mobility, AvgEffT is almost perfect (close to 1.0). We can also observe that in general, AvgEffT drops with increase in the maximum speed of devices for most situations. This is to be expected because higher speeds generally result in more reinstantiations and that results in more ADUs not reaching their intended destinations.

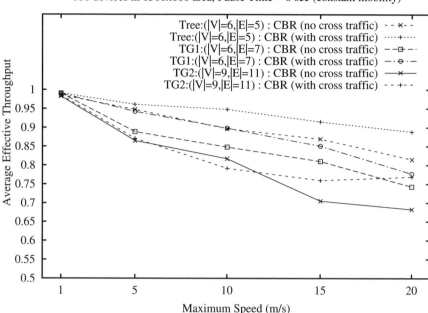

Figure 23.12. **Average Effective Throughput versus Variation of Maximum Speed**

However, AvgEffT rarely drops below 70 percent in the simulated scenarios even under heavy mobility. This demonstrates that our protocols adapt fairly well to mobility and are able to recover from disruptions in task data flow. We can make some more observations from the two figures:

- Exponential traffic pattern occasionally results in a lower throughput than the CBR traffic pattern in scenarios involving nontree task graphs.
- TG-1 usually yields lower throughput than Tree TG.

Exponentially distributed data generation times can occasionally result in large periods without much network activity and this causes the on-demand routing protocols to lose routes to destinations. More route errors cause more frequent TCP back-offs and sometimes result in reinstantiation even if the devices are graph-theoretically reachable from one another. Loss of throughput is greater in the case of nontree TGs than Tree TG, because recovery from the loss of a non-BFS child usually takes more time than a BFS child. On the contrary, in the CBR case, periodic generation of packets keeps routes fresh and hence the task graph suffers less reinstantiation.

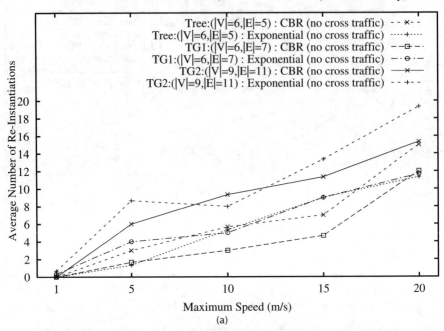

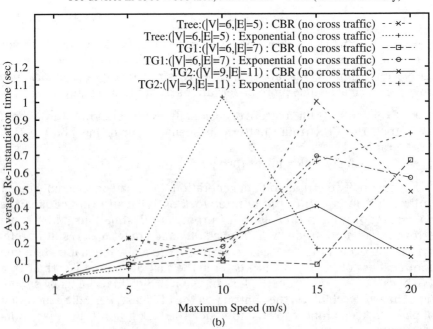

Figure 23.13. (a) Number of Reinstantiations, (b) Average Reinstantiation Time

23.4.4 Number and Time of Reinstantiation

Figure 23.13A shows the average number of reinstantiations underwent during the entire simulation time (400 s). The rate of change in network topology increases with mobility causing more network partitions or route errors. These events in turn prevent HELLO packets from arriving in time and thus triggering more reinstantiations. Because packets caught in transit during the reinstantiation process are dropped (as mentioned earlier, we do not consider application layer buffering in this work), AvgEffT is directly affected by reinstantiations.

Although Tree TG is a subgraph of TG-1, for the CBR data case, TG-1 suffers less reinstantiations because data flow along the non-BFS edges of TG-1 results in the presence of more valid alternate routes (or parts of them). Hence, when a route error happens along a BFS edge (the primary cause of reinstantiations) of TG-1, often these alternate routes come to the rescue before the HELLO timer expires, thus reducing the rate of reinstantiations. TG-2 generally suffers more reinstantiations because it is a larger graph with more depth.

In spite of Tree TG having more reinstantiations than TG-1, it experiences better AvgEffT than TG-1. This is because the data tuples of TG-1 (as well as TG-2) involve flows along non-BFS edges in the graph. Also, the set of reinstantiation events is only a subset of the set of all disruptions. When a non-BFS parent loses a child instance momentarily due to partitions or HELLO time-outs, a reinstantiation will not be triggered because that is the responsibility of the BFS parent of the child instance. Hence, the throughput is affected until a new instance is found by a BFS parent and the non-BFS parent is informed of this event by a one-logical-hop broadcast or a route to the old instance is restored. Also, Tree TG has sinks at all depths, unlike TG-1. Hence, the latter's effective throughput suffers more from a reinstantiation of an intermediate relay node. Exponential traffic generally affects reinstantiations more than CBR traffic especially for the nontree graphs as explained before. The result of that is slightly lower throughput in the respective cases.

Figure 23.13B shows the variation of times taken to reinstantiate a task graph: the times taken to discover a new replacement for a disconnected device which can participate in the task. This time is measured from the time when the rediscovery broadcast is sent out until the time instant when a confirmation is received from the new candidate (this involves two round-trip handshaking steps including the broadcast). Our reinstantiation protocol is able to find a new device nearby within one second. In fact, in most cases, these times are only a few hundred milliseconds. Local network effects are dominant factors in the determination of this metric at higher speeds; hence, there is not much correlation between the values in such cases.

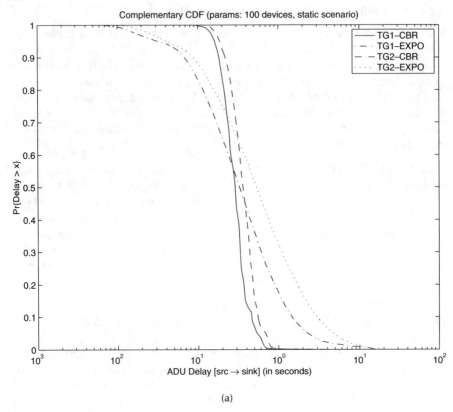

(a)

Figure 23.14. Cumulative Probability Distribution for ADU Delay
(a) Static, (b) Constantly mobile.

23.4.5 Cumulative ADU Delay Distributions

We now examine the nature of the delay distributions that occur as a result of sending task data using CBR and exponential traffic patterns. Figure 23.14 shows the empirical cumulative probability distributions (CDF) of ADU delay samples. A logarithmic scale is used for the delay samples to differentiate between delays at lower and higher ends more effectively. In Figure 23.14A, delays for the static case are plotted. We observe that CBR delay values span a much smaller range than their exponential counterparts. The shape of the task graph does not seem to affect that of the CDF curves. That is primarily because the distribution of sinks in both TG-1 and TG-2 have a common aspect that is a dominant factor in the determination of ADU delays: two sinks each in TG-1 are three and four logical hops away from the source, respectively, but in TG-2, four sinks are three logical hops away and two sinks are four logical hops away from the data source.

Modeling Distributed Applications for Mobile Ad Hoc Networks

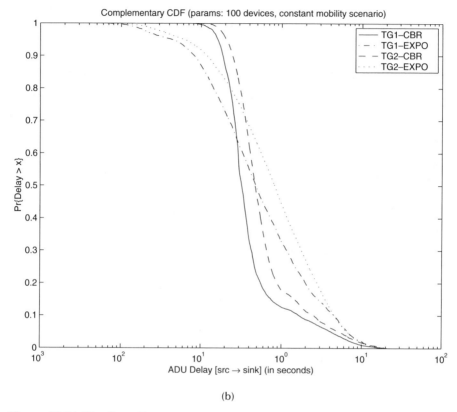

Figure 23.14 (Continued)

CDF curves of delays in the constant mobility scenario have been plotted in Figure 23.14B. We can easily see that although the shapes of the curves are similar at lower values of delay, they become much flatter and somewhat heavy tailed at larger values for both TGs and traffic patterns. These samples correspond to ADUs that had to experience delays due to route errors and expiry of TCP timers. In this work, we do not attempt to investigate the exact statistical nature of the distribution and leave that as a topic of future research.

23.5 Related Work

Service discovery in networks has been a popular topic of research in the industry as exemplified by Service Location Protocol (SLP) [14] and Sun's Jini™ network technology [18]. In both these schemes, a service providing computer registers itself with its attributes at a centralized directory server that the clients can lookup on demand. MOCA is a variation of Jini without any centralized registry [7]. It is specifically designed for mobile computing devices: every device has a service registry component that only the applications running on the local and surrounding devices can benefit from. Our approach is different from these as it operates at a logical layer above service discovery and it can coexist with any of these schemes. Also, it does not depend upon any centralized directory service.

International Naming System (INS) proposes to capture user-intent for discovering appropriate devices suitable to them. The user intent is abstracted into collections of attribute–value pairs that describe the needs of the user. The specific devices that will perform the desired service will be selected by special entities called Intentional Name Resolvers. INS has a feature called Intentional Anycast and late binding, which is somewhat similar to what we call instantiation of task graph nodes. However, INS does not to attempt to systematically use the logical structure of a distributed task for resilient application execution.

Hodes et al. [15] have investigated means of composing services for heterogeneous mobile clients. Their work primarily focuses on controlling office equipment from mobile devices and design of client–device interfaces. They too have not addressed the issues involved in composing complex services from simple devices with specific interaction patterns between them. In general, none of the aforementioned approaches consider scenarios in which multiple specialized devices need to offer their services in a cooperative manner for the provision of a more complex service, a case that we believe will be increasingly common in a ubiquitously networked world.

IBM's Platform-Independent Model for Applications (PIMA) [3] has a vision somewhat similar to ours. Although they argue briefly for the design of applications in terms of subtasks instead of specific devices, they have not mentioned any approach for realizing this vision so far. Our task graph concept on the other hand is a systematic and concrete approach that can help realize this vision.

The concept of a task graph was originally proposed in the parallel computing and scheduling literature for representing tasks that can be split temporally into subtasks and then allocated to different homogeneous processors connected by a fixed high-performance interconnect for reducing the total completion time [10, 17]. Our notion of a task graph is different from this classical one. We are not necessarily concerned with tasks that are distributable among multiple homogeneous processors for speed-up. Rather, most tasks that we are concerned with in this work involve several specialized heterogeneous devices that communicate with each other and are possibly mobile, and there is no notion of minimizing the total completion time. However, if we are interested in solving a large scale distributed computing task on a network of homogeneous mobile devices, then our notion of a task graph will be similar to the classical one. Therefore, our task graph formulation is more general than the one used in the parallel computing context.

23.6 Conclusion

In this chapter, we presented a framework for embedding and executing a distributed application on a network of specialized, potentially mobile devices. We developed a task graph abstraction for applications by taking into account the dependencies induced by the data flows existing between the components of an application. We described the task embedding problem and presented an optimal polynomial time algorithm with respect to an average hop-count measure called dilation, for the special case where the task graph is a tree. We also described how it can be heuristically extended for general graphs. Owing to the unreasonable requirements and time complexity of the aforementioned algorithm, we presented a more practical distributed heuristic algorithm (and protocol) for embedding a given task graph onto a MANET. We also presented a scalable, local disconnection detection and repair mechanism for recovering from task disruptions caused by node mobility and failures.

We showed by simulations that our protocols are able to instantiate and reinstantiate task graphs satisfactorily in constantly mobile scenarios, although the use of a better reliable transport protocol than TCP can yield better performance. As a part of future work, we plan to investigate mechanisms of developing user level applications on top of the task graph layer described in this chapter. These applications will be completely oblivious of the task graph node–physical address mappings during their execution and this can be a major benefit in failure prone mobile networked environments.

Acknowledgment

This work was done when the first author was a Ph.D. candidate at Boston University and was supported by the National Science Foundation (NSF) under Grant no. ANI-0073843. Any opinions, findings, and conclusions or recommendations expressed in this material are those of the author(s) and do not necessarily reflect the views of the National Science Foundation.

Notes

1. Printer, photocopier, digital picture frame, etc., are examples of classes.
2. We assume that the printer Application Programming Interface does not work at the granularity of printing a dot.
3. If a relaying node in the path from source to sink becomes uninstantiated, effective throughput will be affected because some data flows will be discarded and will not reach the data sinks.
4. The Principle of Optimality holds for problems whose structure is such that their optimal solutions contain the same for the smaller subproblems [8].

5. The broadcast is controlled by sending the query packet to all one-hop neighbors, which examine its contents and decide whether to rebroadcast it. A time-to-live (TTL) field in the packet also prevents it from causing a storm.
6. B_j here acts as the local coordinator responsible for instantiation of nodes rooted below it.
7. In an ideal situation, all data originating at the source should reach the instances of the sink nodes in TG (A_i and C_k in the example in Figure 23.6) after having been massaged and relayed by the intermediate devices (B_j).
8. The parent concludes this if it does not get a HELLO-ACK from that child before the expiry of its HELLO timer.
9. Although exchanging HELLO messages with higher frequency could result in the DSR caches having fresher routes.

References

1. W. Adjie-Winoto, E. Schwartz, H. Balakrishnan, and J. Lilley. The Design and Implementation of an Intentional Naming System. In *Proceedings of the 17th ACM Symposium on Operating Systems Principles (SOSP)*, Kiawah Island, SC, December 1999.
2. S. Bajaj, L. Breslau, D. Estrin, K. Fall, S. Floyd, P. Haldar, M. Handley, A. Helmy, J. Heidemann, P. Huang, S. Kumar, S. McCanne, R. Rejaie, P. Sharma, K. Varadhan, Y. Xu, H. Yu, and D. Zappala. Improving Simulation for Network Research. Technical Report 99-702, University of Southern California, Los Angeles, March 1999. http://www.isi.edu/nsnam/ns.
3. G. Banavar, J. Beck, E. Gluzberg, J. Munson, J. Sussman, and D. Zukowski. Challenges: an Application Model for Pervasive Computing. In *Proceedings of the 6th ACM MobiCom Conference*, Boston, August 2000.
4. P. Basu. A Task Based Approach for Modeling Distributed Applications on Mobile Ad Hoc Networks. Ph.D. thesis, Boston University, Boston, May 2003. http://hulk.bu.edu/projects/adhoc/Basu-PhDThesis2003.pdf.
5. P. Basu, W. Ke, and T.D.C. Little. Scalable Service Composition in Mobile Ad hoc Networks Using Hierarchical Task Graphs. In *Proceedings of the 1st Annual Mediterranean Ad Hoc Networking Workshop (Med-Hoc-Net)*, Sardegna, Italy, September 2002.
6. P. Basu, W. Ke, and T.D.C. Little. Dynamic Task Based Anycasting in Mobile Ad Hoc Networks. *ACM/Kluwer Journal for Mobile Networks and Applications (MONET)*, vol. 8, no. 5, October 2003.
7. J. Beck, A. Gefflaut, and N. Islam. MOCA: a Service Framework for Mobile Computing Devices. In *Proceedings of the International Workshop on Data Engineering for Wireless and Mobile Access (MobiDE)*, Seattle, August 1999.
8. D.P. Bertsekas. *Dynamic Programming and Optimal Control*, vol. 1, Belmont, MA: Athena Scientific, 1995.
9. Bluetooth Consortium. http://www.bluetooth.com.
10. S. Bokhari. On the Mapping Problem. *IEEE Transactions on Computers*, vol. 30, no. 3, 1981.
11. K. Chandran, S. Raghunathan, S. Venkatesan, and R. Prakash. A Feedback-Based Scheme for Improving TCP Performance in Ad Hoc Wireless Networks. *IEEE Personal Communications Magazine*, February 2001.
12. B.P. Crow, I. Widjaja, J.G. Kim, and P.T. Sakai. IEEE 802.11 Wireless Local Area Networks. *IEEE Communications Magazine*, vol. 35, no. 9, pp. 116–126, September 1997.
13. M. Esler, J. Hightower, T. Anderson, and G. Borriello. Next Century Challenges: Data-Centric Networking for Invisible Computing The Portolano Project at the University of Washington. In *Proceedings of the 5th ACM MobiCom Conference*, Seattle, August 1999.

14. E. Guttman. Service Location Protocol: Automatic Discovery of IP Network Services. *IEEE Internet Computing,* July 1999.
15. T. Hodes, R. Katz, E. Servan-Screiber, and L. Rowe. Composable Ad-Hoc Mobile Services for Universal Interaction. In *Proceedings of the 3rd ACM MobiCom Conference,* Budapest, Hungary, September 1997.
16. G. Holland and N. Vaidya. Analysis of TCP Performance over Mobile Ad Hoc Networks. In *Proceedings of the 5th ACM MobiCom Conference,* pp. 219–230, Seattle, August 1999.
17. R. Monien and H. Sudborough. Embedding One Interconnection Network in Another. *Computing Suppl.,* vol. 7, pp. 257–282, 1990.
18. S. Oaks and H. Wong. *JINI in a Nutshell.* 1st ed., Sebastopol, CA: O'Reilly and Associates, Inc., March 2000.
19. ZigBee Alliance. http://www.zigbee.org.

Chapter 24
Medium Access Control Mechanisms in Mobile Ad Hoc Networks

Chansu Yu, Ben Lee, Sridhar Kalubandi, and Myungchul Kim

Abstract

Media Access Control (MAC) protocol plays an important role in providing fair and efficient allocation of limited bandwidth in wireless LANs. The basic medium access model in the IEEE® 802.11 standard, known as *distributed coordination function* (DCF), is widely used in wireless LANs. Research efforts in wireless multi-hop networks, where wireless nodes need to forward packets on other's behalf, try to measure up to or improve upon this standard. This chapter presents an in-depth discussion on the problems with IEEE 802.11, especially those relevant in a multi-hop network, and discusses various techniques that have been proposed to enhance the channel utilization of multi-hop wireless networks.

24.1 Introduction

Mobile devices coupled with wireless network interfaces will become an essential part of future computing environment consisting of *infrastructured* and *infrastructureless* wireless LAN networks [1]. Wireless LAN suffers from collisions and interference due to the broadcast nature of radio communication and thus requires special MAC protocols. These protocols employ control packets to avoid such collisions, but the control packets themselves and packet retransmissions due to collisions reduce the available channel bandwidth for successful packet transmissions. At one extreme, aggressive collision control schemes can eliminate the retransmission overhead, but at the cost of large control overhead. At the other extreme, the lack of control over collisions offers zero control overhead, but it may need to expend a large amount of channel bandwidth for retransmissions.

DCF is the basic medium access method in IEEE 802.11 [4], which is the most popular wireless LAN standard, and it makes prudent trade-offs between the two overheads. DCF supports best effort delivery of packets at the link layer and is best described as the *Carrier Sense Multiple Access with Collision Avoidance* (CSMA/CA) protocol. Although DCF works reasonably well in infrastructured wireless LAN environment, this is not necessarily true in a *mobile ad hoc network* (MANET) environment. A MANET is an infrastructureless multi-hop network that consists of autonomous, self-organizing, and self-operating nodes, each of which communicates directly with the nodes within its wireless range or indirectly with other nodes via a dynamically computed, multi-hop route.

Although the multi-hopping technique can potentially maximize the channel use by allowing multiple simultaneous transmissions occurring separated in space [2, 3], all participating nodes must undertake the role of routers engaging in some routing protocol required for deciding and maintaining the routes. In comparison to one-hop wireless networks with base stations, multi-hop networks suffer from more collisions because nodes are not partitioned into a number of disjoint cells, but are overlapped successively in space. Therefore, congestion at one particular area in a MANET may affect the neighboring areas and can propagate to the rest of the network. In addition, multi-hopping effectively increases the total data traffic over the network by a factor of the number of hops. Moreover, it potentially causes self-generating collisions in addition to those from other data streams because each node acts as a router and uses a single network interface to receive a packet as well as to forward the previous packet of the same data stream to the next hop node.

This chapter overviews key elements of DCF, discusses problems of DCF when used in a multi-hop MANET environment, and surveys various mechanisms that balance the above mentioned two overheads to enhance the channel use in the presence of increased chance of collisions. These mechanisms can be broadly classified as *temporal* and *spatial* approaches depending on their focus of optimization on the channel bandwidth. The temporal approaches attempt to better use the channel along the time dimension by optimizing the parameters or improving the *backoff algorithm* of the DCF protocol [5–8]. On the other hand, the spatial approaches try to find more chances of spatial reuse without significantly increasing the chance of collisions. These mechanisms include *busy tone channel* [9], *transmission power control* [10–12], and *directional antenna* [13–17].

Medium Access Control Mechanisms in Mobile Ad Hoc Networks

The organization of the chapter is as follows:

- Section 24.2 provides a general description of MAC algorithms and discusses DCF of IEEE 802.11.
- Section 24.3 and Section 24.4 discuss the temporal and spatial MAC techniques, respectively, to enhance the channel use based on DCF.
- Section 24.5 presents concluding remarks.

24.2 MAC Protocols

A MAC protocol in a multiaccess medium is essentially a distributed scheduling algorithm that allocates the channel to requesting nodes. Two commonly used access principles in wireless networks are *fixed-assignment channel access* and *random access* methods [18]. In the former method, a pair of nodes is statically allocated a certain time slot (frequency band or spread spectrum code), as is the case for most voice-oriented wireless networks. On the other hand, in random access MAC protocols, the sender dynamically competes for a time slot with other nodes. This is a more flexible and efficient method of managing the channel in a fully distributed way, but suffers from collisions and interference. This section provides a general discussion on the random access MAC and then offers an in-depth discussion on DCF of IEEE 802.11.

24.2.1 Random Access MAC

Random access MAC protocol in radio networks has long been an active research area. The throughput of *ALOHA carrier sensing* protocols in the presence of collisions has been analyzed with a wide range of system parameters, such as propagation delay and offered load. A key factor here is the *vulnerable period*, during which for a node to transmit a packet successfully without collisions, other interfering nodes should not attempt to transmit during the node's transmission time [19]. In the pure *ALOHA* scheme, the vulnerable period is twice the packet transmission time as shown in Figure 24.1. This is fairly large and cannot be ignored unless communication traffic is sufficiently light. It has been reported that the maximum achievable channel use is only 18 percent for *pure ALOHA* and 36 percent for *slotted ALOHA* even including retransmissions [19]. The carrier-sensing mechanism reduces this period substantially by sensing the medium before attempting to transmit a packet. The chance of collisions is reduced to the case where a node does not sense the medium correctly due to the propagation delay, which is fairly small compared to the packet transmission time.

Unfortunately, collisions are not completely avoidable in carrier-sensing MAC protocols due to interfering *hidden terminals* [21]. When a mobile node is located near the receiver, but far from the sender, this node may be unaware of the on-going communication and causes collisions at the

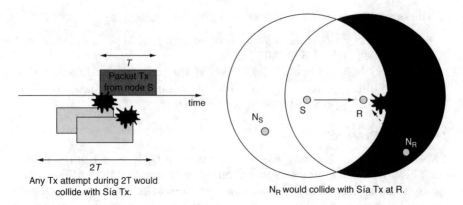

Figure 24.1. Vulnerable Period and Vulnerable Region in Random Access MAC Protocols

receiver by initiating its own data transfers. In Figure 24.1, N_R is an example of a hidden terminal when node S and node R are the sender and the receiver, respectively. Here, the sender S cannot sense N_R's transmission, even though it is strong enough to corrupt the transmission from S to R. The shaded area shown in Figure 24.1, where the hidden terminals can hide, is called the *vulnerable region*.

A *busy tone* is one approach used to avoid the hidden terminal problem in a carrier sensing radio network [20]. Whenever any node detects a packet being transmitted, it starts to send a signal, called a busy tone, in a separate frequency channel. For example, when node S starts to send a packet to node R, node R as well as node N_S will start to send a busy tone. All the nodes that can hear the busy tone will not initiate their own transmission and thus node R will not experience collision. A critical problem with the use of busy tones is that too many nodes (all two-hop neighbors of node S) will be inhibited from transmitting. The number of nodes affected will typically be about four times the number of nodes within the transmission range of the receiver, which is the only set of nodes that should be inhibited. Therefore, even though this approach almost completely eliminates collisions, it is not a very promising approach from a throughput standpoint [20].

24.2.2 DCF of IEEE 802.11 MAC

The IEEE 802.11 wireless LAN standard adopts a dynamic channel allocation scheme based on a carrier-sensing technique, called DCF, as its basic MAC layer algorithm. Four key elements of DCF are *ACK, RTS/CTS* with *NAV, IFS,* and *Backoff algorithm* with *CW*. This subsection introduces these four key elements, which are essential for understanding the utilization enhancing techniques in the following sections.

Medium Access Control Mechanisms in Mobile Ad Hoc Networks

24.2.2.1 ACK for Collision Detection. *Acknowledgment* (ACK) packets enable a mobile node to determine whether its transmission was successful or not because it cannot otherwise detect a collision. The sender is made aware of the collision after it times out waiting for the corresponding ACK for the packet transmitted. If no ACK packet is received or an ACK is received in error, the sender will contend again for the medium to retransmit the data packet until the maximum allowed number of retransmissions has been tried. If all fails, the sender drops the packet, consequently leaving it to a higher level reliability protocol. Note that this sort of link level ACK is not usually used in wired networks because wired links are quite reliable and collisions are easily detected.

24.2.2.2 RTS/CTS and NAV for Solving Hidden Terminal Problem. In DCF, collisions from the nodes hidden in the vulnerable region can be effectively avoided by a four-way handshake based on *request-to-send* (RTS) and clear-to-send (CTS) packets. By exchanging the two short control packets between a sender and a receiver, all neighboring nodes recognize the transmission and back off during the transmission time advertised along with the RTS and CTS packets. Using this information, each node maintains a network allocation vector (NAV), which indicates the remaining time of the on-going communication. Figure 24.2 shows the transmission range of RTS and CTS control packets. Nodes N_S and N_R would receive RTS and CTS, respectively, and set their NAVs accordingly to refrain themselves from accessing the medium during the transmission of node S. Figure 24.2 shows the four-way handshake between S and R, as well as IFS and contention window, which will be described below.

However, as discussed in Section 24.1, the reduction in the chance of collisions occurs at the expense of increased control overhead involved with the exchange of RTS and CTS packets, which can be significant for short frames. For this reason, DCF allows the use of the RTS/CTS mechanism, but does not require it and suggests the use of the `RTSThreshold` parameter to determine the payload size for which RTS/CTS should be used [7]. This parameter is not fixed and has to be set separately by each mobile node.

24.2.2.3 IFS for Prioritized Access to the Channel. *Interframe spacing* (IFS) is the time interval during which each node has to wait before transmitting any packet and is used to provide a prioritized access to the channel. For example, *short IFS* (SIFS) is the shortest and is used after receiving a DATA packet to give the highest priority to an ACK packet. *DCF IFS* (DIFS) is larger than SIFS and is used when initiating a data transfer. When RTS/CTS is used, the RTS packet can be transmitted after waiting for DIFS duration of time. All other frames (CTS, DATA, and ACK) use SIFS before attempting to transmit. Figure 24.2 shows the usage of DIFS and SIFS. Two other IFSs are *point coordination function IFS* (PIFS) and *extended IFS* (EIFS), which will be discussed shortly in this section.

MOBILE COMPUTING HANDBOOK

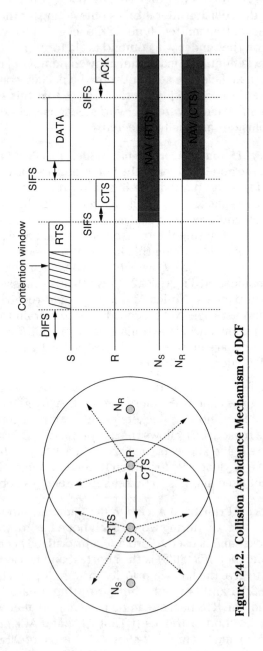

Figure 24.2. Collision Avoidance Mechanism of DCF

Medium Access Control Mechanisms in Mobile Ad Hoc Networks

24.2.2.4 Backoff Algorithm with CW to Provide Fair Access with Congestion Control. The above mentioned IFS is followed by an additional waiting time defined by the backoff algorithm used in DCF. The main purpose of the backoff algorithm is to reduce the probability of collisions when contention is severe. After waiting for the IFS duration, each competing node waits for a backoff time, which is randomly chosen in the interval (0, *CW*), defined as contention window. During the first transmission of a packet, *CW* is set to its minimum preset value, *CWmin*. If the channel continues to be idle during the backoff time, it transmits (winner). Other waiting nodes (losers) become aware of the transmission, freeze their backoff time, and contend again in the next competition cycle after the current transmission completes. Now, the frozen backoff time plays an important role in ensuring fairness. Definition of fairness may differ, but in general all nodes entering the competition for the first time should have on an average equal chance of transmitting and nodes that have lost in the previous competition cycle should have higher priority than newly arrived nodes during the current competition cycle. The losers are given a higher priority by using the remaining frozen backoff time, thereby preserving the first-come, first-serve policy.

The aforementioned access scheme has problems under heavy or light loads. If CW is too small compared to the number of competing nodes, it causes many collisions. On the other hand, if *CW* is too large, it causes unnecessary delays [21]. DCF adopts the *binary exponential backoff* scheme to allow an adaptive solution to this problem. When a node fails to receive an ACK in response to transmission of a DATA packet, it needs to contend in the next competition cycle. However, *CW* is doubled after the collision and this continues until *CW* reaches a preset limit, *CWmax*. It is noted that *CW* is restored to its minimum, *CWmin*, when a node successfully completes a data transmission. Figure 24.3 shows the flow chart of the backoff algorithm used in DCF.

24.2.2.5 EIFS to Protect ACK from Collisions. The RTS/CTS mechanism together with NAV effectively eliminate the vulnerable region introduced in Figure 24.1B. However, some packets are still vulnerable to collisions. For example, consider the coverage area of a radio transmitter, which depends on the power of the transmitted signal and the path loss. Each radio receiver has particular power sensitivity; for example, it can only detect and decode signals with strength larger than this sensitivity [22]. There are two threshold values when receiving radio signals: *receive threshold (RXThresh)* and *carrier sense threshold (CSThresh)*. If the power of the received signal is higher than RXThresh, it is regarded as a valid packet and passed up to the MAC layer. The corresponding distance for two nodes to communicate successfully is called the *transmission range*.

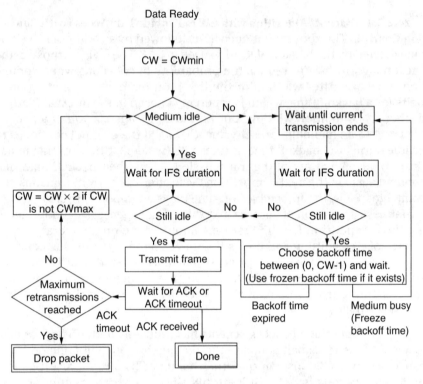

Figure 24.3. Backoff Algorithm Used in DCF of IEEE 802.11 MAC

On the other hand, if the received signal power is lower than *CSThresh*, it is discarded as noise and thus the node can start its own transmission or reception. If the signal power is in between *RXThresh* and *CSThresh*, the node cannot receive the packet intelligibly but acknowledges that some active transmission is going on. The corresponding distance is referred to as the *interference range*. Thus, when node S transmits a data packet to node R, there are four different groups of nodes in the network as shown in Figure 24.4:

1. A node is within the transmission range of S or R (Group I). Thus, it can receive RTS or CTS and sets its NAV accordingly.
2. A node is outside of transmission range of S and R, but is within the interference range of S and R (Group II). Thus, it cannot receive packets intelligently but recognizes the on-going communication.
3. A node is outside of interference range of R but is within the interference range of S (Group III). Thus, it cannot sense CTS and ACK transmission from R.
4. A node is outside of interference range of S but is within the interference range of R (Group IV). Thus, it cannot sense data packet transmission from S.

Medium Access Control Mechanisms in Mobile Ad Hoc Networks

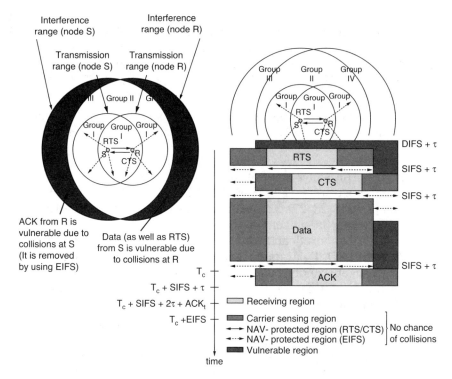

Figure 24.4. Vulnerable Region with Considering the Interference Range (τ: Propagation Delay)

Nodes in Group I correctly set their NAVs when receiving RTS or CTS and defer their transmission until the S–R communication is finished. Nodes in Group II cannot decode the packets and do not know the duration of the packet transmission, but they do sense on-going communications and thus do not cause collisions.

However, ACKs (from R to S) and DATA (from S to R) are vulnerable to collisions due to nodes in Group III and Group IV, respectively. Collisions are critical for any type of packets, but ACK collisions are a more serious problem because an ACK packet forms a vital piece of information as the last step of the four-way handshake. A loss of ACK results in retransmission of long DATA packet and thus significantly degrades the performance. *Extended IFS* (EIFS) is used in DCF to prevent collisions with ACK receptions at the sender. When nodes detect a transmission, but cannot decode it (more specifically, a physical layer header is received correctly, but a MAC layer frame is received in error), they set their NAVs for the EIFS duration. For example, in Figure 24.4, when S completes its data transmission at T_C, nodes in Group II and Group III would set their NAV to $T_C + \textit{EIFS}$. At $T_C + \textit{SIFS} + \tau$, R replies back to S with an ACK and the transmission is completed at T_C + *SIFS* + 2τ + ACK_t, where τ is the propagation delay of the channel and ACK_t is the transmission time for the ACK packet. If EIFS is larger than *SIFS* + 2τ + ACK_t, nodes in Group II and Group III would not corrupt the ACK packet from R to S. These nodes have to wait an additional DIFS to start the competition, thus EIFS is set to *SIFS* + ACK_t + *DIFS* in the IEEE 802.11 MAC standard.

Table 24.1 summarizes the characteristics of a typical radio transceiver and the four key elements of DCF with typical values for the related parameters.

24.2.2.6 Performance Limit of DCF. There has been active research on estimating the performance of IEEE 802.11 MAC, analytically as well as via simulation [7, 8, 18, 23–27]. Among them, Cali et al. have provided a mathematical model for the maximum achievable throughput [8]. According to their results, the theoretical throughput is bounded by around 80 percent when the typical DCF parameters are used (with propagation delay of 1 μsec and packet size of 50 μsec ~ 5 msec). In reality, DCF operates very far from the theoretical limits due to collisions and control overhead associated with RTS/CTS and the backoff algorithm.

In a multihop MANET, the situation becomes worse due to the reasons discussed in Section 24.1. Li et al. showed that the end-to-end throughput is at most one fourth of the channel bandwidth even without any other interfering nodes [28]. In other words, when IEEE 802.11based 2 Mbps wireless network interface is used, a source-destination pair in a MANET cannot support more than 500 kbps. This is mainly due to collisions among intermediate forwarding nodes of the same data stream. In addition, the control overhead of DCF aggravates the situation and the maximum throughput is reduced to about one seventh of the channel bandwidth [28]. When other data traffic exists, the throughput is reduced even further. For example, Xu and Saadawi reported that multiple simultaneous Transmission Control Protocol (TCP) sessions in a MANET result in unreasonably low aggregate throughput and suffers from severe unfairness [23].

24.3 Enhancing Temporal Channel Utilization

As pointed out previously, the performance limitation is mainly due to the limited capability of MAC protocols in a multi-hop communication environment. A key idea for improving DCF for MANET is *adaptivity*. That is, each node should be able to behave adaptively according to traffic intensity in its vicinity. This section discusses the non-adaptive characteristics of DCF and the temporal approaches proposed in the literature [5–8]. These methods attempt to enhance the effective channel use by reconsidering the DCF parameters such as `RTSThreshold` (Section 24.3.1) and the backoff algorithm (Section 24.3.2) to better schedule the channel along the time dimension.

Table 24.1. Radio Transceiver Characteristics and Key Elements of DCF (914 MHz, 1 Mbps Lucent WaveLAN using Direct Sequence Spread Spectrum)

Key Elements	Parameters	Typical Values	Comment
Radio transceiver	Transmission power	0.2818 W	
	RxThresh	3.652×10^{-10} W	Transmission range 250 m (with two-ray ground model)
	CSThresh	1.559×10^{-11} W	Interference range 550 m (with two-ray ground model)
ACK	ACK frame size	376 μsec	184-bit ACK packet with 144 and 48 bits of physical layer preamble and header over 1 Mbps link
RTS/CTS and NAV	RTS frame size	424 μsec	232-bit RTS packet with 144 and 48 bits of physical layer preamble and header over 1 Mbps link
	CTS frame size	376 μsec	184-bit CTS packet with 144 and 48 bits of physical layer preamble and header over 1 Mbps link
	RTSThreshold		Not specified
	Retry limit for a long packet	4	For DATA packet longer than `RTSThreshold`
	Retry limit for a short packet	7	For RTS and shorter DATA packet
IFS	SIFS (Short IFS)	10 μsec	For CTS, DATA, and ACK packet
	DIFS (DCF IFS)	50 μsec	For RTS and short DATA packet
	EIFS (Extended IFS)	436 μsec	SIFS (10) + ACKt (376) + DIFS (50)
Backoff algorithm	Slot time	20 μsec	
	CWmin	32	Equivalent to 640 μsec
	CWmax	1024	Equivalent to 20.48 msec

24.3.1 RTS/CTS Mechanism

24.3.1.1 Optimal Setting of RTSThreshold to Tradeoff between Control and Collision Overhead. As discussed in Section 24.2.2, the parameter RTSThreshold determines whether RTS/CTS is used or not. However, this parameter is not fixed in the DCF standard as discussed previously. Khurana et al. studied the throughput of an IEEE 802.11-based ad hoc network to obtain the optimal parameters for DCF including the RTSThreshold [5]. Assuming that the physical layer uses *direct sequence spread spectrum* (DSSS) and DCF uses typical parameters as in Table 24.1, they recommend a value of 250 bytes for RTSThreshold [5]. In other words, the RTS/CTS exchange is beneficial only when DATA packet size is larger than 250 bytes. Weinmiller et al. performed a similar study and concluded via simulation that the best throughput is obtained when 200 to 500 bytes is used for RTSThreshold [7]. Note that this size should take into account the necessary physical layer preamble and header according to the MAC packet format called *MAC protocol data unit* (MPDU) as noted in Table 24.1.

A better idea is to adjust the parameter depending on the traffic and the collision probability. Even if DATA packet size is large, the RTS/CTS exchange is a waste of bandwidth if the number of hidden terminals is small and collisions are unlikely. Therefore, the optimal value for RTSThreshold would depend on the traffic intensity, which can be estimated indirectly by noting the number of collisions experienced [5, 7].

24.3.2 Exponential Backoff Algorithm

24.3.2.1 Conservative CW Restoration to Reduce Collisions. In DCF of IEEE 802.11, the contention window is reduced to the minimum value (CWmin) for every new packet whether the last packet was successfully delivered or not. Even if the network area is congested with many competing data streams, each packet transmission starts with the minimum window size and thus experiences a large number of collisions before its window size becomes appropriate [8, 24]. In addition, restoration of CW to CWmin makes the backoff algorithm unfair, because it favors the mobile node that has most recently transmitted [23]. In the first part of Figure 24.5, node A wins in the first competition cycle because it chooses the smaller backoff time ($BOFF_A$) than node B and node C ($BOFF_B$ and $BOFF_C$). While node A restores its CW to CWmin in the next competition cycle, node B and node C, being losers, keep the same CW as in the second part of Figure 24.5. Even though node B and node C reduce their backoff time by using the frozen values ($BOFF_B\ BOFF_A$ and $BOFF_C\ BOFF_A$, respectively), node A has a better chance of winning in the next competition cycle again due to the reduced CW size.

To solve the collision and fairness problem, Bharghavan et al. proposed a *Multiplicative Increase and Linear Decrease* (MILD) algorithm where the contention window size increases multiplicatively on collisions but

Medium Access Control Mechanisms in Mobile Ad Hoc Networks

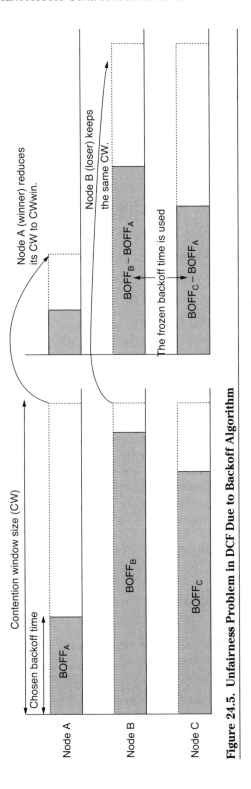

Figure 24.5. Unfairness Problem in DCF Due to Backoff Algorithm

decreases linearly on successful transmission [6]. MILD algorithm works well when the network traffic is high. However, under light traffic condition, it incurs an additional delay to return the CW to CWmin, which is not the case in the original backoff algorithm.

24.3.2.2 Different Treatment of New and Lost Nodes for Fairness.

Weinmiller et al. investigated the effect of CW restoration to CWmin together with the frozen backoff time [7]. In the initial state, the backoff algorithm in DCF results in an equally distributed probability for each slot to be selected. However, in the following competition cycle, the probability is not equally distributed. Consider an example in Figure 24.5. Because $BOFF_A$ is the winner's backoff time in the first competition cycle and the losers use the frozen backoff time in the next competition cycle, the contention window of these nodes is effectively reduced to $(0, CW - BOFF_A)$. Still within this reduced contention window, all slots are selected with the same probability by these nodes. However, newly entering nodes will choose their slot with equally distributed probability within the whole range of the contention window (0, CW). Therefore, slots later than $CW - BOFF_A$ have a significantly lower probability to be chosen compared to the earlier slots. After several competition cycles, the slot selection probability becomes a decreasing staircase function.

As far as the collision probability is concerned, this leads to a high chance of collisions at earlier slots because these slots will most probably be selected two or more times. An equally distributed probability for every slot to be chosen is the favored situation in terms of collision avoidance. Weinmiller et al. suggested two alternative solutions for this fairness problem, both of which attempt to offer the later slots in $(CW - BOFF_A, CW)$ to the newly entering nodes and earlier slots in $(0, CW - BOFF_A)$ to the nodes that have lost the previous competition [7]. These schemes assume that a newly arriving node knows the winning slot of previous competition, which may not be the case under certain conditions.

24.3.2.3 Dynamic Tuning of CW to Minimize the Collision Probability.

Cali et al. observed that the collision probability increases as the number of active nodes increases, but it cannot be dynamically controlled due to the static backoff algorithm of DCF [8]. In other words, the optimal setting of CW, and thus the optimal backoff time, can be achieved by estimating the number of active nodes in its vicinity at runtime. Because each node can estimate the number of empty slots in a virtual transmission time by observing the channel status, the number of active nodes can be computed and exploited to select the appropriate CW without paying the collision costs [8].

Table 24.2 summarizes the channel utilization enhancing techniques discussed in this section.

Table 24.2. Enhancing Temporal Channel Utilization

Key Elements	Parameter	Problem	Solution Technique
RTS/CTS and NAV	`RTSThreshold`	Undetermined or fixed `RTSThreshold`	Optimal preset value: 250 bytes MPDU [5] 200–500 bytes MPDU [7] Adaptive adjustment based on traffic and collision probability [5] collisions experienced [7]
Backoff algorithm	CW restoration to CWmin	Many collisions or large delay	MILD [6]
	Frozen backoff time	Staircase-like slot selection probability and more collisions	Offer later slots to new nodes and earlier slots to old and lost nodes [7]
	Backoff algorithm	CW is not optimal	Dynamic tuning with the estimation of the number of active nodes in its vicinity at runtime [8]

24.4 Enhancing Spatial Channel Utilization

In this section, we discuss MAC protocols that better utilize the channel along the spatial dimension. Although the temporal approaches in Section 24.3 can be applied to single-hop wireless LANs as well as multi-hop MANETs, the spatial approaches discussed in this section focus on multi-hop MANETs and exploit the characteristics unique to the multi-hop communication environment. The *Dual Busy Tone Multiple Access* (DBTMA) protocol [9] employs a busy tone to reserve only the space around the receiver to encourage spatial reuse. Based on the same concept of busy tone, the *Power Controlled Multiple Access* (PCMA) scheme [10] further reduces the interference range by employing the transmission power control. An alternative to these two approaches is the use of directional antenna to transmit or receive data only along a certain direction, which reserves only a fraction of space compared to that of omnidirectional antenna [13–17]. The following three subsections discuss the three approaches, respectively.

24.4.1 Busy Tone to Solve the Exposed Terminal Problem

To avoid interference from other transmissions, a source–destination pair should reserve some spatial area, but the area should be as small as possible to encourage more spatial reuse. One example of excessive space reservation in DCF is the RTS/CTS mechanism: because collisions occur only at the receiver side, it is not necessary to reserve space around the sender. This is known as the *exposed terminal problem* [21], which means that some nodes around the sender are overly exposed to the on-going communication and experience unnecessary delay until the sender completes its data transmission.

The DBTMA protocol [9] uses busy tone with RTS/CTS to solve the exposed terminal problem. A separate control channel is used for both control packets (RTS and CTS) and two busy tones (transmit and receive busy tones, BT_t and BT_r). The main feature of DBTMA is the use of the control channel to completely eliminate the hidden as well as the exposed terminal problem. BT_t and BT_r on the control channel indicate that the node is transmitting and receiving on the data channel, respectively. All other nodes sensing the BT_r signal (hidden terminals) defer their transmissions; nodes sensing the BT_t signal do not attempt to receive. Thus, exposed terminals can sense BT_t, but not BT_r, so that they can safely reuse the space by transmitting their packets. Figure 24.6 shows the DBTMA protocol with two busy tones on the left.

In addition, busy tone can help solve the collision problem due to mobility. The conventional RTS/CTS scheme may not work well in a network with highly mobile nodes. This is because nodes may come within the range of either the sender or receiver after the RTS/CTS exchange. With DBTMA, such hidden terminals do not exist because the receiver continuously sends the BT_r signal to its neighbors.

24.4.2 Transmission Power Control to Reduce Interference Range Radially

When a node's radio transmission power is controllable, its direct communication range, as well as the number of its immediate neighbors, is also adjustable. Although higher transmission power increases the transmission range, lower transmission power reduces the collision probability by reducing the number of competing nodes. In the PCMA protocol [10], a source–destination pair uses *request power to send* (RPTS) and *acceptable power to send* (APTS) control packets to compute the optimal transmission power based on their received signal strength, which will be used when transmitting DATA packets. PCMA also uses the busy tone channel to advertise the noise level the receiver can tolerate. A potential transmitter first senses the busy tone to detect the upper bound of its transmission power for all control and DATA packets. Figure 24.6 shows the PCMA protocol with busy tone on the right.

Transmission power control approach has been actively studied for other purposes, such as energy saving or topology control. For example, Gomez et al. proposed using the maximum power level for RTS and CTS

Medium Access Control Mechanisms in Mobile Ad Hoc Networks

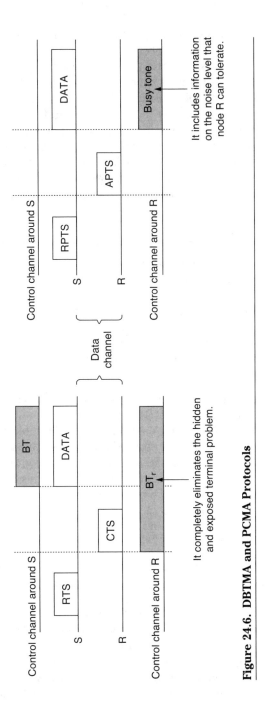

Figure 24.6. DBTMA and PCMA Protocols

packets and lower power levels for DATA packets [11]. This does not increase or decrease the collision probability, but nodes can save a substantial amount of energy by using a low power level for DATA packets. However, this approach has a problem with respect to ACK reception because EIFS (used to protect ACK) is only effective when DATA packets are transmitted at full power as discussed in Section 24.2.2. The *Power Control MAC* (PCM) protocol addresses this problem by transmitting data at a reduced power level most of the time, but periodically transmits at the maximum power level to inform to its neighboring nodes about the current transmission. Another related area of research is routing protocols based on transmission power control [29–31]. We do not discuss these protocols in detail in this chapter because they are designed to save energy rather than improve channel use. For a detailed discussion on this subject, please refer to [32].

24.4.3 *Directional Antenna to Reduce Interference Range Angularly*

Unlike an omnidirectional antenna, a directional antenna has a directional radiation pattern making it possible to transmit to a subset of its neighbors [33]. When it is used for transmission, it can significantly reduce the unwanted interference to nodes outside its directional pattern. Similarly, when it is used for reception, the receiver can eliminate the interference signals from directions other than the signal source [13]. Thus, directional antennas improve spatial reuse and reduce multi-path propagation, which can result in better channel utilization.

With omnidirectional antennas, one-hop neighbors within the range of the sender (S) or the receiver (R) defer their transmission based on RTS/CTS as shown in Figure 24.7A. Although a hidden terminal N_R should defer its transmission to protect node R's reception, an exposed terminal N_S unnecessarily defers its transmission because it would not have interfered with the ongoing S–R communication. This wastes the spatial channel bandwidth around node S. Directional antennas can eliminate this problem by using *directional RTS* (DRTS) and *directional CTS* (DCTS) instead of *omnidirectional RTS* (oRTS) and *omnidirectional CTS* (oCTS) as shown in Figure 24.7B and Figure 24.7C.

A key question then is how can collisions be avoided with DRTS and DCTS packets. For example, in Figure 24.7C, when N_R wishes to transmit directly to R, it simply transmits because N_R did not receive DCTS from node R and thus it is not aware of the S–R communication (*deafness problem* [16]). This may or may not cause collisions at node R depending on the underlying antenna model (*directional hidden terminal problem*). Another important question is how to find the desired direction for the transmission and reception when initiating DRTS or replying with DCTS. This section discusses three representative *directional MAC* (DMAC) algorithms

Medium Access Control Mechanisms in Mobile Ad Hoc Networks

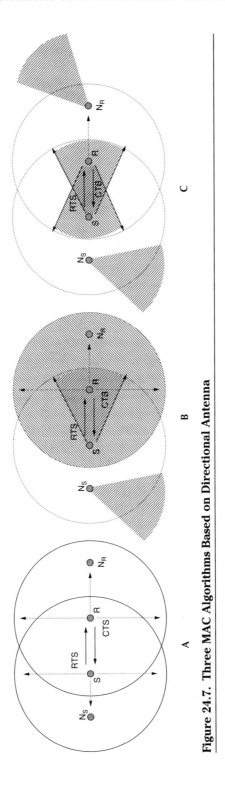

Figure 24.7. Three MAC Algorithms Based on Directional Antenna

based on oRTS/oCTS [13], DRTS/oCTS [14], and DRTS/DCTS [15], respectively, as shown in Figure 24.7.

24.4.3.1 oRTS/oCTS-Based DMAC. Naspuri et al. proposed the oRTS/oCTS-based DMAC protocol [13], where all control packets are transmitted omnidirectionally and only DATA packets are transmitted directionally. Collisions are avoided as in conventional omnidirectional MAC algorithms and the additional benefit is the significant reduction in interference by transmitting and receiving DATA packets over a small angle. The key feature of this scheme is a mechanism to determine the direction of the other party of the communication. Here, the radio transceiver is assumed to have multiple directional antennas and each node is capable of switching any one or all antennas to active or passive modes, known as *directional reception capability*. An idle node listens to on-going transmission on every direction. When it receives an oRTS addressed to itself, it can determine the direction of the sender by noting the antenna that received the maximum power of the oRTS packet[1] [13]. Similarly, the sender estimates the direction of the receiver by receiving the oCTS packet. Thus, a receiver is not influenced by other transmissions from other directions. Figure 24.7A shows the oRTS/oCTS-based DMAC scheme.

24.4.3.2 DRTS/oCTS-Based DMAC. Ko et al. proposed two DMAC schemes based on DRTS [14]. The first scheme trades off between spatial reuse and collision avoidance by using DRTS and oCTS. Although oCTS helps avoid collisions from hidden terminals, such as N_R in Figure 24.7B, DRTS helps improve the spatial channel use by eliminating the exposed terminal problem. (N_S is free to attempt its transmission during the S–R communication.) The second scheme uses both DRTS and oRTS to reduce the probability of collisions of control packets in the sender's vicinity caused by the exposed terminal. The usage rule is if there is no on-going communication in every direction around a sender, then it transmits an oRTS. Otherwise, the sender transmits a DRTS. In both schemes, nodes require external location tracking support such as global positioning service (GPS) to determine the direction of the nodes they would like to communicate with. Based on the location of the receiver, the sender may select an appropriate directional antenna to send packets (DRTS and DATA packets) to the receiver.

24.4.3.3 DRTS/DCTS-Based DMAC. Wang and Garcia-Luna-Aceves observed that the benefit of spatial reuse achieved by a DMAC protocol can outweigh the benefit of a conservative collision avoidance mechanism that sends some omnidirectional control packets to silence potential interfering nodes [15]. Their approach uses both DRTS and DCTS and aggressively reuses the channel along the spatial dimension at the cost of increased chance of collisions. In Figure 24.7C, N_S and N_R can initiate their own transmissions during a *S–R* communication. It is noted that nodes have directional reception capability as discussed previously and thus the transmission from N_R does not cause collisions at node *R*. Location tracking support is required for implementing this scheme.

24.4.3.4 Other DMAC Protocols. Before concluding this section, we introduce two additional DMAC protocols — *Multi-hop RTS MAC* (MMAC) [16] and *Receiver-Oriented Multiple Access* (ROMA) [16]. Choudhury et al. made an important observation that the gain of directional antennas is higher than that of omnidirectional antennas; thus they have a greater transmission/reception range [16]. Even if the receiver is within the sender's transmission range, the receiver may not be able to communicate with the sender if its reception range does not include the sender. This is quite possible when the sender transmits directionally knowing the receiver's location (via GPS), but the receiver tries to receive omnidirectionally because it does not know about the transmission attempt from the sender. Therefore, even though DATA packets can be transmitted over a single hop using directional antenna at both nodes, it is possible for control packets such as DRTS to take more than one hop. MMAC takes into account this fact and uses multi-hop RTS for delivering DRTS to the receiver over a number of hops.

Another recent DMAC protocol proposed by Bao and Garcia-Luna-Aceves is not based on RTS/CTS, but uses a transmission schedule determined statically based on node identifier and time slot number [17]. Although on-demand medium access schemes determine the communicating pair by exchanging short control signals such as RTS/CTS before each transmission session, scheduled medium access schemes prearrange or negotiate a set of timetables for individual nodes or links. ROMA is such a schedule-based MAC protocol where the communicating nodes are paired with the designated time slots based on the schedule; thus the transmissions are collision-free [17].

Table 24.3 summarizes the channel utilization enhancing techniques discussed in this section.

24.5 Conclusions

Mobile ad hoc networks are composed of nodes that are self-organizing and communicate over wireless channels usually in a multi-hop fashion. They exhibit dynamic topology and share limited bandwidth, with most nodes having limited processing abilities, and energy constraints. In this chapter, we have considered some of the techniques in the design of MAC protocols with DCF of IEEE 802.11 as a reference model. Each of these schemes tries to maximize network capacity, reduce congestion at the

Table 24.3. Enhancing Spatial Channel Utilization

Conventional Facility	Problem	Additional Facility	Solution Technique
Single channel for data and control packet	Unnecessary space reservation around the sender by RTS (Exposed terminal problem)	Separate busy tone channel	Advertise the communication over the busy tone channel (DBTMA) [9]
Single power model	Unnecessary interference and space reservation when the communicating distance is short	Transmission power control of radio transceiver	Advertise the tolerable noise level over the busy tone channel (PCMA) [10] Use low power for DATA packets [11] Periodic power adjustment when delivering DATA packets (PCM) [12]
Omni-directional antenna model	Unnecessary interference and space reservation because communication is omnidirectional	Directional antenna	Omnidirectional control packet transfer but directional DATA packet transfer [13] Directional RTS [14] Directional RTS and CTS [15] Multi-hop RTS to take into account the difference in antenna gain [16] Schedule-based directional MAC [17]

MAC layer, and ensure fairness by balancing the control overhead to avoid collisions. Key techniques used to enhance temporal utilization are to optimize the DCF parameters such as RTSThreshold and those associated with the backoff algorithm, which is used to avoid collisions in DCF. Spatial reuse assumes special importance in multi-hop networks. Busy tone method, transmission power control, and directional transmissions are the key techniques in this direction. Among these, the possibilities provided by directional transmissions are most promising because they can reduce interference and collisions considerably and can be used in conjunction with the other two techniques. Transmission power control methods not only help in reducing interference, but also in energy conservation.

Medium Access Control Mechanisms in Mobile Ad Hoc Networks

Note

1. Several directional antenna models have been proposed. A sectored antenna is assumed for the oRTS/oCTS-based scheme. It consists of multiple (M) directional antennas, each of which has a conical radiation pattern spanning an angle of $2\pi/M$ radians. A mobile node can look out simultaneously with all of its M antennas and recognize the direction of arrival by noting the antenna on which the gain is the maximum. A *directional beam-forming antenna* is used for directional transmission or reception by beam-forming toward the intended receiver or sender. Thus, it is usually used along with an omnidirectional antenna for listening on all directions. A *multi-beam adaptive array model* is based on an antenna array, capable of forming multiple beams for several simultaneous receptions or transmissions.

References

1. Forman, G. and Zahorjan, J., The Challenges of Mobile Computing, *IEEE Computer*, vol. 27, no. 4, pp. 38–47, 1994.
2. Frodigh, M., Johansson, P., and Larsson, P., Wireless Ad Hoc Networking — the Art of Networking without a Network, *Ericsson Review*, no. 4, pp. 248–263, 2000.
3. Perkins, C., *Ad Hoc Networking*, Boston: Addison Wesley, 2001.
4. Stallings, W., IEEE 802.11 Wireless LAN Standard, *Wireless Communications and Networks*, Ch. 14, Upper Saddle River, NJ: Prentice Hall, Inc., 2002.
5. Khurana, S., Kahol, A., Gupta, S.K.S. and Srimani, P.K., Performance Evaluation of Distributed Co-Ordination Function for IEEE 802.11 Wireless LAN Protocol in Presence of Mobile and Hidden Terminals, MASCOT'99, pp. 40–47, 1999.
6. Bharghavan, V., Demers, A., Shenker, S., and Zhang, L., MACAW: a Media Access Protocol for Wireless LANs, *ACM SigComm*, pp. 212–225, 1994.
7. Weinmiller, J., Woesner, H., Ebert, J.-P., and Wolisz, A., Analyzing and Tuning the Distributed Coordination Function in the IEEE 802.11 DFWMAC Draft Standard, MASCOT'96, 1996.
8. Cali, F., Conti, M., and Gregori, F., Dynamic Tuning of the IEEE 802.11 Protocol to Achieve a Theoretical Throughput Limit, *IEEE/ACM Tr. Networking*, vol. 8, no. 6, pp. 785–799, December 2000.
9. Deng, J. and Haas, Z.J., Dual Busy Tone Multiple Access (DBTMA): a New Medium Access Control for Packet Radio Networks, *IEEE ICUPC'98*, Florence, Italy, October 5–9, 1998.
10. Monks, J.P., Bharaghavan, V., and Hwu, W.W., A Power Controlled Multiple Access Protocol for Wireless Packet Networks, *IEEE InfoCom*, 2001.
11. Gomez, J., Campbell, A.T., Naghshineh, N., and Bisdikian, C., Conserving Transmission Power in Wireless Ad Hoc Networks, *ICNP'01*, 2001.
12. Jung, E.-S., and Vaidya, N.H., A Power Control MAC Protocol for Ad Hoc Networks, *ACM/IEEE MobiCom'02*, 2002.
13. Nasipuri, A., Ye, S., You, J., and Hiromoto, R.E., A MAC Protocol for Mobile Ad Hoc Networks Using Directional Antennas, IEEE WCNC, 2002.
14. Ko, Y.-B., Shankarkumar, V., and Vaidya, N.H., Medium Access Control Protocols Using Directional Antennas in Ad Hoc Networks, *IEEE InfoCom*, 2000.
15. Wang, Y., and Garcia-Luna-Aceves, J.J., Spatial Reuse and Collision Avoidance in Ad Hoc Networks with Directional Antennas, *IEEE GlobeCom*, 2002.
16. Choudhury, R.R., Yang, X., Ramanathan, R., and Vaidya, N.H., Using Directional Antennas for Medium Access Control in Ad Hoc Networks, *ACM/IEEE MobiCom*, 2002.
17. Bao, L. and Garcia-Luna-Aceves, J.J., Transmission Scheduling in Ad Hoc Networks with Directional Antennas, *ACM/IEEE MobiCom*, 2002.

18. Pahlavan, K. and Krishnamurthy, P., Wireless Medium Access Alternatives, *Principles of Wireless Networks*, Ch 4, Upper Saddle River, NJ, Prentice Hall Inc, 2002.
19. Kleinrock, L. and Tabagi, F.A., Packet Switching in Radio Channels: Part I — Carrier Sense Multiple-Access Models and Their Throughput-Delay Characteristics, *IEEE Tr. Communications*, vol. COM-23, no. 12, December 1975.
20. Bertsekas, D. and Gallager, R., Multiaccess Communication, Data Networks, 2nd ed., Ch. 4, Upper Saddle River, NJ: Prentice Hall Inc, 1992.
21. Schiller, J., Wireless LAN, *Mobile Communications*, Ch. 7, Boston: Addison Wesley, 2000.
22. Pahlavan, K. and Krishnamurthy, P., Characteristics of Wireless Medium, *Principles of Wireless Networks*, Ch. 2, Upper Saddle River, NJ: Prentice Hall, Inc, 2002.
23. Xu, S. and Saadawi, T., Does the IEEE 802.11 MAC Protocol Work Well in Multihop Wireless Ad Hoc Networks?, *IEEE Communications Magazine*, pp. 130–137, June 2001.
24. Chhaya, H.S. and Gupta, S., Performance Modeling of Asynchronous Data Transfer Methods of IEEE 802.11 MAC Protocol, *Wireless Networks*, vol. 3, pp. 217–234, 1997.
25. Bruno, R., Conti, M., and Gregori, E., Optimization of Efficiency and Energy Consumption in p-Persistent CSMA-Based Wireless LANs, *IEEE Tr. Mobile Computing*, vol. 1, no. 1, pp. 10–31, January–March, 2002.
26. Foh, C. and Zukerman, M., Performance Analysis of the IEEE 802.11 MAC Protocol, *European Wireless*, pp. 184–190, 2002.
27. Kwak, B.-J., Song, N.-O., and Miller, L.E., Analysis of the Stability and Performance of Exponential Backoff, *IEEE WCNC*, 2003.
28. Li, J. et al., Capacity of Ad Hoc Wireless Networks, *ACM/IEEE MobiCom'01*, pp. 61–69, 2001.
29. Chang, J.-H. and Tassiulas, L., Energy Conserving Routing in Wireless Ad-hoc Networks, *IEEE InfoCom*, pp. 22–31, 2000.
30. Doshi, S. and Brown, T.X., Minimum Energy Routing Schemes for a Wireless Ad Hoc Network, *IEEE InfoCom*, 2002.
31. Narayanaswamy, S., Kawadia, V., Sreenivas, R.S., and Kumar, P.R., Power Control in Ad-Hoc Networks: Theory, Architecture, Algorithm and Implementation of the COMPOW Protocol, *European Wireless*, pp. 156–162, 2002.
32. Yu, C., Lee, B., and Youn, H.Y., Energy Efficient Routing Protocols for Mobile Ad Hoc Networks, *Wireless Communications and Mobile Computing (WCMC) Journal*, Hoboken, NJ: John Wiley and Sons, Ltd., 2003.
33. Horneffer, M. and Plassmann, D., Directed Antennas in Mobile Broadband System, *IEEE InfoCom*, 1996.

Section VI
Mobile and Ad Hoc Wireless Networks II

Chapter 25
Quality of Service Routing in Mobile Ad Hoc Networks: Past and Future

Jamal N. Al-Karaki, Ahmed E. Kamal

Abstract

In recent years, quality of service (QoS) in mobile ad hoc networks (MANETs) as a research topic has started to receive attention from a growing number of researchers. Future MANETs are expected to support a wide range of real-time multimedia applications. The requirements for timely delivery of multimedia traffic raise new challenges for next generation MANETs. QoS routing is the first step toward achieving guaranteed end-to-end QoS in MANETs. A QoS routing protocol selects network routes with sufficient resources for the satisfaction of the requested QoS parameters. The goal of QoS routing solutions is to satisfy the QoS requirements for each admitted connection, while achieving the global efficiency in resource utilization. In this chapter, we present an overview of QoS routing protocols in MANETs. We discuss several problems that arise in achieving QoS routing and review proposed QoS routing solutions. We present the strengths and the weaknesses of the different QoS routing strategies and point out possible future directions in the QoS routing problem in MANETs.

25.1 Introduction

This chapter is concerned with QoS routing in MANET. QoS refers to satisfying certain requirements of a connection, in terms of a set of constraints, which can be link constraints or path constraints. QoS routing is an essential part of the QoS architecture. Before a connection can be admitted or any resources can be reserved, a feasible path between a source–destination pair

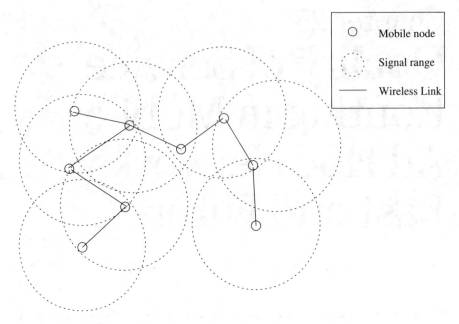

Figure 25.1. Mobile Ad Hoc Network: an Example

(i.e., a path that has sufficient available resources capable of satisfying the QoS requirements) must be found. QoS routing is a routing mechanism under which such paths are determined on the basis of knowledge of resource availability in the network, as well as the QoS requirements of the flows or connections [1]. The objectives of QoS routing are to find a feasible path between a source–destination pair, if one exists, that optimizes the use of network resources and satisfies the required QoS guarantees. Although QoS resource reservation and QoS routing problems are closely related and may be coupled or decoupled in QoS architectures, the two mechanisms have distinct responsibilities. QoS resource reservation is used to reserve resources (e.g., buffer space, bandwidth), set up and maintain virtual connections, release resources, and tear down connections in the network. However, the QoS routing algorithm must first select a feasible path that has a good chance of meeting the QoS requirements [1].

On the other hand, a MANET consists of wireless mobile nodes that may move freely in the absence of a fixed infrastructure resulting in frequent changes in routes, frequent unpredictable topology changes, and link disconnections. The path between any pair of users can traverse multiple wireless links resulting in multi-hop routing in general (Figure 25.1). Mobile nodes can be heterogeneous, thus enabling an assortment of different types of links to be part of the same ad hoc network. The job of the network is to discover the links between the mobile nodes and to build paths for

Quality of Service Routing in Mobile Ad Hoc Networks: Past and Future

user communications. Within the ad hoc network, each node may act as a router and forward packets on behalf of other nodes. Discovering links and building paths across the MANET are challenging problems [11, 22]. Such problems are even exacerbated when these paths have to satisfy some QoS requirements. Traditional routing algorithms designed for wireline networks are not suited for QoS provisioning in MANETs [23]. The main problem is that traditional routing protocols have been designed for fixed networks with infrequent topology changes and typically assume symmetric and fixed capacity links, which makes them unsuitable for MANETs.

The challenge of QoS provisioning in MANETs is still an open research problem due to the unique characteristics of this type of network. To provide a complete QoS solution, the interaction and cooperation of several components (e.g., QoS routing, resource reservation, QoS capable Media Access Control [MAC] layer, and physical layer) must be studied and quantified. This is due to the strong interdependencies between the different layers. To understand the complexity of the problem, consider a routing protocol that computes different routes based on the flows requirements. The set of routes and their flows determine the load on each link in the network. The selection of a link in a certain route also implies the selection of a certain transmission power that guarantees transmission quality without degrading the transmission quality of neighboring links. Because the wireless medium is shared, interference between links exists and a suitable design of the MAC protocol must coordinate among transmissions to provide fair sharing of the wireless link. However, the MAC may not be able to accommodate the offered load alone as the channel behavior introduces errors. This means that the link quality may degrade. When link quality drops as a result of motion, noise, obstacles, etc., the link metrics exhibit new values and the routing protocol has to recompute the routes. This process continues indefinitely among the bottom three layers. One concludes that MAC and routing protocols depend strongly on each other; therefore, their joint performance should be optimized. Note also that the physical layer parameters have a significant impact on the performance of the MAC protocol and hence on the routing layers and the end-to-end network performance. The layer interdependencies may extend all across the protocol stack.

QoS routing in MANETs has only started to receive attention recently. To provide QoS in this kind of network, it is important to discover and react to future effects (e.g., route breakage or topology changes), which will produce service disruption and may result in the violation of the QoS guarantees. QoS routing, in particular, is more difficult in MANETs than in other types of networks for the following reasons:

- The absence of a fixed infrastructure coupled with the ability of nodes to move freely cause frequent route breakage and unpredictable topology changes.

- The overhead of QoS routing is too high for bandwidth limited MANET, because the mobile node should have some mechanisms to store, update, and maintain the precise link state information.
- The limited bandwidth resource is usually shared among adjacent nodes due to the wireless medium.
- The nodes themselves can be heterogeneous, thus enabling an assortment of different types of links to be part of the same network.
- The traditional meaning that the required QoS should be ensured once a feasible path is established is no longer true because of the mobility-caused path breakage or power depletion of the mobile nodes. QoS routing should rapidly find a feasible alternate route, that will guarantee continuous service.

Designers of QoS routing algorithms for MANETs must consider several design issues such as:

- QoS metric selection (e.g., bandwidth, delay)
- QoS path computation methods (e.g., reactive, proactive)
- QoS state propagation and maintenance
- Distributed versus centralized algorithmic design
- The routing architecture (e.g., flat or hierarchical)
- Scalability for large networks

The QoS routing protocol must also deal with imprecise state information due to node (router) movement and topology changes. Furthermore, a QoS routing scheme for ad hoc networks must balance efficiency and adaptivity, while maintaining low control overhead. Therefore, a MANET QoS routing protocol should have the following properties:

- A routing protocol should be distributed to increase reliability. Where all nodes are mobile, it is unacceptable to have a centralized routing protocol. Hence, a distributed operation is a basic requirement in the design of any routing protocols in MANETs.
- A routing protocol should assume routes as unidirectional links. Wireless medium may cause a wireless link to be available in one direction only due to physical factors and it may not be possible to communicate bidirectionally. Thus a routing protocol must be designed considering unidirectional links.
- A routing protocol should be power efficient.
- A protocol should be more reactive than proactive to avoid protocol overhead.

Examples of research addressing the QoS routing problem in MANETs can be found in [2–5]. These protocols differ with regard to the design choices discussed above. Table 25.1 presents a comparison of the design choices for some QoS routing protocol classes presented in this chapter.

Table 25.1. Comparison of QoS Routing Algorithms for MANETs

	Core-Based [3]	Ticket-Based [4]	Bandwidth-Based [7]	Predictive-Based [10]	Position-Based [40]	Power-Based [35]
QoS metric	BW	BW, DLY	BW	BW, DLY	BW	PWR
State maintenance	Local	Local	Global	two-hop information	Local	Local
QoS state propagation	BW changes	Periodic	Periodic	OD	OD	PWR changes
Routing class	Dist	Dist	Dist	Dist	Dist	Dist
Route computation	OD	OD	OD	OD	OD	OD
Routing architecture	Clustered	Flat	Flat	Flat	C, F	Clustered
Single or multiple paths	S	M	M	S	M	M
Power issues	No	No	No	No	No	Yes
Scheduling issues	No	No	No	No	No	No
Channel access	TDMA	TDMA	TDMA	TDMA	TDMA	TDMA

OD — on-demand; C, F — clustered or flat; Dist — distributed; BW — bandwidth, DLY — delay, S — single, M — multiple, PWR — power

The purpose of this chapter is to provide a survey on the recent developments in the area of QoS routing in MANETs. We present the different QoS routing problems and their challenges and the QoS routing strategies. We classify the different approaches and discuss various techniques within this classification. We also compare the existing routing algorithms and then outline the future issues and challenges related to the QoS routing problem in MANETs.

The rest of the chapter is organized as follows:

- Section 25.2, provides the basics of QoS in MANETs as well as the challenges and difficulties associated with the problem of QoS routing in MANETs.
- Section 25.3 presents a general and detailed literature review of the QoS routing schemes in MANETs and groups different approaches and classifies them according to many parameters and contrast similarities and differences among such techniques.
- Section 25.4 presents a perspective of future directions of QoS routing.
- Section 25.5 concludes with a summary of the features of QoS protocols presented and future research directions.

25.2 Quality of Service in MANETs: The Basics

QoS provisioning in MANETs is different from QoS provisioning in other networks, whether wireline or wireless infrastructured networks, in the sense that mobility (e.g., route breakage) may result in service disruption and cause QoS guarantees to be violated. Therefore, alternate routes that are also QoS compliant must be found before the existing data connection over the old route breaks in order to reduce the outage times. QoS support in MANETs encompasses issues at all layers of the protocol stack (i.e., application, transport, network, MAC, and the physical layers). This was motivated by the emergence of applications that require at least statistical QoS guarantees and sometimes deterministic guarantees. QoS routing is the facilitator for QoS provisioning in MANETs and the solution of the issues related to QoS routing is fundamental for enabling QoS in MANETs. The basic problem of QoS routing is to find a path that satisfies the QoS constraints or the required QoS guarantees.

QoS guarantees can be either deterministic or statistical. Although, deterministic guarantees will always be met. Under all circumstances, statistical guarantees allow the guarantee to be met with a certain probability. The characteristics of wireless MANETs preclude any tight bounds on QoS performance measures. For example, it would be useless to reserve sufficient resources via a resource reservation protocol to guarantee a worst case delay for a high priority flow, if we cannot guarantee the delay on wireless links, especially that those links are subject to outage. Instead, a QoS

Quality of Service Routing in Mobile Ad Hoc Networks: Past and Future

model that provides a statistical differentiation fits better in this environment. To illustrate this point, let us focus on the provisioning of deterministic QoS in MANETs, which requires that QoS guarantees be maintained in the presence of failures. For example, consider the following two cases of failures:

1. Transient data corruption or loss — there are two main options to address this problem — request a retransmission or accept the loss. From a QoS point of view, retransmission has the advantage of maintaining the integrity of the information. However, the act of requesting and receiving retransmission may be time consuming, which also impacts the required QoS guarantees.
2. Failure of part of the network — it is the situation where part of the network ceases to provide the required QoS in a sustained manner (e.g., in the case of network congestion or partial network disconnection). Two options are available — void the ongoing transmissions that depend on the failed part or relocate the activities from the failed part elsewhere. The latter option requires redundancy of services and possibly providing for partial renegotiation of QoS during the lifetime of a connection. This may raise considerations of cost in equipment, computations, and network resources.

To support QoS in MANETs, we need to define the QoS metrics that quantify QoS and then fully understand the research issues and difficulties in provisioning QoS in MANETs. This is discussed in detail in the following subsections.

25.2.1 QoS Metrics

QoS is usually specified as a set of service requirements that needs to be met by the network. These service requirements are in terms of end-to-end performance, such as delay, bandwidth, probability of packet loss, delay jitter, etc. Power consumption, security, and service coverage area are other QoS metrics specific to MANETs. Although loss probability, cost, and delay jitter are useful QoS metrics, delay and bandwidth are the two most important QoS metrics. In general, the QoS metrics could be concave, multiplicative, or additive. The following definition illustrates the difference between the three types.

Definition 25.1

Let P be a path that consists of the following k nodes $(s, n_1, n_2, ..., n_{k-3}, n_{k-2}, d)$ where (s, d) are the source and destination nodes, respectively. Let M be a QoS metric on the path P, and let $m_P(i, j)$ be the value of the QoS metric M on the link (i, j) on path P. Let the required value of the QoS metric M on the path P be represented by V(P). If $V(P) = \min(m_P(s, n_1), ..., m_P(n_{k-2}, d))$, then metric M is concave. The QoS metric M is additive if $V(P) = m_P(s, n_1) + m_P(n_1, n_2) + \ldots \ldots m_P(n_{k-2}, d)$. Finally, the QoS metric M is multiplicative if $V(P) = \Pi(m_P(s, n_1) \ldots m_P(n_{k-2}, d))$.

For example, the bandwidth metric is concave (i.e., a certain amount of bandwidth must be available on each link along the path). Delay, delay jitter, and cost are additive and the probability of packet loss can be expressed using a multiplicative relation. The bandwidth considered for making a routing decision is the residual bandwidth available for new traffic. The bandwidth of a path is defined as the minimum of the residual bandwidth of all links on the path or the bottleneck bandwidth. The delay has a number of components — queuing delay, scheduling delay, transmission delay, and propagation delay. Such metrics are not necessarily independent. Queuing delay, for example, is determined by bottleneck bandwidth and traffic characteristics. Note that bottleneck bandwidth and the total delay can be viewed as the width and length of a path. The problem of QoS routing is then to find a path in the network given the constraints on its width and length.

In some recent work, researchers argue that network security should be regarded as a QoS metric [25]. We will not consider security issues in this chapter, because it is an independent and a broad topic by itself. However, it is worth noting that most security protocols increase the overhead in terms of extra messages and increased data; therefore, the required security level may also be subject to a number of trade-offs applied by the QoS scheme.

25.2.2 Challenges of QoS Routing Support in MANETs

Several technical challenges face the design of efficient QoS routing protocols in MANETs mostly related to the constantly changing network topology, the limited bandwidth of a shared wireless medium, and the limited power capacity. Different QoS routing approaches were proposed in the literature in an attempt to overcome various difficulties that prevent the wide deployment of MANETs. These difficulties are due to the following challenges:

- Challenges due to the dynamic topology — the issue of mobility does not exist in fixed networks. Even in infrastructured wireless networks, the mobile nodes move from the domain of one access point to the domain of another access point. In MANETs, there is a high possibility that the topology may vary at a fast rate. The complications imposed by mobility in MANETs may severely degrade the network quality. The frequent route breakage is a natural consequence of mobility, which complicates routing. This problem is exacerbated when paths need to satisfy certain QoS guarantees during the connection lifetime. When the network topology changes frequently enough, it would not be possible for any protocol with reasonable overhead to discover the paths and establish the connections that provide QoS guarantees. As a result, design of QoS routing protocols is challenged by frequent topological changes in MANETs. Such topological dynamics are further complicated by the natural grouping behavior in the mobile user's movement, which leads to frequent network partitioning. Network partitioning poses significant challenges not only to the provisioning of QoS in ad hoc networks, but to connection establishment at large. This is because the partitioning disconnects many mobile users from the rest of the network.

- Challenges due to the scarce resources — the wireless spectrum is a limited resource that must be used efficiently. In addition, the wireless medium is a shared medium where signal attenuation, interference, multipath propagation effects, such as fading, and the unguided nature of the transmitted wave all contribute to wasting the bandwidth resource. Moreover, some overhead is often required to support reliable data transmission. Because bandwidth availability has a direct affect on the QoS routing, effective management of this resource is a key factor in QoS routing.

- Challenges due to the absence of communication infrastructure — standard networks use an infrastructure. In MANETs, there is no preexisting infrastructure, there is no default router, and every mobile node should be able to act as a router and be able to forward packets to other nodes. Therefore, a QoS routing protocol must consider the self-creating and self-organizing features of MANETs.

- Challenges due to power limitations — power is a limited resource in MANETs. Solutions that reduce power consumption will often be favored, all other factors being equal. Mobile nodes need to use their battery limited power supply in a manner that prolongs the lifetime of the battery. If the battery power is used blindly, mobile nodes will fail quickly and this affects the network availability and functionality. Service disruption due to power failure is therefore a problem that needs to be avoided. Power-aware routing schemes, therefore, are designed to provide solutions for this problem. Power-aware routing routes are selected such that nodes with high remaining power are selected.

- Challenges due to heterogeneous nodes and networks — mobile nodes can be heterogeneous, thus enabling an assortment of different types of links to be part of the same ad hoc network. MANETs are typically heterogeneous networks with various types of mobile nodes. In a military application, different military units ranging from soldiers to tanks can come together, hence forming an ad hoc network. Nodes differ in their power capacities and computational powers. Thus, mobile nodes will have different packet generation rates, routing responsibilities, network activities, and power draining

rates. Dealing with node heterogeneity is a key factor for the successful operation of heterogeneous MANETs.
- Challenges due to link quality — the problem of link quality is particularly significant in MANETs. The essential effect on MANETs is that the link quality can become extremely variable, often in a random manner. Although some parts of this effect can be predicted because variations in link quality impact packet delivery and trigger error recovery procedures, the main QoS parameters such as bandwidth availability, latency, reliability, and jitter are all affected. This effect can happen either during or between connections. In the former case, a link quality might become too bad while a connection is in place. In the latter, a new connection with the same requirements as a previously established one is rejected because link status variability increases and, as a result, the link becomes unreliable.
- Challenges due to the maintenance of state information — QoS routing consists of two basic tasks. The first task is to collect the state information and keep it up-to-date. The second task is to find a feasible path for a new connection based on the collected information. The performance of any routing algorithm directly depends on how well the first task is solved. State information can be local or global. In local state, each node is assumed to maintain its up-to-date local state, including the queuing and propagation delay, the residual bandwidth of the outgoing links, and the availability of other resources. The combination of the local states of all nodes is called a global state. Every node is able to maintain the global state by using a link-state based routing protocol, which exchanges the local states among the nodes periodically. The global state kept by a node is always an approximation of the current network state due to the constantly varying network topology and, consequently, link states that encounter a nonnegligible delay for propagating between nodes. In general, as the network size grows the imprecision increases.
- Challenges due to other layers — the choice of the medium access scheme in MANETs is difficult due to the time-varying network topology and the lack of centralized control. Time division multiple access (TDMA) or dynamic time assignment of frequency bands is complex because there is no centralized control. Frequency division multiple access (FDMA) is inefficient in dense networks and code division multiple access (CDMA) is difficult to implement due to node mobility and the subsequent need to keep track of the frequency-hopping patterns or spreading codes for nodes in a time-varying neighborhood. At the MAC layer, we also have link layer reliability problems that are related to the high bit error rate, in addition to the possible packet collision problems. QoS at the MAC layer needs further research (QoS-aware MAC). Note that a QoS-aware MAC can serve as an infrastructure for facilitating the QoS routing.

Quality of Service Routing in Mobile Ad Hoc Networks: Past and Future

- Challenges due to lack of centralized control — the QoS path discovery process may be centralized or distributed. In MANETs, distributed algorithms are preferable due to the lack of a central entity. Although distributed applications are better suited for MANETs due to its peer-to-peer architecture, important network applications and services such as Web servers, location information databases, and network services (Dynamic Host Configuration Protocol [DHCP], Simple Network Management Protocol [SNMP]) are inherently centralized. These services are often critical to the mobile node's operation, such that every node requires constant and guaranteed access to them. Therefore, designing protocols for MANETs requires paying attention to both issues.

In the next section, we review the efforts that have been exerted in the area of QoS routing in MANETs. These efforts attempt to overcome, or at least circumvent, the above challenges to provide possible solutions for the problem of QoS routing in MANETs.

25.3 QoS Routing Protocols in MANETs: Current Trends

In this section, we present a survey of the QoS routing schemes and protocols presented in the literature. Compared to the abundant work on QoS routing for fixed wireline networks, results for QoS routing in MANETs are relatively scarce due to the difficulties mentioned earlier, as well as the relative recency of MANETs. However, some promising work on QoS routing in MANETs has emerged recently. In the remainder of this chapter, we present a survey of this work. To streamline this survey, we use a classification according to the network structure and quality of service constraints. The classification is shown in Figure 25.2.

QoS routing protocols in MANETs are classified into four broad categories:

1. Flat routing schemes, which are further classified, according to their design philosophy, into five classes:
   a. Proactive
   b. Reactive
   c. Predictive
   d. Ticket-based probing
   e. Bandwidth-based
2. Hierarchical routing which can be a multitier network or two-tier network in its simple form
3. Position-based routing that may or may not use an external location determination server (e.g., global positioning system [GPS])
4. Power-based routing that optimizes the use of battery lifetime

In flat routing approaches, each node participating in QoS routing plays an equal role. In contrast, hierarchical routing usually assigns different roles

MOBILE COMPUTING HANDBOOK

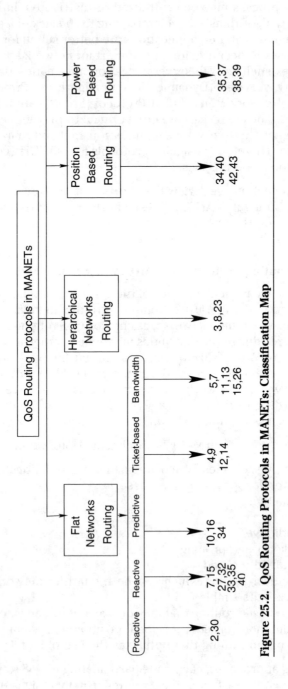

Figure 25.2. QoS Routing Protocols in MANETs: Classification Map

to mobile nodes. QoS routing that needs help from geographic location system (e.g., GPS) requires each node to be equipped with a GPS card. This requirement is quite realistic today because such devices are becoming inexpensive and can provide reasonable precision. However, a GPS-free approach [41] may also be used if GPS use for a certain application is infeasible. With power being a limited resource in mobile ad hoc networks, routing that takes into consideration the power level at each node is a necessity.

25.3.1 QoS Routing in Flat Networks

The protocols we review here fall into five categories:

1. Proactive, table-based routing schemes
2. On-demand or reactive routing
3. Predictive routing
4. Ticket-based routing
5. Bandwidth-based routing

Proactive table-based routing schemes require each of the nodes in the network to maintain and update routing information, which is used to determine the next hop for a packet transmission to reach destination. On-demand routing is an emerging routing philosophy in MANETs. Routes are created when necessary based on a query-reply approach. It differs from the conventional proactive routing protocols in that no permanent routing information is maintained at network nodes, thus providing a scalable routing solution in large MANETs. Predictive routing detects changes in link status and network topology *a priori* and uses this information to build stable routes that have low probability of failing. The basic idea of ticket-based probing is described as follows. A ticket is the permission to search one path. The source node issues a number of tickets based on the available state information. Probes (routing messages) are sent from the source toward the destination to search for a low-cost path that satisfies the QoS requirement. Each probe is required to carry at least one ticket. At an intermediate node, a probe with more than one ticket is allowed to be split into multiple ones, each searching a different downstream subpath. The maximum number of probes at any time is bounded by the total number of tickets. Because each probe searches a path, the maximum number of paths searched is also bounded by the number of tickets. If a QoS routing protocol supports QoS via separate end-to-end bandwidth calculation and allocation mechanisms, it is called bandwidth-based routing. The bandwidth-based routing scheme depends on the use of TDMA medium access scheme in which the wireless channel is time-slotted and the transmission scale is organized as frames (each containing a fixed number of time slots). A global clock or time-synchronization mechanism is used such that the entire network is synchronized on a frame and slot basis.

25.3.1.1 Proactive QoS Routing Protocols.

Proactive protocols have many desirable properties, especially for applications with real-time communications. This type of communication requires QoS guarantees, such as low-latency route establishment, alternate QoS path support, and resource state monitoring. Because the satisfaction of such a requirement is dependent on the accuracy of the routing information stored in the tables, frequent network topology changes may render this information obsolete. In MANETs, the performance of proactive routing protocols deteriorates with frequent node movement.

In [2], the Optimized Link State Routing (OLSR) protocol was proposed. OLSR is a proactive routing protocol that exchanges topology information with other nodes in the network regularly. The protocol inherits the stability of a link state algorithm and has the advantage of having routes immediately available when needed due to its proactive nature. OLSR is an optimization over the classical link state protocol, tailored for mobile ad hoc networks. The key concept used in the protocol is that of multi-point relays (MPRs). MPRs are selected nodes that forward broadcast messages during the flooding process. MPR selection is the key point in OLSR. The smaller the MPR set is, the less overhead the protocol introduces. The idea of MPR is to minimize the overhead of flooding messages in the network by reducing duplicate retransmissions in the same region. Each node in the network selects a set of nodes in its neighborhood that may retransmit its messages. Each node selects its MPR set among its one-hop neighbors. This set is selected such that it covers (in terms of radio range) all nodes that are two hops away. The nodes which are selected as a MPR by some neighbor nodes announce this information periodically in their control messages. Therefore, a node announces to the network, that it has reachability to the nodes which have selected it as MPR. In route calculation, the MPRs are used to form the route from a given node to any destination in the network. The protocol uses the MPRs to facilitate efficient flooding of control messages in the network. OLSR inherits the concept of forwarding and relaying from High Performance Radio Local Area Network (HIPERLAN) (a MAC layer protocol). These MPRs can be used to facilitate the efficient flooding of control messages. The QoS support in the protocol is implemented by extending the routing table to include two parameters — minimum available bandwidth and maximum delay. It is worth mentioning that selecting MPRs is similar to finding the minimum dominating set of an arbitrary graph.

Three algorithms were developed in [29] that allow OLSR to find the maximum bandwidth path. Through simulation, these algorithms were shown to improve OLSR in the static network case. Two of the proposed algorithms, called $OLSR_{R2}$ and $OLSR_{R3}$ and discussed below, were found to perform well (i.e., guarantee that the highest bandwidth path between any two nodes is found). In the first algorithm, called $OLSR_{R1}$, MPR selection is

Quality of Service Routing in Mobile Ad Hoc Networks: Past and Future

almost the same as that of OLSR described earlier. However, when there are more than one one-hop neighbors covering the same number of uncovered two-hop neighbors, the one with the largest bandwidth link to the current node is selected as MPR. The idea behind the second algorithm, called $OLSR_{R2}$, is to select the best bandwidth neighbors as MPRs until all the two-hop neighbors are covered. The third algorithm, called $OLSR_{R3}$, selects the MPRs in a way such that all the two-hop neighbors have the optimal bandwidth path through the MPRs to the current node. Here, optimal bandwidth path means the bottleneck bandwidth path is the largest among all the possible paths. Note that the overhead when using $OLSR_{R3}$ may increase compared with the original OLSR algorithm because we may increase the number of MPRs in the network. This is because $OLSR_{R3}$ may select a different MPR for each two-hop neighbor.

25.3.1.2 Reactive QoS Routing Protocols. A recent approach intended to overcome the scalability and routing overhead problems of proactive protocols is the reactive, on-demand routing. Examples are destination sequenced distance vector (DSDV) [18] and Dynamic Source Routing (DSR) [17], which split routing into discovering a path and maintaining a path. Maintaining a path happens only while the path is in use in order to make sure that it can still be used. Thus, no periodic updates are needed. The key distinguishing feature of DSR is the use of source routing. That is, the sender knows the complete hop-by-hop route to the destination. These routes are stored in a route cache. The data packets carry the source route in the packet header. If any link on a source route is broken, the source node is notified using a route error (RERR) packet. The source removes any route using this link from its cache. A new route discovery process must be initiated by the source if this route is still needed.

The Ad Hoc On-Demand Distance Vector Routing Protocol (AODV) [28] has been proposed for best effort routing in MANETs. When a route to a new destination is needed, the node broadcasts a route request (RREQ) packet to find a route to the destination. Each node that participates in the route acquisition process places in its routing table the reverse route to the source node. A route reply (RREP) packet, which contains the number of hops required to reach the destination node, D, and the most recently seen sequence number for the node D, can be created whenever the RREQ reaches either the destination node or an intermediate mode with a valid route to the destination. To provide QoS support using AODV, a minimal set of QoS extensions has been specified for the RREQ and RREP messages [27]. Specifically, a mobile host may specify one of two services — maximum delay or minimum bandwidth. Before a node can rebroadcast a RREQ or unicast a RREP to the source, it must be capable of meeting the QoS constraints. Upon detecting that the requested QoS can no longer be maintained, a node must send an Internet Control Message Protocol (ICMP) QoS

LOST message back to the source. The specific extensions for the routing table and control packets (e.g., RREQ and RREP messages) are shown in [27].

In [7], a QoS routing protocol that can establish QoS routes with reserved bandwidth in a network employing TDMA is presented. A small network is assumed, where the topology changes at a relatively slow rate, and sessions transmit at constant bit rates. Assuming TDMA is used at the MAC layer, the QoS measure is the amount of bandwidth (given in number of time slots). The authors showed that the path bandwidth calculation problem is NP-complete. Hence, an algorithm for calculating the end-to-end bandwidth on a path is developed and used together with the route discovery mechanism of AODV to set up QoS routes. A session specifies its QoS requirement as the number of transmission time slots it needs on its route from a source to a destination. For each session (flow), the QoS routing protocol will both find the route and the slots for each link on the route. To provide a bandwidth of R slots on a given path P, it is necessary that every node along the path find at least R slots to transmit to its downstream neighbor and these slots do not interfere with other transmissions. Because of these constraints, the end-to-end bandwidth on the path is not simply the bandwidth on the bottleneck link. In addition to building QoS routes, the protocol also builds a best-effort route when it learns of such a route. The best-effort route is used when a QoS route is not available.

A shortcoming of the QoS routing protocol in [7] is that it is designed without considering the situation when multiple QoS routes are being set up simultaneously. A route request is processed under the assumption that it is the only one in the network at the moment. When multiple routes are being set up simultaneously, they reserve their own transmission time slots independently and may interfere with one another. It is possible that two QoS routes will block each other when they are trying to reserve the same time slots simultaneously, which may lead to a deadlock situation. This protocol works well in small networks (or over short routes) and under low network mobility. For a large or highly mobile network, it lacks the scalability and the flexibility to deal with frequent route failures.

In [15], Chen et al. presented an on-demand, link-state-based routing protocol. The proposed protocol can find multiple paths between a source–destination pair. CDMA-over-TDMA channel techniques were used to calculate the end-to-end path bandwidth of a QoS multipath routing. (Figure 25.3 describes a reservation example where in (A) the link state information is collected and in (B) the destination selected two paths to receive data from the source node.) The basic idea of this protocol is to reactively collect link state information from source to destination. The information will be used to construct a flow network, which is a network topology sketched from source to destination. If there are multiple paths between the source–destination pair, the destination node will select the

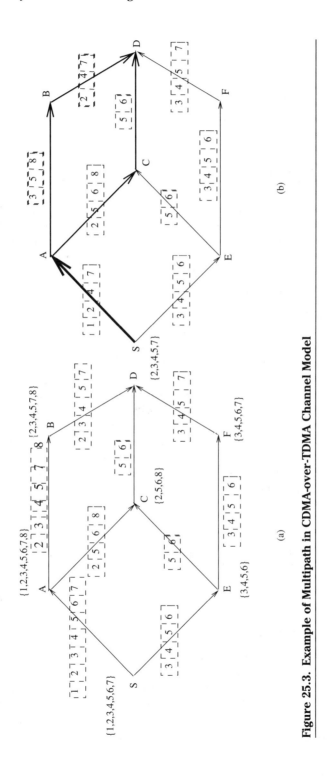

Figure 25.3. Example of Multipath in CDMA-over-TDMA Channel Model

MOBILE COMPUTING HANDBOOK

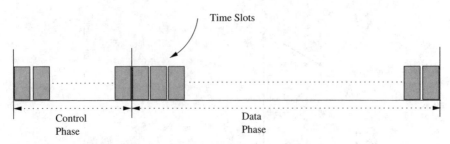

Figure 25.4. TDMA Time Frame Divided into Two Phases

path that satisfies its bandwidth requirement and replies back to the source node. The protocol is able to produce high success rate for the route requests. However, the method used to calculate or reserve bandwidth is not clear. The scalability of the proposed protocol for large networks is questionable. In fact, all protocols that use TDMA to calculate bandwidth are only efficient for small and low mobile networks.

In [31], Lin proposed an on-demand QoS routing protocol that also performs admission control. The QoS metric is bandwidth. Multiple paths are searched at the same time between a source–destination pair. If multiple QoS paths are found, the shortest path will be selected. Because the scheme is pure on-demand, there is no maintenance of any routing tables and there is no exchange of routing information. The path is only discovered upon request. Therefore, in large networks, this path discovery process may incur significant setup times. The network is using a TDMA slot via CDMA channel allocation among different nodes (i.e., CDMA is overlaid on top of the TDMA infrastructure). Hence, multiple sessions can share a common TDMA slot via CDMA. The process is carried through two phases — control phase and data phase — as shown in Figure 25.4. In the control phase, all control functions, such as slot and frame synchronization, power measurement, code assignment, slots request, etc., are performed. The amount of data slots per frame assigned to a virtual circuit (VC) is determined according to bandwidth requirement. In the control phase, each node uses pure TDMA with full power transmission to broadcast its information to all of its neighbors in a predefined slot, such that the network control functions can be performed in a distributed fashion. All nodes take turn in this process. It is assumed that the information can be heard by all of a node's adjacent nodes.

By the end of the control phase, each node should know the channel reservation status and can decide which free slots to request, if any. Therefore, the available path between two nodes is the set of free slots between them. When the destination node receives a RREQ packet from the source node, it returns a RREP packet by unicasting back to the source. If the reservation process fails, then the scheme gives up and sends an appropriate message back to the destination. The control messages are also used to calculate bandwidth hop-by-hop. The success of this scheme is heavily dependent on the reservation procedure and it may incur large setup times in large networks. It also does not provide fairness among different connections, as slots are not distributed among nodes in a fair fashion.

In [32], a QoS routing framework that consists of admission control, resource reservation, and QoS routing were proposed. The shortest path was always used for routing packets. The framework deals with finding shortest paths that satisfy a minimum bandwidth through on-demand channel assignment methods. The framework is heavily dependent on the resource reservation and admission control components where resources are tentatively reserved when a new connection with certain QoS bounds is required. Because nodes are mobile, resource reservation is performed at places where a mobile node is expected to visit. Two on-demand channel assignments proposed by the authors in a different work are employed in the proposed framework. The authors claim that their proposed framework is capable of performing well consistently in all performed experiments. The disadvantage of this approach is that its complexity is high and it is hard to make a resource reservation in MANETs as the nodes are highly mobile. Furthermore, the proposed bandwidth assignment does not really guarantee the bandwidth as links are prone to failure. In addition, the use of shortest path does not always guarantee the required quality, which affects the percentage of accepted calls and could cause congestion on some paths.

25.3.1.3 Predictive QoS Routing Protocols. In this kind of QoS routing, the protocol attempts to predict the link state change or topology change beforehand and tries to avoid using unstable links in calculating a path. This scheme is hence probabilistic in nature and its success is highly dependent on the ability to predict well with high probability.

In [10], Shah and Nahrstedt discussed the use of the updates of geographic location to develop a predictive location-based QoS routing scheme based on a location resource update protocol, which assists a QoS routing protocol. The motivation is that state information in a dynamic environment like MANETs does not remain current for long. The predicted locations are used to build future routes before existing routes break and thus avoid route recomputation delay. The proposed protocol is heavily dependent on the prediction of node locations. The direction of motion of a mobile node is taken into account when attempting to predict its future location. The approach involves an update protocol, a location-delay prediction scheme, and the QoS routing protocol to route multimedia data to the destination. The location-delay prediction scheme is based on a location-resource update protocol, which assists the QoS routing protocol. The update protocol is used to distribute nodes geographical location and resource information (e.g., battery power, queuing space, processor speed, transmission range, etc.). The update packets used by the update protocol contain time stamps, current geometric coordinates, direction of motion, velocity, and also resource information pertaining to the node that is used in the QoS routing protocol. The location-delay prediction scheme is used to predict the new location and the end-to-end delay of a node based on the last update messages received from that node. The scheme assumes that the end-to-end delay for a data packet from, say a to b, is equivalent to the delay experienced by the latest update from b to a. It has been found that the end-to-end delays for packets travelling between a and b usually remain more or less similar for no more than 0.5 seconds. This means that the end-to-end delay to node b will be predicted to be equal to the delay experienced by the most recent update that has arrived from b. The delay prediction for each node is based on the end-to-end propagation delay for that node from a certain node, say a.

In QoS routing, each node maintains two tables — the update table and routing table — to compute paths at the source node. Although this protocol seems promising, the overhead in maintaining and updating tables is high. Moreover, the prediction accuracy is highly dependent on the previous states of mobile nodes, which might be misleading. A QoS-aware admission control is also proposed in [10]. A small mathematical model is used to predict the new location of a mobile node. A mobile node uses the following two equations to determine its new position (x_p, y_p) at a future time t_p:

$$x_p = x_2 + \frac{v(t_p - t2)(x_2 - x_1)}{[(x_2 - x_1)^2 + (y_2 - y_1)^2]^{1/2}}, \quad y_p = y_1 + \frac{(y_2 - y_1)(x_p - x_2)}{x_2 - x_1}.$$

Using the proposed scheme and to be able to do the above calculations, each mobile node has information about the whole topology of the network. It can thus compute a source route from itself to any other node using the information it has and can include this source route in the packet header when it is routed. The prediction scheme proves reduction in the overhead involved in the updates and the QoS routing scheme. However, the prediction scheme keeps track of a lot of information (e.g., tables) and must update these tables frequently when topology or link states change, which results in large computation and communication costs.

In [16], the authors propose a QoS routing scheme that is an extension of the Precomputation-based Selective Probing (PCSP) scheme. PCSP precomputes QoS variations as well as the cost and QoS metrics of the least-cost path (LC path) and the best quality path (BQ path) at each node, taking into account the imprecision of state information. The information is then used with two selective probing methods where one uses the QoS-satisfying LC paths and the other using the least cost of found feasible paths, thus excluding the paths that can never be optimal. To be able to execute the algorithm, the set of links are divided into two different sets — a stationary link set and a transient link set. A routing path is supposed to use stationary links whenever possible to reduce the path failure probability. A newly formed link is regarded as a transient link, but becomes a stationary link if it survives for a certain duration of time. The proposed PCSP scheme can achieve low routing overhead and small route setup time while guaranteeing high success ratio. However, the measurements of the lifetime of the links are based on predictions of the link states, which might change more often as nodes are free to move. The proposed scheme cannot guarantee that by trying to capture the network state in advance, that the paths found are optimal.

In [33], a QoS routing protocol with mobility prediction was proposed. This protocol searches for paths that consist of low mobility nodes. Therefore, the lifetime of this path will be longer than other paths. The protocol uses a metric of path expiration time, which is based on the node's mobility speed and location with respect to other nodes. Although this protocol shows a good performance, the nodes' mobility patterns and speeds in MANETs cannot be enforced or assumed before hand. Hence, the accuracy of the calculated path or the expected lifetime of that path cannot really be measured.

25.3.1.4 Ticket-Based Probing Routing. The basic idea of the ticket-based probing scheme [4] is to utilize tickets to limit the number of paths searched during route discovery. A ticket is the permission to search a single path. When a source wishes to discover an admissible route to a destination, it issues a probe (routing message) to the destination. A probe is required to carry at least one ticket, but may carry more tickets (i.e., connection requests with tighter requirements are issued more tickets). At an intermediate node, a probe with more than one ticket is allowed to split into multiple ones, each searching a different downstream subpath. Hence, when an intermediate node receives a probe, it decides, on the basis of its available state information, whether the received probe should be split and to which neighbors the probes should be forwarded. In the case of route failures, ticket-based probing utilizes three mechanisms — path rerouting, three level path redundancy, and path repairing. Rerouting requires that the source node be informed of a path failure. After which, the source initiates the ticket-based algorithm to locate another admissible route. The path redundancy scheme establishes multiple routes for the same connection. For the highest level of redundancy, resources are reserved along multiple paths and every packet is routed along each path. In the second level of redundancy, resources are reserved along multiple paths; however, only one is used as the primary path and the others serve as backup paths. In the third level of redundancy, multiple paths are

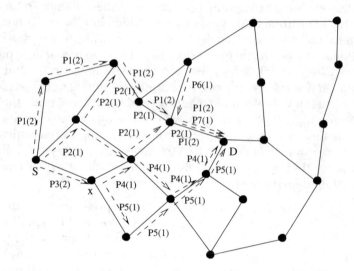

Figure 25.5. An Example of Use of Ticket's in Ticket-Based QoS Rating
(Number in parentheses is number of tickets.)

selected, but resources are only reserved on the primary path. The path-repairing mechanism tries to avoid the cost of rerouting by attempting to repair the route at the point of failure. Ticket-based probing is proposed as a general QoS routing approach for MANETs and can handle different QoS constraints (i.e., bandwidth, delay, packet loss, and jitter). For example, ticket-based QoS routing solutions for the bandwidth and delay-constrained routing problems were presented in [4].

In [4], backup paths for maintenance of the routing paths when nodes move, join, or leave the network are also used. The proposed algorithms require that each node knows (with some imprecision) the delay and bandwidth on the least-delay and largest-bandwidth paths to any possible destination. Thus, in effect, each node should maintain the states of all links in the network, which requires a quadratic communication overhead for the state updates. Figure 25.5 shows an example of how source uses three probes p_1, p_2, and p_3 to find a path to the target D. When one or more probes arrives at the destination node, the hop-by-hop path is known and delay/bandwidth information can be used to perform resource reservation for the QoS satisfying path.

In wireline networks, a probability distribution can be calculated for a path, based on the delay and bandwidth information. In MANETs, however, building such probability distribution is not difficult and not feasible. This is because wireless links are subject to breakage at any time and state information of link and network topology is imprecise in nature. Hence, a simple model that provides information about the state of the network, although still imprecise, is needed. Such a model was proposed in the ticket-based probing algorithm in [4]. It uses history and current estimated delay variations and smoothing formula to find an estimation for the current delay. Moreover, to adapt to the dynamic topology of MANETs, this algorithm allows the different levels of route redundancy mentioned earlier. A route maintenance technique that provides rerouting in case the original path breaks was also proposed. Two routing techniques are considered in [4], both limited to low mobility networks. The first technique is based on the availability of only local state information; the other assumes possibly inaccurate knowledge of global states. For QoS routing using only the local state information, [4] introduces two different distributed routing algorithms, a source initiated routing algorithm and a destination-initiated routing algorithm. Both rely on the use of probe packets (ticket-based probing) for identifying a feasible route where the information for multiple feasible routes is stored in the probes instead of the intermediate routers. A broken route in this scheme is detected by using the beaconing protocol that detects adjacent neighbors. Another modified scheme for finding a QoS route based on the ticket-based probing is presented in [14], where the authors proposed an algorithm with two modified terminating strategies for the original method proposed in [4]. The objective of the modification is to reduce the message overhead per connection. The terminating strategies are designed for delay constrained routing where the routing process ends at the destination nodes either when all tickets reach destination or if a probe time-out occurs. The time-out is needed for an invalidated probe which traverses the network until it reaches the destination. The invalid probe traversal increases the message overhead. Hence, it was proposed to discard all invalid probes instead of sending them to their destinations. To do that, the authors proposed to modify the data structure of the ticket distribution process where a new field called hotness is added. Hence, the termination will depend both on the total number of tickets and the hotness field value. The hotness field represents the degree of importance of the connection: the higher the value, the more important the connection. The first termination strategy deals with the hot value probes and the second termination strategy deals with cold probes. When the probe is hot, destination responds quickly, but if the probe is cold, the destination will wait for more probes to arrive before the time-out. At the end of time-out, the destination selects the path that suits the connection.

Liao et al. have proposed a multipath QoS routing protocol in [12]. The protocol also searches for multiple QoS paths at the same time. The QoS metric used is the available bandwidth. The multiple paths found collectively satisfy the required QoS. In general, the multipath QoS routing algorithm is suitable for MANETs with limited bandwidth. For the duration of the connection and due to mobility, a single path satisfying the QoS requirements is unlikely to exist all the time. Hence, multiple paths can be used in parallel to deliver packets to the destination given that all these paths satisfy the required QoS guarantees. The difference between this protocol and that proposed in [4] is that multiple paths are used at the same time to deliver packets, but in [4], different paths are tested and the most suitable path is selected by the destination to deliver the data packets.

In [9], the authors developed an algorithm that is based on the ticket probing and they used fuzzy logic to model imprecise network information. The proposed algorithm is used to generate a maximum number of probe messages that searches for a path. A hop-by-hop path selection criteria is made to select the best path among the candidate paths. The difference between [4] and [9] is the way that dealing with imprecise state information for the QoS metrics considered. In [9], the number of generated probe messages is based on an inference rule derived by a fuzzy logic system.

25.3.1.5 Bandwidth Calculation Based Routing. Using a TDMA scheme in wireless MANETs allows each node to know about the free slots between itself and its neighbors through a simple broadcasting scheme. Based on this information, bandwidth calculation and assignment can be performed distributively. However, determining slot assignment while searching for the available bandwidth along the path is a NP-complete problem. Bandwidth calculation requires knowledge of the available bandwidth on each link along the path as well as resolving the scheduling of free slots.

Note that in wireline networks, the path bandwidth is the minimum available bandwidth of the links along the path. In time-slotted MANETs, however, bandwidth calculation is much harder. In general, we not only need to know the free slots on the links along the path, but also need to determine how to assign the free slots at each hop.

In [5], a DSDV–based QoS routing scheme was proposed. The routing protocol provides QoS support via separate end-to-end bandwidth calculation and allocation mechanisms, thus called bandwidth-based routing. The proposed bandwidth routing scheme depends on the use of a CDMA over TDMA [26] medium access scheme in which the wireless channel is time-slotted, the transmission scale is organized as frames (each containing a fixed number of time slots), and a global clock or time-synchronization mechanism is used. That is, the entire network is synchronized on a frame and slot basis. The path bandwidth between a source and destination is defined as the number of free or available time slots between them. Bandwidth calculation requires knowledge of the available bandwidth on each link along the path as well as resolving the scheduling of free slots. This problem is NP-complete and thus, requires a heuristic approach. To support fast rerouting during path failures (e.g., a topological change), the bandwidth routing protocol maintains secondary paths. When the primary

Quality of Service Routing in Mobile Ad Hoc Networks: Past and Future

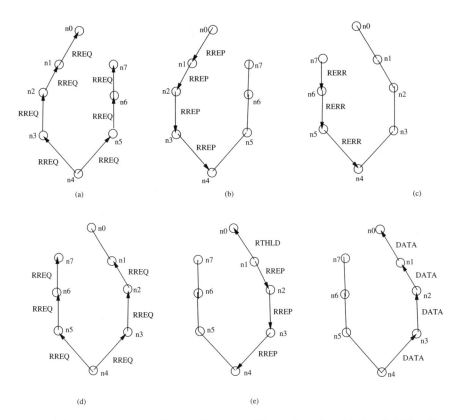

Figure 25.6. An Example of Route Setup and Route Repair with the QoS Routing Protocol
(A similar example appeared in C. Zhu and M.S. Corson, *InfoCom 2002*. vol. 2, pp. 958–967, 2002.)

path fails, the secondary route is used (i.e., becomes the primary route) and another secondary is discovered. A similar algorithm has been proposed by Lin and Liu [5], where an end-to-end bandwidth calculation and bandwidth allocation scheme is also used to assign link bandwidth. The algorithm in [5] also maintains secondary and ternary paths between source and destination for the use in rerouting when the primary path fails.

In [7], an on-demand QoS routing protocol based on AODV for TDMA-based MANETs was proposed. A QoS-aware route reserves bandwidth from source to destination by reserving free time slots. In the route discovery process of AODV, a distributed algorithm is used to calculate the available bandwidth on a hop-by-hop basis and using the RREQ/RREP query cycle as shown in Figure 25.6A and Figure 25.6B, where node n_4 initiates the route discovery process. When a route fails, a RERR message is sent back to the source as illustrated in Figure 25.6C. Only the destination node can reply to a RREQ message that has come along the path with sufficient bandwidth. If

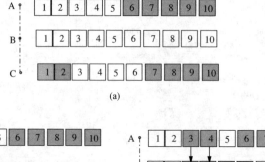

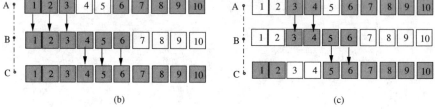

Figure 25.7. TDMA-Based Bandwidth Reservation Protocol

an intermediate node learns of such a path, a RREQ message received by this intermediate node can also be sent back by this intermediate node, hence quicker responses can be obtained. This is explained in Figure 25.6D, Figure 25.6E, and Figure 25.6F where node n_1 replied to the RREQ and data starts flowing on the returned path. If the RREQ is received by a node with inadequate bandwidth, it will be dropped by the intermediate nodes. The protocol can handle limited mobility by restoring broken paths and it is suited for small MANETs with short routes.

In [11], a bandwidth reservation scheme is also used to provide QoS guarantees in MANETs. A route discovery protocol is proposed, which is able to find a route with a given bandwidth (represented by number of slots). The proposed protocol can reserve routes while addressing both the hidden-terminal and exposed-terminal problems. Unlike previous protocols, this protocol addresses the previous two problems by not assuming that the bandwidth of a link can be determined independently of its neighboring links. Recall that a QoS request is represented in terms of number of slots. Figure 25.7 shows an example of how bandwidth reservations are applied in this protocol. The QoS route discovery is an on-demand source routing basis and works similar to the DSR protocol on disseminating route-searching packets. Take the path from A to C going through node B. In Figure 25.7A, the white slots associated with each node are free and the gray slots are busy. Matching the free slots between nodes, we obtain five common free time slots (i.e., 1, 2, 3, 4, 5) between A and B and four common free time slots (i.e., 3, 4, 5, 6) between B and C. In Figure 25.7B, an attempt is made to make a reservation from A to C with three slots. Unfortunately, this is not achievable for obvious reasons. Hence, the possible amount of reservation from A to C is only two slots as shown in Figure 25.7C and the situation cannot be improved, unless we change the assignment for A.

In [13], an AODV variant protocol, called QoS-AODV, is proposed. Similar to [11], QoS-AODV incorporates slot scheduling to find QoS routes over the network. The differences between [13] from [11] is that the protocol in [13] uses an integrated route discovery and bandwidth reservation protocol. Unlike other path finding and route discovery protocols that ignore the impact of the data link layer, QoS-AODV incorporates slot scheduling information to ensure that end-to-end bandwidth is actually reserved. QoS-AODV performs path search simultaneously with time slot scheduling by using simple heuristic algorithms. The protocol creates virtual connections by reserving MAC TDMA time slots along one of the discovered paths. Another protocol proposed in [15], which is a DSDV variant protocol, also performs bandwidth calculation and reservation using TDMA. The CDMA-over-TDMA Channel Model is used in this protocol, where the use of a time slot on a link is only dependent on the status of its one-hop neighboring links. Multipath routes are searched to the destination that satisfies the bandwidth requirement. The destination determines a QoS multipath routing and replies to the source node. This protocol collects link state information from source to destination in a reactive manner. A mobile node knows the available bandwidth to each of its neighbors. When a source node S needs a route to a destination D of bandwidth B, it will send out some RREQ packets, each of which carries the path history and link state information. Each RREQ packet records all link state information from source to destination. The destination collects all possible link state information from different RREQ packets sent from the source. A partial network, which is a flow network, is constructed in the destination node after receiving multiple information packets. An algorithm is applied at the destination to determine a better result for QoS multipath routing. After determining a multipath route, a reply packet is sent from the destination to the source. On the reply's way back to the source, the bandwidths are confirmed and reserved. The protocol works in a number of phases outlined in details in [15].

25.3.2 Hierarchical QoS Routing Protocols

Hierarchical or cluster-based routing, originally proposed in wireline networks, is a well-known technique with advantages related to scalability and efficient communication. As such, the concept of hierarchical routing is also used to perform QoS routing in MANETs.

Using concepts from multilayer adaptive control, Chen et al. proposed in [8] an approach for controlling QoS in large ad hoc networks by using hierarchically structured multi-clustered organizations. In the proposed

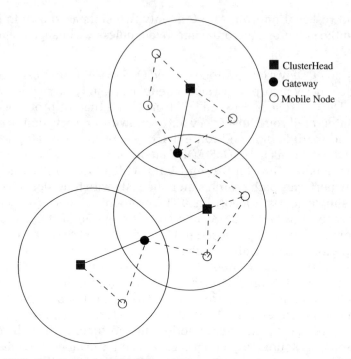

Figure 25.8. Clustering in MANETs: an Example

approach to QoS routing in MANETs, the network can be stand-alone or connected to wireline networks. The proposed scheme uses bandwidth as the QoS metric for nodes grouped in clusters and using a time slotted scheme (i.e., TDMA) inside each cluster. The role of cluster dynamics and mobility management, as well as resource reservation and route repair and router movement on QoS are addressed in detail. First, the network is partitioned into clusters where neighboring clusters use different spreading codes to reduce interference and to enhance spatial reuse of channels. Within each cluster, the MAC is implemented using a time slotted frame, which is divided into a control phase and data phase as described earlier. The QoS metric considered is the bandwidth represented in terms of time slots. The available bandwidth computation is carried out independently at each node. A loop free DSDV scheme is used for routing in this architecture. To avoid loops in DSDV, modifications to the well-known Bellman–Ford algorithm, which is the basis for DSDV, is applied. Although the proposed scheme makes what is called fast reservation in order to adapt to network mobility, it is still hard to guarantee QoS requirement based on reservations only. The protocol is an extension to the DSDV routing algorithm. Figure 25.8 shows an example of clustering used in this protocol where clusterheads maintain the TDMA-based schedules in each cluster.

Two new QoS routing schemes, both based on link state protocols as the underlying mechanism, appear in [9]. Both protocols attempt to reduce routing update overhead: one by selectively adjusting the frequencies of routing table update, and the other by reducing the size of the table. The QoS path is computed locally at each source node based on the routing table. Although the states can never be guaranteed to be accurate, the update selection strategy managed to avail a helpful network state that improves the network performance.

In [23], an extension to the Fisheye State Routing (FSR) and Hierarchical State Routing (HSR) protocols with QoS guarantees were proposed. QoS support is offered by adding an entry to the link bandwidth and channel quality information in the routing tables. However, no specific QoS routing algorithm was discussed in [23]. In [3], a core extraction distributed ad hoc routing (CEDAR) algorithm has been proposed as a QoS routing algorithm for small- to medium-size MANETs, consisting of tens to hundreds of nodes. The core of the network is a subset of nodes selected to perform network management and routing function in the network. CEDAR is an on-demand source routing algorithm, which includes three key components — core extraction, link state propagation, and route computation. In the core extraction phase, the core of the network is extracted by approximating a minimum dominating set (MDS) of the ad hoc network using only local computation and local state information. The MDS is the minimum subset of nodes, such that every node in the network is in the MDS (i.e., is a core node) or is a neighbor of a node in the MDS. Each node in the core then establishes a unicast virtual link with other core nodes over a distance of three hops or less away in the ad hoc network. Each node that is not in the core chooses a core neighbor as its dominator. The core nodes are responsible for collecting local topology information and performing routing on behalf of the nodes in their respective domain (or immediate neighborhood). Figure 25.9 shows a MANET with three nodes selected to act as core nodes (i.e., virtual topology). The core nodes are linked by virtual links in the virtual topology. A virtual link represents one or more links in the physical topology.

A protocol that uses the core (i.e., MDS) of the network is called a core-based routing protocol. CEDAR uses core-based routing mechanisms for two primary reasons:

1. Because of the bandwidth and power constraints, reducing the number of nodes participating in route maintenance (i.e., state propagation and path restoration) is expected to increase network performance and increase network scalability.
2. Because of the hidden terminal and exposed terminal problems, local broadcast may be highly unreliable in MANETs. Using only a subset of nodes should reduce the negative effects of local broadcast.

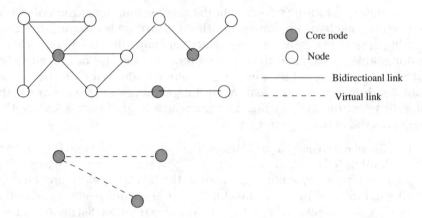

Figure 25.9. Network with Core Nodes and Corresponding Core Graph: an Example

QoS routing in CEDAR is achieved by propagating the bandwidth availability information of stable links in the core subgraph. When a link, (a,b), experiences a significant change (i.e., changes by some threshold value) in available bandwidth, a and b must inform their respective dominators. The dominators are responsible for propagating state information via slow-moving increase waves and fast-moving decrease waves (i.e., messages) to all other core nodes via the core broadcast mechanism.

25.3.3 Position-Based QoS Routing Protocol

Little work has been done in position-based QoS routing. In using position-based routing, the research community usually assumed the use of GPS as an external position determination server. GPS provides location information (latitude, longitude, and possibly altitude) to mobile nodes in MANETs that are equipped with GPS cards. The use of GPS is justified by the fact that GPS cards will be, in the near future, inexpensive and will be deployed in each car, and possibly in every user terminal. Accuracy measurement in GPS can be enhanced by using differential GPS, which offers accuracy up to a few meters. Position-based routing [40] is a new trend in performing routing in MANETs. Routing in this approach requires including the destination's position in the packet, in addition to the position of neighbor nodes that will forward the packet. Therefore, no route establishment and maintenance are needed. As a result, the efficiency of any position-based routing protocol is dependent on both the selection of the position servers and the selection of the forwarding strategy. If the use of GPS is not feasible, a GPS-free approach can also be used to build a local coordinate system based on the exchange of node's relative position [41].

In [39], a DSR-based protocol, which uses GPS for location determination was proposed. By letting each node maintain a table of the position of all other nodes, the mobile node maintains a snapshot for the network connectivity graph and therefore will be able to compute paths locally without the need for route discovery delay. By exploiting location information obtained from GPS, the proposed protocol enhances the end-to-end delay for packet delivery compared to the original DSR. This enhancement is possible because a source node has information about other nodes' locations, hence selecting the path that will satisfy the end-to-end delay with high probability based on its location database. However, the exchange of position information in this protocol may use much of the network resources. Furthermore, if the network topology changes frequently, which is normal in MANETs, the positions maintained in the position tables become stale quickly and need to be updated more often. The update process consumes bandwidth in a limited-bandwidth MANET.

In [33], the authors used GPS to obtain the position and speed of the mobile nodes in the network area. They used this information to build routes that will be valid for a certain period of time and can therefore be used to provide QoS paths, which satisfy certain delay guarantees. Therefore, the proposed protocol is also a predictive protocol in the sense that the motion of the nodes is predicted to build routes in the network. The lifetime of any link between two mobile nodes is defined as a function of their current locations, current speed, and moving directions. Therefore, this information is used on every link to find the expected lifetime of the whole path. The authors claimed that by having a stable path, the path setup time will be reduced, as well as the network control overhead. The authors did not describe how the position information is distributed among nodes and how the prediction information is updated. The GPS-based routing algorithm has two drawbacks. One is that GPS cannot provide the nodes with much information about the physical environment and the second is the power dissipation overhead of the GPS device itself.

In [41], the authors proposed a distributed mechanism for GPS free positioning in MANETs. They used the method of triangulation and coordinate translation to form the coordination system. Being independent of the external position server, the coordinate system provides continuous information about positions of different nodes that can be used to perform QoS routing in MANETs. The disadvantage of this solution is that it is computationally intensive and is also expensive in terms of the number of messages to be exchanged before a coordinate system can be established, because each node individually reorients its coordinates to the reference node's coordinates.

In [42], an approach for integrating QoS in the flooding-based route discovery process was proposed. The proposed positional attribute-based next-hop determination approach (PANDA) discriminates the next-hop node based on its location or capabilities. When a RREQ is broadcast, the intermediate node has two options:

1. It will randomly rebroadcast it to its neighbors and the delay in this case is also random and called random rebroadcast delay.
2. Instead of random rebroadcast, the receivers opt for a delay inversely proportional to their abilities in meeting the QoS requirement of the path.

The decision at the receiver side is made on the basis of a predefined set of rules. Thus, the end-to-end path will be able to satisfy the QoS constraints as long as it is intact. A broken path will initiate a QoS-aware route discovery process, which restarts the whole process again.

In [43], an integrated framework is proposed for performing QoS routing in MANETs. The GPS information[1] was used to divide the network area into fixed and square zones (clusters) and an optimal election algorithm is executed inside each zone to select one node as a clusterhead. Figure 25.10 shows an example of the clustering process in both homogeneous (identical transmission ranges) and heterogeneous (variable transmission ranges) networks by using a location service. The role of the clusterhead is dynamically changed when the network status changes. The set of the selected clusterheads form a rectilinear virtual topology. The virtual topology is used as a wireless virtual backbone to perform QoS routing. An extended version of Open Shortest Path First (OSPF) coupled with an extended version of the weighted fair queueing (WFQ) operate on the virtual topology to provide end-to-end statistical guarantees in terms of bandwidth and maximum delay. The motivation of using OSPF is that OSPF is currently used in the Internet, hence integration between MANETs and the Internet is implemented easily. The extended version of OSPF employs an efficient link update mechanism and the extended version of WFQ to obtain link costs. A hybrid criterion is used to compute QoS routes on the virtual topology. The objective is to maximize the call acceptance rate in MANETs.

25.3.4 *Power-Aware QoS Routing in MANETs*

There is an increasing interest in the power-aware routing in MANETs. Ad hoc wireless networks are power constrained because nodes operate with limited battery energy. To maximize the lifetime of these networks (defined by the condition that a fixed percentage of the nodes in the network will die out due to lack of energy), network-related transactions through each mobile node must be controlled such that the power dissipation rates of all nodes are nearly the same (i.e., avoid overloading a subset of nodes).

There have been some studies on power-aware routing protocols for MANETs. In [34], a source-initiated (on-demand) routing protocol for MANETs that increases the network lifetime was proposed. Simulation

Quality of Service Routing in Mobile Ad Hoc Networks: Past and Future

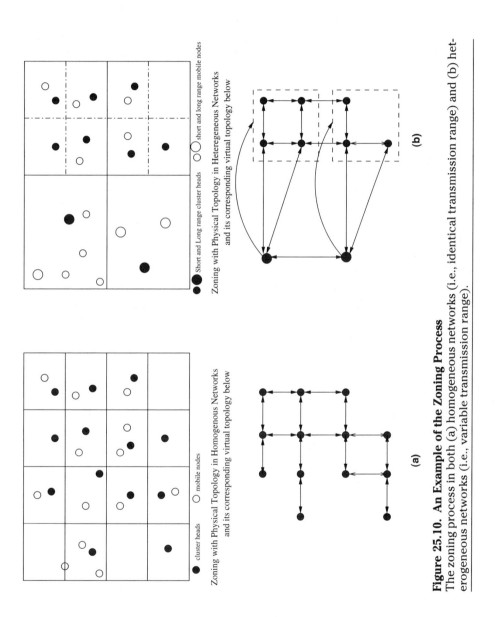

Figure 25.10. An Example of the Zoning Process
The zoning process in both (a) homogeneous networks (i.e., identical transmission range) and (b) heterogeneous networks (i.e., variable transmission range).

results show that the proposed power-aware source routing protocol has a better performance than other source initiated routing protocols in terms of the network lifetime. A greedy policy was applied to the fetched paths from the cache to make sure no path would be overused and also to make sure that each selected path has the minimum battery cost among all possible paths between two nodes. Power-aware source routing solves the problem of finding a route π at route discovery time t such that the following cost function is minimized:

$$C(\pi,t) = \sum_{i \in \pi} C_i(t)$$

where $C_i(t) = \rho_i(F_i/R_i(t))$ and ρ_i is the transmit power of node i, F_i is the full capacity of node i battery, R_i is the remaining battery capacity, and α is a weighting factor. The authors also presented route discovery and maintenance of the routes based on the DSR techniques. When an intermediate node receives a RREQ packet, it starts a timer (T) and keeps the cost in the header of that packet as Min–Cost. If additional RREQs arrive with same destination and sequence number, the cost of the newly arrived RREQ packet is compared to the Min–Cost. If the new packet has a lower cost, Min–Cost is changed to this new value and the new RREQ packet is forwarded. Otherwise, the new RREQ packet is dropped. The destination node waits the threshold number (T) of seconds after the first RREQ packet arrives. During that time, the destination examines the cost of the route of every arrived RREQ packet. When the timer (T) expires, the destination node selects the route with the minimum cost and replies. Subsequently, it will drop any received RREQs. Route maintenance is needed when connections between some nodes on the path are lost due to their movement or the energy resources of some nodes may be depleting quickly. A local approach was adopted for route maintenance because it minimizes control traffic. In the local approach, each intermediate node in the path monitors the decrease in its remaining energy level, which increases its link cost when necessary. When the link cost increase goes beyond a threshold level, the node sends a RERR back to the source and the route is invalidated.

In [37], the authors proposed a routing algorithm based on minimizing the amount of power (or energy per bit) required to deliver a packet from source to destination. The algorithm proposes to use a function f(A) to denote node A's reluctance to forward packets and to choose a path that minimizes the sum of f(A) for nodes on the path. This routing protocol addresses the issue of energy critical nodes. As a particular choice for f, [37] proposes f(A) = 1/g(A) where g(A) denotes the remaining lifetime, normalized to be in the interval [0, 1]. To minimize the total energy consumed per packet, it has been observed in [37] that the routes selected will be identical to routes selected by shortest hop routing, because the energy consumed in transmitting (and receiving) one packet over one hop is considered constant. More precisely, the problem is stated as:

$$Minimize \sum_{i \in R} P(i, i+1) \quad (25.1)$$

where P(i, i + 1) denotes the power needed for transmitting (and receiving) between two consecutive nodes, i and i + 1 (i.e., link cost), in the route R. The link cost can be defined for two cases:

1. Transmission power is fixed.
2. Transmission power is varied dynamically as a function of the distance between the transmitter and intended receiver.

For the first case, energy for each operation (receive, transmit, broadcast, discard, etc.) on a packet is given in [35] by $E_{packet} = b^*packet - size + c$ where b and c are the appropriate coefficients for each operation. Coefficient b denotes the packet size-dependent energy consumption, whereas c is a fixed cost that accounts for acquiring the channel and for MAC layer control negotiation. In the second case, packets are retransmitted via some intermediate nodes that may be available in MANETs. The idea is to vary the transmitted power level (T) in accordance with the distance d such that the received power follows the relation $R = T/d^\alpha$. Several algorithms were proposed for this second case where the value of the required power transmission levels to achieve communication is calculated by making use of the positions of the intermediate nodes relative to the sender node. Assuming that we can set additional nodes in arbitrary positions between the source and destination, the power optimal packet transmissions can be estimated as discussed in [36].

In [36], a localized routing algorithm for the second case is proposed. The authors [36] assume that the power needed for transmission and reception is a linear function of d^α where d is the distance between the two neighboring nodes and α is a parameter that depends on the physical environment. They make use of the GPS position information to transmit packets with the minimum required transmit energy. The key requirement of this technique is that the relative positions of nodes are available to all nodes. However, this information may not be readily available. The main disadvantage of the problem formulation of the previous approach is that it always selects the least-power cost routes. As a result, nodes along these routes tend to die sooner because of the battery energy exhaustion. This is doubly harmful because the nodes that die early are precisely the ones that are needed most to maintain the network connectivity (and hence useful service life). Therefore, it is better to use a higher power cost route if it avoids using nodes that have a small amount of remaining battery energy.

The minimum battery cost routing algorithm, proposed in [38], minimizes the total cost of the route. It minimizes the summation of the inverse of the remaining battery capacity for all nodes on the routing path. A conditional max–min battery capacity routing algorithm is also proposed in [38]. This algorithm chooses the route with minimal total transmission power if all nodes in the route have remaining battery capacities higher than a threshold; otherwise routes including nodes with the lowest remaining battery capacities are avoided. Minimum battery cost routing showed better performance than min–max routing in terms of expiration time of all nodes. Conditional max–min routing showed different behavior that depended on the value of the chosen threshold. However, because there is no guarantee that minimum total transmission power paths will be selected under all circumstances, it can consume more power to transmit user traffic from a source to a destination, which actually reduces the lifetime of all nodes.

25.4 QoS Routing in MANETs: Future Research Directions

MANETs are likely to expand their presence in the future communication infrastructure. The need to support QoS in MANETs thus becomes an important issue. In this section, we discuss some of the most important issues that still need further investigation and more research in the area of QoS provisioning in general and QoS routing in MANETs in particular. Despite various advantages and unlimited application chances, MANETs are still far from being deployed on large-scale commercial basis. This is because some fundamental ad hoc networking problems remain unsolved or need more optimal solutions.

Although it is a difficult issue, designing QoS protocols for MANETs is quite interesting and challenging. In this chapter, we presented an account of such challenges. We also presented a survey of the state-of-the-art QoS routing in this area. Several important research issues and open questions remain to be addressed to facilitate QoS routing support in MANETs. In the following, we list some of those problems pertaining to QoS provisioning and routing in MANETs:

- Multi-class traffic — the issue of accommodating user traffic with multiple classes is difficult, especially in heavy traffic situations (i.e., guaranteeing QoS for lower level classes may be extremely difficult or impossible).
- QoS routing that allows preemption — an open area for further research. The development of QoS routing policies, algorithms, and protocols for handling user data with multiple classes coupled with preemption is also an open area for further research. Differentiated services technology proposed for the Internet might help in this situation.

Quality of Service Routing in Mobile Ad Hoc Networks: Past and Future

- Different operational conditions — the issue of providing QoS under different operational conditions (e.g., failure modes) is worthy of further investigation. The type of QoS provided will be dependent on the type of error and its place. Preservation of QoS guarantees under various failure conditions in ad hoc networks is barely touched.
- Mobile nodes position identification — the provisioning of QoS is dependent on the position of mobile nodes. Each mobile node must know its adjacent neighbors and other mobile nodes. This is difficult because of dynamic changes of the network topology occur frequently in an ad hoc network. This has a direct impact on the QoS routing protocol. The use of external entities (e.g., GPS) is promising despite the need to solve the problems associated with the use of GPS system. A suitable position determination system and position upgrade mechanism is highly required.
- Packet prioritization — packet exchanges should not be treated with equal priority in a QoS network. The exchange of control packets should receive higher priority than user data packets in a network designed for QoS. The QoS policy must allow different priorities to exist even among different flows of user packets.
- Mobility model — guaranteeing QoS in such a network may be impossible if the nodes are too mobile. The challenges increase even more for those ad hoc networks that support both best effort services and those with QoS guarantees and are required to interwork with each other. Most protocols used impractical mobility models for their evaluations. There is a need for a mobility testbed that reflects the actual nodes mobility patterns.
- Layer integration — recent trend of MANETs is to combine the working of the physical layer and the MAC layer with the data link layer and network layer to enhance the network performance. The impact of layer integration of the QoS provisioning and performance of the network is not fully quantified yet.
- Internet-MANET interaction — the interaction between MANETs and the existing global information infrastructure — the Internet — is another major area of research. A form of gateway that provides this kind of interaction is still in its early design stages and needs further investigation. Being able to connect to a wireline network (e.g., Internet) can help in the QoS routing. For example, when MANET becomes partitioned, an alternate routing between the network partitions can be carried out through the wireline network.
- MANET security — security is an important issue that is a desirable feature in any QoS routing protocol. There should be a form of network-level or link-layer security aspects in the protocol such that malicious retransmissions, manipulation, snooping, and redirection of packets are not allowed.

- Network partitioning — in MANET, natural grouping behavior in the mobile user's movement leads to frequent network partitioning. Network partitioning poses significant challenges to the QoS routing protocol because partitioning disconnects many mobile users from each other.
- Nodes cooperation — some mobile nodes may misbehave by agreeing to forward packets and then failing to do so. Obviously, this has a direct impact on QoS in MANETs and may lead to an effectively disconnected network. A node may misbehave because it is overloaded, selfish, malicious, or broken. Hence, node cooperation is essential for the network to perform its duties. Efficient algorithms that ensure or force node cooperation in MANETs are still needed.
- Heterogeneous networks — it is generally assumed that all nodes in MANETs are homogenous both in terms of capacity and functionality. MANETs are typically heterogeneous networks with various types of mobile nodes. In military application, different military units ranging from soldiers to tanks can come together; hence forming an ad hoc network. In conference application, different types of mobile devices such as personal digital assistants (PDAs), smart badges, and laptops may exist in the ad hoc network at the same time. Nodes differ in their power capacities and computational speeds. Thus, mobile nodes will have different packet generation rates, routing responsibilities, network activities, and power draining rates. QoS issues in heterogeneous networks and especially QoS routing should be investigated in more depth. This issue has just started to receive attention in the literature.

We believe that QoS routing should be on-demand as traffic distribution is not uniform throughout the network. Routing should adapt to changes in traffic patterns such that the network and nodal resources are not misused. In short, research still needs to be done in the area of QoS in general and QoS routing in MANETs in particular.

25.5 Conclusion

The key to support QoS in MANETs is QoS routing. QoS routing in MANETs is a growing area of research. Different approaches have been proposed to provide QoS routing in MANET. In this chapter, an attempt to survey QoS routing protocols and paradigms was made. We outlined the challenges that make the QoS routing problem in MANETs difficult. Overall, QoS routing protocols were generalized into flat, hierarchical, position-based, and power-aware routing paradigms. It is clearly evident from the analysis that no particular protocol fulfills all the requirements of QoS routing in MANETs. However each protocol works well under certain scenarios and for some QoS metrics. After presenting an extensive review of the current protocols, we outlined future directions in this field, which is open for further research. In

Quality of Service Routing in Mobile Ad Hoc Networks: Past and Future

conclusion, numerous challenges must be overcome to realize the practical benefits of ad hoc networking. These include effective routing, medium (or channel) access, mobility management, power management, security, and, of principal interest, QoS issues, mainly pertaining to delay and bandwidth management and provisioning.

Note

1. A GPS-free approach is also possible.

References

1. E. Crawley, R. Nair, B. Rajagopalan, and H. Sandrick, A framework for QoS based routing in the Internet, RFC 2386, August 1998.
2. A. Munaretto, H. Badis, K. Al Agha, and G. Pujolle, A link-state QoS routing protocol for ad hoc networks, *Proceedings of the 4th International Workshop on Mobile and Wireless Communications Network,* September 2002, pp. 222–226, 2002.
3. P. Sinha, R. Sivakumar, and V. Bharghavan, CEDAR: a Core-extraction distributed ad hoc routing algorithm. In *Proceedings of IEEE InfoCom 99,* New York, March 1999.
4. S. Chen and K. Nahrstedt, Distributed quality-of-service routing in ad hoc networks. *IEEE Journal on Selected Areas in Communication* [Special Issue on Ad hoc Networks], vol. 17, no. 8, pp. 1488–1505, 1999.
5. C.R. Lin and J.-S. Liu, QoS routing in ad hoc wireless networks, *IEEE Journal on Selected Areas in Communication,* vol. 17, no. 8, pp. 1426–1438, August 1999.
6. C. Perkins, E. Royer, and S.R. Das, Quality of service for ad hoc on-demand distance vector (AODV) routing, Internet Draft, July 2000.
7. C. Zhu and M.S. Corson, QoS routing for mobile ad hoc networks, *InfoCom 2002.* vol. 2, pp. 958–967, 2002.
8. T.W. Chen, J.T. Tsai, and M. Gerla, QoS routing performance in multihop, multimedia, wireless networks, *IEEE on Universal Personal Communications,* vol. 2, pp. 557–561, 1997.
9. G. Raju, G. Hernandez, and Q. Zou, Quality of service routing in ad hoc networks, *Wireless Communications and Networking Conference (WCNC 2000),* vol. 1, pp. 263–265, 2000.
10. S. Shah and K. Nahrstedt, Predictive location-based QoS routing in mobile ad hoc networks, *IEEE International Conference on Communications 2002 (ICC 2002),* vol. 2, pp. 1022–1027, 2002.
11. W. Liao, Y. Tseng, and K. Shih, A TDMA-based bandwidth reservation protocol for QoS routing in a wireless mobile ad hoc network, *ICC 2002,* vol. 5, pp. 3186–3190, 2002.
12. W.H. Liao, Y.C. Tseng, S.L. Wang, and J.P. Sheu, A multi-path QoS routing protocol in a wireless mobile ad hoc network, *IEEE International Conference on Networking (ICN),* 2001.
13. I. Gerasimov and R. Simon, A bandwidth-reservation mechanism for on-demand ad hoc path finding, *Proceedings of the 35th Annual Simulation Symposium,* pp. 20–27, 2002.
14. M. Hashem, M. Hamdy, and S. Ghoniemy, Modified distributed quality-of-service routing in wireless mobile ad-hoc networks, *MELECON 2002,* pp. 368–378, 2002.
15. Y. Chen, Y. Tseng, J. Sheu, and P. Kuo, On-demand, link-state, multi-path QoS routing in a wireless mobile ad-hoc network, *Proceedings of Europian Wireless Conference,* 2002.
16. G. Byeong and W. Lee, Extended precomputation based selective probing (PCSP) scheme for QoS routing in ad-hoc networks, *IEEE 56th Vehicular Technology Conference,* vol. 3, pp. 1342–1346, 2002.

17. D. Johnson and D. Maltz, Dynamic source routing in ad hoc wireless networks, In T. Imielinski and H. Korth, Eds., *Mobile Computing,* Norwell, MA: Kluwer Academic, 1996 .
18. C.E. Perkins and P. Bhagwat, Highly dynamic destination-sequenced distance-vector routing (DSDV) for mobile computers, *Computer Communications Review,* pp. 234–244, October 1994.
19. S. Chen, Routing support for providing guaranteed end-to-end quality-of-service, Ph.D. thesis, University of Illinois at Urbana-Champaign, 1999. http://cairo.cs.uiuc.edu/papers/SCthesis.ps.
20. C. Richard Lin, QoS routing in ad hoc wireless networks, *Proceedings of the 23rd Annual Conference on Local Computer Networks (LCN'98),* pp. 31–40, 1998.
21. A. Ariza, E. Casilari, and F. Sandoval, QoS routing with adaptive updating of link states, *Electronics Letters,* vol. 37, no. 9, pp. 604–606, April 26, 2001.
22. M. Frodigh, S. Parkvall, C. Roobol, P. Johansson, and P. Larsson, Future-generation wireless networks, *IEEE Personal Communications,* vol. 8, no. 5, pp. 10–17, October 2001.
23. A. Iwata, C. Chiang, G. Pei, M. Gerla, and T.W. Chen, Scalable routing strategies for ad hoc wireless networks, *IEEE Journal Selected Areas in Communications,* vol. 17, no. 8, pp. 1369–1379, August 1999.
24. K. Chen, S. Samarth, and K. Nahrstedt, Cross-layer design for data accessibility in mobile ad hoc networks, *Wireless Personal Communications,* vol. 21, pp. 49–76, 2002.
25. S. Holeman, G. Manimaran, and J. Davis, Differentially secure multicasting and its implementation methods, In *Proceedings of ICCCN,* Phoenix, AZ, October 2001, pp. 212–217.
26. J. Kurose and K. Ross, Computer networking: A top-down approach featuring the Internet, Boston: Addison Wesley, 2001.
27. C. Perkins, E. Royer, and S.R. Das, Quality of service for ad hoc on-demand distance vector (AODV) routing, Internet Draft, July 2000.
28. C. Perkins and E. Royer, Ad hoc on demand distance vector (AODV) routing, Internet Draft, November 1998.
29. Y. Ge, T. Kunz, and L. Lamont, Quality of service routing in ad-hoc networks using OLSR, In *Proceedings of the 36th Annual Hawaii International Conference on System Sciences (HICSS'03),* Big Island, HI, 2003.
30. R. Leung, J. Liu, E. Poon, A. Chan, and B. Li, MP-DSR: a QoS-aware multi-path dynamic source routing protocol for wireless ad-hoc networks, In *Proceedings of Local Computer Networks (LCN),* pp. 132–141, 2001.
31. C.R. Lin, On-demand QoS routing in multihop mobile networks, In *Proceedings of the 20th Annual Joint Conference of the IEEE Computer and Communications Societies (InfoCom),* vol. 3, pp. 1735–1744, 2001.
32. G.N. Aggelou and R. Tafazolli, QoS support in 4th generation mobile multimedia ad hoc networks, In *Proceedings of the Second International Conference on 3G Mobile Communication Technologies 2001,* Conf. Publ. no. 477, pp. 412–416, 2001.
33. J. Chen, J. Wang, S. Deng, and Y. Tang, QoS routing with mobility prediction in MANET, *IEEE Pacific Rim Conference on Communications, Computers, and Signal Processing 2001 (PACRIM 2001),* vol. 2, pp. 357–360, 2001.
34. M. Maleki, K. Dantu, and M. Pedram, Power-aware source routing protocol for mobile ad hoc networks, *ACM ISLPED'02,* August 12–14, 2002.
35. S. Lindsey, K. Sivalingam, and C.S. Raghavendra, Power aware routing and MAC protocols for wireless and mobile networks, In Ivan Stojmenivic, Ed., *Wiley Handbook on Wireless Networks and Mobile Computing,* Hoboken, NJ: John Wiley & Sons, 2001.
36. J. Stojmenovic and X. Lin, Power-aware localized routing in wireless networks. In *Proceedings of the IEEE IPDPS,* Cancun, May 2000.
37. S. Singh, M. Woo, and C. Raghavendra, Power-aware routing in mobile ad hoc networks, In *Proceedings of MobiCom '98 Conference,* Dallas, October 1998.

Quality of Service Routing in Mobile Ad Hoc Networks: Past and Future

38. C.K. Toh, Maximum battery life routing to support ubiquitous mobile computing in wireless ad hoc networks, *IEEE Communication Magazine,* June 2001.
39. S. Basagni, I. Chlamtac, and V. Syrotiuk, Dynamic source routing for ad hoc networks using the global positioning system, In *Proceedings of the IEEE Wireless Communications and Networking Conference (WCNC 1999)*, vol. 1, pp. 301–305, 1999.
40. H. Hartenstein, M. Mauve, and A. Widmer, A survey on position-based routing in mobile ad hoc networks, *IEEE Network,* vol. 15, no. 6, pp. 30–39, November/December 2001.
41. S. Capkun, M. Hamdi, and J.-P. Hubaux, GPS-free positioning in mobile ad-hoc networks, In *Proceedings of Hawaii International Conference on System Sciences,* Maui, HI, January 2001, pp. 3481–3490.
42. J. Li and P. Mohapatra, PANDA: A positional attribute-based next-hop determination approach for mobile ad hoc networks, Technical Report, Department of Computer Science, University of California, Davis, 2002.
43. J.N. Al-Karaki and A.E. Kamal, End-to-end support for statistical quality of service in heterogeneous mobile ad hoc networks, *Computer Communications*, in press.

Chapter 26

Issues in Scalable Clustered Network Architecture for Mobile Ad Hoc Networks

Ben Lee, Chansu Yu, and Sangman Moh

Abstract

As large-scale, high-density multi-hop networks become desirable for many applications, there exists a greater demand for scalable mobile ad hoc network (MANET) architecture. Due to the increased route length between two end nodes in a multi-hop MANET, the challenge is in the limited *scalability* despite the improved spatial diversity in a large network area. Common to most existing approaches for a scalable MANET is the *link cluster architecture* (LCA), where mobile nodes are logically partitioned into groups called *clusters*. Clustering algorithms select master nodes and maintain the cluster structure dynamically as nodes move. Routing protocols use the underlying cluster structure to maintain routing and location information in an efficient manner. This chapter discusses the various issues in scalable clustered network architectures for MANETs. This includes a classification of link-clustered architectures, an overview of clustering algorithms focusing on master selection, and a survey of cluster-based routing protocols.

26.1 Introduction

MANET is an infrastructureless multi-hop network where each node communicates with other nodes either directly or indirectly through intermediate nodes. Because MANETs are infrastructureless, self-organizing, rapidly deployable wireless networks, they are highly suitable for applications involving special outdoor events, communications in regions with no wireless infrastructure, emergencies and natural disasters, and military operations. Handling node mobility may be the most critical issue in a MANET; thus previous research efforts have focused mostly on routing or multicasting protocols that result in consistent performance in the presence of wide range of mobility patterns.

As large-scale, high-density multi-hop networks become more desirable for many applications, a greater demand exists for scalable MANET architecture. However, when the network size increases, routing schemes based on the flat network topology (or flat routing protocols) become infeasible because of high protocol overhead and unreliability/interference caused by networkwide flooding of routing-related control packets [1, 2]. Recently, a number of studies have addressed this problem. For example, Li et al. suggested that a large-scale multi-hop network is feasible only when most of communication is local so that the broadcasts of routing-related control packets are restricted to the local areas rather than flooded to the entire network [3]. Morris et al. considered scaling of MANETs to hundreds of thousands nodes, where control packets are not flooded but directed only to some particular locations where the intended destination is most likely to be located [4]. Grossglauser and Tse also proposed an approach where each node localizes its data transfers by buffering the traffic until the destination node is within its radio range [5]. The last solution increases delay and requires a large buffer at each node, while the first two approaches either require a special facility such as *global positioning system* (GPS) to track nodes' locations or assume communication traffic follows a certain pattern.

Recently, more general approaches for a scalable MANET have been explored in the literature [6–18, 32, 33, 36]. A common aspect to these approaches is that the flat network topology is restructured to produce the LCA, which is one of the promising architectural choices for a scalable MANET [6]. Typically, an entire multi-hop MANET is divided into a number of one- or two-hop networks, called *clusters*, and the clusters are independently controlled and dynamically reconfigured as nodes move. Within each cluster, one node is chosen to perform the function of a *master*[1] and some others to perform the function of *gateways* between clusters. The cluster architecture improves the scalability by reducing the number of mobile nodes participating in some routing algorithm, which in turn significantly reduces the routing-related control overhead. Other advantages are less chances of interference via coordination of data transmissions and more robustness in the event of node mobility by judiciously selecting stable nodes as masters.

Issues in Scalable Clustered Network Architecture for Mobile Ad Hoc Networks

This chapter presents a survey of routing protocols for clustered architecture in a large-scale MANET, which can be classified into the following two types:

1. LCA for routing backbone
2. LCA for information infrastructure

The latter type overlays an information infrastructure to support an efficient means of providing routing information; the former type constructs a routing backbone not only to maintain routing information, but also delivers data packets to intended destinations. Master nodes in a cluster architecture-based protocol collectively maintain routing information of all mobile nodes. For nodes in each cluster, a proactive scheme (distance vector or link state) can be used because the network diameter of each cluster is usually small and thus the corresponding control overhead is not high. However, for nodes outside of a cluster, each master node uses either one of the following routing principles as in flat routing protocols:

- Proactive update
- On-demand searching

The chapter is organized as follows:

- Section 26.2 presents the classification of cluster architecture-based routing protocols for MANETs based on the above mentioned cluster architectures and routing principles.
- Section 26.3 and Section 26.4 describe numerous cluster-based routing protocols with the discussion on the cluster type they construct, corresponding control and clustering overheads, and advantages and disadvantages. In particular, Section 26.3 focuses on routing protocols on LCA for routing backbone and Section 26.4 on those based on LCA for information infrastructure.
- Section 26.5 summarizes all the cluster-based protocols with comparisons and draws conclusions.

26.2 Classification of Cluster Architecture-Based Routing Protocols

Before discussing each protocol in detail, this section provides the classification of cluster-based routing protocols. The classification is based on cluster structures these protocols build and routing methods they employ to find the destination node or the destination node's master. Section 26.2.1 briefly overviews flat routing protocols proposed for MANETs. Section 26.2.2 introduces several cluster structures and their characteristics. Section 26.2.3 introduces routing principles used in cluster architecture-based routing protocols and the overall classification.

26.2.1 Flat Routing Protocols and Their Scalability

The routing protocols proposed for MANETs are generally categorized as either *table-driven* or *on-demand* based on the timing of when the routes are updated. With table-driven routing protocols, each node attempts to maintain consistent, up-to-date routing information to every other node in the network. This is done in response to changes in the network by having each node update its routing table and propagate the updates to its neighboring nodes. Thus, it is *proactive* in the sense that when a packet needs to be forwarded, the route is already known and can be immediately used. As in the case for wired networks, the routing table is constructed using either *link state* or *distance vector* algorithms containing a list of all the destinations, the next hop, and the number of hops to each destination. Many routing protocols including *Destination-Sequenced Distance Vector* (DSDV) [19] and *Fisheye State Routing* (FSR) protocol [20] belong to this category; they differ in the number of routing tables manipulated and the methods used to exchange and maintain routing tables.

With on-demand driven routing, routes are discovered only when a source node desires them. *Route discovery* and *route maintenance* are two main procedures. The route discovery process involves sending a route request packet from a source to its neighbor nodes, which then forward the request to their neighbors, and so on until the route request packet reaches the destination node. Once the route is established, some form of route maintenance process maintains the routes in each node's internal data structure. Each node learns the routing paths as time passes not only as a source or an intermediate node, but also as an overhearing neighbor node. In contrast to table-driven routing protocols, not all up-to-date routes are maintained at every node. *Dynamic Source Routing* (DSR) [21] and *Ad Hoc On-Demand Distance Vector* (AODV) [22] are examples of on-demand driven protocols.

Now consider the scalability of these flat routing protocols as network size increases with the number of mobile nodes, n. The total effective bandwidth increases as $O(n)$ because more concurrent transmissions can be supported. However, this advantage of spatial reuse is diminished due to the increased path length ($O(n)$) in a larger network area. For this reason, networkwide end-to-end bandwidth remains the same even though network size increases [23, 24]. Although this scenario holds for data traffic, this is not true for control traffic caused by the underlying routing protocol. The increased path length causes more chance of route failures and results in higher overhead to maintain the routes. More importantly, in a table-driven routing protocol, the size of routing table grows as function of $O(n)$ as network size increases and the control traffic due to the periodic exchange of the routing tables grows as function of $O(n^2)$ because more nodes exchange larger tables. In an on-demand routing protocol such as

Issues in Scalable Clustered Network Architecture for Mobile Ad Hoc Networks

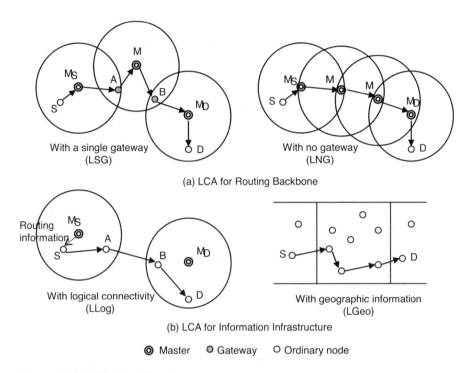

Figure 26.1. LCA Classification

DSR, a route request packet is broadcast to a larger number of nodes with higher frequency and thus the control traffic also increases as function of $O(n^2)$.

In addition to the higher protocol overhead mentioned above, a large-scale MANET suffers from unreliable broadcasts. Unlike unicast communication that usually employs a *four-way handshake* — request-to-send, clear-to-send, data, and acknowledgment packets — [25] to improve link-level reliability, broadcasts are inherently unreliable in wireless ad hoc networks. A large-scale MANET aggravates the problem because such broadcasts are performed in a series, one after the other [1]. Redundant broadcasts and contention/collisions among the broadcasts [26] significantly increase the control overhead in a large-scale MANET.

26.2.2 Cluster Architectures

Cluster architecture is a scalable and efficient solution to the above mentioned problems by providing a hierarchical routing among mobile nodes. Figure 26.1 shows different cluster architectures with different levels of clusters overlapping and different responsibilities imposed on master nodes. As introduced in Section 26.1, they can be broadly categorized into two types — *LCA for routing backbone* and *LCA for information infrastructure* — based on how the master nodes are used. A straightforward difference between the two types is that the former imposes more responsibility on master nodes, but the latter needs to provide an additional mechanism for routing. An important design issue in the information infrastructure approach is to select a set of master nodes that gather and scatter routing information with minimal overhead. On the other hand, in the routing backbone approach, maintaining master-to-master connections and high-level topology among the masters are more important issues to deliver data packets efficiently.

Figure 26.1A shows examples of routing backbones through which data packets are routed. Depending on the number of gateways between two masters, they are called as *LCA for routing backbone with a single gateway* (LSG) and *with no gateway* (LNG), respectively. In LNG, master nodes perform the functions of gateways, and thus, intermediate nodes in a routing path consist only of masters. SPAN [32], *Near Term Digital Radio* (NTDR) networking [33], and *Geographic Adaptive Fidelity* (GAF) [34] are example protocols that construct LNG. *Cluster Gateway Switching Routing* (CGSR) [27], *Hierarchical State Routing* (HSR) [28], *Cluster-Based Routing Protocol* (CBRP) [30], *Adaptive Routing using Clusters* (ARC) [31], *Destination Sequenced Clustered Routing* (DSCR) [27], and *Landmark Ad Hoc Routing* (LANMAR) [29] construct LSG. Note that CBRP and ARC also allow two neighboring masters to contact directly or indirectly via a pair of gateways. This is to avoid frequent changes in masters and prevent network partitioning as will be discussed in Section 26.3.1.

Approaches for constructing routing backbones shown in Figure 26.1A impose high demand on channel bandwidth and require node stability on the backbone nodes to prevent bottlenecks as well as a single point of failure. In addition, they may result in suboptimal routing paths, because every intermediate node must be either a master or a gateway. Therefore, an alternative solution is to construct a virtual infrastructure that serves only as container for routing information as in Figure 26.1B. Routing is carried out based on the flat routing principle without going through masters but route searching is more localized based on the virtual information infrastructure [11]. *Core Extraction Distributed Ad Hoc Routing* (CEDAR) [35], *Zone Routing Protocol* (ZRP) [36], *Zone-based Hierarchical Link State* (ZHLS) [37], and Grid Location Service (GLS) [38] routing protocols fall into this category. It is noted that the last two protocols use geographic location information obtained via GPS to define clusters, which we refer to as *LGeo (LCA for information infrastructure with geographic information)*. Once a destination's physical location is obtained, a more efficient routing scheme can be employed. The first two protocols define clusters based on logical connectivity, which we refer to as *LLog (LCA for information infra-*

structure with logical connectivity). *Distributed Database Coverage Heuristic* (DDCH) [11] and *Max–Min D-Cluster Formation* (MMDF) [13] are also efficient clustering algorithms but not complete routing protocols, which we also include in our discussion.

26.2.3 Cluster Architecture-Based Routing Protocols

The main idea behind constructing a LCA is to reduce the routing-related control overhead involved with searching for the destination node in a large network. Each master node can easily maintain the location information of ordinary nodes in its cluster using local communications. However, to obtain information of a destination node D in a remote cluster, each master has to perform the following tasks: identify the cluster where the destination node D or its master node M_D is located and forward data packets toward M_D and let it deliver the packets to D. Therefore, the node–master association (D, M_D) for all nodes must be maintained. A CBRP updates the association table based on either:

- Proactive update of the association of all nodes
- On-demand searching for M_D corresponding to D

among master nodes over the underlying cluster structure.

Proactive approaches can provide a faster data delivery but a large table containing associations for all mobile nodes needs to be periodically propagated. Notice, however, that the corresponding overhead is far less than that of maintaining link status or distance vector to all nodes because the node–master association changes less frequently than the wireless link status. Moreover, by applying a more stable cluster structuring algorithm, which we will discuss in Section 26.3.1, the update period can be greatly reduced. On the other hand, for on-demand approaches, the master node M_D is searched based on typical *route discovery procedure* as used in on-demand flat routing protocols such as DSR [21] or AODV [22]. The underlying cluster structure is used to relay the route request packet in order to avoid the overhead of a networkwide search. Table 26.1 summarizes cluster routing protocols and their characteristics.

26.3 LCA for Routing Backbone

One important design problem in constructing a LCA for routing backbone is to select master nodes so that they can form an efficient routing infrastructure. Section 26.3.1 discusses the master selection and cluster maintenance algorithms for LSG and LNG in a MANET. Section 26.3.2 and Section 26.3.3 discuss the LSG- and LNG-based routing protocols, respectively.

26.3.1 Clustering Algorithms

Designing a clustering algorithm is not trivial due to the following reasons:

Table 26.1. Cluster-Based Protocols and Their Cluster Architectures

Cluster Architecture	Routing Principle for Nodes Outside of a Cluster	
	Proactive Update	**On-Demand Searching**
LCA for routing backbone	LSG with master-to-gateway routing (Section 26.3.2): CGSR, HSR	LSG (no or two gateways are also allowed) (Section 26.3.2): CBRP, ARC
	LSG with flat routing (Section 26.3.2): DSCR, LANMAR	LNG with master-to-master routing (Section 26.3.3): SPAN, NTDR, GAF
LCA for information infrastructure		LLog (Section 26.4.2): CEDAR, ZRP
		LGeo (Section 26.4.3): ZHLS, GLS

- Electing a master node among a set of directly connected nodes is not straightforward, because each candidate has a different set of nodes depending on the spatial location and the radio transmission range.
- A clustering algorithm must be a distributed algorithm and resolve conflicts when multiple mutually exclusive candidates compete to become a master.
- The clustering algorithm must be able to dynamically reconfigure the cluster structure when either some nodes move or some masters need to be replaced due to overloading.
- In the presence of mobility, it must preserve its cluster structure as much as possible and reduce the communication overhead to reconstruct clusters [7].

Below we will discuss the cluster construction problem involving the first two issues, and then explain the cluster maintenance algorithm that must deal with the last two issues.

26.3.1.1 Master Selection Algorithms for LSG. There are various clustering algorithms used to construct a LSG. In the *identifier-based algorithm* [6], a node elects itself as a master if it has the lowest-numbered identifier in its uncovered neighbors, where any node that has not yet elected its master is said to be uncovered. Figure 26.2A shows the process of master selection based on this algorithm. Node 1 and node 4 elect themselves as masters and node 2 and node 3 are covered by those masters. Among uncovered nodes (5, 6, and 7), node 5 elects itself as a master because it has the lowest identifier. By definition, a master node cannot have another

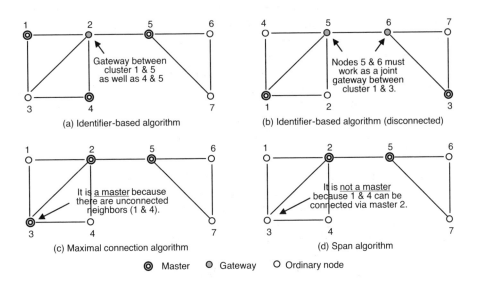

Figure 26.2. Master Selection Algorithms

master as a neighboring node and thus, this algorithm produces a single-gateway structure. The *connectivity-based algorithm* [6] uses the node connectivity instead of node identifier to determine a master because it potentially provides a cluster structure with fewer masters. (When a tie occurs, a node identifier is used to resolve the conflict.)

In the *randomized clustering algorithm* [33], a node elects itself as a master if it does not find any masters within its communication range. Because multiple candidates may compete to become a master, conflicts are resolved by a random delay. That is, when a node detects no neighboring master nodes, it first waits for a randomly selected time. If it still detects no master nodes after the delay, it now becomes the master and immediately announces this information to its neighbors. This algorithm is logically the same as the identifier-based algorithm when the random wait time is translated to the node identifier. The *adaptive clustering algorithm* proposed in [7] forms disjoint clusters, where each cluster is assigned a different communication channel from those in neighboring clusters. Without this assumption, the algorithm is equivalent to identifier-based clustering algorithm and results in a single-gateway network.

Identifier- or connectivity-based algorithms are the basic clustering algorithms used in most of cluster routing protocols. To implement these algorithms, each node periodically broadcasts its identifier or connectivity information to its neighbors and elects a master that has the lowest identifier or the highest connectivity. However, it is important to note that these clustering algorithms may not form a connected cluster structure. This happens when the overlapping area between two adjacent clusters does not contain any single mobile node and thus there is no node assuming the task of a gateway between two clusters. For example, Figure 26.2B shows the same ad hoc network as in Figure 26.2A but with different assignments of node identifiers. Identifier-based clustering algorithm selects node 1 and node 3 as masters but there is not a single node that is included in both clusters.

CBRP [30] and ARC [31] protocols take this problem into account by allowing a pair of gateways (or a *joint gateway* [31]) between two masters. For example, node 5 and node 6 in Figure 26.2B should work as gateways between two clusters. Each of these nodes periodically broadcasts the information on master nodes that it can communicate directly or indirectly via another node. Thus, each master is able to determine other neighboring masters that are two hops as well as three hops away.

26.3.1.2 Cluster Maintenance Algorithms for LSG.
Now, we consider the cluster maintenance procedure. Mobility of ordinary nodes can be simply handled by changing its master node accordingly. Mobility of a master node is a more difficult problem not only because a new master node must be elected, but also because it may affect the entire cluster structure of the network. The identifier-based clustering is more stable than the connectivity-based clustering because connectivity changes frequently as nodes move. In [7], the authors measured the stability of cluster architecture by counting how many nodes migrate from one cluster to another and demonstrated the importance of the stability factor by showing that it directly affects the general network performance.

There are some mechanisms to make the cluster structure more stable. *Least Cluster Change* (LCC) clustering algorithm is the most common denominator, which is used in CGSR [27], CBRP [30], ARC [31], and DSCR [27]. The two LCC rules are as follows:

1. When an ordinary node contacts another master, no change in mastership occurs without reevaluating the basic master selection rule such as lowest-ID or highest-connectivity clustering algorithm.
2. When two masters contact each other, one gives up its mastership based on the basic rule among the two but not among all possible candidates. Some nodes in the loser's cluster should reelect a new master because they are not within the transmission range of the winning master.

However, the problem with the second LCC rule is that it can cause a rippling effect across the network. CBRP [30] modifies the rule a step further to propose the *contention rule* to reduce the frequency of changes in mastership. Unlike the second LCC rule stated above, two masters are allowed to contact each other for less than the predefined contention period. The contention rule is effective when two masters contact temporarily and are separated in a short period of time. ARC protocol [31] adopts the *revocation rule* replacing the second LCC rule. When two masters contact with each other, one master becomes an ordinary node only when its cluster becomes a subset of the other master's cluster. In other words, CBRP and ARC temporarily allow a cluster structure with no gateway.

However, a highly stable structure may easily overload the master nodes. This may produce many undesirable problems because every mobile node is inherently identical in its capability as well as its responsibility in a MANET. Thus, it is necessary to change the master nodes periodically to prevent overloading and to ensure fairness.

26.3.1.3 Master Selection Algorithms for LNG. The *maximal connection algorithm* [10] shown in Figure 26.2C is the most straightforward no-gateway algorithm. A node elects itself as a master if there are two neighbors that are not directly connected. With this clustering algorithm, master nodes collectively provide a routing backbone that always guarantees the shortest path. In other words, intermediate nodes of the shortest path between any two nodes are all master nodes. To see this, consider an intermediate node (for example, node 5 in Figure 26.2C) along a shortest route between node 1 and node 6 (route 1–2–5–6). Node 5 relays packets between the proceeding node (node 2) and the succeeding node (node 6) along the shortest path but, because this node is a part of a shortest route, these two nodes are not directly connected. Therefore, by definition, the intermediate node (node 5) must be a master node because there are two unconnected neighbors.

The SPAN algorithm [32] is a similar scheme but produces fewer master nodes. To select the master nodes, the Span protocol employs a distributed *master eligibility rule* where each node independently checks if it should become a master or not. The rule is *if two of its neighbors cannot reach each other either directly or via one or two masters, it should become a master* [32]. In Figure 26.2D, unlike the maximal connection algorithm, node 3 is not a master node because two of its neighbors, node 1 and node 4, can be connected via a master node 2. A randomized backoff delay is used to resolve contention. By definition, for each pair of nodes that are two hops away, they are directly connected or there is a two-hop or three-hop route where all intermediate nodes are masters. In other words, master nodes connect any two nodes in the network providing the routing backbone. Therefore, the SPAN algorithm produces a no-gateway network, even though the paths are not always the shortest.

Table 26.2. Clustering Algorithms for LCA for Routing Backbone

CBR	Protocol	Clustering Algorithm	Comment
Single gateway	CGSR	Basic + LCC algorithm	"Basic" means the clustering algorithm based on the lowest identifier or the highest connectivity.
	HSR	Basic algorithm	
	DSCR	Basic + LCC algorithm	
	LANMAR	None	Group mobility is assumed so that relative relationship among mobile nodes in a group does not change over time and results in a natural clustering.
	CBRP	Basic + LCC + Contention rule	A pair of gateways is allowed between two clusters.
	ARC	Basic + LCC + Revocation rule	A pair of gateways is allowed between two clusters.
No gateway	NTDR	None	It is assumed that nodes are clustered around a number of geographic locations and they naturally form clusters.
	SPAN	Span algorithm	Master eligibility rule is defined.
	GAF	None	A network area is geographically partitioned into grids and each node can easily associate it with the corresponding cluster.

Master overloading is also a problem in LNG. In the SPAN algorithm, a master node periodically checks if it should withdraw as a master and gives other neighbor nodes a chance to become a master. Ordinary nodes also periodically determine if they should become a master or not based on the master eligibility rule stated above. Table 26.2 summarizes the clustering algorithms for LSG and LNG.

26.3.2 LSG-Based Routing Protocols

For cluster routing protocols, maintaining node–master association (D, M_D) of all mobile nodes in a MANET is the key issue. Routes to local nodes in each cluster are usually updated using a proactive algorithm (i.e., each node broadcasts its link state to all nodes within its cluster). Because they share the same master, their node–master associations are automatically updated. However, node–master association of remote nodes is maintained

Issues in Scalable Clustered Network Architecture for Mobile Ad Hoc Networks

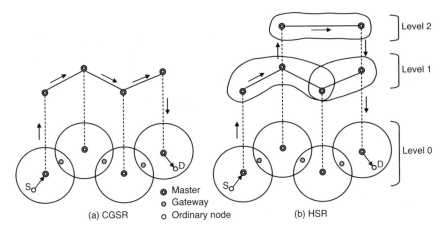

Figure 26.3. CGSR and HSR Protocols

either proactively or reactively. This section discusses six cluster routing protocols, four proactive (CGSR, HSR, DSCR, and LANMAR) and two on-demand protocols (CBRP and ARC). Note that, even though these protocols are all based on single-gateway cluster structure, two protocols (DSCR and LANMAR) use flat routing scheme rather than conventional master-to-gateway routing. Nevertheless, we categorize them as LSG protocols because data packets are routed via ordinary nodes toward M_D, thus one master node plays an important role in routing.

26.3.2.1 CGSR and HSR: Proactive Protocol with Conventional Master-to-Gateway Routing. In CGSR [27], each master node maintains the distance and vector to all other masters based on the DSDV routing principle. The next hop node to each of the neighboring masters should be a gateway shared by the two clusters and thus CGSR offers a hierarchical master-to-gateway routing path. Each node keeps a *cluster member table* where the node–master associations of all mobile nodes in the network are stored; this information is broadcast periodically to other nodes. Upon receiving a packet, a node consults its cluster member table and routing table to determine the nearest master along the route to the destination. Next, the node checks its routing table to determine the particular node that can be used to reach the selected master. It then transmits the packet to this node. Figure 26.3A shows an example of the CGSR routing protocol between S and D.

The HSR protocol [28] combines dynamic, distributed multi-level hierarchical clustering with an efficient location management. It maintains a hierarchical topology, where elected masters at the lowest level become ordinary nodes of the next higher level. The ordinary nodes of a physical cluster (in the lowest hierarchy) broadcast their link information to each

623

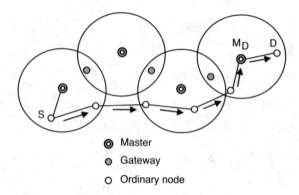

Figure 26.4. DSCR and LANMAR Protocols

other. The master summarizes its cluster's information and sends it to neighboring masters via gateway as it is in CGSR. Figure 26.3B shows an example of the HSR routing protocol with three levels of hierarchy.

In HSR, a new address for each node, *hierarchical ID* (HID), is defined as the sequence of MAC addresses of the nodes on the path from the top hierarchy to the node itself. This hierarchical address is sufficient to deliver a packet to its destination by simply looking at the HID. However, the drawback of HSR also comes from using HID, which requires a longer address and frequent updates of the cluster hierarchy and the hierarchical addresses as nodes move. In a logical sense, this is exactly the same as the *cluster member table* defined in CGSR. However, in case of HSR, the main difference is that the corresponding overhead depends on mobility; it may become zero when nodes do not move and there is no HID change.

26.3.2.2 DSCR and LANMAR: Proactive Protocols with Flat Routing toward M_D.
DSCR [27] is similar to CGSR and HSR in that each node maintains the distance and vector to all masters and has complete information on (D, M_D) association of all mobile nodes. The main difference is that DSCR forwards the data packets to the next hop node, which is not necessarily a master or a gateway. In fact, the concept of gateway is not defined in DSCR and data packets are delivered based on a flat routing scheme. A clear advantage of the DSCR protocol is that the route acquisition time is very small and the routing path is usually the shortest one because it does not need to go through other masters or gateways except the destination's master. Figure 26.4 shows an example of the DSCR routing protocol.

In LANMAR [29], nodes move as inherent groups and there is a master node, called a *landmark*, in each group. As in DSCR, each node periodically exchanges topology information with its immediate neighbors based on FSR routing principle [20] and exchanges distance vector table to all masters. But unlike DSCR, node–master associations do not need to be

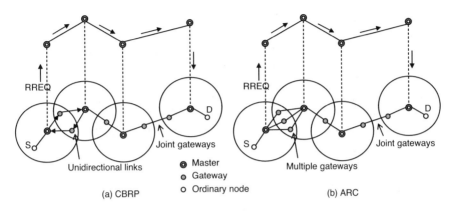

Figure 26.5. CBRP and ARC Protocols

updated because they are known to all the participating nodes. Advantages of LANMAR are small route acquisition time and the shortest routing path. As in DSCR, a routing path does not go through any master nodes, including the destination's master node, M_D. When the packet reaches near the destination cluster, any node that receives the packet may know the destination as one of its neighbors and directly delivers the packet rather than forwarding it to M_D. Figure 26.4 shows an example of the LANMAR routing protocol, which is conceptually the same as DSCR.

26.3.2.3 CBRP and ARC: On-Demand Protocols with Conventional Master-to-Gateway Routing (Allowing No, Single, or Joint Gateways). In CBRP [30] and ARC [31], each node periodically broadcasts its link state to its neighbors as in CGSR and HSR with additional information on neighboring masters that it learns from its neighbors (*neighbor* or *node table*). Therefore, a master is aware of all the ordinary nodes in its cluster and all neighboring masters that are two hops and three hops away (*cluster adjacency* or *cluster master table*), and thus, they support a pair of gateways between two clusters. For each neighboring cluster, the table has entry that contains the gateway through which the cluster can be reached and the master of the cluster.

For (D, M_D) association, CBRP and ARC take an on-demand approach (unlike CGSR and HSR). When a source, S, has to send data to a destination, D, route request packets are flooded only to the neighboring masters. On receiving the request, a master checks to see if D is in its cluster. If so, then the request is sent directly to the destination; otherwise, the request is sent to all its adjacent masters. When the route request reaches D, it replies back to S via the intermediate masters and gateways. Figure 26.5A and Figure 26.5B show examples of the CBRP and ARC routing protocol, respectively.

While the route reply packet goes through the master-to-gateway routing path, intermediate masters can calculate an optimized hop-by-hop route while forwarding the reply packet. Thus, data packets may not follow the master-to-gateway routing path but are delivered along a shorter path [30]. Figure 26.5A shows an example of the CBRP routing protocol. A unique feature to CBRP is that this protocol takes asymmetric links into account, which makes use of unidirectional links and, thus, can significantly reduce network partitions and improve routing performance.

Two new ideas in ARC are:
1. Master revocation rule to preserve the existing cluster structure as long as possible and thus reduce the clustering overhead (see Section 26.3.1)
2. Multiple gateways between clusters for more stable connections

While data packets are forwarded through the hierarchical master-to-gateway routing path, packet header in each data packet contains a source route in the form of master-to-master connections. The benefit of this is that each intermediate master can adaptively choose a gateway when it forwards the data packet to the next hop master and thus provide better packet delivery capability.

26.3.3 LNG-Based Routing Protocols (On-Demand Protocols with Master-to-Master Routing)

One of main benefits of building a no-gateway structure is energy conservation in addition to the routing efficiency. Each node can save energy by switching its mode of operation into *sleep mode* when it has no data to send or receive. SPAN [32] and GAF [34] adopt this approach. In NTDR [33], each node saves power by reducing its transmission power just enough to reach local nodes, but a master should have a large transmission power to reach nodes in remote clusters. In either case, LCA is essentially used, where a master node coordinates the communication on behalf of ordinary nodes in its cluster.

One clear difference between SPAN and NTDR is the power model each assumes. The cluster architecture in Figure 26.6A is based on symmetric power model as used in the SPAN protocol, where master nodes have the same radio power and thus the same transmission range as ordinary nodes. On the other hand, Figure 26.6B shows the asymmetric power model used in the NTDR protocol, where master nodes have longer transmission range. SPAN uses a distributed clustering algorithm discussed in Section 26.3.1, but NTDR does not use any specific clustering algorithm because it is assumed that nodes are naturally clustered in a special environment such as a military setting. On-demand routing principle is used in SPAN and NTDR and route request packets and data packets follow a master-to-master routing path.

Issues in Scalable Clustered Network Architecture for Mobile Ad Hoc Networks

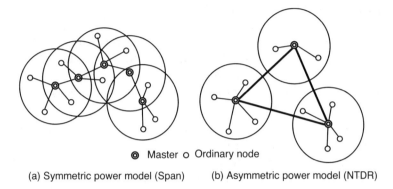

(a) Symmetric power model (Span) (b) Asymmetric power model (NTDR)

Figure 26.6. LNG Architecture with Different Power Models

Routing and energy-efficient operation in GAF protocol [34] are similar to SPAN, but the clustering algorithm is fundamentally different. In GAF, each node uses location information based on GPS to associate itself with a *virtual grid* so that the entire area is divided into several square grids; the node with the highest residual energy within each grid becomes the master of the grid. Other nodes in the same grid can be regarded as redundant with respect to forwarding packets; thus they can be safely put to sleep without sacrificing the *routing fidelity* (or routing efficiency).

26.4 Cluster Architecture for Information Infrastructures

For a large network with many nodes and frequent topology changes, mobility and location management of all mobile nodes pose a high demand of network traffic. The main objective of a LCA for information infrastructure is to select a set of master nodes, which possess routing information of all nodes, so that every ordinary node can reach at least one master within a certain bounded number of hops (e.g., k hops). Searching for the destination node's location and the corresponding routing path is localized within a k-hop cluster rather than an expensive networkwide search. As discussed in Section 26.2.2, the cluster structure is based either on logical connectivity (LLog) or geographic information (LGeo). Section 26.4.1 discusses the master selection algorithms that use these types of LCAs. Section 26.4.2 and Section 26.4.3 discuss LLog- and LGeo-based routing protocols, respectively.

26.4.1 Clustering Algorithms

The clustering algorithms for LCA for information infrastructure turns out to be the *minimum set covering* (MSC) problem, or called a *minimum dominating set* (MDS) problem over a graph representing the ad hoc network. It finds a smallest number of masters such that every node in the network is covered within k hops [1, 11, 13, 35]. The MSC or MDS problem is a well-known NP-hard problem [1, 11, 13]. A number of heuristic clustering algorithms have been proposed to select master nodes that approximate a MDS without resorting to global computation. Note that the basic idea of the heuristics is to select lowest-ID or highest-connectivity node as discussed in Section 26.3.1 with the competition extended to k-hop neighbors rather than just direct (one-hop) neighbors.

The CEDAR protocol [35] is a connectivity-based algorithm with $k = 1$. To provide stability to the master selection algorithm, it gives preference to master nodes already present in its neighbors. Among those master nodes, the one that has more nodes in its cluster is given a higher priority. DDCH [11] is another connectivity-based master selection algorithm for the MSC problem. A link state algorithm is employed with the range of link update limited to k hops. A node is either a master or an ordinary node. An ordinary node can be in one of three states such as *normal, panic,* and *samaritan*. A node enters the panic state if there is no master within k-hop cluster. It sends and receives state packets within $2k$ hops. If it has the maximum number of panic nodes within its k-hop cluster, it becomes a master node.

MMDF [13] provides another heuristic algorithm for the same MSC problem in the context of ad hoc networks. Unlike CEDAR and DDCH, it is an identifier-based algorithm also extended to k-hop cluster. Although identifier-based algorithms are more stable than connectivity-based algorithms (see Section 26.3.1), they may have a balance problem because every ordinary node in the overlapping area of two nearby clusters selects the higher-ID master. Because the overlapping can be quite large in a k-hop cluster structure, cluster sizes tend to be very different and unbalanced. MMDF addresses this problem by using two k rounds of information exchange (*floodmax* and *floodmin*). During the first k rounds, each node selects the highest-id node in each node's k-hop cluster and then, during the second k rounds, it selects the smallest-ID node among the survivals in the first k rounds. One of the features of the MMDF heuristic is that it tends to reelect existing masters even when the network configuration changes. There is also a tendency to evenly distribute the mobile nodes among the masters and evenly distribute the responsibility of acting as masters among all nodes.

The clustering approach of ZRP [36] is unique in that every node is regarded as a master. Each node defines its own k-hop cluster and maintains a set of *border* nodes as gateways to neighboring clusters. Thus, it does not require a specific master selection algorithm.

In ZHLS [37] and GLS [38], constructing a cluster structure is straightforward based on GPS-like location facility. The network area is geographically partitioned into clusters (*grids*) and each node can easily associate it with the corresponding cluster based on its physical coordinates. In ZHLS,

Table 26.3. Clustering Algorithms for LCA for Information Infrastructure (k-Hop Clustering)

LCA	Protocol	Cluster Structure	Clustering Algorithm
LLog	CEDAR	No gateways	Connectivity-based algorithm with $k = 1$. Preference is given to a master that has a larger number of ordinary nodes in its cluster.
	ZRP	Every node is a master	Every node maintains neighbors within its k-hop cluster and *border nodes* as gateways.
LGeo	ZHLS	No masters Multiple gateways between clusters	Gateways link to neighboring grids and maintain information of the nodes within its grid.
	GLS	Every node has a different set of masters (location servers)	Grid hierarchy is formed where each node is located exactly one grid of each size.

there are no masters but gateways are defined as the ones that have links to neighboring grids. Note that a gateway in this case is included in just one cluster. While exchanging link state information between neighbors, each node recognizes itself as a gateway and it uses the stored routing information when relaying packets to neighboring grids.

In GLS [38], the grid structure has more than one level hierarchy as in the HSR protocol discussed in Section 26.3.2. For example, four small sized grids are combined to become a higher level grid. Each node is located in exactly one grid of each size and one master for each grid maintains the location information of the node. This means master nodes for a node are relatively dense near the node but sparse further away from the node. A unique feature to GLS is that there is a set of master nodes for each ordinary node, determined by *consistent hashing*, but the set is totally different from node to node. The rule to select the master of node D is: *A node with the least identifier greater than D's identifier among the candidates becomes a master of D, where ID space is considered to be circular.* In short, for a given ID and a set of candidates, the master node can be deterministically determined. A set of masters for a destination node is used when searching for the location of the node, which we will explain in detail later in Section 26.4.3. Table 26.3 summarizes clustering algorithms used in LLog and LGeo.

26.4.2 LLog-Based Routing Protocols

As discussed previously in Section 26.3, maintaining node–master association (D, M_D) of all nodes is the key design issue in a large-scale MANET. In this section, we discuss two routing protocols (CEDAR and ZRP) that use

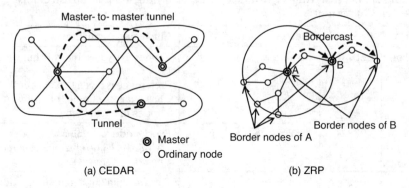

Figure 26.7. CEDAR and ZRP Protocols

cluster architecture as information infrastructure. They employ an on-demand routing principle when searching for the location of a destination node.

26.4.2.1 CEDAR Protocol. CEDAR [35] has three components — master selection (*core extraction*), link state propagation, and route computation. Master nodes are dynamically selected using a connectivity-based algorithm discussed in the last section. When S wants to send the packet to D, it informs its master M_S. Then, M_S finds the path to M_D using DSR-like on-demand probing. Two unique features in CEDAR are QoS routing and *core broadcast* mechanism. In CEDAR, each node can request a communication path to D with a bandwidth requirement. To support this, stable high-bandwidth links are advertised further away while relatively unstable low-bandwidth links are known only to its local neighbors.

Core broadcast mechanism is used to discover D or M_D and to propagate link state information of stable links. Because broadcast is inherently unreliable in a wireless environment (see Section 26.2.1), CEDAR maintains an explicit tunnel between two neighboring master nodes. When a master receives a *core broadcast message*, the master uses the tunnels to unicast the message to all its nearby master nodes. A more recent work combines CEDAR with DSR and AODV to propose *DSRCEDAR* and *AODVCEDAR* [1]. Figure 26.7A shows the CEDAR protocol with three clusters and master-to-master tunnels.

26.4.2.2 Zone Routing Protocol. In ZRP [36], each node has a predefined *zone* (*k*-hop cluster) centered at itself in terms of a number of hops. It consists of three components. Within the zone, proactive *Intra-Zone Routing Protocol* (IARP) is used to maintain routing information. IARP can be any link state or distance vector algorithm. For nodes outside of the zone, reactive

Issues in Scalable Clustered Network Architecture for Mobile Ad Hoc Networks

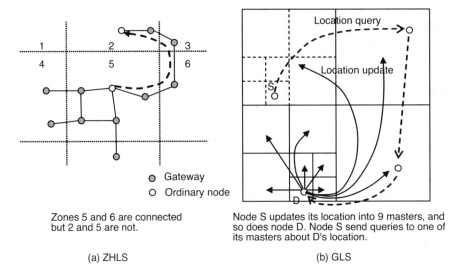

(a) ZHLS

Zones 5 and 6 are connected but 2 and 5 are not.

(b) GLS

Node S updates its location into 9 masters, and so does node D. Node S send queries to one of its masters about D's location.

Figure 26.8. ZHLS and GLS Protocols

Inter-Zone Routing Protocol (IERP) is performed. IERP uses the conventional route request packets to discover a route. It is broadcast via the nodes on the border of the zone (called *border nodes*) and such a route request broadcast is called *Bordercast Resolution Protocol* (BRP). Figure 26.7B shows ZRP with $k = 2$.

26.4.3 LGeo-Based Routing Protocols

This section discusses ZHLS and GLS where cluster structure is simply given based on physical locations obtained via GPS. Routing principle in ZHLS is on-demand searching for the destination cluster. (Note that it does not search for M_D because masters are not defined in ZHLS.) In GLS, location information of a node is distributed to a number of masters and the routing principle is a hybrid of on-demand searching and proactive update.

26.4.3.1 ZHLS Routing Protocol. In ZHLS [37], the network is divided into nonoverlapping clusters (*zones*) without any masters (*zone-heads*) as shown in Figure 26.8A. A node knows its physical location by geographic location techniques such as GPS. Thus, it can determine its *zone id* by mapping its physical location to a zone map, which has to be worked out at design stage. Each node periodically exchanges link state information, called node LSP (*link state packet*), with its neighbors and thus knows the local topology of its zone. For intrazone routing, a shortest path algorithm is used for routing. For interzone routing, zone LSP is propagated globally throughout the network so that each node knows the zone-level topology and the next hop node toward every zone.

Given the zone id and the node id of a destination, the packet is routed based on the zone id until it reaches the correct zone. Then, in that zone, it is routed based on node id. Because the zone id of D changes due to mobility, the association of (D, zone id of D) can be obtained based on on-demand searching through the zone-level topology via gateway nodes. As discussed in Section 26.4.1, there are no masters in ZHLS but gateways may exist between two zones. In Figure 26.8A, zone 4 and zone 5 have two pairs of gateways and zone 5 and zone 6 have a pair of gateways. However, it is possible for two nearby zones to have no gateways, such as zone 2 and zone 5 in Figure 26.8A. In this case, the routing path consists of a number of interzone connections.

26.4.3.2 GLS Protocol. As in ZHLS, the GLS [38] protocol provides a grid network based on physical locations. The basic routing principle used in GLS is *geographic forwarding*. The source S forwards packets toward the destination's physical location meaning that any intermediate node can determine whether it is along the direction between S and D by knowing the locations of S, D, and itself and decides whether to forward or not [38]. Therefore, routing is essentially a two-step process: find the destination node's location and perform geographic forwarding toward that location. In fact, geographic forwarding is used not only to route data packets but also to route location queries to masters that have location information of the destination.

As discussed in Section 26.4.1, GLS replicates the location information of a node at a small set of master nodes (*location servers*), where the set is different from node to node. For example, in Figure 26.8B, node D's location information is maintained at nine masters. Node D periodically updates its location into those masters: three in order-1 squares, three in order-2 squares, and another three in order-3 squares. (This in turn means that node D knows the locations of the nine master nodes and the location update is based on geographic forwarding.) When node S wishes to send data packets to D, it can query one of the nine masters about D's location. Although node S does not know master nodes of D, it can query its masters, especially the most promising master which has the least id greater than node D's id, hoping that it happen to have D's location. Eventually, the query will reach a location server of D, which will forward the query to node D itself. Because the query contains node S's location, it can respond directly using geographic forwarding.

26.5 Summary and Conclusion

Due to the increased path length between two end nodes in a multi-hop MANET, scalability is a challenging issue. A large-scale MANET is feasible only when the task of route search is localized so that the corresponding

Issues in Scalable Clustered Network Architecture for Mobile Ad Hoc Networks

overhead does not increase as the network grows. As one of the promising architectural choices for a scalable MANET, the LCA was discussed, where mobile nodes are logically partitioned into clusters that are independently controlled and dynamically reconfigured with node mobility. By exploiting the spatial locality of communication in MANET applications, the clustered network architecture associated with hierarchical (inter- and intracluster) routing is more scalable compared to nonhierarchical ones. This chapter classified and surveyed LCAs for MANET in terms of clustering algorithms and routing protocols.

Table 26.4 summarizes the CBRPs with their routing principles and unique features.

Table 26.4. Comparison of Cluster-Based Routing Protocols

LCA	Cluster-Based Protocol	Features	Route Pattern	Intercluster Routing Principle
LCA for Routing Backbone				
LSG	CGSR		S, M_S, G, ... G, M_D, D	Proactive update
	HSR	Multilevel clusters		
	DSCR		S $\Rightarrow M_D$, D	
	LANMAR	Group mobility assumed for all nodes within a cluster		
	CBRP	Joint gateways for better connectivity Unidirectional links considered	S, M_S, G, ... G, M_D, D (Route request packets follow a master-to-gateway routing path although actual data packets use a flat routing scheme toward M_D.)	On-demand searching
	ARC	Multiple gateways between two masters for improved robustness		
LNG	SPAN	LNG structure with small number of master nodes	S, M_S, ... M_D, D	
	NTDR	Asymmetric Power Model		
	GAF	GPS-based clustering		

MOBILE COMPUTING HANDBOOK

Table 26.4. Comparison of Cluster-Based Routing Protocols (Continued)

LCA	Cluster-Based Protocol	Features	Route Pattern	Intercluster Routing Principle
	LCA for Information Infrastructure			
LLog	CEDAR	QoS routing Unicast-based *core broadcast* for reliability	Flat routing principle	
	ZRP	Every node being a master *Border-cast* through border nodes		
LGeo	ZHLS	Zone-level routing via gateways		
	GLS	A set of masters (location servers) for each node	Geographic forwarding	Hybrid

Note

1. Master nodes are alternatively called as clusterheads [9], coordinators [32], core [35], leader [31], or a member of dominating set [10] or a backbone network [11].

References

1. P. Sinha, R. Sivakumar, and B. Vaduvur, Enhancing Ad Hoc Routing with Dynamic Virtual Infrastructures, *Proc. of the 20th Annual Joint Conference of the IEEE Computer and Communications Societies (InfoCom 2001),* April 2001.
2. X. Hong, K. Xu, and M. Gerla, Scalable Routing Protocols for Mobile Ad Hoc Networks, *IEEE Network Magazine,* pp. 11–21, July–August 2002.
3. Li, J. et al., Capacity of Ad Hoc Wireless Networks, *Proc. of ACM/IEEE MobiCom'2001,* pp. 61–69, 2001.
4. R. Morris, J. Jannotti, F. Kaashoek, J. Li, and D. Decouto, CarNet: A Scalable Ad Hoc Wireless Network System, *Proc. of 9th ACM SIGOPS European Workshop,* Kolding, Denmark, September 2000.
5. M. Grossglauser and D. Tse, Mobility Increases the Capacity of Ad-hoc Wireless Networks, *Proc. of InfoCom'01,* pp. 1360–1369, 2001.
6. M. Steenstrup, Cluster-Based Networks, in C.E. Perkins, Ed., *Ad Hoc Networking,* Ch. 4, Boston: Addison Wesley, 2001.
7. C.R. Lin and M. Gerla, Adaptive Clustering for Mobile Wireless Networks, *IEEE Journal on Selected Areas in Communications,* vol. 15, no. 7, pp. 1265–1275, September, 1997.
8. T.J. Kwon and M. Gerla, Clustering with Power Control, *Proc. of IEEE Military Communications Conference (MILCOM'99),* 1999.
9. Z.J. Haas and S. Tabrizi, On Some Challenges and Design Choices in Ad-Hoc Communications, *Proc. of IEEE MILCOM'98,* 1998.

10. J. Wu, F. Dai, M. Gao, and I. Stojmenovic, On Calculating Power-Aware Connected Dominating Sets for Efficient Routing in Ad Hoc Wireless Networks, *IEEE/KICS Journal of Communication Networks*, vol. 4, no. 1, pp. 59–70, March 2002.
11. B. Liang and Z.J. Haas, Virtual Backbone Generation and Maintenance in Ad Hoc Network Mobility Management, *Proc. of IEEE InfoCom'00*, 2000.
12. T. Salonidis, P. Bhagwat, L. Tassiulas, and R. LaMaire, Distributed Topology Construction of Bluetooth Personal Area Networks, *Proc. of IEEE InfoCom'01*, 2001.
13. A.D. Amis, R. Prakash, T.H.P. Vuong, and D.T. Huynh, Max-Min D-Cluster Formation in Wireless Ad Hoc Networks, *Proc. of IEEE InfoCom'00*, 2000.
14. S. Chakrabarti and A. Mishra, QoS Issues in Ad Hoc Wireless Networks, *IEEE Communications Magazine*, pp. 142–148, February 2001.
15. E.M. Royer and C.-K. Toh, A Review of Current Routing Protocols for Ad Hoc Mobile Wireless Networks, *IEEE Personal Communications*, April 1999.
16. Z.J. Haas and B. Liang, Ad Hoc Location Management Using Quorum Systems, *ACM/IEEE Trans. on Networking*, April 1999.
17. N. Malpani, J.L. Welch, and N. Vaidya, Leader Election Algorithms for Mobile Ad Hoc Networks, *Proc. of 4th Int'l Workshop on Discrete Algorithms and Methods for Mobile Computing, and Communications*, August 2000.
18. F. Xue and P.R. Kumar, The Number of Neighbors Needed for Connectivity of Wireless Networks, *Wireless Networks*, vol. 10, pp. 169–181, 2004.
19. C. Perkins and P. Bhagwat, Highly Dynamic Destination-Sequenced Distance-Vector Routing (DSDV) for Mobile Computers, *Computer Communications Review*, pp. 234–244, 1994.
20. G. Pei, M. Gerla, and T.-W. Chen, Fisheye State Routing: a Routing Scheme for Ad Hoc Wireless Networks, *Proc. of IEEE Int'l Conf. on Communications (ICC'00)*, pp. 70–74, 2000.
21. D. Johnson and D. Maltz, Dynamic Source Routing in Ad Hoc Wireless Networks, In T. Imielinski and H. Korth, Eds., *Mobile Computing*, pp. 153–181, Norwell, MA: Kluwer Academic, 1996.
22. C. Perkins and E. Royer, Ad-Hoc On-Demand Distance Vector Routing, *Proc. of 2nd IEEE Workshop on Mobile Computing Systems and Applications*, pp. 90–100, 1999.
23. P. Gupta and P.R. Kumar, The Capacity of Wireless Networks, *IEEE Trans. on Information Theory*, vol. 46, no. 2, pp. 388–404, March 2000.
24. T.J. Shepard, A Channel Access Scheme for Large Dense Packet Radio Networks, *Proc. of ACM SIGCOMM'96*, 1996.
25. W. Stallings, IEEE 802.11 Wireless LAN Standard, *Wireless Communications and Networks*, Ch. 14, Upper Saddle River, NJ: Prentice Hall, Inc., 2002.
26. S.Y. Ni, Y.C. Tseng, Y.S. Chen, and J.P. Sheu, The Broadcast Storm Problem in a Mobile Ad Hoc Network, *Proc. of the 5th Annual ACM/IEEE International Conference on Mobile Computing and Networking (MobiCom'99)*, pp. 152–162, 1999.
27. C.-C. Chiang, H.-K. Wu, W. Liu, and M. Gerla, Routing in Clustered Multihop, Mobile Wireless Networks with Fading Channel, *Proc. of the IEEE Singapore International Conference on Networks*, pp. 197–211, 1997.
28. A. Iwata, C. Chiang, G. Pei, M. Gerla, and T. Chen, Scalable Routing Strategies for Ad-hoc Wireless Networks, *Proc. of IEEE JSAC'99*, August 1999.
29. G. Pei, M. Gerla, and X. Hong, LANMAR: Landmark Routing for Large Scale Wireless Ad Hoc Networks with Group Mobility, *Proc. of IEEE/ACM MobiHOC 2000*, pp. 11–18, August 2000.
30. M. Jiang, J. Li, and Y.C. Tay, Cluster Based Routing Protocol (CBRP), IETF Internet Draft, August 1999. http://www.ietf.org/proceedings/99nov/I-D/draft-ietf-manet-cbrp-spec-01.txt.
31. E.M. Belding-Royer, Hierarchical Routing in Ad Hoc Mobile Networks, *Wireless Communication and Mobile Computing*, vol. 2, no. 5, pp. 515–532, 2002.

32. B. Chen, K. Jamieson, R. Morris, and H. Balakrishnan, Span: An Energy-Efficient Coordination Algorithm for Topology Maintenance in Ad Hoc Wireless Networks, *Proc. of 7th ACM International Conference on Mobile Computing and Networking (MobiCom'01)*, Rome, July 2001.
33. Near Term Digital Radio (NTDR), http://www.gordon.army.mil/tsmtr/ntdr.htm.
34. Y. Xu, J. Heidemann, and D. Estrin, Geography-Informed Energy Conservation for Ad Hoc Routing, *Proc. of Int'l Conf. on Mobile Computing and Networking (MobiCom'2001)*, pp. 70–84, 2001.
35. R. Sivakumar, P. Sinha, and V. Bharghavan, CEDAR: a Core-Extraction Distributed Ad Hoc Routing Algorithm, *IEEE Journal on Selected Areas in Communication*, vol. 17, no. 8, pp. 1–12, August 1999.
36. M.R. Pearlman and Z.J. Haas, Determining the Optimal Configuration for the Zone Routing Protocol, *IEEE Journal of Selected Areas in Communications*, vol. 7, no. 8, pp. 1395–1414, August 1999.
37. M. Joa-Ng and I.-T. Lu, A Peer-to-Peer Zone-Based Two-Level Link State Routing for Mobile Ad Hoc Networks, *IEEE Journal on Selected Areas in Communication*, vol. 17, no. 8, pp. 1415–1425, August 1999
38. J. Li, J. Jannotti, D.S.J. De Couto, D.R. Karger, and R. Morris, A Scalable Location Service for Geographic Ad Hoc Routing, *Proc. of the 6th Int'l. Conf. on Mobile Computing and Networking (MobiCom'00)*, pp. 120–130, August 2000.

Chapter 27
Routing and Mobility Management in Wireless Ad Hoc Networks

Ravi Sankar

27.1 Introduction

The past decade has experienced an exponential growth in the use of wireless networks because they enable mobility — a characteristic trait that separates them from other wired networks. This is the result of many different factors including the availability of an assortment of portable devices — such as wireless phones, personal digital assistants (PDAs), and affordable laptop computers — and the accessibility of new services that offer high-speed connectivity and increased communication quality for advance mobile wireless computing.

In general, mobility can be classified as user mobility and device mobility. A mobile and wireless network such as cellular network uses both user and device mobility at the same time. In contrast, a fixed and wireless network such as a wireless local area network (WLAN) provides wireless access but supporting mobility is not an issue. Further, a network can be with or without a fixed infrastructure. In a fixed infrastructure network such as cellular network [1], base stations (BSs) in each cell that provide the radio coverage are fixed and the mobile switching centers (MSCs) act as gateways to the wired telecommunication infrastructure — public switched telephone network (PSTN) — whereas in an ad hoc network, there is no infrastructure in place (i.e., there are no fixed BSs or routers). In other words, the mobile hosts[1] (or nodes) form their own network dynamically by discovering and maintaining wireless connectivity to other hosts by functioning as routers.

27.2 Ad Hoc Network: Definition, Characteristics, and Applications

An ad hoc network is a collection of wireless mobile hosts or terminals forming a distributed reconfigurable network topology. They can operate without the aid of any fixed infrastructure or centralized control and can be rapidly deployed and reconfigured. Ad hoc networks can be compared to fixed infrastructure networks by a number attributes such as scalability, flexibility, controllability, routing complexity, coverage, and reliability [2]. The concept of ad hoc was initially developed for military applications but currently is being considered for many commercial applications including home networking, nomadic computing, networking for disaster relief, search-and-rescue operations, and for large public events such as conventions and conferences [3–5]. These networks, due to their capability of handling node failures and fast topology changes, provide users with ubiquitous communication, computing capability, and information access regardless of location.

Ad hoc networks use peer-to-peer architecture that can be either single-hop or multiple-hop network topology depending on the location of different users, which changes over time. The IEEE® 802.11 WLAN standard [2, 6–10] supports single-hop peer-to-peer ad hoc networking. In such a scenario, when a mobile host is turned on for the first time, it listens for a beacon signal informing the presence of an ad hoc network either from the access point (AP) or from another host. If it does not exist, the host then takes the responsibility of establishing the ad hoc network and informing others who join at a later time of its existence. In some applications, like in the wide area network (WAN), users may be spread over a wide area and a mobile may not be able reach some hosts in the network due to the transmitter signal power limitations. In this situation, it has to enlist other mobiles to aid in forwarding the packets to its final destination. Thus, end-to-end connection between any two mobile hosts may consist of multiple hops. The European Telecommunications Standards Institute (ETSI) broadband radio access networks (BRANs) High Performance Radio Local Area Network (HIPERLAN) wireless LAN standard [2, 11] support multi-hop peer-to-peer ad hoc networking. This multi-hop network configuration is also used in many military tactical networks requiring reliable communications under unpredictable propagation conditions, over rapidly and widely changing geographic environments, and data gathering in inhospitable remote terrains.

A working group (WG) in the protocol area of the Internet Engineering Task Force (IETF) has been studying development and experimentation with mobile ad hoc network (MANET) approaches. The primary focus of the WG is to develop and evolve MANET routing protocols, to stimulate research and address issues on modeling, simulation, performance, implementation, and quality of service (QoS). According to the WG, a MANET is

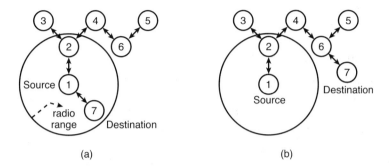

Figure 27.1. An Ad Hoc Network with Seven Wireless Mobile Nodes

an autonomous system of mobile routers (and associated hosts) connected by wireless links. The routers are free to move randomly and organize themselves arbitrarily; thus, the network's wireless topology may change rapidly and unpredictably. Such a network may operate in a standalone fashion or may be connected to the larger Internet or a fixed wired network [4, 12].

MANET hosts are equipped with wireless transceivers using an omnidirectional (broadcast) antenna or directional (point-to-point) antenna or both. The radio transmission range of the mobile hosts is limited and so the connection between two hosts has to be established through multi-hop links that are either bidirectional or unidirectional. All nodes in a MANET need to be within the radio transmission range of one another so as to maintain the network connectivity. Figure 27.1 illustrates a simple example of an ad hoc network comprising seven mobile hosts. Initially, node 1 and node 7 are closely located to each other as shown in Figure 27.1A. The radio transmission range encompasses both of these nodes and therefore establishes a communication link between the nodes that requires no routing. Suppose node 7 moves away from node 1's range as shown in Figure 27.1B; in this case, direct transmission between the hosts is not possible. Communications between the hosts can only be accomplished with the help of routing, where node 2, node 4, and node 6 serve as intermediate routers as shown in Figure 27.1B and the packets are forwarded between node 1 and node 7 as in the case of a typical packet-switched network using a conventional routing protocol. Every mobile host in an ad hoc network must operate as a router to maintain wireless connectivity and forward packets from other hosts. The challenge is to find an optimal route to establish communication to the farthest node in the network when there are hundreds of mobile hosts. To accomplish this task, one needs to design a multi-hop routing mechanism that is capable of adapting to changing network topologies for robust communication between power-constrained mobile hosts using bandwidth-constrained wireless channels (i.e., the routing protocols should minimize the usage of valuable resources such as power, memory, and bandwidth) [4, 5].

27.3 Desired Characteristics of Routing Protocols for MANETs

It is a significant challenge to provide reliable high-speed end-to-end communication in ad hoc networks given their dynamic network topology due to the mobility (random movement) of the hosts, decentralized control, power and bandwidth limitations, multi-hop connections, and nonuniform characteristics of signal propagation in wireless channels. As mentioned above, the routing protocols designed for traditional high-speed networks are not efficient and hence inappropriate for ad hoc networks because they cannot quickly adapt (update the routing tables) to the dynamically changing topology and the network performance drops drastically due to excessive overhead produced by periodic routing updates. Due to the nature of the ad hoc networks, a routing protocol should be distributed enough to increase the reliability and assume the routes as unidirectional links. Each node should be intelligent enough to make routing decisions with the aid of other nodes. Mobile hosts will have limited resources in terms of power, memory, and bandwidth. For optimal use, the protocol should be power efficient and should not broadcast routing updates unnecessarily [4].

27.4 Conventional Routing Protocols

In a packet-switched network, the purpose of the routing protocol (algorithm) is to determine an optimal route to deliver a packet from source to destination using multiple hops: a number of intermediate nodes or routers that the packet traverses along that route. The selection of the route is based on minimizing some measurable performance criterion such as number of hops, delay, cost, throughput, or a combination of some of these factors and in general minimization of the usage of network resources [13–16]. Routing protocols can be categorized as centralized or distributed, adaptive (dynamic) or fixed (static), and reactive or proactive or combinatorial (hybrid). In a centralized routing, all decisions are made at a designated node such as a network control center, where as in a distributed routing protocol, each node shares the responsibility in making the routing decision. In adaptive routing protocols, such as the distance vector routing and link state routing, the routing decision may change as the network condition changes, such as congestion on a link or change in topology (link or node failure). In fixed or nonadaptive routing protocols, such as the shortest path routing and flooding, in contrast, the decision is not based on the measurements or estimates of network traffic or topology. A reactive or on-demand routing protocol takes required actions, such as discovering routes, only when needed, but a proactive protocol discovers the routes before they are needed. Hybrid methods combine the best of both to form a more efficient one. Two of the most commonly used routing algorithms, distance

Routing and Mobility Management in Wireless Ad Hoc Networks

vector routing and link state routing, are based on the adaptive distributed routing:

1. *Distance vector routing* — requires each router to maintain a routing table with distance to all possible destinations from itself. These routing tables will be frequently broadcasted to all the neighbors and all the routers in the network will use this to update their routing tables periodically. This algorithm computes the shortest path from source to destination. While forwarding a packet, each router compares distances received from each destination to each of its neighbors to determine the next hop on the shortest path.
2. *Link state routing* — requires each router to maintain a partial map of the network. Periodical router broadcast updates called link state advertisements (LSAs) regarding the link status and topology changes are flooded throughout the network. All the routers note the change and recompute their routes accordingly.

27.4.1 Problems with Conventional Routing

The follwoing factors must be considered when designing appropriate routing protocols for MANETs:

- Due to the nonuniform propagation characteristics, transmission between two mobile hosts in a wireless network may not be the same in both directions and in the wireless environment, some routes discovered by conventional routing protocols will not work. All conventional routing protocols assume that routes are bidirectional and of the same quality, which is not always the case in ad hoc networks.
- A conventional routing protocol also calculates the redundant routes and updates the routing tables. This increases the size of the routing table, the processing time, and power consumption.
- Periodic updates without a change in the topology waste precious network bandwidth in the wireless networks. In the process of sending and receiving the routing update packets, the battery power is also consumed.
- Routing protocols periodically send control or signaling messages for making measurements and broadcasting routing tables to other nodes. In a large network with long routes, this would result in significant overhead with adverse effect on the battery power consumption.

27.5 Review of Ad Hoc Routing Protocols

There are a number of routing protocols for MANETs that have been either designed or are works-in-progress and many are under consideration within the WG of the IETF. Their implementations now exist and are in various stages of maturity, yet there is no comprehensive study of performance evaluation and comparison of all the routing protocols that are available. For an excellent survey on ad hoc routing protocols, readers are referred to [17, 18]. There are a few performance comparisons between selective protocols that have been reported [19–21]. Based on those studies and from general consensus, there are three or four major routing protocols that appear to be in contention if at any time a routing protocol for MANET is standardized by the IETF. In this section, we will describe the operation of each of these protocols and then compare their various characteristics. In general, MANET routing protocols can be categorized as table-driven, on-demand, and hybrid.

27.5.1 Table-Driven Routing Protocols

Each node in the network maintains routing tables that contain the routing information between that node and every other node in the network. Any change in the network topology and link status will require the nodes to propagate updates throughout the network, thereby maintaining consistent and up-to-date routing information about the network. Table-driven routing protocols address the problems associated with the big routing tables and unnecessary routing updates as they are designed to minimize the routing table size and updates required. Some of the existing table-driven routing protocols include Destination-Sequenced Distance Vector (DSDV) routing, Wireless Routing Protocol (WRP), Link State Routing (LSR), Clusterhead Gateway Switch Routing (CGSR), and Hierarchical Routing (HR). They differ in the way they propagate the changes in the network structure and in the number of routing tables used to maintain the network routing information at each node.

27.5.1.1 Destination-Sequenced Distance Vector. Each node in DSDV [22], based on the classical Bellman–Ford routing algorithm [13–15], maintains a routing table that lists all possible destinations and the number of hops required to reach each destination. Routing tables will be periodically broadcasted throughout the network to update the routing information. For example in the ad hoc network shown in Figure 27.1B if node 4 needs to send a packet to node 7, it first examines its routing table for the next hop. From the routing table shown in Table 27.1 for this example, node 4 needs a minimum of two hops to reach node 7 with the next hop being node 6. Each routing entry in the table is marked with a sequence number assigned by the destination node to distinguish between stale routes from new ones and to avoid creating deadlock loops. The sequence number is subject to change upon any changes in the neighboring nodes. When routing a packet, recent route with the highest sequence number is used. If two routes have the same sequence number, then the route with the best metric (shortest path) will be used (Figure 27.2).

Table 27.1. Node 4 Routing Table

Destination	Next Hop	Number of Hops	Sequence Number
1	2	2	S406_1
2	2	1	S128_2
3	2	2	S564_3
4	4	0	**S710_4**
5	6	2	S392_5
6	6	1	S076_6
7	6	2	S128_7

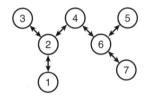

Figure 27.2. DSDV Routing Operation Example

Routing updates can be either *event driven* or *time driven*. To avoid wasting of bandwidth, routing updates can be done in two ways — *full dump* or an *incremental update*. In the full dump type of update, the full routing table is broadcasted and this must be done infrequently, whereas in an incremental update, packets are small in size and are used to convey the information regarding the changes in the routing table after the last full dump. When a network is stable, incremental updates are used to avoid the bandwidth waste. However in a fast changing network, the number of incremental updates can become large, thereby requiring more frequent full dumps. Routing entries are marked with sequence numbers to distinguish between stale routes from new ones, thereby avoiding deadlock loops. When routing a packet, a recent route with the highest sequence number is used. If two routes have the same sequence number then the route with the best metric (shortest path) will be used.

DSDV suffers from congestion control and from bandwidth-limitations. To overcome these problems extended versions of the DSDV protocol such as R-DSDV [23] and M-DSDV [24] have been proposed. R-DSDV is a randomized version of the DSDV routing protocol. The randomization consists of advertising the routing table according to probabilities, which determines independent advertisement rates for different nodes in the network rather than advertising whenever there is a change in the routing table. M-DSDV is a multipath routing protocol. It aims to discover and maintain multiple quasi-shortest paths to the destinations. The advantage of M-DSDV is that the traffic load is distributed to the multiple paths at the same time without forwarding loops.

27.5.1.2 Wireless Routing Protocol. WRP [25] has a unique feature of checking the consistency of all its neighbors every time it detects a change in the links of any of its neighbors. This helps to eliminate loops and enables fast convergence to changes in the topology. Each node in the network maintains a distance table, a routing table, a link cost table, and a message retransmission list (MRL). Distance table of node N_s contains the distance of each destination N_d via each neighbor N_n of the node. It also contains the downstream neighbor of N_n through which this path is realized. The routing table of N_s contains the distance of each destination node N_d and the predecessor, and the successor of node N_s on the path. It also contains a tag to identify whether the entry is a simple path, a loop, or an invalid path. The link cost table contains cost to link to each neighbor of the node and the number of time-outs since an error free message was received from that neighbor. The MRL consists of information to let a node know which of its neighbors has not acknowledged its updated message and to retransmit the updated message to that neighbor. Mobile nodes send update messages, which consist of destination identification, distance to the destination, and predecessor of the destination only to neighboring nodes. Nodes in the MRL should acknowledge the updates. Upon receiving the updates, the node updates its distance table and looks for a new path that is better for sending the route information. Any new path so found is relayed back to the original nodes so that they can update their tables. The nodes also update their routing table if the new path is better than the existing path. If there is no change in the routing table, the node is required to send a HELLO message to ensure the connectivity. Otherwise the lack of messages from the node indicates the failure of the link, which may cause a false alarm. On receiving the acknowledgment message (ACK), the sender updates its MRL.

27.5.1.3 Link State Routing Protocols. In Global State Routing (GSR) [26], each node keeps just the neighbor's link state in its topology table. It also maintains the next hop and distance tables. When there is a link change, a routing message is forwarded so that nodes can update their tables when the sequence number of the message is different from the one stored in the topology table.

Fisheye State Routing (FSR) [27, 28] is an improvement to GSR in that it attempts to reduce the routing update overhead. It is well-suited for large MANETs. In FSR, a node stores the link state update of a destination to its neighbors with a frequency that depends on the hop distance to

Routing and Mobility Management in Wireless Ad Hoc Networks

that destination. It retains the routing entry for each destination. State updates corresponding to the farthest destinations are propagated with lower frequency than those for close by destinations. From the state tables, nodes construct the topology map of the entire network and compute efficient routes. The route becomes progressively accurate as the update packet approaches its destination. In contrast to most LSR protocols where the link state updates are flooded from each source, here the entire link state information is exchanged only with the neighbors. The updates are propagated periodically as aggregates based on the scope relative to the destination. Scope is defined in terms of the nodes that can be reached in a certain number of hops.

Optimized Link State Routing (OLSR) [29] provides an optimization of the classical link state algorithm tailored to fit the requirements of mobile networks. It is a proactive, table-driven protocol that exchanges topology information periodically with other nodes. OLSR inherits the concept of forwarding and relaying from HIPERLAN. Each node selects a set of its neighbor nodes as multipoint relays (MPRs) and these MPRs announce this information periodically in their control messages. The routes from source to destination will consist of only MPRs as intermediate nodes. OLSR provides optimal routes in terms of number of hops. The number of message and control overheads is minimized by the use of MPRs, which facilitate efficient flooding as opposed to the classical flooding mechanism.

Topology Broadcast Based on Reverse-Path Forwarding (TBRPF) [30] is also a proactive, table-driven protocol designed for MANET. This provides hop-by-hop routing along the shortest paths to each destination. Using a modified Dijkstra's algorithm [13–15], each node computes a source tree based on partial topology information stored in its table. TBRPF uses a combination of periodic incremental updates of the partial source tree to neighbors. It performs neighbor discovery using differential control messages to report changes in the status of neighbors.

27.5.1.4 Clusterhead Gateway Switch Routing. CGSR [31] is a clustered multi-hop network with several heuristic routing algorithms. The nodes form a cluster and a clusterhead is selected using a distributed clusterhead selection algorithm. All the nodes in the transmission range of the clusterhead belong to that cluster. Nodes that can communicate with two or more clusterheads are gateway nodes and they are used for intercluster communication. To avoid performance degradation due to frequent changes to the clusterheads in response to dynamic changes in the network topology, CGSR uses a least cluster change (LCC) algorithm. Clusterhead change occurs only if a change in network causes two clusterheads to come into one cluster or one of the nodes move out of the range of all the clusterheads.

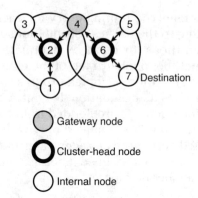

Figure 27.3. CGSR Routing Showing a Data Path from Source to Destination

Each node must keep a *cluster member table,* which consists of destination clusterheads of all the nodes in the network. Nodes broadcast cluster member table updates periodically throughout the network using DSDV. Neighboring nodes will update their cluster member tables with these updates. In addition to the cluster routing table, each node also maintains a routing table to determine the next hop in routing packets to destination. A packet sent by the node is first routed to the clusterhead of that particular cluster. Clusterheads will look into the cluster member table and route the packet to the gateway in the destination cluster, which will send the packet to the destination via its clusterhead. For the ad hoc network shown in Figure 27.3, if node 3 is the source and node 7 is the destination, then any packet sent by node 3 is first routed to its clusterhead node 2. Clusterhead node 2 will look into its cluster member table and route the packet to the gateway node 4 in the destination cluster, which will route the packet to the destination node 7 via its clusterhead node 6.

If the destination cluster gateway is not in the range of the sending cluster gateway, the clusterhead will forward the packet to the next hop by looking at the routing table. Due to the cluster division of the network topology and cluster election methods, CGSR is best suited for networks with continuous changes in the network topology. The clustering approach of CGSR limits the number of messages that need to be sent resulting in low latency. Hence, it is suitable for real-time traffic and also it is scalable. It is important to note that frequent clusterhead changes can adversely affect routing protocol performance because nodes are busy in clusterhead selection rather than packet relaying.

27.5.1.5 Hierarchical Routing Protocols. Hierarchical State Routing (HSR) [32] is based on the concept of organizing nodes in a logical hierarchical structure using recursive clustering. As in the case of CGSR, nodes that are close to each other within a certain radio transmission range are grouped as clusters. Special nodes for each cluster are selected, clusterhead and gateway for any intracluster and intercluster communication, respectively. Clusterheads in turn organize themselves by forming higher level clusters and so on forming a three-level hierarchy. The main goal of hierarchical clustering is to reduce the overall routing overhead. This has to be optimized at each level.

Landmark Routing Protocol (LANMAR) [33] uses similar multilevel hierarchical addressing, which identifies the position of a node in the hierarchy and helps to find a route to it. Each node in the network uses a scoped routing as in the case of FSR to learn about routes within a given scope (maximum number of hops). To route a packet to a destination outside its scope, a node will have to direct the packet to the destination via its corresponding landmark, which is similar to the clusterhead in CGSR.

27.5.1.6 Summary of Table-Driven Protocols. Table 27.2 summarizes the characteristics or properties of the table-driven routing protocols and compares in terms of their performance parameters.

27.5.2 On-Demand Routing Protocols

In this type of routing there are no periodic routing updates and routes are discovered only when required (i.e., on-demand). The routing process consists of two steps — *route discovery* and *route maintenance*. When a source node requires the route to a destination, it initiates a route discovery process, which will discover the route from source to destination. The route maintenance process monitors the validity of the route. When a route maintenance process detects problems with the existing route, it initiates the route discovery process to detect a new route.

Even for the most efficient proactive table-driven routing protocols, continuous tracking of the network topology changes in a practical MANET produces an overwhelming amount of control traffic. Further, the bandwidth used for this is wasted in a high mobility environment when the route information acquired may become outdated even before it is used. On the contrary, on-demand protocols initiate route discovery only when needed by the source node and so the amount of overhead traffic is much lower, which comes at the expense of the delay in routing. So, reactive, on-demand routing protocols are generally viewed as being more suitable than proactive, table-driven protocols for power and bandwidth constrained MANETs. Some of the existing on-demand routing protocols include Dynamic Source Routing (DSR), Ad Hoc On-Demand Distance Vector Routing (AODV), and Associativity Based Routing (ABR).

27.5.2.1 Dynamic Source Routing. DSR [34, 35] is based on the source routing scheme used in the wired networks. Each packet carries the address of each host in the network through which the packet should be

Table 27.2. Summary of Table-Driven Routing Protocol Characteristics

Characteristic/ Parameter	DSDV	WRP	LSR	CGSR	HR
Network route topology	Flat	Flat or Hierarchical	Flat	Hierarchical	Hierarchical
Loop-free using sequencing	Yes				
Number of required tables	Two	Four	Varies	Two	Varies
Multiple routes possibility	No	No	No	No	Yes
QoS and multicast capability	No (extensions to support multicasting available)				
Periodic broadcasting (uses update messages)	Yes				
Frequency of update	Periodically				
Uses special nodes	No	No	No	Yes (clusterhead or landmark)	
Routing metric	Shortest path				
Time complexity[1] (Number of steps/operation)	O(d)	O(h)	≤ O(d)	O(d)	O(h)
Communication complexity[2] (Number of messages/operation)	O(n)	O(n)	≤ O(2n)	O(n)	O(n)

[1] Time complexity provides a measure of delay computed from the number of steps needed to perform the routing protocol operation.

[2] Communication complexity provides a measure of utilization computed from the overhead needed to perform the routing protocol operation.

Notations used in the Tables:

n — Number of nodes in network

d — Network distance (between the farthest source and destination)

h — Height of the routing tree

n_c — Number of nodes affected by topology change

forwarded to reach the destination host. All the mobile hosts in the network participating in the routing maintain a route cache to store the information about the source routes they have learned. Before sending the packet, the sender will check the route cache for a source route to destination. If a route is found, the sender uses the route to transmit the packet to the destination or else the sender will start the route discovery process to find the route to the destination. To reduce the size of the route cache, each entry will be associated with an expiration period. DSR eliminates the need for every node in the network to do periodic route discovery advertisements because the route will be explicitly specified in each packet from the route discovery. This considerably reduces the network bandwidth overhead and battery power consumption. DSR works well with unidirectional asymmetric links.

27.5.2.1.1 Route Discovery. From Figure 27.2, if node 4 is a source initiating a route discovery process for node 7 as the destination, it floods a route request (RREQ) packet with a unique request ID. All the nodes in the range of the source node 4 (node 3, node 2, node 6) will receive this packet. A RREQ packet identifies the destination node 7 as the target of the route discovery. A RREQ packet consists of the address of the initiator (node 3) target of the request (node 7) and the route record. In the route record the sequence of hops taken by the RREQ packet to the target is recorded. Request ID is used to eliminate any duplicate RREQs. Any node, for example node 6, receiving the RREQ packet performs the following:

- If it has already received the request, it drops the packet.
- If it is the target node, it returns a route reply to the source node.
- Otherwise, the node appends its address to the route record and rebroadcasts the updated route request. The target node 7 identifies itself as the destination from the rebroadcast RREQ and sends the route reply packet to the source node 3.

The route reply packet consists of a complete list of the route record from source to destination. For a target node to return the route reply, it will examine its own route cache for a route back to the source node. It will send the route reply using this route if found, otherwise the target node will invoke route discovery to the source node. To prevent infinite recursion of route discoveries, it must piggyback the route reply on the packet containing the route request. When the source node receives the route reply, it caches this route in its route cache and uses the route to send the packets to the target node.

27.5.2.1.2 Route Maintenance. DSR uses two types of packets for route maintenance — route error and passive acknowledgment. After sending a packet to the next hop, the sender may be able to listen to the transmission of that particular packet from that node to another node. Using this passive acknowledgment, the validity of the existing route is monitored. If an explicit acknowledgment mechanism is not available, the node transmitting the packet can request DSR-specific software acknowledgment to be returned by the next node along the route. When a node fails to receive a passive acknowledgment, a route error packet will be sent to the original sender to invoke a new route discovery process. Nodes that receive the error packet will delete the route entry associated with the broken link. DSR mainly relies on Media Access Control (MAC) for notification of link failures.

27.5.2.2 Ad Hoc On-Demand Distance Vector Routing. AODV [36, 37] routing algorithm is designed for ad hoc mobile networks. AODV is an improvement on the DSDV algorithm because it minimizes the required routing updates. Routes will be discovered only on-demand and maintained as along as they are needed by the source as opposed to DSDV that maintains the list of all the routes. It is capable of supporting both unicast and multicast routing. When a source node requires sending a message to a destination, it checks its routing table for previous entries. If the route is listed then the packet will be forwarded in that route. Otherwise, the path discovery process will be initiated. The route will be monitored by the route maintenance process as long as it is required.

27.5.2.2.1 Route Discovery. When a source node requires the route to a destination, it broadcasts a RREQ packet containing the source number, its current sequence number for that destination, and broadcast ID to its neighbors. Nodes receiving this packet check for the destination number. If it is not the destination, then it will be rebroadcast to its neighbors and the node will set up backward pointers in the routing table. If the destination number matches with its own number, then a route reply packet, which includes the number of hops between the source and destination and its sequence number, will be unicasted to the source. The source receiving the reply message sets forward pointers to the destination. Once the pointers are set to the destination, the route will be established and the source will begin to transmit packets. If the source receives a sequence number higher than the present sequence number of a particular destination, it updates the routing information of the destination and begins using the better route.

27.5.2.2.2 Route Maintenance. If a node along the route moves, then the upstream neighbor notices the move and informs each of its upstream neighbors by sending the link failure notice message. The source will finally get the message and initiate the route discovery process again to find another route to the destination. In the case of source movement, it initiates the route discovery process to find a new route to the destination.

A routing protocol called AODV with path accumulation (AODV-PA) has been proposed to improve the efficiency of AODV and DSR [38]. Here,

AODV is modified to enable the path accumulation feature of DSR during the route discovery process to get routing information. This has been shown to perform better under conditions of increased load and mobility.

27.5.2.3 Associativity Based Routing. ABR [39] defines a new metric for routing known as the degree of association stability. This is free from loops, deadlocks, and packet duplicates. Routes are selected based upon associativity state of nodes. All the nodes in the network periodically generate *beacons* to signify their existence. When a node receives a beacon from its neighbor, it updates the associativity table by incrementing the associativity tick to that particular neighbor. Tick is a variable counter, which will be updated with response. High tick value represents low mobility and low tick value represents high mobility of the nodes. When a neighboring node moves out of the range its associative ticks will be reset. The ultimate aim of this routing technique is to find the long-lived routes for an ad hoc network. ABR routing consists of three different phases — route discovery, route reconstruction, and route deletion.

27.5.2.3.1 Route Discovery. This phase consists of broadcast query and await-reply. Source node broadcasts a query message in search of route to destination. When an intermediate node receives the query message it will append its address and associativity tick to the query packet. The next node erases the upstream node neighbor's associativity and retains only the entries pertaining to it and its upstream node. Packets arriving at the destination will contain the nodes along the path from source to destination and their associativity ticks. The destination will select the long-lived path by examining the number of node ticks in each path. If multiple paths have the same level of ticks, then the shortest path will be selected. The destination will send an await-reply to the source node on the selected path.

27.5.2.3.2 Route Reconstruction. This phase consists of partial route discovery, invalid route erasure, valid route updates, and new route discovery. If a source node moves, then the await-reply process will be initiated and a new route will be discovered from the source to the destination. The route notification message is used to erase the routes associated with the downstream nodes. When the destination or intermediate node moves, its immediate upstream node erases the route and a localized query process will generate a query packet to find if the node is still reachable. It selects the best route depending upon the replies. This process is called partial route discovery. Invalid routes will be erased using the route notification.

27.5.2.3.3 Route Delete. When a route is no longer needed, the source node initiates a route delete broadcast. All the nodes along that route delete their route entries in routing tables.

Table 27.3. Summary of On-Demand Routing Protocol Characteristics

Characteristic/Parameter	DSR	AODV	ABR
Network route topology	Flat	Flat	Flat
Loop-free using sequencing		Yes	
Multiple routes possibility	Yes	No	No
QoS and multicast capability	No	Yes	No
Periodic broadcasting (uses update messages)	None	None	Beacons
Frequency of update	None	None	Periodically
Uses HELLO message or beacons	No	Yes	Yes
Routing metric	Shortest path	Newest and shortest path	Degree of association stability
Time complexity (Number of steps/operation)	O(2d)	O(2d)	O(2d)
Communication complexity (Number of messages/operation)	O(2n)	O(2n)	≤ O(2n)

27.5.2.4 Summary of On-Demand Protocols. Table 27.3 summarizes the characteristics or properties and compares the performance parameters of the on-demand routing protocols discussed.

27.5.3 Hybrid Routing Protocols

As mentioned earlier, hybrid protocols combine the best features of reactive, on-demand, and proactive, table-driven routing. They include Zone Routing Protocol (ZRP) and Temporally Ordered Routing Algorithm (TORA).

27.5.3.1 Zone Routing Protocol. ZRP [40–42] is a hybrid protocol combining both the principles of table-driven and on-demand protocols. It consists of two protocols: Intra-Zone Routing Protocol (IARP) and Inter-Zone Routing Protocol (IERP), which are the proactive and reactive components of the ZRP, respectively. The network is divided into nonoverlapping routing zones and runs independent protocols for inter- and intra-zones. Different zones can operate with different protocols. IARP operates within a zone (hop distance that route updates are relayed) and supports route acquisition and maintenance. It tracks all the possible routes, so all nodes within a zone learn about its local network connectivity. Traditional proactive link state protocols or any other table-driven protocol can be modified to serve

Routing and Mobility Management in Wireless Ad Hoc Networks

as an IARP. In IERP, a source node finds a destination node not located within the same zone by sending control messages to all border nodes until it is found. Optimal routing zone diameter should be chosen for a scaled topology. Any existing reactive protocol can be adapted to IERP.

27.5.3.2 Temporally Ordered Routing Algorithm. TORA [43, 44] is based on neither link state nor distance-vector algorithms, but it is a member of the link-reversal class of algorithms. This distributed, loop-free routing protocol is designed to minimize communication overhead by localizing the reaction to topological changes. This adaptive protocol can simultaneously support both source-initiated, on-demand routing for some destinations and destination-initiated, proactive routing for others. Because TORA control messages are localized to a small set of routers near the occurrence of a topological change, nodes must maintain routing information about adjacent routers. TORA maintains route entries for each destination similar to the distance-vector routing approach, however the distance metric is not used for routing because it does not continuously compute the shortest path. So rather than providing optimal routing, it is designed to achieve low time and communication complexities by supporting a mix of reactive and proactive routing on a per destination basis.

TORA assigns direction (upstream and downstream) to the link with the neighboring routers based on a relative *height* metric associated with each router. Conceptually it can be considered as the router's height (i.e., links are directed from the higher to the lower router such that directional assignment for the forward flow of packets to reach destination is downstream). TORA performs four basic functions — route creation, maintenance, erasure, and optimization. Route creation is by generating a directed sequence of links based on the height metric. Maintenance of the routes involves adapting the structure in response to topological changes such as when there is a link failure, it will trigger link reversals to form new directed paths to the destination. In cases when the network becomes partitioned, links that are partitioned from the destination are marked undirected to delete invalid routes during the route erasure process. TORA include secondary mechanism to optimize the routes in which routers reselect their heights to improve the routing structure by means of four control messages — query (QRY), update (UPD), clear (CLR), and optimization (OPT).

27.5.4 Comparison of the Routing Protocols

Table 27.1 and Table 27.2 highlight the similarities and differences between the table-driven and on-demand routing protocols, respectively. Table 27.4 briefly provides the overall comparison review of table-driven, on-demand, and hybrid protocols.

Table 27.4. Overall Comparison of Ad Hoc Routing Protocol Types

Characteristic/Property	Table-Driven	On-Demand	Hybrid
Network route topology	Mostly flat except CGSR and HR	Flat	Flat
Routing strategy	Proactive	Reactive	Uses both
Route discovery/updates	Periodically	As needed	Periodically and as needed
Routing information availability	Always	As needed	Uses both
QoS support	Shortest-path (number of hops) or height as QoS metric		
Issues	Optimizes route delay at the expense of bandwidth and power consumption	Optimizes bandwidth usage at the expense of delay and power consumption	Optimizes both delay and bandwidth usage

27.5.5 Other Protocols

In addition to the routing protocols discussed earlier, currently there is significant interest in the development of two other classes of routing protocols — power-aware and QoS-aware. Other routing protocols that are based on location and those that support multicasting and multipath routing have been reported.

27.5.5.1 Power-Aware and QoS-Aware Routing. The objective of an ad hoc protocol is to either optimize for power-efficiency or certain defined QoS parameters, which may include the following metrics: minimize per packet power consumption (shortest-hop path) and cost, maximize load sharing and distribution among the routers to maximize the life of the network [45, 46]. For battery power consumption at nodes, these metrics have been used for determining only broadcast routes in wireless ad hoc networks [47]. Other reported studies include modification of MAC and physical (PHY) layers of IEEE 802.11 standard for power management to extend battery life [48] and development of distributed MAC scheme that provides QoS real-time access to ad hoc Carrier Sense Multiple Access (CSMA)-based wireless networks [49]. In applications such as wireless ad hoc sensor networks, nodes are expected to be low cost, small and lightweight (<100 grams), and consume low power (<100 microwatts). Current research focus has been in the design of protocols that are energy efficient,

Routing and Mobility Management in Wireless Ad Hoc Networks

as well as the study of issues in network scalability and survivability [50–54].

There are also QoS related studies on ad hoc networks such as the bandwidth reservation proposal [55–57]. The proposed routing with bandwidth as QoS parameter includes bandwidth calculation and reservation methods. Another related study reports on QoS routing performance and requirements in wireless ad hoc networks [58]. In core extraction distributed ad hoc routing algorithm (CEDAR) [59], QoS information is calculated at specific core nodes and propagated as waves for overall QoS awareness in the network.

27.5.5.2 Location-Based Routing. The geographical location information if available can be used effectively to improve the routing performance of ad hoc networks. This information can be obtained using global positioning system (GPS) or other location positioning technologies. Location-Aided Routing (LAR), Geographic Addressing and Routing (GeoCast), Distance Routing Effect Algorithm for Mobility (DREAM), and Greedy Perimeter Stateless Routing (GPSR) are some of the location-based ad hoc routing protocols that can be found in the literature. An excellent review and comparative analysis of these protocols can be found in [18]. Location Trace Aided Routing (LOTAR) is another location-aided protocol that uses LAR and GSR methods to predict when a route will break and then attempts to establish a better route [60]. Each node on the active path monitors the connectivity of its neighbors. When the hand-off condition is met, it starts a local route discovery algorithm, and hands off traffic through the new route. If discovery fails, it sends global discovery message to the source. The route is also optimized locally. When a node discovers that there is no need for it to bypass the data flow, because both neighboring nodes can communicate to each other, it leaves the active path.

For effective location-based routing, the routing update processing must be done at a rate faster than the network mobility because the node's location might change quickly in an ad hoc network with dynamic rapidly changing topology. It is more challenging if the nodes move at different rates.

27.5.5.3 Flooding and Multicasting. Flooding method is useful in MANET for the route discovery that is required for on-demand protocols. The broadcast addressing mechanisms available for IPv4 (Internet Protocol, version 4) are not suitable because ad hoc network nodes require forwarding of flooded packets. New multicast groups for flooding to all nodes are specified in this protocol, which can be adopted for both IPv4 and IPv6 [61]. Every node maintains a list that includes the flooded packet identifier (FPI) to keep track of the received and retransmitted flooded packets. When a node receives a flooded packet, it checks its list for the FPI. If there is a matching FPI in the list entry, then it discards the packet because it has been already received and forwarded. Otherwise, the node retransmits the packet. Each node in the network is enabled for forwarding and retransmission of any distinct flooded packet it receives.

The protocol designed to address the broadcast and multicast functionality in MANET uses the route discovery mechanism defined in DSR to flood the data packets throughout the network. This can be considered as an extension to DSR [62]. Another protocol [63] introduces flooding for on-demand routing using directional antennas to reduce flooding overhead. By knowing the direction range of the source and destination, each node floods only a fraction of the network with queries. Flow Oriented Routing Protocol (FORP) for real-time IPv6 flows, such as voice and data for highly mobile ad hoc wireless networks using multi-hop hand-off is presented in [64].

27.5.5.4 Multipath Routing. Link failures in MANET occur due to node mobility resulting in frequent changes in topology and wireless channel conditions. The use of multiple routes circumvents this problem of link failures and reduces the effect of congestion. Hybrid routing protocols such as ZRP and TORA support multipath routing and multipath extensions to on-demand routing such as DSR and AODV have been developed where alternate routes are discovered, maintained, and used when needed. Traffic is distributed through only one path. The multipath property of routing protocols is also essential for distribution of load in the network. Multiple source routing is one such method that achieves this by a weighted round-robin heuristic-based scheduling strategy among multiple paths. Split multipath routing aims on maintaining maximally disjointed paths, but the traffic is distributed only on two routes per session. Another distributed multipath routing algorithm (MPATH) builds multiple paths for each destination in the network such that they are loop-free at all times [65].

27.6 Performance Issues and Challenges

Study of these protocols and performance comparisons reveals that table-driven protocols yield inferior performance when compared to on-demand protocols due to their regular routing updates, which consume large amounts of the available bandwidth and computational overhead in the presence of mobility and dynamic channel conditions. The protocol performance is generally measured in terms of network size specified by the number of nodes, network connectivity specified by the average number of neighboring nodes, rate of change of network topology, link capacity specified by the link data transfer rate in bits/second, effectiveness to adapt to varying traffic patterns, channel conditions, and node mobility [4]. The number of metrics to assess the performance includes end-to-end throughput and delay, time-to-route discovery and establishment, routing efficiency measured in terms of overall utilization (measure of packet loss rate), and transmission efficiency (measure of control or overhead wastage). Finally, the overall assessment would be based on how well the protocol satisfies the design criteria set forth and the QoS constraints met in terms of bandwidth, power or energy, shortest-path (number of hops), or some combination thereof.

The MANET performance depends on how well it copes with the effect of its different characteristics: MAC and routing protocols must take into account the dynamic multi-hop network topology that changes rapidly and randomly at unpredictable times due to host mobility and the wireless communication channel conditions. Protocols must adapt to the bandwidth-constrained variable capacity wireless links and account for the effect of frequency-selective and multipath fading, noise, and cochannel interference conditions. Another issue is the energy or power conservation. In applications such as wireless sensor networks, the batteries powering the sensor nodes cannot be replenished or replaced and so it needs to be minimally used and the protocol design criteria for optimization needs to include low power drainage when in use (i.e., reduced time and communication complexities). In addition, security limitations and the need for scalability for large (hundreds or thousands of nodes) MANETs as in the case of sensor networks are open issues that have to be addressed.

27.7 Mobility Management in Ad Hoc Networks

Mobility management, which encompasses location management, resource allocation, and hand-offs, is one of the most important research issues in wireless networks. In wireless cellular networks and wireless Internet that operates in a single-hop wireless environment, only the end nodes (source and destination) are mobile. In contrast, in multi-hop ad hoc wireless networks the intermediate nodes also exhibit mobility and so the mobility management aspects are quite different in that sense.

For wireless cellular (or personal communication system [PCS]) systems such as IS54/IS136/TDMA, IS95/CDMA, and GSM, mobility management functions such as hand-off, roaming, registration/authentication, and location tracking are handled by the signaling protocols IS-41 and GSM MAP (Global System for Communication — Mobile Application Part) [66], respectively. The strategies proposed in these standards involve partitioning into location areas through a two-tier system of home and visited databases called home location register (HLR) and visitor location register (VLR). Similarly the mobility management signaling protocol Mobile IP standardized by the IETF for Wireless Internet has two-tier system of permanent IP address on a home network and care-of-address (CoA) in the foreign network while visiting away from home.

In ad hoc networks, if the topology and network address is flat then the mobility management is unnecessary and can be handled simply by the routing algorithm. In general, for the table-driven routing protocols such as DSDV, WRP, and LSR, the routing tables have entries to all destinations and can deliver a packet without any mobility management. This is based on the assumption that there is only one ad hoc network domain. Similarly for the on-demand protocols such as DSR and AODV or hybrid protocols such as ZRP and TORA, the routes are discovered as needed and there is no need for any mobility management, specifically roaming or location management. However, for any hierarchically structured ad hoc networks using routing protocols such as CGSR and HR, there exist a number of mobility management techniques that are similar to the one adopted for hierarchical infrastructure network (cellular network and wireless Internet). In [67], mobility management for HSR protocol has been proposed, where the wireless ad hoc network is hierarchically organized into multi-level clustering structure. The key concept of logical subnets is used to achieve mobility management.

27.8 Conclusion

The main goal of this chapter is to provide a review of mobility aspects in ad hoc networks, specifically addressing its impact on current routing protocols, performance issues, and design challenges due to user or node mobility. The chapter starts with an overview of MANETs by defining their characteristics and applications. Then it describes the problems with conventional routing and the desired characteristics of routing protocols for ad hoc networks. This is followed by an extensive review of different protocols used in ad hoc routing, classified according to the routing strategy. A comparison of the protocols, highlighting their features and characteristics for the table-driven, on-demand, and hybrid protocols are provided. Recent research focuses on ad hoc routing protocols that support QoS requirements and are power- or energy-efficient are outlined. Aspects of mobility management in the context of routing for ad hoc networks are discussed.

Note

1. The host and node are interchangeable. It can also be called a station, device, or sensor if it is a wireless sensor network.

References

1. T.S. Rappaport, *Wireless Communications: Principles and Practice,* 2nd ed., Upper Saddle River, NJ: Prentice Hall, 2002.
2. K. Pahlavan and P. Krishnamurthy, *Principles of Wireless Networks,* Upper Saddle River, NJ: Prentice Hall, 2002.

3. M.S. Corson and A. Ephremides, A Distributed Routing Algorithm for Mobile Wireless Networks, *ACM/Baltzer Wireless Networks Journal,* vol. 1, no. 1, pp. 61–81, February 1995.
4. M.S. Corson and J. Macker, Mobile Ad hoc Networking (MANET): Routing Protocol Performance Issues and Evaluation Considerations, RFC 2501, January 1999.
5. C.E. Perkins, Ed., *Ad Hoc Networking,* Boston: Addison Wesley, 2000.
6. W. Stallings, *Wireless Communications and Networks,* Upper Saddle River, NJ: Prentice Hall, 2002.
7. B. Ohara and A. Petrick, *IEEE 802.11 Handbook: A Designer's Companion,* Piscataway, NJ: IEEE Press, 1999.
8. IEEE Computer Society LAN/MAN Standards Committee, Wireless LAN Medium Access Control (MAC) and Physical Layer (PHY) Specifications, IEEE Std. 802.11-1997, 1997.
9. IEEE Computer Society LAN/MAN Standards Committee, Wireless LAN Medium Access Control (MAC) and Physical Layer (PHY) Specifications: High-Speed Physical Layer in the 5 GHz Band, IEEE Std. 802.11a-1999, 1999.
10. IEEE Computer Society LAN/MAN Standards Committee, Wireless LAN Medium Access Control (MAC) and Physical Layer (PHY) Specifications: Higher-Speed Physical Layer Extension in the 2.4 GHz Band, IEEE Std. 802.11b-1999/Cor 1-2001, 1999–2001.
11. HIPERLAN2, http://www.hiperlan2.com.
12. IETF Mobile Ad Hoc Networks (MANET) WG Charter, 1997. http://www.ietf.org/html.charters/manet-charter.html.
13. A.S. Tanenbaum, *Computer Networks,* 3rd ed., Upper Saddle River, NJ: Prentice Hall, 1996.
14. D. Bertsekas and R. Gallager, *Data Networks,* 2nd ed., Upper Saddle River, NJ: Prentice Hall, 1992.
15. W. Stallings, *Data and Computer Communications,* 6th ed., Upper Saddle River, NJ: Prentice Hall, 2000.
16. Introduction to Routing, http://www.cisco.com/univercd/cc/td/doc/cisintwk/ito_doc/routing.htm.
17. E.M. Royer and C.-K. Toh, A Review of Current Routing Protocols for Ad Hoc Wireless Networks, *IEEE Personal Communications,* vol. 6, no. 2, pp. 46–55, April 1999.
18. X. Hong, K. Xu, and M. Gerla, Scalable Routing Protocols for Mobile Ad Hoc Networks, *IEEE Network,* pp. 11–21, July/August 2002.
19. S.-J. Lee, M. Gerla, and C.-K. Toh, A Simulation Study of Table-Driven and On-Demand Routing Protocols for Mobile Ad Hoc Networks, IEEE Network, pp. 48–54, July/August 1999.
20. E. Celebi, Performance Evaluation of Wireless Mobile Ad Hoc Network Routing Protocols, M.S. Thesis, Bogazici University, Turkey, February 2001, http://cis.poly.edu/ece/ebi/esim.pdf.
21. S.R. Das, C.E. Perkins, E.M. Royer, and M.K. Marina, Performance Comparison of Two On-Demand Routing Protocols for Ad Hoc Networks, *IEEE Personal Communications,* vol. 8, no. 1, pp. 16-28, February 2001.
22. C.E. Perkins and P. Bhagwat, Highly Dynamic Destination-Sequenced Distance-Vector Routing (DSDV) for Mobile Computers, *Computer Communications Review,* pp. 234–244, October 1994.
23. A. Bourkerche, A. Fabbri, and S.K. Das, Analysis of Randomized Congestion Control in DSDV Routing, *Proc. 8th International Symposium on Modeling, Analysis, and Simulation of Computer and Telecommunication Systems,* pp. 65–72, September 2000.
24. X. Dong and A. Puri, A DSDV-Based Multipath Routing Protocol for Ad-Hoc Mobile Networks, In *Proc. Int. Conf. Wireless Networks (ICWN'02),* Las Vegas, June 2002.
25. S. Murthy and J.J. Garcia-Luna-Aceves, An Efficient Routing Protocol for Wireless Networks, ACM Mobile Networks and App. J., pp. 183–197, October 1996.

26. T.-W. Chen and M. Gerla, Global State Routing: A New Routing Scheme for Ad-Hoc Wireless Networks, *Proc. IEEE Int'l. Conf. on Communications (ICC)*, Atlanta, pp. 171–175, June 1998.
27. M. Gerla, X. Hong, and G. Pei, Fisheye State Routing Protocol (FSR) for Ad Hoc Networks, IETF Draft, draft-ietf-manet-fsr-02.txt, December 2001.
28. G. Pei, M. Gerla, and T.-W. Chen, Fisheye State Routing: A Routing Scheme for Ad Hoc Wireless Networks, Proc. IEEE Int'l. Conf. on Communications (ICC), New Orleans, June 2000.
29. T. Clausen, P. Jacquet, A. Laouiti, and P. Minet et al., Optimized Link State Routing Protocol, IETF Draft, draft-ietf-manet-olsr-06.txt, September 2001.
30. R.G. Ogier, F.L. Templin, B. Bellur, and M.G. Lewis, Topology Broadcast Based on Reverse-Path Forwarding (TBRPF), IETF Draft, draft-ietf-manet-tbrpf-05.txt, March 2002.
31. C.-C. Chiang, Routing in Clustered Multihop, Mobile Wireless Networks with Fading Channel, *Proc. IEEE SICON*, pp. 197–211, April 1997.
32. A. Iwata, C.-C. Chiang, G. Pei, M. Gerla, and T.-W. Chen, Scalable Routing Strategies for Ad Hoc Wireless Networks, *IEEE J. Selected Areas in Communications,* vol. 17, no. 8, pp. 1369–1379, August 1999.
33. M. Gerla, X. Hong, and L. Ma, Landmark Routing Protocol (LANMAR) for Large Scale Ad Hoc Networks, IETF Draft, draft-ietf-manet-lanmar-03.txt, December 2001.
34. D.B. Johnson and D.A. Maltz, Dynamic Source Routing Protocol in Ad Hoc Wireless Networks, In T. Imielinski and H. Korth, Eds., *Mobile Computing,* pp. 153–181, Norwell, MA: Kluwer Academic Publishers, 1996.
35. D.B. Johnson, D.A. Maltz, Y.-C. Hu, and J.G. Jetcheva, The Dynamic Source Routing Protocol for Mobile Ad Hoc Networks (DSR), IETF Draft, draft-ietf-manet-dsr-07.txt, February 2002.
36. C.E. Perkins and E.M. Royer, Ad-Hoc On Demand Distance Vector Routing, In *Proc. 2nd IEEE Workshop on Mobile Computing Systems and Applications,* New Orleans, pp. 90–100, February 1999.
37. C.E. Perkins, E.M. Belding-Royer, and S.R. Das, Ad Hoc On Demand Distance Vector (AODV) Routing, IETF Draft, draft-ietf-manet-aodv-10.txt, January 2002.
38. S. Gwalani, E.M. Belding-Royer, and C.E. Perkins, AODV-PA: AODV with Path Accumulation, In *Proc. IEEE International Conference on Communications (ICC),* pp. 527–531, 2003.
39. C.-K. Toh, A Novel Distributed Routing Protocol to Support Ad Hoc Mobile Computing, In *Proc. IEEE 15th Annual Int'l. Phoenix Conf. on Computers and Communications,* pp. 480–486, March 1996.
40. Z.J. Haas, M.R. Pearlman, and P. Samar, The Intrazone Routing Protocol (IARP) for Ad Hoc Networks, IETF Draft, draft-ietf-manet-iarp-01.txt, June 2001.
41. Z.J. Haas, M.R. Pearlman, and P. Samar, The Interzone Routing Protocol (IERP) for Ad Hoc Networks, IETF Draft, draft-ietf-manet-ierp-01.txt, June 2001.
42. M.R. Pearlman and Z.J. Haas, Determining the Optimal Configuration for the Zone Routing Protocol, *IEEE J. Selected Areas in Communications,* vol. 17, no. 8, pp. 1395–1414, August 1999.
43. V.D. Park and M.S. Corson, A Highly Adaptive Distributed Routing Algorithm for Mobile Wireless Networks, In *Proc. IEEE InfoCom,* Kobe, Japan, April 1997.
44. V.D. Park and M.S. Corson, Temporally-Ordered Routing Algorithm (TORA) Version 1 Functional Specification, IETF Draft, draft-ietf-manet.tora-spec-04.txt, July 2001.
45. V. Rodoplu and T.H. Meng, Minimum Energy Mobile Wireless Networks, *IEEE J. Selected Areas in Communications,* vol. 17, no. 8, pp. 1333–1344, August 1999.
46. S. Singh, M. Woo, and C.S. Raghavendra, Power-Aware Routing in Mobile Ad Hoc Networks, In *Proc. 4th Annual ACM/IEEE Int'l. Conf. on Mobile Computing and Networking (MobiCom),* Dallas, pp. 181–190, October 1998.
47. S. Singh, C.S. Raghavendra, and J. Stepanek, Power-Aware Broadcasting in Mobile Ad Hoc Networks, In *Proc. IEEE Int'l. Symposium on Personal, Indoor, and Mobile Radio Communications (PIMRC),* Osaka, Japan, September 1999.

48. R. Kravets and P. Krishnan, Power Management Techniques for Mobile Communication, In *Proc. 4th Annual ACM/IEEE Int'l. Conf. on Mobile Computing and Networking (MobiCom)*, Dallas, pp. 157–168, October 1998.
49. J.L. Sobrinho and A.S. Krishnakumar, Quality-of-Service in Carrier Sense Multiple Access Ad Hoc Wireless Networks, *IEEE J. Selected Areas in Communications*, vol. 17, no. 8, pp. 1353–1368, August 1999.
50. C.F. Chiasserini and R.R. Rao, Routing Protocols to Maximize Battery Efficiency, In *Proc. IEEE Military Communications Conf. (MILCOM)*, Los Angeles, pp. 496–500, October 2000.
51. J. Chang and L. Tassiulas, Energy Conserving Routing in Wireless Ad Hoc Networks, *IEEE InfoCom*, Tel Aviv, pp. 22–31, March 2000.
52. C.-K. Toh, Maximum Battery Life Routing to Support Ubiquitous Mobile Computing in Wireless Ad Hoc Networks, *IEEE Comm. Mag.*, pp. 138–147, June 2001.
53. R.C. Shah and J.M. Rabaey, Energy Aware Routing for Low Energy Ad Hoc Sensor Networks, *IEEE Wireless Communications and Networking Conf. (WCNC)*, Orlando, FL, pp. 350–355, March 2002.
54. M.B. Pursley, H.B. Russell, and J.S. Wysocarski, Tradeoffs in the Design of Routing Metrics for Frequency-Hop Wireless Networks, In *Proc. IEEE Military Communications Conf. (MILCOM)*, Los Angeles, pp. 65–69, October 2000.
55. S. Chen and K. Nahrstedt, Distributed Quality-of-Service Routing in Ad Hoc Networks, *IEEE J. Selected Areas in Communications*, vol. 17, no. 8, pp. 1488–1505, August 1999.
56. C.R. Lin and J.-S, Liu, QoS Routing in Ad Hoc Wireless Networks, *IEEE J. Selected Areas in Communications*, vol. 17, no. 8, pp. 1426–1438, August 1999.
57. C.R. Lin and J.-S, Liu, On-Demand QoS Routing Protocol for Mobile Ad Hoc Networks, In *Proc. IEEE InfoCom*, Anchorage, pp. 1735–1744, April 2001.
58. T.W. Chen, J.T. Tsai, and M. Gerla, QoS Routing Performance in Multihop Wireless Networks, *IEEE 6th Int'l. Conf. on Universal Personal Communciations (ICUPC)*, October 1997.
59. R. Sivakumar, P. Sinha, and V. Bharghavan, CEDAR: A Core-Extraction Distributed Ad Hoc Routing Algorithm, *IEEE J. Selected Areas in Communications*, vol. 17, no. 8, pp. 1454–1465, August 1999.
60. K. Wu and J. Harms, Location Trace Aided Routing in Mobile Ad Hoc Networks, In *Proc. IEEE Int'l. Conf. on Computer Communications and Networks (ICCCN)*, Las Vegas, October 2000.
61. C.E. Perkins, E.M. Belding-Royer, and S.R. Das, IP Flooding in Ad-Hoc Mobile Networks, draft-ietf-manet-bcast-00.txt, IETF Draft, November 2001.
62. J.G. Jetcheva et al., A Simple Protocol for Multicast and Broadcast in Mobile Ad Hoc Networks, draft-ietf-manet-mbcast-01.txt, IETF Draft, July 2001.
63. A. Nasipuri, R.E. Hiromoto, and J. Mandava, On-Demand Routing Using Directional Antennas in Mobile Ad Hoc Networks, In *Proc. IEEE Int'l. Conf. on Computer Communications and Networks (ICCCN)*, Las Vegas, October 2000.
64. W. Su and M. Gerla, IPv6 Flow Handoff in Ad Hoc Wireless Networks Using Mobility Prediction, In *Proc. IEEE Global Telecommunications Conference (Globecom)*, Rio de Janeiro, pp. 271–275, December 1999.
65. S. Vutukury and J.J. Garcia-Luna-Aceves, A Distributed Algorithm for Multipath Computation, In *Proc. IEEE Global Telecommunications Conference (Globecom)*, Rio de Janeiro, pp. 1689–1693, December 1999.
66. Y.-B. Lin and I. Chlamtac, *Wireless and Mobile Network Architectures*, Hoboken, NJ: Wiley, 2001.
67. G. Pei and M. Gerla, Mobility Management in Hierarchical Multi-Hop Mobile Wireless Networks, In *Proc. IEEE 8th Int'l. Conf. on Computer Communications and Networks (ICCCN)*, Boston, pp. 324–329, October 1999.

Chapter 28
Localized Broadcasting in Mobile Ad Hoc Networks Using Neighbor Designation

Wei Lou and Jie Wu

28.1 Introduction

A mobile ad hoc network (MANET or ad hoc network) [13] enables wireless communications between participating mobile nodes without the assistance of any base station. Two nodes that are out of one another's transmission range need the support of intermediate nodes that relay messages to set up a communication between each other. The broadcast operation is the most fundamental role in ad hoc networks because of the broadcasting nature of radio transmission: When a sender transmits a packet, all nodes within the sender's transmission range will be affected by this transmission. This is usually referred to as the promiscuous receive mode. The advantage is that one packet can be received by all neighbors; the disadvantage is that it interferes with the sending and receiving of other transmissions, creating an exposed terminal problem — an outgoing transmission collides with an incoming transmission — and a hidden terminal problem — two incoming transmissions collide with each other. Broadcast operation has extensive applications, such as when used in the route query process in routing protocols [20, 33, 37], when sending error messages to

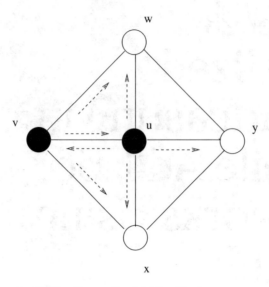

Figure 28.1. Sample Ad Hoc Network with Five Nodes

erase invalid routes [32], or when used as an efficient mechanism for reliable multicast in highly dynamic wireless networks [17].

In general, broadcasting refers to a process of transmitting a packet so that each node in a network receives a copy of this packet. Flooding is the simplest approach for broadcasting: every node in the network forwards the packet exactly once. Flooding ensures the full coverage of all the network, that is, the broadcast packet is guaranteed to be sent to every node in the network, providing the network is static and connected and the Media Access Control (MAC) layer of the communication channel is error-free during the broadcast process. However, flooding generates many redundant transmissions. Figure 28.1 shows a sample network with five nodes. When node v broadcasts a packet, node u, node w, and node x will receive the packet. Node u, node w, and node x will then forward the packet and finally node y will also broadcast the packet. Apparently, there is much broadcast redundancy for blind flooding in this case. Transmitting the broadcast packet only by node v and node u is enough for a broadcast operation. When the size of the network increases and the network becomes dense, more transmission redundancy will be introduced and these transmissions are likely to trigger considerable transmission collision and contention. This is a serious broadcast storm problem [31] that finally collapses the whole network.

The broadcast storm problem can be avoided by reducing the number of nodes that retransmit the broadcast packet. Ni et al. classified the broadcast algorithms into two categories — probabilistic approach and deterministic

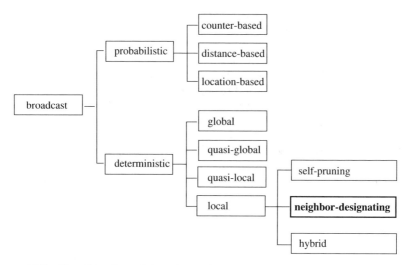

Figure 28.2. Classification of Broadcast Algorithms

approach [31]. Counter-based, distance-based, and location-based schemes belong to probabilistic approaches. For the deterministic approach, Wu and Lou [52] further classified it into four classes — global, quasi-global, quasi-local, and local algorithms. In [49], Wu and Dai classified the local broadcast algorithms into three classes — self-pruning, neighbor-designating, and hybrid algorithms (Figure 28.2).

The rest of this chapter is organized as follows:

- Section 28.2 classifies the basic broadcast algorithms.
- Section 28.3 describes several neighbor-designating-based broadcast algorithms.
- Section 28.4 discusses three extensions of the neighbor-designating-based broadcast algorithms.
- Section 28.5 summarizes the topic of neighbor-designating-based broadcast algorithms.

28.2 Classification

This section discusses in detail three levels of classification of broadcast algorithms.

28.2.1 Probabilistic Algorithms

The probabilistic approach for a broadcast operation is as follows. Upon receiving a broadcast packet, each node forwards the packet with probability p. The value p is determined by relevant information gathered at each node. Although the probabilistic approach provides a good stochastic result, it cannot guarantee the full coverage. In [31], other probabilistic approaches were also discussed.

28.2.1.1 Counter-Based Scheme. Upon receiving a previously unknown packet, the node initiates a waiting timer and a counter. The counter increases one for each received redundant packet. When the waiting timer expires, if the counter is larger than a threshold value, the node will not rebroadcast the packet; otherwise, the node will broadcast it.

28.2.1.2 Distance-Based Scheme. Upon receiving a previously unknown packet, the node initiates a waiting timer. Before the waiting timer expires, the node checks the location of the senders of each received packet. If any sender is closer than a threshold distance value, the node will not rebroadcast the packet. Otherwise, the node rebroadcasts it when the waiting timer expires.

28.2.1.3 Location-Based Scheme. Upon receiving a previously unknown packet, the node initiates a waiting timer and accumulates the coverage area that has been covered by the arrived packet. When the waiting timer expires, if the accumulated coverage area is larger than a threshold value, the node will not rebroadcast the packet. Otherwise, the node will broadcast it.

Other enhancements to the above probabilistic algorithms are discussed in [5, 15, 44].

28.2.2 Deterministic Algorithms

The deterministic approaches provide full coverage of the network for a broadcast operation. In other words, only a subset of nodes forward the broadcast packet and the remaining nodes are adjacent to the nodes that forward the packet. The nodes that forward the broadcast packet form a forward node set for a particular broadcast operation. Basically, a forward node set is a connected dominating set (CDS). A dominating set (DS) is a subset of nodes such that every node in the graph is either in the set or is adjacent to a node in the set. If the subgraph induced from a DS of the network is connected, the DS is a CDS. Finding a minimum connected dominating set (MCDS) in a given graph is NP-complete; in a unit disk graph, it has also been proved to be NP-complete [30]. There are, in general, two models of neighbor set information — neighbor set without node positions and neighbor set with node positions (obtained through global positioning system [GPS] or other means). The latter model trivializes the approximation problem of CDS. That is, approximation algorithms with a constant approximation ratio can be easily derived. On the other hand, finding an approximation algorithm with a small constant approximation ratio is still a challenging issue in the absence of global network information [4]. Heuristic methods are

Localized Broadcasting in Mobile Ad Hoc Networks

normally used to balance cost (in collecting network information and in decision making) and effectiveness (in deriving a small forward node set).

The CDS of a graph can be constructed with global or local information. The distinction between global and local is not a clear-cut situation. Through several rounds of sequential information exchanges, global information can be assembled based on local information only. However, sequential information propagation can be costly. There are four types of broadcast protocols [52] — global, quasi-global, quasi-local, and local.

28.2.2.1 Global. Broadcast protocols — centralized or distributed — are based on global state information. The most widely used global broadcast protocol is based on Guha and Khuller's approximation algorithm [14] and has been used in protocol design by Das et al. [10]. All nodes are initially colored white. The node with the maximum node degree is selected and colored black and all its neighbors are colored gray. An iterative selection process runs until there are no white nodes left. Select a gray node that has the maximum number of white neighbors, color the selected node black and its white neighbors gray. The resultant set of black nodes is a CDS. This algorithm is centralized and works well except for some extreme cases. A modified algorithm selects the gray node u and its neighbor v if they can cause the maximum number of white nodes to change color to gray when both u and v are changed to black. The modified algorithm guarantees an approximation ratio $O(\ln \Delta)$ under any random graph, where Δ is the maximum node degree of the network. Therefore, this algorithm can be used as a lower bound of the MCDS.

28.2.2.2 Quasi-Global. Distributed broadcast protocols are based on partial global state information. Unlike the global broadcast protocol, the quasi-global broadcast protocol does not need to collect the whole global state. Only partial global state information is collected, typically with the help of a global infrastructure such as a spanning tree. The protocol proposed in [3, 45] fits into this category. A spanning tree is first constructed starting from the selected root (through an election process), a maximal independent set (MIS) is constructed level by level down the tree. An independent set (IS) is a set in which no two nodes are neighbors. An MIS is an IS in which any other node in the network is a neighbor of a node in the set. Nodes in the MIS are colored black. Clearly, an MIS is a DS. Specifically, nodes are labelled according to a topological sorting order of the tree. Then nodes are labelled based on their positions in the order starting from the root v. All nodes are white initially. The root v is marked black first and other nodes are marked black unless there is a black neighbor. Each parent of a black node in the tree acts as a connector by marking gray. Black and gray nodes form a CDS. This spanning-tree-based CDS (STCDS) broadcasting generates a CDS with a constant approximation ratio of 8 (i.e., the size of the CDS is at most eight times the size of the MCDS). Also, other than the tree-level information needed to determine the topological sorting order of each node, no other global state information is distributed. However, like the MCDS, the STCDS requires O(diam) sequential rounds, because both the spanning tree construction and status marking process are serialized. In addition, the STCDS does not support locality of maintenance. Movements of hosts may trigger the reconstruction of the whole spanning tree.

28.2.2.3 Quasi-Local. Distributed broadcast protocols are based on mainly local state information and occasional partial global state information. The cluster approach falls into the Quasi-Local Model. Cluster structure is a two-level hierarchical structure and it is formed by first electing a clusterhead and, then, its neighbors joining in the cluster as nonclusterhead members. There are many clustering approaches [7, 8, 11, 12, 23]. A simple one, called the lowest-ID (LID) cluster algorithm, initializes all nodes white. When a white node finds itself having the lowest ID among all its white neighbors, it becomes a clusterhead and colors itself black. All its white neighbors join in the cluster and change their colors to gray. This iterative process continues until there are no white nodes left. The black nodes form the set of clusterheads. Each gray node belongs to one and only one clusterhead. That clusterhead is called the dominator of the gray node. The clusterhead and its dominated gray neighbors form a cluster. The LID may exhibit sequential propagation, as happens when the network is a linear chain with decreasing IDs from one end to the other end (this is the reason this approach is called quasi-local), resulting in O(diam) rounds of information exchanges. However, this situation rarely happens. In the average case, the cluster formation can be considered as a localized process. Clusterheads form a DS, but not a CDS. Clusterheads and gateways together form a CDS.

Another variation of cluster approach proposed by Sinha, Sivakumar, and Bharghavan [41], called core broadcast (CB), also includes the selection process of forward nodes: Initially all nodes are white. A white node determines its dominator by selecting its black neighbor that has the maximum number of nodes that regard this black node as their dominators. In case there is no black neighbor, the white node selects the node (white or gray) with the maximum node degree within its one-hop neighborhood (including itself) as its dominator. After the white node has chosen its dominator, it colors itself gray if it is not selected as a dominator by itself or by its neighbors; otherwise, it marks itself black if it has been selected as a dominator. The coloring process continues until there are no white nodes left. Eventually, all the black nodes become cores. In the core broadcast, each node computes its forward node set. A node's forward node set includes all its black neighbors. It also includes those gray neighbors that either have a black neighbor that is not covered by the forward node set or have a gray neighbor whose dominator is not covered by the forward node set.

The core broadcast requires only the nodes in the forward node set relay the broadcast so it reduces the broadcast redundancy. The set of cores, like the set of clusterheads, is a DS of the network. Although the set of clusterheads is also an IS, the set of cores does not have this property because two cores may be neighbors.

28.2.2.4 Local. Distributed broadcast protocols are based solely on local state information. The local broadcast protocol is based solely on local information without exhibiting any sequential propagation of state information. It also supports locality of maintenance. However, although this approach is competitive in the average case, it does not guarantee performance in the worst case such as a constant approximation ratio. Wu and Li's [51] marking process is a local broadcast protocol. All nodes are initially white. A node marks itself black only when it has two unconnected neighbors. After the marking process, the black nodes form a CDS. Rule 1 and rule 2 aim to remove redundant nodes from the CDS. Rule 1 allows a black node u to change its color from black to white if it can find another black node v, with $id(u) < id(v)$, to cover all u's neighbors. For rule 2, a black node u changes itself to white if there exist two connected nodes, v and w, with $id(u) = \min\{id(u), id(v), id(w)\}$, that can collectively cover all u's neighbors. Recently, Dai and Wu [9] have extended rule 1 and rule 2 to rule k to further reduce nodes in the CDS without increasing the computational complexity. If a black node u can be covered by k connected black nodes and $id(u)$ is smaller than any ID of these k nodes, then u can change itself to white. A constant number of rounds (two or three depending on the implementation) are needed to construct a CDS and its maintenance. Wu and Li's approach has been applied to broadcasting in [43] where only black nodes (besides the source) forward the broadcast packet.

28.2.3 Local Algorithms

Wu and Dai [49] proposed a deterministic generic distributed broadcast scheme and classified the local broadcast algorithms into three categories — self-pruning, neighbor-designating, and hybrid broadcasting approaches. In all these schemes, the status of each node is determined in a decentralized manner based on node's current local view. A view is a snapshot of network state, including network topology and broadcast state, along time. A node can utilize its K-hop neighborhood information to build its local view. One-hop and two-hop neighborhood information are the most common cases. Also, the broadcast packet can carry some broadcast state information, such as the next selected node to forward the packet, the recently visited nodes and their neighbor sets. The status of each node can be determined by itself or by its neighbor. For a specific node, the upstream node that has sent a broadcast packet to this node is viewed as a forwarded node; the downstream node selected by this node to forward the broadcast packet is viewed as forward node; the downstream node that is designated not to forward the packet is viewed as a nonforward node. Notice that the node status under the current view will be changed in the next view; that is, a forward node in current view will be a forwarded node in the next view.

The generic distributed broadcast scheme constructs a CDS for a particular broadcast that depends on the location of the source and the progress of the broadcast process. Each node v determines its status and the status of some of its neighbors under a current local view. Each node has the forwarding status by default and the status can be changed to nonforwarding if the coverage condition is met. Coverage condition is described as follows:

> Node v has a nonforwarding status if for any two neighbors u and w, a replacement path exists that connects u and w via several intermediate nodes (if any) with higher priorities than the one of v.

The coverage condition indicates that when every pair of neighbors of v can be connected through other nodes, node v, as the connecting node for its neighbors, can be replaced (i.e., can take the nonforwarding status). Note that replacement can be applied iteratively. To avoid possible cyclic dependency situations, a total priority order needs to be defined among all nodes, such as node ID or a pair of node degree and node ID.

In self-pruning approaches, a node will resign its role of forwarding the broadcast packet by itself if the replacement path from the source can be found for each of its neighbors. Nodes in each replacement can be either forwarded nodes or nodes with higher priorities. In the neighbor-designating broadcasting approaches, a node can determine its neighbor's forwarding/nonforwarding status; that is, a forward node selected by its upstream sender updates its view when it receives the packet and determines its status and its neighbors' status consequently. In the hybrid approaches, both self-pruning and neighbor-designating methods are applied to determine a node's status. The marking process discussed in the last section is an example of self-pruning and multi-point relay (MPR) is an example of neighbor-designating.

Figure 28.3 shows the coverage condition that applies to the self-pruning and neighbor-designating approaches. In Figure 28.3A, suppose v has received a packet from u. If v's one-hop neighbors are all covered by a set of forwarded nodes (black) or some white nodes with higher priorities, v can drop its role to relay the packet. Figure 28.3B shows how the neighbor-designating approach works. Suppose u is the current node and node v is

Localized Broadcasting in Mobile Ad Hoc Networks

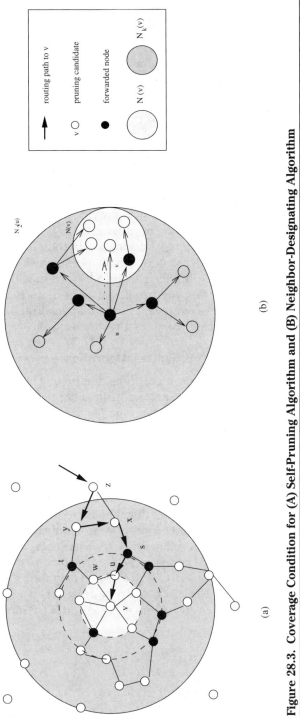

Figure 28.3. Coverage Condition for (A) Self-Pruning Algorithm and (B) Neighbor-Designating Algorithm

any neighbor that is not selected as a forward neighbor. Because u and the selected designated forward neighbors cover all the two-hop neighbors of u which include one-hop neighbors of v, node v is covered by a set of (connected) coverage nodes.

In the subsequent discussion, we only consider the neighbor-designating algorithms that utilize local information. The following assumptions are also used:

1. The transition is error free; that is, each message (broadcast packet or network state message) sent from a node will eventually reach its neighbors.
2. Location information of each node is not available. Location-based broadcasting has been extensively studied as in [34, 42, 43].
3. Network topology is a connected graph without unidirectional links. A sublayer can be added [40, 46] to provide a bidirectional abstraction for unidirectional ad hoc networks.
4. All nodes have fresh topology information in their local views at the beginning of the broadcast period and the network topology does not change during the broadcast period. Note that if the network topology changes during the broadcast period, no broadcast algorithm can ensure full coverage, except a special mobility management mechanism is used, such as the one in [50].

28.3 Neighbor-Designating Broadcast Algorithms

In this section, we describe some algorithms that belong to the category of neighbor-designating broadcast. Each algorithm adopts the heuristic strategy where a minimum number of forward nodes are selected so that other neighbors can take the nonforward status.

28.3.1 Forward Node Selection Process

Theoretically, a MANET is represented as a unit disk graph $G(t) = (V, E)$, where the node set V represents a set of wireless mobile nodes and the edge set E represents a set of bidirectional links between the neighboring nodes. Two nodes are considered neighbors if and only if their geographic distance is less than the transmission range r. We use $N_k(v)$ to represent the k-hop neighbor set of v, where nodes in the set are no more than k hops further from v. $N_k(v)$ includes v itself. ($N_1(v)$, one-hop neighbor set, can be simply represented as $N(v)$.) Node u's k-hop node set $H_k(u)$ consists of all nodes that are exactly k hops away from u. $N_k(u)$ and $H_k(u)$ have the following relationships:

$$N_k(u) = N_{k-1}(u) \cup H_k(u)$$

$$N_{k-1}(u) \cap H_k(u) = \phi$$

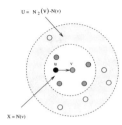

Figure 28.4. Multiple Relays

If S is a node set, N(S) is the union of the neighbor sets of every node in S, that is, $N(S) = \bigcup_{\forall w \in S} N(w)$.

For the instance a node u broadcasts a packet, u selects a subset from X to cover U by using the greedy algorithm in the set coverage problem [29]. The greedy algorithm, called forward node set selection process (FNSSP), works as follows:

1. Each node w in X calculates its effective node degree $deg_e(w) = |N(w) \cap U|$.
2. A node w_1 with the maximum $deg_e(w_1)$ is first selected, w_1 is removed from X and $N(w_1)$ is removed from U. A tie is broken by using node ID.
3. If U is not empty, each node recomputes its effective node degree and another node w_2 with the maximum $deg_e(w_2)$ is selected.
4. Repeat step 2 and step 3 until U becomes empty.
5. The node set $\{w_1, w_2, ...\}$ forms a forward node set.

28.3.2 Multi-Point Relays

Qayyum et al. [39] proposed selected MPRs as forward nodes to propagate link state messages in their Optimized Link State Routing (OLSR) protocol. The MPRs are selected from one-hop neighbors to cover the entire set of two-hop neighbors (Figure 28.4). The MPRs are selected as follows: If u intends to forward a packet, u uses the forward node set selection process to select its forward node set from $X = N(v)$ to cover two-hop neighbors in $U = N_2(v) - N(v)$. The selected forward nodes become MPRs.

In the sample network shown in Figure 28.5, when node 3 uses MPR algorithm, $U = N_2(3) - N(3) = \{1, 5, 8\}$ and $X = N(3) = \{2, 3, 4, 6, 7\}$, node 3 selects node 2, node 4, and node 6 as its forward nodes.

When sending a broadcast packet, each selected MPR runs a restricted forward node selection process applied in [19]: if an MPR v first receives a broadcast packet from a neighbor that does not designate v as a forward node, v does not forward this packet even if v may be selected as a forward node later by another neighbor.

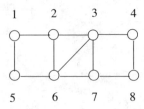

Figure 28.5. Sample Network

In Figure 28.5, node 3 selects node 2, node 4, and node 6 as MPRs; node 6 selects node 2, node 3, and node 7 as MPRs. Suppose node 2 starts a broadcast, node 7 may receive the broadcast packet from both node 3 and node 6. If node 7 first receives a packet from node 3 and then receives a duplicated broadcast packet from node 6, but because node 3 does not designate node 7 to forward the broadcast, node 7 will not forward the broadcast packet later when it receives the same broadcast packet from node 6 and has been designated to forward the packet by node 6 as a MPR. On the other hand, if node 7 first receives a packet from node 6, node 7 will forward the broadcast packet.

In [1], the requirement for a MPR to forward the packet is more restricted. A MPR will become a forward node if its node ID is smaller than all its neighbors or if it is selected as a MPR by its neighbor with the minimal node ID.

In the same network shown in Figure 28.5, when applying the rule proposed in [1], node 7 will never forward a broadcast packet whether the packet is from node 3 or node 6 because node 3, which is the neighbor of node 7 with the smallest node ID, does not select node 7 as a MPR.

In [47], Wu further extends the algorithm in [1] by introducing the concept of *free neighbor*. Node u is a free neighbor of v if v is not the neighbor of u with the smallest node ID. The neighbors of free neighbors can be removed from the two-hop neighbors of a node before it uses the forward node selection algorithm to designate its MPRs. Another extension rule proposed in [47] is that a node that has a smaller ID than all its neighbors and also has two unconnected neighbors becomes a forward node. This extension rule will be more effective in relatively sparse networks.

In Figure 28.5, we investigate node 3's neighbors. Node 2 and node 6 are node 3's free neighbors, therefore, $N(2) = \{1, 2, 3, 6\}$ and $N(6) = \{2, 3, 5, 6, 7\}$ can be removed from $N_2(3) - N(3)$. After that, node 6 only selects node 4 as its forward node. For node 7, because its free neighbors are node 6 and node 8 and $N_2(7) - N(7) - N(6) - N(8)$ are empty, node 7 will not select any forward node.

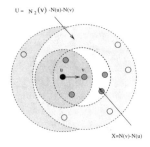

Figure 28.6. Dominant Pruning

28.3.3 Dominant Pruning

Lim and Kim [22] provided a dominant pruning (DP) algorithm. Unlike the MPR, the DP excludes the neighbors of the upstream forwarded node from the current node's two-hop neighbor set. In Figure 28.6, suppose u sends a packet and u selects v as its forward node. Because the neighbor set of u, $N(u)$, has been covered by u and the neighbor set of v, $N(v)$, will be covered by v, v does not need to select other forward nodes to cover them again. Therefore, v can determine its forward node set from $X = N(v)$ to cover $U = N_2(v) - N(v) - N(u)$. The forward node selection process from X to cover U is the same as above.

For the sample network shown in Figure 28.5, suppose node 3 receives a packet from node 6. When node 3 uses DP algorithm, $U = N_2(3) - N(6) - N(3) = \{1, 8\}$ and $X = N(2) = \{2, 4, 6, 7\}$. Node 3 selects node 2 and node 4 as its forward nodes.

Peng and Lu [36] proposed an Ad Hoc Broadcast Protocol (AHBP) algorithm similar to the DP. In their algorithm, forward nodes are called Broadcast Relay Gateways (BRGs). BRGs, using the same forward node selection process to determine their downstream BRGs, will forward the broadcast packet and inform their designated BRGs. The AHBP considers the case of the mobility of the node. When v receives a packet from u that is not listed in its neighbor set, v assumes itself to be a designated BRG and rebroadcasts the packet.

28.3.4 Total Dominant Pruning and Partial Dominant Pruning

Lou and Wu [25] proposed total dominant pruning (TDP) and partial dominant pruning (PDP) to extend the DP by further reducing the number of two-hop neighbors that need to be covered by one-hop neighbors. The TDP requires the upstream forwarded node u to piggyback $N_2(u)$ along with the broadcast packet. With this information, the selected forward

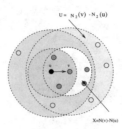

Figure 28.7. Total Dominant Pruning

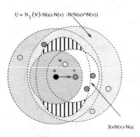

Figure 28.8. Partial Dominant Pruning

node v can remove $N_2(u)$, instead of $N(u)$ in DP, from $N_2(v)$ (Figure 28.7); that is, $U = N_2(v) - N_2(u)$. X can also update to $U(u) - N(v)$. In the PDP, the broadcast packet does not attach the upstream forwarded node's two-hop neighbor set. When receiving the packet from node u, node v extracts the neighbors of the common neighbors of u and v (i.e., neighbors of nodes in $N(u) \cap N(v)$) from $N_2(v)$ because these nodes are covered by u's forward node set $F(u)$. Therefore, the uncovered two-hop neighbor set U becomes $N_2(v) - N(u) - N(v) - N(N(u) \cap N(v))$ (Figure 28.8). The selection process for both TDP and PDP is the same as before: selecting forward nodes from X to cover U.

For the sample network shown in Figure 28.5, suppose node 3 receives a packet from node 6. When node 3 uses the TDP algorithm, $U = N_2(3) - N_2(6) = \phi$ and $X = N(3) - N(6) = \{4\}$. Node 3 does not select any node as its forward node. When node 3 uses the PDP algorithm, $U = N_2(3) - N(6) - N(3) - N(N(3) \cap N(6)) = \{8\}$ and $X = N(3) - N(2) = \{4\}$. Node 3 selects node 4 as its forward node.

28.3.5 CDS-Based Broadcast Algorithm

Peng and Lu [35] proposed a CDS-based broadcast (CDSB) algorithm. It considers not only the sender of the broadcast packet, but also the forward nodes with lower node IDs that are selected by the sender to determine a

Localized Broadcasting in Mobile Ad Hoc Networks

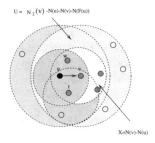

Figure 28.9. CDS-Based Broadcasting

selected forward node's forward node set. For a sender u, suppose u selects nodes t, v, w (id(t)<id(v)<id(w)) as its forward nodes. When nodes t, v, w receive the packet, t updates its uncovered two-hop neighbor set $U(t) = N_2(t) - N(t) - N(u)$; v updates its uncovered two-hop neighbor set $U(v) = N_2(v) - N(v) - N(u) - N(t)$ because $N(t)$ is covered by t. Likewise, w's uncovered two-hop neighbor set is $U(w) = N_2(w) - N(w) - N(u) - N(t) - N(v)$ (Figure 28.9). Notice that v may not forward the packet if $U(v)$ is empty.

For the sample network shown in Figure 28.5, suppose node 6 is the source, it selects node 2, node 3, and node 7 as forward nodes. The forward nodes are piggybacked with the broadcast packet. When node 2 receives a packet from node 6, it updates $U(2) = N_2(2) - N(6) - N(2) = \{4\}$ and $X(3) = N(2) = \{1, 2, 3, 6\}$. Node 2 selects node 3 as its forward node. When node 3 receives the packet from node 6, it updates $U(3) = N_2(3) - N(6) - N(3) - N(2) = \{8\}$ and $X(3) = N(3) = \{2, 3, 4, 6, 7\}$. Node 3 selects node 4 as its forward node. When node 7 receives the packet from node 6, it updates $U(7) = N_2(7) - N(6) - N(7) - N(2) - N(3) = \phi$ and $X(7) = N(7) = \{3, 6, 7, 8\}$. Node 7 does not select any node as its forward node.

28.4 Other Extensions

We describe three extensions that also applied the neighbor-designating approach for broadcasting. The first broadcast approach is based on the cluster structure, the second is a generic K-hop zone-based algorithm, and the third considers the reliability issue.

28.4.1 *Cluster-Based Broadcast Algorithm*

Although cluster-based broadcast algorithms are not local algorithms, they usually work well with local state information and low time delay. Basically, the clustered network converts any dense network to a sparse one consisting of clusterheads only because clusterheads form a DS of the network. Moreover, clusterheads and gateways form a CDS of the network.

Therefore, this is enough to fulfill a broadcast operation when all clusterheads and gateways forward a broadcast packet.

Alzoubi et al. [2] proposed a cluster-based message-optimal CDS algorithm. In this algorithm, a CDS is constructed locally in a constant approximation ratio with message complexity in O(n), where n is the size of the network. This message complexity reaches the lower bound in the network with n nodes. It was formed with two steps:

1. Clusterheads are determined by the lowest-ID clustering algorithm. A clusterhead knows all its two-hop and three-hop clusterheads with two rounds of neighborhood information exchanges.
2. Each clusterhead selects a node to connect each two-hop clusterhead and a pair of nodes to connect each three-hop clusterhead. All the clusterheads and selected nodes form a CDS of the network.

A cluster-based broadcast algorithm is proposed in [24] based on the coverage of the neighbor set. A clusterhead v's coverage set C(v) is a set of clusterheads that are in a specific coverage area of v. It can be a three-hop coverage set, which includes all the clusterheads in $N_3(v)$, or a two-and-a-half-hop coverage set, which includes all the clusterheads in $N_2(v)$ and the clusterheads that have members in $N_2(v)$. Figure 28.10 illustrates a clusterhead v's three-hop and two-and-one-half-hop coverage sets. In this network, the clusterhead of c is in v's three-hop coverage set, but not in v's two-and-one-half-hop coverage set. In general, the size of a clusterhead's two-and-one-half-hop coverage set is less than that of its three-hop coverage set. Therefore, the cost of maintaining the two-and-one-half-hop coverage set can be less than that of the three-hop coverage set.

Each clusterhead gathers the information of its coverage set by exchanging neighborhood information with its neighbors. Figure 28.11A illustrates construction of a coverage set with three-hop and two-and-one-half-hop coverage set. For three-hop coverage set, node 6 sends a message M1: $CH\_HOP1(6) = \{2^*\}$, which contains its one-hop clusterhead neighbors. Node 8 sends a message M2: $CH\_HOP1(8) = \{2^*, 3\}$. Here, * indicates the clusterhead of the cluster that the node belongs to. Likewise, node 5 and node 7 send messages M3 and M5: $CH\_HOP1(5) = \{1^*\}$ and $CH\_HOP1(7) = \{1, 4^*\}$. After receiving M1 and M2, node 7 may form a message M6, which contains its two-hop clusterhead neighbors and associated gateways: if node 6 is selected, $CH\_HOP2(7) = \{2[6], 3[8]\}$; if node 8 is selected, $CH\_HOP2(7) = \{2[8], 3[8]\}$. Here, $CH\_HOP2(u) = \{v[w], \ldots\}$ means that clusterhead u connects to clusterhead v via w. Node 5 also sends message M4: $CH\_HOP2(5) = \{2[6]\}$. By receiving M3, M4, M5, and M6, node 1 can build its local view as Figure 28.11B or Figure 28.11C. For the two-and-one-half-hop coverage set, node 1 builds its local view as Figure 28.11D. Here, clusterhead 3 is unknown to node 1.

Localized Broadcasting in Mobile Ad Hoc Networks

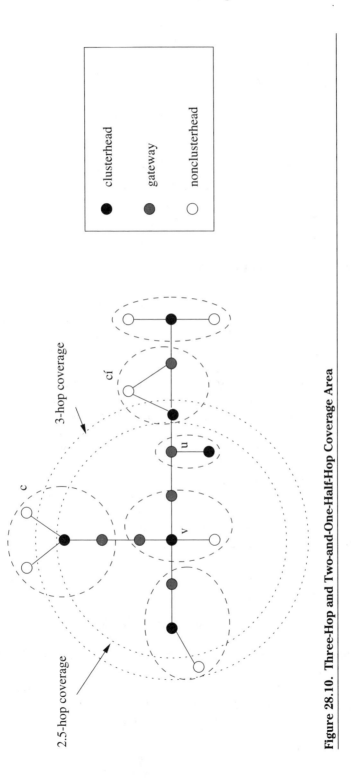

Figure 28.10. Three-Hop and Two-and-One-Half-Hop Coverage Area

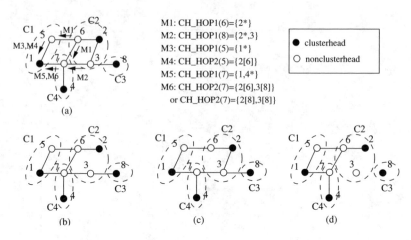

Figure 28.11. Constructing a Coverage Set

The clusterhead needs to select forward nodes to connect each clusterhead in its coverage set. In [2], each clusterhead randomly selects one or two nodes to connect its adjacent clusterheads. In [26], a greedy algorithm similar to forward node selection process is applied when a clusterhead v receives a broadcast packet p from its upstream clusterhead u. Suppose p is a new packet for v and p also attaches u's forward node set $F(u)$ and u's coverage set $C(u)$. v can update $C(v) = C(v) − C(u) − \{u\}$ because all the clusterheads in $C(u) \cup \{u\}$ are covered by F(u) and they do not need to be covered again. Clusterhead v selects the gateway that connected the maximum number of clusterheads in C(v), puts the gateway into F(v), and removes the connected clusterheads from C(v). This process repeats until C(v) becomes empty. The selection process works for both three-hop and two-and-one-half-hop coverage sets. When the two-and-one-half-hop coverage set is used, F(u) may cover some extra clusterheads in addition to $C(u) \cup \{u\}$; that is, if clusterhead v is three hops away from u and u uses a path (u, f, r, v) to deliver the broadcast packet to v, clusterheads in N(r) also receive the broadcast packet. These clusterheads can also be excluded from C(v). Therefore, the updated $C(v) = C(v) − C(u) − \{u\} − N(r)$.

The cluster-based broadcast algorithm consists of the following steps:

1. If the source is not a clusterhead, it sends the broadcast packet p to its clusterhead.
2. When a clusterhead receives p from its upstream clusterhead sender for the first time, it uses the above forward node selection process to choose forward nodes that cover all the clusterheads in its coverage set. The coverage set of this clusterhead, together with its selected forward nodes, are piggybacked with p for the forwarding purpose.

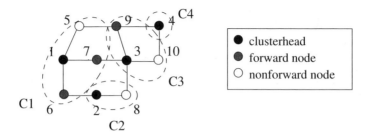

Figure 28.12. Cluster-Based Broadcast with Two-and-One-Half-Hop Coverage Set

3. For a nonclusterhead node that receives p for the first time, if it is a forward node, it relays p; otherwise, it does nothing.

Figure 28.12 illustrates the broadcast process in a cluster-based CDS with two-and-one-half-hop coverage set. Suppose the source is node 1. Node 1's two-and-one-half-hop coverage set $C(1)$ is $\{2, 3\}$, it selects node 6 and node 7 to forward the packet to clusterhead 2 and clusterhead 3. The broadcast packet piggybacks the forward node set $F(1) = \{6, 7\}$ and the two-and-one-half-hop coverage set $C(1) = \{2, 3\}$. When clusterhead 2 receives the broadcast packet from clusterhead 1, it updates $C(2) = C(2) - C(1) - \{1\} = \{1, 3\} - \{2, 3\} - \{1\} = \phi$; then, it only locally broadcasts the packet. When clusterhead 3 receives the packet from clusterhead 1, it updates $C(3) = C(3) - C(1) - \{1\} = \{1, 2, 4\} - \{2, 3\} - \{1\} = \{4\}$; therefore, clusterhead 3 selects node 9 to forward the packet to clusterhead 4. $F(3) = \{9\}$ and $C(3) = \{1, 2, 4\}$ are piggybacked with the packet. After clusterhead 4 receives the packet, it only locally broadcasts the packet because all clusterheads in $C(4)$ have received the packet. In total, 7 nodes (1, 2, 3, 4, 6, 7, and 9) will forward the packets.

28.4.2 K-hop Zone-Based Algorithm

A node's K-hop zone consists of all the nodes within K hops from the given node. For a given network, K can be set from 0 to the diameter of the network. One extreme case is K = 0; that is, nodes in the network have no neighbor set information or the neighbor set information is most likely out-of-date most of the time because of the high mobility of the nodes. The only possible strategy for routing is flooding. The other extreme case is that K is equal to the diameter of the network; that is, each node knows the global information of the network. Therefore, the optimum solution can be determined in this circumstance.

A K-hop cluster-based algorithm is proposed in [21]. Each node gets its K-hop neighbor set information. A cluster is composed of all nodes within K hops from a given node. Each node belongs to one cluster. When a broadcast occurs, only border nodes, which are exactly K hops away from the sender, will relay the broadcast. A similar connectivity-based K-hop clustering algorithm is proposed in [6]. One main disadvantage of these algorithms is that the overlapped area of two neighboring sender's K-hop neighbor sets cause much redundancy when K is large.

Lou and Wu [27] proposed a generic K-hop zone-based broadcast protocol. A generic K-hop zone-based broadcast protocol broadcasts a packet in four steps:

1. The sender uses the forward node set selection process to select its forward node set to cover its K-hop zone. It broadcasts the packet, which attaches the forward node set.
2. A forward node (excluding the border nodes) forwards the packet when it first receives the packet.
3. A nonforward node receives the packet but does not forward it.
4. A border node, which is a selected forward node exactly K − 1 hops away from the source, after receiving the packet, becomes a new sender to disseminate the packet.

Nodes that are within K − 1 hops of the sender can be excluded from the K-hop zone of the border node. In addition, nodes that are within K − 1 hops of a border node with lower ID can also be excluded from the *K*-hop zone of a border node with higher ID if the two border nodes are within K − 1 hops. Figure 28.13 illustrates the K-hop zone-based broadcast protocol. The source *u* selects forward nodes to cover its K-hop zone. Gray nodes are forward nodes that just relay the broadcast packet. Black nodes are the border nodes that have their own K-hop zones. White nodes are the nonforward nodes that only receive the packet.

The forward node set selection process is described as follows. A sender u computes its forward node set to cover all the nodes in its K-hop zone. In each iteration, u selects some nodes in $H_k(u)$ to cover all the nodes in $H_{k+1}(u)$, where $0 \le k \le K-1$. Specifically, u itself covers all nodes in $H_1(u)$; some selected nodes in $H_1(u)$ cover all nodes in $H_2(u)$, ..., until some selected nodes in $H_{K-1}(u)$ cover all nodes in $H_K(u)$. In each iteration, the selection criterion is that the node with the maximum number of uncovered neighbors is selected first. A tie is broken by node ID. All selected nodes form u's forward node set F(u).

For the sample network shown in Figure 28.5, suppose node 1 is the source and it has three-hop neighbor set information: $H_0(1) = \{1\}$, $H_1(1) =$ (1) = $\{2, 5\}$, $H_2(1) = \{3, 6\}$ and $H_3(1) = \{4, 7\}$. Node 1 covers $N_1(1)$. Node 2 and node 5 are first selected from $H_1(1)$ to cover all nodes in $H_2(1)$; then node 3 is selected to cover nodes in $H_3(1)$. Therefore, F(1) = {2, 3, 5} and node 3 is a border node. Then node 3 becomes a new sender. Among nodes in node 3's three-hop zone, node 4 will be selected to cover node 8.

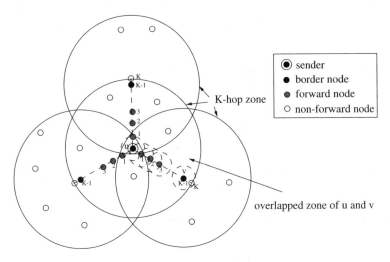

Figure 28.13. K-Hop Zone-Based Broadcast Protocol

28.4.3 Reliable Broadcast Algorithm

The traditional reliable broadcast protocols can be classified into two categories. The first category enforces strong reliability guarantees which provide an atomic operation for the successful delivery of a message to all the nodes [16]. The disadvantage is its poor scalability even in a very stable network. The second category is based on the feedback mechanisms of acknowledgment (ACK) or negative acknowledgment (NACK). It can also be classified as sender initiated and receiver initiated approaches [38]. In the sender initiated approach, the receiver acknowledges each message it receives. The ACKs are unicasted to the sender who maintains all the records for all receivers to confirm the success of the delivery. Only missing packets are retransmitted, either to individual requested receivers, or to all receivers. However, the requirement of sending ACKs in response to the reception of a packet for all receivers may cause channel congestion and packet collision, which is called ACK implosion [18]. Moreover, the amount of records that the sender must maintain to track the receiver set may also grow large. In the receiver initiated approach, the receiver is responsible for reliable delivery. Each receiver maintains reception records and requests repairs via a NACK when errors happen. Several strategies can be applied for the receiver initiated approaches, such as sender-oriented, flat-receiver-oriented, and hierarchical-receiver-oriented approaches. The problem of the receiver initiated approach is the long end-to-end delay brcause the sender must wait for the next broadcast packet to determine if the previous one is successfully delivered or not. Therefore, it can be applied only when the sender has many packets to be sent.

Lou and Wu [28] proposed a reliable broadcast algorithm, called double-covered broadcast (DCB) algorithm, which aims to reduce broadcast redundancy by decreasing the number of forward nodes but still providing high delivery ratio for each broadcast packet in a dynamic environment. The algorithm uses the method in which the sender overhears the retransmission of the forward nodes to avoid the ACK implosion problem. Also, the algorithm guarantees that each node is covered by at least two transmissions so that it can avoid a single error due to the transmission collision. Moreover, the algorithm does not suffer the disadvantage of the receiver-initiated approach, which needs a much longer delay to detect a missed packet.

The double-covered broadcast algorithm works as follows. When a sender broadcasts a packet, it selects a subset of one-hop neighbors as its forward nodes to forward the broadcast based on a greedy approach. The selected forward nodes satisfy two requirements:

1. They cover all the nodes within two hops of the sender.
2. The sender's one-hop neighbors are either forward nodes or nonforward nodes but are covered by at least two neighbors, once by the sender itself and once by one of the selected forward nodes.

After receiving the broadcast packet, each forward node records the packet, computes its forward nodes and rebroadcasts the packet as a new sender. The retransmissions of the forward nodes are received by the sender as the acknowledgment of receiving the packet. The nonforward one-hop neighbors of the sender do not acknowledge receipt of the broadcast. The sender waits for a predefined duration to overhear the rebroadcasting from its forward nodes. If the sender fails to detect all its forward nodes retransmitting during this duration, it assumes that a transmission failure has occurred for this broadcast because of the transmission error or because the missed forward nodes are out of its transmission range. The sender then resends the packet until all forward nodes are retransmitted or the maximum number of retries is reached.

Like most of the other neighbor designating algorithms, the DCB supposes each node knows its one-hop and two-hop neighbor sets $N(v)$ and $N_2(v)$. The forward node set selection process executes at each forward node to determine its own forward node set. We consider the two cases where a node v determines its forward node set $F(v)$:

1. v is the source of the broadcast: v uses FNSSP algorithm to find $F(v)$ in $X = N(v)$ to cover $U = N_2(v)$.
2. v is a selected forward node to relay the broadcast packet: Suppose v has already received the packet from a node set $V(v)$ and each node w in $V(v)$ has its own forward node set $F(w)$. v uses FNSSP algorithm to find $F(v)$ in $X = N(v) - V(v) - \cup_{\forall w \in V(v)} F(w)$ to cover $U = N_2(v) - N(V(v)) - \cup_{\forall w \in V(v)} N(F(w))$.

Localized Broadcasting in Mobile Ad Hoc Networks

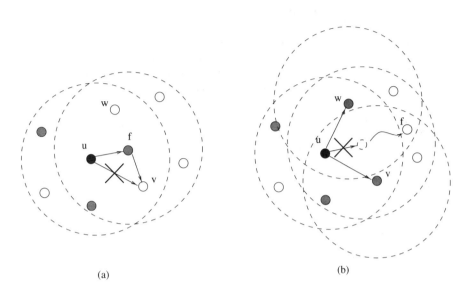

Figure 28.14. Transmission Errors
(A) A transmission error occurs at a nonforward node n. (B) Alternative forward node m and node n are selected to cover the area that is supposed to be covered by the missed forward node f.

For the sample network in Figure 28.5, we use the DCB algorithm to select each node's forward node set. In case 1, we suppose node 3 is the source, then node 3 selects node 2, node 4, and node 6 as its forward nodes. In case 2, we consider node 6 when it receives a packet from node 3. $U(6) = N_2(6) - N(3) - N(2) - N(4) = \{5\}$, node 6 selects node 5 as its forward node.

A node may fail to receive the broadcast packet from its neighbors because of a transmission collision with other neighbors, the high transmission error rate of the radio channel, or the out-of-range movement of the node. Each nonforward node has been at least covered by two forwarding nodes; even if the nonforward node missed the packet from one, it still has a second chance to receive the packet from the other one (Figure 28.14A). The reception of a forward node needs to be confirmed because forward nodes are the key nodes in the network that need to relay the broadcast packet. The loss of the reception may cause the transmission error to propagate. If the sender does not detect the forward node's retransmission signal, the sender will select alternative forward nodes to cover the coverage area of the missed forward node and resend the broadcast packet until it receives the confirmation from its forward nodes or the maximum times of retries is reached (Figure 28.14B).

28.5 Summary

In this chapter, we described several broadcast algorithms that use the neighbor-designating approach. More complicated algorithms based on neighbor designating approach are also introduced. The forward node selection process is the basis for all the broadcast algorithms discussed here. The optimal solution is NP-complete and more theoretical discussion about the optimal MCDS solution can be found in [2, 14, 30, 45]. The algorithms discussed above are all heuristic algorithms. Except for the cluster-based broadcast algorithm, all other algorithms mentioned here have no constant approximate ratio to the optimal solution. Therefore, they will show poor performance for broadcast operation when the network becomes extremely dense. Recently, some efforts, such as [48], have been made to extend various local algorithms to be effectively used in dense networks.

Acknowledgment

This work was supported in part by NSF Grants CCR 9900646, ANI 0073736, and EIA 0130806.

References

1. C. Adjih, P. Jacquet, and L. Viennot. Computing connected dominated sets with multipoint relays. http://www.inria.fr/rrrt/rr-4597.html, 2002.
2. K.M. Alzoubi, P.J. Wan, and O. Frieder. Message-optimal connected dominating sets in mobile ad hoc networks. *Proc. of 3rd ACM Int'l Symposium on Mobile Ad Hoc Networking and Computing (MobiHoc'2002)*, pp. 157–164, 2002.
3. K.M. Alzoubi, P.J. Wan, and O. Frieder. New distributed algorithm for connected dominating set in wireless ad hoc networks. *Proc. of 35th Hawaii Int'l Conf. on System Sciences (HICSS-35)*, pp. 3881–3887, January 2002.
4. G. Calinescu, I. Mandoiu, P.J. Wan, and A. Zelikovsky. Selecting forwarding neighbors in wireless ad hoc networks. *Proc. of ACM DIALM'2001*, pp. 34–43, December 2001.
5. J. Cartigny and D. Simplot. Border node retransmission based probabilistic broadcast protocols in ad-hoc networks. *Proc. of HICSS-36*, January 2003.
6. G. Chen, F.G. Nocetti, J.S. Gonzalez, and I. Stojmenovic. Connectivity based k-hop clustering in wireless networks. *Proc. of 35th Hawaii Int'l Conf. on System Sciences (HICSS-35)*, January 2002.
7. C.C. Chiang and M. Gerla. Routing and multicast in multihop, mobile wireless networks. *Proc. of IEEE Int'l Conf. on Universal Personal Communications*, vol. 2, pp. 546–551, 1997.
8. C.C. Chiang, H.K. Wu, W. Liu, and M. Gerla. Routing in clustered multihop, mobile wireless networks with fading channel. *Proc. of IEEE Singapore Int'l Conf. on Networks*, pp. 197–211, 1996.
9. F. Dai and J. Wu. Distributed dominant pruning in ad hoc wireless networks. *Proc. of IEEE 2003 International Conference on Communications (ICC 2003)*, vol. 1, pp. 353–357, May 2003.
10. B. Das, R. Sivakumar, and V. Bharghavan. Routing in ad-hoc networks using a spine. *Proc. of the 6th Int'l Conf. on Computer Communications and Networks (ICCCN'97)*, pp. 1–20, September 1997.

Localized Broadcasting in Mobile Ad Hoc Networks

11. A. Ephremides, J.E. Wieselthier, and D.J. Baker. A design concept for reliable mobile radio networks with frequency hopping signaling. *Proc. of the IEEE*, vol. 75, no. 1, pp. 56–73, 1987.
12. M. Gerla and J. Tsai. Multicluster, mobile, multimedia radio network. *ACM/Baltzer Wireless Networks*, vol. 1, pp. 255–265, 1995.
13. IEFT MANET Working Group. http://www.ietf.org/html.charters/manet-charter.html.
14. S. Guha and S. Khuller. Approximation algorithms for connected dominating sets. *Algorithmica*, vol. 20, no. 4, pp. 374–387, 1998.
15. Z. J. Haas, J. Y. Halpern, and L. Li. Gossip-based ad hoc routing. *Proc. of InfoCom 2002*, vol. 3, pp. 1707–1716, June 2002.
16. V. Hadzilacos and S. Toueg, Fault-tolerant broadcasts and related problems, In S. Mullender, Ed., *Distributed Systems*, 2 ed., pp. 97–145, Boston: Addison Wesley, 1993.
17. C. Ho, K. Obraczka, G. Tsudik, and K. Viswanath. Flooding for reliable multicast in multi-hop ad hoc networks. *Proc. of ACM DIALM'99*, pp. 64–71, August 1999.
18. M. Impett, M.S. Corson, and V. Park. A receiver-oriented approach to reliable broadcast ad hoc networks. *Proc. of Wireless Communications and Networking Conference (WCNC'2000)*, vol. 1, pp. 117–122, 2000.
19. P. Jacquet, A. Laouiti, P. Minet, P. Muhlethaler, A. Qayyum, and L. Viennot. Optimized link state routing protocol. draft-ietf-manet-olsr-07.txt, November 2002.
20. D.B. Johnson and D.A. Maltz. Dynamic source routing in ad-hoc wireless networks, In T. Imielinski and H. Korth, Eds., *Mobile Computing*, pp. 153–181. Boston: Kluwer Academic Publishers, 1996.
21. D. Kim, S. Ha, and Y. Choi. K-hop cluster-based dynamic source routing in wireless ad-hoc packet radio network. *Proc. of VTC'98*, pp. 224–228, 1998.
22. H. Lim and C. Kim. Flooding in wireless ad hoc networks. *Computer Communications Journal*, vol. 24, Nos. 3–4, pp. 353–363, 2001.
23. C. Lin and M. Gerla. Adaptive clustering for mobile wireless networks. *IEEE Journal on Selected Areas in Communications*, vol. 15, pp. 151–162, 1999.
24. W. Lou and J. Wu. Efficient broadcast with forward node set in clustered mobile ad hoc networks. *Proc. of the 11th Int'l Conf. on Computer Communications and Networks (ICCCN'2002)*, pp. 398–403, October 2002.
25. W. Lou and J. Wu. On reducing broadcast redundancy in ad hoc wireless networks. *IEEE Trans. on Mobile Computing*, vol. 1, no. 2, pp. 111–123, April–June 2002.
26. W. Lou and J. Wu. A cluster-based backbone infrastructure for broadcasting in MANETs. *Proc. of IPDPS'2003, Workshop of WMAN*, April 2003.
27. W. Lou and J. Wu. A k-hop zone-based broadcast protocol in mobile ad hoc networks. 2003. *Proc. of IEEE Globecom 2004*, (in press), 2004.
28. W. Lou and J. Wu. Double-covered broadcast (DCB): A simple reliable broadcast algorithm in mobile ad hoc networks. *Proc. of IEEE InfoCom 2004*, March, 2004.
29. L. Lovasz. On the ratio of optimal integral and fractional covers. *Discrete Mathematics*, vol. 13, pp. 383–390, 1975.
30. M.V. Marathe, H. Breu, H.B. Hunt III, S.S. Ravi, and D.J. Rosenkrantz. Simple heuristics for unit disk graphs. *Networks*, vol. 25, pp. 59–68, 1995.
31. S. Ni, Y. Tseng, Y. Chen, and J. Sheu. The broadcast storm problem in a mobile ad hoc network. *Proc. of ACM/IEEE MobiCom'99*, pp. 151–162, August 1999.
32. V.D. Park and M.S. Corson. Temporally-ordered routing algorithm (TORA) version 1: Functional specification. *Internet Draft*, 1997.
33. M.R. Pearlman and Z.J. Haas. Determining the optimal configuration of the zone routing protocol. *IEEE Journal on Selected Areas in Communications*, vol. 17, no. 8, pp. 1395–1414, February 1999.
34. A. Pelc. Broadcasting in Radio Networks. In I. Stojmenovic, Ed., *Handbook of Wireless Networks and Mobile Computing*, Hoboken, NJ: John Wiley & Sons, Inc, 2002.
35. W. Peng and X. Lu. Efficient broadcast in mobile ad hoc networks using connected dominating sets. *Journal of Software*, 1999.

36. W. Peng and X. Lu. AHBP: An efficient broadcast protocol for mobile and hoc networks. *Journal of Science and Technology*, 2002.
37. C. Perkins and E.M. Royer. Ad-hoc on-demand distance vector routing. *Proc. of 2nd IEEE Workshop on Mobile Computing Systems and Applications (WMCSA)*, pp. 90–100, New Orleans, February 1999.
38. D.G. Petitt. Reliable multicast protocol design choices. *Proc. of MILCOM'97*, vol. 1, pp. 242–246, 1997.
39. A. Qayyum, L. Viennot, and A. Laouiti. Multipoint relaying for flooding broadcast message in mobile wireless networks. *Proc. of 35th Hawaii Int'l Conf. on System Sciences (HICSS-35)*, pp. 3898–3907, January 2002.
40. V. Ramasubramanian, R. Chandra, and D. Mosse. Providing a bidirectional abstraction for unidirectional ad hoc networks. *Proc. of 21st Annual Joint Conference of the IEEE Computer and Communications Societies (InfoCom)*, vol. 3, pp. 1258–1267, June 2002.
41. P. Sinha, R. Sivakumar, and V. Bharghavan. Enhancing ad hoc routing with dynamic virtual infrastructures. *Proc. of IEEE InfoCom'2001*, vol. 3, pp. 1763–1772, April 2001.
42. I. Stojmenovic. Location updates for efficient routing in ad hoc networks, in I. Stojmenovic, Ed., *Handbook of Wireless Networks and Mobile Computing*, pp. 451–471, Hoboken, NJ: John Wiley & Sons, Inc., 2002.
43. I. Stojmenovic, S. Seddigh, and J. Zunic. Dominating sets and neighbor elimination based broadcasting algorithms in wireless networks. *IEEE Trans. on Parallel and Distributed Systems*, vol. 13, no. 1, pp. 14–25, January 2002.
44. Y.-C. Tseng, S.-Y. Ni, and E.-Y. Shih. Adaptive approaches to relieving broadcast storms in a wireless multihop mobile ad hoc network. *IEEE Trans. on Computers*, vol. 52, no. 5, pp. 545–557, May 2003.
45. P.J. Wan, K. Alzoubi, and O. Frieder. Distributed construction of connected dominating set in wireless ad hoc networks. *Proc. of IEEE InfoCom'2002*, vol. 3, pp. 1597–1604, June 2002.
46. J. Wu. Extended dominating-set-based routing in ad hoc wireless networks with unidirectional links. *IEEE Transactions on Parallel and Distributed Systems*, vol. 13, no. 9, pp. 866–881, September 2002.
47. J. Wu. An enhanced approach to determine a small forward node set based on multipoint relay. *VTC 2003*, October 2003.
48. J. Wu and F. Dai. Distributed formation of a virtual backbone in ad hoc networks using adjustable transmission ranges. *Proc. of IEEE ICDCS 2004*, pp. 372–379, March 2004.
49. J. Wu and F. Dai. A generic distributed broadcast scheme in ad hoc wireless networks. *Proc. of ICDCS 2003*, pp. 460–468, May 2003.
50. J. Wu and F. Dai. Mobility management and its applications in efficient broadcasting in mobile ad hoc networks. *Proc. of IEEE InfoCom 2004*, March 2004.
51. J. Wu and H. Li. On calculating connected dominating sets for efficient routing in ad hoc wireless networks. *Proc. of ACM DIALM'99*, pp. 7–14, August 1999.
52. J. Wu and W. Lou. Forward-node-set-based broadcast in clustered mobile ad hoc networks. *Wireless Networks and Mobile Computing*, [Special Issue on Algorithmic, Geometric, Graph, Combinatorial, and Vector Aspects], vol. 3, no. 2, pp. 155–173, 2003.

Chapter 29
Energy-Efficient Wireless Networks

Ionut Cardei and Ding-Zhu Du

Abstract

A key issue for applications with wireless networks is efficient energy utilization. In most cases, mobile wireless nodes are powered by electrochemical batteries with limited capacity, which imposes a strict limit on the application lifetime. There is high interest in optimizing power consumption and considerable research effort has been invested in this area. In this chapter, we survey techniques for power management at different layers of the network protocol stack.

29.1 Introduction

Wireless networks have a dramatic impact on the way people interact and on how information has become readily available. Miniaturization and developments in radio frequency (RF) communications technology, protocols, and computing platforms made possible new classes of applications where rapid deployment and support for mobility are crucial, such as cellular communications, law enforcement, civilian search and rescue, and a variety of military applications, where wireless networks provide services for situational awareness, personal communication, command/control, information, and reconnaissance.

A general classification for wireless networks considers the nature of the network topology:

- Infrastructure-based networks, where every host can directly communicate with a base station (BS) that coordinates communication
- Ad hoc networks, where nodes are self-configurable and able to forward packets so that end-to-end communication is achieved with multi-hop paths

For many applications employing mobile terminals, energy-efficient operation remains a critical requirement, at least until new technologies based on fuel cells and portable renewable sources become widely available. Wireless stations are in general powered by electrochemical batteries, having a rather limited capacity, caused mainly by the low energy density, limited in part by constraints in size and weight. As servicing or replacing batteries may not be feasible for some applications (e.g., wireless sensor networks), extending network lifetime by applying power-aware communications and computing has received considerable attention. Energy efficiency has been addressed through low power design at the hardware level as well as through power-aware mechanisms at all layers of the network protocol stack.

In this chapter, we present recent research in the area of energy efficiency in wireless communications. We survey mechanisms for power-aware communications and energy management applied at the wireless link layer and medium access control sublayer. We also review energy-efficient techniques for ad hoc routing and topology control, as well as an innovative approach for energy harvesting.

Within the context of network operation, energy is consumed for communications and processing. Computation energy is used for executing protocols and algorithms within network adapters or host processing platforms, while energy spent for communications relates to network data transfer. Mobile stations (or wireless terminals) have a digital radio — an RF transceiver — on board that provides the physical and link layer services. A radio can be in one of the following states (modes) — transmit, receive, idle, and sleep. In the idle state, the host node does not transmit or receive packets, although the receiver is active, listening for incoming transmissions. Thus, in the idle state, the receiver pipeline is powered on, with power consumption nearly as high as in receiving mode, when the radio actually is receiving a transmission. In many studies, from an energy utilization viewpoint, researchers do not differentiate the idle state from the receiving state. The highest power consumption is in transmit mode and the lowest in the sleep mode, when a node turns off both its transmitter and receiver. As an example, consider the power consumption for the `transmit:receive:sleep` states for the 11 megabits per second (Mbps) Lucent® IEEE® 802.11 WaveLAN personal computer (PC) [1]: `1.4W:0.9W:0.05W`.

The work in [1] describes a series of experiments on energy consumption with these network interface cards (NICs) in sending and receiving broadcast versus point-to-point data packets. Considering these characteristics, it is important to design protocols and algorithms that reduce transceiver energy consumption.

We identify several mechanisms used to reduce energy consumption and to extend network lifetime:

Energy-Efficient Wireless Networks

- Have nodes enter sleep state as often and for as long as possible
- Choose routing paths that minimize energy consumption
- Selectively use nodes based on their energy status
- Construct communication and data delivery structures (e.g., broadcast/multicast tree, underlying topology) that minimize energy consumption
- Energy harvesting
- Reduce networking overhead
- Power-aware wireless link adaptation

Scheduling node activity to switch to the sleep state is an important and efficient method to conserve energy resources. Radios would be switched off whenever a node is not expected to send or receive packets, so that it avoids overhearing transmissions intended for other nodes that would otherwise waste energy for reception and decoding.

Energy-efficient routing protocols use power-aware criteria in choosing the routing paths. If a few nodes are overused, they can deplete their power earlier, triggering network partition. Therefore, to prolong the network lifetime, it is also imperative to select routing nodes also based on current residual battery power level. Underlying communication and data delivery structures are implicitly or explicitly used by the routing protocol and their maintenance should aim at minimizing the maximum energy consumption per node.

A straightforward way to further improve energy efficiency is to reduce the communication overhead of the various protocol layers. For instance, at higher load, collisions at medium access for contention-based Media Access Control (MAC) protocols triggers retransmissions, thereby increasing the energy consumed per useful bit. Additionally, the size and periodicity of background control packets need to be considered. The paper [2] presents a survey of works that consider energy-efficient techniques for each network layer.

Wireless sensor networks (WSN) are an emerging class of wireless networks that promise to have a significant impact on a broad range of applications relating to surveillance, health care, and environmental monitoring, etc. WSNs consist of a large number of miniaturized devices, each having communication, computation, and sensing ability, all collaborating to perform a common sensing task. Wireless sensor nodes are battery powered and mechanisms for energy savings can effectively extend their operational time. The specific architecture of WSNs and the characteristic constraints on resource availability pose significant challenges to rapid development and wide deployment. In this chapter, we will present energy-efficient techniques specifically designed for WSNs.

The rest of this chapter is structured as follows:

- Section 29.2 describes a power-aware link adaptation technique.
- Section 29.3 describes a novel energy harvesting application for wireless networks.
- Section 29.4 presents power-saving methods based on node activity scheduling.
- Section 29.5 presents routing related techniques.
- Section 29.6 discusses power-aware topology control.
- Section 29.7 concludes with some final remarks.

29.2 Power-Aware Link Layer Adaptation

Wireless links undergo dynamic channel conditions in time and space, caused by path loss, interference, multipath fading. In addition, network mobility introduces another variation factor that can further affect signal quality, so that, the bit error rate at the receiver can vary over several orders of magnitude in a brief time interval. To mitigate against channel impairment, wireless networks employ various mechanisms of forward error correction (FEC) and automatic repeat request (ARQ) to provide the required level of quality of service (QoS). For applications with energy-constrained battery-power nodes, such as WSNs, significant energy savings can be achieved by implementing an adaptive error control strategy that adapts various link layer parameters depending on channel quality to minimize power utilization, trading off some degree of QoS offered to the higher layers.

FEC inserts redundant bits in each user frame that allow receivers to repair, to a certain degree, frames damaged by burst errors. The overhead cost of FEC is constant, typically a fraction of the total available bit rate. FEC is most effective when the channel conditions are bad. With a clean channel, the redundant bits are not used and waste useful network capacity. With ARQ, the receiver informs the sender on the status of received frames, so that the sender can retransmit lost frames. ARQ works best when the channel is good and frames are lost only occasionally. With deteriorating conditions (high bit error rate), the number of errors grows and so does the number of retransmissions, so that a considerable fraction of time is spent doing retransmissions, resulting in a sharp drop in effective data rate.

Hybrid FEC–ARQ schemes employ a lower level of FEC, with less redundancy, relying on retransmissions to deal with poor channel conditions when FEC cannot compensate for extended burst errors.

The work in [3] presents an approach for energy efficiency in wireless networks employing adaptive error control. Their link layer changes dynamically with the channel condition, so that energy consumed per useful bit transmitted is optimized, while still maintaining QoS. In addition to the energy cost of transmission, their approach also considers the cost of

Energy-Efficient Wireless Networks

performing error control, so that the energy cost per bit metric reflects realistically the overall energy efficiency of the link layer. The proposed link layer adapts both the error control FEC/ARQ mix and the maximum frame length.

The authors model the wireless link with a Discrete Time Markov Chain (DTMC) [4, 5]. Each state in the chain corresponds to a different fading level of a binary symmetric channel, with a specific bit error rate (BER), dependent on the signal to noise ratio (SNR) in that state. The analysis further applies a Rayleigh fading channel on a DTMC with a good state and a bad state. The derived probabilities of being in the good state and in the bad state depend only on the Rayleigh fading envelope. From the mean bit error rate derivation, BER = P(Good) BER_G + P(Bad) BER_B, the authors conclude that the mean BER only depends on the terminal's speed. Two main parameters are therefore used for the characterization of the channel dynamics. First, the good state BER is varied between 10^{-2} and 10^{-8}. The second main parameter is the speed of the terminal, which also dictates error burstiness. The bad state BER is set constant at 0.5.

The total energy consumed per useful bit (E_t) depends on the transmission energy ($E_{t, tx}$) and the energy spent on error correction ($E_{t, comp}$): E_t = $E_{t, comp}$ + $E_{t, tx}$. When the channel quality drops or when the code rate increases to 1.0, the mean number of retransmissions and the transmission energy increases, too. In contrast, $E_{t, comp}$, the energy spent on FEC encoding/decoding decreases. Conversely, analysis of Reed Solomon coding for different block sizes and code rates applied to the Rayleigh channel indicates that with less redundancy or with improved channels, the mean number of retransmissions decreases while the computational cost increases. Both effects combined yield a minimum energy per bit that can be achieved with an adaptive link layer with adjustable code rate.

29.3 Energy Harvesting

An innovative approach for improving the lifetime of WSNs is to coordinate sensing task allocation with the spatio-temporal characteristics of energy availability and to employ energy harvesting. The work in [6] describes a distributed framework for WSNs that enables nodes to discover the energy environment and to use availability information to schedule and share sensing tasks accordingly. This framework can be integrated with other energy efficiency mechanisms, such as sleep scheduling and power-aware routing.

Simple and uncoordinated use of harvesting energy sources, such as solar cells, microbial fuel cells, or acoustic/vibration transducers, do not bring out the full potential for network lifetime improvement. Sensing tasks and communications, in general, require cooperation between multiple nodes, and, as energy resources may not be distributed uniformly, task allocation adapted to energy availability will extend network lifetime more than a simple nonadaptive approach.

The proposed Environmental Energy Harvesting Framework (EEHF) plays two roles. First, it uses local measurements to learn adaptively the energy properties of the environment and about the renewal opportunities. For each node, a single number C is computed from various energy availability parameters to reflect the cost of using that node for scheduling. Second, EEHF exports this information to energy-aware task assignment functions, such as load balancing, leader elections, clustering, and power-aware communication. Scheduling decisions based on node costs would inherently select nodes with higher environmental energy availability.

A prototype architecture of EEHF is illustrated with an energy harvesting network with solar cells. The main EEHF architecture components are:

- Spectral estimation
- Prediction filter
- Stochastic consumption predictor
- Parameterize block
- Scalability-friendly information exchange

The first component, spectral estimation, tries to detect a pattern for energy availability and determines the time parameter T, over which the energy availability is predicted, from the energy availability waveform spectral analysis, for instance, using the Fast Fourier Transform.

The prediction filter uses the T parameter from the spectral estimator to predict E_m, the energy expected to be harvested within the next period of time T. Another parameter derived from the prediction filter is the prediction confidence, η.

The stochastic consumption predictor records or estimates for every time period T the average energy consumption in the WSN. The Parameterize component uses the available battery energy E_b and the other parameters to compute the cost C: $E = w_1(E_m - E_{cm})\eta + w_2 E_b$, and $C = 1/E$, where E_{cm} is the energy consumed in every T period by tasks out of scheduler's control and w_1 and w_2 are weighting coefficients.

The previous framework components are responsible for the local characterization of energy availability. Lifetime optimal scheduling needs global information and scalability-friendly information exchange provides spatial characterization of energy availability in a scalable way, both in terms of network density and size. The first method for sharing local information proposed is network averaging. The average value of E and the maximum value are computed distributively and then each node offers to accept a workload proportional to $L = (E - E_{av})/(E_{max} - E_{av})$ if $L > 0$ and to go to sleep if $L < 0$. The second method has distributed scheduler benefit directly from local node energy availability information. For instance, a power-aware route discovery based on Bellman–Ford algorithm picks the minimum cost route without having to share all information on the entire network.

The EEHF framework has been applied to the routing problem and evaluated through simulation with a solar cell application, achieving more than 100 percent lifetime improvement for scenarios with strong nonuniform energy distribution.

29.4 Scheduling Node and Radio Activity

Commonly used in energy-efficient MAC protocols, the concept of conserving energy by scheduling wireless node activity to alternate between active state (transmit or receive) and low-energy sleep state (also called dozing state) is effective. A node in sleep state is not aware of network activity and cannot participate in traffic exchange, but, on the other hand, consumes very little power compared to the idle state. Spending more time with the receiver deactivated extends the battery lifetime. As the transition time between the sleep mode and the active mode can be as high as 800 microseconds [7], to reduce the number of transitions, it is recommended that the nodes schedule traffic into bursts in which a station can continuously transmit or receive data.

In this section, we present research in MAC protocols for wireless networks and WSNs that implement energy efficiency by scheduling nodes to alternate between active and low-energy sleep states.

29.4.1 Power-Aware Medium Access Control

The IEEE 802.11 wireless standard [8] covers three functional areas of the MAC sublayer — reliable data delivery, medium access control, and security. The data transfer function uses two control frame handshakes: after every unicast data frame is received successfully, the receiver replies with an acknowledgment (ACK) frame. To reduce contention in a collision domain, a four frame exchange protocol may be used for unicast frames:

1. The sender first sends a request to send (RTS) frame to the destination
2. The destination responds with a clear to send (CTS) frame
3. The sender sends the data frame
4. The receiver replies with an ACK

The RTS/CTS scheme avoids the hidden terminal problem and is required by the 802.11 standard, but may be disabled. The basic medium access protocol is the Distributed Coordination Function (DCF), which implements medium sharing through the use of Carrier Sense Multiple Access with Collision Avoidance (CSMA/CA). Collision avoidance uses a binary exponential backoff procedure, which is invoked when a station that wants to transmit a frame detects a busy medium. Point Coordination Function (PCF) is an alternative access method built on top of DCF with medium access controlled by a central point coordinator through polling.

The IEEE 802.11 MAC provides a power management technique. The basic concept is to have all stations that operate in power-save (PS) mode to synchronize and wake up at the same time. If a station receives an announcement about data to be delivered, it stays awake until it receives the frame, otherwise it may return to the sleep state. This mechanism is easily accomplished in infrastructure networks where the point coordinator synchronizes all mobile stations and buffers the frames for stations in sleep state. This is done by periodically sending a beacon, which contains both a time stamp as well as a traffic indication map (TIM), which announces all u packets for stations in doze mode. The mobile stations that wake up to receive the beacon will determine if there is any pending traffic, in which case they will stay wake until the transmission is over. Power management for the DCF is accomplished in a distributed fashion. After the beacon interval, all mobile stations compete for sending the beacon, using the standard backoff algorithm. Packets for a station in sleep state are buffered by the sender and announced using ad hoc TIMs (ATIMs), which are sent after the beacon during the ATIM window. All stations are awake during the ATIM window; therefore the announced receiver knows to stay awake until data transmission takes place. Both ATIMs and data packets are acknowledged and are sent using the standard backoff algorithm.

There are many published research works that explore further and look at different ways to improve the energy conservation mechanism proposed by IEEE 802.11. The Energy Conserving MAC (EC-MAC) protocol [9] was designed for a centrally controlled asynchronous transfer mode (ATM) network, having energy efficiency as the primary goal. After the schedule is transmitted by the BS, the nodes in the network know to be awake only during the periods when they exchange traffic. This improves the energy consumption compared with 802.11, where the destination remains awake until the data transfer is completed. Also, by scheduling the exact order of data transfer, collisions are avoided and the number of retransmissions will be decreased. Further energy improvements are achieved when frames from or to the same station are allocated contiguous slots, thereby saving the transition energy between active–sleep states.

In [10], authors proposed PAMAS (Power-Aware Multi-Access Protocol with Signaling), a medium access control protocol for ad hoc networks, based on the Multiple Access with Collision Avoidance (MACA) protocol, with the addition of a separate signaling channel. The signaling channel is used to send RTS/CTS control messages and busy tones, which are sent while a node receives a packet (at the beginning of message reception or when a receiving node gets a RTS from another station). The main way to save energy comes from nodes shutting off their transceiver when they overhear transmissions which were not directed to them. A node powers off when it cannot send or receive data: either it has no data to transmit and one neighbor starts transmitting or it has data to transmit, but at least one neighbor is transmitting and another is receiving. For a station to determine the time to power off, the proposed probe protocol determines the longest transfer time for a sender or receiver neighbor. Experiments conducted on a random network topology, line topology, and fully connected network topology, show power savings of 10 to 70 percent.

29.4.2 Energy-Efficient MAC Protocols for WSN

Wireless sensors are devices equipped with sensing, processing, and communication capabilities. The sensor nodes may communicate or forward data through multi-hop paths. This section presents approaches for power savings that apply to sensor networks.

In [11], authors proposed S-MAC, a MAC protocol for wireless sensor networks, having energy efficiency as the primary design objective. The modeled network has many sensor nodes capable of multi-hop communication, deployed in an ad hoc manner, which are inactive for long periods of time and become active when something is detected. This protocol assumes nodes collaborate for a common application and adapts to changes to network size, density, and topology.

The S-MAC protocol assumes the principal sources of energy waste are:

- Collisions
- Message overhearing — receiving packets addressed to other nodes
- Control packet overhead
- Idle listening

S-MAC uses the following mechanisms to reduce the energy waste from all the above sources. First, it lets nodes sleep periodically if they are in the idle listening mode. Neighboring nodes synchronize their sleep schedule, forming virtual clusters. To avoid collisions, S-MAC follows a mechanism similar to 802.11, using `RTS/CTS/DATA/ACK` packet sequence. To avoid overhearing, S-MAC employs a mechanism similar to PAMAS, by putting the neighbors of both a sender and a receiver to sleep after they hear a RTS or CTS packet, for the duration of the current transmission. Still, S-MAC uses only one channel for data and signaling. The last mechanism used is message passing, where a long message is divided into many small packets, sent in a burst. Only one RTS/CTS packet sequence is used, but every data fragment is acknowledged to avoid the hidden terminal problem.

These mechanisms may affect per hop fairness and latency. The performance of this protocol is evaluated using Rene Motes, developed at University of California — Berkeley, and shows energy savings of S-MAC compared with 802.11 DCF.

29.4.3 Node Activity Scheduling

This section presents two approaches for energy-efficient node scheduling mechanisms for WSNs:

1. A centralized solution that works when transmissions from a coordinator node can be broadcast to the entire network
2. A fully distributed approach that uses locally available information to employ a fairly elaborate schedule coordination protocol

In [12], the authors address the problem of energy efficiency in wireless sensor applications for surveillance of a set of targets with known locations. When ground access in that area is prohibited, one method is to deploy remotely (e.g., from an aircraft) a large sensor population, in the targets' proximity. The sensors send the monitored information to a central processing node. Every target must be monitored at all times and every sensor is able to monitor all targets within its operational range. Energy savings are obtained by scheduling the sensor node transmission such that they are in the sleep mode as much as possible. The proposed method consists of dividing the set of sensors into disjoint sets, such that every set of sensors completely covers all targets. Then, every set responsible for target monitoring is activated in a round-robin fashion. The nodes from the active set are in the active state, whereas all other nodes are in a low-energy sleep state. This method also ensures balanced energy consumption among all sensor nodes. The lifetime of the network is extended proportionally to the number of sets found. The paper presents an efficient heuristic for computing the maximum number of disjoint sets. Once the sensors are deployed, they send their location information to the central node, which computes the disjoint sets and sends back the membership information. Knowing the set it belongs to and the total number of covers, every sensor is able to identify the time periods when it will in active or sleep states. This mechanism can be implemented at the MAC layer. In this case, the node synchronization is accomplished with periodic beacon messages transmitted by the central node. On the other hand, the medium access for the sensors within the same set can be implemented using any MAC protocol suitable for sensor networks.

Another approach for maximizing lifetime of an ad hoc network by coordinating node activity is presented in [13]. The metric for network lifetime used in this research is the time before the network becomes partitioned. Similar to other research presented in this chapter, their approach for extending network lifetime consists of scheduling sleep intervals, when the transceiver is deactivated and power consumption is very low. The remaining active nodes in the network must be connected so that between every two nodes there must be a communication path available formed by other active nodes. The complexity of this problem can be proven to be NP-complete by reducing from an instance of the minimum connected dominating set problem. The authors have developed a heuristic and a distributed protocol that uses only information available locally within a configurable number of hops.

At the core of the solution to the sleep coordination problem is the Care-Free Sleep theorem, stating necessary and sufficient conditions for a node to enter the sleep state without partitioning the network. According to the theorem, a node can go to sleep if and only if after the node has entered sleep state, there is a communication path between any of its active neighbors and any of its sleeping neighbors has at least one active neighbor. *if*: after the node (A) is deactivated, a path S – D interrupted by the sleeping node A at two of its neighbors, B and C, can be repaired by plugging in path B – C, that must exist, so that new S – D path becomes S – B – C – D. Both nodes B and C would have at least an active neighbor. *only if*: provided the two conditions are not met, the network becomes disconnected, with one node isolated, or a path between two active neighbors breaks up.

The authors propose a distributed protocol to coordinate sleep schedules, using local information and following the concept from the theorem. A current computation node (CCN) determines the nodes in its neighborhood that can be safely switched to the sleep mode. Multiple CCNs are active in the network and a token-based protocol helps electing CCNs. A CCN evaluates all its neighbors (and itself) for the best candidate for the sleep state. Routing information up to k hops away helps the CCN to select the node with the minimum energy on the shortest path between any two neighbors. Then, the CCN passes the token to its active neighbor that has not been CCN for the longest time and the process continues.

All nodes sleep for the same time T_s. T_s determines the amount of power savings and must be adapted to three factors:

1. Energy of the alternative path to the node
2. Expected traffic energy utilization
3. Length of alternative paths to the node

For simulation purposes, the sleep time was determined with respect to node density, overall network lifetime and number of tokens.

With the sleeping approach, the network lifetime was extended with more than 200 percent for higher network densities and with 50 to 100 percent, for average to medium densities.

The main contributions of this work consist in the insights of the Care-Free Sleep theorem and the token-based distributed sleep coordination protocol, which can be easily applied to WSNs.

In [14], the authors propose a network topology management technique for WSNs which exploits the time dimensions instead of the density dimension. In an event-based sensing application, transmissions are infrequent during the most prevalent monitoring state. Keeping nodes in sleep state increases the link setup time when data transmissions to the sink node are necessary. The proposed Sparse Topology and Energy Management (STEM) scheme addresses this problem by trading off energy utilization against transition latency to the data transfer state. The concept for STEM is fairly simple. Each node periodically turns on its radio for a short period to listen for incoming transmissions. The node that wants to transmit (called initiator) transmits beacons with the node ID that it is trying to wake up (target node). After receiving a beacon, the target node replies and both radios now have an open communication link. To transmit the data further over a multi-hop path, the initiator role is passed to the current target until a path is established. To avoid interference with the wakeup protocol, STEM uses a separate radio on a different frequency to transmit the actual data. Systems from Sensoria Corporation provide such a dual transceiver radio. During the data transmission time, participating nodes continue to execute the wakeup protocol. To handle collision cases in the wakeup plane, a node activates its data radio when there is a collision in the wakeup plane. The radio on the data channel can detect transmissions, providing a sort of carrier sensing. When a signal is detected on the data plane, the node will not reply on the wakeup plane, because its acknowledgment will likely collide with other transmissions. Because STEM is orthogonal to network density, it can be further integrated with Geographic Adaptive Fidelity (GAF) [15], a location-based topology management scheme, to improve energy efficiency.

29.5 Energy Conservation in Ad Hoc Routing

Ad hoc wireless networks are infrastructureless, with nodes establishing connections on-the-fly, without a centralized coordinator. All nodes have the ability to route packets, so that when the nodes are not within radio range of each other, they achieve end-to-end connectivity through multi-hop routes. This feature allows ad hoc wireless networks to be easy to deploy and to heal rapidly following node failure and link impairment.

One difficulty is raised by the node mobility. Mobility triggers frequent topology updates and therefore, generates higher control message overhead at the network layer. Methods to reduce the energy consumption include:

- Considering battery resources when selecting the route
- Reducing the communication overhead of control messages

- Efficient route reconfiguration mechanisms (as effect of topology change)

29.5.1 Energy-Efficient Routing Protocols

Traditional routing metrics, such as minimum hop-count are not appropriate for ad hoc wireless networking, as they may overuse the energy resources of a small set of nodes in favor of others. To increase node and network lifetime, the work in [16] introduces five new power-aware metrics for determining routes in ad hoc wireless networks. Intuitively, it is desirable to route packets through lightly loaded nodes, with sufficient power resources. The new metrics are:

1. Minimize energy consumed per packet — for any packet, the goal of this metric is to minimize the sum of energies consumed in every node involved in forwarding the packet from source to destination. One disadvantage is early energy consumption for some nodes, with impact on the network lifetime.
2. Maximize time to network partition — this can be accomplished by load balancing among the critical nodes. Optimizing this metric is difficult if application requires low delay and high throughput.
3. Minimize variance in node power levels — this is similar to load sharing, by keeping the amount of unfinished work in all nodes the same.
4. Minimize cost per packet — the cost of sending a packet along some path is defined as the sum of the costs of all nodes in path, where the cost of a node denotes the reluctance to forward packets. By incorporating the battery characteristics in defining the node cost, a different goal can be achieved such as avoiding nodes with depleted energy resources or increasing the time to network partition.
5. Minimize maximum node cost — where the cost of a node is the cost of routing a packet. The goal is to increase the time until the first node fails.

The authors conducted simulations to validate the benefits of using these power-aware metrics. The authors compared the performance of shortest-hop routing versus shortest-cost routing as defined by the fourth metric, using three performance metrics — end-to-end packet delay, average cost/packet, and average maximum node cost. The results indicated no difference in the end-to-end delay between the two routing methods. By using shortest-cost routing, some packets may have a longer delay when they avoid high cost nodes, but others have shorter delays by avoiding congested nodes. The experiments indicate higher cost savings for larger networks, for moderate network loads, and in denser networks.

Performing routing such that the energy consumed along the selected path is minimized can result in overusing and depleting the power of a small set of nodes. A better approach is to select a routing path with the objective of maximizing the network lifetime by balancing energy consumption among the nodes. This idea is explored in [17], where authors address the problem of routing in a static wireless network, having a set of source nodes generating packets that must reach a set of designated nodes. An instance of such a network is a sensor network, with sensors generating data that is sent to more powerful nodes for processing. Every node can participate in data forwarding and can adjust its power within a range, resulting in a set of possible one-hop away neighbors. Considering the objective of maximizing the system lifetime, the authors proposed a class of flow augmentation algorithms and a flow redirection algorithm, that balance the energy consumption among network nodes, proportionally with their energy reserves. Performance evaluations with simulations show that the system lifetime is improved in average by 60 percent compared to minimum transmitted energy routing.

29.5.2 Power-Aware Broadcast and Multicast Tree Construction

Broadcast and multicast are important functions for a routing protocol. An important issue in ad hoc wireless networks is the broadcast/multicast delivery structure. Traditional routing protocols for ad hoc wireless networks used flooding for broadcast. In mobile ad hoc wireless networks broadcast by flooding is usually costly and results in serious redundancy, contention, and collisions; this is referred to as the broadcast storm problem [18]. Alternative methods need to be designed for delivery of broadcast/multicast traffic considering the limitations in resource availability for a mobile node — energy, bandwidth, transceivers, etc.

Compared with wired networks where network links and their capacity are known *a priori*, in ad hoc wireless networks links depend on factors such as the distance between nodes, transmission power, and interference. The wireless channel is characterized by the broadcast property and when omnidirectional antennas are used, all nodes within the sender's transmission range receive the message. The total power required to reach a set of nodes is the maximum power necessary to reach each individual node, whereas in wired networks, the cost for sending a message to a set of nodes is the sum of the costs of the individual transmissions. Therefore it is important to design algorithms and protocols that reflect the node-based model of wireless communications, compared with the link-based model of wired networks.

These characteristics are pointed out in [19], where authors address the problem of constructing the minimum energy-source-based broadcast and multicast trees, by determining which nodes belong to the tree and the power used by these nodes for transmission. The problem is addressed for a stationary network, when bandwidth and transceiver resources are considered unlimited.

Every node, equipped with an omnidirectional antenna, can choose the transmission power from a range of values and has several transceivers, thus being able to support several multicast sessions simultaneously.

There is a tradeoff in choosing the transmission power at a node. Higher transmission power results in a higher connectivity, with more nodes being reached in one hop at the cost of higher interference and higher energy usage. The authors design three algorithms for the broadcast tree construction — Broadcast Incremental Power (BIP), which uses a node-based cost; Broadcast Least-Unicast-cost (BLU) and Broadcast Link-based MST (BLiMST) that both use a link-based cost. BIP starts the broadcast tree construction from the source node and at every step adds a new node such that the additional power cost is minimum, in a manner similar to Prim's algorithm. BLU is constructed by superpositioning a minimum-cost path from the source node to every other node. BLiMST uses the standard minimum spanning tree (MST) to build the tree, assuming a link cost between every two nodes. The transmission range of each node is then set to the distance to the farthest neighbor in the broadcast tree. A sweep operation can be applied to improve the broadcast trees performance by removing unnecessary transmissions. The solutions proposed for constructing the multicast tree use BIP or BLiMST by pruning transmissions not intended for multicast group members or use BLU by considering only the unicast paths to the desired destinations. Simulations show that BIP produces better results than the other two link-based algorithms.

Theoretical models and performance analysis of these protocols are further studied ratio of BIP is between 13/3 and 12 and the approximation ratio of MST is between 6 and 12. For the shortest path tree, the authors proved the approximation ratio to be at least $n/2$, where n is the number of nodes.

29.6 Energy-Aware Connected Network Topology

The topology of a multi-hop wireless network is defined by the set of communication links between node pairs and is used mainly by the routing mechanism. Topology depends both on uncontrollable parameters such as node mobility, interference, or weather, as well as on controllable parameters such as transmission power. Using a wrong topology may impact the network capacity and packet delay and may decrease robustness to node failure. A sparse topology may cause network partitioning and affects the end-to-end delay, whereas too dense a topology reduces bandwidth and spatial reuse with effect on aggregated network capacity. Because energy resources are limited, to extend the network lifetime, one important strategy is to control the node's transmit power to achieve the desired topology qualities.

The work in [20] addresses the problem of controlling the network topology by adjusting the transmission power at each node to minimize the maximum node transmission power in the entire network, subject to the network being connected or biconnected. Biconnectivity provides two different paths between every pair of nodes, improving fault tolerance and load balancing. Two optimal centralized algorithms are proposed for static networks — CONNECT and BICONN-AUGMENT. CONNECT builds the topology first as a MST and then applies a post processing phase. In the second phase, every node is considered and its power is decreased to the maximum possible extent, which does not disconnect the induced graph. BICONN-AUGMENT algorithm employs a greedy technique, which starts from a connected network and adds links until the resulting topology is biconnected. A similar post processing phase is used to ensure per node minimality. For mobile networks, the authors propose two distributed heuristics that adaptively adjust the transmission power to maintain the desired topology considering the effect of mobility. Both heuristics do not require any special control messages. In the first heuristic, Local Information No Topology (LINT), every node gets the current neighbor information from the routing protocol and attempts to keep the number of neighbors bounded by increasing or decreasing its operational power.

In the second heuristic, Local Information Link-State Topology (LILT), there are two mechanisms to control connectivity — the Neighbor Reduction Protocol (NRP) and the Neighbor Addition Protocol (NAP). When a node receives a routing update, it determines the status of the topology. If it is biconnected, no action is taken. If it is disconnected, the power is set to maximum. If it is connected, a timer is set proportional to the distance from the node to the first articulation point (a node whose removal partitions the network). After that time, if the network is not biconnected, the power is set to maximum. It is possible that the network overreacts by having more nodes increasing their power, but this effect will be regulated by the NRP. Experiments are performed with a system that uses a flat link-state routing mechanism. The radio used is the Utilicom Longranger, which has transmission power control and uses the CSMA channel access protocol. The performance metrics considered are throughput, end-to-end delay, maximum transmission power across all nodes (for static networks), and average transmission power (for static networks). The results show that for static networks, BICONN gives the best throughput and adapts well to a changing node density.

CONNECT suffers from congestion hot spots at low densities. BICONN uses significantly more power than CONNECT at lower densities, triggered by the isolated nodes. The conclusion is that at high densities it is better to use BICONN, whereas at lower densities the choice depends on which is more important — power or throughput. In mobile environments, LINT has a better throughput but a higher delay than LILT. This occurs mainly because the link-state database is often out of date, causing false alarms and power increases in LILT. Nevertheless, the experiments show that the performance of multi-hop wireless networks, in practice, can be substantially improved with topology control.

29.7 Conclusion

In this chapter, we reviewed recent research in the area of energy efficient mechanisms for wireless networks and WSNs. The techniques presented addressed energy efficiency at the data link layer and above. Extending the operation time of wireless networks is a particularly important issue in cases when terminals have size, weight, and cost constraints, as well as when battery replacement or recharging is not feasible. One emerging solution for extending the network lifetime involves harvesting energy from the environment and integrating information on renewable energy availability with node task scheduling. Other research results presented here can be broadly classified for using one of four main mechanisms for energy efficiency:

1. Power-aware link layer
2. Scheduling node activity
3. Power-aware routing
4. Topology control

Some of these approaches provide benefits for particular environments and can be implemented at various layers in the protocol stack. They provide a solid basis for future developments, but are not definitive answers to the problem of optimal energy usage in wireless networks.

References

1. Feeney, L.M. and Nilsson, M., Investigating the Energy Consumption of a Wireless Network Interface in an Ad Hoc Networking Environment, in *Proc. of IEEE InfoCom 2001*.
2. Jones, C.E. and Sivalingam, K.M., A Survey of Energy Efficient Network Protocols for Wireless Networks, *ACM/Baltzer Journal on Wireless Networks*, vol. 7, no. 4, pp. 343–358, 2001.
3. Lettieri, P., Schurgers, C., and Srivastava, M., Adaptive Link Layer Strategies for Energy Efficient Wireless Networking, *ACM/Baltzer Jourlal of Wireless Networks*, vol. 5, no. 5, pp. 339–355, 1999.
4. Swarts, F. and Ferreira, H.C., Markov Characterization of Digital Fading Mobile VHF Channels, *IEEE Transactions on Vehicular Technology*, vol. 43, no. 4, pp. 977–985, November 1994.
5. Wang, H.S. and Moayeri, N., Finite-State Markov channel — a Useful Model for Radio Communication Channels, *IEEE Transactions on Vehicular Technology*, vol. 44, no. 1, pp. 163–171, February 1995.
6. Kansal, A. and Srivastava, M.B., An Environmental Energy Harvesting Framework for Sensor Networks, *Proceedings of the 2003 International Symposium on Low Power Electronics and Design*, Seoul, Korea, 2003.
7. Havinga, P.J.M. and Smit, G., Energy-Efficient TDMA Medium Access Control Protocol Scheduling, *Proceedings Asian International Mobile Computing Conference (AMOC 2000)*, November 2000.

8. ANSI/IEEE, Standard 802.11, Wireless LAN Medium Access Control (MAC) and Physical Layer (PHY) Specifications, 1999.
9. Sivalingam, K.M. et al., Design and Analysis of Low-Power Access Protocols for Wireless and Mobile ATM Networks, *ACM/Baltzer Journal of Wireless Networks,* vol. 6, no. 1, 2000.
10. Singh, S. and Raghavendra, C.S., PAMAS-Power Aware Multi-Access protocol with Signaling for Ad Hoc Networks, *Computer Communications Review,* vol. 28, no. 3, July 1998.
11. Ye, W., Heidemann, J., and Estrin, D., An Energy-Efficient MAC Protocol for Wireless Sensor Networks, *Proc. InfoCom 2002,* June 2002.
12. Cardei, M. and Du, D.Z., Improving Wireless Sensor Network Lifetime through Power Aware Organization, *ACM Journal of Wireless Networks,* in press.
13. Koushanfar, A.F. et al., Low Power Coordination in Wireless Ad-Hoc Networks, *Proceedings of the 2003 International Symposium on Low Power Electronics and Design,* Seoul, Korea, 2003
14. Schurgers, C. et al., Topology Management for Sensor Networks: Exploiting Latency and Density, *Proceedings of the 3rd ACM International Symposium on Mobile Ad Hoc Networking and Computing,* Lausanne, Switzerland, 2002.
15. Xu, Y., Heidemann, J., and Estrin, D., Geography-Informed Energy Conservation for Ad Hoc Routing, *Proc. of MobiCom 2001,* Rome, pp. 70–84, July 2001.
16. Woo, M., Singh, S., and Raghavendra, C.S., Power-Aware Routing in Mobile Ad Hoc Networks, *Proceedings of ACM MobiCom,* 1998.
17. Chang, J.-H. and Tassiulas, L., Energy Conserving Routing in Wireless Ad-Hoc Networks, *Proceedings IEEE InfoCom,* 2000.
18. Tseng, Y.C. et al., The Broadcast Storm Problem in a Mobile Ad Hoc Network, *Journal of Wireless Networks,* vol. 8, pp. 153–167, 2002.
19. Wieselthier, J.E., Nguyen, G.D., and Ephremides, A., On the Construction of Energy-Efficient Broadcast and Multicast Trees in Wireless Networks, *Proceedings of IEEE InfoCom,* 2000.
20. Ramanathan, R. and Rosales-Hain, R., Topology Control of Multihop Wireless Networks Using Transmit Power Adjustment, *Proc. of IEEE InfoCom,* 2000.

Section VII
Power Management

Chapter 30
Power Management for Mobile Computers

*Thomas L. Martin, Daniel P. Siewiorek,
Asim Smailagic, and Jolin Warren*

Abstract

Energy consumption is a first class design constraint for mobile computers. Having a long battery life with adequate performance is an important criterion for users, as is having a system that is small and lightweight. The energy consumed by a system determines the size, weight, and volume of the battery it must have to achieve a given battery life. The battery is a significant fraction of the overall size and weight, so reducing energy consumption is a key step in making a system that is small and lightweight.

This chapter describes the important considerations in managing energy for mobile computing systems. There is no magic bullet for achieving energy efficiency. The whole system must be taken into account, including both the power sources (batteries) and power consumers (hardware and software). Every aspect of the system has an impact on the energy efficiency and the impact of a given subsystem is usually dependent on its interactions with other subsystems. The goals of energy-aware computing for battery-powered devices are to meet the performance expectations of the user and to maximize the amount of work the user can do before the battery becomes discharged [27]. For some energy-aware techniques, these goals are consistent with each other. For most, however, these goals are conflicting. Consequently, a designer must carefully balance performance and the work that can be completed in a battery life.

Because energy efficiency is a system problem, this chapter takes a hierarchical approach. After a brief introduction of the units of power and energy, the chapter begins with a description of batteries and their behavior. It then describes the impact of power supply efficiency on power consumption, followed by an overview of hardware and software power management techniques. The chapter concludes with general guidelines on the relationship between battery life and power consumption and the evaluation of power management options.

30.1 The Relationship between Power and Energy

One must be careful to distinguish between power and energy. These two terms are often (mistakenly) used interchangeably. Although they are closely related, they are distinctly different from one another. As will be explained below, a system with lower power consumption than another may actually consume more energy.

Power is the time derivative of energy,

$$P = \frac{dE}{dt} \tag{30.1}$$

Power is typically expressed in units of Watts (W) and energy in units of Joules (J). The relationship between Watts and Joules is:

$$1\,J = 1\,W \times 1\,s \tag{30.2}$$

There are a number of equivalent relationships that arise frequently in electronics, allowing the units of energy and power to be related to units of charge, voltage, and current:

$$1\,W = 1\,V \times 1\,A \tag{30.3}$$

$$1\,J = 1\,V \times 1\,A \times 1\,s = 1\,V \times 1\,C \tag{30.4}$$

From the relationship of power and energy, it can be shown that a low power consumption mobile computer may not be a low energy consumption mobile computer if the computational performance of the system is low (i.e., the computer takes a long time to execute a program). A typical problem faced by a designer implementing a power management system is choosing between a lower power state that has a longer execution time and a higher power state that has a shorter execution time. All other things being equal, it is better to pick the state with the lowest energy consumption.

30.2 Batteries

Batteries are the most common power source for mobile computers. For designers of mobile computers, the most important parameters of a battery are:

- Primary (disposable) or secondary (rechargeable)
- Nominal voltage
- Nominal charge capacity

Power Management for Mobile Computers

Table 30.1. Theoretical Properties of Common Rechargeable Battery Chemistries

Chemistry	Theoretical Voltage, V	Theoretical Charge Capacity, C/kg	Theoretical Energy Capacity, Wh/kg
Li-ion	3.9	100	390
NiCd	1.3	180	240
NiMH	1.3	210	270

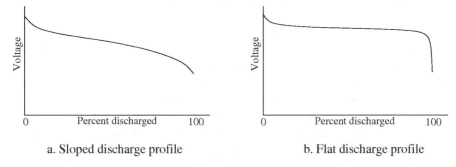

a. Sloped discharge profile b. Flat discharge profile

Figure 30.1. Typical Voltage Discharge Profiles

The materials used for the battery's electrodes determine these parameters. Lithium ion (Li-ion) batteries are the most widely used batteries in mobile computing today. Nickel metal hydride (NiMH) and nickel cadmium (NiCd) are used in some low-cost devices and disposable alkaline batteries are used in others. Table 30.1 lists the theoretical properties of several rechargeable battery families. For more information on battery chemistries please refer to [5].

The charge capacity of a battery is typically specified in Ampere-hours (Ah), where 1 Ah = 3600 C. The energy capacity of a battery is the nominal voltage multiplied by the nominal charge capacity and is typically given in Watt-hours (Wh). Li-ion batteries are currently the most common choice for mobile computing because of their superior energy capacity.

The properties listed in Table 30.1 are theoretical values that consider only the electrode materials. Actual values for voltage and capacity are lower in practice and are affected by the details of the batteries manufacture. Voltage and capacity are also affected by electrochemical phenomena during the charge/discharge cycle. Figure 30.1 shows typical voltage discharge profiles for batteries. Figure 30.1A shows a sloped discharge that is typical of Li-ion batteries and Figure 30.1B shows a flat discharge profile

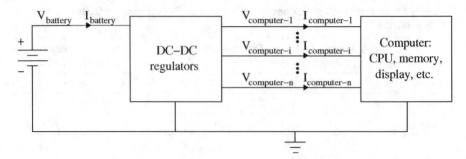

Figure 30.2. Typical Mobile Computer Power Supply

that is typical of NiCd and NiMH batteries. The sloped discharge profile is a nearly linear function of the percent of capacity for the middle portion of the discharge and the flat discharge profile is nearly constant for the middle part of the discharge. For a battery with a flat discharge profile, it is difficult to estimate its state of charge by the output voltage because the voltage will vary by only a few tens of millivolts over a large fraction of the battery life. The range of voltages from fully charged to fully discharged will also determine the input voltage parameters of the DC–DC regulators for a mobile computing system, as will be explained in Section 30.3. Finally, because of the battery's internal resistance and electrochemical polarization effects, the battery voltage is also an instantaneous function of the load current.

The energy capacity is a function of the load as well [5]. As the load power increases, the energy capacity decreases, as will be shown in Section 30.6. This energy is not lost; it is available at lower load power. Because energy capacity depends upon load power, two loads with the same average power may have different battery lives [27]. Consequently, power management decisions should account for changes in energy capacity to ensure that the battery life is improved.

30.3 Power Supplies

The efficiency of the DC-DC regulators of a mobile computing device has a major impact on its power consumption. An inefficient regulator can easily consume 20 to 50 percent of the total power of the system. Consequently, the regulator must be considered to ensure an energy efficient system.

A block diagram of the hardware in a typical mobile computer is shown in Figure 30.2. One or more DC–DC regulators sit between the battery and the computing hardware to provide regulated voltages from the battery voltage. The DC–DC regulators are typically referred to as the power supply. Modern computing hardware requires several different voltages, so Figure

Power Management for Mobile Computers

30.2 shows n voltages being supplied by the DC–DC regulators, $V_{computer\text{-}i}$. From a given voltage $V_{computer\text{-}i}$ the computer draws a current $I_{computer\text{-}i}$.

There are two types of DC–DC regulators commonly used in mobile computing — linear regulators and switching regulators. The features of concern are that linear regulators require fewer components to implement than switching regulators but typically are less efficient. Thus a designer is faced with a trade-off between component count (which tends to increase PCB area and cost) and power consumption.

The efficiency of a power supply is the ratio of its output power to its input power or P_{output}/P_{input}. If the power supply were 100 percent efficient then the amount of power drawn from the battery would be equal to the amount of power consumed by the computing hardware. In practice, however, power supply efficiencies are less than 100 percent, so that the amount of power drawn from the battery is greater than that consumed by the computing hardware. The difference between the two is consumed by the power supply. With several DC–DC regulators as shown in Figure 30.2, the power drawn from the battery is:

$$P_{battery} = \sum_{i=1}^{n} V_{computer-i} \times I_{computer-i} \times \frac{1}{E_i} \qquad (30.5)$$

where E_i is the efficiency of the ith regulator.

For linear regulators, the input current is equal to the output current, so that the efficiency becomes:

$$Efficiency = \frac{P_{output}}{P_{input}} = \frac{V_{output} \times I_{output}}{V_{input} \times I_{input}} = \frac{V_{output}}{V_{input}} \qquad (30.6)$$

One parameter of concern with a linear regulator is its dropout voltage, the minimum difference between input and output voltage for which the output voltage is still regulated, so the relationship between the input voltage and the output voltage is:

$$V_{input} \geq V_{output} + V_{dropout} \qquad (30.7)$$

Dropout voltages are typically a few hundred millivolts and increase as the load current increases. As an example, consider a linear regulator where the $V_{dropout}$ = 0.5 V, V_{input} = 3.5 V, and V_{output} = 2.5 V. Then using equation (30.5), the efficiency of the regulator would be:

$$Efficiency = \frac{V_{output}}{V_{input}} = \frac{2.5V}{3.5V} = 0.71$$

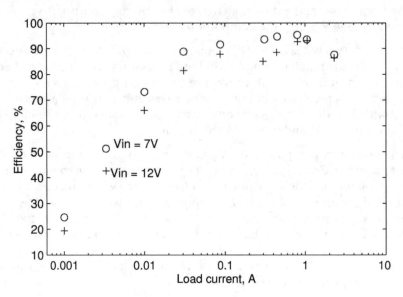

Figure 30.3. Efficiency versus Load for a Switching Regulator

Consequently, for every 1 W consumed by the computing hardware, the battery would need to supply 1 W/0.71 = 1.4 W. If V_{input} is lowered to the minimum voltage, $V_{output} + V_{dropout}$, the efficiency increases to 2.5V/3.0V = 0.83, and every 1 W consumed by the computing hardware draws 1 W/0.83 = 1.2 W from the battery.

For switching regulators, finding the efficiency is more complicated, as it is a function of the input voltage and the output current. Typically the efficiency for a switching regulator is found from a plot such as Figure 30.3, which shows the efficiency versus load current of a switching regulator. The x-axis is the load current on a logarithmic scale and the y-axis is the efficiency on a linear scale. Each curve on the diagram shows the efficiency for a particular input voltage. The general characteristics of this plot are typical of many switching regulators, with an efficiency of 90 to 95 percent over a couple of order of magnitudes of load currents, but much lower efficiency at lighter loads. There is also some dependence on the input voltage, but generally the change of efficiency is small over the range of battery voltage from the beginning to the end of discharge.

The system will usually be operating in the region where efficiency is high, such as greater than 90 percent. As an example, consider the regulator in Figure 30.3. Using the 7 V V_{in} curve, then a load power of 1 W at 5 V (which is the output voltage for this particular regulator) would be a load current of 0.2 A. The efficiency at this operating point is approximately 95

Power Management for Mobile Computers

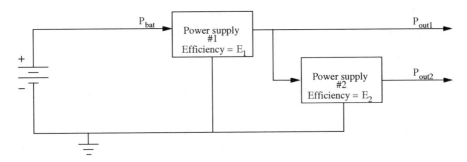

Figure 30.4. Cascaded Power Supplies, Usually a Bad Idea

percent, so the 1 W consumed by the computing hardware would require 1 W/0.95 = 1.05 W from the battery.

An implication of the effect of power supply efficiency is that it is generally a bad idea to cascade supplies in the manner shown in Figure 30.4. The input voltage to power supply 1 is the battery, but the input to power supply 2 is the output of power supply 1. This situation typically arises when the battery voltage range is outside the input voltage range of power supply 2. The efficiency of power supply 1 is E_1 and the efficiency of power supply 2 is E_2. P_{bat}, P_{out1}, and P_{out2} are the power drawn from the battery, supply 1 and supply 2, respectively. Given this, then:

$$P_{bat} = \frac{\left(P_{out1} + \frac{P_{out2}}{E_2}\right)}{E_1} = \frac{P_{out1}}{E_1} + \frac{P_{out2}}{(E_1 \times E_2)} \tag{30.8}$$

If P_{out2} dominates, then the overall efficiency will be close to $E_1 \times E_2$. Even if both supplies are efficient, then the overall efficiency will suffer. For example, if both supplies are 90 percent efficient and $P_{out2} \gg P_{out1}$, the overall efficiency will be about 81 percent.

Finally, a major difference between linear regulators and switching regulators with respect to estimating battery life is that a linear regulator is a constant current load and a switching regulator is a constant power load. For a given load, a linear regulator will draw the same current no matter what the input voltage is and thus the efficiency of the linear regulator will increase from the beginning of the discharge to the end of the discharge as the battery voltage decreases. A switching regulator, on the other hand, will have essentially constant efficiency during the discharge. For a given load, it will draw the same power from the battery over the course of a discharge. Thus the current drawn from the battery will increase as the battery voltage decreases during the discharge.

30.4 Hardware Power Management States

For the computing hardware, there are two main choices for saving power:

1. Putting inactive subsystems into low power idle, standby, and sleep modes
2. Decreasing the performance and power of a subsystem while it is active

The decision to put a subsystem into a low power mode or to decrease its performance is usually made in software, typically by the operating system (OS) with some assistance from applications. OS control of power management states will be discussed in Section 30.5.

In the early 1990s, a hardware-only method changing to a low power mode based upon inactivity timers was common. A hardware countdown timer was set to a given value every time an event occurred. If the timer reached zero then the subsystem in question was put into a low power mode. An example of this was to control the central processing unit (CPU) power management state using a timer that was reset every time a key on the keyboard was pressed. Inactivity timer-based power management was found to be ineffective because the timer control was too coarse and simple events did not adequately convey the state of the system. For example, CPU inactivity timers based upon keyboard presses fail when the user is performing a CPU intensive, noninteractive task, such as compiling a large program or updating a spreadsheet. In these cases, the CPU would be halted or set to run at a lower frequency because the user had not pressed a key for some time, with the result that the computation would take much longer to finish.

The decision to transition a subsystem into a low power state must be based upon the cost of making the transition to and from the low power state, the relative power consumption of the active state and the low power state, and the amount of time until the device will be needed again. The energy saved by being in the low power state must outweigh the energy cost of making the transition to and from the low power state.

The hard drive is a subsystem where the energy cost of transitioning from one power state to another is high. A typical hard drive has three main power states:

1. Active — the disk is spinning and either being read or written
2. Idle — the disk is spinning but not being read or written
3. Standby — the disk is not spinning

Going from standby to idle or active takes seconds while using roughly the same power as active mode. Table 30.2 shows the characteristics of the IBM® Microdrive [1] and the Hitachi Travelstar® 80GN [2].

Power Management for Mobile Computers

Table 30.2. IBM Microdrive [1] and Hitachi Travelstar 80GN [2] Characteristics

	IBM Microdrive	Hitachi Travelstar 80GN
Active power:		
Read	0.73 W	2.1 W
Write	0.83 W	2.2 W
Seek	0.66 W	2.3 W
Idle power	0.50 W	1.85 W
Standby power	0.07 W	0.25 W
Spin-up power	0.66 W	4.7 W
Standby to idle transition time	0.5 s–0.7 s	4.5 s–9.5 s

To save energy, the energy saved by being in standby mode must be greater than the energy required to spin up the drive (i.e., to go from standby to idle mode) or:

$$P_{idle} \times t_{idle} - P_{stdby} \times t_{stdby} \geq P_{spinup} \times t_{spinup} \qquad (30.9)$$

where P_x is the power consumed while in state x and t_x is the time spent in state x. In this case, t_{idle} is equal to t_{stdby}; this is the time between disk accesses. Then:

$$t_{idle} \geq \frac{P_{spinup} \times t_{spinup}}{P_{idle} - P_{stdby}} \qquad (30.10)$$

is the condition for saving energy by going into standby mode. In the case of the IBM Microdrive, this means that power can be saved by putting the drive in standby mode when the idle time is greater than 0.76 s to 1.07 s. For the Hitachi Travelstar, however, the idle time must be greater than 13.2 s to 27.9 s.

This example is oversimplified because it ignores the amount of power consumed in the rest of the system while the drive is spinning up from standby to idle mode. Section 30.7 will revisit this example by considering the effect on the whole system.

Besides low power idle, standby, and sleep states, some devices have the flexibility to remain active but at a reduced power and performance, trading off performance for power consumption. The goal of a power–performance trade-off should be to meet the performance expectations of the user while increasing the amount work that can be completed before the battery becomes exhausted.

The most widely studied power–performance trade-off to date has been the problem of CPU speed-setting, dynamically adjusting the CPU's clock frequency to meet the computational workload while saving power. Weiser et al. first pointed out that it is feasible to build systems that can use CPU speed-setting to reduce energy consumption [3]. The speed would be set by the OS to reduce the energy consumption of the CPU while still meeting the performance requirements of the user.

CPU speed-setting makes the following ideal assumptions:

- Performance is proportional to clock frequency f.
- Power is proportional to fCV^2, where C is capacitance and V is voltage.

Given these two assumptions, if V is held constant while f is changed, then the energy per operation is constant for all frequencies. Therefore, reducing the CPU frequency will not save energy if the voltage is held constant while the frequency is changed. For speed-setting to be useful, both the CPU voltage and frequency must be decreased [3]. (The voltage determines the maximum frequency, so it is not possible to reduce the voltage without also reducing the frequency.) If both are reduced by a factor of s, then the power of the CPU will change by a factor of s^3, the time per operation will change by a factor of $1/s$, and consequently the energy per operation will change by a factor of s^2. Thus running as slowly as possible will minimize the energy per operation and if the battery capacity is ideal, maximize the work that can be completed before the battery is exhausted.

In practice, however, these two assumptions do not hold true. First, performance may not be proportional to the CPU frequency. For example, in applications with a high cache miss ratio, performance can be limited by memory bandwidth rather than CPU frequency. Once an application becomes limited by memory bandwidth, increasing the CPU frequency will have little effect on the application performance and will increase the energy per operation. This nonideal speedup could be avoided by designing the memory subsystem such that speed of the memory is matched to the CPU's maximum frequency. However, using faster memories may increase system cost unacceptably, especially in price-sensitive systems such as personal digital assistants (PDAs) and notebook computers.

Second, the power consumed by the CPU in most systems is only one of many factors in the overall consumption, so that the system power is not proportional the CPU frequency. Figure 30.5 illustrates this, showing the power consumption of the Itsy handheld computer versus CPU frequency. Even if it were to have a variable-voltage CPU, the total system power would be nearly linear with respect to frequency. If the Intel® StrongARM™ SA-1100 were a variable-voltage CPU, only its 1.5V supply would be able to vary. The 3.3V supply powers the CPU's input/output (I/O) pins and must

Power Management for Mobile Computers

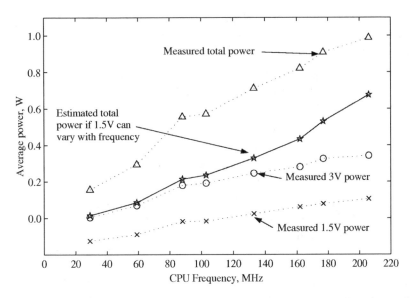

Figure 30.5. Measured System Power and Estimated Itsy System Power If 1.5V CPU Supply Could Scale with Frequency
(From T. Martin and D. Siewiorek, *IEEE Transactions on Very Large Scale Integrated Systems,* vol. 9, no. 1, Feb 2001, pp. 29–34. ©2001 IEEE. With permission.)

remain fixed. Figure 30.5 shows that if only the 1.5V CPU voltage were allowed to vary, then the power of the system would be dominated by the 3.3V supply, which supplies the CPU I/O pins and other subsystems (e.g., memory, display, and interface ports). The system power would be more linear than cubic. Consequently, even if system performance and battery capacity were ideal, the assumption that power is proportional to s^3 would be invalid and the goal of "running as slowly as possible" this assumption leads to would be invalid as well.

Finally, even if the assumption were to hold true, in order to save energy while maintaining the expected performance, the system must be able to predict how much performance will be necessary in the future. If the performance is not predicted correctly, then either the CPU frequency will be set too high, wasting energy, or it will be set too low, not meeting performance expectations. It turns out that predicting future performance in a general purpose mobile computer is a difficult problem [11].

30.5 Software

The work in the area of low power software includes power-savings and awareness in applications, compilation, and operating systems. The consensus is that the potential for power savings in software is greater than the potential for savings in hardware, but that the software savings are more difficult to achieve [6]. Hence, what is evident in the following paragraphs is that more work has been done in saving power at the lower levels of the software (i.e., at the OS and compilation levels).

The potential for algorithmic changes to save power is considered to be large. The choice of algorithm constrains the hardware and power savings achieved in hardware in two ways:

1. If one algorithm takes longer to execute than another, then the system will consume more energy because it is active longer.
2. If an algorithm blindly uses resources, those resources cannot be put into low power idle modes.

A trivial example of this is a routine that polls a device rather than using interrupts. Polling keeps the device active even when it does not have any work to do. Interrupts, on the other hand, make the device active only when there is an event. The related work at the algorithmic level thus looks to reduce performance and resource requirements. At this time, researchers are still trying to gain an understanding of how application software affects power consumption. The following is intended to be a representative list of the current work in low power software, but it is by no means exhaustive.

Most of the work in low power software has been in the OS, specifically utilizing power management features of the underlying hardware. The OS is in a better position to judge whether a device should be put into a low power mode than the device itself is, because it has a view of the overall state of the system. The OS is also in a better position to judge than an application because it can balance the needs of several applications. Furthermore, if the OS makes the power management decisions then the applications do not need to be modified.

The Advanced Configuration and Power Interface (ACPI) standard was developed to create a standard interface between hardware and the OS for power management on laptop computers [20]. ACPI defines several *global states* for the whole system (G0 — working, G1 — sleeping, G2 — soft off, and G3 — mechanical off), *device power states* for individual components (D0 — fully on, D1 — device dependent, D2 — device dependent, and D3 — off), and *processor power states* for the microprocessor (C0 — executing instructions, C1, C2, and C3 lower power states). Within the C0/D0 states, devices and processors may also have *performance states* (P0 to Pn, in which n < 16), where they are fully on but executing at reduced power and performance. ACPI allows the OS to interact with the components of the system to find their power states and parameters and then to use that information to make transitions between power states of the system and devices. Deciding which transitions to make and when to make them has been an area of much research, generally called *dynamic power management* [21].

Power Management for Mobile Computers

Another area of low power software is compilation for reduced power consumption. Tiwari et al. examine instruction-level power consumption for both the Intel x86 architecture and a Fujitsu RISC (reduced instruction set computing) architecture [8, 9]. The power consumption for each instruction is measured, as are the interinstruction and data contributions. The compiler is modified to take into account the power consumption as well as the timing cost of each instruction. The major energy savings come from reducing the time to complete a computation, not from using lower power instructions. The energy savings was nearly identical to the execution time savings in most cases. For example, their biggest time savings was 36 percent with a corresponding energy savings of 41 percent. The peak power for their test cases typically increases, but the cycle count for the programs is reduced by a much larger amount, so that the overall energy consumption is reduced. Simunic et al. examined compiler optimizations and source modifications for reducing power on the StrongARM SA-1100; their results also showed a high correlation between execution time and energy consumption [10]. Their most effective changes reduced execution time by 35 percent with an energy savings of 32 percent. Source code modifications that were particularly effective in reducing energy consumption included using table look-ups instead of switch statements, replacing character variables with integers, passing values to functions in registers instead of on the stack, and using unsigned integer division instead of signed. Although these techniques were effective for their particular processor and compiler, they may not be effective for all processors or compilers. However, exploring similar changes for given processors and compilers may lead to effective techniques for them.

Flinn and Satyanarayanan describe a tool called PowerScope for profiling the energy usage of applications for mobile computing and correlating dynamic power traces to procedure call traces [4]. PowerScope permits the measurement of energy consumption on a function-by-function basis. Given a tool like PowerScope, a software developer can find which functions are consuming the most energy and target them first for the compilation and source code modifications described in previous paragraphs.

30.6 Case Study

This section is a brief case study of system level power management on mobile systems [22]. Other interesting case studies include the IBM Linux® Wristwatch, an (ARM)-based (no pun intended) wearable computer [23, 24], and the Compaq Itsy, a StrongARM™-based personal digital assistant [25, 26].

Some experiments described in the next section use the Itsy platform. Itsy is a very small mobile device, with a clock-throttleable StrongARM processor with 8 megabytes (MB) of flash memory, 64 MB of random access

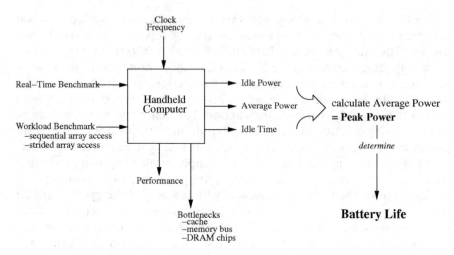

Figure 30.6. Inputs and Outputs of a Handheld/Wearable Computer System under Power Consumption Tests

memory, 3 serial ports, an infrared port, a speaker, a microphone, a gray-scale display, and a touch screen [26]. The processor clock can be varied from 30 megahertz (MHz) up to 200 MHz. At the highest clock frequency, the processor is rated at approximately 90 Dhrystone MIPS (millions of instructions per second). The Itsy runs the Linux OS, weighs about 120 grams, and consumes less than 1 W. We have ported the Sphinx speech recognition system to the Itsy as well as GeoPlex™ Peer software.

The inputs shown in Figure 30.6 are two benchmarks and the CPU frequency. These inputs affect the performance and whether or not bottlenecks are experienced. More importantly, the inputs affect the system's idle time, average power consumption, and idle power consumption. These three output parameters are used to calculate the system's active or peak power, which is used to determine battery life.

30.6.1 Memory Bottleneck and Dynamic CPU Speed-Setting

As described in Section 30.4, CPU speed-setting is an effective method for reducing power consumption. The typical assumption is that performance is proportional to frequency [3, 12]. Such an assumption ignores the memory subsystem. For example, in applications with a high cache miss ratio, performance can be limited by memory bandwidth rather than CPU frequency. Once an application becomes limited by memory bandwidth, increasing the CPU frequency will have little effect on the application performance and will increase the energy per operation. This nonideal speedup could be avoided by designing the memory subsystem such that speed of the memory is matched to the CPU's maximum frequency. However, using faster

Power Management for Mobile Computers

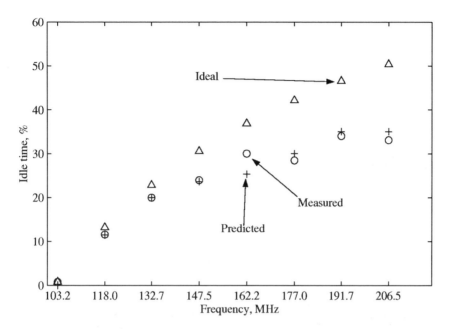

Figure 30.7. Ideal, Measured, and Predicted Idle Time as a Function of CPU Frequency
(From T. Martin et al., *ACM Transactions on Embedded Computing Systems,* vol. 2, no. 3, August 2003, pp. 255–276. © 2003, ACM. With permission.)

memories may increase system cost unacceptably, especially in price-sensitive systems such as PDAs and notebook computers. In addition, it may be desirable to implement a CPU speed-setting scheme on an existing hardware platform that cannot be redesigned. Finally, acknowledging that the memory subsystem can have a large effect on CPU speed-setting will show system designers that it is an area of prime concern.

The code chosen for the experiments shown here was the Sox audio player. The justification for this choice is twofold. First, we expect an audio player to be one of the typical applications for a mobile computer. Second, we wanted code that adequately exercised the memory hierarchy and OS [13, 14]. Measuring the energy of an application that fits into the cache or that uses no OS resources would not be representative of a mobile computer capable of running a variety of applications.

Figure 30.7 shows the ideal, measured, and predicted idle time versus CPU frequency for an audio playback application, which was measured to spend 49 percent of time accessing main memory at the base frequency of 103 MHz. At 103 MHz, the processor has almost no idle time; frequencies below 103 MHz are not shown because the application cannot be run in real-time.

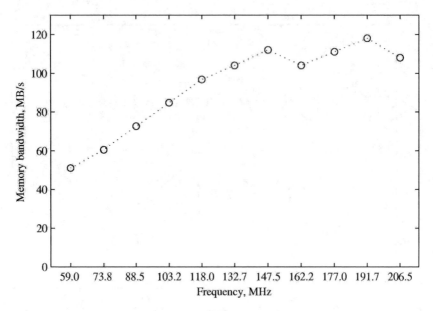

Figure 30.8. Itsy Main Memory Bandwidth versus CPU Frequency
(From T. Martin et al., *ACM Transactions on Embedded Computing Systems,* vol. 2, no. 3, August 2003, pp. 255–276. © 2003, ACM. With permission.)

The results show that assuming that performance scales with CPU frequency (labeled *ideal* in Figure 30.7) will overestimate the idle time by a factor of 1.5 at the maximum CPU frequency. But by assuming that main memory bandwidth scales at a different rate than the CPU frequency, as shown in Figure 30.8, and by using Amdahl's Law [15], the idle time can be accurately predicted (labeled *predicted* in Figure 30.7). The worst-case error in the predicted value was less than 5 percent, compared to an error of nearly 20 percent using ideal assumptions. Consequently, a realistic CPU speed-setting policy must account for nonideal memory behavior.

Another reason it is necessary to accurately predict the idle time is for estimating the active power of a system from a measurement of its average power. In [27], it was shown that due to nonideal battery behavior, the peak power consumed by a system, the power consumed while it is active, is a better indicator of the battery's energy capacity than the average power.

It can be shown that the predicted active power of a system, $P_{activep}$, is given by:

$$P_{activep} = \frac{P_{ave} - (\%t_{idle} \times P_{idle})}{\%t_{active}} \qquad (30.11)$$

where P_{ave} is the measured average power, t_{idle} is the idle time, and P_{idle} is the power while the system is idle.

Power Management for Mobile Computers

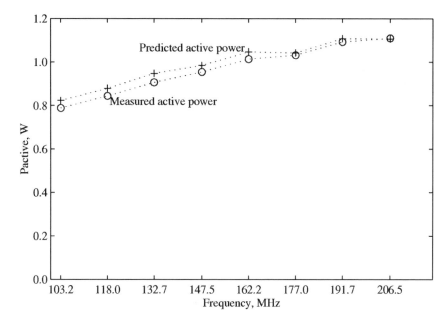

Figure 30.9. Comparison of Predicted and Actual Power versus CPU Frequency
(From T. Martin et al., *ACM Transactions on Embedded Computing Systems*, vol. 2, no. 3, August 2003, pp. 255–276. © 2003, ACM. With permission.)

Figure 30.9 shows the measured (P_{active}) and predicted active power ($P_{activep}$) versus the CPU frequency for the Itsy as it executes Sox. The predicted active power is within 4.5 percent in all cases. Thus by accounting for the main memory behavior, it is possible to accurately predict the active power by measuring the idle power and the average power at each CPU frequency.

30.6.2 Dependence of Battery Capacity on Load Power

In addition to reducing the energy usage of the power consumers, it is also necessary to consider the energy available from the power source, typically a battery. As described in Section 30.2, if a battery were ideal, it would provide the same amount of energy no matter how quickly the battery is discharged. In reality, however, the amount of energy the battery delivers depends on the power of the load. As the power of the load increases, the battery's energy capacity decreases.

For a typical battery, if the power consumption of the load varies, the energy capacity of the battery is determined by the peak power consumption sustained over roughly 1 second. If two different loads have the same average power but different dynamic power, then they may have a much different battery life because the energy capacity will vary, as illustrated by Figure 30.10, created using a simulation model of a Li-ion battery. The solid

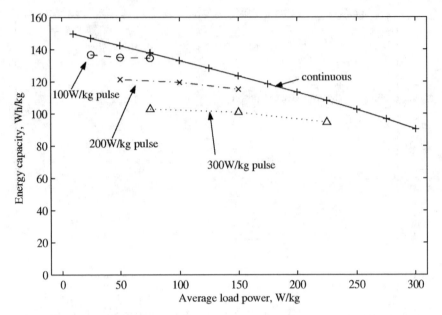

Figure 30.10. Battery Capacity versus Load Power for a Continuous Load and for Intermittent Loads of Different Peak Powers and Duty Cycles.
(From T. Martin and D. Siewiorek, *IEEE Transactions on Very Large Scale Integrated Systems,* vol. 9, no. 1., February 2001, pp. 29–34. © 2001, IEEE.)

line indicates the energy capacity of the battery for a load of a constant value and the triangles connected by dashed lines represent pulsed loads of 25 percent, 50 percent, and 75 percent duty cycles with a peak value of 300 W/kg and a low value of 0 W/kg. (The simulator used to generate this result uses the specific energy (energy per kilogram [kg]) and specific power (power per kilogram) in its calculations.) The solid line shows the decrease in the energy capacity as the load power increases, approximately 40 percent. The triangles show that the peak power is a dominant factor in the energy capacity delivered, as the capacity for the pulsed loads with a 300 W/kg peak have nearly the same capacity as the 300 W/kg continuous load. In addition, the battery life for the pulsed loads will be less than that of a continuous load of the same average power. Thus using the nominal energy capacity for a battery to estimate battery life will result in an overestimate if the peak power is large. If there is a large peak power, a better estimate can be obtained by using the energy capacity at that peak power.

Figure 30.11 illustrates three different methods for predicting the battery life as the CPU frequency is varied for the Sox application described previously. The first method is to measure the average power of the system at each frequency and then to divide the battery rated capacity by the average power. This is illustrated in the figure by the points labeled measured average

Power Management for Mobile Computers

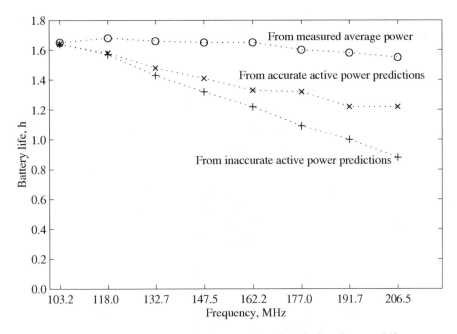

Figure 30.11. Comparison of Three Methods for Predicting Battery Life over a Range of Frequencies
(From T. Martin et al., *ACM Transactions on Embedded Computing Systems,* vol. 2, no. 3, August 2003, pp. 255–276. © 2003, ACM. With permission.)

power. This method ignores the variation in the energy capacity with the load power and leads to an overestimate of the battery life. The second method illustrated in Figure 30.11 is to take into account the variation in energy capacity but to predict the active power from the average power without accounting for the affect nonideal memory behavior has on peak power, as described earlier in this section. In this case, peak power will be overestimated, leading to an underestimate of the battery life, shown in the figure as by the points labeled inaccurate active power predictions. The third, correct method is to account for both the battery capacity and the memory behavior, as illustrated by the points labeled accurate active power predictions.

30.7 General Guidelines

This section provides some general guidelines for power management of mobile computing systems.

30.7.1 An "Amdahl's Law" for Power Management

Assuming that a battery's energy capacity is constant, then power management has an equivalent to Amdahl's Law for computer performance [19].

The increase in battery life due to a given power management state is constrained by how much power is saved in that state and what fraction of time is spent in that state. The battery life is given by:

$$BatteryLife = \frac{BatteryEnergy}{(1 - F_{idle}) \times P_{active} + F_{idle} \times P_{idle}} \quad (30.12)$$

where F_{idle} is the fraction of time that can be spent in a low power state, P_{idle} is the power consumption of the system while in the low power state, and P_{active} is the power consumption of the system while active. Thus a large improvement in power consumption in some state will not translate to a large improvement in battery life if that state is used infrequently. Similarly, because both P_{idle} and P_{active} are the power of the system, if the power savings is achieved in a subsystem that is a small portion of the overall system power, then the battery life increases will be limited.

30.7.2 Evaluating Power Management Options

Lorch and Smith provide a number of principles for evaluating modifications to save power [7]. These principles are not specific to any particular modification, but instead are generally applicable.

First, a modification to some subsystem should be evaluated based upon its impact on the power consumption of the system, not on its impact on the power consumption of that particular subsystem. This is related to the Amdahl's Law for power management. If a component consumes only a small fraction of the overall system power, then even a total reduction of its power consumption will only decrease the overall system power by a small amount.

Second, maximizing the battery life does not necessarily mean maximizing the amount of work that can be performed in the battery life [16]. The power savings of a modification must be balanced against the performance impact. As an example, the Navigator 1 wearable computer at CMU could save power by slowing its CPU [19]. Although the battery life was increased by a factor of 1.3, the performance was decreased by a factor of 0.32. Thus the work completed per battery life was 1.3 × 0.32 = 0.42 of the work that could be completed running at maximum CPU speed.

Third, the effects of the modification on the rest of the system should be considered. For example, consider the hard drive example of Section 30.4. During the spin-up time, instead of accounting only for the power of the hard drive itself, the proper approach is to account for the power of the whole system while it waits for the hard drive to spin up. Thus if the IBM Microdrive is in a system where the system power is 10 W while the drive is spinning up, then the time that the drive should be idle before putting it in standby saves power becomes 11.5 s to 16.2 s instead of 0.76 s to 1.07 s.

Finally, the modifications should be evaluated against the typical behavior of the system, not the worst case. There is often a large difference between typical and maximum power consumption and comparing a modification against the maximum power consumption may be misleading.

30.8 Conclusions

Power management is a system problem. All parts of the system must be taken into account, from batteries and power supplies to hardware components, OS, and applications. Greedy approaches to energy efficiency that consider subsystems in isolation will not be optimal. Second-order effects, such as variations in battery capacity and limited main memory bandwidth, adversely affect the ability to save power and must be factored into. A system-level approach, one that considers power consumers and power sources, is the proper method for examining energy efficiency in battery-powered computing systems.

References

1. IBM, Hard disk drive specifications IBM Microdrive with CF+ Type II interface. December 2001.
2. Hitachi, Travelstar 80GN OEM Specification v1.3, June 2003.
3. M. Weiser, B. Welch, A. Demers, and S. Shenker, Scheduling for Reduced CPU Energy, *Proceedings of the 1st USENIX Symposium on Operating Systems Design and Implementation,* November 1994, pp. 13–23.
4. J. Flinn and M. Satyanarayanan, PowerScope: A Tool for Profiling the Energy Usage of Mobile Applications, *Proceedings of the 2nd IEEE Workshop on Mobile Computing Systems and Applications,* New Orleans, February 25–26, 1999, pp. 2–10.
5. D. Linden and T. Reddy, *Handbook of Batteries, Third Edition,* New York: McGraw-Hill, 2002.
6. S. Kiaei and S. Devadas, Panel session organizers, Which Has Greater Potential Power Impact: High-Level Design and Algorithms or Innovative Low Power Technology?, *Proceedings of the 1996 International Symposium on Low Power Electronics and Design,* August 1996, P. 175.
7. J. Lorch and A. Smith, Software Strategies for Portable Computer Energy Management, *IEEE Personal Communications Magazine,* vol. 5, no. 3, June 1998, pp. 60–73.
8. V. Tiwari, S. Malik, and A. Wolfe, Compilation Techniques for Low Energy: an Overview, *Proceedings of the 1994 Symposium on Low Power Electronics,* October 1994, pp. 38–39.
9. V. Tiwari, S. Malik, and A. Wolfe, Power Analysis of Embedded Software: a First Step Toward Software Power Minimization, *IEEE Transactions on Very Large Scale Integration Systems,* vol. 2, no. 4, December 1994, pp. 437–445.
10. T. Simunic, L. Benini, and G. de Micheli, Energy-Efficient Design of Battery-Powered Embedded Systems, *Proceedings of the 1999 International Symposium on Low Power Electronics and Design,* August 16–17, 1999, pp. 212–217.
11. D. Grunwald, P. Levis, K. Farkas, C. Morrey, and M. Neufeld, Policies for Dynamic Clock Scheduling, *Proceedings of OSDI 2000,* October 2000, pp. 73–86.
12. K. Govil, E. Chan, and H. Wasserman, Comparing Algorithms for Dynamic Speed-Setting of a Low Power CPU, *Proceedings of the 1st ACM International Conference on Mobile Computing and Networking,* 1995, pp. 13–25.
13. A. Agarwal, *Analysis of Cache Performance for Operating Systems and Multiprogramming.* Boston: Kluwer Academic Publishers, 1989.

14. R. Uhlig, D. Nagle, T. Mudge, S. Sechrest, and J. Emer, Instruction Fetching: Coping with Code Bloat, *Proceedings of the 22nd International Symposium on Computer Architecture,* July 1995, pp. 345–356.
15. J. Hennesey and D. Patterson, *Computer Architecture: A Quantitative Approach,* 2nd ed., San Mateo, CA: Morgan Kaufmann Publishers, Inc., 1996.
16. T. Martin and D. Siewiorek, A Power Metric for Mobile Systems, *Proceedings of the 1996 International Symposium on Low Power Electronics and Design,* August 1996, pp. 37–42.
17. J. Warren, Interaction between Architecture Performance and Power Consumption in Mobile Systems, Master's Project Report, Electrical and Computer Engineering Department, Carnegie Mellon University, 2000.
18. M. Ettus, Power Consumption and Latency as a Function of Path Loss and Routing in Multihop SS-CDMA Wireless Networks, Master's Project Report, Electrical and Computer Engineering Department, Carnegie Mellon University, 1999.
19. T. Martin, Evaluation and Reduction of Power Consumption in the Navigator Wearable Computer, Master's Project Report, Electrical and Computer Engineering Department, Carnegie Mellon University, 1994.
20. Intel Corporation, Instantly Available Technology — ACPI, http://developer.intel.com/technology/iapc/acpi.
21. E. Macii, Dynamic Power Management of Electronic Systems, *IEEE Design and Test of Computers,* vol. 18, no. 2, March–April 2001, pp. 6–9.
22. T. Martin, D. Siewiorek, A. Smailagic, M. Bosworth, M. Ettus, and J. Warren, A Case Study of a System-Level Approach to Power-Aware Computing, *ACM Transactions on Embedded Computing Systems,* vol. 2, no. 3, August 2003, pp. 255–276.
23. C. Narayanaswami, N. Kamijoh, M. Raghunath, T. Inoue, T. Cipolla, J. Sanford, E. Schlig, S. Venkiteswaran, D. Guniguntala, V. Kulkarni, and K. Yamazaki, IBM's Linux Watch, the Challenge of Miniaturization, *IEEE Computer,* vol. 35, no. 1, January 2002, pp. 33–41.
24. N. Kamijoh, T. Inoue, C.M. Olsen, M.T. Raghunath, and C. Narayanaswami, Energy Trade-Offs in the IBM Wristwatch Computer, *Proceedings of the 5th International Symposium on Wearable Computers,* October 8–9, 2001, pp. 133–140.
25. M.A. Viredaz and D.A. Wallach, Power Evaluation of a Handheld Computer, *IEEE Micro,* vol. 23, no. 1, January–February 2003, pp. 66–74.
26. W.R. Hamburgen, D.A. Wallach, M.A. Viredaz, L.S. Brakmo, C.A. Waldspurger, J.F. Bartlett, T. Mann, and K. Farkas, Itsy: Stretching the Bounds of Mobile Computing, *IEEE Computer,* vol. 34, no. 4, April 2001, pp. 28–36.
27. T. Martin and D. Siewiorek, Nonideal Battery Properties and Their Impact on Software Design for Wearable Computers, *IEEE Transactions on Computers,* August 2003, pp. 79–84.
28. T. Martin and D. Siewiorek, Nonideal Battery and Main Memory Effects on CPU Speed-Setting for Low Power, *IEEE Transactions on Very Large Scale Integrated Systems,* vol. 9, no. 1, February 2001, pp. 29–34.

Chapter 31
Power Awareness and Management Techniques

Ahmed Abukmail and Abdelsalam (Sumi) Helal

31.1 Introduction

When talking about power sources for mobile devices, we must come to the realization that the power source is also mobile. Therefore, it is a finite power source that has a limited lifespan. This power source is the battery of the mobile device and not the A/C adapter, because the A/C adapter provides an infinite power source, but it is plugged in the wall, which makes the device immobile.

31.1.1 Motivation

This limitation of the battery of a mobile device creates the need for extending its lifetime, simply because of the fact that a mobile device without power is useless irrespective of how many capabilities it could have or how nice and slick it looks. Additionally, mobile devices that do not regulate power consumption seem to be thermally ineffective. In other words, the machines tend to get hotter with time, which ultimately results in somewhat of a hazardous environment. Notice how hot a laptop computer gets when you actually have it directly on your lap. Some commercial laptops have actually caused small fires.

Therefore, the need for developing techniques, mechanisms, and hardware components to extend the lifetime of a mobile device's battery, and ultimately result in better thermal management, whether it is a laptop computer, a personal digital assistant (PDA), or a cellular telephone, has come to the forefront of academic research as well as becoming an industry concern. Additionally, with the advancement and wide use of wireless networks, network connectivity will be a foregone conclusion. Therefore it will be a waste if a network is available but the mobile device has no juice left in it to connect to that network and be able to compute and communicate with the rest of the world. That means the idea of computing anywhere, anytime will have a lesser meaning and will be less useful. From that we can conclude that the availability power/energy in a mobile device complements the availability of a wireless network to which the device can connect.

31.1.2 Taxonomy of Research and Industry Solutions

In [1], it has been determined that within a mobile device there are several components that consume most of the power within the device such as the processor, the memory, network adapter, the screen, and the video card, as well as hard drives. Additionally, it was concluded that there are three levels of opportunity to reduce power consumption in a computer system. These levels are the main layers of any computer system:

1. The architectural and hardware layer
2. The operating system layer, which includes the communication sublayer
3. The applications layer, including compilers

As far as hardware solutions are concerned, the industry targeted battery systems by the introduction of smart batteries, which became supported by operating system power management subcomponents such as advance power management (APM), and advanced configuration and power interface (ACPI). The industry also targeted the processor with respect to power management: processors like Intel® PCA (Personal Client Architecture) processors, ARM's family of processors, and Transmeta's Crusoe™ line of processors. At the hardware level, reducing the capacitance load, as well as frequency and voltage scaling, aid in reducing the power consumed.

In addition to hardware, we can move up to the next layer in a computer system, which is the operating system, in order to complement managing/reducing power at the hardware layer. Solutions include energy management subsystems such as APM and ACPI. Additional methods to reduce power at the operating system include memory and input/output (I/O) management, communication techniques, and process scheduling.

The next layer in which we can mange power is the application layer. In this layer, power management is handled specific to the application. In other words, developers are aware of the power consumed by their applications and they develop accordingly. Another way to reduce power at this level is to make the developer unaware of power requirements and that can be done at the compiler level. So, compilers will generate code that is optimized for power reduction.

The rest of this chapter will be organized as follows:

- Section 31.2 discusses some of the prominent industry and research contributions to reducing power at the hardware layer.
- Section 31.3 discusses those contributions made to the operating system layer.
- Section 31.4 discusses reducing power at the application layer.
- Section 31.5 introduces some of the tools used for estimating and measuring power consumption.
- Section 31.6 concludes the chapter.

31.2 Hardware and Architecture Techniques

The first thing that research and development attacks, when discovering that devices consume lots of power, is the hardware itself or the architecture of the computer system. The results were better batteries, better processors, as well as software and architectural solutions to reduce the power consumed by the complementary metal-oxide semiconductor (CMOS) circuitry components.

31.2.1 Smart Batteries

31.2.1.1 Battery Basics. Every battery has two terminals — positive and negative. Electrons form on the negative terminal. Once a connection is made between the positive and the negative terminal (battery is installed), electrons will start to flow from the negative terminal to the positive terminal. Inside of the battery certain chemical reactions occur to produce electrons. This reaction only occurs when the electrons are traveling from one terminal to the other: meaning, once you install the battery. When talking about rechargeable batteries (which are the batteries that we are concerned about in mobile devices), these devices have suffered from the problem of memory effect. The memory effect is when the battery does not get recharged fully due to the fact that it has not drained completely prior to recharging, however the lithium ion batteries have solved this problem. However we still have the issue of once the battery is built, and as long as it is connected, the reaction is going to happen, draining the battery power and eventually rendering it useless until it gets either recharged or replaced.

31.2.1.2 Intelligent Power Drainage. The shortcoming of regular batteries is that, as mentioned, they do not intelligently drain power. Once connected to a device, the power drainage is going to happen no matter what. This motivation led to the development of the smart battery system (SBS) [2]. This battery is different than the other nonsmart ones in that it provides information and system status to the host machine through the system management bus (SMBus), as well as it has its own recharger. This information will and can be used by the system to handle power management within the mobile device. Operating systems use standards and specifications such as APM [3] and ACPI [4]. Currently, most laptop computers are shipping with smart batteries.

There are a number of companies in the industry that produce products for SBSs. These companies include, but are not limited to, Samsung, Motorola, and Hitachi as battery vendors. Semiconductor vendors that produce chips for smart batteries include companies like Powersmart, Acer, and Adaptec.

31.2.2 Energy-Aware Processors

Another way to tackle power consumption in a computer system is to produce processors that do not drain as much power. Numerous microprocessor companies in the industry have products that are power-efficient as well as thermally efficient. Intel's Xscale® architecture is designed to optimize low power consumption [5]. Xscale represents an integral building block of Intel's PCA architecture [6]. Intel also utilized the SpeedStep® technology in building the Mobile Pentium® III processor. SpeedStep provides the advantage of the processor dropping to a lower frequency and lower voltage when powered by battery, which results in conserving battery life while maintaining a high level of performance [7]. Transmeta Corporation has the Crusoe family of processors [8]. The Crusoe is an x86 compatible family of processors that is lightweight, high-performance as well as much less power consuming than other compatible mobile processors (70 percent less). In addition, the ARM family of processors is a widely popular family of processors and is geared toward reduction in power consumption while maintaining a high level of performance [9]. In the next section, a more theoretical approach to reducing power in processors is introduced.

31.2.3 Reducing Power through CMOS Circuitry Components

31.2.3.1 The Power Consumption Equation.
Power can be measured by a simple equation: $P = V \times I$, where V is the voltage and I is the current in amps. Additionally, the electric current equation is $I = Q \div t$, where Q is the electrical charge and t is the time in seconds, which is the reciprocal of the frequency, which means that $I = Q \times f$. Because $Q = V \times C$, where V is the voltage and C is the capacitance load in farads, the final power equation will be:

$$P = CfV^2$$

where C is the capacitance load, f is the clock frequency, and V is the voltage. Therefore, reducing any of the variables of this equation results in power reduction. This can only be done at the architectural level of a computer system.

31.2.3.2 Voltage and Frequency Scaling.
The equation given in the previous section suggests that if the voltage or the frequency are reduced, then the power consumption will be reduced due to the direct relationship between these values. Therefore, researchers targeted voltage and frequency scaling techniques to achieve that result. Smit and Havinga in [1] gave the argument that a reduction in voltage will mean reduction in performance, therefore additional hardware will be needed to balance that difference to keep the same throughput.

An approach for dynamic frequency and voltage scaling was given in [10]. In this work, the claim, via simulated results, was that they achieved power dissipation similar to that of the Crusoe TM5400 by Transmeta. Their solution, at compile-time, assigns a clock frequency, and voltage levels for input loop executions. The main idea for their work is to assign deadlines to tasks and use these deadlines to allow for voltage and frequency scaling. Simply, if a task finishes at time 0.5 t, and its deadline is at time t, then it will not hurt the execution if the task takes t time to complete. So, at compile-time they identify these tasks and generate code that will allow for lower voltage and lower frequency to be used in the execution.

The work in [10] supports the argument that additional hardware will be needed to achieve voltage and frequency scaling, which was mentioned in [1].

31.2.3.3 Capacitance Load Reduction.
Reduction in the capacitance load is another method to reduce the power consumed by a computer, again due to the aforementioned equation. Two main mechanisms can be used to handle this goal. As mentioned in [1], memory capacitance is greater than that of the processor. Therefore reducing memory operations and keeping as much work on the processor and in registers is extremely important. This kind of work has been done in [11] via reducing the off-chip activity. The solution is basically to use a minimum cost network flow to find minimum energy solutions by simultaneous partitioning of variable into memory and possibly allocate them into registers. This work resulted in 1.4 to 2.5 times energy improvement.

Another mechanism to reduce this variable in the power equation is to reduce the logic state transition when handling memory addresses. In any processor, the program counters tend to change multiple bits when accessing the next instruction or data from memory locations and that depends on the size of the instruction and if that instruction was a jump instruction or a sequential instruction. That transition increases the capacitance load. To minimize the transition, a Gray Code solution can be used to facilitate memory addressing. This solution along with a compiler solution was done in [12]. Sequential Gray codes vary only in one bit at a time, therefore, aside from jump instructions, there will only be a variation of one bit, which is the minimum that can be achieved when accessing memory addresses.

31.2.4 Power Reduction through Architectural Design

Architectural design can always be one of the most attractive solutions to the power problem and that is due to the fact that you can always leave the architecture the same but add to it another component that will aid in reducing power consumption. In [13], Bellas et al. introduced a compilation methodology and supported it with an addition of a second level cache between the processor and the first level cache. This cache is a smaller cache and is much simpler to use than that of the first level cache.

Another architectural solution was introduced in [14]. They introduced direct addressed caches that will allow the software to access any cached data without the hardware doing any tag check. That means that the power consumed by the tag check will be saved. In this work, they also introduced a compiler solution that will complement this hardware solution.

31.3 Operating Systems and Communication Techniques

Along with the architecture comes the operating system that will run on this architecture. Therefore, we can optimize this component to help in consuming less power. Most operating systems now come with energy management solutions that are configurable by the user. However, additional methodologies have been researched to handle such issues.

31.3.1 Energy Management Solutions

Operating systems now implement certain, user-configurable, energy management solutions. APM is one of these solutions [3]. In APM, a user can define the power scheme that he or she desires to use. Some of the variables that a user can control are the screen, the hard drive, and suspend/sleep modes. The user can instruct the screen and the hard drive to stop/sleep when the system has been idle for a while. The user can even specify if the system is idle or, in the case of a laptop, if the lid is closed, the whole computer can save its state on the hard drive and then go to sleep. When it wakes up it will restore itself back to the previous state.

Intel and other partners contributed to the ACPI specification mentioned before in [4]. Intel discussed some shortcomings of APM such as not considering future capabilities of personal computers (PCs) that include their communication capabilities because APM drops the communication link. Therefore, they came up with their own solution in Instantly Available PC, which uses the ACPI specification to manage power while keeping the system online via either the modem or a LAN card.

31.3.2 Memory and I/O Management

We mentioned before that memory access is expensive when it comes to power consumption. The desire here is to try to have data access on the chip via registers or cache as much as possible. In addition to the memory, accessing secondary storage is even more expensive, so we would like to access the disk as little as possible. In [15], a method was introduced to make fewer incorrect file predictions as a way to save energy. They accomplished this task by observing the probability of file access as well as the repeated history of file access to avoid unnecessary access to each file. They claim that programs access more or less the same files in the same order when they execute every time, giving themselves a good knowledge for determining what they call program-specific last-successors for each file.

31.3.3 Communication Techniques

Power-saving communication techniques are increasingly important in this day and age. We can claim with a great deal of certainty that everyone who has a computer now has access to the Internet and is using e-mail and the Web in one form or another. Therefore, if we find ways to save power on communication, we will definitely prolong the battery life of any communicating mobile device.

Any communication system has three states — sending, receiving, and idle. Switching between these three states is power consuming [1]. Therefore, reducing the transition from one state to another will definitely reduce power loss. Therefore, instead of serving the communication request of sending, we can buffer the data to send in large chunks. The same thing can be done on the receiving end. We can only accept data that is of a certain size optimized for our power needs. Otherwise, our system needs to go into an idle mode. In addition, with the availability of solutions that conform the ACPI specification, there is no need any more to modify network protocols to support full connectivity while the system is in sleep mode.

Kravets and Krishnan in [16] have developed a method for managing power in mobile communication. They introduce a method for managing the communication's device on a host through suspend and resume operations. Simply, they claim that a communication device will continue to draw power unless it is suspended. Their solution takes into account the need for the host to know the communications patterns, so when it is suspended there is no communication happening, otherwise problems such as buffer overflow will occur on the host and other hosts. So, the functionality of the suspend and resume is adaptable such that it will avoid such overflow.

Additionally, in [17], Loy and Helal have introduced an application-based solution that can easily be used in developing energy management techniques in the operating system. A discussion of this work is given in Section 31.4.2.

31.3.4 Scheduling

Energy-aware scheduling in the system requires *a priori* knowledge of the activity inside that system [18]. Monitors can be incorporated within the operating system to handle this. In [18], Bellosa introduced an event-driven energy accounting as a way to manage power in an operating system. This is done by online analysis of energy-usage patterns that are fed back into a scheduler to control the CPU's clock speed. The work here was done with real hardware as opposed to simulation.

31.4 Software Application Techniques

Software solutions, as a high-level power management technique, are becoming increasingly important for power management. In [19], Ellis made a case for the need for higher level power management. Simply, the gains from higher level power management are great at the application level, as well as the operating system level, and that stems from people's interests. People can enhance on hardware advances in power management, but also when hardware solutions are not available, software developers can handle this issue to either complement hardware solutions or obtain better results without newer, more advanced hardware. In the software and application techniques there are two important sections — compiler development and application-specific techniques.

31.4.1 Compilation Techniques

Research and experiments have shown that compilation for performance does not mean compiling for power [20]. That created the need to come up with different optimization techniques than those that are used for optimization for performance and rightfully so, because compilation for power is extremely important. It is important and that is due to the fact that it is an attractive solution to programmers who do not need to worry about power consumption and wants to let the compiler or the operating system handle it transparently. Research has been done to optimize compilers for power consumption and these techniques used methods mentioned before such as Gray Code addressing, memory operand reduction, and register allocation techniques. However, a lot of work that was done at the compiler level showed that a hardware/architectural modification would be necessary to get significant results. In this section, we will discuss some techniques that showed promise and some that did not when dealing with compiler optimization.

31.4.1.1 Reordering Instructions.
It was shown in [1] that energy consumption is directly proportional to the frequency of switching the signal from 0 to 1 (i.e., logical state transitions). In [21], it was also mentioned that switching is a function of present input and previous state. So, the previous instruction is a factor in the function. Therefore, reordering instructions

can be a factor in reducing switching activity, which therefore reduce energy consumption. In utilizing this technique in the 486DX2 architecture, it did not show much favorable results, it only showed little improvement. However, a proposal to investigate this technique on other architectures was suggested.

31.4.1.2 Reduction of Memory Operands.
In [21] as well, it was shown via experiments on the 486DX2 architecture that instructions with memory operands have higher energy consumption than instructions with register operands. Pipeline stalls, misaligned accesses, and cache misses add to the cost energywise. Compiler optimizations achieve reducing the number of memory operands. In this research, the authors claim that the most efficient way to reduce the memory operands is via register allocation for temporary and frequently used variables and that also leads to potential reduction in pipeline stalls and cache misses. However there are some issues with register allocation optimization techniques, such as more complex compilers, longer compile time, and register allocation algorithms need to be modified to optimize for low power. Additionally, larger caches, which will result in lower miss rates will result in this technique being a little less significant.

31.4.1.3 Code Generation through Pattern Matching.
Code generation through pattern matching and dynamic programming was introduced by Aho in [22] and was later utilized in [23] for generating code that optimizes for performance. The idea is to find a cover for intermediate representation directed acyclic graph (DAG) for each basic block of code. To find that cover, dynamic programming and tree matching is used so that the overall cost is minimized. The cost function introduced in [22, 23], take performance under consideration. Therefore, to optimize for power, the metric used is power usage as opposed to the number of clock cycles used for the instructions. It was suggested that further investigation of this technique needs to be done.

31.4.1.4 Remote Task Execution.
This is believed to be one of the most attractive compiler optimization techniques because it takes into account the idea that although communication might be a drain on the battery, it can also be an opportunity to save power by shipping the execution elsewhere. The idea of compiler-based remote task execution has been proposed in other research. Kremer et al. have done some work that uses compiler-based remote execution of certain tasks in image processing applications [24]. They basically utilized what they called checkpoints in the code to determine the task's delimiters. Additionally in [25], Rudenko et al. introduced remote task execution mechanisms. However, this approach migrates entire processes to remote servers and then the client remote machine waits to receive the results back from the server. As far as

Java™ applications are concerned, the work done in [26] introduces the possibility of migrating the compilation (both optimizing and nonoptimizing) step of running a Java program to a server and that is due to the power consumption cost of compiling Java programs.

31.4.2 Application-Level Techniques

Attacking the power problem at the application level has been done quite extensively. However, the solutions provided are only useful in certain applications and most of the time cannot be generalized to allow for use in other applications. Most of the work seems to have been done in the signal and image processing applications due to the amount of drainage that they inflect on the battery, because they use computation power as well as screen power.

Loy and Helal have introduced an active mode power management (AM/PM) where they increase the amount of information available through the use of a power–aware Application Programming Interface (API) [17]. This saved up to 62 times the power used when compared to using no power management. They examined the amount power used by a connected wireless PC card when the connection is not needed. At that point, they take advantage of stealing back as much of the power as possible. The targeted applications in this work were an e-mail client as well as a Web browser.

In [27], a collaborative relationship between operating systems and applications is demonstrated to meet user-specified goals for battery life. In this work, applications can dynamically modify their behavior to conserve energy. The way this is done is by the operating system monitoring energy supply and demand. It is as simple as if the supply of energy is plentiful, applications perform best, otherwise they will be biased toward conserving energy. PowerScope [30] is used here to validate the measurement of energy consumption for accurate estimation.

In [28], an excellent application is developed with energy awareness in mind. MP3 (Motion Picture Experts Group Audio Layer-3) players are some of the most popular devices in the market at the present time. This work presents designing an energy-aware MP3 player. They not only tackle the hardware side of the device, but also the software side. They talked about the two technologies that influenced the design of wearable computers, which are system-on-chip (SOC) and system-in-package (SIP). These two technologies are the reason that both hardware and software solutions will need to be developed for energy-aware wearable devices. In this work, they analyze a single-chip multimedia system to be used in the wearable device. They take into account the entire computing environment such as hardware, packaging, and software design. They achieve low energy consumption by the use of detailed statistical analysis of the energy consumption. They use an in-house designed run-time energy estimation tool.

Power Awareness and Management Techniques

Additionally, in [29] another multimedia application is investigated with respect to power awareness. They present a middleware framework for coordinating the adaptation of a multimedia application to the hardware resources. They have three goals in mind when designing their system:

1. To have a soft real-time guarantee of multimedia application deadline
2. To have sufficient energy for multimedia applications to finish their task
3. To waste as little energy as possible

To meet the three goals, the framework presented makes three useful contributions:

1. Making multimedia application make energy-aware processor reservation
2. Formally modeling adaptability of hardware, software, and user preference
3. Delivering sufficient processor and energy resources to the application and operate the processor as slow as possible to save energy

31.5 Tools and Packages for Low-Power Design and Measurement

When it comes to developing power-aware techniques, developing and ultimately using tools and methods that measure and estimate the power consumed by applications is essential. Without these tools and techniques, it is near impossible to gather accurate information about applications to be able to pinpoint the sections of programs that dominate power consumption.

31.5.1 PowerScope

PowerScope [30] is an extremely useful tool to estimate power consumption of applications via profiling. This tool maps energy consumption to program structures. It can be determined, using this tool, what specific processes consumed energy during a specific time period. Furthermore, energy consumed by procedures/subroutines within a process can be determined. By providing this fine granularity of feedback, PowerScope allows one to focus system components that are responsible for the largest energy consumption. Experiments by the developers of PowerScope yielded 46 percent reduction in total energy consumed.

The architecture of this system is quite simple. It is composed of a data collection phase and an offline analysis phase. The data collection phase uses a multi-meter [34] connected by a Hewlett-Packard interface bus (HPIB) connection to a data collection computer running an energy monitor component. A profiling computer running an application and a system monitor component are getting power by being connected to the multi-meter for measurement. The profiling computer also provides the trigger for the multi-meter to start measuring.

The offline analysis phase uses a third component called the energy analyzer, which will take the results produced in the data collection phase by the system monitor and the energy monitor and correlate the two to produce the energy profile of the application.

It can be concluded that by providing a profile for energy consumption, pinpointing the sections of code that consume lots of power can be achieved and dealt with. This is analogous to using CPU profilers to discover components of code that waste processor cycles.

31.5.2 Derivatives of SimpleScalar

SimpleScalar [31] is a tool set used for building modeling applications for analyzing program performance and hardware and software coverification. By using it, users can develop modeling applications to simulate real programs running on various modern microprocessors and operating systems. SimpleScalar has simulators with different granularities. In other words, it has high-level functional simulators as well as low-level detailed simulators. SimpleScalar is widely used in the research community. The SimpleScalar Web site mentioned that in 2000 more than a third of all research papers published in top architecture conferences used the tool set for design evaluation.

The toolset also includes performance visualization tools, statistical analysis resources, as well as debug and verification infrastructure.

The simulators included in SimpleScalar can emulate instructions sets for Alpha, Pendulum Instruction Set Architecture (PISA), advanced RISC machine (ARM), and x86. It includes machine definition infrastructure to separate architectural details from simulator implementation.

The licensing model for SimpleScalar allows users to extend the tool set. As a result the power management community took advantage of that to extend the tool set to develop projects for measuring and estimating power consumed by applications. These projects include the Power Analyzer Project at University of Michigan [32] and the Wattch project at Princeton University [33].

31.5.2.1 The Power Analyzer. This project is a joint effort between University of Michigan and University of Colorado. The goal of this project is to create an early power estimator that will allow for examining power/performance trade-offs. This analyzer is a power-aware cycle-level simulator. The target processors that they are after are generally commercially used Pocket PCs and handheld computers. This system provides dynamic as well as static power consumption profile.

31.5.2.2 The Wattch Project. The Wattch project included developing a power measurement tool on the Web. The group called the project CASTLE —

Power Measurement on the Web. In this application, a user would connect to the system and estimate the power consumed by an application program. It also provides multi-meter results for the power consumed by the CPU.

31.5.3 Other Power Estimation Techniques

In addition to PowerScope and SimpleScalar derivatives, other research has been done to estimate energy for certain applications, systems, and devices. In [35], an energy model for the Palm OS™ is described. In this work, they targeted challenges faced by the programmers when developing energy efficient code. So, they built their simulators to handle these challenges. In [36], a hardware/software SOC design's framework for power estimation is introduced. It is based on concurrent and synchronized execution of a hardware simulator and an instruction set simulator. In [37], a survey of techniques to develop tools for automatic design of low-power VLSI systems.

31.6 Conclusion

Rapid advances in mobile technology, as well as wider and wider use of these devices will continue to increase the need for the continuation of research and development targeting reduction in power usage. The aforementioned techniques and opportunities have yet to be complete. New methodologies and techniques continue and will continue to be developed to save as much power as possible. Along with advances in battery technology as well as energy-aware CPUs and memory systems, software developers need to realize that there is a need for a high-level (application level) power awareness. As always, hardware and software computer components do complement each other. Just because one has become more power-efficient, does not mean that the other has to stay put. With advances in both, we will be able to achieve much better results.

References

1. G. Smit and P. Havinga, A Survey of Energy Saving Techniques for Mobile Computers, Internal Report, University of Twente, 1997.
2. SBS IF Specifications, http://www.sbs-forum.org/specs/index.html.
3. Advanced Power Management v. 1.2, http://www.microsoft.com/whdc/archive/amp_12.mspx.
4. ACPI — Advanced Configuration and Power Interface, http://www.acpi.info/.
5. Intel Xscale Technology, http://www.intel.com/design/intelxscale/.
6. Intel PCA Developer Network, http://www.intel.com/pca/developernetwork/.
7. Mobile Pentium III Processors, Intel SpeedStep Technology, http://www.intel.com/support/processors/mobile/pentiumiii/sb/CS-007509.htm.
8. Simply Revolutionary, http://www.transmeta.com/technology/index.html.
9. Processor Core Overview, http://www.arm.com/armtech/CPUs.

10. C.-H. Hsu, U. Kremer, and M. Hsiao, Compiler-Directed Dynamic Frequency and Voltage Scheduling. In *Proceedings of the Workshop on Power-Aware Computer Systems (PACS),* November 2000.
11. C.H. Gebotys, Low Energy Memory and Register Allocation Using Network Flow, In *Proceedings of DAC '97,* Anaheim, CA.
12. C.-L. Su, C.-Y. Tsui, and A. Despain, Low Power Architecture Design and Compilation Techniques for High-Performance Processors.Technical Report ACAL-TR-94-01 February 15, 1994, University of Southern California.
13. N. Bellas, I. Hajj, C. Plychronopoulos, and G. Stamoulis, Architectural and Compiler Support for Energy Reduction in the Memory Hierarchy of Hi Performance Microprocessors, In *Proceedings of ISLPED '98,* Monterey, CA, pp. 70–75.
14. E. Witchel, S. Larsen, C.S. Ananian, and K. Asanovic, Direct Addressed Caches for Reduced Power Consumption, In *Proceedings of the 34th Annual International Symposium on Microarchitecture, MICRO-34,* Austin, TX, December 2001.
15. T. Yeh, D. Long, and S. Brandt, Conserving Battery Energy through Making Fewer Incorrect File Predictions. In *Proceedings of the IEEE Workshop on Power Management for Real-Time and Embedded Systems,* Taipei, Taiwan, May 2001.
16. R. Kravets and P. Krishnan, Power Management Techniques for Mobile Communication. In *Proceedings of MOBICOM '98,* Dallas, TX, pp. 157–168.
17. R. Loy and A. Helal, Active Mode Power Management in Mobile Devices, In *Proceedings of the 5th World Multi-Conference on Systematics, Cybernetics, and Informatics,* Orlando, FL, July 2001.
18. F. Bellosa, The Benefits of Event-Driven Energy Accounting in Power-Sensitive Systems, In *Proceedings of the 9th ACM SIGOPS European Workshop,* Kolding, Denmark, September 2000.
19. C.S. Ellis, The Case for Higher-Level Power Management. In *Proceedings of the 7th IEEE Workshop on Hot Topics in Operating Systems,* Rio Rico, AZ, March 1999.
20. M. Valluri and L. John, Is Compiling for Performance == Compiling for Power? In *Proceedings of the 5th International Workshop on Interaction between Compilers and Computer Architecture (INTERACT-5),* Monterrey, Mexico, January 20, 2001.
21. V. Tiwari, S. Malik, and A. Wolfe, Compilation Techniques for Low Energy: An Overview. In *Proceedings of the Symposium on Low-Power Electronics,* San Diego, CA, October 1994.
22. A. Aho, M. Ganapathi, and S. Tjiang, Code Generation Using Tree Matching and Dynamic Programming, *ACM Transactions on Programming Languages and Systems,* vol. 11, no. 4, October 1989, pp. 491–516.
23. C. Fraser, D. Hanson, and T. Proebsting, Engineering Efficient Code Generators using Tree Matching and Dynamic Programming, Technical Report CS-TR-386-92, Princeton University, August 1992.
24. U. Kremer, J. Hicks, and J.M., Rehg, Compiler-Directed Remote Task Execution for Power Management. In *Proceedings of the Workshop on Compilers and Operating Systems for Low Power (COLP '00),* Philadelphia, PA, October 2000.
25. A. Rudenko, P. Reiher, G. Popek, and G. Kuenning, The Remote Processing Framework for Portable Computer Power Saving. *ACM Symp. Appl. Comp.,* San Antonio, TX, February 1999.
26. J. Palm, J. Eliot, and B. Moss, When to Use a Compilation Service? In *LCTES'02-SCOPES'02,* Berlin, Germany, June 19–21, 2002.
27. J. Flinn and M. Satyanarayanan, Energy-Aware Adaptation for Mobile Applications. *SOSP-17 12/1999,* Kiawah Island, SC.
28. J. Haid, W. Schogler, and M. Manninger, Design of an Energy-Aware MP3-Player for Wearable Computing. In *Proceedings of the Workshop on Mobile Computing (TCMC 2003),* Graz, Austria, March 2003.

29. W. Yuan, K. Nahrstedt, and X. Gu, Coordinating Energy-Aware Adaptation of Multimedia Applications and Hardware Resources. In *Proceedings of the 9th ACM Multimedia (Multimedia Middleware Workshop)*, Ottawa, Canada, October 2003.
30. J. Flinn and M. Satyanarayanan, PowerScope: A Tool for Profiling the Energy Usage of Mobile Applications. In *Proceedings of the 2nd IEEE Workshop on Mobile Computing Systems and Applications*, 1999.
31. SimpleScaler LLC, http://www.simplescalar.com.
32. The SimpleScalar-Arm Power Modeling Project, http://www.eecs.umich.edu/~tnm/power.
33. MRM Research Group, http://parapet.ee.princeton.edu.
34. Agilent Technologies, http://www.agilent.com.
35. T. Cignetti, K. Komarov, and C. Schlatter Ellis. Energy Estimation Tools for the Palm. In *Proceedings of the 3rd ACM International Workshop on Modeling, Analysis, and Simulation of Wireless and Mobile Systems (ACM MSWiM 2000)*, Boston, MA, 2000.
36. M. Lajolo, A. Raghunathan, S. Dey, L. Lavagno, and A. Sangiovanni-Vicenetelli, Efficient Power Estimation Techniques for HW/SW Systems. In *Proceedings of the IEEE VOLTA'99 International Workshop on Low Power Design*, pp. 191–199, Como, Italy, March 4–5, 1999.
37. E. Macii, M. Pedram, and F. Somenzi, High-Level Power Modeling, Estimation, and Optimization. In *Proceedings of DAC '97*, Anaheim, CA.

Chapter 32
Adaptive Algorithmic Power Optimization for Multimedia Workload in Mobile Environments

Luca Benini and Andrea Acquaviva

32.1 Introduction

Resource usage of multimedia applications depends on the multimedia workload characteristics and the desired quality of service (QoS). Multimedia workload is strongly variable due to the heterogeneous nature of the information content, whereas quality of service depends on user requirements. In addition, both can be affected by wireless link quality. For instance, consider a video streaming application composed by a video capture device that grabs, encodes, and transmits video data through a wireless network to a remote video decoding and playback device. User tuneable network parameters, such as transmission power and convolutional code rate, lead to variations in error probability and latency affecting the PSNR (peak signal to noise ratio) and the perceived quality of the video sequence. On the other side, the decoder may experience a variable input data rate due to channel quality variations such as noise level and fading.

Algorithmic power optimization policies in this context aim at reducing the power consumption of the mobile device either by tuning algorithmic and network parameters or by appropriately configuring resource power states to adapt to workload characteristics, network conditions, and desired quality of service.

Depending on the targeted application, different algorithmic parameters can be adjusted to trade off quality of service for system power. For example, quantization parameters in MPEG-4 (MPEG — Motion Picture Experts Group) encoding algorithms may be tuned to reduce communication energy (spent by the network interface card) with little loss in visual quality. The corresponding impact in computational energy (spent by the processor) has been shown to be very small [Zhao02]. This suggests that power optimization policies may nonuniformly affect different parts of the system. Hence, effective techniques should be aware of the power consumption of all components of a wearable device.

In an energy constrained environment, a user can decide to trade off quality of service with power to increase battery lifetime and thus service duration. As an example, consider a video telephone application. The user may decide to lower the image quality to extend battery lifetime and conversation duration. Algorithms gradually trade off QoS/performance for energy can be defined as energy scalable [Sinha02]. Not all the algorithms are energy scalable and different algorithms have different energy-quality behavior. Algorithms can be modified to achieve better energy scalability [Sinha00, Sinha02, Bhardwaj01].

Communication or transmission power can also be reduced by performing an efficient resource management through explicit power state configuration of processor and devices. In general purpose architectures such as those used in modern mobile devices, resources are redundant and as they must accommodate the peak of computational, memory, and input/output (I/O) requirements. For this reason, resources can be put in low-power modes characterized by lower performance adapting to the multimedia workload and the required QoS. For instance, processor speed and voltage can be scaled down if applications provide information on their requirements, such as processor utilization [Yuan01a].

Applying a power optimization policy means making power-related decisions that affect the use of hardware resources. Some policies try to adapt resource utilization by exploiting information about the workload, system state (e.g., battery level, wireless channel conditions), and required QoS so that the power consumption is minimized. Closed loop policies may be adopted as they exploit output information to adjust power-related parameters. For instance, PSNR feedback can be used to avoid strong variations in video quality.

At this level, decisions depend only on the single application. However, a further level of power optimization exploits information from the surrounding environment, represented by local or remote entities like the operating system or a remote content provider interacting with the application. In this case, techniques are said to be collaborative. For instance, a

Adaptive Algorithmic Power Optimization for Multimedia Workload

remote server may provide workload information to aid a dynamic processor voltage scaling (DVS) algorithm [Chung02] or perform energy efficient traffic reshaping to save power spent by a wireless network interface [Acquaviva03].

OS-collaborative techniques are mandatory in a multiprocessing environment to perform explicit management of system resources like processors, network interfaces, and video displays. For instance, scaling the processor speed affects all the applications running on the system. As a consequence, the scaling decision cannot be made by a single application, but must be coordinated by the OS. In this case, applications may give information about their resource usage or QoS requirements to an underlying power manager acting at the OS-level [Yuan01a, Min02].

In this chapter, we describe algorithmic techniques aimed to minimize power consumption in mobile devices running multimedia applications. First, we focus on energy scalability of multimedia algorithms and we describe application level power optimization techniques to improve energy–quality behavior by exploiting scalability to trade off power with QoS. Then, we analyze policies that perform explicit resource management (processors, peripherals) in both a stand-alone and collaborative way.

32.2 Scalability and Energy Optimization

In this section, we first introduce the concept of scalability in multimedia applications by analyzing scalability properties of multimedia algorithms that can be exploited to trade off QoS for power consumption. Then we analyze its relationship to energy and power consumption through Energy–Quality curves (E-Q). We show how E-Q behavior can be improved through algorithmic transformations.

32.2.1 Scalability in Modern Multimedia Applications

Multimedia services must reach several users having different terminals characterized by diverse local resources (display size, storage and processing capabilities, interconnection bandwidth) and located in a highly mobile scenario, where environmental conditions are strongly variable. A high adaptation capability is required in such a context. Adaptability is obtained by exploiting the concept of scalability. An application of this concept can be found, for example, in encoding-transmission-decoding applications. The data stream is packetized according to the content of the data to enable fast transmission of low-resolution but critical information, followed by progressive transmission of additional details carried by additional data packets. The described mechanism provides means for recovering the audiovisual information at its highest quality under the imposed system resource constraints.

Different scalability options can be available depending on the algorithm characteristics and implementations like spatial resolution, quality level, and temporal resolution of video and audio sequences. This enables a variety of trade-offs between QoS and resource costs, such as memory size, processing requirements, and power consumption, which we will explore in this chapter.

32.2.1.1 Scalable Source Coding with Wavelets. To analyze the adaptation opportunities allowed by modern multimedia applications, we give a detailed example of scalable source coding using wavelets, which has been extensively analyzed in [Chirila-Rus02]. Source coding is usually performed as a preprocessing step before channel coding and modulation of data to be transmitted to the network. Source coding is usually performed by a complex compression algorithm characterized by frequency and timing domain decorrelation and successive coding.

Discrete cosine transform (DCT) is one of the most used decorrelators, on which widespread adopted compression algorithms like MPEG-2 and MPEG-4 are based. However, discrete wavelet transform (DWT) is gaining importance because of its good intrinsic features like spatial and quality scalability. In fact, it is currently adopted in multimedia standards like MPEG-4 (in the area of visual texture coding) and JPEG2000 (JPEG — Joint Photographic Experts Group).

Due to the high demand for scalable coding techniques that support adaptation to varying environmental conditions, several solutions for scalable DCT coding have been proposed, but they are recognized to be unsatisfactory. These include MPEG-4 fine grained scalable (FGS) video coding that provides only quality scalability at a cost of a bit rate overhead from 30 to 80 percent. Clearly, this leads to a higher transmission power, so that the scalability in this case does not help. In contrast, wavelet transform has intrinsic spatial scalability. The lowest resolution representation of the original image can be successively refined until the original image size is obtained without any cost in additional bandwidth requirements.

The wavelet transform (WT) provides a multi-resolution/multilevel data representation of the original image, therefore supporting spatial scalability. Each level of the WT is obtained by recursively decomposing the previous (Lowpass–Lowpass) resolution level into four subimages — the Lowpass–Lowpass (LL), the Lowpass–Highpass (LH), the Highpass–Lowpass (HL) and the Highpass–Highpass (HH) through associated Low- and Highpass filters in conjunction with downsampling. The LL-subimages are often referred to as DC images, whereas the LH-, HL-, and the HH-subimages are called AC or detail subimages. The lowest resolution DC image, grouped together with all AC subimages, creates the actual WT data structure, which is as large as the original image. Starting from the DC image, successive

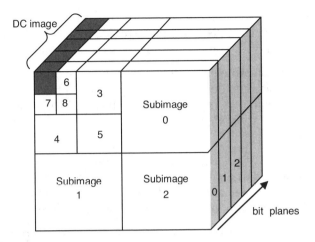

Figure 32.1. Wavelet Bitplanes

higher resolutions of the image can be obtained by injecting the detail images into the DC image.

Quality scalability is obtained through the bit plane structure. A single bit plane collects all the digits of the same weight of different wavelet coefficients. In essence, the digits of the radix M of the wavelet coefficients are transmitted one by one, so that each coefficient (only the dominant ones, see [Chirila-Rus02]) is separately and progressively transmitted through successive quantization/approximation phases. Quality scalability is obtained by decoding the wavelet data structure from the most significant bitplane down to the least significant bitplane (Figure 32.1). As shown in Figure 32.1, each bitplane contains information related to all the subimages, so that quality and spatial scalability are independent.

A wavelet-based image encoder can generate a set of substreams that can be used for a scalability target. The scalability can be spatial scalability (targeting users with different display sizes), image quality scalability (inherent to reduced processing power), or a combination of these two. To better explain the use of these properties, let us consider the examples taken in [Chirila-Rus02]. The authors use three different application scenarios. The first is characterized by two clients with different bandwidth connections, one limited to 100 kilobits per second (Kbps), the other to 50 Kbps. If the goal is to transmit to both one image per second, a normal approach consists in transmitting two separated bit streams, yielding a total bit rate of 150 Kbps. However, using scalability, it can be transmitted as only one stream of 100 Kbps, the first client decoding the complete bit stream, while the second one only decodes half of the bit stream. To maximize image quality, DC data are put first in the stream, followed by the data belonging to higher significant bitplanes down to lower significant bitplanes. In this way, the PSNR of the image is constantly growing, but the most important information is transmitted at the beginning of the stream.

In a second application scenario, the authors consider two clients with different display size and decoding processing power to show the effectiveness of spatial scalability. The user with the bigger display decodes the full bit stream and the other one decodes only a portion of the stream depending on the capabilities.

As a third application, the authors consider a broadcasting scenario where both spatial and quality scalability can be exploited, because a single bit stream can be sent and each client can be served depending on their resources like bandwidth, display size, and processing power.

It can be seen from the previous discussion that several parameters can be tuned to adapt the information produced by modern multimedia algorithms to user terminal capabilities. In the following subsection, we will discuss some of the scalability opportunities available in MPEG-4 algorithm.

32.2.1.2 Scalable Source Coding in MPEG-4. MPEG-4 is an ISO/IEC standard being developed by MPEG. Although MPEG-1 standard was mainly targeted to CD-ROM applications and the MPEG-2 for digital and high definition television, with higher quality as well as bandwidth requirements (2 megabits per second (Mbps) to 30 Mbps), the MPEG-4 standard primarily focuses on interactivity, higher compression, universal accessibility and portability of video content, with rates between 5 Kbps and 64 Kbps for mobile applications and up to 2 Mbps for television and film applications.

MPEG-4 video supports all the functionalities provided by MPEG-1 and MPEG-2, compressing rectangular-sized images at varying levels of input formats, frame rates, and bit rate while providing better visual quality at comparable bit rates. Furthermore, the MPEG-4 image and video coding algorithms give efficient representation of visual object of arbitrary shapes with the goal to support content-based functionalities. Each frame of an input video sequence can be segmented into a number of arbitrarily shaped image regions (called video objects [VOs]), each representing a particular physical object or content within scenes [Zhao02].

MPEG-4 employs block-based motion estimation and compensation techniques to efficiently exploit temporal redundancies of the video content in image sequences. The texture information is coded using the DCT similar to previous MPEG standards.

MPEG-4 supports scalable coding, the technique allowing access or transmission of VOs at various spatial and temporal resolutions. This allows it to support receivers with different bandwidths or to provide a layered video bit stream amenable to prioritized transmission. Receivers can choose not to reconstruct the full resolution VOs by decoding subsets of the layered bit stream to display the VOs at lower spatial or temporal resolution or with lower quality [Zhao02]. Both spatial and temporal scalability are supported by MPEG-4. Here we focus on temporal scalability; later we will survey a technique that exploits this property to improve energy efficiency of the coding process.

Temporal scalability involves partitioning the video object planes (VOPs) that can be defined as the VO in a determined time instant. VOPs are partitioned into layers, where the lower layer is coded by itself to provide the basic temporal rate and the enhancement layer is coded with temporal prediction on the lower layer. Similar to spatial scalability, temporal scalability has an additional advantage in that it provides resilience to transmission errors. Object-based temporal scalability can also be exploited to allow control of picture quality by controlling the temporal rate of each VO under the constraint of a given bit-budget [Zhao02].

32.2.2 Energy Scalability

Scalability properties of modern multimedia algorithms allow for adaptation of multimedia content to characteristics of mobile devices. These characteristics can be intended as computational and bandwidth capabilities, but also as energy requirements. Hence, scalability can be exploited to achieve a good trade-off between battery lifetime and QoS. Thus, energy consumption can be reduced at the cost of a degradation of quality of the multimedia content. However, not all algorithms scale well with respect to energy.

This observation lead to the concept of energy scalability, that has been introduced in [Sinha02] as the properties of algorithms to trade off computational accuracy (or quality, Q) with energy requirement (E). More precisely, an algorithm is said to be energy scalable if, when the available computational energy is reduced, the impact on quality gradually reduces. The concept can be extended to the total system energy.

Algorithms evidencing this property have a good E–Q behavior. The E–Q behavior of multimedia applications can be modified to increase energy scalability through algorithmic transformations. Clearly, energy overhead due to these transformations must be small with respect to total energy consumption.

The formalization of the concept of desirable E–Q behavior can be easily introduced through the E–Q graph, which represents the function Q(E). Here Q represents some quality metric (e.g., PNSR, mean square error) as a function of the system energy.

Consider now two algorithms (I and II) that perform the same function. II would be more scalable compared to I if:

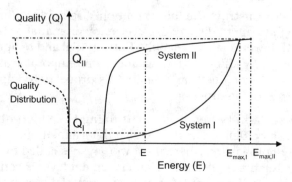

Figure 32.2. Energy Scalability

$$Q_{II}(E) > Q_I(E), \forall E. \quad (32.1)$$

The desired E–Q behavior described above can be easily expressed through the Q(E) function. In fact, we would like a curve to maximally concave downward (with respect to the energy axis), so that:

$$\frac{d^2Q(E)}{dE^2} \leq 0 \quad (32.2)$$

The E–Q behavior suggested by (32.2) is not always obtainable globally (i.e., across 0 E E_{max}), however, in an average case, for a given energy availability, E, we would like the obtainable quality Q(E) to be as high as possible (Figure 32.2).

As an example of energy scalable application design, we can consider a filtering application as reported in [Sinha02]. Finite impulse response (FIR) filtering is one of the most commonly used digital (digital signal processor [DSP]) operations, frequently used in multimedia applications. An N-tap FIR filter is defined by:

$$y[n] = \sum_{k=0}^{N-1} h[k]x[n-k]. \quad (32.3)$$

The filtering operation involves N to multiply and accumulate (Media Access Control [MAC]) cycles. For desired E–Q behavior, the MAC cycles that contribute most significantly to the output y[n] should be done first (i.e., the larger in magnitude). A simple rule could be to sort the impulse response in decreasing order of magnitude of its coefficients. Even if the data sample multiplied to the coefficient might be so small as to mitigate the effect of the partial sum, nevertheless, in an average case, the coefficient reordering by magnitude yields a better E–Q performance than the original scheme [Sinha02]. The authors illustrate the scalability results for

Adaptive Algorithmic Power Optimization for Multimedia Workload

a low pass filtering of speech data sampled at 10 kilohertz (kHz) using a 128-tap FIR filter. Results on about average energy consumption per output sample have been obtained on an Intel® StrongARM SA-1100 [Strongarm] operating at 1.5 V voltage supply and 206 megahertz (MHz) clock frequency. Comparing the original scheme to the reordered one, an improvement of the E–Q behavior has been observed, even if input data fluctuations lead to deviations from the ideal behavior suggested by [Acquaviva 03].

32.3 Adaptive Algorithmic Power Optimization

Based on algorithmic degrees of freedom discussed in the previous sections, several power optimization strategies can be found in literature. They aim to trade off QoS for computational or communication power by tuning power-related parameters. Computational power is saved by reducing the number of operations to be performed, while communication power is reduced by acting on the bit rate. The rationale behind these techniques is to adapt to characteristics of the multimedia workload to reduce the quality penalty caused by parameter tuning. In essence, they try to act on those parameters that show good energy scalability. Some of these policies may perform a resource management by configuring the underlying hardware (both processor and peripherals) to adapt to the new computational and communication requirements. For example, processor frequency and voltage can be adjusted to save system energy. Resource power management requires hardware knobs allowing reconfiguration via software. Recent processors for embedded systems have been designed in a power-conscious way. As a result, several of today's cores support different software-tuneable frequency and voltage levels. Moreover, peripherals are often characterized by multiple power states. For example, wireless network interfaces can be configured in power save mode and their transmission energy can be modulated upon software commands. Recent studies showed that also liquid crystal displays (LCDs) are power manageable [Choi02, Gatti02].

Power optimization policies can exploit information from the environment to improve their adaptability. Thus, a multimedia application can be seen as a component of a more complex system. It typically interacts locally with the OS and remotely with a content provider (i.e., a remote server). A typical example is given by a video capture system where, on the server side, video frames grabbed by a camera are encoded before transmission over a wireless network. At the client side, a decoding application running on a palmtop plays back the decoded video sequence. A collaborative power optimization technique can try for example to exploit server knowledge of workload to reduce power consumption of the client system. Examples of these policies have been presented in literature [Acquaviva03, Chung02]. Other types of collaborative policies exploit the OS knowledge of total system workload to perform a power efficient resource management.

Such collaboration is mandatory in a multiprocessing system where resources are shared among several applications.

Noncollaborative or stand-alone techniques are described below and collaborative techniques are analyzed later in this section.

32.3.1 Stand-Alone Power Management

Adaptive power optimization techniques can be classified based on the quantity they try to adapt to and the software or hardware knobs they exploit to achieve adaptation. Adaptation can be performed on workload (image size, resolution, video/audio content), network conditions (distance range, fading, multipath), user terminal characteristics (battery level, display size, processing capabilities), and required QoS. Once an adaptation parameter is established, achieving power efficiency through adaptation is not straightforward because of:

- The intrinsic difficulty to choose an adaptation step (it is obviously not possible to continuously look at the adaptation parameter)
- A fine tuning of the regulation variable is not always possible
- The effects of the regulation on power consumption may be contrasting

To clarify the last point, let us consider a compression algorithm. Improving communication energy efficiency by reducing the amount of data to be transmitted requires more computational energy spent by the processor to perform the compression task. The total energy balance must be evaluated in this case if we are interested in increasing the battery lifetime of our mobile device.

Workload adaptation is usually exploited to achieve a better usage of communication or computational resources while matching some constraints on the output (bandwidth or QoS requirements). For instance, several adaptive source coding algorithms look at the variable nature of input data (image and audio characteristics, required QoS) to reduce the amount of processing (which saves computational power) or to reduce the bit rate (which saves communication power) while keeping the required QoS level. Closed loop techniques may be used to this purpose. In the following, we will summarize optimization techniques in literature in this area and we will outline possible future research directions.

32.3.1.1 Adaptive Encoding Algorithms. Multimedia encoding algorithms are exploited in mobile devices to perform data compression before transmission to a remote machine and are aimed at reducing information's bandwidth requirements by reducing the bit rate. Even if lower bit rate means low energy required for data transmission, the total system's energy may increase due to higher computational power.

In this section, we first analyze a technique aimed at reducing processor power through optimization of a computation intensive function of modern encoders, then we discuss techniques that try to reduce communication power by adaptation of source coding and transmission rate.

32.3.1.1.1 Adaptive Motion Estimation. In [He97a], a low-power motion estimation algorithm that exploits adaptive pixel truncation is presented. Motion estimation is a computational intensive task in widely used digital pulse code modulation/discrete cosine transform (DPCM/DCT) compression schemes to estimate the motion vector of a particular macro-block. Instead of transmitting the compressed pixel data, only the motion vector and the prediction error are coded and transmitted. This greatly reduces the temporal–spatial redundancy and the number of bits required to represent the macro-block. To reduce the complexity and the power consumption of the motion estimation, a bit truncation technique can be used. In fact, in [He97b], it has been shown that the incoming image pixel value can be truncated by four without significant degradation in the algorithm performance, which is measured by the PSNR of the images. The adaptive approach proposed in [He97a] consists of determining the number of truncated bits in real-time depending on the quality of the picture. If the accuracy is too low, the prediction error will accumulate and the magnitude of data feed to the quantizer is large. To maintain a constant bit rate, a large quantization step size is required, which degrades the picture quality. In the proposed adaptive scheme, the number of truncated bits will then be reduced by masking fewer least significant bit (LSB) bits. To determine the image quality, instead of directly measuring the PSNR, the quantization step size is observed. This scheme tries to achieve a better trade-off between power consumption and picture quality. A different approach has been presented by [Minocha99], where the motion estimation algorithm is dynamically adapted to input data in order to save memory accesses and arithmetic operations.

32.3.1.1.2 Adaptive Source Coding and Transmission. Several techniques to reduce the transmission energy have been presented by researchers [Eisenberg02, Lan97]. In general, these techniques try to achieve an optimal trade-off between energy consumption and quality of the resulting media information, especially in the video domain. In a recent work, Eisenberg et al. proposes an adaptation approach where the source coding parameters are adjusted jointly with transmission rate and transmission power. The authors considered the problem of compressing a video sequence for transmission over a wireless channel and formulate a general optimization problem for minimizing the energy required for transmitting the video under distortion and delay constraints [Eisenberg02]. At the source coding level, the authors act on error resilience and concealment techniques and at the communication level they consider transmission power and transmission rate adaptation. In the formulation of the optimization problem, the following assumptions are made:

- Delay constraints translate in constraints on the bit rate for each packet k (B_k), which depends on coding parameters for that packet (μ_k) and on transmission rate (R_k)
- The frame distortion (D_f) depends both on how the video is encoded (quantization) and transmitted (transmission rate, power)
- The frame delay (T_f) depends on the number of bits used to transmit each packet of the frame (B_k) and the packet transmission rate (R_k)
- The distortion can be measured at the transmitter
- The transmission power can be adjusted at run-time

Formally, the objective can be written as:

$$\underset{(\mu_k, P_k)}{\text{minimize }} E_{TOT} = \sum_{k=1}^{K} \frac{B_k(\mu_k)}{R_k} P_k, \qquad (32.4)$$

subject to:
$$D^f(\{\mu_k, P_k, R_k\}) \leq D^0,$$

and
$$T^f(\{\mu_k, R_k\}) \leq T^0.$$

The packet loss adaptation (variable packet loss [VPL]) is based on jointly controlling transmission power and source-coding parameters by adapting to the characteristic of the video sequence to achieve energy efficiency. The expected distortion is computed. If this distortion is tolerable when a packet is lost and concealed, then the mobile device can save time and energy by not transmitting the packet. The encoder forces the decoder to conceal a packet using information from neighboring packets. The constraints on the distortion are met because the mobile device is able to vary the transmission energy in response to variations in the source content. During periods of high visual activity (such as scene changes) the receiver has more difficulty concealing lost packets. However, by simultaneously adapting the source coding and transmission power, the mobile device can provide more protection to packets during these periods.

Transmission rate adaptation (variable rate [VR]) takes place by simultaneously adapting transmission rate and source coding. Less transmission energy is used by decreasing the transmission rate when channel fading is large, while the transmission rate is increased when the channel fading is small. To provide high video quality, more bits are required to encode each frame. Therefore, to meet delay constraints, higher transmission rates are needed to decrease the expected distortion.

Comparative results (based on simulations) with respect to fixed packet loss (FPL) and fixed rate approaches show that the adaptive approach can significantly improve the energy efficiency of the mobile device.

32.3.1.1.3 Adaptive MPEG-4 Encoding Parameter's Tuning. In the previous section, we described how the MPEG-4 coding standard allows for scalable coding. Now we show how this property can be exploited to optimize power consumption. In [Zhao02] several factors and their impact on power consumption are studied. In particular, the focus is on quantization level, number of bidirectional video object planes (B-VOPs), the error resilience technique used, content-based coding, and spatial and temporal scalability.

As described in previous section, the MPEG-4 algorithm supports three coding modes (I-VOP, P-VOP, B-VOP). B-VOP gives the smaller average bit number per VOP, because of the bidirectional prediction. As reported in [Zhao02], the quantization parameter (QP) related to B-VOP affects the PSNR less than I-VOP and P-VOP. In addition, B-VOPs are the least important among the three, because no other VOPs depend on them. From these considerations, B frames can be coarsely quantized to improve the compression ratio with little loss in visual quality.

By implementing this policy, 79 percent, 86 percent, and 88 percent of energy savings can be obtained on average for transmitting I-, P-, and B–VOPs, respectively [Zhao02].

Another characteristic of MPEG-4 that can be exploited to reduce power consumption is that a video frame can be segmented into a number of VOs and encoded separately. Separate decoding and reconstruction of the objects allow the interoperability and manipulation of content of the original scene by simple operation on the bit stream. This is called content-based functionality. Because a video object has an arbitrary shape, the shape information must also be encoded and transmitted. Even in [Zhao02], authors compare the communication energy spent by decoding an image sequence encoded as a single object with respect to the energy spent by decoding the same sequence where the image is decomposed into two objects. The results show that the communication energy (proportional to the total number of bits in the bit stream) and the computational energy increase significantly when multiple video objects are encoded. This can be exploited for adaptive power minimization. In fact, when the time-varying residual battery energy of a mobile device attains a certain threshold, which could trigger the power saving mode, it is possible to opt for a low quality video stream to reduce computational energy. This can be achieved by sacrificing the quality of less important objects when multiple video objects are coded and transmitted. This is better than sacrificing the quality of the total frame when only one video object is coded. For instance, if the background information does not change substantially, it can be encoded in a lower frame rate and with less detail. Results are reported related to a case in which fewer details of the background of a test video sequence are encoded and transmitted. Savings in both communication and computation energy can be obtained.

Temporal scalability in MPEG-4 as described above can also be exploited for low-power communication needs. For example, the base layer can be encoded with a frame rate half of the frame rate used for the enhancement layer. The latter can be the same frame rate as with no scalability. Results presented in [Zhao02] based on simulations point out that scalable encoding enables the control of total energy subject to quality constraints.

32.3.1.2 Adaptive Decoding Algorithms. Mobile devices run decoding algorithms to play back audio/video information either stored on a local memory or transmitted by a remote server in a compressed format. To achieve a continuous playback to guarantee the quality of the video and audio reproduction, the decoding process must periodically produce a determined amount of data, depending on the output data rate and the characteristics of the playback sequence (sample rate for audio, frame rate, image size, and resolution for video information). Energy efficiency of the decoding process can be enhanced by performing a power-aware computational resource management. Adaptive techniques can try to reduce the processor speed (and, consequently, the voltage) while matching real-time constraints (just-in-time computation). Buffering techniques may be exploited when designing multimedia applications to further relax time constraints.

In the rest of this section, we present an adaptive speed setting algorithm for streaming applications and a buffer insertion technique to enhance dynamic processor clock and voltage scaling.

32.3.1.2.1 Adaptive Clock/Voltage Setting. To reduce the computational power at the decoder side, multimedia algorithms can exploit speed/voltage scaling capabilities of modern microprocessors. Clock speed reduction by a factor of s allows scaling down voltage as well, thus leading to energy reduction by a factor of s^3 [Chandrakasan92]. However, slowing down the processor speed lead to a performance drop and for this reason it must be done in a workload adaptive way to avoid deadline misses that may cause degradation of the playback quality.

Several DVS techniques suitable for multimedia workloads have been presented in the past. In [Simunic01], the authors propose a method to estimate the interarrival time of MPEG audio and video frames from the network that are used to apply a dynamic voltage scaling policy. In [Delaney02], a DVS policy has been exploited to reduce the energy consumption of a speech recognition front-end running on a wearable device.

It has also been stated by Acquaviva et al. that for real-time multimedia streaming applications, a speed-setting approach with no voltage scaling can also lead to consistent energy reductions [Acquaviva01]. This is in contrast with the common assumption that speed-setting is effective only accompanied by an adequate voltage-setting policy. Of course, if voltage is scaled with frequency, more power can be saved, but the point here is that this is not a forced choice.

Speed-setting effectiveness depends on the workload characteristics and the system's architecture (both hardware and software). It reduces the costs of memory latency in terms of CPU wait states, hence, in execution dominated by memory access (high miss rate) and where memory latency is higher, this technique is more effective [Bellosa99]. In addition, from a system energy perspective, because the CPU clock often feeds other on-chip components, additional system power can be saved by reducing useless work on these as well (even if in some cases they implement power down and gated clock strategies).

Idleness can be classified as implicit idleness and explicit idleness. The first identifies CPU idleness dispersed among useful operations (mainly during memory wait cycles on cache misses). This term varies with frequency: because memory access time is fixed, adjusting the frequency involves variations in number of wait states in a bus cycle. This happens when (as usual) the CPU is not the speed limiting element. The second is due to coarsely clustered idle cycles. Explicit idleness is quite common in practice. When the execution time is fixed, as in the case of real-time constrained algorithms, making a computation faster involves the need of storing the results of computation in a buffer waiting for some event external at the CPU. During that time, the CPU experiences idleness, which can be eliminated without affecting the algorithm effectiveness by increasing the time spent in useful operations, that is, by lowering the CPU frequency. Explicit idleness can be reduced also by putting the processor in a low-power state while waiting and restoring the running state when the external event arrives (i.e., an external interrupt), but in this case, it is needed to account for the time and energy overhead needed to shut down and wake up the CPU.

An implementation that exploits this concept has been used in a low-power MPEG-layer3 audio decoding algorithm capable of adapting to the bit rate and sample rate of the input audio stream. Based on these two parameters, the algorithm chooses the lowest possible frequency that allows matching real-time constraints imposed by the audio playback rate [Acquaviva01].

In practice, the real-time constraints are represented by a frame rate (FR) requirements, which is the number of MPEG Audio Layer-3 (MP3)

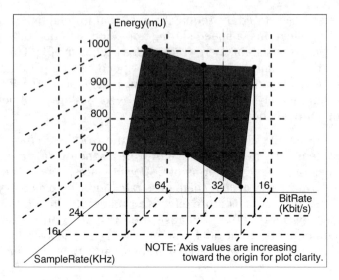

Figure 32.3. Frequency Setting Curves

frames that must be elaborated by the processor in the time unit to guarantee continuous audio play back. The minimum frequency that guarantees to match real-time constraints is found by looking at a FR(f) curve. For a given clock frequency, the achievable FR strongly depends on the bit rate (br) and sample rate (sr) of the compressed stream. This makes it possible to identify experimentally a set of three FR(f) curves (FRA, FRB, FRW) that represent, for a given br and sr, the best case, average case, worst case FR curves. By intersecting these curves with the required FR, a range of frequencies is determined (fmin to fmax) that can be used to clock the processor. The lower frequency value in the range allows for large energy reductions and the higher value guarantees the required FR. The effectiveness of the policy is given by the fact that the obtained frequency range is short, as shown in Figure 32.3.

It is important to stress that the FR(f) curves are not linear in general. This is because the memory system and interfaces do not speed up like the processor with increasing clock frequency. The slower the speed of the external hardware (e.g., memory access time), with respect to the processor, the flatter the performance curve, and the greater can be the effectiveness of the speed-setting policy.

Other than the hardware characteristics, the shape of the curve depends on the ratio between the computation time spent inside the CPU and that spent outside the CPU. Considering the nonideality of the external memory, this can be expressed also as the ratio between the external accesses and the total memory accesses, which in turn is equal to the cache miss rate.

Adaptive Algorithmic Power Optimization for Multimedia Workload

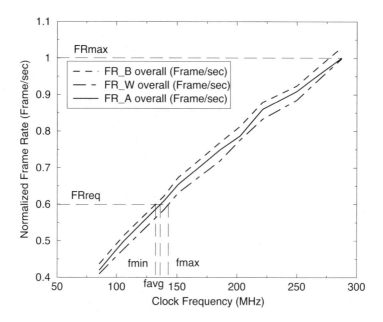

Figure 32.4. Energy as a Function of Bit Rate and Simple Rate

Results of the implementation of this policy on a wearable computer prototype show the total energy cost of decoding a compressed audio stream with different levels of compression and bit rate. In the x–y plane are the points corresponding to different versions of the audio stream, while on the z-axis is the energy consumption when the policy is applied. The results show how our algorithm adapts to workload, consuming less energy when computational load decreases. This behavior is in contrast with the one of the unutilized algorithm, which consumes less energy when bit rate and sample rate increase. This behavior is explained considering that in idle intervals the CPU spends a lot of power polling a synchronization variable. When the workload is higher, the CPU spends more time in decoding instructions, which are less power-expensive (Figure 32.4).

32.3.1.2.2 Improving Energy Scalability through Buffer Insertion. Applications can be modified to better exploit clock scaling capabilities of underlying hardware. Actual energy consumption can be reduced with a negligible hit in performance, thus improving energy scalability. As an example on how this concept can be exploited in decoding algorithms, we examine the approach taken by Lu et al. They presented a design approach for a multimedia application that requires constant output rates and sporadic jobs that need prompt responses. The method is based on splitting it into stages and inserting data buffers between them [Lu02]. Data buffering has three purposes:

1. To support constant output rates
2. To allow frequency scaling for energy reduction
3. To shorten the response times of sporadic jobs

The authors construct a frequency assignment graph where each vertex represents the current state of the buffers and the frequencies of the processor. The authors develop an efficient graph-walk algorithm that assigns frequencies to reduce energy. The same method can be applied to perform voltage scaling and the combination of frequency and voltage scaling.

The proposed technique is aimed at reducing power by dynamic frequency scaling on processors that have only finite frequencies through data buffer insertion in multimedia programs. Data is processed and stored in the buffers when the processor runs at higher frequency. Later, the processor runs at a lower frequency to reduce power and data is taken from the buffer to maintain the same output rate. Before the buffers become empty, the processor begins to run at a higher frequency again.

Buffering can also shorten the response time of a sporadic job, if there is enough data in the buffers. In fact, the processor can handle a sporadic job without affecting the output rate of the multimedia program.

The authors' experimental results on a StrongARM-based computer show that four discrete frequencies are sufficient to achieve nearly maximum energy saving. The method reduces the power consumption of a MPEG program by 46 percent. The authors also demonstrate a case that shortens the response time of a sporadic job by 55 percent.

32.3.2 *Collaborative Power Management*

Even if multimedia algorithms can be designed to address low-power requirements, often a stand-alone approach is not enough to satisfy tight power constraints imposed by battery-operated devices. Collaborative techniques may be used in this case to develop more aggressive power management techniques. In addition, collaborative techniques are mandatory in some cases. In fact, considering for example, algorithms that perform dynamic voltage/clock processor scaling, they are supposed to run alone in the system. Furthermore multimedia applications often run in a multiprocessing environment, where hardware resources are shared among different processes. In such a context, resources management must be coordinated by the OS that knows the needs of all active applications in the system.

Collaboration can be used also to enhance the adaptation by exploiting workload knowledge provided by surrounding systems interacting with the device running the application. In particular, often mobile devices communicate with a remote machine to download video and audio content for real-time streaming purposes. The remote server application can provide

Adaptive Algorithmic Power Optimization for Multimedia Workload

workload information allowing for effective power management decision at the client side. In a multiclient environment, the server can also implement a power-aware scheduling strategy. These kinds of techniques are well-suited for hot spot servers that must handle connections with several clients providing them heterogeneous data streams.

Collaborative policies may target the reduction of both communication and computational power, as we will describe in the rest of this section.

32.3.2.1 Operating System Collaborative Techniques. Multimedia processing in portable and embedded devices such as cell phones, wireless terminals, handhelds, and PDAs is often done under the control of OSs (either general purpose or real-time). An OS coordinates resource access (peripherals, CPUs, and memories) for all the applications. Because a multimedia workload translates in application resource requirements, the OS can perform adaptation by monitoring the usage of resources and by selectively configuring their power states. Current OSs implement simple policies based on time-out triggered by user interactions or on CPU current utilization. More advanced policies have been proposed from researchers aimed at exploiting characteristics of the multimedia workload and applications [Chandrasena00, Pouwelse01, Qu00]. In the rest of this section, we examine two techniques. The first is a predictive technique that exploits characteristics of multimedia workloads without performing any explicit information exchange between the application and the OS. The second technique is based on communication of resource requirements from applications.

32.3.2.1.1 Predictive Technique for Multimedia Workload. Kumar et al. proposed a modification of a real-time operating system (RTOS) kernel to perform power-aware scheduling of multimedia tasks that exploit the inherent tolerance of many multimedia applications to lost data samples due to factors like communication noise or network congestion [Kumar01]. This tolerance is used as an immunity noise margin that mitigates the effects of a wrong adaptation. In fact, the proposed strategy is based on a DVS technique that uses a history of the actual computation requirements of the previous instances of a task to predict the computation required by the next instance. The prediction may have an error that may result in an underestimation that may lead to a deadline miss. The tolerance to a small percentage of missed deadlines can be exploited to do an aggressive DVS.

Results of the application of the prediction strategy on a MPEG player are shown in Table 32.1. Power reduction achieved by two prediction strategies is shown. The number of frames missing their deadlines for the second strategy (row II) is much smaller compared to those of the first strategy (row I). This enhancement has been obtained by considering that I frames in MPEG are important parts of the sequence, so the prediction has

Table 32.1. Power Reduction Obtained with the Predictive Policy

Row Number	Power Reduction Compared to Full Power Mode	Power Reduction Compared to Low Mode without Prediction	Number of Deadlines Missed
I	95%	60%	33%
II	90%	30%	10%

been restricted to when P frames are about to be decoded. I frames are assumed to take always their worst case time to decode. Refer to [Kumar01] for a more detailed explanation of the prediction strategies. Summarizing, in both cases, significant power reduction can be obtained using prediction strategies in the OS scheduler.

32.3.2.1.2 Application-Assisted Technique. The technique described above does not involve any explicit information exchange between applications and OSs, because it tries to exploit their intrinsic features. A different approach is taken by Yuan and Nahrsted that proposes a middleware framework in which applications explicitly provide resource needs and receive feedback from the middleware [Yuan01a]. The middleware coordinates the processor/power resource management (PPRM) and has four major characteristics:

1. Provides a power-aware resource reservation mechanism, where admission control is based on the processor utilization and power availability
2. Adjusts the speed and corresponding power consumption of the processor upon events, triggered by the change of the system workload or power availability
3. Updates reservation contracts of multimedia applications to maintain their resource requirements while adjusting the processor speed
4. Notifies applications about the change of resource status to enable them to adapt their behavior and complete tasks before power runs out

The architecture of the middleware is shown in Figure 32.5. The OS exports hardware resources (such as processor and power) status to the middleware layer, which receives resources requests from applications. The middleware layer consists of three major components — the Dynamic Soft Real-Time (DSRT) processor scheduler [Yavatkar95], the power manager, and the coordinating PPRM framework.

The DSRT scheduler allows multimedia applications to reserve processor resources and corresponding power resources and monitors the system workload. The power manager monitors the power availability (i.e.,

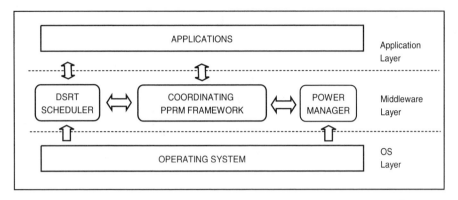

Figure 32.5. Middleware Architecture of PPRM Framework

remaining battery lifetime) and the processor power consumption. The coordinating PPRM framework:

- Determines polices on how to adjust reservations according to power availability
- Uses the corresponding polices to differentiate applications in case of low power availability
- Adjusts the processor speed to achieve minimum wasted energy
- Notifies application, if it cannot extend the battery life and meet the processor resource requirements of applications under low power availability

The main characteristic of the DSRT scheduler is that it provides a power-aware resource reservation mechanism that separates soft real-time multimedia applications from best effort applications and statistically multiplexes processor resources between them. Each real-time application reserves a certain amount of processor resource, required capacity, CRE (as explained below), and the real-time workload is the sum of the required capacity of all admitted real-time applications in the system. The best-effort workload is limited by the available unreserved processor resource. On the other side, multimedia applications make processor resource reservations through the power-aware reservation specification in [Yuan01b], as shown in Table 32.2.

It must be noted that the parameter duration is used to determine if the power is enough for the application to finish its task. The other important component of this approach is the coordinating algorithm. It dynamically adjusts the speed of the processor to meet the following goals:

- Ensuring enough power availability for all admitted multimedia applications (that is, the battery can last for the maximum duration of all multimedia applications)

Table 32.2. Power-Aware Reservation Specification

Parameter	Meaning
class	Periodic constant processing time (PCPT) or periodic variable processing time (PVPT) [PARSEC97].
period	Inform the scheduler when to release a new job, which must be finished before a deadline.
utilization	How much percentage of processor resource to reserve.
speed	Context for utilization (request utilization percentage when the processor runs at speed).
duration	How long the application lasts.
weight	Importance of the application under low power availability.

- Allocating the required capacity of processor resource to each multimedia application under high power availability
- Reducing the required capacity C_{RE} to weight*C_{RE} for each multimedia application and notifying applications under low-power availability
- Running the processor as slowly as possible to save energy while meeting the above goals

Meeting the processor/power resource requirement of a soft real-time application means that the required capacity of processor resource can be allocated to the application and the power availability is enough for the duration of the application. Note that a certain amount of processor resource is shared among all best-effort applications to protect starvation for best-effort applications.

Here we report the experiment presented in [Yuan01a] to better illustrate how the proposed policy works. The experiment has the following steps:

1. For the first 16 seconds, there is no real-time workload, so the processor runs at the slowest speed 300 MHz and best-effort applications can use all processor resources.
2. At time 17, the mpegplay program requests to reserve (utilization 60 percent, speed 300 MHz, weight 0.9, duration 600 seconds) 4 and the required capacity is 60%*300 MHZ/600 MHZ = 0.3. This reservation request can be admitted without adjusting the processor speed.
3. At time 41, the math program requests to reserve (utilization 80 percent, speed 300 MHz, weight 0.6, duration 100 seconds) and the required capacity is 80%*300 MHZ/600 MHZ = 0.4. The coordinating policy adjusts the processor speed to 500 MHz to admit this reservation request with new utilization 48 percent. It also

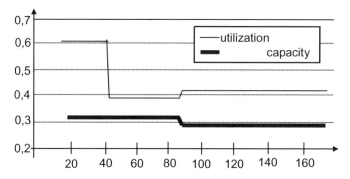

Figure 32.6. Example of Application of PPRM Framework

updates the reservation contract of the mpegplay with new utilization 36 percent to maintain its required capacity.

4. At time 89, the power manager finds that the actual power availability is 400 seconds and not enough for the mpegplay program. Therefore, the coordinating policy adjusts the required capacity of mpegplay to 0.3*0.9 = 0.27, and the required capacity of math to 0.4*0.6 = 0.24. Then it slows down the processor to 400 MHz, and allocates new utilization 40.5 percent and 36 percent to mpegplay and math, respectively. The predicted power availability is 400 seconds*1.5W/0.9W = 666.6 seconds, which is enough for both programs.

Figure 32.6, adapted by [Yuan01a], reports the utilization and the required capacity as a function of time for the proposed example. Summarizing, the authors show that a percentage of 39.5 percent of energy can be saved.

The presented technique presents good evidence that when applications provide their resource requirements to power-aware OSs, better power efficiency can be achieved with respect to predictive techniques.

32.3.2.2 Server-Assisted Collaborative Techniques

32.3.2.2.1 Communication Energy Optimization. Mobile devices are often used as playback clients connected by a wireless link to a remote content provider. The energy spent by the wireless network interface card to receive streaming data can strongly affect battery lifetime. Other than for actually downloading useful data, a consistent part of this energy is spent by the card in idle state while listening for the channel or to receive broadcast traffic.

The application server, based on its knowledge of workload and traffic shape, may provide information to the clients to selectively shut down their network interfaces. The effectiveness of such an approach has been recently explored in a wireless local area network (WLAN) environment [Acquaviva03].

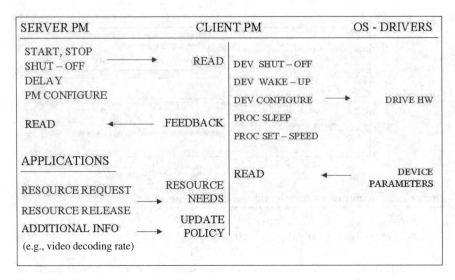

Figure 32.7. Communication Framework

In this work, the authors present a power management infrastructure composed by a couple of power manager modules, local and remote. The local power manager provides an Application Programming Interface (API) to power related parameters of the system. In particular, it can control WLAN power states (off, doze, active), CPU speed, and power states (active, idle, sleep).

This interface can be exploited to implement a server controlled power management strategy. The proposed technique exploits the server knowledge of the workload, traffic conditions, and feedback information from the client to minimize WLAN power consumption. Two entities are defined — a server power manager (server PM) and a client power manager (client PM). Both are implemented as a part of the Linux® OS and they provide the power control interface to the applications. Server PM uses the information obtained from the client and the network to perform energy efficient traffic reshaping so that WLAN can be turned off. Client PM communicates through a dedicated low-bandwidth link with the server PM and implements the power controls by interfacing with the device drivers. It also provides a set of APIs that client applications can use to provide extra information back to the server.

The communication protocol between server and client is shown in Figure 32.7. Power control commands are issued by the server PM, interpreted by the client PM, and translated in the appropriate device driver's function calls. Upon request from the server PM, device specific information is fetched by the client PM through device driver's calls. In addition,

Adaptive Algorithmic Power Optimization for Multimedia Workload

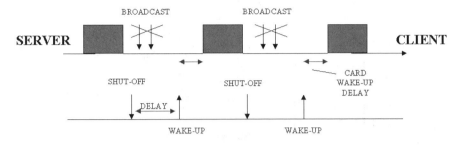

Figure 32.8. Example of Application of the Server Assisted Policy

the application specific information can be retrieved by the client PM via application API calls (Figure 32.8).

The server PM achieves energy reduction at the wireless interface card of the client by means of traffic reshaping and controlling power states of the card. It schedules the transmission to the client in bursts in order to compensate for the client's performance and energy overheads during the transitions between the card's on and off states. The client's WLAN card is switched off once the server has sent a burst of data that will keep the client application busy until the next communication burst. Burst size and delay between bursts must be precomputed at the server. The goal is to have a delay large enough to almost empty the client input buffer and small enough burst size to avoid overflow while keeping the buffer sufficiently filled. An illustration of the shut-off policy is shown in Figure 32.9. The horizontal axes represent the time. In the upper axis, the network traffic is represented; in the lower axis, the power management decisions made by the server are represented.

When the server decides to transmit (gray boxes) the WLAN is switched on; when the server stops the transmission the WLAN is switched off (down arrow in the lower axis), thus discarding the broadcast traffic (down arrows in the upper axis). Before beginning a new transmission time interval, the server wakes up the WLAN (up arrow in the lower axis).

The only parameters needed by the server to implement this policy are the burst size and time between bursts, which determines the time the card is in off state. The burst size has been established as the largest value as possible to avoid overflow conditions in the access point (AP) buffer or the application buffer, whichever is smaller. The burst delay can be computed by exploiting the server's knowledge of MPEG-4 frames composition (i.e., the number of packets needed to compose a video frame). This number may be strongly variable depending on the characteristics of the video sequence.

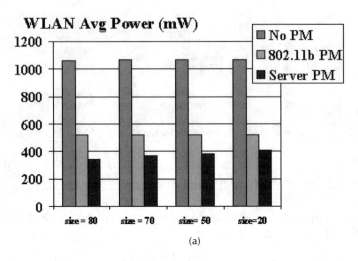

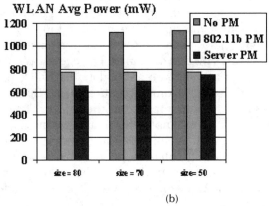

Figure 32.9. Results of the Remote Power Control Policy

Experimental results performed on two benchmarks are presented in Figure 32.9. The y-axis presents the average power consumed by the WLAN card when receiving the MPEG-4 stream for different burst sizes and for medium broadcast traffic conditions. The experiment has been performed in three different situations:

1. WLAN always on
2. WLAN with only IEEE® 802.11b PM
3. WLAN controlled by the server

As shown in Figure 32.9A, for the first benchmark the server controlled approach (darker plot) saves 67 percent of average power compared to leaving WLAN always on (grey plot) and 50 percent compared to the default 802.11b PM (light plot). The average power savings increase as the

Adaptive Algorithmic Power Optimization for Multimedia Workload

burst size increases because that enables longer times between bursts and thus better compensation for the transition delay between the WLAN's on and off states. Note that in all three cases the video plays back in the same amount of time, as it continues to be real-time. Thus the reported average power savings directly correspond to energy savings. For the second benchmark (Figure 32.9B), the savings are smaller because the delays are very short and in some cases the card cannot be switched off. However, we are still able to save 41 percent of power with respect to leaving the card always on and 15 percent with respect to the standard 802.11b power management protocol with bursts of 80 packets. It can be observed that for this benchmark a burst size smaller than 50 packets cannot be used, because the computed delays are too short and the card can be never switched off.

32.3.2.2.2 Computational Energy Optimization. Server knowledge of workload can be exploited also to reduce computational energy of the client. Chung et al. implemented a collaborative technique enabling adaptive dynamic voltage scaling in MPEG decoders [Chung02]. The author's approach is similar to [Acquaviva01] (described in Section 32.3.1.2) and it is based on the fact that many multimedia applications have a periodic property, but each period shows large variation in terms of its execution time. Exact estimation of such variation is a crucial factor for low-energy execution with DVS technique. Although noncollaborative DVS techniques focus only on client sites and their quality heavily depends on the accurateness of worst execution time estimation, in a server-assisted approach, the contents provider supplies the information of the execution time variations in addition to the content itself. This makes it possible to be DVS independent from worst case execution time estimation. The extra work required for the contents provider for this purpose is fully compensated by the benefits for the end users because single content is often provided to many users.

The policy is based on the observation that decoding costs computed on different architectures show similar behavior.

The decoding cost of each frame is its decoding time normalized to that of the first frame of the sequence. Figure 32.10A shows the decoding cost of the first 30 frames of a video clip (cindy, 80x60). It can be observed that the decoding cost in each system varies in the similar direction. In other words, the decoding time ratio between two different frames is much less sensitive to system architecture compared to the absolute decoding time. Thus, decoding cost is more appropriate information rather than decoding time to consider various client system architectures for DVS.

Suppose that the reference system of the contents provider is SimpleScalar and the client systems are ST200 and Strong ARM. Then the decoding costs of two client systems (*actual decoding costs*) can be obtained by

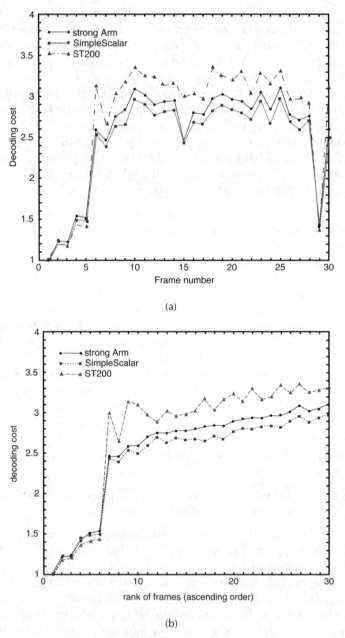

Figure 32.10. (A) Decoding Costs (B) Ordered Decoding Costs

scaling the decoding cost of SimpleScalar. The scaling factor can be computed by measuring the decoding times of a few frames on each client system. Although decoding costs show similar variations, the difference at

Adaptive Algorithmic Power Optimization for Multimedia Workload

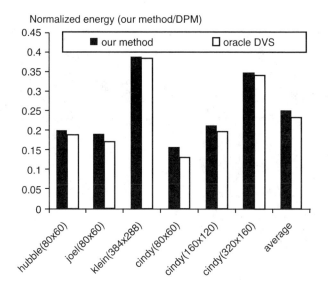

Figure 32.11. Results for the Server Assisted Policy

each frame is not a constant. Thus, decoding cost translation should consider the nonlinear relation between decoding cost and actual decoding cost. This must be handled by the clients as explained later.

The construction of a decoding cost table is needed to translate in the decoding cost to the actual decoding cost. Due to the nonlinear relationship between the decoding cost and the actual decoding cost, single scaling factor is not sufficient and a piecewise linear model is appropriate for the translation, as shown in Figure 32.11B. The important observation is that similar decoding costs can use the same scaling factor to approximate their actual decoding costs with acceptable error ratio.

The developed DVS algorithm exploits all these observations by constructing a DVS table to map each clock frequency and supply voltage pair to the appropriate actual decoding cost, as explained in [Chung02]. Summarizing, the algorithm works as follows. The first frame is decoded at full speed, then for the other frames, the MPEG decoder computes the actual cost of the current frame using the scaling factor selected from the cost scaling table. If the scaling factor is not yet computed, the decoder goes to a learning phase where the current frame is decoded at full speed and its decoding time is measured and compared to the decoding time of the first frame to find out the appropriate scaling factor. Next, the voltage and clock frequency pair is selected from the DVS table.

Experimental results show the effectiveness of this policy by means of comparison with the oracle DVS. The oracle is able to predict the exact

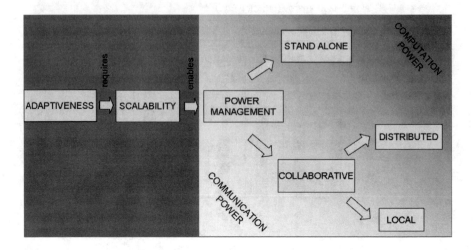

Figure 32.12. Scheme of the Chapter

decoding time for each frame thus providing an upper bound to the DVS performance.

32.4 Conclusion

In this chapter, we focused on adaptive techniques to improve energy efficiency of mobile devices handling multimedia workloads. By analyzing current multimedia algorithms and trends for their development, we described available software degrees of freedom and those actually exploited by state-of-the-art optimization techniques. The conceptual flow we follow for description and classification purposes can be summarized in Figure 32.12.

To achieve adaptiveness to multimedia workload, algorithms are required to be scalable. Scalability involves the possibility to trade off quality of media content with utilization of computational and communication resources. This enables the development of power management strategies that try to save power by selectively configuring the underlying hardware to match the utilization profile. To perform this matching, power management strategies can exploit information at different layers. Stand-alone techniques rely on information given by the algorithm itself; collaborative techniques exploit information from the environment represented by the OS (local management) or by a remote application like a media content provider (distributed management). All of these approaches can be exploited to achieve both computational and communication power optimization as shown in this chapter.

References

[Acquaviva01] A. Acquaviva, L. Benini, and B. Riccò, Software Controlled Processor Speed-Setting for Low-Power Streaming Multimedia, *Transaction on CAD*, November 2001.

[Acquaviva03] A. Acquaviva, T. Simunic, V. Deolalikar, and S. Roy, Remote Power Control of Wireless Network Interfaces, *Proceedings of PATMOS*, Turin, Italy, September 2003.

[Ji01] Z. Ji, Q. Zhang, W. Zhu, and Y.Q. Zhang, End-to-End Power-Optimized Video Communication over Wireless Channels, *Proceedings of the IEEE Workshop on Multimedia Signal Processing*, October 2001.

[Minocha99] J. Minocha and N. Shanbhag, A Low-Power Data-Adaptive Motion Estimation Algorithm, *Proceedings of IEEE Workshop on Multimedia Signal Processing*, September 1999.

[Chirila-Rus02] A. Chirila-Rus, G. Lafruit, and B. Masschelein, Scalability and Error Protection, in T. Basten, M. Geilen, and H. De Groot, Eds., *Ambient Intelligence, Impact on Embedded System Design*, Norwell, MA: Kluwer Academic Publisher, 2003.

[Yuan01a] W. Yuan and K. Nahrstedt, A Middleware Framework Coordinating Processor/Power Resource Management for Multimedia Applications, *Proceedings of IEEE GlobeCom*, November 2001.

[Lu02] Y.H. Lu, L. Benini, and G. De Micheli, Dynamic Frequency Scaling with Buffer Instertion for Mixed Workloads, *IEEE Transaction on CAD*, November 2002.

[Zhao02] J. Zhao, R. Chandramouli, N. Vijaykrishnan, M.J. Irwin, B. Kang, and S. Somasundaram, Influence of MPEG-4 Parameters on System Energy, *Proceedings of IEEE ASIC/SOC*, 2002.

[Chung02] E.Y. Chung, L. Benini, and G. De Micheli, Contents Provider-Assisted Dynamic Voltage Scaling for Low Energy Multimedia Applications, *Proceedings of IEEE ISLPED*, August 2002.

[Sinha02] A. Sinha, A. Wang, and A. Chandrakasan, Energy Scalable System Design, *IEEE Transaction on VLSI*, Vol 10, no. 2, April 2002.

[He97a] Z.L. He, K.K. Chan, C.Y. Tsui, and M.L. Liou, Low-Power Motion Estimation Design Using Adaptive Pixel Truncation, *IEEE Proceedings of ISLPED*, 1997.

[He97b] Z.L. He and M.L. Liou, Reducing Hardware Complexity of Motion Estimation Algorithms using Truncated Pixels, *Proceedings of IEEE ISCAS*, June 1997.

[Yuan01b] W. Yuan, K. Nahrstedt, and K. Kim, R-EDF: a Reservation Based EDF Scheduling Algorithm for Multiple Multimedia Task Classes, *IEEE Real-Time Technology and Applications Symposium*, May 2001.

[Kumar01] P. Kumar and M. Srivastava, Power Aware Multimedia Systems using Run-Time Prediction, *Proceedings of IEEE VLSI Design*, January 2001.

[Yavatkar95] R. Yavatkar and K. Laksman, A CPU Scheduling Algorithm for Continuous Media Applications, *Proceedings of the Workshop on Network and OS Support for Digital Audio and Video*, April 1995.

[Gatti02] F. Gatti, A. Acquaviva, L. Benini, and B. Riccò, Power Control Techniques for TFT LCD Displays, *Proceedings of ACM CASES*, Grenoble, France, 2002.

[Min02] R. Min and A. Chandrakasan, A Framework for Energy-Scalable Communication in High Density Wireless Networks, Proceedings of IEEE ISLPED, August 2002.

[Sinha00] A. Sinha, A. Wang, and A. Chandrakasan, Algorithmic Transforms for Efficient Energy Scalable Computation, *Proceedings of IEEE ISLPED*, August 2000.

[Bhardwaj01] M. Bhardwaj, R. Min, and A. Chandrakasan, Quantifying and Enhancing Power Awareness of VLSI Systems, *IEEE Transactions on VLSI*, vol. 9, no. 6, December 2001.

[Strongarm] Intel, Intel StrongARM SA-1110 Microprocessor Advanced Developer's Manual, June 2000.

[Simunic01] T. Simunic, L. Benini, A. Acquaviva, P. Glynn, and G. de Micheli, Dynamic Voltage Scaling and Power Management for Portable Systems, *IEEE Proceedings of DAC*, June 2001.

[Chandrasena00] L.H. Chandrasena and M.J. Liebelt, A Comprehensive Analysis of Energy Savings in Dynamic Supply Voltage Scaling Systems Using Data Dependent Voltage Level Selection, *Proceedings of IEEE International Conference on Multimedia and Expo,* July–August 2000.

[Pouwelse01] J. Pouwelse, K. Langendoen, and H. Sips, Energy Priority Scheduling for Variable Voltage Processors, *IEEE Proceedings of ISLPED,* August 2001.

[Chandrakasan92] A.P. Chandrakasan, S. Sheng, and R.W. Brodersen, Low Power CMOS Digital Design, *IEEE Journal of Solid State Circuits,* vol. 27, no. 4, April 1992.

[Choi02] I. Choi, H. Shim, and N. Chang, Low-Power Color TFT LCD Display for Hand-Held Embedded Systems, *IEEE Proceedings of ISLPED,* August 2002.

[Lan97] T. Lan and A.H. Tewfik, Adaptive Low-Power Multimedia Communications, *IEEE Workshop on Multimedia Signal Processing,* June 1997.

[Eisenberg02] Y. Eisenberg, C.E. Luna, T.N. Pappas, R. Berry, and A.K. Katsaggelos, Energy Efficient Wireless Video Communications for the Digital Set-Top Box, *Proceedings of International Conference on Image Processing,* September 2002.

[Delaney02] B. Delaney, N. Jayant, M. Hans, T. Simunic, and A. Acquaviva, A Low-Power, Fixed-Point, Front-End Feature Extraction for a Distributed Speech Recognition System, *IEEE Proceedings of ICASSP,* May 2002.

[Qu00] G. Qu and M. Potkonjak, Energy Minimization with Guaranteed Quality of Service, *Proceedings of IEEE ISLPED,* July 2000.

[Bellosa99] F. Bellosa, OS-Directed Throttling of Processor Activity for Dynamic Power Management, TR-I4-99-03, June 1999.

Chapter 33
Energy-Aware Web Caching over Hybrid Networks

Françoise Sailhan and Valérie Issarny

Abstract

A terminal's latency, connectivity, energy, and memory are the main characteristics of today's mobile environments whose performance may be improved by caching. In this chapter, we present an adaptive scheme for mobile Web data caching, which accounts for congestion of the wireless network and energy limitation of mobile terminals. Our main design objective is to minimize the energy cost of peer-to-peer communication among mobile terminals so as to allow for inexpensive Web access when a fixed access point is not available in the communication range of the mobile terminal. We propose a collaborative cache management strategy among mobile terminals interacting via an ad hoc network. We further provide evaluation of the proposed solution in terms of energy consumption on mobile devices.

33.1 Introduction

The limited capabilities of handheld mobile terminals, including smaller display and limited processing power and memory, complicate direct access to content. A solution to this issue lies in the use of proxy agents that adapt the content according to the capacities of the mobile terminals, implementing techniques like data compression and filtering, and format conversion. The proxy architecture is based on a centralized scheme that is well-suited to infrastructure-based wireless networks. Infrastructure-based networks are characterized by the deployment of base stations (also referred to as infrastructures), which handle traffic to and from mobile terminals in their transmission range. In the context of a proxy-based architecture, the base station forwards messages for mobile terminals to the proxy. Then, the proxy becomes a bottleneck when there are a large number of mobile terminals. In addition, the centralized approach implemented by the proxy architecture is not suitable for highly dynamic networks, such as ad hoc networks that do not involve any infrastructure in the communication scheme. Instead, mobile terminals communicate directly using peer-to-peer communication. Ad hoc networks enable mobile terminals to cooperatively form a dynamic and temporary network without any preexisting infrastructure. Hence, this is a cheap solution. In general, ad hoc and infrastructure-based networking should be seen as complementary rather than as competitive. Ad hoc networking is more convenient for accessing information available in the local area and possibly reaching a wireless local area network (WLAN) base station, which comes at no cost for users. Ultimately, the user may decide to pay for communication using wireless global networking facility, if the connectivity using the WLAN happens to be bad.

The issue that we are addressing is setting up an ad hoc network of mobile terminals that cooperate to exchange data and more specifically Web data. Hence, this enables Web access at no financial cost for mobile users. Our solution lies in implementing a cooperative caching strategy among mobile terminals over the ad hoc network. The proposed caching strategy has the potential to significantly reduce latency and hence user-perceived delays and to further reduce the bandwidth utilization, by not requiring systematic Internet access. In that context, it is crucial to account for the specifics of mobile terminals. Mobile terminals that will soon be available will embed powerful hardware (e.g., liquid crystal display [LCD] screens, accelerated three-dimensional [3D] graphics, high-performance processors), and an increasing number of devices (e.g., DVD, CD). However, the capacity of batteries goes up slowly and all these powerful components reduce battery life. Thus, it is compulsory to devise adequate solutions to energy saving on terminals for all the constituents of the mobile environment (i.e., application software, network operating system, and hardware) [18]. In particular, communications being one of the major sources of energy consumption [21], a number of wireless communication protocols reduce energy consumption. It is further mandatory for these protocols to be coupled with distributed application software that are designed so as to minimize energy consumption, including the one associated with communication.

This chapter introduces such a distributed application software, which implements ad hoc cooperative Web caching among mobile terminals. The proposed solution aims at improving the Web latency on mobile terminals while optimizing associated energy consumption. As a result, it masks network downtimes and makes mobile terminals more autonomous and less dependent upon the network. Our solution accounts for both the capacities of mobile terminals and the network features; it includes:

- A cooperative caching protocol among mobile terminals
- An associated local caching strategy for mobile terminals

The latter is coupled with a prefetching strategy, which aims at reducing latency while accounting for the added network traffic and the terminal's limited resources.

The rest of the chapter is organized as follows:

- Section 33.2 proposes an overview of power-aware solutions provided at the network level for wireless devices.
- Section 33.3 introduces the ad hoc cooperative Web caching protocol.
- Section 33.4 presents the local cache management and prefetching strategies.
- Section 33.5 gives an evaluation of our proposal, focusing on the energy consumption associated with ad hoc cooperative caching.
- Section 33.6 concludes with a summary of our contribution.

33.2 Power-Aware Communication

Personal digital assistants (PDAs) and cell phones gain more memory, new features, and more options for connecting anytime, anywhere, but battery power is still an obstacle to true mobile freedom. Energy consumption is indeed a crucial factor for designing battery-operated systems. Consequently, energy consumption has been largely considered in the hardware design of mobile terminals. In the following, we study more specifically energy consumption associated with communication, which represents more than 50 percent of the total system power [31]. Increased mobility is a common trend; people now want to have reliable access to a corporate network and the Internet, anytime, anywhere. According to Frost and Sullivan,[1] the market of wireless technologies will grow from $300 million to $1.6 billion in 2005 for the total worldwide WLAN. Many enterprises fully launch out into WLAN technology, largely because it provides a flexible and cost-effective access to the Internet Protocol (IP) core network without being physically connected. Compared to wired local area networks (LANs), WLAN systems are now faster to deploy and cheaper to operate. As a result, mobile devices are being networked and an increasingly larger share of their energy budget is attributed to communication. This has led to investigating energy-efficient solutions for wireless networks at all the layers of the protocol stack. In the following, we survey solutions for the Media Access Control (MAC) (Section 33.2.1), routing (Section 33.2.2) and transport (Section 33.2.3) layers.

33.2.1 Energy Saving at the MAC Layer

One of the major sources of energy waste in wireless communication is attributed to packet loss due to, for example, signal attenuation from large obstacles, reflection in the indoor environment, and drastic reduction of power signal with the distance. To improve transmission performance, adaptive transmission techniques make a trade-off between the reduction of interference and the power consumption associated with a transmission [30].

Depending on the mode of transmission (i.e., infrastructure-based or ad-hoc-based), the sources of energy consumption differ. Hence, distinct power-aware mechanisms have been developed for each mode. An infrastructure-based wireless network uses a base station with which mobile terminals interact for communication (i.e., base stations forward messages that are sent and received by mobile terminals). The main feature of power-saving mechanisms for infrastructure-based networks is to have mobile terminals remaining in sleep power-saving mode most of the time and to have them waking up periodically to receive beacon frame messages from the base station. The base station then buffers information intended for mobile terminals and indicates as a part of beacon frame messages that information is being stored, waiting for delivery. Thus, mobile terminals awake only during the amount of time necessary for receiving information from the base station. The ad hoc mode enables mobile terminals to cooperatively form a dynamic network without a preexisting infrastructure because mobile terminals communicate with each other using peer-to-peer links. In this context, the major source of energy waste comes from:

- Collision overhearing induced by multiple transmitters and receivers communicating over the same channel
- Idle listening resulting from the continuous monitoring of the channel
- Control overhead induced by the exchange of control packets

The third cost lies in the reception of useless packets (i.e., control and coordination packets) to be discarded. However, these messages are inherent to the control of transmission and thus cannot be avoided. On the other hand, a number of solutions have been proposed to limit the impact of the two other sources of energy waste. These are surveyed below.

33.2.1.1 Minimizing Collisions. To minimize collisions in point-to-point traffic, and thus the resulting energy consumption, the IEEE® 802.11 and SMAC (Sensor-MAC) [16] protocols provide the same collision avoidance mechanism. Before transmitting a packet, a mobile terminal first listens to determine whether a terminal is transmitting. If the channel is busy, the terminal defers the transmission. In addition, to reduce the probability of two mobile terminals transmitting at the same time and colliding because they cannot hear each other, the IEEE 802.11 and the SMAC protocols provide the virtual carrier sensing mechanism. Prior to any point-to-point transmission, the sender broadcasts a request to send (RTS) control message, which specifies the destination node and the data size (for duration estimate). The sender then waits for a clear to send (CTS) message

Table 33.1. Energy Consumption on Nodes for Point-to-Point Communication

Mobile	Energy Consumption (μW.sec)	m (μW.sec/byte)	p (μW.sec)
Sender X	$\varepsilon_{send} = m_{send} \times size + p_{send}$	$m_{send} = 1.9$	$p_{send} = 454$
Destination A	$\varepsilon_{dest} = m_{dest} \times size + p_{dest}$	$m_{dest} = 0.5$	$p_{dest} = 356$
Nondestination nodes			
In range of sender X and destination A	$\varepsilon_{AX} = m_{AX} \times size + p_{AX}$	$m_{AX} = -0.22$	$p_{AX} = 210$
In range of sender X	$\varepsilon_X = m_X \times size + p_X$	$m_X = -0.04$	$p_X = 90$
In range of destination A	$\varepsilon_A = m_A \times size + p_A$	$m_A = 0$	$p_A = 119$

from the destination node. Once it receives the CTS, the sender sends the data message. Finally, the destination node sends an acknowledgment (ACK) message upon receiving the data message. Table 33.1 gives the energy cost, relative to the size of the message, for the sender and the receiver. With IEEE 802.11, the energy consumed for sending and receiving is given by the linear equation [12]: $\varepsilon = m \times size + p$, where size is the message size, and m (respectively p) denotes the incremental (respectively fixed) energy cost associated with the message. For the sender X, the high value of the incremental cost m_{send} is due to the emission of the data message. The fixed cost p_{send} results from the reception of two control messages (i.e., CTS and ACK) and from the emission of the RTS message. For the destination node A, the fixed cost p_{dest} is due to the emission of two control messages (i.e., CTS and ACK) and to the reception of the RTS message. The value of m_{dest} for A follows from the reception of the data message.

Unlike the aforementioned protocols, the Power-Aware Multi-Access protocol with Signaling (PAMAS) uses two separate channels to exchange control messages (i.e., RTS and CTS messages) and data messages, respectively [22]. The use of a separate control channel implies that there is no collision between data and control packets. In addition, the collision overhead is reduced: when a mobile terminal receives or transmits packets, it can still forbid other terminals to use the data channel. When a mobile A is willing to transmit data to a mobile terminal B, A sends a RTS message to B. But, if a neighbor of A or B is receiving or transmitting data, it transmits a busy tone that collides with the CTS message sent in response. As a result, A is not going to transmit a packet. To reduce the number of collisions resulting from simultaneous transmissions, the Energy Conserving MAC (EC-MAC) [17] protocol elects a coordinator, which acts as a base station. From the standpoint of transmission scheduling, the coordinator runs periodic frames that subdivide into a number phases. At the start phase of a frame, the coordinator broadcasts a frame synchronization message.

This message contains multiple types of information (e.g., framing, synchronization information, uplink transmission orders). In the subsequent request/update phase, mobile terminals transmit requests for connection. Then, the new user phase allows new mobile terminals to register to the coordinator. The following schedule phase allows the coordinator to broadcast a schedule message containing the slot permissions for the final data phase of the given frame; each permission specifies the mobile terminal that should transmit or receive. The main idea of the protocol is that collisions resulting from simultaneous transmissions are avoided by the centralized coordinator, which distributes the time allocated for transmission and reception among mobile terminals.

33.2.1.2 Minimizing Channel Listening. Time devoted to listening to the channel has to be kept to a minimum to reduce energy consumption. Although the IEEE 802.11 protocol addresses energy efficiency, it was not the central issue in the protocol design. With IEEE 802.11, mobile terminals keep track of the channel status most of the time through quasi-constant monitoring in the idle mode. Consequently, mobile terminals are in the listening mode, except those that are in the transmission range of a sender or receiver, which can switch off the wireless interface during the time allocated to the transmission. Table 33.1 gives the energy cost, relative to the size of the message, for nondestination nodes for the IEEE 802.11 protocol. Nondestination nodes in the range of the sender receive RTS messages and thus enter a reduced energy consumption mode during data emission. This leads to have a negative value for m_{AX} and m_X because the energy consumption is less than the one in the idle mode. The fixed cost p_{AX} for nondestination nodes in the range of both the sender and the destination node is greater than the fixed cost p_X for nondestination nodes in the range of the sender only, because the latter do not receive the CTS and ACK messages. Finally, nondestination nodes in the range of the destination node but not the sender receive the CTS and ACK messages. On the other hand, they do not receive the RTS message and thus cannot enter in the reduced energy consumption mode; this leads to having the incremental cost m_A be equal to 0. The PAMAS protocol additionally includes the packet length in both the CTS and RTS messages [22]. This allows all nodes in the transmission range of the sender or destination node to switch off during transmission (or reception). Unlike IEEE 802.11, the EC-MAC protocol is based on a combination of reservation and scheduling mechanisms. As a consequence, terminals do not have to monitor continuously; mobile terminals are awake during the period of time necessary to transmit and receive data and to coordinate with the coordinator. To limit the time slot dedicated to channel listening on a mobile terminal, the SMAC protocol synchronizes neighboring terminals so that they are in the sleep mode most of the time and regularly synchronized to listen [16]. The protocol tries to reduce the waste of energy by minimizing the listening time, while increasing latency because any sender must wait for its receiver to wake up (i.e., next synchronization phase). The SMAC protocol targets applications for which communication delays are not important. Such applications include those over sensor networks, which are characterized by long periods of inactivity during which there are no sensing events. Compared to SMAC, the noteworthy aspect of the PAMAS and IEEE 802.11 protocols is that their power saving mechanisms do not affect the delay or the throughput behavior and the one of EC-MAC is that it supports quality of service (QoS) requirements for multimedia traffic.

33.2.2 Energy Saving at the Routing Layer

In the infrastructure-based mode, mobile terminals communicate directly with a base station providing access to the wired IP network. Ad hoc networks are multi-hop networks where every mobile terminal potentially acts as a router and forwards packets for them to reach the final destination. Thus, mobile terminals cooperatively maintain network connectivity without relying on any infrastructure or centralized administration. Multi-hop protocols build upon single-hop packet transmission controlled by the MAC layer. Then, the main issue addressed in the design of an ad hoc (network) routing protocol is the computation of the communication path between any two mobile terminals.

33.2.2.1 Ad Hoc Routing Protocols. Basic ad hoc routing protocols compute the shortest path and use a cost metric based on the number of hops, assuming that the energy consumption increases proportionally with the number of hops. There exist two base types of ad hoc routing protocols — proactive and reactive (also referred to as on-demand). Proactive protocols (e.g., Optimized Link State Routing [OLSR] [9], Destination-Sequenced Distance-Vector [DSDV] [26]) update their routing tables periodically. According to [6], the energy consumed by proactive protocols is stable regardless of the traffic load. Unfortunately, if the network is inactive in terms of data traffic, proactive protocols induce a continuous energy consumption. Compared to proactive protocols, reactive protocols (e.g., Ad-Hoc On-Demand Distance Vector [AODV] [25], Dynamic Source Routing [DSR] [20], Temporally Ordered Routing Algorithm [TORA] [23]) compute the communication path between any two mobile terminals only when a communication is requested between the two. For instance, DSR maintains routing tables as follows. Upon a request on the mobile terminal for a communication path that is either missing or no longer valid in the routing table, DSR returns several possible paths toward the terminal and the shortest one is selected for communication. In addition, DSR embeds in each message sent, the IP addresses of the mobile terminals forwarding the message, hence enabling each mobile terminal involved in the message routing to both forward the message and update its table. A comparison of reactive routing protocols with respect to energy consumption is proposed

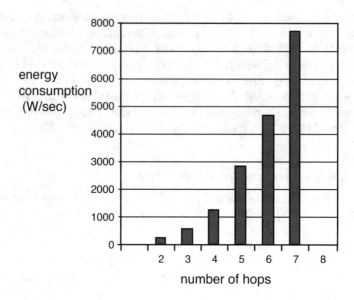

Figure 33.1. Energy Consumption Generated by the Reception of ZRP Traffic in a Zone per second

in [6] under several conditions (e.g., rate of data exchange, number of mobiles, network area). Results show that an increasing number of sources leads to an increase in the number of routing packets. On the other hand, the energy consumed follows a slower shape compared to the increase of the number of sources. This behavior is mainly attributed to the fact that reactive protocols learn new routes from previously sent packets. Zone Routing Protocol (ZRP) [14] is a hybrid protocol that combines the reactive and proactive modes. The design rationale of ZRP is that it is considered advantageous to accurately know the neighbors of any mobile terminal (i.e., mobile terminals that are accessible in a fixed number of hops), because they are close. Therefore, communicating with neighbors is less expensive and neighbors are most likely to take part in the routing of the messages sent from the terminal. As a result, ZRP implements:

- A proactive protocol for communication with mobile terminals in the neighborhood
- A reactive protocol for communication with the other terminals

With respect to a given mobile terminal, its neighborhood is referred to as its zone. Performance of ZRP has been evaluated in [14, 15] using event-driven simulation; this evaluation is gauged by considering the control traffic generated by ZRP. Figure 33.1 gives the energy consumption generated by the reception of ZRP traffic. Messages are used to maintain the zone routing tables with a beacon broadcast period of 0.2 s. To obtain the path to a mobile terminal out of the zone, additional control messages are exchanged only on-demand.

33.2.2.2 Energy-Aware Ad Hoc Routing. A number of solutions for reducing the energy consumption associated with ad hoc routing protocols have been investigated. In particular, it has been recognized that selecting the shortest path is not the best choice with respect to minimizing energy consumption, because if the transmission power is dynamically set, a number of short hops results in less energy consumption than fewer longer hops. Thus, to gain the maximum energy saving, one solution is to transmit datagrams at the minimum transmission power that is required for successful reception [11]. The chosen route is then the one that consumes the minimum amount of energy to get the packet to the destination. This solution is based on existing reactive routing protocols such as DSR or AODV. The minimum transmission power given by the MAC layer and required to get to the next hop is included along with the terminal ID in the route request packet, which is further broadcasted by its receiver. Finally, the destination node inserts this power information for each hop in the routing header of the route that it returns to the requesting source node. Thus, the source node gets the power value for each hop from the route reply, from which it computes the total energy cost of the route. A similar approach is undertaken in [28], which assumes that each mobile terminal has a low-power global positioning system (GPS) receiver on board. The transmission power is then dynamically set according to the distance separating two nodes, which is computed according to the location data provided by the GPS. Because this computation requires knowledge of the position of the nearby nodes, each node broadcasts its position. Ultimately, the sender node is able to identify the route that minimizes energy consumption, based on the energy cost of each link of the path. The use of GPS further allows for synchronization among mobile nodes based on the absolute time information provided by GPS. Each node then wakes up periodically to listen for changes and goes back to the sleep mode to conserve energy. Still, thanks to the GPS, a mobile terminal uses a localized search for computing a route and takes into consideration only the links in the immediate neighborhood of the destination or those that are potential candidates. Thus, the mechanism reduces the flood of the network to a localized part of it. The distributed and randomized algorithm introduced in [8] attempts to minimize the number of coordinators (i.e., nodes that forward packets) and to put in the sleep mode noncoordinator terminals. Thus, the number of mobile terminals involved in communication is kept to a minimum and the overall network lifetime is extended. In the proposed algorithm, each mobile terminal broadcasts periodically a `Hello` message containing the terminal state with respect to its possible coordination role. From these messages, each node builds a list of the neighbor coordinators. Meanwhile, the procedure for selecting a route based on the use of the optimal link (either in terms of the number of hops or overall energy consumption) has a negative impact on the network lifetime. The use of a restricted set of mobile terminals (i.e., the coordinators) to route packets leads to the depletion of the energy of the coordinators and thus to their eventual failure. To ensure that all mobile terminals share the task of providing connectivity and hence that energy consumption induced by routing is fairly distributed over all nodes, nodes acting as coordinators are changed over time. Periodically, any noncoordinator node checks whether it should become a coordinator as follows. If the node discovers that two of its neighbors cannot reach each other, either directly or via one or two coordinators, it changes its state to the coordinator state. As illustrated above, sharing the cost of routing among mobile terminals calls for selecting different routes over time for communication between any two given nodes. The protocol proposed in [29] finds multiple paths between a source and a destination and assigns to each path a probability of being chosen. The protocol then achieves a rotation among the chosen routes and leads to nonsystematic use of the same terminals for routing. To reduce the wide disparity of energy level of the mobile terminals, and to maximize the time until the first battery drains out, nodes are classified according to three levels [13]: normal, warning, and danger, corresponding to a remaining energy respectively greater than 20 percent, between 10 and 20 percent, and less than 10 percent. The cost of routing may further be weighed according to the energy level of the routing nodes, so as to route packets through terminals that have plenty of remaining battery. For example, the cost assigned to a route involving a nonpowerful node (i.e., danger or warning level) is higher than the one corresponding to the transmission through a normal node.

33.2.3 *Energy Saving at the Transport Layer*

Once a route is discovered by the routing protocol, data transfer is achieved by a transport protocol such as Transport Control Protocol (TCP), which is the most commonly used. TCP is responsible for verifying correct data delivery from the source to the destination. TCP is tuned for wired networks and performs well in this environment. In wireless networks, TCP exhibits much poorer performance, mainly because wireless networks suffer from packet losses due to high error state and hand-offs. TCP reacts by using congestion control whereas errors are not congestion-related, leading to impact performance. Further investigations of energy consumption under different implementations of the TCP layer has been undertaken, through both simulation (e.g., [4, 24]) and experiment (e.g., [1]). In particular, results presented in [1] show that the frequency of route failures, the routing overhead, and the delay in route establishment are the key parameters affecting TCP throughput in multi-hop ad hoc networks and that the stability of the route is crucial for the performance of TCP. Furthermore, performance can change drastically, depending on the ad hoc routing protocol [2]. One solution is thus to suspend packet transmission when conditions become adverse and resume transmission when they improve [10]. To cope with the problem of using TCP in wireless networks, two major approaches have been proposed. The first one assumes that problems associated with transmission are local. Then, message loss that is not congestion-related is hidden to the sender. The second approach assumes that the sender can distinguish between packet loss due to congestion and packet loss due to other reasons, enabling accurate invocation of congestion control algorithms.

This section has provided an overview of network-level solutions toward minimizing energy consumption associated with communication over ad hoc networks. It is further crucial to develop distributed application software that also accounts for the minimization of the overall energy cost and, in particular, the one of communication. The remainder of this chapter introduces such an application, which allows for energy-aware Web access from mobile terminals.

33.3 Web Caching in Ad Hoc Networks

We have not yet reached the point where anywhere, anytime Internet access is actually offered. The proposed collaborative Web architecture addresses the above issue, concentrating more specifically on Web caching in a mobile environment to allow for Web access, without requiring availability of any infrastructure in the nearby environment. More specifically, our solution lies in supporting ad hoc collaborative Web caching among mobile terminals, exploiting both the ad hoc and infrastructure-based modes of the network for enhanced connectivity. We use ZRP over IEEE 802.11 as the base ad hoc routing protocol for achieving ad hoc cooperative caching among mobile terminals. Mobile terminals belonging to the zone of a given terminal then form a cooperative cache system for the reference terminal because the cost for communicating with them is low both in terms of energy consumption and message exchanges. However, cooperative caching must not be restricted to the mobile terminals belonging to the zone: low-cost reachability of a base station must be accounted for as well as knowledge of a terminal that does not belong to the zone but that is likely to store a requested Web document given commonalties in performed Web accesses. We first introduce the ad hoc communication protocol over ZRP for the retrieval of Web documents (Section 33.3.1) and then detail the ad hoc cooperative caching protocol (Section 33.3.2).

33.3.1 Ad Hoc Communication for Cooperative Web Caching

A mobile terminal may get Web data that is not cached locally through two communication paths:

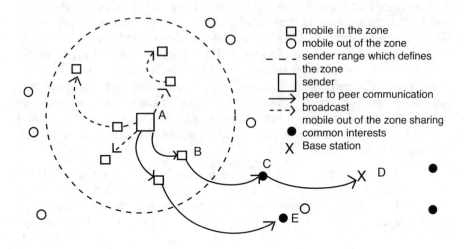

Figure 33.2. Getting Web Data

1. Using the infrastructure-based mode, the terminal may interact with the nearby base station, which forwards the request to the Web.
2. Using the *ad-hoc*-based mode, the terminal may submit a request for the data to the mobile terminals in its transmission range (i.e., accessible in one hop in a base WLAN or in a number of hops using some ad hoc routing protocol).

The former is the most efficient in terms of energy consumption; the mobile terminal sends a single message to a fixed station that has an infinite energy budget. However, the *ad-hoc*-based mode must be enabled for the case where a base station is not reachable in one hop. In this case, a base station can still be reachable in a number of hops, thanks to mobile terminals forwarding the requests. The number of hops that are necessary to access a base station depends on the terminal's location. Let N be this number, then any mobile terminal that is at a distance greater or equal to N is not contacted to get a document. Figure 33.2 depicts the case where the mobile terminal A reaches the base station D in three hops, using the mobile terminals B and C for routing the request. Then, if either a mobile terminal in the zone of A (e.g., B belonging to the path leading to the base station or any other terminal in the zone) or a known mobile terminal located outside the zone but at a lesser distance than the base station D (e.g., C that is in the path leading to D or E that does not belong to the path) holds the requested document in its cache, it returns it to A. Otherwise, the request reaches D and D forwards it to the Web.

We get the following ad hoc communication protocol over ZRP to retrieve a remote Web object W with respect to a given mobile terminal A:

- In-zone communication:
  - If a base station is in the zone of A, then A requests for W through the base station only.
  - Otherwise, A broadcasts the request message for W to the mobile terminals in the zone of A.
- Peer-to-peer communication — if there is no base station in the zone of A, then:
  - If W is not cached by any of the mobile terminals in the zone of A, then a peer-to-peer communication scheme is achieved with mobile terminals that share interests with A (see 33.3.2) and that are at a distance that is less than the one between A and the nearest base station. Mobile terminals outside the zone of A are basically known through two ways:
    - They belong to the path used to reach the nearest base station.
    - They were previously either in the zone or in the path used to reach the base station.
  - The request for W is ultimately forwarded to the nearest base station.

Based on the above, the communication cost and hence the energy cost associated with getting a Web object is kept to a minimum: broadcast is within a zone and peer-to-peer communication occurs only with mobile terminals that are both the most likely to store a requested object and closer than a base station.

33.3.2 Ad Hoc Cooperative Caching

Web data are distributed/cached among the mobile terminals according to Web accesses performed by their user. Without a proxy-type architecture that centralizes requests, local statistics are relied on for a mobile terminal A to identify mobile terminals that are likely to store a Web object requested on A. Such statistics are maintained on A through a terminal profile for every mobile terminal with which A interacts. For a mobile terminal T, its profile is characterized by a value that counts the number of times T is either known to cache an object requested by A or requested for an object to A that A had in its cache. The values of the terminal profiles is used to identify the mobile terminals with which peer-to-peer communication is undertaken. Specifically, known mobile terminals outside the zone and that are at a distance less than a base station, are weighed according to the value of F = terminal profile/hops, where the number of hops, hops, is obtained from the routing table. In addition to the management of terminal profiles to identify mobile terminals that share common interests, we must account for the heterogeneity of the terminals' capacities (i.e., battery, processing, storage, communication). In particular, for two mobile terminals that are equally likely to store a requested object, it is better to contact the one that has the greatest capacity. Ignoring the case where a base station is accessible in the zone and given the ad hoc communication protocol aimed at cooperative Web caching that was discussed in the previous section, the request for a Web object W that is not cached locally, from a mobile terminal A is handled as follows:

1. A first broadcasts the requests for W within the zone.
2. If W is not retrieved then the retrieval protocol iterates on sending the request for W to known mobile terminals outside the zone and at a distance less than the base station according to the maximization of F.
3. The base station is ultimately contacted.

The processing of requests is further handled as follows. A mobile terminal that receives the request for W and caches it, increments its local value of A's terminal profile. If the terminal is further willing to cooperate (e.g., absence of energy safeguarding or of security policy enforcement), it returns a hit message, which embeds:

- TTL that gives the time-to-live field of the document.
- Capacity that characterizes the capacity of the terminal to handle requests, whose value is in the range [0–1], 1 denoting the highest capacity. Currently, we use a simple scheme to set the value of Capacity; it is equal to the percentage of the energy budget that is left. It is part of our future work to investigate a more accurate way of computing Capacity, in particular accounting for the various terminal resources.

For every hit message that it receives, A increments the terminal profile of the sender. Among the mobile terminals that replied by a hit message, A selects the terminal from which W should be obtained, that is the one that maximizes the following function: R = Capacity ($\lambda \times$ TTL + $\mu \times$ hops).

More precisely, the value of R is computed as follows. The metrics used for consistency (i.e., TTL) and for communication cost (i.e., hops that gives the number of hops) are distinct and should be made comparable. As a dispersion measure, we use standard deviation. Thus, considering that A received n hit messages for W, A computes the following mean m_{TTL} (resp. m_{hop}) and standard deviation σ_{TTL} (resp. σ_{hop}) of TTL (resp. hops):

$$m_{TTL} = \frac{1}{N} \sum_{i=1}^{N} TTL_{mobile_i} \quad \sigma_{TTL} = \sqrt{\frac{1}{N} \sum_{i=1}^{n} \left(TTL_{mobile_i} - m_{TTL}\right)^2}$$

$$m_{hop} = \frac{1}{N} \sum_{i=1}^{N} hop_{mobile_i} \quad \sigma_{hop} = \sqrt{\frac{1}{N} \sum_{i=1}^{n} \left(hop_{mobile_i} - m_{hop}\right)^2}$$

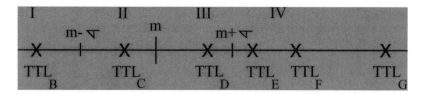

Figure 33.3. TTL Distribution

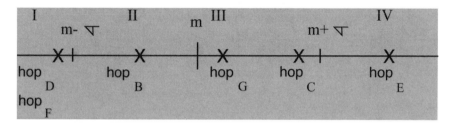

Figure 33.4. Hops Distribution

Figure 33.3 and Figure 33.4, respectively, give a set of TTL and hops values considering reception of hit messages from terminals [B–G]. More precisely, we distinguish four intervals for the values of TTL and hops — Level $I_1 =]-\infty, m - \sigma]$, Level $I_2 = [m - \sigma, m]$, Level $I_3 = [m, m + \sigma]$, Level $I_4 = [m + \sigma, +\infty]$. Then, the values of TTL are mapped into one of the ranges [0–1], [1–2], [2–3], and [3–4], if they belong to I_1, I_2, I_3, and I_4, respectively. The value taken in the target range is then set proportionally to the value in the initial range and is the one used in the computation of R. The same applies for the value of *hops*. For illustration, Table 33.2 gives the initial TTL and *hops* values and associated TTL and *hops* in the targeted unified metric, considering the distributions given in Figure 33.3 and Figure 33.4.

Then, the values of λ and μ are set so that $\lambda + \mu = 1$ and will be chosen according to the respective weighing factors for consistency and communication cost. Figure 33.5 gives the value of R/capacity for mobile terminals [B–G] considered in Figure 33.3 and Figure 33.4, according to the value of λ, ranging from 1 (i.e., $\mu = 0$ and the communication energy cost is ignored) to 0 (i.e., $\mu = 1$ and the communication energy cost is the only selective factor). Note that we do not use `miss` messages for mobile terminals to notify that they do not cache a requested object.

This is to minimize both network load and energy consumption. Hence, we need to use time-outs to detect the absence of a requested object. The value of the time-out is set according to the greatest number of hops that are involved to interact with the mobile terminals to which the object is

Table 33.2. TTL and Hop Metric

Mobile	TTL	TTL Metric	Hop	Hop Metric
B	$(m-\sigma) - \frac{1}{2}\sigma = \frac{1}{2}$	$1 - \frac{1}{2}$	4	$1 + \frac{1}{\sqrt{5}}$
C	$m - \frac{3}{7}\sigma = \frac{4}{7}$	$2 - \frac{3}{7} = \frac{11}{7}$	8	$\frac{25}{7} + \frac{2}{\sqrt{5}}$
D	$(m+\sigma) - \frac{3}{10}\sigma = \frac{27}{10}$	$3 - \frac{3}{10} = \frac{27}{10}$	2	$\frac{27}{10}$
E	$(m+\sigma) + \frac{1}{4}\sigma = \frac{13}{4}$	$\frac{17}{8}$	10	$\frac{49}{8}$
F	$(m+\sigma) + \frac{5}{6}\sigma = \frac{23}{6}$	$\frac{29}{12}$	2	$\frac{29}{12}$
G	$(m+\sigma) + 2\sigma = 5$	4	6	$2 + \frac{1}{2\sqrt{5}}$

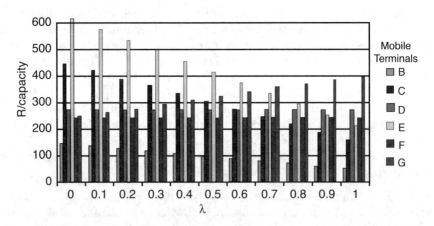

Figure 33.5. R/capacity Function

requested, together with the current network load. Upon expiration of the time-out, if `hit` messages have been received, the Web object will be requested to the mobile terminal that maximizes R. Otherwise, the next iteration of the cooperative caching protocol is processed. For the case where a `hit` message is received after time-out expiration, although the object is still not retrieved, the message is accounted for in the current step of the protocol.

33.4 Local Caching

The ad hoc cooperative caching protocol introduced in the previous section is complemented with a local caching strategy, which is adaptive according to the current capacity of the terminal. In particular, the local cache is managed in a way that accounts for the available energy and the network connectivity (Section 33.4.1). To reduce the user perceived delay, the caching strategy is further coupled with a prefetching strategy (Section 33.4.2).

33.4.1 Cache Management

Local cache management mainly amounts to the implementation of a replacement algorithm that is run when the cache gets full. Ideally, the algorithm must remove from the cache, data that will not be accessed in the future. A number of replacement algorithms have been proposed in the literature for Web caches. However, these are aimed at stationary hosts that have an infinite energy budget and are strongly connected. In the case of caching on a mobile terminal, it should be considered that the mobile terminal must save energy and that the network connectivity is not reliable. We thus weigh every cached document according to both its probability of being accessed in the future and the energy cost associated with getting the document remotely. Then, documents with the lowest weights are those that are removed from the cache. The document weight is computed according to the following criteria:

- Popularity — the Popularity value serves to approximate the probability of future access, both on the terminal and from remote terminals, as enabled by the cooperative caching protocol. The probability is approximated according to the number of times the document has been requested since it has been cached.
- AccessCost — the AccessCost value gives an estimate of the energy cost associated with getting the document remotely if it is to be removed from the cache. This cost varies depending on whether a base station is accessible in the zone of the terminal, the document is cached on a mobile in the zone of the terminal, or communication out of the zone is required to retrieve the document. In the first two cases, the cost is quite low, but it may be quite high in the third case. The value of the access cost is computed according to the energy consumption associated with intrazone and interzone communication. Intrazone communication holds if a base station is known to be in the zone. It is further assumed if the document was obtained from a terminal that is still in the zone, as identified using the routing table.
- Coherency — a document is valid for a limited lifetime, which is known using the TTL field. However, when the energy remaining on

the terminal is low, it is better to favor energy saving over the accuracy of the document. Hence, the value of Coherency is equal to v_{TTL} where v is initially set to the actual document TTL and increases as the available energy decreases. We get the following function to compute the document weight:

$$Q = \alpha \times Popularity + \beta \times AccessCost + \gamma \times Coherency + \delta \times size$$

The values of α, β, γ, and δ are set so as make the values of AccessCost, Popularity, Size, and Coherency, decreasingly prominent factors for deciding whether a document should be kept in the cache. The metric used to compare the different parameters is similar to the one described in Section 33.3.2, for the function R, and α, β, γ, and δ are adjusted as for the λ and μ factors. Notice that our Q function offers similarities with the one used by hybrid replacement algorithms that were proposed in the literature for Web caching on stationary hosts (e.g., [19]). In the same way, we use a function that accounts for a combination of criteria in the replacement decision. However, our function differs in that energy savings is a prominent criterion used by our algorithm.

33.4.2 Prefetching

Although prefetching is at the expense of bandwidth consumption, it is a useful mechanism in a mobile environment because it enables masking disconnection occurrences to the user. We propose an online prefetching strategy that accounts for the specifics of mobile terminals in terms of resource availability and network load. Online prefetching strategy can conveniently be complemented with offline strategy when the mobile terminal is being powered or when a disconnection is anticipated. Due to the lack of space, we do not present the offline strategy in the remainder of this discussion, although it may be inferred from the online strategy coupled with a graphical user interface as the one of the ARTour Web Express software [7]. The prefetching strategy is enabled only if the energy remaining on the terminal exceeds a given threshold and the network load is low, which is detected according to the number of error messages generated by the underlying network protocol that are received on the mobile terminal. Our online prefetching strategy relies on a prediction algorithm that exploits the history of past Web accesses, while minimizing the memory space consumed for maintaining the history. The prediction algorithm selects the page to be prefetched, which is the one that has the highest probability to be accessed next given the page that is being visualized by the user. Given the visualization of a page W, the page W_g is prefetched if it was previously accessed more than once and it has the highest probability of being accessed subsequently to the visualization of W. The latter condition is checked for using a dependence tree that is maintained for each page that has been accessed in the past (over a given period of time). The

Energy-Aware Web Caching over Hybrid Networks

dependence tree associated to a page W gives the sequence of consecutive accesses from W, which corresponds to the root node. Each directed arc of the tree is weighed by the probability of accessing the page denoted by the destination node, from the page denoted by the source node (i.e., the number of times this specific path was followed divided by the total number of accesses from the source node). However, to keep the size of the trees small, their maximum depth is set to two, which has further shown to be sufficient with respect to prediction accuracy in the Web [3].

When a page W is downloaded, all the links $l_{i=1...n}$ embedded in W that provide access to pages $W_{i=1...n}$ are examined. The dependence tree associated with W is used to identify the respective probability of each $W_{i \in \{1,...,n\}}$ to be accessed next, noted $P(W_i/W)$. Consider first the use of a tree of depth one, qualified as level-one tree, which minimizes memory consumption. Consider further that the user requests access to the page $W_{g \in \{1,...,n\}}$, from W. The level-one tree associated with W_g (i.e., the tree whose root is W_g) gives the probability of access for every page $W_j = _{1...m}$ that is denoted by a link $l_{j=1...m}$ embedded in W_g. However, the probability of accessing $W_{h \in \{1,...,m\}}$ from W_g given in the level-one tree does not account for the fact that the access follows from an access to W, which can be ignored if $P(W \cap W_g \cap W_h) = 0$. Maintaining dependence trees of depth two, qualified as level-two trees, is thus more accurate because it gives the exact value of $P(W_h/W_g/W)$ (Figure 33.6). In general, maintaining level-two trees is optimal from the standpoint

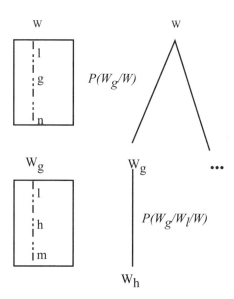

Figure 33.6. Level-Two Dependence Tree

of prediction accuracy in the Web. However, maintaining such trees consumes a significant memory space, which is not affordable by mobile terminals having a small storage budget. Hence, level-two trees are maintained on mobile terminals by default and level-one trees are used on mobile terminals having a low memory budget. In addition, when the space used for maintaining trees exceeds a given limit, trees associated with pages that have the lowest weights W are removed. For selecting the page to be prefetched, we weigh the probabilities of access given by the dependence trees with additional criteria that have been proven useful in the Web for better prediction accuracy [5].

These criteria relate to:

1. The page site because it has been observed that a page located on the same site as the one of the page being visualized has a higher probability to be accessed next than a page located on a distinct site
2. The type of the link because it has been observed that the probability of selecting a link differs depending on whether it is a button, an image, or an hypertext link
3. The presence of the page in the user's bookmarks

Hence, for a page W being visualized and the set of pages $W_{i=1...n}$ that are denoted by the links $l_{i=1...n}$ embedded in W, the page $W_{g \in \{1,...,n\}}$ that is selected for prefetching is the one that maximizes the following value:

$$P = l \times t \times b \times P(W_j/W)$$

where $P(W_j/W)$ actually denotes $P(W_j/W)$ or $P(W_j/W/W')$ depending on whether level-one or level-two trees are used and l, t, and b are weighing factors that are respectively associated with the page location, the type of the link, and the presence of the page in the user's bookmarks.

33.5 Evaluation

Control messages generated by the ZRP routing protocol (see Section 33.2.2) and messages associated with ad hoc cooperative caching affect the network traffic and energy consumption on mobile terminals. Given the energy consumption generated by ZRP, we evaluate energy consumption associated with ad *hoc* cooperative caching, as the sum of the energy consumption induced for the various mobile terminals that are involved in the cooperation, which adds to the energy cost induced by ZRP. In the following, we first examine the energy consumption induced by ad hoc (multihop) networking (Section 33.5.1), and then the energy cost of ad hoc cooperative caching (Section 33.5.2). The latter includes the energy cost due to both communication and computation. However, we do not consider the energy cost associated with computation because it is induced by any local cache management and it is negligible compared to the energy cost of communication: the energy cost of transmitting 1 kilobyte (KB) is approximately

Table 33.3. Energy Consumption in a One-Hop Zone

Mobile Terminal in Range of	Surface	Number of Mobile Terminals	Total Energy for a Zone
The sender X and the receiver A	$S_{AX} = \theta\,(r^2 - a_1^2) + \beta\,(R^2 - a_2^2)$	$n_{AX} = \dfrac{NS_{AX}}{S}$	$N_{AX}\,\varepsilon_{AX}$
The sender X	$S_X = \pi r^2 - S_{AX}$	$N_X = \dfrac{NS_X}{S} = \pi r^2 - \theta r^2 \sin^2\theta + \beta\,R^2 \sin^2\beta$	$N_X\,\varepsilon_X$
The receiver A	$S_A = \pi R^2 - S_{AX}$	$N_A = \dfrac{NS_A}{S} = \pi R^2 - \theta r^2 \sin^2\theta + \beta\,R^2\sin^2\beta$	$N_A\,\varepsilon_A$

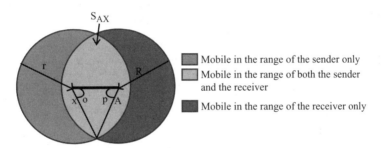

Figure 33.7. Mobile Terminals in the Transmission Range of the Sender and of the Destination Node

the same as the energy consumed to execute three million instructions [27]. We further use the following wireless interface for our evaluation: 2.4 Ghz direct sequence spread spectrum (DSSS) Lucent® IEEE 802.11 WaveLan PC (Bronze) 2 Mbps.

33.5.1 Energy Consumption of Ad Hoc Networking

Consider a network of N (= 500) mobile terminals whose transmission range is of about 250 meters (m), that is such that the mobile terminals are uniformly distributed in the network surface S with S = 4000 [m] 4000 [m]². The overall energy consumption associated with the emission of one message within a one-hop zone is the sum of the energy consumed by every mobile terminal in the one-hop zone (see Table 33.3 where notations follow from Figure 33.7) and is given by $\varepsilon = n_{AX} \times \varepsilon_{AX} + n_X \times \varepsilon_X + n_A \times \varepsilon_A$. Then, the energy consumption of the overall network is the sum of the energy consumed per one-hop zone that is traversed by this message.

MOBILE COMPUTING HANDBOOK

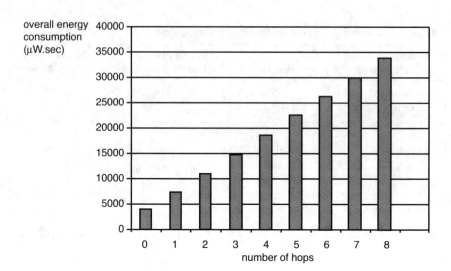

Figure 33.8. Energy Consumption in the Network for the Retrieval of Web Data

Table 33.4. Energy Consumption According to the Network Density

Number of mobile terminals in the network surface	500	600	700
Energy consumption of nondestination mobile terminals ($\mu W.sec$)	1389	18959	2136
Ratio	9	7	6.11

Figure 33.8 gives the energy consumption associated with the delivery of a message of 1 Kb to the destination node according to the number of hops necessary to reach the destination, which is the sum of the energy consumed on all the nodes involved in the communication.

The figure clearly demonstrates that the energy consumption increases with the number of hops. This directly follows from the resulting increase of both mobile terminals forwarding the message and nondestination nodes receiving control messages. Table 33.4 further evaluates the impact of the network density on energy consumption. For a constant number of hops (= 3), we see that increased density of terminals results in additional energy consumption for nondestination nodes. But, the ratio:

$$\frac{\varepsilon_{send} + \varepsilon_{dest} + \varepsilon_{forwarding\_mobile\_terminals}}{\varepsilon_{overall\_non\_destination\_mobile\_terminals}}$$

highlights the weak impact of message reception on nondestination nodes, on the overall energy consumption. Indeed, for 600 mobile terminals and a

Energy-Aware Web Caching over Hybrid Networks

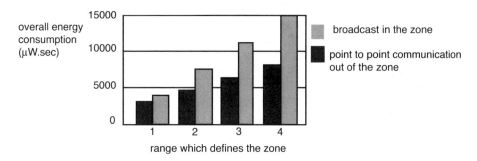

Figure 33.9. Energy Consumption in the Network for the Broadcast and the Peer-to-Peer Communication

destination node at 4 hops of the sender, the energy consumed by nondestination nodes is about 6 times less than the energy consumed by the sender, the receiver and the 3 forwarding nodes.

33.5.2 Energy Consumption of Ad Hoc Cooperative Caching

Having examined the energy consumption associated with data delivery over the ad hoc network, we now focus on the energy consumption induced by our ad hoc cooperative caching protocol. A request message for a Web page induces broadcasting (toward mobile terminals in the requester's zone) and peer-to-peer communication (with mobile terminals sharing the same interests). Before broadcasting a message, the sender listens to the channel; if no traffic is detected, the message is broadcasted. Table 33.5 gives the energy consumption induced by in-zone broadcasting and the one associated with peer-to-peer communication is given in Table 33.1 (Section 33.3.2.1.1). The difference between the energy consumption associated with broadcasting and the energy consumption associated with peer-to-peer communication is due to the emission of control messages (RTS, CTS, and ACK messages) in the case of peer-to-peer communication. Figure 33.9 gives the energy consumed by cooperative caching according to the number of hops that defines the zone. Precisely, the figure gives the cost associated with broadcasting plus peer-to-peer communication with a

Table 33.5. Energy Consumption on the Sender and Destination Nodes for Broadcast

Mobile	Energy consumption (μW.sec)	m (μW.sec/byte)	P (feeney μW.sec)
Sender X	$\varepsilon_{brsend} = m_{brsend} \times size + p_{brsend}$	$M_{brsend} = 1.9$	$P_{brsnd} = 266$
Destination A	$\varepsilon_{brdest} = m_{brdest} \times size + p_{bdest}$	$m_{bdest} = 0.5$	$P_{brdest} = 56$

node at a distance that adds one hop to the number of hops that defines the zone.

As expected, the overall energy consumption of both broadcast and peer-to-peer communication increases with the number of hops that defines the zone (because more mobiles are in the zone). Comparing energy consumption of broadcasting with the one of peer-to-peer communication, we find that the cost added by broadcasting is weak compared to the number of mobile terminals contacted, and hence supports our approach.

33.6 Conclusion

Embedded systems are now becoming ubiquitous. In particular, they have large applicability in mobile, battery-operated personal communication systems, for which energy is a major constraint. However, effective ubiquitous usage of mobile systems remains limited by available energy. There has been extensive study on minimizing the energy cost of mobile systems at the hardware, operating system, and protocol levels. On the other hand, the design of mobile applications accounting for energy saving has received little attention. This chapter has presented such an application, which aims at providing energy-efficient Web access from mobile terminals while enhancing connectivity, through cooperative caching among mobile terminals over ad hoc networks. Our solution combines two key features:

1. Combined exploitation of multi-hop and infrastructure-based communication for enhanced, energy-efficient connectivity
2. Collaborative caching among mobile terminals up to the reachability of a base station for increased, energy-efficient data availability

In particular, we provide a collaborative scheme that increases performances by selecting the most suitable terminal to retrieve information, resulting in a higher hit ratio while limiting the network overflow. This selection process relies on user's profile defining a common center of interests and also investigates different metrics such as bandwidth availability, connectivity, terminal capacity (e.g., battery resources, memory size, and processing speed) to define with which mobile terminal communication is going to be undertaken. Our solution further integrates a prefetching strategy to mask disconnection to the user and to reduce the user's perceived delay, while limiting additional network traffic and being adaptive according to the storage capacity of the terminal. To validate our approach targeting a minimization of the energy consumption, we have provided a detailed evaluation of the energy consumption generated by our cooperative caching strategy. This evaluation takes into account the energy consumption of all terminals involved in the communication, including the cost for sender, forwarding terminals, and nondestination nodes.

Notes

1. http://www.frost.com/prod/servlet/frost-home.pag.
2. This specific configuration example is the one used in [15] for the evaluation of ZRP.

References

1. Agrawal, S. and Singh, S., An experimental study of TCP's energy consumption over a wireless link, in *Proc. EPMCC*, 2001.
2. Ahuja, A. et al., Performance of TCP over different routing protocols in mobile ad hoc networks, in *Proc. IEEE VTC*, 2000.
3. Aumann, Y. et al., Predicting event sequences: Data mining for prefetching Web pages, in *Proc. KDD*, 1998.
4. Bansal, S. et al., Energy efficiency ad throughput for TCP traffic in multi-hop wireless networks, in *Proc. IEEE InfoCom*, 2002.
5. Brewington, B.E. and Cybenko, G., How dynamic is the Web?, in *Proc. WWW9*, 2000.
6. Cano, J.C. and Manzoni, P., A performance comparison of energy consumption for mobile ad hoc routing protocols, in *Proc. IEEE MASCOTS*, 2000.
7. Chang, H. et al., Web browsing in a wireless environment: disconnected and asynchronous operation in ARTour Web Express, in *Proc. ACM/IEEE MobiCom*, 1997.
8. Chen, B. et al., Span: An energy-efficient algorithm for topology maintenance in ad hoc wireless networks, in *Proc. ACM/IEEE MobiCom*, 2001.
9. Clausen, T. et al., Optimized link state routing protocol, in *Proc. IEEE INMIC*, 2001.
10. Zorzi, M. and Rao, R.R., Energy constrained error control for wireless channels, *IEEE Personal Communication Magazine*, vol. 4, 1997.
11. Doshi, S., Bhandare, S., and Brown, T.X., An on-demand minimum energy routing protocol for a wireless ad hoc network, *ACM SIGMOBILE*, 2002.
12. Feeney, L. and Nilsson, M., Investigating the energy consumption of a wireless network interface in an ad hoc networking environment, in *Proc. IEEE InfoCom*, 2001.
13. Gupta, N. and Das, S.R., Energy-aware on-demand routing for mobile ad hoc networks, in *Proc. IEEE IWDC*, 2002.
14. Haas, Z., A new routing protocol for the reconfigurable wireless networks, in *Proc. IEEE ICUP*, 1997.
15. Haas, Z. and Pearlman, M.R., The performance of query control schemes for the zone routing protocol, *ACM/IEEE Transactions on Networking*, vol. 9, no. 4, 2001.
16. Ye, W., Heidemann, J., and Estrin, D., An energy-efficient MAC protocol for wireless sensor networks, in *Proc. IEEE InfoCom*, 2002.
17. Sivalingam, K.M. et al., Design and analysis of low-power access protocols for wireless and mobile ATM networks, *Wireless Networks*, vol. 6, no. 1, 2000.
18. Intel. Mobile power guide v1.00. http://developer.intel.com/design/mobile, 2000.
19. Jinag, Z. and Kleinrock, L., Web prefetching in a mobile environment. *IEEE Personal Communications*, vol. 5, no. 2, 1998.
20. Johnson, D. and Maltz, D., Dynamic source routing in ad hoc wireless networks, In T. Imielinski and H. Korth, Eds., *Mobile Computing*, 1996.
21. Jones, C.E. et al., A survey of energy efficient network protocols for wireless networks, *Wireless Networks*, vol. 7, no. 4, 2001.
22. Singh, S. and Raghavendra, C., PAMAS: Power aware multi-access protocol with signalling for ad hoc networks, *ACM Computer Communications Review*, 1999.
23. Park, V. and Corson, M., A highly adaptive distributed routing algorithm for mobile wireless networks, in *Proc. IEEE InfoCom*, 1997.
24. Zorzi, M. and Rao, R.R., Throughput and energy performance of TCP on a wideband CDMA air interface, *Wireless Communications and Mobile Computing*, 2002.
25. Perkins, C.E., Ad hoc on-demand distance vector routing, in *Proc. WMCSA*, 1999.

26. Perkins, C.E. and Bhagwat, P., Highly dynamic destination-sequenced distance-vector routing (DSDV) for mobile computers. *Computer Communications Review*, 1994.
27. Pottie, G.J. and Kaiser, W.J., Wireless integrated network sensors, *Communications of the ACM*, May 2000.
28. Roduplu, V. and Meng, T., Minimum energy mobile wireless networks, in *Proc. IEEE JSAC*, 1999.
29. Shah, R.C. and Rabaey, J.M., Energy-aware routing for low energy ad hoc sensor networks, in *Proc. IEEE WCNC*, 2002.
30. Sheu, S.T., Tsai, Y., and Chen, J., Mr2 rp: The multi-rate and multi-range routing protocol for IEEE 802.11 ad hoc wireless networks, *Wireless Networks*, vol. 9, no. 3, 2003.
31. Simunic, T. et al., Dynamic power management for portable systems, *Mobile Computing and Networking*, 2000.

Chapter 34
Transmitter Power Control in Wireless Computing

Savvas Gitzenis and Nicholas Bambos

34.1 Introduction

Following the explosive growth of wireless telephony in the 1990s, wireless mobile computing is rapidly expanding in the current decade. Indeed, thanks to the digitization of the networks and the miniaturization of the mobile devices, second generation (2G) cellular networks have enjoyed a phenomenal success, offering mobile telephony services. One decade later, the same networks upgraded to 2.5G, started offering data/packet services enabling continuous Internet access on the move at rates comparable to dial-up modems. In the forthcoming years, the 3G networks are expected to increase the access data rates, which will enable a suite of new applications previous networks could not support, such as video-telephony, music on demand, etc.

In the unlicensed spectrum bands, a similar course has taken place. The initial burst of sales regarding cordless telephony is now being succeeded by wireless local area network (WLAN) and personal area network (PAN) devices — prominent examples of them being the IEEE® 802.11 suite and the Bluetooth® standards, respectively. With the infrastructure being deployed in both licensed and unlicensed bands at a quick pace, computing rapidly transitions from the desktop to mobile.

34.1.1 Power Control Issues in Wireless Packet Communication

A key issue in wireless network design is that of transmitter power control or how each radio transmitter should choose the power level to transmit at to its intended receiver on its wireless communication link. It is obviously of critical importance because power transmitted on each link is seen as interference by other links sharing the same channel, that is, links operating over the same set of wireless resources (frequency bands in frequency division multiple access [FDMA], time slots in time division multiple access [TDMA], or spreading codes in code division multiple access [CDMA] systems).

Early wireless cellular networks employed rudimentary power control schemes (i.e., only CDMA systems mandated power control at the uplink to mitigate the near–far effect) and relied instead on the use of high spatial separation between cochannel links to confine interference, usually considering the worst case interference scenario. Such an approach has serious drawbacks, resulting in low spectrum reuse density and high transmitter energy consumption. To mitigate these shortcomings, network designers can increase the infrastructure density, deploying more base stations (BSs) or access points (APs) in a given area. Thus, the BSs and APs come closer to the wireless terminals and the radio links get shorter, permitting substantial reduction of transmission power levels; moreover, as each link uses the same wireless resources over shorter distances, the spatial reuse increases. Unfortunately, this solution can be rather expensive in terms of the infrastructure cost. Moreover, the strain on network control overhead increases due to higher signaling and control information exchange; indeed, for given user mobility levels, the rate of hand-offs between the terminals and the BSs and APs increases, making it difficult to maintain an adequate quality of service (QoS) grade on each link.

As next generation wireless networks are expected to support significantly higher numbers of users, it will be advantageous to employ — in conjunction with denser infrastructure — smart power control schemes allowing for more efficient use and reuse of the wireless resources. Minimizing the transmission power of each individual link (given its required QoS grade) lowers the overall interference in the channel, allowing for denser packing of links through higher spectrum reuse. Moreover, the judicious selection of transmitter power can help each wireless terminal minimize its energy consumption, a key issue for battery-operated mobile devices.

From an individual link's perspective, it may initially seem advantageous to use higher transmitter power to compensate for higher radio signal propagation losses, channel interference, etc. However, higher power will result in increased interference on other links, which would raise their own powers to compensate for it, potentially resulting into a vicious cycle of power increases. It is, therefore, in the interest of all links to cooperate in order to coexist in the channel efficiently, minimally disturbing each other. Yet, the coordination between them need not be centrally managed. Indeed, although the interference can be be detrimental when unconstrained, it can be quite helpful in providing a collective feedback signal about the congestion state of the network. Namely, by measuring the interference at the receiver, the individual links can coordinate their transmissions and avoid destructive collisions of transmitted packets in a fully distributed and autonomous (hence, scalable) manner, to be seen later.

Power control may be considered to operate at the physical and Media Access Control (MAC) layers and naturally couples with modulation/coding or mode control. In particular, a transmission mode is a combination of modulation and coding schemes; for example, the transmitter may choose from different modulation schemes (binary phase shift keying (BPSK), 4-QAM, [QAM — quadrature amplitude modulation] or 16-QAM, etc.) and different error-correcting codes (by adjusting the number of redundant bits padded to the payload). By choosing a mode, the transmitter effectively adjusts the frame transmission rate and the robustness of the transmitted data against errors. Thus, it can opt for a high data-rate modulation/coding mode, at the cost of an increased bit error rate, or be more conservative and choose a low-rate mode that provides higher immunity against errors. For optimum wireless link performance, the two decisions of power and mode control is best to be taken jointly in a single step to leverage important synergies, as we will see later.

Aiming to understand the power control and bandwidth allocation issues in wireless computing, we investigate the following two core canonical problems:

1. Power-controlled packet forwarding over a wireless link — that is, packets arrive at the transmitter buffer of each link, which autonomously decides what power and mode to use to transmit, given its packet backlog and channel interference.
2. Power-controlled data prefetching over a wireless link — that is, wireless mobile terminals issue requests to download data items residing on a server over a wireless link. The requests are issued by the terminal's user throughout some computing process (e.g., Web browsing, database query). To enhance the performance of the system, the terminal may prefetch data items in anticipation of future requests. The choice of transmitter power and mode to download data is critical, given the candidate items for request and the channel interference.

The first problem is a basic one in wireless data communications. It is similar to the classical queuing model of a server with a queue, except that the service rate is adjusted by the transmission power level and the channel interference. The goal is to minimize the average transmission power (energy) spent on the wireless link to serve its traffic load, over an acceptable delay that buffered packets can tolerate, which reflects the link's QoS grade. The wireless link is time-varying due to transmission activity of other links in the channel which induce interference, as well as terminal mobility resulting in varying propagation losses, fading, shadowing, etc.

At first-cut, an obvious strategy for the transmitter would be to increase or decrease the transmission power in response to the changes sensed in the channel, trying to maintain a constant throughput over the wireless link. Indeed, this is the right strategy when the traffic is unelastic to buffering, as in the case of voice. However, with data packets, the transmitter has an alternate option: instead of trying to compensate for the wireless channel deficiencies, it may choose to ride on the fluctuations. Hence, when the channel dips to low quality, the transmitter can buffer the incoming data — data packets can usually be buffered with minimal degradation in the QoS. Thus, upon a packet arrival, the transmitter is faced with a dilemma: should it attempt to transmit the packet immediately or should it better buffer the packet and wait for the wireless link to improve? The latter choice essentially takes a bet on the channel improving shortly later on, so that the packet can be transmitted with less power (energy). The price for postponing the transmission is the increase in the packet delay, which affects the QoS. Moreover, it requires increasing the buffer space to store all the packets that may arrive during the period that the transmitter stays idle.

The above considerations essentially define the fundamental trade-off between the average power cost and the buffering/delay cost. The optimal operating point is not unique, instead there is a multitude of them depending on the different priorities that may be assigned to the power over the delay cost. These priorities can be assigned at the level of each individual link, thus some links may employ more power-sensitive and delay-tolerant policies in contrast to the more power-aggressive and delay-sensitive ones others may choose.

The second problem can be regarded as the reverse of the first. Instead of the transmitter sending packets to the receiver, it is the mobile terminal (receiver) that places requests to download data from the former server (transmitter) over the wireless link. To shorten the download/access delay time, the technique of prefetching may be employed: if the system can predict the future user requests to an adequate degree of confidence, it can download the anticipated data in advance, that is to prefetch them.

Prefetching has been originally proposed to shorten the access delay in slow communications links. However, in wireless networking, if exercised jointly with power control, it can also help minimize the power cost by selecting to transmit only during periods when the channel is in a good state.

Similar to the previous problem, a trade-off between power and delay cost is induced in the system. Namely, a delay-sensitive system tends to prefetch as much as possible and keep the transmitter busy all the time, thus spending lots of power (energy) and risking downloading items that may never be requested. On the contrary, a power-sensitive system limits itself by transmitting only:

- When the wireless link quality is adequate
- Data that has a relatively high probability of being accessed in the future

Unlike packet forwarding, which is a classic MAC/physical layer problem, observe that in prefetching, the power used for transmitting each particular data item is coupled to the probability that the item will be accessed in the future. In other words, it is a problem involving cross-layer design, as the data requested is jointly decided in the MAC/physical and the application layer.

In the rest of the chapter, we study these two problems and some interesting extensions of them. Our presentation emphasizes the fundamental principles, we deliberately avoid engaging in architectural or implementation issues.

34.2 Packet Forwarding over Single-Mode Wireless Links

Let us assume that the time is slotted and indexed by $n \in \{1, 2,...\}$. Each slot corresponds to the time duration required to transmit one packet/frame over the wireless link. The transmitter is equipped with a buffer, which stores packets that are stored while waiting to be transmitted. Let b_n be the number of packets in the buffer at the beginning of time slot n. Let i_n be the interference/stress state of the wireless channel during time slot n, which incorporates the channel interference, propagation losses, scattering, fading, shadowing, and all other effects that are potentially detrimental to the quality of communication. It takes values in I, the set of all possible channel stress states, and is assumed to remain constant throughout the duration of each time slot. It may switch randomly, though, between consecutive ones.

Observing the channel state i_n and the backlog size b_n at the beginning of time slot n, the transmitter decides whether to transmit the head packet of the queue and if so, at power level p. The latter is chosen from a set $\mathcal{P}$ of all possible power levels the transmitter may use. For our purposes, deciding not to transmit is equivalent to setting the power to $0 \in \mathcal{P}$. Each packet transmitted and successfully decoded at the receiver is removed from the buffer at the end of time slot n, or equivalently, the beginning of $n + 1$; otherwise (if not successfully transmitted/decoded) it remains in the buffer to be again retransmitted in a future time slot. For simplicity, we assume that the receiver immediately notifies the transmitter whether the packet was successfully received or not, possibly using an acknowledgment/negative acknowledgment (ACK/NACK) message sent over a highly reliable control channel in negligible time. Moreover, along with the ACK/NACK, the received packet may also notify the transmitter about the channel state i_n observed at the beginning of time slot n.

Let $s(p, i)$ be the probability that a transmitted packet is successfully received and decoded, given that power $p \in \mathcal{P}$ was used to transmit it during a time slot when the channel interference state was $i \in \mathcal{I}$. Successful packet transmissions in different time slots are assumed to be statistically independent. As a baseline model, we can typically use:

$$s(p,i) = s\left(\frac{p}{i}\right),$$

where p/i reflects the signal to interference (plus receiver noise) ratio (SIR) and $s(\cdot)$ is any increasing function. Actually, any function $s(p, i)$ can be used in our analysis below, as long as it is increasing in the power p and decreasing in the interference i, which is indeed expected in any transmission system.

To spotlight the fundamental power versus delay trade-off in wireless packet transmission and for other methodological purposes which are to become clear later, we assume in the analysis that the channel interference i_n is nonresponsive to the transmitter power p_n: that is, it is driven by an agent extraneous to the system. In practice, interference on the link under consideration from other links sharing the same channel will be responsive, as links interact and their dynamics are coupled/entangled. Indeed, if a link raises its power, the others will see higher interference and may also raise theirs, resulting in increased interference on the primary one. However, the nonresponsive channel assumption is a good methodological vehicle, allowing for studying the link in virtual isolation, yet in a fluctuating channel. It can be viewed as a first step toward the analysis of the complete coupled link system and is model reduction, which proves to be very useful. Indeed, its solution allows us to gain important insights regarding the power versus delay trade-off in wireless packet transmission, which proves to be relevant for the general case of the responsive channel as well. We can also see *ex post facto* that Power Controlled Multiple Access (PCMA) algorithms designed under the nonresponsive channel assumption will indeed perform surprisingly well even in the case of responsive channel stress [7, 8].

Summing up the above discussion, we assume here that the channel interference/stress process i_n evolves as an irreducible Markov chain with transition probabilities:

$$r_{ii'} = P[i_{n+1} = i' \mid i_n = i]$$

and is statistically independent from other random events (potential packet arrivals and successful packet transmissions).

Given this set of assumptions regarding the operation of the system, we proceed to discuss the core of the power control problem, expanding that later in several steps.

34.2.1 Optimally Emptying the Transmitter Buffer

In the simplest form of the problem, we would like to optimally empty the transmitter buffer, when the system starts with $b_0 = b$ packets in it. In every time slot n, the following two key costs are incurred:

1. A power cost equal to the power $p_n \in \mathcal{P}$ used for transmission during slot n
2. A delay/buffering cost $D(b_n)$, where $D(\cdot)$ is some increasing function assigned by the system operator or manager.

The sum of the two costs is the total cost incurred by the the link in each time slot. The goal is to find a power control policy p_n that minimizes the total cost incurred until the buffer is cleared and all packets have been successfully transmitted.

Note that the system evolves as a controlled Markov chain with state (b_n, i_n) and the optimal power action p_n^* depends on the current system state (b_n, i_n). In order to specify it, we formulate the problem in the context of dynamic programming [1], as follows. Let $V(b, i)$ be the cost-to-go, that is, the minimal expected total cost that will be incurred until the buffer clears, given that initially there are b packets in it and the channel interference is i. Then, the standard dynamic programming recursion is simply (for $b \neq 0$):

$$V(b,i) = \inf_{p \in \mathcal{P}} \left\{ p + D(b) + s(p,i) \left[\sum_{i' \in I} r_{ii'} V(b-1,i') \right] + \left[1 - s(p,i) \right] \sum_{i' \in I} r_{ii'} V(b,i') \right\} \quad (34.1)$$

with boundary condition $V(0, i) = 0$. Observe that the first two minimization terms in (34.1) correspond to the cost incurred in the current time slot and the last two terms express the future cost-to-go. The latter is the sum of the various costs-to-go starting from all system states (b', i') reachable from (b, i), weighted by the probability of transitioning to (b', i') in the next time slot. The transmitter has no control over i' — as the channel is nonresponsive and evolves autonomously — however, it can control the probability that $b' = b - 1$ or $b' = b$ in the next time slot by adjusting p, influencing thus the $s(p, i)$ transmission success probability.

If the function $s(p, i)$ and the channel transition probabilities $\{r_{ii'}\}$ are known *a priori*, the recursion (34.1) can be used to compute the optimal power control p^* as a function of the current backlog b and interference i or $p^* = p^*(b, i)$. In fact, the solution can be carried out offline and the computed values $p^*(b, i)$ can be stored into a look-up table. Thus, when in time slot n the buffer has b_n packets and the channel state is i_n, the power level chosen will be simply $p^*_n = p^*(b_n, i_n)$.

In practice, both $s(p, i)$ and $r_{ii'}$ might be estimated via statistical profiling or learning methods, observing the transmission events and channel transitions over some long-enough period of time and doing a best-fit model approximation under the nonresponsive Markovian Channel Model assumption using statistical classification techniques for a given acceptable level of model complexity (i.e., number of distinct states of the channel). In some idealized cases (for channel noise) of certain transmission technologies, the function $s(p, i)$ is analytically computable from digital communications theory.

In any case, even if $s(p, i)$ and $\{r_{ii'}\}$ are unknown, we can still design efficient power control schemes based on a few simple observations explained below. They simply leverage key structural properties of the optimal power control solutions without requiring full specification of those solutions.

34.2.1.1 The Simple Case of Independent Channel Interference — Power Phases.
Let us rearrange the terms of (34.1) and rewrite it as:

$$V(b,i) = \inf_{p \in \mathcal{P}} \{p - s(p,i)X(b,i) + Y(b,i)\} \tag{34.2}$$

where:

$$X(b,i) = \sum_{i' \in I} r_{ii'} \left[V(b,i') - V(b-1,i') \right]. \tag{34.3}$$

$$Y(b,i) = D(b) + \sum_{i' \in I} r_{ii'} V(b,i') \tag{34.4}$$

Observe now that in the special case where the channel state i is independent from time slot to time slot, the transition probabilities $\{r_{ii'}\}$ reduce to the stationary distribution $\pi_{i'}$ of the channel state, that is $\pi_{ii'}$ for every $i, i' \in I$. Thus, X and Y lose their dependence on i and become functions of b only (note though that X and Y are still dependent on the distribution $\pi_{i'}$ of the channel state, but not on the current value i). Hence, assuming that the channel interference/stress i_n is independent in different time slots, we get:

$$X(b) = \sum_{i \in I} \pi_i \left[V(b,i) - V(b-1,i) \right] \tag{34.5}$$

$$Y(b) = D(b) + \sum_{i \in I} \pi_i V(b,i). \tag{34.6}$$

Consider now for simplicity the set of power levels $\mathcal{P} = [0, p_{\max}]$. If $s(p, i)$ is concave in p, then the minimization (34.2) can be performed analytically, by solving the equation:

$$\frac{\partial s(p,i)}{\partial p} = \frac{1}{X(b)}. \tag{34.7}$$

Then, if the root $\hat{p}$ of (34.7) lies in $(0, p_{max})$, the optimal transmission power level p^* is equal to the root $\hat{p}$ of (34.7), $p^* = \hat{p}$ (otherwise $\hat{p}$ has to be truncated to 0 or p_{max} to get p^*).

To better understand the structural properties of $p^*(b, i)$, let us consider specific examples of $s(p, i)$. In particular, two typical functional forms of $s(p, i)$ are the following:

$$s_1(p,i) = \frac{p}{\alpha p + \beta i} \tag{34.8}$$

$$s_2(p,i) = 1 - e^{-\delta \frac{p}{i}} \tag{34.9}$$

where $\alpha \geq 1$, $\beta > 0$, and $\delta > 0$. For these functional forms, we correspondingly get from (34.7) that:

$$p_1^*(b,i) = \begin{cases} \min\left\{\frac{1}{\alpha}\left[\sqrt{\beta X(b) i} - \beta i\right], p_{max}\right\}, & i < \frac{X(b)}{\beta} \\ 0, & i \geq \frac{X(b)}{\beta} \end{cases} \tag{34.10}$$

$$p_2^*(b,i) = \begin{cases} \min\left\{\frac{i}{\delta}\ln\frac{\delta X(b)}{i}, p_{max}\right\}, & i < \delta X(b) \\ 0, & i \geq \delta X(b) \end{cases} \tag{34.11}$$

The two power-control policies (34.10) and (34.11) are plotted in Figure 34.1 (for simplicity we remove the power ceiling, by setting $p_{max} = \infty$). Note that although the success probability functions (34.8) and (34.9), as well as the corresponding solutions of (34.7), (34.10) and (34.11), respectively, have very different analytic forms, the plots are very similar. This is quite interesting and indicates that the analysis that follows could hold true for any reasonable functional form $s(p, i)$ — increasing in p and decreasing in i — that could describe a communication link. This is indeed intuitively possible, but can also be verified by simulation and testing.

Observe that for any fixed $X(b)$ — hence, fixed b — both p_1^* and p_2^* go through three distinct phases with respect to i:

1. Aggressive (increasing on i)
2. Soft back-off (decreasing on i)
3. Hard back-off (constant and equal to zero)

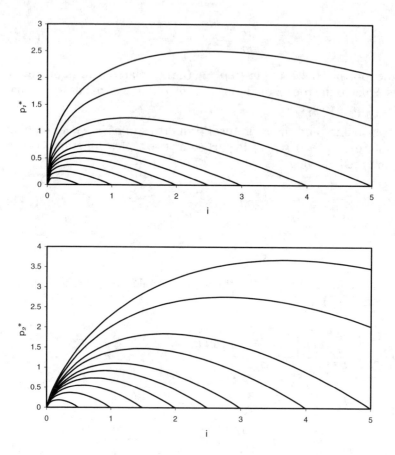

Figure 34.1. Optimal Power versus Interference Functions
Plots of optimal power versus interference function for various fixed backlog pressures $X(b)$ = 0.5, 1, 1.2, 2, 2.5, 3, 4, 5, 7.5, 10 and p_{max} = ∞. On the top, (34.10) is plotted; on the bottom, (34.11) is plotted. In both plots, the family of curves are ordered, the lowest being for $X(b)$ = 0.5 and the highest for $X(b)$ = 10.

In the aggressive phase, the channel stress i is low, thus the optimal strategy for the transmitter is to actively transmit with some positive power $p^*(i)$ that maintains the probability of successful reception at a high level. Hence, for small increases in i, the optimal power $p^*(i)$ increases as well trying to match the channel. However, as i takes even higher values, this effort gradually relaxes, due to the increased transmission cost required to compensate for the channel. As a result, $p^*(i)$ has a concave form. Eventually, at some i, $p^*(i)$ attains a maximum value signaling the end of the aggressive phase.

In the soft back-off phase, the channel stress i is high, turning inefficient costwise to compensate for the channel. Instead, as the channel degrades (i increases), the transmitter reduces its power and gradually withdraws (softly backs off). Ultimately the power reaches zero and the hard back-off phase begins.

In the hard back-off phase, the power remains equal to zero, as the channel stress i is prohibitively high to attempt any transmission. In essence, the transmitter prefers to incur the delay/buffering cost, awaiting for the channel to improve, instead of spending enormous amounts of power (energy) to successfully push a packet through the degraded link.

Regarding the dependence of p^* on the buffer backlog b, it is indirectly expressed through $X(b)$. $X(b)$ is:

- Positive — from (34.5), it is equal to a weighted sum of positive differences $V(b, i) - V(b - 1, i)$ as one more packet in the queue increases the cost-to-go
- Increasing on b, as $D(b)$ is an increasing function[1]

Observe now, that as $X(b)$ increases, not only do the values of p_1^* and p_2^* increase, but the length of the first two phases (where p_1^* or p_2^* are positive) expands as well. Thus, as more packets accumulate in the queue, $X(b)$ becomes higher, forcing the transmitter to be more aggressive. In essence, $X(b)$ corresponds to a backlog pressure, that is a measure of the stimulus of the buffered packets forcing the transmitter to be active.

Last, regarding the dependence on p_{max}, although the plots do not depict the effect of the power ceiling p_{max}, it is trivial to see from the formula that any finite value of p_{max} simply truncates p_1^* and p_2^*.

34.2.2 Incorporating Packet Arrivals and Buffer Overflows

In the system presented so far, we considered a simplified version of the problem which precluded the arrival of new packets at the transmitter. In most cases however, the packets do not arrive at the transmitter all together or in batches, but at random times, before the transmitter has the chance to empty its buffer. Extending the model to incorporate packet arrivals, let $\lambda \in (0, 1)$ be the probability of a packet arriving at the transmitter at the beginning of each time slot (thus $1-\lambda$ is the probability that no packet arrives). The packet arrival process is assumed to be Poisson; that is the event of each packet arrival is considered to be independent of the other arrival events and the other random events taking place in the system, such as the channel stress state transitions, successful or erroneous receptions, etc. The arriving packets join the buffer at its tail, provided that the buffer is not full. To formally express this, the buffer has a finite capacity equal to Q packets; any packet that arrives and finds it full ($b_n = Q$) is blocked and discarded from the system.

The new model is very much alike to the previous one in Section 34.2.1, albeit somewhat more complicated. We can again write a dynamic programming equation, incorporating the packet arrival process with probability λ

and an extra cost term L. L will be the penalty incurred in the event of a packet being rejected whenever the buffer is full. Its purpose is to be an extra incentive to the transmitter to empty its queue whenever the latter fills up, so as to lower the probability of packet loss.

The study of this enhanced model does not reveal significant additional intuition than the one already gleaned in Section 34.2.1. Hence instead of pursuing it further, we directly jump into designing power control algorithms and evaluate them in the presence of packet arrivals.

34.2.3 Design of PCMA Algorithms — Responsive Channel

With the above remarks, we are ready to design efficient PCMA algorithms, even in the absence of knowledge of the channel dynamics. In the above analysis, we did not compute $X(b)$, because this is not possible when $r_{ii'}$ are unknown. Instead, we get around this dead-end and specify directly $X(b)$ to be an increasing function, such as:

$$X(b) = \theta b + \psi, \qquad (34.12)$$

where θ and ψ are positive constants. In essence, this is an implicit switch in our formulation; up to now, $D(b)$ was assumed given, therefore $X(b)$ would be expressed based on $D(b)$. Now, $X(b)$ is given and thus $D(b)$ is a quantity that can be expressed from the given $X(b)$. By changing the function $X(b)$ (the slope θ in [34.12] in particular), we can make the transmitter more or less aggressive, in accordance to the power versus delay trade-off.

Although the above analysis was based on an extreme operating point for the system, that has:

- Nonresponsive and per slot independent channel
- No packet arrivals
- Knowledge of i_n in the beginning of time slot n at the transmitter

we will relax the above and simply apply the findings in a system with packet arrivals and responsive (and hence dependent) channel i. Moreover, to make the protocol operation more realistic, we allow the transmitter at the beginning of time slot n to know only slot channel state i_{n-1} at the previous time slot, not i_n as in the model analyzed. Thus, the PCMA algorithms of (34.10) or (34.11), become:

$$P_{1,n} = \begin{cases} \min\left\{\frac{1}{\alpha}\left[\sqrt{\beta(\theta b_n + \psi)i_{n-1}} - \beta i_{n-1}\right], P_{\max}\right\}, & i_{n-1} < \frac{\theta b_n + \psi}{\beta} \\ 0, & i_{n-1} \geq \frac{\theta b_n + \psi}{\beta} \end{cases} \qquad (34.13)$$

$$p_{2,n} = \begin{cases} \min\left\{\dfrac{i_{n-1}}{\delta}\ln\left[\dfrac{\delta}{i_{n-1}}(\theta b_n + \psi)\right], p_{\max}\right\}, & i_{n-1} < \delta(\theta b_n + \psi) \\ 0, & i_{n-1} \geq \delta(\theta b_n + \psi) \end{cases} \quad (34.14)$$

where we have substituted $X(b)$ from (34.12).

The above PCMA algorithms are compared against the dynamic power control (DPC) algorithm [2], which is a power control algorithm designed for voice communications. DPC updates the transmission power according to (when $b > 0$):

$$p_{n+1} = \frac{\gamma}{G} i_n \quad (34.15)$$

where γ is the desired signal to interference plus noise ration (SINR) at the receiver and G is the link gain (actually path loss). Observe that it does not take into account the packet backlog in the buffer or any power cost — it blindly tries to maintain a constant SINR p/i equal to γ.

As reported in [7, 8], both the two PCMA algorithms (34.13) and (34.14) outperform DPC (34.15) by a large margin in all operational scenarios considered. We do not elaborate further on the performance in the limited scope of this chapter, the reader is encouraged to read the literature on the subject.

34.2.4 The Multi-Transmitter/Multi-Receiver Case

The model presented assumes one wireless link, consisting of a single transmitter and a single receiver. However, the same model can be applied with minor modifications in the case that there are more than one independent transmitters sharing the same packet queue or more than one independent receivers that packets can be forwarded to (not to be confused with links with multiple receive or transmit antennae). The most prominent example of such networks are cellular or wireless LAN architectures regarding:

- The uplink — the transmitter is the mobile terminal that can transmit its packets to one of the many receivers (BSs or APs) in its vicinity
- The downlink — the transmitter is the fixed network that chooses from which BS/AP to send packets to the mobile terminal. In this case, we assume that the system has a common packet queue for all its transmitters and in each time slot decides which one to use for transmission.

Recursion (34.1) still applies in the above models with the difference that now the action p and the channel stress i become vectors instead of scalars. Assuming that there are K receivers (multi-receiver case) or K transmitters (multi-transmitter case), it is:

$$i = (i^1, i^2, \cdots, i^K),$$

where i^k describes the channel stress state of each individual link connecting the mobile terminal and the kth BS/AP of the network. Observe that the model is general enough to cover the cases that the K different wireless links operate over the same wireless resources — thus the stress states i^k will be correlated, or over orthogonal channels (i.e., different frequency/time slot/orthogonal spreading codes in FDMA/TDMA/CDMA systems, respectively), thus i_k may be independent. The channel stress state transition probabilities $r_{ii'}$ describe the vector channel stress as a single Markov process; in the important case that the K channels are independent, each channel stress i^k is an independent Markov process and thus $r_{ii'}$ can be decomposed into the product of the transition probabilities of the states of each individual link.

The control action decision p will include, apart from power transmission level component $p^t \in \mathcal{P}^t$, an index $k \in \{1, 2, ..., K\}$ identifying which of the K links will be used for transmission (that is to which receiver shall the transmitter transmit in the case of multi-receiver or from which transmitter the network will transmit in the case of multi-transmitter). Thus p = (k, p^t).

With the above extensions, the success probability becomes equal to:

$$s(\mathrm{p},\mathrm{i}) = s(k, p^t, i^1, i^2, \cdots, i^K) = s(p^t, i^k), \quad (34.16)$$

where $s(p^t, i^k)$ is the probability that the packet is successfully received if transmitted at power level p^t on link i^k. Thus (34.1) can be rewritten as:

$$V(b,\mathrm{i}) = V(b, i^1, i^2, \ldots, i^K)$$

$$= \inf_{k \in \{1,2,\ldots,K\}, p^t \in \mathcal{P}^t} \left\{ p^t + D(b) + s(p^t, i^k) \left[\sum_{i' \in I} r_{ii'} V(b-1, i') \right] \right. \quad (34.17)$$

$$\left. + \left[1 - s(p^t, i^k) \right] \sum_{i' \in I} r_{ii'} V(b-1, i') \right\}$$

with boundary condition $V(0, \mathrm{i}) = 0$.

Equation (34.17) can be solved using the techniques for the single transmitter/receiver case. The new component in the control action k shall trivially be selected as the one minimizing i^k:

$$k^* = \underset{k \in \{1,2,\cdots,K\}}{\operatorname{argmin}} \{i^k\}. \quad (34.18)$$

Then, (34.7) and the resulting PCMA policies of Section 34.2.3 are directly applicable to decide on the transmission power level p^t.

34.3 Packet Forwarding over Multimode Transmission Links

Toward the more efficient utilization of the wireless resources, we study the case of multiple transmission modes. We assume that the transmitter has the capability of switching the modulation scheme, the coding scheme or other transmission parameters. This extra degree of freedom allows the more efficient use of the wireless channel. Thus, if the channel quality level is high, the transmitter can choose a high-order mode and bundle multiple packets in a single transmission (e.g., switching from BPSK to 4-QAM). Conversely, if the channel is sensed to be of low quality, the transmitter may choose to add more error-correcting bits to protect the transmitted frames against a higher number of bit errors.

As we already have seen in the previous section, the channel state is one of the key factors in deciding what the transmission power should be. Because both the transmission mode and the power level decision depend on the quality of the wireless link, it is natural to take these decisions jointly in one step.

For our formulation, a mode m will be loosely defined as a transmission scheme that allows the transmission of L_m packets in a single time slot. As each mode packs, more or less densely, packets in a time slot, it will be characterized by its own effective throughput: that is, the packet success rate. Thus, it will be assumed that:

$$s(k,m;p,i)$$

is the probability that k packets out of L_m transmitted are successfully received when mode $m \in \mathcal{M}$ is utilized with transmission power level $p \in \mathcal{P}$ and when the channel stress state is $i \in I$. $\mathcal{M} = \{1, 2, \ldots, M\}$ will be the set of all available transmission modes. Without loss of generality, we will order the modes according to L_m, hence $m_1 < m_2$ if $L_{m_1} < L_{m_2}$.

As in the previous section, our modeling will be general enough to encompass all functions $s(k, m; p, i)$. However, we will make some assumptions regarding its structure. Let us consider:

$$S(k,m;p,i) = \sum_{k'=k}^{L_m} s(k';m,p,i), \qquad (34.19)$$

which is the probability that at least k packets are successfully received when transmitting at mode m and power p under channel stress i. Then, keeping the other parameters fixed, $S(k, m; p, i)$ is assumed to be:

- Increasing (actually nondecreasing) in p, which loosely expresses the intuition that by increasing the power p, the average throughput should rise.
- Decreasing (actually nonincreasing) in i, which expresses the intuition that as the channel deteriorates, the average throughput should fall.
- Bell-shaped on m, that is to initially rise with m, attain a maximum at some m and then decline as m gets higher. This is justified by the mode structure, for low values of m too few packets are transmitted, thus increasing m should be beneficiary, whereas for high values of m too many packets are squeezed in a single time slot, thus they become easily corrupted.

34.3.1 Optimally Emptying the Transmitter Buffer

Incorporating the multiple modes in the dynamic programming formulation of Section 34.2.1, (34.1) takes the new form:

$$V(b,i) = \inf_{\substack{m \in \mathcal{M} \\ p \in \mathcal{P}}} \left\{ p + D(b) + E(m) + \sum_{k=0}^{L_m} s(k,m;p,i) \sum_{i' \in I} r_{ii'} V(b-k,i') \right\}. \quad (34.20)$$

for $b \geq L_M$. Special boundary equations need to be written for low values of b, when the packets in the buffer may be less than the number of packets L_m some modes may attempt to transmit; however we suppress them as they are not necessary for our study.

$E(m)$ is a new term in the cost structure, expressing any overheads that may be associated with the choice of the transmission mode m. In particular, $E(m)$ may be associated to the different signal processing overhead at each mode m and is expected to be increasing on the complexity of the transmission mode.

Similar to (34.1), (34.20) can, in theory, be solved precisely to find the optimal policy $(m^*(b, i), p^*(b, i))$ if the channel dynamics $r_{ii'}$ are known to the transmitter. However, the analytic study (34.20) will provide significant insight into the structure of the problem, which will then allow us to design efficient policies even under unknown $r_{ii'}$.

34.3.2 Structural Properties — The Independent Channel Stress Case

Equation (34.20) can be rewritten as:

$$V(b,i) = \inf_{\substack{m \in \mathcal{M} \\ p \in \mathcal{P}}} \left\{ Y(b,i) - \sum_{k=1}^{L_m} s(k,m;p,i) \sum_{k'=0}^{k-1} X(b-k',i) \right\} \quad (34.21)$$

where:

$$X(b,i) = \sum_{i' \in I} r_{ii'}[V(b,i') - V(b-1,i')] \tag{34.22}$$

$$Y(b,i) = p + D(b) + E(m) + \sum_{i' \in I} r_{ii'} V(b,i') \tag{34.23}$$

using the identity $s(0;m,p,i) = 1 - \sum_{k=1}^{L_m} s(k,m;p,i)$.

Observe that the minimization in (34.21) can be performed in two steps. First compute for all $m \in \mathcal{M}$ the power $\tilde{p}(m;b,i)$ that minimizes the right-hand side. Then, after substituting p with the computed $\tilde{p}(m;b,i)$ in (34.21), we can compute the optimal transmission mode m^* in a second minimizing step with respect to m.

Elaborating further in the above, as in Section 34.2.1.2, $\tilde{p}(m;b,i)$ can be computed analytically via the root $\hat{p}(m;b,i)$ of:

$$\frac{\partial}{\partial p} \sum_{k=1}^{L_m} s(k,m;p,i) \sum_{k'=0}^{k} X(b-k',i) = 1. \tag{34.24}$$

provided that the differentiated quantity is concave on p. The optimal transmission mode is given then by:

$$m^*(b,i) = \underset{m \in \mathcal{M}}{\mathrm{argmin}} \left\{ E(m) - \sum_{k=0}^{k} L_m s(k,m;\tilde{p}(m;b,i),i) \sum_{k'=0}^{k} X(b-k',i) \right\} \tag{34.25}$$

and therefore, the optimal power level is:

$$p^*(b,i) = \tilde{p}(m^*(b,i);b,i). \tag{34.26}$$

Similar to Section 34.2.1.2, assuming that the channel stress is independent on different time slots, $r_{ii'} = \pi_{i'}$, and therefore $X(b, i)$ becomes independent on i. Before proceeding to study the solution of (34.24) and (34.25), note that (34.24) can be hard to solve as it involves differentiating a sum of probabilities weighted by the sums of the terms $X(b-k, i)$. However, it takes a simpler form, if the event of each packet transmitted in mode m being successfully received is independent. If this is the case, then:

$$s(k,m;p,i) = \binom{L_m}{k} s^k(m;p,i)[1-s(m;p,i)]^{L_m - k} \tag{34.27}$$

where $s(m; p, i)$ is the probability that each one of the L_m transmitted packets is successfully received. Note that the effective throughput of mode m

is equal to the expected number of packets successfully received, that is $L_m s^k(m; p, i)$.

Consider now the more interesting case of the backlog being much larger than the number of transmitted packets $b \gg L_m$ for all modes m. As we have seen in the previous section, $X(b)$ is increasing on b, however for $b \gg L_m$, the relative difference between $X(b)$ and $X(b - L_m)$ is small, that is $X(b) \gg X(b) - X(b - L_m)$. Then, we can approximate:

$$X(b) \approx X(b-1) \approx \cdots \approx X(b - L_m).$$

and hence,

$$\sum_{k'=0}^{k-1} X(b-k') \approx kX(b). \qquad (34.28)$$

Thus, from (34.28) and (34.27), (34.24) can be rewritten as:

$$\frac{\partial s(m;p,i)}{\partial p} = \frac{1}{L_m X(b)}, \qquad (34.29)$$

which is of the same form as (34.7).

If the transmitter knows what the function $s(m; p, i)$ is, it can solve (34.29) for each available mode and decide from (34.25) the optimal transmission action (μ^*, p^*). To better study the solution, let us study the following example of a three-mode transmitter:

1. $L_1 = 1$ and $s(1;p,i) = 1 - \dfrac{i}{p+i}$,

2. $L_2 = 2$ and $s(2;p,i) = 1 - \dfrac{2i}{p+2i}$,

3. $L_3 = 4$ and $s(3;p,i) = 1 - \dfrac{10i}{p+10i}$

The above choice results in:

1. $\hat{p}(1;b,i) = \max\left\{\sqrt{X(b)i} - i, 0\right\}$

2. $\hat{p}(2;b,i) = \max\left\{\sqrt{4X(b)i} - 2i, 0\right\}$

3. $\hat{p}(3;b,i) = \max\left\{\sqrt{40X(b)i} - 10i, 0\right\}$

Transmitter Power Control in Wireless Computing

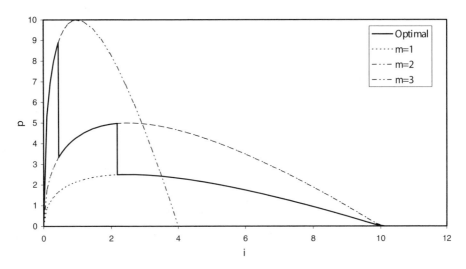

Figure 34.2. Plots of $\tilde{p}$ for $m = 1, 2, 3$ and p^* (Labeled Optimal) for Backlog Pressure $X(b) = 10$

Note that $p(m; b, i)$ is the optimal power level used for single-mode transmission with mode m.

Figure 34.2 plots $p(m; b, i)$ and the optimal $p^*(m; b, i)$ for the above modes. The overhead cost $E(m)$ is taken to be zero for all modes. Observe that the optimal transmission scheme jumps between the different modes. Specifically, for low channel stress, it begins with the highest order mode. As the channel stress increases, it becomes advantageous to switch to a lower order mode, as the power cost of maintaining a high-order transmission mode becomes prohibitive. Instead, it is better to downgrade to a lower mode that requires less power. After downgrading to a lower mode, the transmitter initiates a new aggressive phase. For large values of the channel stress, at the lowest order mode, the transmitter goes through a soft back-off phase before ceasing to transmit (hard back-off).

Regarding the dependence of m^* and p^* on b, it is indirectly expressed through the backlog pressure $X(b)$. We already know from Section 34.2 — the arguments are similar — that $X(b)$ is an increasing function of b. Figure 34.3 plots p^* for increasing values of $X(b)$ corresponding to increasing values of backlog b — the jumps in p* corresponds to mode changes. Observe that as the backlog pressure $X(b)$ increases, p^* takes higher values, and the interval of the channel stress for each transmission mode (at each jump in p^*, m^* changes to a lower order mode) expands. Thus, the transmitter becomes more aggressive both in the power level and the transmission mode utilized.

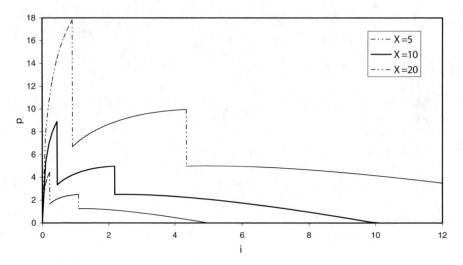

Figure 34.3. Plots of p* for Values of Backlog Pressure $X(b)$ Equal to 5, 10, 20

34.3.3 Design of Multimode PCMA Algorithms for Responsive Channels

Although the multimode transmission scheme analysis assumed that a channel was nonresponsive and independent in different time slots and without packet arrivals, we nonetheless apply the findings of the analysis in the general case.

Specifically, we use (34.25) and (34.29) to create an efficient transmission mode $m(b, i)$ and power level $p(b, i)$ policy. Similar to the previous section, we directly specify $X(b)$ for example equal to (34.12).

As reported in [13], the multimode PCMA algorithms outperform the single-mode PCMA algorithms of the previous section. In particular, they demonstrate important reduction in the average packet delay and the average transmitted power per packet and moreover they significantly increase the capacity of the system.

34.4 Data Prefetching over Single-Mode Transmission Links

The model presented so far is a classical queuing model where data packets are sent only if and when needed. Upon arrival at the transmitter they are subject to queuing if the latter is busy transmitting other packets. When the transmitter's buffer empties, the wireless link remains idle and thus its capacity gets unexploited. Switching to a new paradigm of operation, we examine the case of a wireless terminal requesting data from a server over a wireless link. Taking advantage of the idle time between the downloads, the system may try to fetch data before an actual request for them is placed. Thus, the link utilization increases and the access delay is mitigated, as data already prefetched is readily available at the receiver.

To exercise prefetching though, it is necessary to augment the receiver with a cache/buffer where prefetched data is stored. Moreover, compared to the problem of the previous section, the system has to make buffer decisions managing the contents of the terminal's cache regarding what to prefetch or fetch and what to evict from the buffer on top of the transmission power ones. The two decisions are entangled in each other, as the latter is related to the importance of the former. Thus, the problem is essentially one of joint buffer and power control.

The above model captures various wireless computing paradigms such as Web browsing or database queries. In these scenarios, the transmitted data are Web pages or database information items, respectively, transmitted from a Web/database server to a wireless terminal.

34.4.1 System Model

Let $\mathcal{D}$ be set of all data items available at the server and $B \subseteq \mathcal{D}$ be the contents of the buffer/cache of the terminal. The buffer B has a finite capacity of b data items. In each time slot, the system has to decide:

- What data item $f \in \mathcal{D}$ to request to download from the server.
- What power level $p \in \mathcal{P}$ to use for transmission.
- What data item $p \in B$ to evict from the buffer. This applies in the case that the buffer is full to its capacity, that is $|B|=b$.

Regarding the power level decision p and the wireless link operations in general, we reuse the model of Section 34.2. Namely, we assume that a data item has the size of a packet, hence in each time slot, only one data item can be transmitted. The data item will be successfully received with probability $s(p, i)$, where $p \in \mathcal{P}$ is the transmission power (energy) level, and $i \in I$ is the channel stress state.

For the terminal to send the requests for data to the server, a reverse link from the terminal to the server is needed. As in Section 34.2, this link is assumed to be reliable because it is of low bandwidth and carries only control information. In this system, the reverse link communicates information regarding:

- The data requests of the terminal
- The stress state i of the wireless channel as measured at the receiver
- The acknowledgments notifying the transmitter (server) whether the reception of the transmitted data was successful or not

Before describing the buffer management control actions f and e, let us first define the cost structure of the problem. The goal is to minimize the total cost, which consists of the power and the delay components. Namely, in each time slot, the system incurs:

- A power cost p, equal to the power (energy) used for transmission at the server, which may also incorporate other considerations such as implicit or explicit charges for the use of the wireless resources.
- An access delay cost D, whenever the user requests a data item d that is not in the buffer B. D is a cost related to the degradation in the perceived QoS resulting from the delay in serving the user delay, but may include other considerations too (e.g., the lock-up of computing resources).

To describe more accurately when and how D is incurred, let u_n be a variable describing the operational state of the user or the computing process at time n, taking values in the set $\mathcal{U}$. When at state $u_n = u$, the user tries to access (requests) the data item $d = d(u)$. Thus, if the requested item is not in the buffer, $d \notin B$, the system incurs the access delay cost D and the user does not switch state, but remains at u, $u_{n+1} = u_n = u$, waiting for its request to be satisfied. However, if the requested item can be found in the buffer, $d \in B$, the user can immediately access it. Hence, no delay cost is incurred in the system and at the next time slot, the user may transition to a new state $u_{n+1} = u'$ (or remain at u). Specifically, let:

$$q_{uu'} = P[u_{n+1} = u' | u_n = u]$$

be the transition probability[2] from u to u'. Thus, the time the user will remain at u consists of a number of time slots in delay if, when transitioning at u, $d(u)$ was not in the buffer, plus $1/(1 - q_{uu})$ time slots on average before spontaneously switching to a new state.

In both cases, the terminal can use the current time slot n to fetch a data item $f \in \mathcal{D}$ from the server. Clearly, in the former case, there is no point in delaying fetching d as in each time slot that $d \notin B$, the system keeps incurring a delay cost D. Instead, in the latter case, the system has the opportunity to exercise prefetching and download any data item $f \notin B$.

Although not immediately clear from the above description, the model presented is quite general in describing the data and how they are requested by the user/computing process. In particular, regarding:

- Variable-size data items — we have assumed that all data items are of unit size: that is, they can be transmitted in a single time slot. The case of variable-size data items can be still captured in our formulation by splitting any large data item that spans over multiple packets into multiple data items. Then, we can enforce that the terminal requests all the component data items at a time by appropriately modifying the user state space.[3]

- Users with finite memory — in the simplest user modeling, $d(\cdot)$ can be chosen to be a one-to-one function (or without loss of generality, $u \equiv d$); with this choice, the next requested data item depends only on the last requested item d_{-1}. To have the next user request depend on the last N request $d_{-1}\ d_{-2},\ldots,\ d_{-N}$ let $u \equiv (d_{-1},\ d_{-2},\ldots,\ d_{-N})$. Clearly, $d(u)$ will no longer be one-to-one, as the number of user states will be greater than the number of data items. Note though that this more refined scheme of the user model increases the complexity of the user state space, $\mathcal{U} = \mathcal{D}^N$.

As it should be clear, for the system to exercise prefetching efficiently, it has to have reliable estimates of the future user requests. For simplicity, it is assumed that the system has full knowledge of the user transition probabilities $q_{uu'}$. In practice, $q_{uu'}$ can be estimated by methods of statistical profiling, either at the terminal observing a single user or at the server by observing a large user population. In the limited scope of this chapter, we do not engage in the problem of how such a model is built, instead we assume that it exists and is readily available to the system.

Before proceeding to the dynamic programming formulation of the problem, let us complete the model details. The random events of:

- The user transitions $u \to u'$
- The channel stress transitions $i \to i'$
- The successful or not reception of the transmitted data (with probability $s(p, i)$)

are considered to be statistically independent and moreover to be independent in different time slots. With these assumptions, the system evolves as a controlled Markov chain. The above framework models the channel as nonresponsive, a simplification that keeps the complexity of the model low. As in packet forwarding, the resulting power/buffer policies are expected to perform well, even in the presence of responsive channel (e.g., responsive interference).

34.4.1 The Dynamic Programming Formulation

First we define the global system state $S_n = (u_n, B_n, i_n)$ at time slot n. It comprises:

- The user state $u_n \in \mathcal{U}$
- The buffer contents $B_n \in \mathcal{B}_b$, where $\mathcal{B}_b$ is the set of subsets of $\mathcal{D}$ of up to b elements
- The channel stress state $i_n \in I$

Thus, S_n takes values in $\mathcal{S} = \mathcal{U} \times \mathcal{B}_b \times I$.

In each time slot n, the system has to decide the control action μ_n, which comprises:

Table 34.1. Allowed Transitions

Case	S'	$\Phi_S^{S'}(\hat{B},p)$	$C_s(p)$
$d(u) \in B$	(u',B,i')	$q_{uu'}[1-s(p,i)]r_{ii'}$	p
	$(u',\hat{B},i')$	$q_{uu'}s(p,i)r_{ii'}$	
$d(u) \notin B$	(u,B,i')	$[1-s(p,i)]r_{ii'}$	$p+D$
	$(u,\hat{B},i')$	$s(p,i)r_{ii'}$	

The allowed transitions from state $S = (u, B, i)$ when control action $\mu = (\hat{B},p) \in \mathcal{B}_{b,B} \times \mathcal{P}$ is exercised. The probability of transitioning to any other $S' \in \mathcal{S}$ is zero.

- The desired next buffer state $\hat{B}_n$, the terminal aims to reach at the next time slot $n + 1$ by trying to prefetch $\{f_n\} = \hat{B}_n - B_n$. If $|B_n| = b$, then the buffer/cache is full to its capacity; therefore, the system has to decide also which item $\{e_n\} = B_n - \hat{B}_n$ to evict from the buffer, if f_n is successfully received. Hence, $\hat{B}_n$ takes values in the set $\mathcal{B}_{b,B} = \{\hat{B} \subseteq \mathcal{B}_b : |\hat{B} - B| = 1, |\hat{B}| \leq b\}$.
- The power level $p_n \in \mathcal{P}$ that will be used to transmit f_n.

Table 34.1 summarizes the possible next states $S = (u,B,i) \rightarrow S' = (u',B',i')$ along with the transition probabilities $\Phi_S^{S'}(\mu)$, and the cost $C_S(\mu)$ incurred per time slot, where $\mu = (\hat{B},p)$ is the control action exercised. Indeed, starting from $S = (u, B, i)$:

- If $d(u) \in B$, the user state may switch to u' and the channel switch independently to i'. Thus, $S' = (u', B', i')$ where B' is either $\hat{B}$ or B depending on whether the transmission was successful or not. The total cost per time slot is equal to the power cost p.
- If $d(u) \notin B$, then the user state cannot switch to a new state, but the channel can change to i'. Thus, $S' = (u, B', i')$ where B' is either $\hat{B}$ or B depending on whether the transmission was successful or not. The total cost per time slot includes the delay penalty D on top of the power cost p.

Note that the p component of μ affects directly the probability of the next buffer state B' ($\hat{B}$ versus B), whereas the choice of $\hat{B}$ is indirect, interslot, aiming at the data item that is most likely to induce delay cost in the future.

The goal is to minimize the total cost incurred throughout the system evolution. Let $J(S)$ be the cost-to-go: that is, the average minimum cost until it ceases evolving. It is assumed that the user eventually stops requesting data items (for example, the user switches off the terminal). In the user state space, this translates to the user transitioning to an absorbing terminal state $u^o \in \mathcal{U}$, which the user never exits from, that is $q_{u^o u^o} = 1$.

Thus the system ceases its evolution upon entering any state S in the set $\mathcal{S}^o = \{(u^o, B, i) : B \in \mathcal{B}, i \in I\}$, hence:

$$J(S) = 0 \text{ for any } S \in \mathcal{S}^o.$$

For any other $S \in \mathcal{S} - \mathcal{S}^o$, the dynamic programming recursion applies:

$$J(S) = \inf_{\substack{\hat{B} \in \mathcal{B}_{b,B} \\ p \in \mathcal{P}}} \left\{ C_S(p) + \sum_{S' \in \mathcal{S}} \Phi_S^{S'}(\hat{B}, p) J(S') \right\} \quad (34.30)$$

The first term in the right-hand side of (34.30) corresponds to the cost incurred in the current time slot n, whereas the sum conditions on what the system state S will be at next time slot $n + 1$ to express the future cost.

In theory, (34.30) is possible to solve if the function of successful reception $s(p, i)$ and the transition probabilities of the user $q_{uu'}$ and the channel $r_{ii'}$ are known. The optimal power control $\mu^* = (\hat{B}^*, p^*)$ is the minimizing argument of the right-hand side of the solution of (34.30) and depends on the system state $S = (u, B, i)$. However, in any practical system:

- $r_{ii'}$ are difficult to estimate as the wireless channel is highly volatile, as argued in the previous section.
- The state space is too complex for (34.30) to be solved directly. Indeed, there are about $b|\mathcal{D}|$ choices for $\hat{B}$ to consider and the buffer can have a number of states in the order of $|\mathcal{D}|b$. Thus, (34.30) is practical only for tiny datasets $\mathcal{D}$.

Therefore, because (34.30) is not directly applicable, we have to devise suitable heuristics to approximate it. In the rest of the section, we study the structure of the problem, which will help us design efficient prefetching control policies $\mu(u, B, i) = (\hat{B}(u, B, i), p(u, B, i))$.

34.4.2 The Structural Properties of the Power Decision p

Let us first attack the power control component p of the control action μ. The analysis is almost identical to the one of Section 34.2.1.2, so we briefly restate the most important points. Indeed, if $\hat{B}^*(u, B, i)$ was known, then for any $s(p, i)$ concave on p and $\mathcal{P} = [0, p_{\max}]$, the minimizing power $p^*(S)$ could be computed analytically by:

$$\frac{\partial s}{\partial p}(p, i) = \frac{1}{X(S)} \quad (34.31)$$

where:

$$X(S) = X(u,B,i) = \begin{cases} \sum_{\substack{u' \in \mathcal{U} \\ i' \in I}} q_{uu'} T_{ii'} \left[J(u',B,i') - J(u', \hat{B}^*(u,B,i), i') \right], & d(u) \in B \\ \sum_{i' \in I} r_{ii'} \left[J(u,B,i') - J(u, \hat{B}^*(u,B,i), i') \right], & d(u) \notin B \end{cases} \quad (34.32)$$

Observe that $X(S)$ in (34.32) is very similar to (34.3). Indeed:

- $X(S)$ is equal to the difference in the expected cost-to-go in the next time slots if the reception of the data transmitted in the current slot is erroneous over to being successful, as in Section 34.2.1.2.
- If the channel stress is independent in each time slot, $X(S)$ loses its dependence on i and becomes a function of u, B only.
- If $s(p, i)$ is equal to (34.8) or (34.9), (34.31) results to the solutions (34.10) and (34.11), respectively. Thus, regarding the dependence of the optimal power p^* on i, it goes through the aggressive, the soft back-off and the hard back-off phase as depicted in Figure 34.1.
- As far as the u, B components of the state space are concerned, their dependence is indirect through $X(u, B)$. In packet forwarding, X depended on b and corresponded to the backlog pressure: the pressure resulting from the buffering cost of the queued packets. Instead, in the case of the data fetching or prefetching X corresponds to the **prefetch pressure**, which arises from the value to the system of the data items not in the buffer B in comparison to the items already in B. The value of the data is essentially measured by the potential delay cost they may incur to the system: the more close probabilisticwise to be requested given u, the more pressure they add to $X(u, B)$.
- Higher values of the per slot delay cost D, increase the cost-to-go J and thus amplify the cost-to-go differences in $X(u, B)$, making the transmitter more aggressive. As shown in Figure 34.1, increasing X, increases both the power level p^* used at a given channel stress state i and the range of values of the channel stress i, where the transmitter is active ($p^* > 0$). This essentially reflects the trade-off between the delay versus the power.

34.4.3 Online Look-Ahead Heuristics for Efficient Buffer Control

Although the above study can help us create a power policy that dispenses with the need of knowing $r_{ii'}$ (similar to what we did in Section 34.2.3), it does not help with the problem of the state complexity arising from the user and buffer components. To reduce the complexity of the user/buffer space and facilitate the buffer control decision we employ the technique of the look-ahead.

The idea behind look-ahead is that beginning from the current state S_n of the system, there exists a small set of states that the system can reach over a limited time horizon, namely the look-ahead horizon. Instead of considering the whole system evolution, the optimization is conducted with respect to the look-ahead horizon only. Therefore, (34.30) will be much easier to solve as it involves a much smaller set of states. In fact, the computation could even be performed online by the system (at either the terminal or the server). The resulting decisions $\mu(S)$ will clearly be suboptimal, but they can be quite efficient as in prefetching the buffer has to be filled with data likely to be requested in the near-to-medium term rather than in the far future. Indeed, [16] demonstrated experimentally that even for small look-ahead horizons, the resulting policies can yield significant performance gains over the case of no prefetching at all. Moreover, if we expand the horizon to include more steps in the future, the decisions $\mu(S)$ are expected to improve and converge to the optimal policy $\mu^*(S)$.

In our model, the look-ahead technique will be applied in the user component of the system state space. Hence, beginning from the current user state u_n, we define the look-ahead horizon to be a set of user states $\mathcal{U}_{u_n} \subset \mathcal{U}$ in the vicinity of u_n. Then (34.30) applies by equating any any user state u outside the look-ahead horizon $\mathcal{U}_{u_n}$ to the terminal state u^o:

$$J(u,B,i) = 0, \text{ for any } (u,B,i) \in \mathcal{S} - \mathcal{U}_{u_n} \times \mathcal{B}_b \times I.$$

The above truncation in the user space induces a similar truncation in the data and thus in the buffer space. Namely, the only relevant items regarding the buffer lie in the set $\mathcal{D}_{u_n} = d(\mathcal{U}_{u_n})$. Any data item $d \notin \mathcal{D}_{u_n}$ is virtually of no use, as it cannot be requested in the set look-ahead horizon $\mathcal{U}_{u_n}$. In other words, any $d \notin \mathcal{D}_{u_n}$ is not a candidate for prefetching and moreover is not an item that needs to be preserved in the buffer when deciding which item e to evict.

34.4.3.1 No Prefetching — Efficient Data Downloading. Let us first consider the case where no prefetching is exercised. This corresponds to the minimal choice of the look-ahead horizon $\mathcal{U}_{u_n} = \{u_n\}$. The only relevant data item is $d = d(u_n)$; thus, the problem essentially reduces to identifying a power policy $p(i)$ to download d if it is not in the buffer. Observe that this is how a system would operate if there is no user profile available: when $q_{uu'}$ are unknown. There is hardly any point in blindly downloading data items at random (if $\mathcal{D}$ is not a small set), thus the best the terminal can do is to download optimally only the requested items.

Let $J^0(u, B, i)$ be the cost-to-go with respect to the $\{u_n\}$ look-ahead horizon. Then, (34.30) takes the form:

$$J^0(u_n,B,i) = \begin{cases} \inf_{p \in P} \left\{ D + p + [1-s(p,i)] \sum_{i' \in I} r_{ii'} J^0(u_n,B,i') \right\} & d(u_n) \notin B \quad (34.33A) \\ 0 & d(u_n) \in B \quad (33.33B) \end{cases}$$

Note that in (34.33A) the decision to fetch $f = d(u_n)$ is hard-coded into the equation (i.e., $d(u) \in \hat{B}$). Moreover, observe that if the buffer is full, it does not matter which data item e will be evicted, as the look-ahead horizon contains $d(u)$ only. Therefore, in this setup, any choice of e is equally good or bad (e.g., choose at random from any $d \in B$).

If the channel transition probabilities $r_{ii'}$ are known, (34.33A) can be used to compute numerically the optimal download power $p(i)$. However, because this is rarely the case, we will revert to the analysis of Section 34.4.2, to design a power policy $p(i)$ for the case that the channel dynamics are not available to the system. Equation (34.31) becomes:

$$\frac{\partial s}{\partial p}(p,i) = \frac{1}{X^0(u_n,B,i)} \qquad (34.34)$$

where:

$$X^0(u_n,B,i) = \sum_{i \in I} r_{ii'} [J^0(u_n,B,i') - J^0(u_n,\hat{B},i')] = \sum_{i' \in I} r_{ii'} J^0(u_n,B,i') \qquad (34.35)$$

(the last step comes from the fact that $J^0(u_n,\hat{B},i) = 0$). Recycling the heuristic from the independent channel stress analysis, we set $X^0(u_n, B, i)$ equal to a constant:

$$X^0_{u_n}(B,i) = \Theta \qquad (34.36)$$

for $B \not\ni d(u_n)$, where Θ is a positive constant controlling the priority of delay versus power cost.

Note that $X^0(u_n, B, i)$ is constant on the current user state u_n and the buffer contents $B \not\ni d(u_n)$, as it does not make any difference which data item is fetched. The only approximation (34.36) introduces is with respect to i — (34.36) assumes that the channel is memoryless.

Then (34.34) and (34.36) can be used to compute the power policy $p(i)$. In particular, if $s(p, i)$ is equal to (34.38) or (34.39), then the resulting power policies $p_1(i)$ and $p_2(i)$ are respectively equal to:

$$p_1(i) = \begin{cases} \min\left\{\frac{1}{\alpha}\left[\sqrt{\beta\Theta i} - \beta i\right], p_{max}\right\}, & i < \frac{\Theta}{\beta} \\ 0, & i \geq \frac{\Theta}{\beta} \end{cases} \qquad (34.37)$$

$$p_2(i) = \begin{cases} \min\left\{\dfrac{i}{\delta}\ln\dfrac{\delta\Theta}{i}, p_{\max}\right\}, & i < \delta\Theta \\ 0, & i \geq \delta\Theta \end{cases} \qquad (34.38)$$

Note that the solution of (34.31) or the computation of (34.37) or (34.38) are simple enough to be computed online.

Regarding the dependence on Θ, it should be clear that higher values of Θ result in power-aggressive/delay-tolerant policies, whereas low values of Θ result in power-sensitive/delay-tolerant policies.

Although the above scheme essentially models the channel as memoryless (independent on different time slots), it has been shown experimentally that it performs surprisingly close to the optimal (that is if $r_{ii'}$ are known) even when the channel has memory [18].

34.4.3.2 Neighbor Prefetching — Depth-1 Look-Ahead. To introduce prefetching to the system, we expand the look-ahead horizon to include the user states reachable in one step from the current user state u_n, $\mathcal{U}_{u_n} = \{u \in \mathcal{U} : u = u_n \text{ or } q_{u_n,u} > 0\}$. By looking at the neighbor user states, the system may decide to prefetch a data item that may be used next. The decision on what and how to prefetch will be taken by solving the 1-step ahead dynamic programming equation:

$$J^1(u_n,B,i) = \inf_{\hat{B}\in\mathcal{B}_{b,B}, p\in\mathcal{P}} \left\{ C_{(u_n,B,i)}(\hat{B},p) + \left[\sum_{\substack{B'\in\{B,\hat{B}\},\\ i'\in I}} \Phi^{\{u_n,B',i'\}}_{\{u_n,B,i\}}(\hat{B},p) J^1(u_n,B',i') \right] \right. $$
$$\left. + \sum_{\substack{u'\in\mathcal{U}_{u_n}-\{u_n\},\\ B'\in\{B,\hat{B}\},\\ i'\in I}} \Phi^{\{u',B',i'\}}_{\{u',B,i\}}(\hat{B},p) J^0(u',B',i') \right\} \qquad (34.39)$$

Besides the cost incurred in the current time slot, the cost-to-go $J^1(u_n, B, i)$ at current user state u_n is expressed through the cost-to-go of the same user state u_n, but potentially different buffer state B (second term) and the cost-to-go of the other user states u (third term). Observe that the former is equal to $J^1(u, B, i)$ as the user does not change state, whereas the latter is equal to the cost $J^0(u, B, i)$ of optimally fetching $d(u)$ as the user has advanced to the next user state.

Although (34.39) may look complicated at first glance, in reality it is quite straightforward to solve:

- If $d(u_n) \notin B$, as already argued, there is no point in delaying to fetch $d(u_n)$, hence $f = d(u_n)$.

- If $d(u_n) \in B$, then the requested data item f should be equal to the item $d(u') \notin B$, of the neighbor user state u' that maximizes $q_{u_n u'}$.

Regarding the item to evict, it is clear that it should be chosen as one that does not appear in the neighbors $d(u')$ or if such an item does not exist, the $d(u') \in B$ that minimizes $q_{u_n u'}$.

Therefore, there is no need to consider all the possible buffer configurations $B' \subseteq \mathcal{D}$, but only up to $N + 1$ of them, where N is equal to the number of neighbors. Indeed, if the user remains at u_n long enough so that the system has adequate time to download all of $\mathcal{D}_{u_n}$, there is an ordering of how the data items should be prefetched. This ordering is quite easy to find by inspecting $q_{u_n u'}$; the higher $q_{u_n u'}$ is, the higher in the list $d(u)$ should be placed. Thus, we have effectively computed the buffer decisions $\hat{B}(u_n, B, i)$.

What remains is to compute the power decisions $p(u_n, B, i)$: beginning from the last item to the first of the list, we can numerically solve the recursion of (34.39) — note that if $B \supseteq D_{u_n}$, then $J^1(u_n, B, i) = 0$. Observe that the equations have the same form as in the previous case of no prefetching (with the exception of some additional terms, which do not complicate the solution), therefore the computation is essentially the same.

In the important case that the channel stress dynamics $r_{ii'}$ are unknown, we will assume:

- That the cost-to-go is reduced by Θ if $d(u_n) \in B$ over the case that $d(u_n) \notin B$, as in (34.35).
- The channel is static. Thus we set $r_{ii'} = 1$ in (34.39).

In the light of the above, (34.39) can be rewritten as:

$$J^1(u_n, B, i) = \begin{cases} \inf_{\substack{d(u_n) \in \hat{B}, \\ \hat{B} \in \mathcal{B}_{b,B}, \\ p \in \mathcal{P}}} \{[1 - s(p,i)]\Theta + p + J^1(u_n, \hat{B}, i)\}, & d(u_n) \notin B \quad (34.40A) \\ \inf_{\substack{\hat{B} \in \mathcal{B}_{b,B}, \\ p \in \mathcal{P}}} \{p + q_{u_n u_n}[s(p,i)J^1(u_n, \hat{B}, i) + (1 - s(p,i))J^1(u_n, B, i)] + \\ + \sum_{\substack{u \in \mathcal{U}_{u_n}, \\ u \neq u_n}} \{q_{u_n u'}[s(p,i)J^0(u', \hat{B}, i) + (1 - s(p,i))J^0(u', G, i)]\}, & d(u_n) \in B \quad (34.40B) \end{cases}$$

where from the previous subsection:

Transmitter Power Control in Wireless Computing

$$J^0(u,B,i) = \begin{cases} \Theta, & d(u) \in B \\ 0, & d(u) \notin B \end{cases}.$$

As in the no-prefetching case, it has been observed experimentally that the decisions taken with the above heuristics are very close to the decisions obtained by (34.39) when $r_{ii'}$ are known [18].

34.4.3.3 Deep Prefetching. The technique used to derive (34.39) can be also applied considering deep look-ahead user graph fragments (e.g., up to h steps ahead). However, in deep prefetching, it is not always easy to decide what the buffer decision B should be, except in the case that all user states in the look-ahead horizon map to different data items and the look-ahead horizon graph is acyclic (excluding the self-loops).

In the general and more interesting case of many user states in the look-ahead horizon mapping to the same data item, one cannot solve (34.39) without considering all the configurations B of the buffer unless some other heuristics/approximations are introduced for dealing with the complexity of the buffer space. Due to the limited space and scope of this chapter, we do not elaborate any further. However, the topic of prefetching and caching in general is vast and calls for more research.

Notes

1. Intuitively this can argued from the fact that the difference $V(b, i) - V(b-1, i)$ is increasing on $D(b)$. Indeed, a delay/buffering cost equal to $D(b)$ will be incurred for a number of time slots until a packet is successfully received. Given that $D(b)$ is increasing on b, it follows that $V(b + 1, i) - V(b, i) > V(b, i) - V(b - 1, i)$, therefore $X(b)$ is increasing on b.
2. In essence, the user is modeled as a time-homogeneous and irreducible Markov process.
3. If d_m is a single data item that needs N_m time slots to be transmitted, it can be divided into N_m unit-sized data items d_m^ℓ, $\ell \in \{1, ..., N_m\}$. Then, any user state u_k such that $d_m = d(u_k)$ should be replaced by a chain $\{u_k^\ell\}$ of N_m states such that $d_m^\ell = d(u_k^\ell)$ for $\ell \in \{1, ..., N_m\}$. Correspondingly, the user Markov process should be modified so that transitions to u_k point to u_k^1 and transitions originating from u_k originate from $u_k^{N_m}$ instead. Last, the $\{u_k^\ell\}$ states should be connected in a chain, that is $q_{u_k^\ell u_k^{\ell+1}} = 1$ for $\ell \in \{1, ..., N_m - 1\}$. Thus, the user upon visiting u_k^1 will request all the d_m^ℓ, effectively downloading all the components of d_m.

References

1. D. Bertsekas, *Dynamic Programming: Deterministic and Stochastic Models*. Upper Saddle River, NJ: Prentice Hall, 1987.
2. G. Foschini and Z. Miljanic, A simple distributed autonomous power control algorithm and its convergence, *IEEE Tran. on Vehicular Technology*, vol. 42, no. 4, pp. 641–646, 1993.

3. K. Pahlavan and A.H. Levesque, *Wireless Information Networks*, New York: John Wiley & Sons, 1995.
4. M.D. Yacoub, *Foundations of Mobile Radio Engineering*, Boca Raton, FL: CRC Press, 1993.
5. J. Rulnick and N. Bambos, Mobile power management for maximum battery life in wireless communication networks, in *Proceedings of InfoCom 1996*, vol. 2, pp. 443–450, 1996.
6. J. Rulnick and N. Bambos, Performance evaluation of power-managed mobile communication devices, *IEEE Int'l. Conference on Communications 1996*, vol. 3, pp. 1477–1481, 1996.
7. N. Bambos and S. Kandukuri, Power controlled multiple access (PCMA) in wireless communication networks, in *Proceedings of InfoCom 2000*, vol. 2, pp. 386–395, 2000.
8. N. Bambos and S. Kandukuri, Power-controlled multiple access schemes for next-generation wireless packet networks, *IEEE Wireless Communications*, vol. 9, no. 3, pp. 58–64, June 2002.
9. L. Fan, Pei Cao, Wei Lin, and Quinn Jacobson, Web prefetching between low-bandwidth clients and proxies: Potential and performance, in *Proceedings of the International ACM Conference on Measurement and Modeling of Computer Systems*, pp. 178–187, 1999.
10. Z. Jiang and L. Kleinrock, Prefetching links on the WWW, in *Proceedings of the IEEE Int'l. Conference on Communications*, vol. 1, pp. 483–489, 1997.
11. Z. Jiang and L. Kleinrock, An adaptive network prefetch scheme, *IEEE Journal on Selected Areas in Communications*, vol. 16, no. 3, pp. 358–368, 1998.
12. Z. Jiang and L. Kleinrock, Web prefetching in a mobile environment, *IEEE Personal Communications*, vol. 5, no. 5, pp. 25–34, 1998.
13. S. Kandukuri and N. Bambos, Multimodal dynamic multiple access (MDMA) in power controlled wireless packet networks, in *Proceedings of InfoCom 2001*.
14. N. Tuah, M. Kumar, and S. Venkatesh, Investigation of a prefetch model for low bandwidth networks, in *Proceedings of ACM International Workshop on Wireless Mobile Multimedia*, pp. 38–47, 1998.
15. N. Tuah, M. Kumar, and S. Venkatesh, Performance modelling of speculative prefetching for compound requests in low bandwidth networks, in *Proceedings of ACM International Workshop on Wireless Mobile Multimedia*, pp. 83–92, 2000.
16. S. Gitzenis and N. Bambos, Power-controlled data prefetching/caching in wireless packet networks, in *Proceedings of InfoCom 2002*, vol. 3, pp. 1405–1414, 2002.
17. S. Gitzenis and N. Bambos, Power-controlled data prefetching and caching in wireless packet networks, Technical Report, SU NETLAB-2002-05-01, Engineering Library, Stanford University.
18. S. Gitzenis and N. Bambos, Efficient data prefetching for power-controlled wireless packet networks in *Proceedings of ACM/IEEE Mobile and Ubiquitous Networking Conferences Mobiquitious 2004*, pp. 64–73, 2004.

Section VIII
Performance and Modeling

Chapter 35
A Survey on Mobile Transaction Models

Abdelsalam (Sumi) Helal, Santosh Balakrishnan, Margaret H. Dunham, and Youzhong Liu

Abstract

With the availability of inexpensive portable computers and with the growth in mobile and satellite communication technologies, mobile computing has attracted great attention from both research and industry. In many ways mobile computing can be compared to distributed computing, but constraints unique to mobile computing, such as low bandwidth of the wireless links, user and host mobility, and high vulnerability, have thrown up many challenges to researchers. Transaction models, data consistency, and cache management have been the focus for database researchers in mobile computing. Mobile transaction models are especially of interest, both academically and industrially, because it is envisaged that a major component of applications in mobile environments will involve database access or activities modeled as transactions. In this chapter, we present a survey on the various mobile transaction models proposed in the literature. We present a brief description of all the models and outline the various issues addressed by the mobile transaction model designers. We also compare mobile transactions on these issues and present a discussion on the issues yet to be addressed.

35.1 Introduction

With the availability of powerful portable computing systems and with the tremendous growth of the wireless communication technology, mobile computing is being projected as the future growth area in both academia and industry. Mobile computing systems when available will allow users to be in constant touch with their office computing resources enhancing productivity by efficiently utilizing the commuting and travel time. Growth of the Internet and the popularity and effectiveness of the World Wide Web is another factor that has spurred the interest in mobile computing systems.

The availability of the World Wide Web on mobile computing systems is expected to open up a new class of location sensitive applications.

In this chapter, we present a survey of mobile transaction models proposed in the literature. These transaction models are aimed at maximizing concurrency and maintaining data consistency in a failure prone and low bandwidth mobile environment. Our main thrust in this chapter is in comparing the transaction models on parameters like scalability, additional infrastructure, communication costs, extensions required for commercial databases, etc. These parameters are indicative of the cost of executing a transaction under a particular model as well as the cost of the deployment of a transaction model. Both issues are crucial to the success of any model in the practical world.

The major factors that differentiate mobile computing from conventional computing are movement, low bandwidth, and frequent disconnections. A mobile user will typically be connected to the fixed network through a cellular network link or a radio or infrared link. In all these cases, the bandwidth available is low, typically in the range of 10 kilobits per second (Kbps) in case of cellular links and 2 megabits per second (Mbps) in the case of an infrared link. As the mobile user moves, the current link may get disconnected, and the user may have to acquire a new link to retain connection to the fixed network. The mobile user may also stray off the service area of wireless providers or lose the connection for extended periods of time. These factors along with the issues arising out of mobility itself, such as how to uniquely identify a mobile system in the internetworked world or how to service location sensitive queries like "Where is the nearest restaurant?" have thrown up a number of challenges and opened up a new field of research.

The research effort in mobile computing has been mainly concentrated in the following major areas:

- Networking
- Operating systems
- Database
- System architectures

The mobile networking research has mainly concentrated on developing protocols for seamless access of a mobile computer in the internetworked world. The Mobile-IP Protocol [PerkIP] and other Internet Protocol (IP)-based internetworking protocols [Ioannidis91, Johnson93a, Teraoka] attempt to modify the IP for use in mobile environments. These protocols define mechanisms that allow a mobile computer to access resources in an internetwork irrespective of its current location as well as allow other systems to access the mobile computer with the same IP address irrespective of the point of attachment of the mobile system. Operating systems research in mobility has mainly concentrated on file systems mechanisms [Kistler91] and event delivery mechanisms [BadriWelling]. Architecture issues and protocol mechanisms have been addressed in [CheGros, Liu-MarMag]. Database issues have been another focal area in mobile computing research. The studied database related issues in mobile computing research include transactions [Chry93a, Helal96a, Niel95, PitouraBhargavaTr, WalChry95, Zas94], lock management [JJ95], and data consistency [PitouraBhargava-Ag95].

The chapter is organized as follows:

- Section 35.2 describes the reference model for mobile environments.
- Section 35.3 presents the characteristics and issues related to mobile transactions.
- Section 35.4 describes the transaction models on which the various mobile transaction models are based.
- Section 35.5 presents the various transaction models.
- Section 35.6 presents a comparison of the models.
- Section 35.7 outlines the issues still to be resolved.

35.2 Reference Model

The generally accepted reference model for the mobile computing environment is depicted in Figure 35.1. The model has a set of hosts on a fixed network, some of which serve as base stations (or mobile support stations [MSS], as some authors refer to them). Each base station serves a number of mobile hosts (MH) (or mobile nodes), which are currently in its cell. MHs as well as the base station have wireless interfaces. The base station communicates with the MH in its cell by broadcasting. As a MH moves across a cell boundary, the base station that was serving it prior to its movement hands over the MH to the base station in the cell that the MH enters. A handoff protocol defines the actions that occur during the handoff. The model proposes the use of location databases that are used to locate MHs. A discussion of location management can be found in [Badrinath92b].

MHs are expected to be laptops or palmtops, which have batteries with a short life span. Further, the wireless communication channel between the MH and the base station is constrained by bandwidth. So, protocols have to be designed so that the load on the MH as well as the amount of communication between the MH and the base station is minimized.

MHs are expected to disconnect from the fixed network frequently. This may happen either due to physical damage suffered by the MHs or due to a voluntary disconnection by the mobile user (to conserve power). The user may later connect to the fixed network at a different location. The network layer protocol has to be capable of handling these disconnections and providing a transparent interface to the upper layers. The Mobile IP

MOBILE COMPUTING HANDBOOK

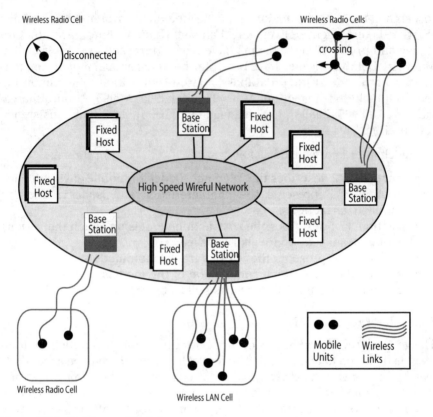

Figure 35.1. Reference Model

developed by IETF [PerkIP] is accepted as the standard for the network layer.

To summarize briefly, the mobile computing environment poses many challenges to hardware, software, and communications. Most of the issues have already been addressed and solutions have been proposed. These are still under investigation and no formal standards have been yet adopted.

35.3 Mobile Transactions: Definition, Characteristics, and Issues

According to the classical definition a transaction is described by its ACID properties — atomicity, consistency, isolation, durability. It has been long recognized that the ACID properties are too restrictive for many applications, which can be modeled as transactions [Elmar91]. Many models have been proposed to extend the traditional transaction model by relaxing the atomicity, concurrency, and isolation requirements [Elmar91]. The motivation for a new model for transactions in mobile environments can be illustrated by the following examples.

A Survey on Mobile Transaction Models

Consider a mobile user ordering takeaway food from a fast food restaurant chain. The transaction is initiated by contacting the nearest restaurant in the chain address of which is obtained from a Yellow Pages server. As the order is being processed, it is possible that the mobile user is already past the service area of the first franchise and has entered the service area of the second franchise. The execution of the transaction may now have to be relocated to the new serving franchise of the mobile user to seek performance. This process may continue until the order is completed and the user is ready to be served. The system may then direct the mobile user to the particular franchise where the customer will be served. This example illustrates an important characteristic of a mobile transaction, the ability to execute the transaction at a geographical proximity to the mobile user who has issued the transaction.

Another example is adapted from [CheGros] to illustrate disconnected or asynchronous operations. Consider a mobile employee with a rather flexible travel plan. The employee may finalize travel plans at some point on the trip and may initiate a transaction to book the most economical and convenient ticket to the destination. Once the transaction is initiated, the user may disconnect from the system. This sets in motion some agent processes that determine the available choices. These agent processes report back the results when the user connects to the system again. Depending on the results, the user may either decide to continue the transaction, initiate a new transaction, or terminate the transaction. This example illustrates another important characteristic of mobile transactions, nondeterministic duration and asynchronous operation.

These examples illustrate the need to relax the strict ACID properties of the traditional transactions for mobile transactions. Transactions executed in mobile environments could be of nondeterministic duration and could get aborted due to failures on the mobile unit (MU) end like power or connection failures. Thus even though transaction models have been designed for long duration transactions [Elmar91, GarciaMolina87], none of the existing transaction models is fully suited for operation in mobile environment.

35.3.1 Characteristics

From the above discussion the following characteristics of mobile transactions can be summarized:

- Nondeterministic lifetime — the MUs from which a mobile transaction is issued will normally be attached to the fixed internetwork through a low bandwidth wireless link. The routing delay in the mobile environment is far higher than in the fixed network. Mobile systems are prone to failures such as battery power loss and wireless link loss. These factors coupled with other factors like disconnection

from the fixed network either due to lack of a communication link or for economic reasons make the duration of a mobile transaction nondeterministic.
- Relocation — To minimize the response time and efficiently use the limited bandwidth available, it is necessary to have the component of the mobile transaction executing on the fixed network to be as close to the MU as possible. Thus it is also necessary to relocate the fixed network component of the mobile transaction as the mobile node moves.

35.3.2 *Definition*

A mobile transaction is a transaction of nondeterministic lifetime submitted from a mobile capable node in a mobile environment. The mobile transaction, in general, can be considered to consist of two components — a MU component and a fixed network component. The fixed network component of the mobile transaction may have to be partially or completely relocated as the MU moves.

35.3.3 *Issues*

Mobile environments can be considered to be similar to highly distributed environments in many respects. But unlike in distributed environments, locations of some hosts are not permanent in mobile environments. This along with the low communication bandwidth, frequent disconnections, and high vulnerability throws up many challenges to researchers. In this section, we outline the issues addressed by researchers in mobile transaction design:

- Data consistency and concurrency control — in mobile environments, data could be replicated on a number of servers throughout the network. Some of these servers could be MUs. Moreover, a MH might operate on cached data while being disconnected from the fixed network. The data conflicts arising in mobile environments could partly be due to the locality of the users accessing the data [Imielinski93b]. The execution of a mobile transaction could also be distributed and relocated among fixed hosts and the mobile nodes. The nondeterministic lifetime of a mobile transaction and the low bandwidth of communication links are other factors that affect concurrency control and cache management.
- Infrastructure requirements — for any model to be successful, it is important that it be moved from the research labs and deployed in the real world. Assuming a wireless communication infrastructure to be well in place, it is important to determine the additional resources required for having a mobile transaction system in place. These resources could range from protocols for location sensitive

service access to mechanisms for optimized query management and controlled query release mechanisms.
- Communication costs — the high cost of the communication links is one of the major constraints in mobile environments. Efficient utilization of bandwidth is thus an important factor on evaluating a transaction model.
- Relocation mechanisms and user profiles — mobile agents are processes or set of processes that perform an activity on the fixed network on behalf of the MU. These agents will typically be a transaction activity that access several databases and report some results to the mobile node. Relocation of transaction execution or mobile agents is necessary to improve response times in mobile environments. Effectiveness of mobile agent relocation has been studied in [LiuMarMag]. Performance can still be improved if the user profiles or user directives can be used to effect anticipatory relocation or to avoid unnecessary relocation.
- Scalability — as mobile computing grows to be more affordable and popular, the number of MUs handled by every base station could be large. Hence, it is very important that a mobile transaction model scale up efficiently.

35.4 Applicable Transaction Models

The mobile transaction models presented in this chapter are based on the extended transaction models developed for open-ended long duration transactions. The extended transaction models, which form the basis for mobile transaction models are:

- Open Nested Transactions [Elmar91]
- Split transactions [PU88a]
- Saga — compensating transactions [GarciaMolina87]

The applicability of these transaction models can be easily explained from the characteristics of mobile transactions. Due to the nondeterministic lifetime of a mobile transaction, it is best characterized as a long-lived transaction. The execution of a mobile transaction could be migrated or relocated as the MU moves after it issues the mobile transaction. The high vulnerability of mobile transactions may warrant arbitrary rollback as well.

35.4.1 Open Nested Transactions

Open Nested Transaction Model is designed for long duration activities. These transactions typically consist of a set of subtransactions, which can be structured as a transaction tree. The Open Nested Transaction Model provides better support for long duration activities. These are also ideally suited for conversational transactions.

35.4.1.1 Properties of Open Nested Transactions.
An open nested transaction is a generalization of multilevel transactions. In a multilevel transaction model, the subtransactions are divided into layers and the nesting depth is the same among all subtransactions. In the case of open nested transactions, the restriction on nesting depth has been eliminated, allowing different nesting depths in different subtransactions trees.

The classical ACID paradigm of transactions is too restrictive for long duration activities, which can be easily modeled as transactions. The Open Nested Transaction Model accommodates these extended transaction activities by relaxing the ACID properties:

- Atomicity — the classical definition requires transactions to be atomic at the lower details. Atomicity of a transaction or a subtransaction can still be enforced if the system ensures that the effect or existence of an aborted transaction is hidden from other transactions and subtransactions. A completed open nested transaction cannot be rolled back, because the results are already visible to other transactions. Open nested transactions are undone by executing compensating transactions, which reverse the effect of the transactions.
- Isolation — in the Open Nested Transaction Model, semantics of the transaction operations are used to relax the isolation of transactions. The operations are defined to be either commutative or compatible if the order of the execution of the operations is insignificant for the success of the application. The result of an operation can be made available to its commuting or compatible operations. Serializability is not compromised in this case, because the order of execution is immaterial in this case.
- Durability — in some applications with long duration transactions, the complete undo of a transaction may not always be acceptable. In the Open Nested Transaction Model, programmers are allowed to tag subtransactions with a `persistent` attribute. The updates of a persistent subtransaction are made persistent as soon as it completes. Compensating subtransactions are not allowed for persistent subtransactions.

35.4.2 Split Transactions

Split transactions [PU88a] or dynamically restructured transactions split a transaction into two independent serializable transactions. The transactions could be committed or aborted independent of each other. The operation `split-transaction` can be used to split a transaction into two. The split transactions are combined together by an inverse operation termed `Join Transaction`. Split transactions are mainly designed for user controlled open-ended transactions.

35.4.2.1 Split Transaction Semantics

- Split transaction — the `split-transaction` operation is used to split a transaction into a set of independent entities. It takes the read and write sets as the input and produces a split:

  `split-transaction(Read(A), Write(A), Read(B), Write(B))`

- Join transaction — the `join-transaction` operation is the inverse of the `split-transaction` operation. It takes the target transaction to which the current transaction is to be joined as the input. Let T be the current transaction to be joined with S. The operation `join-transaction(S)` joins T with S. All data items of T are available to S after this operation. T may be committed or aborted depending on the fate (`commit` or `abort`) of S. The join transaction can also be extended so that the join is done only if the target agrees on the operation too.

35.4.2.2 Properties of Split Transactions. Let a transaction T be split into two transactions A and B. Let the read and write sets of a transaction be denoted as Read(T) and Write(T) respectively. Then:

$$Read(T) = Read(A) \bigcup Read(B)$$

$$Write(T) = Write(A) \bigcup Write(B)$$

The instructions in T are also split into instructions of A and B:

$$Instr(T) = Instr(A) \bigcup Instr(B)$$

The transaction is split if and only if the following conditions hold:

$$Read(A) \bigcup Write(B) = \phi$$

$$Write(A) \bigcup Write(B) = \phi$$

$$Read(B) \bigcup Write(A) = \phi$$

From these properties it is evident that there are no dependencies between transactions A and B. They are entirely independent of each other. There are no serializability constraints between the transactions A and B, moreover there are no data conflicts either. The above restriction can be relaxed; if we assume that the transaction A is committed as soon as T is split and only B is executed after that, then the properties can be written as:

$$Write(A) \cap Write(B) \subseteq Writelast(T)$$

$$Read(A) \cap Write(B) = \phi$$

$$Read(B) \cap Write(A) = Share(A,B)$$

The set WriteLast(T) is the set of data items written last by transaction T. Share(A, B) is the data set shared by transactions A and B. The first property allows transaction B to overwrite the values written by A, but not vice versa. The second property ensures that A precedes B. The third property allows transaction B to use values written by A.

35.4.3 Sagas

Sagas [GarciaMolina87] are defined as a set of relatively independent transactions. The transactions in a saga are termed as component transactions. Each component transaction has a dual termed compensating transaction. A predefined order can be defined for the execution of the saga. The transactions belonging to different sagas can be interleaved in any fashion.

35.4.3.1 Properties of Sagas. As mentioned above, a saga consists of a set of component transactions and corresponding compensating transactions. A compensating transaction can semantically undo the effects of the component transaction. A component transaction can have ACID properties. A component transaction is not allowed to make any changes directly onto the database until it is ready to commit.

Let $T\_1, T\_2, ..., T\_n$ constitute a saga S. The saga S is said to commit if all the transactions $T\_1, T\_2, ..., T\_n$ belonging to S have executed and committed. If the saga has aborted after the commitment of the transaction T_k where $1 \leq k \leq n$, then the correct execution of the saga is:

$$T\_1, T\_2, T\_3, ..., T\_k, CT\_k, CT\_\{k-1\}, ..., CT\_1$$

where CT_k is the compensating transaction for transaction T_k.

35.4.3.2 Limitations of Sagas. The commitment of a saga is dependent on the commitment of all its components. But a saga by itself has no notion of commitment. This introduces a certain amount of inflexibility. Moreover, some activities cannot to be modeled as saga transactions, because they inherently cannot be compensated. The Saga Model has been extended to take care of these limitations.

35.4.3.3 Extensions of Saga Model

- Vital and nonvital components — component transactions are distinguished as vital and nonvital components. A saga can commit if

and only if all vital components commit. A saga need not be aborted if any of the nonvital components abort.
- Nested saga — a nested saga is a saga transaction with sagas as its components. A nested saga can be considered to be a set of nonvital components. Thus a nested saga can commit even if some of its components abort.

35.4.3.4 Noncompensating Transactions. Sagas are designed such that every component transaction will have a compensating transaction. But some transactions inherently cannot be compensated. Different techniques are used to accommodate this. In one method, noncompensating transactions are executed concurrently. In another technique, noncompensating transactions are set up as nested transactions. In yet another technique, additional semantics are used to specify dependencies between transactions.

35.5 Approaches to Mobile Transaction Models

In this section, we describe selected approaches to mobile transaction modeling. We will consider the following models in this section:

- Reporting and cotransaction [Chry93a]
- Clustering [PitouraBhargava95]
- MDSTPM [Zas94]
- Pro-Motion [WalChry95, WalChry99]
- Prewrite [MadriaBhargava]
- Semantic [WalChry95]
- Kangaroo [Helal96]
- Time-based Mobile Transaction [Niel95]
- Two-Tier Replication [Gray96]
- IOT [Saty]
- Bayou [Terry]
- New Transaction Management System [Nagi]

Most of the approaches that use the Mobile Computing Reference Model were described in Section 35.2. If a special reference model is used, we will point it out when we explain the corresponding transaction model.

Based on different taxonomy, the selected mobile transaction models can be classified in different ways.

Mobile transaction models can be classified based on their emphasis on challenges presented in a mobile computing environment. [Chry93a, Helal96, MadriaBhargava, PitouraBhargava95, WalChry95, WalChry99, Zas94] focus on providing solutions to improve the availability of local resources on mobile devices. [Gray96, WalChry95, Zas94] provide solutions for frequent disconnection. [Helal96] addresses the issue of movement. [Niel95] focuses on real-time transaction management in a mobile environment. Mobile transaction models can be classified based on their communication assumptions. [Chry93a, Gray96, Helal96, MadriaBhargava, PitouraBhargava95, WalChry95, WalChry99, Zas94] assume the existence of MSS. [Nagi] on the other hand utilizes the broadcast feature of the downlink of the server.

Mobile transaction models can be classified based on the research approaches. [Chry93a, Helal96, MadriaBhargava, PitouraBhargava95, WalChry95, WalChry99, Zas94] focus on transaction management in a multi-database system. [Gray96, Saty, Terry, WalChry95] use a data replication and coherence approach.

35.5.1 Reporting and Cotransactions

This model [Chry93a] is based on the Open Nested Transaction Model. A computation in a mobile environment is considered to consist of a set of transactions, some of which may execute on the mobile node and others may execute on the fixed host. The model attempts to address the following two issues in particular:

1. Sharing of partial results while in execution
2. Maintaining computation state in a fixed node so that the communication cost is minimal

The model allows the sharing of partial results and transaction relocation.

The model proposes to modify reporting and cotransactions [Chry91, Chry93b] to suit the mobile environment. The model defines a mobile transaction to be a set of relatively independent transactions, which interleave with other mobile transactions. A component transaction can be further decomposed into other component transactions with an arbitrary level of nesting.

Component transactions are allowed to commit or abort independently. If a transaction aborts, all components that have not committed yet may abort. Some of the transactions may have a compensating dual and may be compensated.

The model classifies mobile transactions into the following four types:

1. Atomic transactions — atomic transactions are normal components and may be compensated with atomic compensating duals.
2. Compensatable transaction — this is an atomic transaction whose effects cannot be undone at all. When ready to commit, the transaction delegates all operations to its parent. The parent has the responsibility to commit or abort the transaction later on.
3. Reporting transactions — a reporting transaction can make its results available to the parent at any point of its execution. It could be a compensating or a noncompensating transaction.

A Survey on Mobile Transaction Models

4. Cotransactions — cotransactions behave in a manner similar to the coroutine construct in programming languages. Cotransactions retain their current status across executions. Hence, they cannot be executed concurrently.

35.5.1.1 Properties of Reporting Transactions. A reporting transaction reports its results to other transactions by delegating the results. A reporting transaction can have only one recipient at any given point of time. The changes made by a reporting transaction are made permanent only when the receiving transaction commits. If the receiving transaction aborts, the reporting transaction aborts as well.

35.5.1.2 Properties of Cotransactions. A cotransaction reports its results in a way similar to a reporting transaction. But upon delegation, the transaction stops execution and is resumed from the point where it left off. For any pair of cotransactions, either both commit or both abort.

35.5.2 The Clustering Model

This model [PitouraBhargava95] assumes a fully distributed system. It is designed to maintain consistency of databases. A database is divided into clusters. Each of them defines a set of mutually consistent data. Bounded inconsistencies are allowed to exist between clusters. These inconsistencies are finally reconciled by merging the clusters. This model is based on the *Open Nested Transaction Model* and is extended for mobile computing. A transaction submitted from a MH is composed of a set of weak and strict transactions. Transaction proxies are used to mirror the transactions on individual machines as they are relocated from one machine to another.

35.5.2.1 Clusters. A cluster is defined as "a unit of consistency in that all data items inside it are required to be fully consistent, although data items residing in different clusters may exhibit bounded inconsistency." Clusters can be defined either statically or dynamically. A wide set of parameters can be used for defining clusters. This could include the physical location of data, data semantics, and user definitions.

Consistency between clusters can be defined by using an m-degree relation, in which the clusters are said to be m-degree consistent. The m-degree relation can be used to define the amount of deviation allowed between clusters.

35.5.2.2 Weak and Strict Transactions. A mobile transaction is decomposed into a set of weak and strict transactions. The decomposition is done based on the consistency requirement. The read and write operations are also classified as weak and strict. The weak operations are allowed to access only data elements belonging to the same cluster, whereas strict operations are allowed to access the whole database. For every data item, two copies can be maintained — one strict and the other weak. As mentioned above, a weak operation can access only the local copies of a data item. Weak operations are initially committed in their local clusters. They are once again committed when the clusters are finally merged. The weakly committed values are available only to other weak transactions in the same cluster.

35.5.2.3 Transaction Migration and Proxying. Transaction migration is used for relocating transactions to avoid long network delays, as well as to represent the user mobility. Relocation is denoted by $T_{i \to j}$, which indicates that transaction T was partially executed at site i before it was migrated to site j. Transaction proxies are used to support relocation. As the transaction/host moves, the proxy is relocated.

35.5.3 The Multi-Database Transaction Processing Manager

The MDSTPM is a model proposed by [Zas94]. The model visualizes an environment where MHs submit transactions to a coordinator on the fixed network. The MHs may disconnect from the network. They might reconnect at a later time to query the result of the transaction. The system builds on existing heterogeneous, autonomous database management systems (DBMSs) and defines a new layer residing on top of them.

35.5.3.1 Architecture. The model assumes the MDSTPM to be running on top of each of the DBMSs. When a MH wants to connect to the fixed network, it sends a message to a coordinator on the fixed network to request a connection. The coordinator sends back an acknowledge message to the MH. Similarly, the MH sends a disconnect request message to the coordinator when a voluntary disconnect is desired. The coordinator handles these and other messages asynchronously. It does not notice failures of MHs until the failed host recovers and tries to reconnect. At that moment, the coordinator finds that the MH had disconnected abnormally by checking on a status table that it maintains.

The MDSTPM has the following components:
- Global Communication Manager (GCM) — this is responsible for handling message passing for the local site. It exchanges messages with MHs as well as other sites on the fixed network.
- Global Transaction Manager (GTM) — this module manages the global transactions submitted by MHs. The site to which the global transaction is submitted is designated as the coordinator for that transaction. The other participants in that global transaction are termed Global Transaction Manager Participants (GTMPs). GTM has components for scheduling global transactions (Global Scheduling Submanager [GSS]) and for concurrency control of global subtransactions (Global Concurrency Submanager [GCS]).

A Survey on Mobile Transaction Models

- Global Recovery Manager (GRM) — this is responsible for recovery after a global transaction failure.
- Global Interface Manager (GIM) — this acts as the interface between the MDSTPM and the local DBMS.

35.5.3.2 Transaction Model. The global transaction submitted by the MH to the coordinator is scheduled and executed by the coordinator on behalf of the MH. When a transaction is submitted, it is put into an input queue by the GCM. The transaction undergoes a state transition when it moves from one queue to another. The queues (apart from the input queue) are the allocate queue, the active queue, the suspend queue, and the output queue. The GSS schedules the execution of the transactions in the input queue and moves them to the allocate queue. The transaction gets the locks it needs from the GCS and moves to the active queue. The global transaction is broken down into subtransactions and dispatched to other sites by the GCM. When the global transaction has completed the first phase of the two-phase commit, it is put on a suspend queue. On completion, it is put on the output queue and handed over to the MH that initiated it, when that host connects to the network.

One of the significant features of this model is that once the MH has submitted the transaction to the coordinator, further communication is not required until the MH wants the results of the transaction. There is no mobility or migration of transactions.

35.5.4 Pro-Motion

The goal of Pro-Motion [WalChry95, WalChry99] is to devise methods to allow remote database access and update by mobile computers regardless of connection status and despite various limitations of mobility and portability.

Pro-Motion is a flexible and adaptive infrastructure to support transaction processing in a multitier, mobile client–server operating environment. It allows mobile clients to continue executing competing transactions on data items cached locally while they are moving and not connected to the network, incorporating the modified data back into the database when reconnection occurs.

The architecture consists of a compact manager at the database server; a compact agent at the MH to negotiate, manage compacts, and provide local transaction management for the MU; and a mobility manager at the MSS to help manage the flow of updates and data between the other components in the system.

The fundamental building block is called compact (object) and is the basic unit of caching and consistency. A compact is an object that encapsulate the cached data, methods to access the cached data, information about the current state of the compact (such as name, data type, version, cache status outstanding transaction IDs, and amount of storage), consistency rules to guarantee global consistency, obligations (such as a deadline that creates a bound on the time for which the rights to a resource are held by the MH) and methods that provide an interface with which the MU can manage the compact.

The MH requests compacts from the server when a real or anticipated data demand is created. If the data is available, the server creates a compact (with the help of the compact manager), records it in the compact store, and transmits it to the MU. The MU records the compact in compact registry, which is used by the compact agent to track the location and status of all active compacts. Some of the basic methods of compact are inquire(), notify(), dispatch(), commit(), abort(), and checkpoint(). Other application specific methods can be included as needed.

For a transaction a shared data item is given to all requesters (MU) with an expiration time. MUs with read access are free to read the item until the expiration time, but must obtain permission from all other MUs holding a copy. To do these specific methods read() and modify() are used. The read() method checks for expiration of data and returns the value of the compact data if valid. With the modify() method, the compact communicates with the server to obtain write access. The server in turn obtains permission from other MUs holding the data value and communicates permission (or refusal) to the MU that wants to modify the data. Compacts can also be used to get data to be read and written exclusively. Data can also be stored as fragments.

The compact manager acts as a front-end to a database (DB) server and may execute on the same or independent host from the DB server. The system utilizes an Open Nested Transaction Model as the basis for concurrency control and recovery for mobile transactions processed against the DB server. To the DB server, the compact manager appears to be an ordinary DB client, executing large, long-lived transactions. These transactions become the root transactions of the Nested Transaction Model. Resources needed to create compacts are obtained by these transactions through normal DB operations (reads and writes). Mobile transactions appear as children in the open nested transactions. The transactions processed on the MH appear as siblings. Each sibling transaction may commit or abort independently as long as the consistency constraints expressed in the compacts with which they have shared access are not violated. It should be noted that a mobile transaction may invoke multiple compacts and a compact may support the execution of multiple transactions. The responsibility for the correct execution of mobile transactions is assumed by the MH and accomplished by utilizing the methods encapsulated in the compacts. The root transactions are managed by DB server and committed by the compact manager.

A Survey on Mobile Transaction Models

On each MH, a compact agent is responsible for processing requests on behalf of transactions executing on the MH. The compact agent handles disconnections and manages storage on a MH. It monitors activity and interacts with the user and applications to maintain lists of items, which are candidates for caching. The compact agent acts as the transaction manager for transactions executing on the MH and is responsible for concurrency control, logging, and recovery.

Each of the interactions between the compact agent and the compact manager are processed via the MSS. When an update is sent to MSS, the mobility manager (MM) in the MSS, functions on behalf of the compact agent to complete delivery of update messages. This update by proxy ensures that the compact manager receives the updates in a timely fashion. The MM maintains mobility tables in which each mobile control block (MCB) contains location and database access information that pertains to a single MH.

35.5.5 Prewrite

Prewrite [MadriaBhargava] tries to increase data availability on MH by introducing a `prewrite` operation in addition to standard writes. A `prewrite` makes data value visible at precommit before the `commit` of the mobile transaction (MT). Permanent updates on the database are performed later by the `write` operation at commitment. Two variants of the data are maintained — the `prewrite` and the `write`. Prewrite variant reflects the future state of the data, but may be structurally slightly different from the corresponding write value.

The main idea is to divide the transaction execution between the MH and the DB server. Three operations (`prereads`, `prewrites`, and `precommits`) that will be executed by the transaction manager (TM) are proposed. Ordinary `reads` and permanent `writes` are made by a data manager (DM) at the DB server. The MSS has logging capacities and maintains close relationship with the DM. The transaction execution is divided in two parts, first, the TM executes local transaction that finishes with a `precommit`. In the second part, the DM makes `prewrites` permanent and `commits` the MT. This model considers that MT are long-lived and implementation can be made with nested and split transactions.

The transaction validation is done in two steps. The first one is realized on MH (`local commit`) and the second one (`commit`) at the MSS/DB server. Prewrite does not differentiate connected and disconnected mode. `Local commit` is performed using an atomic commit protocol. At the second step of the validation process, locally committed transactions execute `commit` to make updates permanent on the DB server. Transaction commitment can involve a reconciliation mechanism or transactions reexecution. Neither reconciliation nor reexecution is made. By the transaction processing algorithm and locking protocol, Prewrite ensures that locally committed transactions will commit at the DB server.

Prewrite ensures that the transaction processing algorithm along with the lock-based protocol produces only serializable histories. This serializability is based on the `precommit` order of MT. In Prewrite, objects can have two variants (`write`/`prewrite` value) as design objects (the `prewrite` represents a model of the design) or document objects. In these object types, `prewrites` are different from `writes` and availability is improved with two variations of the same object. Otherwise, using simple objects, `prewrites` are identical to `writes` and the algorithm behaves as using relaxed two-phase locking. If an MT makes a `local commit`, it is sure to commit. Prewrite does not permit a local committed transaction to abort.

35.5.6 Semantic Transaction Processing

The semantics-based mobile transaction processing scheme [WalChry95] views mobile transactions as a concurrency and cache coherence problem. It introduces the concepts of fragmentable and reorderable objects to maximize concurrency and cache efficiency exploiting semantics of operations defined on the data objects. The model assumes a mobile transaction to be a long-lived one characterized by long network delays and unpredictable disconnections.

35.5.6.1 Exploiting Semantics for Concurrency and Caching. Traditional definitions of concurrency and serializability are too strict for most operations [Elmar91]. Semantics of operations defined on an object can be utilized to define correctness criteria so as to maximize the concurrent operations on the object [ChryRaghuRama91]. Both application-dependent and application-independent semantics can be utilized for this purpose.

Commutativity of operations is an important property to allow concurrent operations on an object. If certain operations on an object are commutative, then the DB server can schedule these operations in an arbitrary manner. Recovery also becomes quite simplified. Operations may be commutative either for all states or only for some states of the objects. The input/output (I/O) values of the operations can be used to redefine serial dependencies of the operations. Though this may improve concurrency, it may require more complex recovery mechanisms than the normal schemes. Organization of the object can be used for selective caching of the object fragments, necessary for continuing the operation during the disconnected state. This approach reduces the pressure on the limited wireless bandwidth as well as better utilizes the cache space available on the MH.

Application semantics can be utilized to define the degree of inconsistency, degree of isolation, and degree of transaction autonomy [Chry91,

Elmar91]. Techniques like epsilon serializability and quasi-copies [AlonsoBarbaraMolina90, PuLeff91] can be used to specify allowable inconsistencies in the systems.

35.5.6.2 Fragmentable and Reorderable Objects. This approach utilizes the object organization to split large and complex objects into smaller easily manageable pieces. The semantic information is utilized to obtain better granularity in caching and concurrency. These fragments are cached or operated upon by the MHs and later merged back to form a whole object. Thus the object fragments form the basic unit of consistency. A stationary server dishes out the fragments of an object on request from MUs. The objects are fragmented by a `split` operation. The split is done using a selection criteria and a set of consistency conditions. The consistency conditions include the set of allowable operations on the object and the conditions of the possible state of the object. On completion of the transaction, the MHs return the fragments to the server. These fragments are put together again by the `merge` operation at the server. If the fragments can be recombined in any order then the objects are termed reorderable objects. Aggregate items, sets, and data structures such as stacks and queues are examples of fragmentable objects.

Formally an object O represented as (S, C) where S is the state of the object and C is the set of consistency conditions, is said to be fragmentable if it can be split into fragments $(O\_1, C\_1), (O\_2, C\_2), \ldots (O\_n, C\_n)$ such that each of the fragments supports the same set of operations as object O. The transactions can operate asynchronously on the object fragments. The modified objects when merged still satisfy the consistency constraints of the object O. The object O is reorderable if the fragments, $O\_1, O\_2, \ldots O\_n$, can be merged in any order.

35.5.7 The Kangaroo Transaction Model

This model [Helal96] is based on the global transaction and the split transaction models. In this model, transaction relocation is achieved by splitting the transactions. A mobile transaction is considered as a global transaction in a multi-database environment.

35.5.7.1 Reference Model. The mobile computing environment assumed has a small enhancement compared to the model described in Section 35.2. It consists of three layers. The innermost layer is the DBMS running on the source system. The outermost layer has the mobile nodes, which initiate mobile transactions. The middle layer consists of a data access agent (DAA), which acts as a gateway between mobile nodes and the source system. DAA is assumed to be present in every base system and acts as a router for the data. In general, DAA is a TM for mobile transactions.

35.5.7.2 Transaction Model. In the Transaction Model, the mobile transaction is termed as kangaroo transaction (KT). A KT is a global transaction identified with the user who has issued it by a unique ID. It consists of a set of joey transactions (JT). A JT is associated with the base station or the cell in which it executes. When the MU moves to a new cell, the JT in the previous cell is split into two JTs and one of them is moved to the current cell of the MU. Each JT may consist of a set of local and global transactions. The model is built upon the existing databases. The transactions are micro-managed by the individual database TM.

35.5.7.3 Properties

- Joey transactions — a JT consists of a set of global and local transactions. Each JT should terminate in an `abort`, `commit`, or `split`. A handoff of a mobile node from one cell to another will result in a split of the JT associated with it, if any. Each JT is assigned with a unique ID when it is created.
- Kangaroo transactions — a KT consists of a set of JT. For a KT to be successful, the last JT in the order should end in a commit or abort whereas all other JTs should be split. KT captures the movement behavior of the transaction.

35.5.8 Time-Based Consistency Model

In [Niel95], a time-based model is proposed to maintain consistency in a mobile environment. All objects in the system are associated with a set of time parameters. These time parameters are used to determine whether the object is currently consistent or not. A modification time (MT) is associated with every object on the server. When the object is cached on the MU the modification time of the object on the MU is set to the modification time of the object on the server. The object on the MU is also associated with a consistency time (CT) and a consistency flag. These parameters are used to represent the consistency state of the object on the MU:

- `Read` parameters — when a MU reads an object, a parameter consistency time bound (`CTB`) is associated with it. The object is considered to be consistent only for that time period. The `CTB` can be used to specify an optimistic or pessimistic approach. This model uses time locks for controlling access to objects as well. When a host acquires a read lock on an object a read expiration time is set. The read lock is valid only for that duration.
- `Write` parameters — when a MU writes a cached object a parameter modification time bound (`MTB`) is set on the object. The MU is required to update the server copy within the `MTB`. Otherwise, the modification is invalidated. When the server copy is updated, the time parameters are appropriately updated on the server. Write locks are timed out after the write expiration time (WET). It is necessary for a

MU to establish a connection within this period and update the copy on the server.

Time-out locks provide a flexible and secure mechanism to lock data objects. But this mechanism also requires that a MU correctly estimate the time-out period. Performance could be seriously affected if the time for the lock is either too high or too low.

35.5.9 Two-Tier Replication

This scheme [Gray96] assumes two types of nodes:

1. The mobile nodes that store a replica of the database and may originate tentative transactions
2. Base nodes that are always connected and have replica of the database

At the mobile node two versions of the replicated data item are kept:

1. The master version, which is the most recent value received from the object master
2. The tentative version, which has the most recent value due to local updates

This object is updated by tentative transactions.

There are two kinds of transactions:

1. The base transactions, which work only on master data and produce new master data and involve at most one connected mobile node and may involve several base nodes.
2. Tentative transactions work on local tentative data and also produce a base transaction to be run at a later time on the base nodes.

With the two-tier replication scheme mobile nodes may make tentative database updates. When the mobile node connects to a base station, the mobile node:

- Sends replica updates for any objects mastered at the mobile node to the base node.
- Sends all its tentative transactions to the base node to be executed in the order in which they committed on the mobile node.
- Accepts replica updates from the base node.
- Accepts notice of the success or failure of each tentative transaction.

The mobile node connects to the base, the host base node:

- Sends delayed replica update transactions to the mobile node.
- Accepts delayed update transactions for mobile-mastered objects from the mobile node.
- Accepts a list of tentative transactions, their input messages, and their acceptance criteria. Reruns each tentative transaction in the

order it committed on the mobile node. During this process, the base transactions access object master copies. (The scope rule ensures that base transactions only access data mastered by originating mobile nodes and base nodes.) If the base transaction fails its acceptance criteria, the base transaction is aborted and a diagnostic message is returned to the mobile node.
- When a base transaction commits, the base node sends updates to all other replica nodes.

35.5.10 Isolation-Only Transactions

35.5.10.1 The Approach Taken. The model [Saty] is enabled by the Coda file system, which provides continuous file access to mobile clients using a special form of client disk caching and optimistic replica control. This model hides the mobility of clients from application and users. It also allows disconnected operation. The transactions are IOT. This transaction model is an optional file system facility of the Coda file system. Thus, application writers can select the IOT option to wrap around applications for better consistency protection when the application is used in a mobile environment. Some background information on the Coda file system and the concept of IOT is described in the following.

35.5.10.2 The Coda File System. The Coda file system is a distributed UNIX file system. It addresses the issue of hiding mobility from applications and users. In other words, it provides continuous file access to mobile clients even when they are disconnected from the servers. This is possible by using a technology called disconnected operation, which is a special form of client disk caching employing optimistic replica control. When disconnected, the local cache of the Coda client solely serves file access requests. Updates are performed locally, logged and later reintegrated to servers upon reconnection.

35.5.10.3 What Is an IOT? The IOT is a flat sequence of file access operations bracketed by `begin_iot` and `end_iot`. It guarantees consistency — actually, various levels of serializability, depending on the connection status — for transactions in a mobile UNIX environment. However, it does not guarantee atomicity and only conditionally guarantees durability.

35.5.10.4 IOT Execution Model. The execution of the IOT is carried out with consideration of disconnection. When a user invokes a transaction T, the entire execution of T is first carried out on the client's local cache. Remote accesses also go through the local cache. During T's execution, if no disconnection has occurred while accessing a file, T can safely commit and its result is made visible on the servers. Otherwise, if a disconnection has occurred while accessing a file (i.e., partitioned file access) in T, T enters a pending state and a validation step (i.e., checking if T satisfies global serializability, explained later) must be carried out to ensure consistency. This validation step can only be carried out after the connection is regained. If the validation succeeds, the result can be reintegrated and committed. Otherwise, if the validation does not succeed, a resolution step is needed. The resolution can be carried out automatically or manually and afterwards, the result is committed.

35.5.10.5 Why Isolation Only?

- IOT provides lightweight operation and high efficiency.
- Atomicity is not always desirable.
- Atomicity is not supported due to high resource cost. Undoing a transaction needs a large amount of space and space is a limited resource in mobile clients.
- Durability is guaranteed only when the transaction does not contain partitioned file access, when it is successfully reintegrated, or when it is resolved. In the IOT Model, the result of a pending transaction is visible to subsequent transactions running on the same client, but this result is subject to change based on future validation.

35.5.10.6 IOT Consistency Guarantees. The IOT Model offers stronger consistency guarantees than the isolation of traditional transactions. Transactions are classified into two categories: a first class transaction does not contain any partitioned file access, whereas a second class transaction does. The following are the consistency guarantees provided by IOT:

- First class transactions:
  - Serializability (SR) — serializable with all committed transactions
- Second class transactions:
  - Local serializability (LSR) — serializable with other second class transactions executed on the same client.
  - Global serializability (GSR) — globally serializable with all committed transactions. This needs to be checked at the validation time of the pending transaction T. This is automatically guaranteed when the pending transaction T succeeds in the validation. If the validation fails, a resolution method (i.e., specify what to do) should be provided to guarantee GSR. Some resolution options are:
    - Reexecuting the transaction
    - Aborting the transaction
    - Notifying the users
    - Invoking the transaction's application specific resolver (ASR) — using application-specific knowledge, the transaction writer can attach an ASR to a transaction
  - Global certification order (GCO) — a stronger consistency level may be required for disconnected clients and can be enforced at

the validation of a pending transaction. GCO requires a pending transaction to be serializable not only with but also after all the committed transactions. This makes the isolated transaction running on the mobile client compatible with the most recent system state.

35.5.11 Bayou

Bayou [Terry] is a replicated, weakly consistent storage system designed for a mobile environment. Bayou supports portable computers with limited resources using a flexible client–server architecture in which any MU can act as a server or client for any data item. Bayou takes a two-tier weakly consistent replication scheme combined with asynchronous, epidemic information flow to propagate updates in the system.

35.5.11.1 Two-Tier Replication and Weak Consistency. In Bayou, any data item has a primary and multiple copies, where the primary is used to commit data to a stable value and set the order in which data is committed. To be highly available for `reads` and `writes`, multiple copies are provided at other servers to get read-any/write-any access. The tentative updates made by any copies should contact the primary to get committed.

All Bayou servers move toward *eventual consistency in a* lazy manner. That is, the Bayou system guarantees that all servers *eventually* receive all `writes` via the pairwise antientropy process and that two servers holding the same set of `writes` will have the *same* data contents.

35.5.11.2 Antientropy. Eventual consistency is achieved by antientropy protocol: each server periodically selects another server (or triggered by system or manually) with which to perform a pairwise exchange of `writes`. At the end of this process, both servers have identical copies of the database, with the same `writes` effectively performed in the same order. Version vectors compactly represent the set of updates known to a server and a server with more knowledge will bring another server up-to-date.

Bayou's design has focused on supporting application-specific mechanisms to detect and resolve update conflicts, ensuring that replicas move toward eventual consistency and defining a protocol by which the resolution of update conflicts stabilizes. It provides per write dependency checks for application-specific conflict detection and per write merge procedures for application-specific conflict resolution.

35.5.11.3 Session Guarantees. To present individual applications with a view of the database that is consistent with their own actions, session guarantees are provided. A session is an abstraction for the sequence of `reads` and `writes` performed on a database during the execution of an application.

A Survey on Mobile Transaction Models

Bayou supports four session guarantees for choice of data consistency:
1. Read your writes
2. Monotonic reads
3. Writes follow reads
4. Monotonic writes

One or more of the four guarantees can be requested on a per session basis to get an appropriate trade-off between availability and consistency.

Session guarantees do not address the problem of isolation between concurrent applications.

35.5.12 A New Transaction Management Scheme

This scheme [Nagi] is based on IOT and uses optimistic concurrency control. The main objective here is to minimize wireless communication in the uplink direction by utilizing the broadcasting features of the MSS and by operating the MU-client at four different modes:

1. Fully connected (FC) mode — the MU is fully connected to the fixed network and can listen to broadcasts from MSS as well as transmit to the MSS.
2. Partially disconnected (PD) mode — the MU can listen to the MSS but not transmit to the MSS. This happens when the bandwidth becomes scarce.
3. Fully disconnected (FD) mode — the MU will neither be able to establish an up- nor downlink connection. This happens when the MU is out of range of any MSS cell range.
4. Doze mode — virtually no computation is done at MU to conserve energy.

The MSS broadcasts on the air periodically either data requested by MUs or the whole database. The MSS also broadcasts periodically invalidation and update reports.

The invalidation reports identify data items (not the data itself but an identifier) that were updated in the MSS since the last invalidation report. The MUs can listen to the invalidation reports in PD mode and invalidate the corresponding local cache or commit a transaction locally without contacting the MSS. The main advantage of invalidation reports is that the MU can kill the transactions that accessed invalid data immediately after getting the invalidation report and start resolution rather than the transaction being invalidated at the server thus avoiding useless work.

The update reports are sent at a lower periodic frequency than invalidation reports. An update report has the data items that were changed from the last report: data items requested by MUs while they were in FC mode. The advantage of update reports are that they can be received by MUs in

PD mode and transactions that were aborted due to invalidation can be restarted upon receiving the new data. Update reports also allow read-only transactions to be committed locally instantaneously.

A MU caches frequently accessed data either by listening to broadcasts or by sending specific requests to the MSS while in the FC mode and receiving the reply in FC or PD mode. A transaction at a MU accesses the cached data. Upon completion of the transaction, the MU sends the read/write sets of the transaction to the MSS for remote validation. Validation is done by certifying the read and write of the transaction in all the involved servers. If validation succeeds, the result is committed to the server; otherwise, the transaction is aborted. If the MU cannot establish an uplink with the MSS, the transaction goes to a pending state and later on sent to the MSS upon reconnection.

35.6 Comparative Analysis of Transaction Models

In this section, we present a comparative analysis of the transaction models presented in Section 35.5. We describe how each model addresses the issues outlined in Section 35.3. Table 35.1 and Table 35.2 present the comparison in a table format.

35.6.1 Consistency and Concurrency

The issues of consistency and concurrency have been addressed in the models in a widely varying manner. In the Reporting and Cotransactions Model [Chry93a], compensating transactions are used to maintain the consistency of data, whereas in the Kangaroo Transaction Model [Helal96], the underlying database is relied upon to maintain consistency. In the Clustered Data Model [PitouraBhargava95], the entire data model is designed around maintaining data consistency in a distributed environment. To improve concurrency, the Reporting and Cotransactions Model delegates or reports its operations to other transactions, thus making its results available to other transactions.

- Reporting and Cotransactions Model — in this model, compensating transactions and delegation are used to maintain data consistency. In case of noncompensatable transactions, only the local buffers are operated upon. Delegation transfers the responsibility of committing or aborting a transaction to the delegate rather than on the transaction, which conducted the operation. Delegation also allows intermediate results to be used by other transactions.
- Clustered Data Model — the design of the Clustered Data Model revolves around maintaining data consistency in a fully distributed environment. The data in individual clusters are consistent and there can be bounded inconsistency between clusters. The transaction operations are classified depending on the type of data they access.

A Survey on Mobile Transaction Models

Table 35.1. Comparison of Mobile Transaction Models

	Consistency and Concurrency	Database System Model	Additional Infrastructure
Reporting			
Cotransactions		Multi-database.	None required.
Clustering Model	Bounded intercluster consistency.	Fully distributed database.	None required.
MDSPTM	Relies on underlying database.	Heterogeneous multi-database systems.	None required.
Pro-Motion	Based on object (compact) semantics.	Traditional database systems.	Extended software.
Prewrite	Precommit on MH and guarantees commitment of the precommit result.	Traditional database systems.	Extended software.
Semantics-Based Model	Based on object semantics.	Distributed multi-database.	None required.
Kangaroo Model	Relies on underlying database.	Heterogeneous multi-database.	Requires data access. Assumes extended software interfaces.
Two-Tier	Base transaction uses underlying database.	Distributed database systems.	None required.
IOT	Relies on the file system and strong isolation. Optimistic concurrency control.	Traditional database systems.	None required.
Bayou	Default-weak but user selectable.	Replicated and distributed database systems.	None required.
NTMS	Optimistic concurrency control.	Distributed database.	Extended software at MSS.

Table 35.2. Comparison of Mobile Transaction Models

	Net Management Communications Cost	User Profile	Extensions Required for Commercial Database	Scalability
Reporting and Cotransaction	Handoff information required. Reporting Cotransaction information exchanged.	Can be used for relocating transactions.	TM will have to be extended to handle new transaction types.	Will require high bandwidth.
Clustering Model	Handoff information required.	Used to define clusters and for transaction migration.	TM should be enhanced to handle weak, strict transactions, and clusters definitions.	Large number of clusters or large databases could lead to cluster management.
MDSTPM	No handoff information required. Involves only submission and querying of results.	Can be used for priority queuing.	Requires the TM layer above the database system.	Transaction queuing could create a bottleneck.
Pro-Motion	Handoff information required.	To relocate transaction.	Compact manager acts as front-end of DB server.	Highly scalable.
Prewrite	Handoff information required.	Used for transaction migration	TM should be enhanced to guarantee the commitment of precommitted result.	Enforcing commitment may add loads to the database.
Semantics-Based Model	Handoff information required.	Not required.	Objects should be fragmentable or reorderable. Object managers will be required.	

A Survey on Mobile Transaction Models

Kangaroo Model	Handoff information required. Assumes that each base station can handle transactions	Can be used to relocate transactions.	TM should be able to handle split transactions and recovery mechanisms will have to be enhanced.	Splitting with frequent commits might load the database.
Two-Tier	No handoff information required. Involves submission of results.	Not needed.	No extensions are needed.	Provide high availability and scalability through replication, while avoiding instability.
IOT	No handoff information required. Involves submission of results.	Not required.	The underlying file system provides the IOT features. SO transaction manager need not have special capabilities.	Not a problem unless there are many GSRs that involve reexecution.
Bayou	No handoff information required. Involves submission of results.	Not required for weak consistency.	TM should be able to handle application-specific resolution and session guarantees.	Highly scalable but too many MU and less frequent peer-to-peer contact can delay consistency.
NTMS	No handoff information required. Involves submission of results.	Not needed.	TM should be able to kill and restart transaction based on invalidation reports.	Highly scalable for read-only transactions. Improvement over IOT.

This mechanism of allowing bounded inconsistencies of data elements to increase concurrency is similar to epsilon serializability [RamaPu]. An epsilon transaction is similar to a normal transaction, but the allowable inconsistency is also specified along with the transaction. Consistency of the database system is maintained through divergence control algorithms.

- MDSTPM — in this model, GCS is used to maintain the concurrency of the global transactions. The model assumes that the underlying DBMSs implement their own mechanisms for concurrency control and consistency of local transactions. This model is similar to the Kangaroo Transaction Model in that the mobile transaction is dependent on the underlying database and TM for concurrency and consistency control. But unlike in KT, user mobility does not affect the execution of transaction in this model.
- Pro-Motion Model — based on objects called compacts which have data and associated methods. All operations on the object fragments obey the consistency constraints specified in the compact when the compact is created and sent by the server.
- Prewrite Model — Prewrite ensures that the transaction processing algorithm along with the lock-based protocol produce only serializable histories. This serializability is based on the precommit order of MT. If an MT makes a local commit, it is sure to commit. Prewrite does not permit a local committed transaction to abort.
- Semantics-Based Mobile Transaction Model — the use of fragmentable and reorderable objects maximizes concurrency as well as reduces the cost of caching. The MUs operate on fragments of the data object, which is later pieced together again. All operations on the object fragments obey the consistency constraints specified in the consistency conditions specified when the fragment is dished out by the server.
- Kangaroo Transactions Model — the Kangaroo Transaction Model relies on the underlying transaction model to enforce data integrity. In this model, the TM splits a KT into a set of JTs, which are then executed on the underlying databases. A transaction is split when the mobile node moves and the first transaction of the split gets committed immediately. This releases some of the data items leading to higher concurrency.
- Time-Based Transaction Model — this model uses time as a consistency management mechanism. Every data element is associated with a time bound: duration for which the data element is considered to be consistent. The locks are also allocated for specific time durations. The success of this scheme is highly dependent on the user's choice of the lock duration. If the duration for which the lock is acquired is too low, then it may have to be reacquired. If the duration

A Survey on Mobile Transaction Models

for which the lock is acquired is too high, then it will reduce the allowable concurrency.
- Two-Tier Replication Model — any transaction run on the MH is rerun on the correspondent base node to validate a transaction. Thus the use of base transaction ensures the consistency of the database.
- Isolation-Only Transactions Model — optimistic concurrency control is used to enforce serializability for first class transactions. GSR for second class transactions is provided by validation and reintegration or by several resolution methods. This model guarantees stronger consistency than traditional transactional model.
- Bayou Model — Bayou uses read-any/write-any weakly consistent replication. Eventual consistency for updates is done by peer-to-peer antientropy protocol. Bayou can be operated on an ad hoc network.
- New Transaction Management Scheme — this scheme is based on IOT and uses optimistic concurrency control. It is to reduce uplink communication. With the information from invalidation reports, useless processing is avoided.

35.6.2 Additional Infrastructure Requirements and Compatibility with Commercial Databases

In this section, we discuss the additional infrastructure assumed in each model. We assume that the reference model discussed in previous sections is available. The requirements discussed are apart from this. We also discuss whether the models can utilize the available databases to make available a database system for mobile users. All these models assume that the physical movement information is available to the upper layers.

- Reporting and Cotransactions Model — this model does not assume any network interface other than the mobile environment discussed in the previous sections. The model adapts reporting and cotransactions for a mobile environment. Transaction relocation is also achieved using reporting and cotransactions. The TM will have to be modified to handle reporting and cotransactions. Because concepts like delegation and cotransactions are involved, it will be difficult to implement this model on an existing database and a classical transactional facility.
- Clustered Data Model — although this model assumes no additional network facilities, extensive database modification is required on the DBMS. New operations like Weak and Strict operations are defined. The data objects also have additional attributes reflecting their consistency requirements. Moreover other database operations like clustering of data and cluster merging also will have to be implemented.

- MDSTPM — this model is implemented by defining a MDSTPM layer over the existing DBMSs. This layer acts as an interface between the MH and the underlying multi-database system. The components of this interface include the GCM, GTM, GRM, and GIM, described in Section 35.5.3.1 [gtm_arch]. The queuing mechanism for transactions has to be implemented.
- Pro-Motion Model — software for compact manager at the DB server, a compact agent at the MH to negotiate and manage compacts, and a MM at the MSS to help manage the flow of updates and data between the other components in the system are needed.
- Prewrite Model — software for prewrite on MH is needed. Data manager on fixed network has to be extended to guarantee the commitment of precommit by MH.
- Semantics-Based Mobile Transaction Model — this model requires additional capability at both the server and the MU end to split, operate, and merge objects. Moreover the model also assumes that the database operates on objects so structured that fragmentation and merging is possible. This model attempts to improve consistency and concurrency, whereas the Reporting and Cotransactions Model attempts to model the mobile transaction behavior.
- Kangaroo Transactions Model — the Kangaroo Transaction Model introduces the concept of DAA, which acts as a router of transaction requests. The DAA is assumed to exist on a base station. Thus, a base station will have to be enhanced to provide this facility because the resources available on a base station will be limited. In this model, a split is affected as soon as the mobile node hops from one cell to another, thus every cell should have a DAA ready to serve the transaction. This model can be implemented over an existing database by implementing a global mobile transaction manager (GMTM). The GMTM can accept transaction requests from a mobile capable node and assign it to individual database systems.
- Time-Based Transaction Model — this model requires the database system to use time-based locks. No more enhancements are proposed in the model.
- Two-Tier Replication Model — base nodes need to be always connected to the network. Additional software is needed for coordinating the updates to all other replica nodes.
- Isolation-Only Transactions Model — this model should include additional software to handle GSR of a second class transaction. The under lying distributed UNIX operating system should support the IOT file system.
- Bayou Model — no additional infrastructure is needed. Bayou can operate on an ad hoc network. Server should be available to handle application-specific resolution and session guarantees.

- New Transaction Management Scheme — additional software is needed in the MSS as they need to broadcast update and invalidation reports. The MU should also identify its operating mode and perform appropriately.

35.6.3 Communication Cost and Scalability

As mentioned previously, communication cost forms a significant factor in mobile transaction execution. The contributing factors to this are the cost of communication between the mobile node and the fixed server and the cost of transaction relocation.

- Reporting and Cotransactions — in this model, communication between a mobile node and the fixed server takes place through cotransaction pairs or between a reporting transaction and cotransaction pair. Depending on the transaction mobility and transaction behavior, there could be several rounds of transfer. The results are communicated between the transactions. Apart from this, housekeeping information will also have to be communicated between the nodes. This communication can form a bottleneck that can limit scalability.
- Clustered Data Model — as in the Kangaroo Transaction Model, the Clustered Data Model is required to communicate only the end results and the housekeeping information once the transaction is fired from the mobile node. As discussed previously, the model divides the data into clusters and allows bounded inconsistency between clusters. Thus on every operation, the cluster has to be maintained consistently and the inconsistency between the clusters also will have to be maintained within bounds. The cost of this increases as the size and number of clusters increases. The cost is also dependent on the consistency bounds set. In some cases, the amount of messaging required to maintain consistency could be very high. Thus, this is an important parameter determining the scalability of this model.
- MDSTPM — the communication costs of this model are significantly lower compared to the other models, because communication takes place only for submission and for transferring the results.
- Pro-Motion Model — communication takes place once when the compact is sent to the MU and later when the compact is returned as all the processing is done locally at the mobile client.
- Prewrite Model — communication cost is needed to transfer the precommit results back to MSS.
- Semantics-Based Mobile Transaction Model — the Semantics-Based Mobile Transaction Model attempts to reduce the communication costs by caching only those parts of an object required in a disconnected operation. This reduces the pressure on the low bandwidth

wireless network and utilizes the MU cache space better. As the number of transactions in the system increases, the load on the server may increase correspondingly.
- Kangaroo Transactions Model — in the Kangaroo Transaction Model, once the transaction is fired from a mobile node, only the final results and the housekeeping information need to be transmitted between the mobile node and the stationary server. The housekeeping information will include log and the handoff information. The model requires a transaction to be split and relocated when its originating mobile node moves from one cell to another. This could be a costly operation, because the data items to be committed have to be determined for each transaction split and the transactions will have to relocated as well. These factors can affect scalability.
- Two-Tier Replication Model — communication between MU and base node is high, as MU sends all its tentative transactions to base node.
- Isolation-Only Transactions Model — the communication costs of first class transactions of this model are significantly lower compared to the other models, because all processing is done at the client and communication takes place only for submission and for transferring the results. But in the case of GSR for second class transactions, the communication cost will increase if the transactions are reexecuted due to inconsistencies.
- Bayou Model — communication takes place between two servers during peer-to-peer entropy. Scalability is not a problem.
- New Transaction Management Scheme — limited communication is required in the uplink direction. But downlink is high because of periodic broadcast of several reports.

35.7 Open Issues in Mobile Transactions

In this section, we discuss the issues that we feel are open for research in mobile transactions:

- Network transparency — the transaction models described in Section 35.5 assume that movement information of the mobile node is available to the application layer. Transaction managers are typically implemented in the session or application layer of the Open Systems Interconnection (OSI) reference model. The node movement information may not be available to the upper layers, because a network layer protocol like Mobile-IP [PerkIP] attempts to provide a transparent interface to the upper layers. The issues of network transparency provided by the lower layers and the mobility information required by the upper layers have been addressed for general programming environments [BadriWelling, WellBadriEcoop]. The amount of transparency and movement information expected by the mobile TMs is an important research issue.

- Transaction relocation — transaction relocation is the mechanism by which the mobile transaction models attempt to improve response times and performance, by executing the transaction at a server near to the current location of the mobile node. The transaction relocation process can be generalized to agent relocation, where the mobile agent could be performing a transaction on behalf of the mobile node. The transaction models described in Section 35.5 relocate a transaction as soon as the mobile node moves from one cell to another. The major cost components in transaction relocation are the cost of metadata movement, cost of splitting a transaction, or cost of blocking execution. Other cost components include system usage, network utilization, etc. The naive transaction relocation approach adopted by the mobile transaction models described in Section 35.5 could be costly, if a mobile node was present in a cell for too small a duration. Some heuristics can be applied to improve transaction relocation. These may include programmer directives, network and system loads, etc. Another mechanism for improving efficiency could be doing an anticipatory relocation depending on user directives or the direction of motion. This mechanism is similar to the anticipatory caching mechanism described in [Helal95].
- Programming language support and location-sensitive transaction operations — mobile databases have to deal with the mobile computing issues like relocation of transactions, data inconsistencies, disconnections, and low bandwidth links. Efficiency of the operations can be improved if these events are anticipated and responded to. Structure Query Language extensions can be proposed that will allow a developer to take care of these issues in the design itself. Semantics are also required to deal with the location-sensitive queries that could arise in a database. New transaction operations may be required that allow manipulation of location information.
- Performance evaluation of mobile transaction — for a thorough comparison of various mobile transaction models, it is important that their performance be evaluated. There could be various measures of evaluation like response time, throughput, relocation costs, communication costs, etc. The exact parameters and performance criteria for mobile transactions is not clear and is an open issue.

35.8 Summary

Designing a transaction model for a mobile computing environment poses many challenges to researchers. In this chapter, we introduced the issues involved in transaction processing in a mobile computing environment. We presented three advanced transaction models — Open Nested, Split, and Saga — that have been adapted in mobile transaction models. We presented a comparative analysis of four mobile transaction models — Reporting and Cotransactions, Clustered Data, Kangaroo Transactions, and the Multi-Database Transaction Processing Manager. We also enumerated some topics that are open for research.

References

[AlonsoBarbaraMolina90] R. Alonso, D. Barbara, and H. Garcia-Molina, Data caching issues in an informational retrieval system, *ACM Transactions on Database Systems*, vol. 15, no. 3, pp. 359–384, September 1990.

[Badrinath92b] B.R. Badrinath, T. Imielinski, and A. Virmani, Locating strategies for personal communication networks, in *Proceedings of IEEE GlobeCom '92 Workshop on Networking for Personal Communications Applications*, December 1992.

[BadriWelling] Girish Welling and B.R. Badrinath, Event delivery abstractions for mobile computing, Technical Report LCSR-TR-242, Rutgers University, 1995.

[CheGros] David Chess, Benjamin Grosof, Colin Harrison, David Levine, Colin Parris, and Gene Tsudik, Itinerant agents for mobile computing. *IEEE Personal Communications*, October 1995.

[Chry91] Panos Chrysanthis, ACTA, a framework for modeling and reasoning about extended transactions, Ph.D. dissertation, University of Massachusetts, Amherst, 1991.

[Chry93a] Panos Chrysanthis, Transaction processing in mobile computing environment, in *Proceedings of IEEE Workshop on Advances in Parallel and Distributed Systems*, pp. 77–82, October 1993.

[Chry93b] Panos Chrysanthis and K. Ramamritham, Synthesis of extended transaction models using ACTA, Technical report, University of Pittsburgh, 1993.

[ChryRaghuRama91] P.K. Chrysanthis, S. Raghuram, and K. Ramamritham, Extracting concurrency form objects: A methodology, in *Proceedings of ACM SIGMOD Conf.*, pp. 108–117, May 1991.

[Elmar91] Ahmed K. Elmagarmid, *Database Transaction Models for Advanced Applications*, San Francisco: Morgan Kaufman, 1991.

[GarciaMolina87] H. Garcia-Molina and K. Salem, Sagas, in *Proceedings of ACM Conference of Management of Data*, pp. 249–259, May 1987.

[Gray96] J. Gray, P. Helland, P. O'Neill, and D. Shasha, The Dangers of Replication and a Solution, *Proc. ACM SIGMOD Conference*, New York, NY, pp. 173–182, 1996.

[Helal95] Abdelsalam Helal and Margeret Eich, Supporting mobile transaction processing in database systems, Technical report, University of Texas at Arlington, Southern Methodist University, submitted to the *1st International Conference on Mobile Computing and Networking*, 1995.

[Helal96] Margeret Dunham and Abdelsalam Helal, A mobile transaction model that captures both the data and movement behavior, submitted to *12th International Conference in Data Engineering*, 1996.

[Imielinski93b] T. Imielinski and B.R. Badrinath, Data management for mobile computing, *SIGMOD Record*, vol. 22, no. 1, pp. 34–39, March 1993.

[Ioannidis91] John Ioannidis, Dan Duchamp, and Gerhald Q. Maguire Jr, IP-based protocols for mobile internetworking, in *Proceedings of SIGCOMM'91*, pp. 235–245, September 1991.

[JJ95] Jin Jing, Omran Bukhres, and Ahmed Elmagarmid, Distributed lock management for mobile transactions, *IEEE*, pp. 118–125, 1995.

[Johnson93a] David B. Johnson, Mobile host internetworking using IP loose source routing, Technical report CMU-CS-93-128, School of Computer Science, Carnegie Mellon University, Pittsburgh, PA, February 1993.

A Survey on Mobile Transaction Models

[Kistler91] James J. Kistler and M. Satyanarayanan, Disconnected operation in the Coda file system, in *Proceedings of the 13th ACM Symposium on Operating Systems Principles*, pp. 213–225, 1991.

[LiuMarMag] George Y. Liu, Alexander Marlevi, and Gerald Q. Maguire Jr, A mobile virtual-distributed system architecture for supporting wireless mobile computing and communications, *Wireless Networks*, vol. 2, no. 1, 1996.

[MadriaBhargava] S.K. Madria and B. Bhargava, A transaction model for improving data availability in mobile computing, *Distributed and Parallel Database*, 2001.

[Nagi] Khalil M. Ahmed, Mohamed A. Ismail, Nagwa M. El-Makky, and Khaled M. Nagi, A new transaction management scheme for mobile computing environments, Technical report, Alexandria University, Alexandria, Egypt.

[Niel95] J.D. Nielsen, Transactions in mobile computing, Ph.D. dissertation, DIKU, 1995.

[PerkIP] Charles Perkins, Ed., draft-ietf-mobileip-protocol-15.txt, Technical report, IETF, 1996.

[PitouraBhargava95] Pitoura Evaggelia and Bhargava Bharat, Maintaining consistency of data in mobile distributed environments, in *Proceedings of 15th International Conference on Distributed Computing Systems*, 1995.

[PitouraBhargava-Ag95] Pitoura Evaggelia and Bhargava Bharat, Consistent and recoverable agent-based access to mobile heterogeneous databases, Technical report, Purdue University, 1995.

[PitouraBhargavaTr] Pitoura Evaggelia and Bhargava Bharat, Revising transaction concepts for mobile computing, in *Proceedings of the 1st IEEE Workshop on Mobile Computing Systems and Applications*, pp. 164–168, December 1994.

[PU88a] C. Pu, G. Kaiser, and N. Hutchinson, Split-transactions for open-ended activities, in *Proceedings of the 14th VLDB Conference*, 1988.

[PuLeff91] C. Pu and A. Leff, Replica control in distributed systems: An asynchronous approach, in *Proceedings of the ACM SIGMOD Conference*, pp. 377–386, May 1991.

[QS94] Q. Lu and M. Satyanarayanan, Isolation-only transactions for mobile computing, *ACM SIGOPS Operating Systems Review*, vol. 28, no. 2, April 1994, pp. 81–87.

[RamaPu] K. Ramamritham and C. Pu, A formal characterization of epsilon serializability, *IEEE Transactions on Knowledge and Data Engineering*, 1994.

[Serrano-Alvarado] P. Serrano-Alvarado, C.L. Roncancio, and M. Adiba, Analyzing mobile transactions support for DBMS, in *Proceedings of the 12th International Workshop on Database and Expert Systems Applications (DEXA2001)*, Munich, Germany, pp. 595–600, September 2001.

[Teraoka] F. Teraoka and M. Tokoro, Host migration transparency in IP networks: the VIP approach, *Computer Communication Review*, vol. 23, no. 1, pp. 45–65.

[WalChry95] Gary D. Walborn and Panos K. Chrysanthis, Supporting semantics-based transaction processing in mobile database applications, in *Proceedings of the 14th IEEE Symposium on Reliable Distributed Systems*, September 1995.

[WalChry99] Gary D. Walborn and Panos K. Chrysanthis, Transaction processing in Pro-Motion, in *Proceedings of the 14th ACM Annual Symposium on Applied Computing*, San Antonio, TX, February 1999.

[WellBadriEcoop] Girish Welling and B.R. Badrinath, Mobjects: Programming support for environment directed application policies in mobile computing, in *Proceedings of ECOOP-95*, 1995.

[Zas94] L.H. Yeo and A. Zaslavsky, Submission of transactions from mobile workstations in a cooperative multidatabase processing environment, in *Proceedings of the 14th International Conference on Distributed Computing Systems*, pp. 372–379, June 1994.

Chapter 36
Analytic Mobility Models of PCS Networks

Chien-Hsing Wu

Abstract

Consider a mobile station (MS) moving in a personal communication service (PCS) network that contains fixed base stations (BS). The MS informs the network of its whereabouts by updating its location regularly, while the network uses this location update information to page and connect it to an incoming call. The MS performs a location update based on its distance, movement, direction, or elapsed time [1, 2] and the network is supposed to find the location of the MS in a limited delay period. A delicate balance is maintained between location updates by the MS and paging by the network, whereas the location manager of the network wishes to minimize the total cost due to both by conducting performance analysis or simulations.

An elegant analytic mobility model is essential for efficient performance evaluation of location management schemes in a PCS network. A simple one-dimensional Markov Walk Model was used to characterize the intercell movements of a MS in a linear array of cells [3]. The authors in [4] employed a continuous-time Markov Mobility Model with a huge state space covering all the cells in the network. A conventional two-dimensional (2D) Random Walk Model was commonly used in [5–8] based on the assumption that the directions of the MS are independent and identically distributed (IID).

In the real world, the MS traveling across various landscapes can exhibit disparate mobility patterns. For example, the MS on a highway tends to move along a straight line and the MS on local drives inside a city is likely to move around. The conventional IID Random Walk Model is insufficient to distinguish these mobility patterns due to its oversimplified assumption.

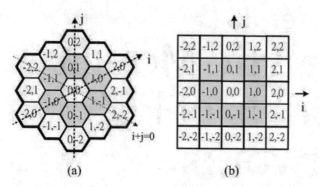

Figure 36.1. A Cluster of Cells, $\mathcal{A}_2$, in a PCS Network, where the Shaded Region Represents Ring $\mathcal{R}_1$
(A) hexagonal cells; (B) rectangular cells.

In this chapter, we present a two-dimensional Markov Walk Model [9] based on the theory of hexagonal or rectangular cellular patterns [10]. A broad class of isotropic processes having circulant transition probability matrices [11] is proposed. By examining the extreme mobility patterns for hexagonal cellular networks, we identify six isotropic processes — IID, directional, turning, ping-pong, h-spin, and t-spin. The Markov Walk Model is then applied to the performance assessment of various location management schemes based on recursive Markov analysis. This analytic framework captures the dynamic behavior of a MS and successfully resolves the subtle differences among various dynamic mobility management schemes.

36.1 System Models

36.1.2 Cellular Systems

As shown in Figure 36.1, a PCS network can be partitioned into hexagonal cells or rectangular cells, each of which represents the radio coverage of a base station. A cell is labeled by the coordinate of its centroid, (i, j), where i and j are integers. Denote the rth ring of cells by $\mathcal{R}_r$ and the cluster of $(k + 1)$ rings by $\mathcal{A}_k = \bigcup_{r=0}^{k} \mathcal{R}_r$. Let the numbers of cells in $\mathcal{R}_r$ and $\mathcal{A}_k$ be denoted by $|\mathcal{R}_r|$ and $|\mathcal{A}_k|$, respectively. We have $\mathcal{R}_0 = \mathcal{A}_0 = \{(0, 0)\}$ and $|\mathcal{R}_r| = |\mathcal{A}_r| - |\mathcal{A}_{r-1}|, r \geq 0$, where $|\mathcal{A}_{-1}| = 0$. A cell $c = (i, j)$ is said to be at distance r away from a cell $\hat{c}$ if $c - \hat{c} \in \mathcal{R}_r$.

For the hexagonal cellular configuration (Figure 36.1A), it is convenient to define the i and j axes of the coordinate system with their positive portions crossing at a 60-degree angle. We have:

$$\mathcal{R}_r = \{(i,j) : \max(|i|, |j|, |i + j|) = r\}, \quad r \geq 0, \qquad (36.1)$$

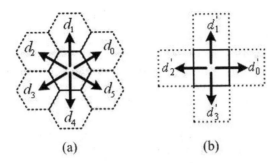

Figure 36.2. Directions for the Mobile Station to Move from a Cell to Its Neighboring Cells
(A) six directions for hexagonal cells; (B) four directions for rectangular cells.

$|\mathcal{R}_r| = 6r + \delta_r$ and $|\mathcal{A}_k| = 3k(k+1) + 1$, where $\delta_0 = 1$ and $\delta_r = 0$ for $r \neq 0$. For example, $\mathcal{R}_1 = \{(1, 0), (0, 1), (-1, 1), (-1, 0), (0, -1), (1, -1)\}$, $|\mathcal{R}_1| = 6$ and $|\mathcal{A}_1| = 7$.

For the rectangular cellular configuration (Figure 36.1B), the i and j axes are usually set to be mutually orthogonal. We have:

$$\mathcal{R}_r = \{(i, j) : \max(|i|, |j|) = r\}, \quad r \geq 0, \tag{36.2}$$

$|\mathcal{R}_r| = 8r + \delta_r$ and $|\mathcal{A}_k| = (2k+1)^2$. For example, $\mathcal{R}_1 = \{(1, 0), (1, 1), (0, 1), (-1, 1), (-1, 0), (-1, -1), (0, -1), (1, -1)\}$, $|\mathcal{R}_1| = 8$ and $|\mathcal{A}_1| = 9$.

36.1.2 Markov Walk Models

When the MS moves into a cell, it resides in the cell for a random period of time and then moves out to one of the neighboring cells. Let $c^{(n)}$ denote the cell where the MS resides immediately after the nth cell boundary crossing instant. Denote the nth direction (displacement or intercell movement) of the MS by $d^{(n)}$. The locations of the MS can then be described by $c^{(0)} = (0, 0)$ and:

$$c^{(n)} = c^{(n-1)} + d^{(n)}, \quad n \geq 1, \quad d^{(n)} \in \mathcal{X}, \tag{35.3}$$

where $\mathcal{X}$ denotes the set of directions for the MS to move from a cell to its neighboring cells, as shown in Figure 36.2. Let $|\mathcal{X}|$ denote the cardinality of $\mathcal{X}$. For the hexagonal cellular configuration, we have $\mathcal{X} = \mathcal{R}_1 = \{d_i, i = 0, \ldots, 5\}$ and $|\mathcal{X}| = 6$, where $d_0, \ldots, d_5$ are the six directions defined in a counterclockwise manner: $d_0 = (1, 0)$, $d_1 = (0, 1)$, $d_2 = (-1, 1)$, $d_3 = (-1, 0)$, $d_4 = (0, -1)$, and $d_5 = (1, -1)$. For the rectangular cellular configuration, we have $\mathcal{X} = \{d'_i, i = 0, \ldots, 3\}$ and $|\mathcal{X}| = 4$, where $d'_0 = (1, 0)$, $d'_1 = (0, 1)$, $d'_2 = (-1, 0)$, and $d'_3 = (0, -1)$.

Three assumptions are made in our model:

1. We assume that the cell residence times of the MS are independent and identically distributed random variables having a probability distribution function $F_m(t)$ and the average cell residence time λ_m^{-1}.
2. The call arrival process is assumed to be a Poisson process with parameter λ_c. The call-to-mobility ratio (CMR) is defined as $\rho = \lambda_c/\lambda_m$.
3. To characterize the dynamic mobility pattern of the MS, we make the special assumption that the direction process $\{d^{(n)}, n \geq 0\}$ constitutes a discrete-time Markov chain over state space $\mathcal{X}$.

The transition probability matrix of the direction process is defined by $\mathbf{P} = [p_{i,j}]$, where $p_{i,j} = P(d^{(n+1)} = d_j | d^{(n)} = d_i)$, $d_i, d_j \in \mathcal{X}$. The steady-state distribution of the direction d_j, denoted by π_j, satisfies the balance equation $\pi_j = \Sigma_{i \in Z_{|\mathcal{X}|}} \pi_i p_{i,j}$ and $\Sigma_{j \in Z_{|\mathcal{X}|}} \pi_j = 1$, where $Z_k = \{0, 1, \ldots, k-1\}$. When $\pi_i = 1/|\mathcal{X}|, i \in Z_{|\mathcal{X}|}$, the direction process is said to be isotropic in terms of steady state distributions (ISS).

The class of ISS processes having circulant transition probability matrices is of special interest. For the hexagonal cellular system, the circulant matrix P is given by:

$$\mathbf{P} = \begin{bmatrix} p_0 & p_1 & p_2 & p_3 & p_4 & p_5 \\ p_5 & p_0 & p_1 & p_2 & p_3 & p_4 \\ p_4 & p_5 & p_0 & p_1 & p_2 & p_3 \\ p_3 & p_4 & p_5 & p_0 & p_1 & p_2 \\ p_2 & p_3 & p_4 & p_5 & p_0 & p_1 \\ p_1 & p_2 & p_3 & p_4 & p_5 & p_0 \end{bmatrix}. \quad (35.4)$$

If $p_1 + p_5 > 0$, then the Markov chain $\{d^{(n)}, n \geq 0\}$ is irreducible and has a unique steady-state distribution $\pi_i = 1/6, i \in Z_6$.

For the rectangular cellular system, the circulant matrix P is given as:

$$\mathbf{P} = \begin{bmatrix} p_0 & p_1 & p_2 & p_3 \\ p_3 & p_0 & p_1 & p_2 \\ p_2 & p_3 & p_0 & p_1 \\ p_1 & p_2 & p_3 & p_0 \end{bmatrix}. \quad (35.5)$$

If $p_1 + p_3 > 0$, then the Markov chain $\{d^{(n)}, n \geq 0\}$ is irreducible and has a unique steady-state distribution $\pi_i = 1/4, i \in Z_4$.

Note that the 2D IID isotropic Random Walk Model with $p_i = 1/|\mathcal{X}|, i \in Z_{|\mathcal{X}|}$, is a special case of the ISS class, and is said to be isotropic in terms of state transition probabilities (IST).

Analytic Mobility Models of PCS Networks

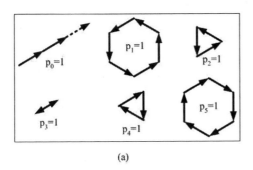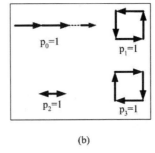

(a)   (b)

Figure 36.3. Extreme Mobility Patterns
(Assume $d^{(0)} = (1, 0)$): (A) hexagonal cells; (B) rectangular cells.
(Source: Wu, C.H., Lin, H.B., and Lan, L.S., A new analytic framework for dynamic mobility management of PCS networks, *IEEE Trans. Mobile Computing*, vol. 1, no. 3, pp. 208–220, July–September 2002. [© 2002 IEEE]. Used with permission.)

We illustrate the extreme mobility patterns in Figure 36.3 with the initial direction $d^{(0)} = (1, 0)$. The mobility pattern can be determined by adequately selecting the mobility control vector $\mathbf{p} = [p_0, p_1, \ldots, p_{|cx|/2}]$ as the first row of the circulant transition probability matrix P. If $p_0 = 1$, the MS always moves along the direction $d^{(0)}$ and its mobility pattern is a directed line. If $p_{|cx|/2} = 1$, the MS always returns to the cell that it just came from so that it bounces back and forth (ping-pong) between two cells. In the hexagonal configuration of Figure 36.3A, while the MS spins counterclockwise (clockwise) on a hexagon for $p_1 = 1$ ($p_4 = 1$), it circulates counterclockwise (clockwise) on a triangle for $p_2 = 1$ ($p_5 = 1$). In the rectangular configuration of Figure 36.3B, the MS rotates counterclockwise (clockwise) on a square for $p_1 = 1$ ($p_3 = 1$).

The ISS process can accurately represent the mobility patterns exhibited by the MS moving across various landscapes. In the real world, the MS tends to move along a straight line on a highway, change its directions on local drives, and move back upon encountering a dead end or making a U-turn. By adjusting the respective dominating probabilities $\{p_0\}$, $\{p_1, p_2, p_4, p_5\}$, and $\{p_3\}$, we can easily construct Markov Walk Models to characterize the specific mobility patterns of the MS.

36.2 Analysis for Location Update

A MS moves across the cells and updates its locations in the network based on a predetermined scheme. In response to an incoming call destined to a MS, the network pages (polls) the cells to find the called MS in accordance with some paging strategy. From the perspective of location management, it is essential to achieve the delicate balance between location updates and paging so that the overall cost is minimized.

36.2.1 Location Tracking and Updates

A PCS network tracks the locations of a MS at the instants of call arrivals and location updates. We define the origin of the (dynamic) coordinate system as the cell where the MS resides at the most recent location tracking instant. The following dynamic location update schemes with the threshold D are considered:

- Distance-based — the MS performs a location update when it moves to a cell in $\mathcal{R}_{D+1}$ with respect to the current origin.
- Enhanced movement-based — The MS uses a counter $b^{(n)}$ to count the number of cell boundaries crossed since the most recent location update and performs a location update when $b^{(n)} = D$. When the MS moves back to the current origin, it resets the counter $b^{(n)}$ to zero and does not perform a location update.
- Direction-based — the MS performs a location update when it changes the direction, $d^{(n+1)} \neq \pm d^{(n)}$, or when it moves to any cell at distance D + 1 from the current origin.

Let $\mathcal{L}^\theta$ denote the residing area of the MS when a call arrival occurs, $\theta \in \{d,m,r\}$. For distance-based ($\theta = d$) and movement-based ($\theta = m$) location update schemes, $\mathcal{L}^d = \mathcal{L}^m = \mathcal{A}_D$. For the direction-based ($\theta = r$) scheme, $\mathcal{L}^r$ is a line segment of length 2D + 1 centered at (0, 0) in some direction $\pm d_j$, $j \in Z_{\frac{|X|}{2}}$.

36.2.2 Two Renewal Processes and α_k

The proposed analysis is based on the theory of Markov chains. The first step to construct an embedded Markov chain is to identify the appropriate sampling instants. In our analysis, the systems are sampled at the call arrival and cell boundary crossing instants. Denote the ith intercall arrival time and intercell boundary crossing time, respectively, by $T_c^{(i)}$ and $T_m^{(i)}, i \geq 1$. Let $A_c^{(i)}(A_m^{(i)})$ denote the ith call arrival (cell boundary crossing) instant. We have $A_c^{(i)} = \Sigma_{n=1}^i T_c^{(n)}$ and $A_m^{(i)} = \Sigma_{n=1}^i T_m^{(n)}$. From the independent assumptions on the interarrival processes, we know that $\{A_c^{(i)}, i \geq 0\}$ and $\{A_m^{(i)}, i \geq 0\}$ are two renewal point processes [12].

Consider the ith call arrival period $[A_c^{(i-1)}, A_c^{(i)}]$ and suppose $A_m^{(j)} < A_c^{(i-1)} < A_m^{(j+1)}$. Let the sampling instants in the ith call arrival period, denoted as $\tau_k^{(i)}, k \geq 0$, be given by:

$$\tau_k^{(i)} = \begin{cases} 0, & k = 0 \\ A_m^{(j+k)} - A_c^{(i-1)}, & k \geq 1. \end{cases}$$

Let $N_m(T_c^{(i)})$ denote the number of cell boundaries crossed by the MS during $[A_c^{(i-1)}, A_c^{(i)}, i \geq 1]$. The timing diagram is illustrated in Figure 36.4, where $N_m(T_c^{(i)}) = K$.

Analytic Mobility Models of PCS Networks

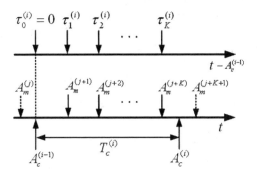

Figure 36.4. Timing Diagram of Call Arrival and Cell Boundary Crossing Processes
(Source: Wu, C.H., Lin, H.B., and Lan, L.S., *IEEE Trans. Mobile Computing*, vol. 1, no. 3, pp. 208–220, July–September 2002. [© 2002 IEEE]. Used with permission.)

By omitting the index i of $\tau_k^{(i)}$ at steady state, we construct the process $\{\tau_k, k \geq 0\}$ in the call arrival period. An essential quantity for the analysis of mobility management is $\alpha_k = P(N_m(T_c) = k)$, the probability that the MS crosses k cell boundaries in a call arrival period [13]. We have shown in [9] that α_k can be succinctly derived as follows.

The process $\{\tau_k, k \geq 0\}$ is a delayed renewal process (Figure 36.4) with the initial delay being the residual cell residence time. From [12], we have:

$$P(N_m(t) = k) = \begin{cases} 1 - \hat{F}_m(t), & k = 0, \\ \hat{F}_m(t) * \left[F_m^{(k-1)}(t) - F_m^{(k)}(t) \right], & k \geq 1, \end{cases} \quad (36.6)$$

where $\hat{F}_m(t) = \lambda_m \int_0^t [1 - F_m(u)] du$, and $F_m^{(k)}(t)$ denotes the k-fold convolution of $F_m(t)$ with itself. Noting that $F_c(t) = 1 - e^{-\lambda_c t}$, $t \geq 0$, we have:

$$\alpha_k = \int_0^\infty P(N_m(t) = k) dF_c(t) = \begin{cases} 1 - \frac{1-\eta}{\rho}, & k = 0, \\ \frac{1}{\rho}(1-\eta)^2 \eta^{k-1}, & k \geq 1, \end{cases} \quad (36.7)$$

where $\eta = F_m^*(\lambda_c)$. The average number of cell boundaries crossed by the MS per call arrival, $E[N_m(T_c)]$, is equal to $1/\rho$.

36.2.3 Recursive Markov Analysis

Consider a tagged call arrival period. Let u_k denote the number of location updates during $[\tau_0, \tau_k]$. Let (c_k, j_k) indicate that at τ_k, the MS moves along direction d_{jk} into cell c_k, where $j_k \in Z_{|\mathcal{X}|}$.

For the proposed Markov Walk Model, $u_k = 0$, $\forall k$, the state process $\{S_k = (c_k, j_k), k \geq 0\}$ forms a Markov chain, where $c_0 = (0, 0)$ denotes the origin set at the most recent call arrival instant. Define $h_k(c, j) = P(S_k = (c, j))$. We can compute $h_k(c, j)$ recursively as follows:

$$h_0(c, j) = \begin{cases} \pi_j, & c = (0,0), \ j \in Z_{|\mathcal{X}|} \\ 0, & \text{otherwise.} \end{cases}$$

$$h_k(c, j) = \sum_{i \in Z_{|\mathcal{X}|}} p_{i,j} h_{k-1}(c - d_j, i), \quad k \geq 1, \ c \in \mathcal{A}_k, \ j \in Z_{|\mathcal{X}|}.$$

The recursive computations aforementioned for the Markov Walk Model can be extended for various location update schemes ($\theta = d, m, r$) by properly defining the state, S_k^θ, and then computing recursively the steady-state distributions $\{h_k^\theta(s) = P(S_k^\theta = s)\}$, as shown in the following:

- Distance-based: $\theta = d$, $S_k^d = (u_k, c_k, j_k)$.

$$h_0^d(u, c, j) = \begin{cases} \pi_j, & u = 0, \ c = (0,0), \ j \in Z_{|\mathcal{X}|} \\ 0, & \text{otherwise.} \end{cases}$$

$$h_k^d(u, c, j) = \begin{cases} \sum_{i \in Z_{|\mathcal{X}|}} p_{i,j} h_{k-1}^d(u, -d_j, i) + f_k^d(u, j), & c = (0,0); \\ \sum_{i \in Z_{|\mathcal{X}|}} p_{i,j} h_{k-1}^d(u, c - d_j, i), & c \in \mathcal{A}_{\langle k \rangle} - \mathcal{R}_0, \ c - d_j \in \mathcal{A}_{\langle k-1 \rangle}; \\ 0, & \text{otherwise}; \end{cases}$$

$$k \geq 1, \ u = 0, 1, \ldots, \left\lfloor \frac{k}{D+1} \right\rfloor, \ j \in Z_{|\mathcal{X}|},$$

where:

$$f_k^d(u, j) = \begin{cases} 0, & u = 0 \text{ or } k \leq D; \\ \sum_{\substack{c: \ c \in R_{D+1} \\ c - d_j \in R_D}} \sum_{i \in Z_{|\mathcal{X}|}} p_{i,j} h_{k-1}^d(u - 1, c - d_j, i), & \text{otherwise.} \end{cases}$$

- Enhanced movement-based: $\theta = m$, $S_k^m = (u_k, b_k, c_k, j_k)$.

$$h_0^m(u, b, c, j) = \begin{cases} \pi_j, & (u, b) = c = (0,0), \ j \in Z_{|\mathcal{X}|} \\ 0, & \text{otherwise.} \end{cases}$$

$$h_k^m(u,b,c,j) = \begin{cases} \sum_{b'=1}^{D}\sum_{i \in Z_{|\mathcal{X}|}} p_{i,j} h_{k-1}^m(u,b',-d_j,i) \\ \quad + f_k^m(u,j), & b=0, c=(0,0); \\ \sum_{i \in Z_{|\mathcal{X}|}} p_{i,j} h_{k-1}^m(u,b-1,c-d_j,i), & b=1,\ldots,D, c \in \mathcal{A}_{(k)}-\mathcal{R}_0, \\ & c-d_j \in \mathcal{A}_{(k-1)}; \\ 0, & \text{otherwise}; \end{cases}$$

$$k \geq 1, \quad u = 0,1,\ldots,\left\lfloor \frac{k}{D+1} \right\rfloor, \quad j \in Z_{|\mathcal{X}|},$$

where:

$$f_k^m(u,j) = \begin{cases} 0, & u=0 \text{ or } k \leq D; \\ \sum_{\substack{c:\, c \in \mathcal{A}_D \\ c+d_j \neq (0,0)}} \sum_{i \in Z_{|\mathcal{X}|}} p_{i,j} h_{k-1}^m(u-1,D,c,i), & \text{otherwise.} \end{cases}$$

- Direction-based: $\theta = r$, $S_k^r = (u_k, c_k', j_k)$, where c_k' is a scalar defined as follows: $c_k' = 0$ if $d_{jk} \neq \pm d_{jk-1}$; $c_k = c_k' d_{jk}$ if $d_{jk} = \pm d_{jk-1}$.

$$h_0^r(u,c',j) = \begin{cases} \pi_j, & u=c'=0, j \in Z_{|\mathcal{X}|} \\ 0, & \text{otherwise.} \end{cases}$$

$$h_k^r(u,c',j) = \begin{cases} \sum_{\substack{n \in Z_{|\mathcal{X}|} \\ n \neq j, j^-}} \sum_{e=-D+1}^{D} p_{n,j} h_{k-1}^r(u-1,e,n) \\ \quad + p_{j,j}\left[h_{k-1}^r(u,-1,j) + h_{k-1}^r(u-1,D,j)\right] \\ \quad + p_{j^-,j} h_{k-1}^r(u,1,j^-), & c'=0; \\ p_{j,j} h_{k-1}^r(u,c'-1,j) + p_{j^-,j} h_{k-1}^r(u,1-c',j^-), & c'=-D+2,\ldots,-1,1,\ldots,D; \\ p_{j^-,j} h_{k-1}^r(u,D,j^-), & c'=-D+1, D>1; \\ 0, & c'=-D; \end{cases}$$

$$k \geq 1, \quad u=0,1,\ldots,k, \quad j \in Z_{|\mathcal{X}|},$$

where we define $j^- = \left(j + \frac{|\mathcal{X}|}{2}\right) \bmod |\mathcal{X}|$ because $d_{j^-} = -d_j$.

885

36.2.4 Distributions µ(u) and ϕ(c)

Denote the distribution that the MS performs location updates u times in a call arrival period by $\mu^\theta(u)$, $\theta \in \{d,m,r\}$. Let $\phi^\theta(c)$ denote the location profile of the MS (i.e., the distribution that the MS resides in cell c when a call arrival occurs). Let $\mu_k^\theta(u)$ and $\phi_k^\theta(c)$ be the (conditional) marginal distributions obtained by summing $h_k^\theta(s)$ over all the states s containing u and c, respectively. Thus, we have:

$$\mu^\theta(u) = \sum_{k=0}^{\infty} \alpha_k \mu_k^\theta(u), \tag{36.8}$$

and the average number of location updates per call arrival is then given by:

$$U^\theta = \sum_{u=0}^{\infty} u\mu^\theta(u) \tag{36.9}$$

We can compute the location profile $\{\phi^\theta(c)\}$ of the MS as:

$$\phi^\theta(c) = \sum_{k=0}^{\infty} \alpha_k \phi_k^\theta(c) \tag{36.10}$$

36.3 Paging and Cost

Although the location area (LA) $\mathcal{L}^\theta$ of a MS depends solely on the location update scheme, the location profile of the MS in $\mathcal{L}^\theta$ depends on both the location update scheme and mobility pattern. Given a location profile $\{\phi^\theta(c), c \in \mathcal{L}^\theta\}$, we can devise the optimal paging strategy to minimize the paging cost.

Consider the probabilistic selective paging (PSP) strategy with a delay constraint of N polling cycles. Let $\mathbf{v}_c = (v_{c,1}, \ldots, v_{c,N})$ be the page control vector for cell $c \in \mathcal{L}^\theta$, where $v_{c,n}$ is the probability that cell c is paged in the nth polling cycle. Note that $\Sigma_{n=1}^{N} v_{c,n} = 1$. When a call arrival occurs, the network performs the following tasks:

- It randomly selects the polling cycle for each cell $c \in \mathcal{L}^\theta$ according to $\mathbf{v}_c$. Thus, the residing area $\mathcal{L}^\theta$ is effectively partitioned into N disjoint LAs, $\mathcal{L}_1^\theta, \ldots, \mathcal{L}_N^\theta$, for each call.
- It polls all the cells in the first LA simultaneously to locate the MS. If the MS does not reside in this area, then it proceeds to page the other LAs sequentially until the MS is found.

A special case of PSP having $\mathbf{v}_c = \mathbf{e}_n$ for $c \in \mathcal{L}_n^\theta$ is the selective paging (SP) strategy that employs a fixed partition of LAs, where $\mathbf{e}_n$ denotes a $1 \times N$ unit vector with unity at the nth position and zeros elsewhere.

For the PSP strategy, the average number of cells polled per call arrival, denoted by G^θ, is shown in [9] to be given as:

$$G^\theta = \left[\sum_{n=1}^{N}\sum_{c\in\mathcal{L}_n^\theta}\phi^\theta(c)v_{c,n}\sum_{j=1}^{n}\sum_{c\in\mathcal{L}^\theta}v_{c,j}\right] + \left[\sum_{c\in\mathcal{L}^\theta}\phi^\theta(c)\sum_{n=1}^{N-1}\sum_{j=n+1}^{N}v_{c,n}v_{c,j}\right], \quad (36.11)$$

which can be simplified for the SP strategy [6] to be:

$$G^\theta = \sum_{n=1}^{N}\sum_{c\in\mathcal{L}_n^\theta}\phi^\theta(c)\sum_{j=1}^{n}\left|\mathcal{L}_j^\theta\right| \quad (36.12)$$

It can be observed that if $\phi^\theta(c) > \phi^\theta(\hat{c})$, $c \in \mathcal{L}_n^\theta$, and $\hat{c} \in \mathcal{L}_{\hat{n}}^\theta$, then we should assign $n \geq \hat{n}$ to reduce the paging cost. In general, we can optimize the page control vectors $\mathbf{v}_c$ for PSP by solving the constrained quadratic programming problem, given a location profile $\{\phi^\theta(c)\}$. For SP, a fast algorithm has been proposed in [14]. We note the following:

- The solutions to these optimization problems are locally optimal and depend on the initial conditions.
- Because SP is a special case of PSP, the optimal solution obtained by PSP is at least as good as that of SP. Both SP and PSP yield very close results in practice.

Let C_u^θ and C_g^θ, respectively, denote the unit location update and per cell paging costs. The total cost of mobility tracking per call arrival, denoted as C^θ, is therefore given as:

$$C^\theta = C_u U^\theta + C_g G^\theta \quad (36.13)$$

36.4 Performance Evaluation

We employ the ISS processes of the 2D Markov Walk Model to assess the performance of various location management schemes in hexagonal cellular PCS networks.

In Table 36.1, we designate six irreducible ISS processes for the hexagonal configuration, including IID, directional, turning, ping-pong, h-spin, and t-spin processes. When p_i is a dominating (diminishing) factor in the mobility control vector $\mathbf{p}$, the corresponding extreme mobility pattern in Figure 36.3 becomes prominent (attenuate). Figure 36.5 depicts the sample paths for these ISS processes.

We set up the parameters in the experiments as follows:

- The call arrival rate is normalized to $\lambda_c = 1$. The cell residence time is exponentially distributed with parameter $\lambda_m = 10$. Thus, the CMR is given by $\rho = 0.1$.

Table 36.1. Six Irreducible ISS Processes for Hexagonal Configuration

ISS Process	p					
	p_0	p_1	p_2	p_3	p_4	p_5
IID	1/6	1/6	1/6	1/6	1/6	1/6
Directional	0.8	0.025	0.025	0.1	0.025	0.025
Turning	0	0.25	0.25	0	0.25	0.25
Ping-Pong	0.1	0.1	0.1	0.5	0.1	0.1
h-spin	0.1	0.8	0.1	0	0	0
t-spin	0	0	0	0.1	0.8	0.1

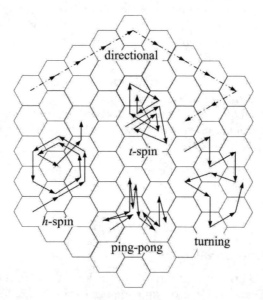

Figure 36.5. Sample Paths of Various ISS Processes
(Source: Wu, C.H., Lin, H.B., and Lan, L.S., *IEEE Trans. Mobile Computing*, vol. 1, no. 3, pp. 208–220, July–September 2002. [© 2002 IEEE]. Used with permission.)

- The unit location update and per cell paging costs, respectively, are selected as $C_u = 10$ and $C_g = 1$.

In Figure 36.6, we illustrate the variations of total costs with respect to various thresholds by conducting the recursive Markov analysis and solving the quadratic optimization problem for PSP. In distance-based and enhanced movement-based schemes (Figure 36.6A and Figure 36.6B), the curves of total costs are bowl-shaped and their bottoms occur at the optimum thresholds of D = 2 or D = 3. In the direction-based scheme, as the

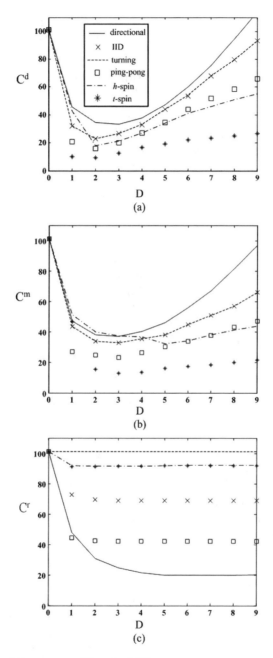

Figure 36.6. Total Cost versus Threshold D
For (A) distance, (B) enhanced movement, and (C) direction-based location update schemes at $\rho = 0.1$.
(Source: Wu, C.H., Lin, H.B., and Lan, L.S., *IEEE Trans. Mobile Computing*, vol. 1, no. 3, pp. 208–220, July–September 2002. [© 2002 IEEE]. Used with permission.)

threshold value D increases, the total cost decreases to reach a saturation point, beyond which the MS will almost surely change its direction before exceeding the threshold.

We make the following observations with respect to the six ISS processes:

1. IID — the IID process corresponds to the conventional 2D IID isotropic Random Walk Model that yields medium performance among the six processes because there is no dominating or diminishing factor. For distance-based ($\theta = d$) and enhanced movement-based ($\theta = m$) schemes, the IID and turning processes produce similar total costs, indicating that both schemes are robust against undirected movements.
2. Directional — with $p_0 = 0.8$, the MS is most likely to stay in its current direction and its location profile concentrates on straight lines. Extreme performance differences are observed. The total cost caused by the directional process is the highest for distance-based and enhanced movement-based schemes, but the lowest for the direction-based ($\theta = r$) scheme.
3. Turning — because $p_0 = p_3 = 0$, the MS constantly changes its direction. The worst-case scenario is the direction-based scheme that behaves like the zero-threshold case (D = 0) because the MS always performs a location update whenever it moves across a cell boundary.
4. Ping-pong — for $p_3 = 0.5$, the MS tends to stay around the origin because it usually moves back and forth between two cells several times before entering into other cells. The overall cost for location management remains low for all location update schemes.
5. h-spin — for $p_1 = 0.8$, the MS is most probable to shift its direction 60 degrees to the left so as to spin counterclockwise in hexagonal patterns. The h-spin process has low costs for the distance-based scheme because the MS frequently rotates inside the LA $\mathcal{A}_D$ and rarely moves beyond the threshold $\mathcal{R}_D$. Because $p_0 = 0.1$ and $p_3 = 0$, the MS seldom maintains its direction and never bounces back and forth between two cells. Thus, the total cost is relatively high for the direction-based scheme.
6. t-spin — for $p_4 = 0.8$ and $p_3 = 0.1$, the MS usually shifts its direction 120 degrees to the right and sometimes bounces back and forth so as to produce clockwise staggered triangular patterns. The t-spin process offers the lowest cost for distance-based and enhanced movement-based scheme. For the direction-based scheme, the total cost rendered by the t-spin process is as high as that by the h-spin process because the MS has a low probability of $p_0 + p_3 = 0.1$ to keep its direction unchanged.

Analytic Mobility Models of PCS Networks

In summary, the six ISS processes are indistinguishable at steady state by possessing identical steady-state probabilities for all directions, but they exhibit disparate mobility patterns that lead to subtle performance variations of dynamic location update schemes. We have demonstrated that the 2D Markov Walk Model with circulant transition probability matrices provides an elegant analytic framework for better understanding of dynamic mobility management schemes in PCS networks.

References

1. Wong, V.W.S. and Leung, V.C.M., Location management for next-generation personal communications networks, *IEEE Network*, pp. 18–24, September/October 2000.
2. Ho, J.S.M. and Akyildiz, I.F., On location management for personal communications networks, *IEEE Communications Magazine*, pp. 138–145, September 1996.
3. Bar-Nor, A., Kessler, I., and Sidi, M., Mobile users: To update or not to update, *ACM-Baltzer Wireless Networks*, vol. 1, no. 2, pp. 187–196, 1994.
4. Lui, J.C.S., Fong, C.C.F., and Chan, H.W., Location updates and probabilistic tracking algorithms for mobile cellular networks, *International Symposium on Parallel Architectures, Algorithms, and Networks*, pp. 432–437, 1999.
5. Ho, J.S.M. and Akyildiz, I.F., Mobile user location update and paging under delay constraints, *ACM-Baltzer Wireless Networks*, vol. 1, no. 4, pp. 413–425, 1995.
6. Akyildiz, I.F., Ho, J.S.M., and Lin, Y.B., Movement-based location update and selective paging for PCS networks, *IEEE/ACM Trans. Networking*, vol. 4, pp. 629–638, August 1996.
7. Casares-Giner, V. and Mataix-Oltra, J., On movement-based mobility tracking strategy — An enhanced version, *IEEE Communication Letters*, vol. 2, no. 2, pp. 45–47, February 1998.
8. Hwang, H.W., Chang, M.F., and Tseng, C.C., A direction-based location update scheme with a line-paging strategy for PCS networks, *IEEE Communication Letters*, pp. 149–151, May 2000.
9. Wu, C.H., Lin, H.B., and Lan, L.S., A new analytic framework for dynamic mobility management of PCS networks, *IEEE Trans. Mobile Computing*, vol. 1, no. 3, pp. 208–220, July–September 2002.
10. MacDonald, V.H., The cellular concept, *AT&T Bell System Technical Journal*, vol. 58, no. 1, pp. 15–41, January 1979.
11. Golub, G.H. and Van Loan, C.F., *Matrix Computations*, Baltimore, MD: Johns Hopkins Press, 1996.
12. Ross, S.M., *Stochastic Processes*, New York: John Wiley & Sons, 1983.
13. Lin, Y.B., Reducing location update cost in a PCS network, *IEEE/ACM Trans. Networking*, pp. 25–33, February 1997.
14. Wang, W., Akyildiz, I.F., and Stuber, G.L., An optimal paging scheme for minimizing signaling costs under delay bounds, *IEEE Communications Letters*, vol. 5, no. 2, pp. 43–45, February 2001.

Chapter 37
Battery Power Management in Portable Devices

Vinod Sharma and A. Chockalingam

Abstract

This chapter provides a tutorial-cum-survey of performance analysis of battery power management schemes in wireless mobile devices. In particular, it focuses on the performance analysis of schemes that exploit the relaxation phenomenon in batteries whereby a battery can rebuild its charge if left idle. With an intent to exploit the relaxation phenomenon, the battery can be allowed to go on intentional vacations during which the battery can recharge itself. The recharge thus built up can effectively increase the number of packets transmitted (in other words, battery life can be extended). Such improved battery life performance would, however, come at the expense of increased packet delay performance if there is only one battery in the mobile device. To analyze the performance of such systems, the battery is modeled as a server with finite capacity and the data packets as customers to serve. The battery life gain versus delay performance trade-off is quantified through analysis and simulations. We also consider systems when there is more than one battery in the device. Then, by suitably choosing the battery to use at a time, one can increase the system life without incurring extra delays. We also discuss the optimal policies to use for choosing the battery at any time.

37.1 Introduction

Wireless mobile devices, like cellular phones and personal digital assistants (PDAs), have become a part of our daily lives. These devices have certain desired characteristics that have made them useful. They are lightweight, small in size, portable and have the ability to perform a variety of tasks including mobile Internet access. These devices need to be powered by batteries, which are limited in their ability to keep the devices working, making the battery energy a valuable resource. To make optimum use of this resource, we need to look at ways to make the devices more energy conscious. Electronic devices mainly comprise hardware (processors, digital signal processor [DSP] chips, etc.) and software (set of instructions). Chips in electronic devices dissipate heat and consume energy while performing operations. The battery life in a device thus depends on how efficiently the on-board chips perform their functions. Battery technology has been unable to match the pace of development taking place in the field of wireless communications and device technology. Chips have become smaller and more computationally powerful whereas batteries have been unable to keep up with the requirement of packing more joules in less area with minimum weight. Therefore, efforts to design chips that consume less energy to perform similar computations have become important. This means not only a rethink of the chip architecture, but also taking a look into new materials that offer better efficiency at submicron range. Furthermore, algorithms and programs need to be written in such a way so as to perform the tasks in an energy efficient manner. For devices that are a part of a wireless network, some of the energy consuming components include the radio frequency (RF) circuit and the wireless interface card on the hardware side and modulation and coding schemes, Media Access Control (MAC) protocol, and routing protocol on the software side [1].

Growing demand for computationally intensive applications for mobile devices has motivated research into ways for improving the battery utilization [2]. It is important that the design of the communication system as a whole (not only at the device or circuit level) should follow energy conservation principles. In particular, energy savings can be sought at different layers of the wireless protocol stack. In the following, first we briefly survey the work that has been going on at the different layers to improve the battery utilization. Then we discuss the literature on performance analysis of battery management schemes.

On the hardware side, dynamic voltage scaling is a technique used in microprocessors to reduce energy consumption [3]. At low speeds, the energy consumed by a processor is less than the consumption at higher speeds. In [4], an energy efficient real-time scheduling algorithm, called Slacked Earliest Deadline First (SEDF) algorithm, that exploits this concept in processors has been analyzed. Also, energy can be saved by reducing the speed of rotation of the hard disk or by completely stopping it. In [5], algorithms to control the hard disk access are presented. The trade-off in the hard disk access time to the energy savings achieved by adapting the speed of the hard disk has been analyzed.

Battery Power Management in Portable Devices

The physical layer is the lowest layer of the protocol stack. Among other things, it deals with modulation schemes and transmitter power control. Energy conscious physical layer design ideas have been considered by several authors. In [6], Shih et al. present a physical layer-aware design of applications, algorithms, and protocols, which minimizes the energy consumption. A power-mode scheduling algorithm that manages the active and sleep states of the node is presented. The cost of forward error correction (FEC) coding and determining the convolutional code to use to minimize the energy consumption for a given probability of error has also been studied. Transmitter power control is yet another physical layer feature that can be used to save battery power in wireless devices. For example, most cellular standards, including Global System for Mobile Communications (GSM), IS-95 code division multiple access (CDMA) etc., use transmitter power control to meet the twin objectives of reducing interference as well as saving battery power [7, 8]. In wireless ad hoc networks like Bluetooth® [9], transmitter power control in discrete power steps is allowed. In [10], algorithms are developed to minimize the transmit power, based on the channel state, to achieve certain bit error rate.

The data link layer controls the access to the shared wireless medium. It also provides a certain degree of data reliability by using automatic repeat request (ARQ) and FEC. Several energy saving ideas have been proposed in the data link layer design as well. In [11], the authors present a protocol, called Power-Aware Multi-Access Protocol with Signaling (PAMAS) protocol, in which terminals using this protocol switch off their wireless interface when they are not sending or receiving packets. This allows the mobiles to save energy as they do not have to listen to packets that are not meant for them. In [12], Sivalingam et al. suggest a centralized scheduling policy for MAC that reduces collisions and thereby reduces the overall energy consumption. Energy efficient MAC protocols for correlated fading channels are proposed and analyzed in [13]. Lettieri and Srivastava, in [14], consider the technique of adapting the size of the MAC layer frame along with adaptive error correcting techniques. The authors optimize the battery energy consumed per useful bit while maintaining certain quality of service (QoS) constraints. In [15], the authors obtain scheduling policies to provide QoS to different users within a certain power constraint. Energy efficient ARQ strategies have been studied in [16]. In [17] the authors propose and analyze energy efficient backoff algorithms on point-to-point wireless links resulting in energy savings.

The network layer protocols provide the functionalities including routing the packets and maintenance of the topology in networks. In traditional wireline networks, routing is often based on shortest path, throughput, and delay as cost metrics in choosing routes. However, for wireless networks various energy-efficiency-based metrics have been considered [18].

In [19], Woo et al. investigate five different metrics — energy consumed per packet, time to network partition, variance in power levels across mobiles, cost per packet, and maximum mobile cost — for achieving energy efficiency in routing. These metrics are used in conjunction with existing routing protocols like Dynamic Source Routing (DSR), Destination-Sequenced Distance-Vector (DSDV), etc. [20]. They show that the cost savings are higher in large networks and at moderate loads. In [21], a routing protocol to maximize the time to network partitioning is presented. They propose flow augmentation and flow redirection algorithms that balance the flows based on energy reserves. This can increase on an average the system life time by 60 percent. Energy efficient routing protocols for Bluetooth are provided in [22]. Energy efficient topology construction and maintenance schemes are presented in [23–25].

The transport layer provides end-to-end reliability and congestion control. Transmission Control Protocol (TCP) operates at this layer of the stack. Zorzi and Rao, in [26], analyze the energy consumption of performance Tahoe, Reno, and New Tahoe versions of TCP. The use of backoff in congestion control mechanisms during error bursts helps in energy savings. In [27], the energy and throughput efficiency of TCP error control strategies for Tahoe, Reno, and New Tahoe are presented. The different error recovery strategies in the variations of TCP result in different energy efficiencies. Based on the results, Tahoe is found to be more-or-less the most energy conserving of the three. A similar conclusion is obtained in [28] when FEC is also used. A TCP-Probing protocol is proposed in [29]. It is shown that a higher throughput with relatively low energy expenditure is achieved by the use of the probing protocol.

The application layer is the interface between the user and the system. Various applications like Web browsing, remote login, etc., reside in this layer. In [30], Agarwal et al. propose decreasing the number of bits and discarding packets as means for reducing the power consumption for encoded video applications. Another possibility is to lower the source coding rate as the battery gets weaker.

37.1.1 Relaxation Phenomenon in Batteries

In addition to energy saving possibilities at various layers of the protocol stack discussed above, yet another way to improve battery life is to exploit the relaxation phenomenon in batteries, whereby a battery can rebuild its charge if left idle [31]. In the following, we describe the relaxation phenomenon. We are concerned about the ways to exploit this phenomenon and modeling and analyzing the resulting gains in system lifetime.

Several studies characterizing the battery discharge behavior have shown that pulsed discharge [32–35] (i.e., discharge with in-between nondischarging, idle periods) performs better than continuous discharge. Particularly,

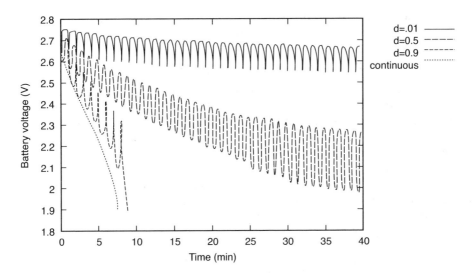

Figure 37.1. Battery Potential versus Time for Different Duty Cycles of Discharge (Current density = 5A/M², cut-off voltage = 1.9V.)

the battery can recharge itself (i.e., recover the potential) if left idle after discharge. This relaxation phenomenon is illustrated in Figure 37.1. Figure 37.1 shows the battery potential (in volts) as a function of time with continuous discharge and pulsed discharge. The parameter d represents the duty cycle of the pulsed discharge pattern (i.e., d = 0.1 means 10 percent discharge time followed by 90 percent relaxation time and so on). The plots in Figure 37.1 are generated using the battery simulation program developed by the Chemical Engineering department at University of California — Berkeley [36]. A 2.76 V Lithium ion battery with a cutoff voltage of 1.9 V is considered. It is assumed that the battery ceases to deliver power once the voltage drops below the cutoff voltage. In other words, the time taken for the battery to fall below the cutoff voltage is the battery life. The slope of battery discharge is determined by the discharge current density. The larger the value of this current density, the steeper will be the discharge slope, and hence lesser will be the time taken to reach the cutoff voltage. In Figure 37.1, the discharge current density is taken to be 5A/m². From Figure 37.1, it can be seen that if the battery is discharged continuously it takes about 7.5 minutes to reach the cutoff voltage. On the other hand, if the battery is discharged in pulsed mode, the battery recovers the voltage during the relaxation periods and it takes more time to reach the cutoff voltage. For example, if the duty cycle of the pulsed discharge is 50 percent (i.e., d = 0.5) it takes more than 40 minutes to reach the cutoff voltage. Also, because the duty cycle is 0.5, the total on-time of the battery is more than 20 minutes, which is about three times the on-time in continuous discharge mode. As the discharge current density is taken to be the same in both continuous and pulsed discharge modes, this essentially means that the battery can deliver energy for a longer duration. This recharge effect advantage in pulsed mode can be exploited for improved energy efficiency in packet communications in wireless mobile devices.

This chapter surveys the methods to exploit the relaxation phenomenon in batteries and the analysis of battery life gain and packet delay performance using a queueing theory approach. Because transmission of packets on a wireless link consumes a significant amount of power, this can be an important way to increase the overall battery life in the wireless mobile terminal. Recently, in [37], Chiasserini and Rao proposed a probabilistic recharge model for the battery. Through simulations, they showed that the recharge phenomenon can be exploited for battery life gain through suitable traffic shaping algorithms and battery level sensed routing strategies [38, 39].

To analyze the performance of schemes that exploit the relaxation phenomenon, the battery can be modeled as a server with finite capacity and the arriving packets at the mobile terminal as the customers to be served. Each transmitted packet consumes energy proportional to the packet size, transmission bit rate, wireless link design, etc. It is possible to intentionally allow the server (battery) to go on vacation for a calibrated amount of time, essentially allowing idle times for the battery to recharge itself. By doing so, the number of customers served can be increased (in other words, battery life can be increased). Expressions for the number of customers served and the average delay for a M/GI/1 queueing system without and with such intentional server vacations can be derived. In [40], it has been shown that allowing forced vacations during busy periods helps to increase battery life. However, because the customers (packets) have to wait in the queue when the server goes on vacation, the battery life gain will come at the expense of increased delay performance of the packets. Use of multiple batteries can improve system lifetime without incurring large delays.

Consequently, a system where there is more than one battery to transmit the packet provides an attractive option. In [41], a simple strategy to share two batteries to increase the overall lifetime of the batteries has been analyzed. In the rest of this chapter, we focus on the performance analysis of schemes that exploit the relaxation phenomenon.

The rest of this chapter is organized as follows:

- Section 37.2 explains the basic model for a single battery system.
- Section 37.3 analyzes this base system and provides several extensions to this base system — system with vacations and multiple battery system.

- Section 37.4 provides analytical and simulation results of the system studied in Section 37.3.
- Section 37.5 considers the optimal battery scheduling problem in the framework of Markov decision theory.

37.2 System Model

Consider data packet transmission from a mobile terminal that draws power from a battery of finite capacity. We model the system as a queueing system with the battery as the server and packets as customers. The packet interarrival time is assumed to be exponential (i.e., Poisson arrivals with rate λ) and service times are assumed to be independent and idenitically distributed (IID) with a general distribution.

We assume that the battery has a nominal capacity of N charge units. That is, if the transmission of a unit length packet consumes one charge unit and if the battery is discharged continuously, then the battery can serve N such packets until it gets completely discharged.

37.2.1 Battery Discharge/Recharge Model

In [37, 38], the battery discharge behavior is assumed to be discrete: each packet occupies a fixed slot size and each packet transmission consumes one charge unit (or integer number of charge units). Also, in [37, 38] the battery recharge behavior (when the battery remains idle) too is modeled as a discrete process: the battery recharges by one charge unit with some probability if left idle for one slot duration. The recharge probability has been assumed to follow an exponential function that decreases with increasing discharged capacity. In other words, the recharge capability of the battery at any time is made proportional to the charge available in the battery at that time (i.e., the greater the available charge, the greater is the ability to recharge). In addition to the nominal capacity N, the battery is assumed to be characterized by another parameter called the theoretical capacity, T charge units [37]. The battery is considered to be fully drained if either the available charge goes to zero or T charge units of capacity has been delivered, whichever occurs first. Typically, the theoretical capacity of the battery is taken to be one or two orders higher than the nominal capacity [37].

Here (until Section 37.4), we consider both the discharge as well as the recharge behavior of the battery as continuous phenomena, as illustrated in Figure 37.2. While serving packets in busy periods, the battery looses charge linearly at a constant slope of unity. During idle periods, the battery recharges linearly with varying slopes depending on the battery level[1] at the beginning of the idle period. The recharge model is more clearly explained as follows. We divide the range of charge from 0 to N using P

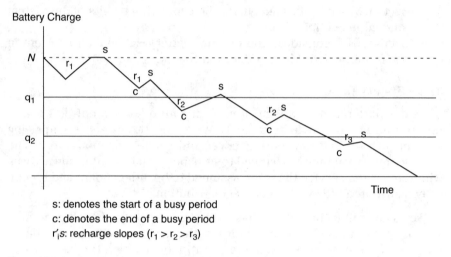

Figure 37.2. Continuous Discharge/Recharge Model

threshold values, $\theta_1, \theta_2, \ldots, \theta_P$. The recharge slope is taken to be $r_1, r_2, \ldots, r_{P+1}$, respectively, when the battery level at the beginning of the idle period is in the range θ_1 to N, θ_2 to $\theta_1, \ldots, 0$ to θ_P. By choosing $r_1 > r_2 > \ldots > r_{P+1}$, the model ensures that the ability to recharge reduces with decreasing battery level, which is more realistic [36]. Thus, the parameters P, θ_i's, and r_is characterize the recharge behavior of the battery, which can be used in the mathematical analysis of the battery life gain due to recharge during idle or vacation periods. Actually a more general recharge function can be easily incorporated in our analysis, as we will comment in the following.

37.3 Analysis

We are interested in analyzing the battery performance in terms of mean number of packets served and mean packet delay when the battery's recharge behavior is exploited by intentionally allowing the battery to go on vacations. We first analyze and quantify the battery life gain due to idle periods (i.e., when buffer is empty) inherent in a M/GI/1 queue. We then analyze a system where the server exhaustively serves all packets in the queue and takes intentional vacations before the start of the next busy period. We call this a system with exhaustive service with vacations. We also propose a nonexhaustive service system with vacations where the server takes vacations during the busy period. First consider a simple M/GI/1 queue. Define a cycle as the time from start of one idle period to the start of the next idle period. A cycle consists of one idle and one busy period as shown in Figure 37.3. Note that the first cycle starts after the first busy period. To obtain the expected number of customers served, we obtain the distribution of the amount of charge left at the end of the first

Battery Power Management in Portable Devices

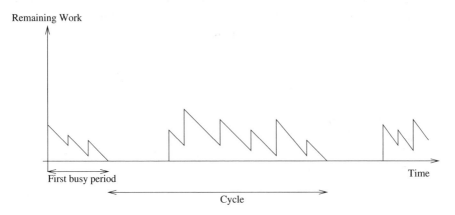

Figure 37.3. Definition of a Cycle in a Single Battery Scheme

busy period, obtain the expected number of cycles after the first busy period until the charge goes to zero, and obtain the expected number of customers served in each cycle, as follows.

For the M/GI/1 queue the idle period is distributed as exp(λ). Let:

$$Z_i = Y_i - B_i, \qquad (37.1)$$

where B_i is the charge consumed during the busy period and Y_i is the amount of recharge during the idle period in the ith cycle, respectively. Then, Z_i is the total charge lost ($Y_i < B_i$) or gained ($Y_i > B_i$) in the ith cycle. Let T_i be the random variable denoting the number of cycles until the charge goes to zero and let z be a random variable that denotes the charge at the end of the first busy period. We need to find:

$$\tau = \inf\left\{n : \sum_{i=1}^{n} Z_i \leq -z\right\}. \qquad (37.2)$$

Let $E_z[\tau]$ denote the expected number of cycles given that the charge after the first busy period is z. (The event that the battery gets discharged in the first busy period itself will be taken care of in (37.9) below.) We can then write the integral equation:

$$E_z[\tau] = P[Z_1 \leq -z] + \int_{-z}^{N-z} (1 + E_{z_1}[\tau]) \cdot dF_{Z_1}(z_1) \qquad (37.3)$$

where on the right side, the first expression corresponds to the event that the battery will get discharged in the first busy cycle and the second that there will be a left over charge. $F_{Z1}(z_1)$ is the cumulative density function (CDF) of Z_1. The right side can be further simplified to:

$$= P[Z_1 \leq N - z] + \int_{-z}^{N-z} E_{z_1}[\tau] \cdot dF_{Z_1}(z_1) = 1 + \int_{-z}^{N-z} E_{z_1}[\tau] \cdot dF_{Z_1}(z_1). \quad (37.4)$$

To obtain the distribution of Z_i, we need to obtain the distributions of Y_i and B_i. The distribution of Y_i is obtained as follows. Let T_i denote the duration of the idle period of the ith cycle and z' denote the charge at the beginning of the ith cycle. Then:

$$Y_i = min(r_k T_i, N - z'), \quad (37.5)$$

and

$$F_{Y_i}(y) = \begin{cases} F_{T_i}(y/r_k) & y \leq N - z' \\ 1 & y > N - z', \end{cases} \quad (37.6)$$

where $k \in \{1, 2, \ldots, P+1\}$ is determined by the charge threshold values between which z' lies. Because T_i is distributed as $\exp(\lambda)$, the CDF of Y_i is given by:

$$F_{Y_i}(y) = \begin{cases} 1 - e^{-\lambda y'}, & y \leq N - z' \\ 1, & y > N - z' \end{cases} \quad (37.7)$$

where $y' = y/rk$. Observe that the distribution of Y_i depends upon z'.

It is easy to see now how a more general recharging function can be incorporated. For example if the recharge slope is $r(z)$ when the charge in the battery is z, then the distribution of Y_i is again (37.6) but now y and y' are related as follows. Let $z(t)$ be the solution of the differential equation:

$$\frac{dz(t)}{dt} = r(z(t)), XZ(0) = Z'$$

Then $y = z(y')$.

Computing the distribution of B_i for a general M/GI/1 queue may not be so easy. Here, we limit ourselves to exponential service times with rate μ. Then, B_i has density [43]:

$$f_{B_i}(b) = \frac{I_1(2b\sqrt{\lambda\mu}) \cdot e^{-b(\lambda+\mu)}}{b\sqrt{\rho}}, \quad (37.8)$$

where $I_1(\cdot)$ is the modified Bessel function of the first order and $\rho = \dfrac{\lambda}{\mu}$. The CDF of Z_1, $F_{Z1}(z_1)$ can then be written as:

$$F_{Z_1}(z_1) = \int_{max(0,-z_1)}^{\infty} F_{Y_1}(z_1+b)f_{B_1}(b)db$$

$$= \int_{max(0,-z_1)}^{N-z-z_1} F_{Y_1}(z_1+b)f_{B_1}(b)db + \int_{N-z-z_1}^{\infty} f_{B_1}(b)db. \qquad (37.9)$$

Equation (37.4) can be numerically solved to obtain $E_z[\tau]$. To obtain $E[\tau]$ we average $E_z[\tau]$ over z, as:

$$E[\tau] = \int_0^N E_{N-z}[\tau]dF_{B_0}(z), \qquad (37.10)$$

where F_{B0} is the distribution of the first busy period. The expected number of customers served can be obtained as the product of the expected number of cycles and the expected number of customers served in a busy period as follows:[2]

$$C = (1+E[\tau])\dfrac{1}{1-\rho}. \qquad (37.11)$$

Note that the expected number of customers served is upper bounded by the theoretical capacity of the battery, μT. The mean delay of packets in the M/M/1 queue is given by [42]:

$$\overline{W} = \dfrac{\lambda}{\mu^2(1-\rho)}. \qquad (37.12)$$

It is possible to handle the general service times of an M/GI/1 queue also, even though it may be computationally more complex. For example, given the Laplace Transform (L.T.) of the service times, one can compute the L.T. $E[e^{sB_i}]$ of a busy period B_i. Then of course $E[e^{-sZ}] = E[e^{-sY_i}]E[e^{sB_i}]$ provides the L.T. of Z_i. Inverting the L.T. of Z_i will provide F_{Zi}, which can be used in (37.3).

37.3.1 Extensions and Generalizations

The analysis provided above can be extended to other systems. In the following, we consider some such extensions along with the necessary details.

37.3.1.1 Exhaustive Service with Vacations.

In this system, the server takes intentional vacations. The server exhaustively serves all packets in the queue and takes a vacation. If there is still no packet in the queue after completion of a vacation, it takes another vacation, and so on. We assume the vacation time distribution to be $\exp(\delta)$, where δ can be a system parameter.

The analysis of this system can be carried out as for the system without vacations with the following modifications. The idle time now is exponentially distributed with mean $1/p\delta$ where $p = \lambda/(\lambda+\delta)$. Also the busy period can start with more than one packet in the queue. Now it can be shown to have the density (see [40] for details):

$$f_{B'}(b) = \frac{4(\lambda+\delta)\sqrt{\lambda\mu}\,\delta}{\lambda} \cdot \frac{e^{-(\lambda+\mu)b}}{\sqrt{b}} \cdot \int_0^\infty \frac{e^{-2(\lambda+\delta)u}\, I_1\!\left(2\sqrt{\lambda\mu b(b+2u)}\right) u}{\sqrt{b+2u}}\, du. \quad (37.13)$$

The mean delay for this system with vacations can be written as [42]:

$$\overline{W} = \frac{\lambda}{\mu^2(1-\rho)} + \frac{1}{\delta}. \quad (37.14)$$

37.3.1.2 Nonexhaustive System with Vacations.

Another way to improve battery life is to interrupt service during busy period and allow vacations. Specifically, we allow the server to go on vacation after serving K packets in a row during a busy period. We assume the duration of this vacation to be IID and distributed as $\exp(\delta)$. Both δ and K are parameters that can be chosen to allow desired vacation times.

To analyze this scheme, consider an approximate system that models the nonexhaustive system with vacations at high arrival rates. Assume the arrival rate is high so that there are always packets to send in the queue. In this case, we can take the busy period to be the sum of K service times. Thus, for the case of $\exp(\mu)$ service times the probability density function (PDF) of the busy period can be written as:

$$f_{B_i}(b) = \frac{\mu^K b^{(K-1)} e^{-\mu b}}{(K-1)!}. \quad (37.15)$$

The vacation period is distributed as $\exp(\delta)$. We then use (37.4), (37.6), and (37.10) to obtain $E[\tau]$ the expected number of cycles to discharge completely. The number of packets served C is then given by:

$$C = \frac{K E[\tau]}{\mu}. \quad (37.16)$$

Battery Power Management in Portable Devices

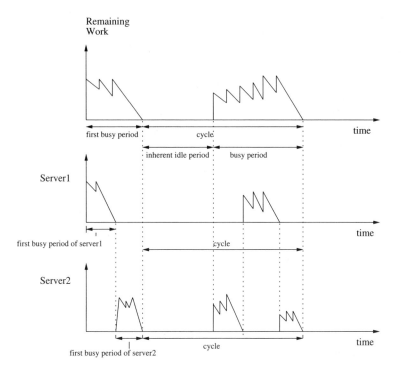

Figure 37.4. Definition of a Cycle in a Dual Battery Scheme

37.3.1.3 Multi-Battery System. In this section, we consider the system with multiple batteries. Each battery has N charge units. A packet can be served by any of the batteries. When one battery serves a packet, the other batteries remain idle (i.e., take a natural vacation). The batteries are assumed to discharge (during packet transmissions) and recharge (during idle periods) as per the discharge/recharge model described in Section 37.2. In this section, we analyze a system with two batteries using a specific scheduling policy: a packet can be served either by the first or by the second battery with probabilities p and (1 − p), respectively. The problem of optimal scheduling of batteries will be considered in the next section.

We are interested in analyzing the performance of the dual battery scheme, in terms of mean number of packets served and mean packet delay. We define a cycle as shown in Figure 37.4.

To find the expected number of packets served by the system, we find the expected number of packets served by both the batteries until one of the batteries expires (i.e., battery charge goes to zero) and the expected number of packets served by the remaining live battery until it expires. To do that, we carry out the following analytical steps:

1. Obtain the distribution of the amount of charge left in each battery at the end of its first busy period.
2. Obtain the expected number of cycles after the first busy period of each battery until its charge goes to zero.
3. Obtain the expected number of packets served in each cycle in each battery until any one battery expires.
4. After any one battery expires, using the charge left in the remaining live battery, evaluate the expected number of packets served by it until it expires, by considering the system as a single battery system without vacations.

Consider the busy period of a M/M/1 queue. Let X be a random variable (r.v.) denoting the number of packets served in a M/M/1 busy period. Let X = n packets are served in this busy period. Let k, $k = 1, 2,\ldots,n$, be the number of packets served by the first battery during a M/M/1 busy period. Then (n – k) packets will be served by the second battery in the same busy period. We use the following notations:

- T_b, T_i: r.v.s denoting busy and idle periods of the M/M/1 queue respectively
- T_{1b}, T_{1i}: r.v.s denoting busy and idle periods of the first battery
- T_{2b}, T_{2i}: r.v.s denoting busy and idle periods of the second battery
- S_i: exponentially distributed r.v. with parameter μ

We are interested in obtaining the distribution of the busy periods of the two batteries. Exact PDF expressions for these busy periods are difficult to obtain. Hence, to facilitate the analysis, we assume that the first battery serves k out of n packets in a M/M/1 busy period continuously and the second battery serves the remaining n – k packets continuously. Note that in the actual system, the service of k packets in a M/M/1 busy period by the first battery can be discontinuous (i.e., service of packets can alternate randomly between the two batteries depending on the scheduling probability, p). Hence this assumption is expected to give approximate results. Later, we will compare the results obtained through this approximate analysis with exact simulation results.

With the above assumption, the CDF of the busy period of the first battery can be written as:

$$P(T_{1b} \leq t) = Pr\{n \text{ packets are served in } M/M/1 \text{ busy period,}$$

$$\text{and } k \text{ out of } n \text{ packets are served by 1st battery,} \quad (37.17)$$

$$\text{and } \sum_{i=1}^{k} S_i \leq t\}.$$

The probability of having n packets served in the M/M/1 busy period is given by [43]:

$$P(X = n) = \frac{1}{n}\binom{2n-2}{n-1}\rho^{n-1}(1+\rho)^{1-2n}, \quad (37.18)$$

where $\rho = \lambda/\mu$. Equation (37.17) can then be written as:

$$P(T_{1b} \le t) = \sum_{n=1}^{N}\sum_{k=1}^{n}\Pr\left(\sum_{i=1}^{k}S_i \le t \mid X = n, k \text{ packets are served by 1st battery}\right)$$
$$\cdot P(X = n) \cdot \Pr(k \text{ out of } n \text{ packets are served by 1st battery}) \quad (37.19)$$
$$= \sum_{n=1}^{N}\sum_{k=1}^{n}\Pr\left(\sum_{i=1}^{k}S_i \le t\right)\frac{1}{n}\binom{2n-2}{n-1}\rho^{n-1}(1+\rho)^{1-2n} \cdot \binom{n}{k}p^k(1-p)^{n-k}.$$

Because S_i's are $\sim \exp(\mu), S = \sum_{i=1}^{k}S_i$ has Erlang distribution. The PDF of the busy period of the first battery can then be obtained from (37.19), as:

$$f_{T_{1b}}(t) = \sum_{n=1}^{N}\sum_{k=1}^{n}\frac{1}{n}\binom{2n-2}{n-1}\rho^{n-1}(1+\rho)^{1-2n}\binom{n}{k}p^k(1-p)^{n-k}f_S(t) \quad (37.20)$$

where f_S is given by the Erlang distribution, $f_S(t) = \dfrac{\mu(\mu t)^{k-1}\exp(-\mu t)}{(k-1)!}$. Similarly, the CDF and PDF of the busy period of the second battery can be obtained as:

$$P(T_{2b} \le t) = \sum_{n=1}^{N}\sum_{k=1}^{n}\Pr\left(\sum_{i=1}^{n-k}S_i \le t\right)\frac{1}{n}\binom{2n-2}{n-1}\rho^{n-1}(1+\rho)^{1-2n}$$
$$\cdot \binom{n}{n-k}p^k(1-p)^{n-k}, \quad (37.21)$$

$$f_{T_{2b}}(t) = \sum_{n=1}^{N}\sum_{k=1}^{n}\frac{1}{n}\binom{2n-2}{n-1}\binom{n}{n-k}\left(\frac{\lambda+\mu}{\lambda}\right)\left(\frac{\rho(1-p)}{(1+\rho)^2}\right)^n\left(\frac{p}{1-p}\right)^k$$
$$\cdot \left(\frac{\mu(\mu t)^{n-k-1}\exp(-\mu t)}{(n-k-1)!}\right) \quad (37.22)$$

Next, we are interested in the distribution of the idle periods of the two batteries. The idle period of the first battery in a cycle consists of two components, namely, the inherent idle period in a M/M/1 queue and the busy period of the second battery. Hence, the PDF of the idle period of the first battery can be written as:

$$f_{T_{1i}}(t) = f_{T_{2b}}(t) * f_{T_i}(t), \tag{37.23}$$

where $*$ denotes convolution operation and $f_{T_i}(t) = \lambda e^{-\lambda t}$. Equation (37.23), in transform domain, can be written as:

$$f_{T_{1i}}(s) = f_{T_{2b}}(s).f_{T_i}(s), \tag{37.24}$$

where $f_{T_i}(s)$ and $f_{T_{2b}}(s)$ are given by [44]:

$$f_{T_i}(s) = \frac{\lambda}{(s+\lambda)}, \tag{37.25}$$

$$f_{T_{2b}}(s) = \sum_{n=1}^{N}\sum_{k=1}^{n} \frac{1}{n}\binom{2n-2}{n-1}\binom{n}{n-k}\left(\frac{\lambda+\mu}{\lambda}\right)\left(\frac{p(1-p)}{(1+\rho)^2}\right)^n \cdot \left(\frac{p}{1-p}\right)^k \left(\frac{\mu}{s+\mu}\right)^{n-k}.$$

Also, we can obtain the following Laplace Transform relation:

$$\frac{\lambda}{s+\lambda}\left(\frac{\mu}{s+\mu}\right)^{n-k} \rightleftharpoons \frac{e^{-\lambda t}\mu^{n-k}(\mu-\lambda)^{-(n-k)}}{\Gamma(n-k)} \cdot \frac{\lambda\bigl(\Gamma(n-k)-\Gamma(k,(\mu-\lambda)t)\bigr)}{\Gamma(n-k)}, \tag{37.26}$$

where $\Gamma(k)$ and $\Gamma(a,z)$ are, respectively, the Euler Gamma function and the incomplete Gamma function, given by $\Gamma(k) = (k-1)!$, k: integer, and $\Gamma(a,z) = \int_{t=z}^{\infty} t^{a-1}e^{-t}dt$. From the above, $f_{T_{1i}}(t)$ can be written as:

$$f_{T_{1i}}(t) = \sum_{n=1}^{N}\sum_{k=1}^{n} \frac{1}{n}\binom{2n-2}{n-1}\binom{n}{n-k}\left(\frac{\lambda+\mu}{\lambda}\right)\left(\frac{p(1-p)}{(1+\rho)^2}\right)^n \left(\frac{p}{1-p}\right)^k$$

$$\cdot \left(\frac{e^{-\lambda t}\mu^{n-k}(\mu-\lambda)^{-(n-k)}\lambda\bigl(\Gamma(n-k)-\Gamma(n-k,(\mu-\lambda)t)\bigr)}{\Gamma(n-k)}\right). \tag{37.27}$$

Similarly, because the busy period of the first battery contributes to the idle period of the second battery, the PDF of the idle period of the second battery, $f_{T_{2i}}(t)$, can be written as:

$$f_{T_{2i}}(t) = f_{T_{1b}}(t) * f_{T_i}(t), \tag{37.28}$$

which can be obtained as:

$$f_{T_{2i}}(t) = \sum_{n=1}^{N}\sum_{k=1}^{n} \frac{1}{n}\binom{2n-2}{n-1}\binom{n}{k}\left(\frac{\lambda+\mu}{\lambda}\right)\left(\frac{\rho(1-p)}{(1+\rho)^2}\right)^n \left(\frac{p}{1-p}\right)^k$$
$$\cdot\left(\frac{e^{-\lambda t}\mu^k(\mu-\lambda)^{-k}\lambda\left(\Gamma(k)-\Gamma(k,(\mu-\lambda)t)\right)}{\Gamma(k)}\right). \qquad (37.29)$$

We are now interested in obtaining the expected number of cycles after the first busy period of each battery until its charge goes to zero. Let τ_1 be a r.v. denoting the number of cycles until the charge of the first battery goes to zero and let τ_2 be a r.v. denoting the number of cycles until the charge of the second battery goes to zero. We first find $E(\tau_1)$ and $E[\tau_2]$. However now, these can be obtained as in the case of a single battery.

Let $x = \min\{E[\tau_i], i = 1,2\}$. Then the average number of packets served by both the batteries until one of the batteries expires is given by:

$$C_1 = (1+x)\frac{p}{(1-\rho)} + (1+x)\frac{1-p}{(1-\rho)} = (1+x)\frac{1}{(1-\rho)}. \qquad (37.30)$$

Now, we need to determine the remaining charge available in the battery which is alive (not expired yet) at the expiry time of the other battery. Let $\min\{E[\tau_i], i = 1, 2\}$ correspond to $i = j$ and $\bar{j}$ denote the complement of j (i.e., when $j = 1$, $\bar{j} = 2$, and vice versa). Then we have:

$$x = \int_0^{N_1} E_{N_1-Z_{\bar{j}}}[\tau]dF_{T_{\bar{j}b}}(z_{\bar{j}}), \qquad (37.31)$$

where $N - N_1$ is the charge left in the live battery at the expiry time of the other battery. Note that, after one battery expires, the live battery serves packets like as a single battery system without vacations. Hence, the expected number of cycles elapsed until $N - N_1$ charge units available in the live battery are exhausted can be obtained as:

$$E[\tau] = \int_0^{N-N_1} E_{N-N_1-Z}[\tau]F_{T_b}(z), \qquad (37.32)$$

where $f_{Tb}(z)$ is the PDF of the busy period in a single battery scheme. Therefore, the expected number of packets served, C_2, by the live battery can be evaluated as described in Section 37.3. The expected number of packets served by the system until both the batteries expire is then given by $C = C_1 + C_2$.

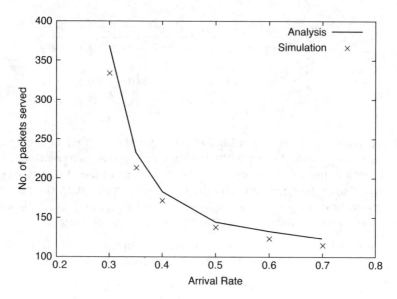

Figure 37.5. Number of Packets Served versus Arrival Rate in the System without Vacation in a Single Battery Scheme with $\mu = 1$

37.4 Performance Results and Discussion

In this section, we present the numerical results for the analysis we carried out in the previous section. We consider the following system parameters:

- Nominal capacity N = 100 charge units
- Number of thresholds P = 3
- Threshold values $\theta_1 = 75$, $\theta_2 = 50$, $\theta_3 = 25$
- Recharge slopes $r_1 = 0.4$, $r_2 = 0.3$, $r_3 = 0.2$, $r_4 = 0.1$
- Service time parameter $\mu = 1$.

First, from (37.11), we compute the mean number of packets served in a simple M/M/1 queue with battery recharge due to inherent idle periods. Figure 37.5 shows the number of packets served for different arrival rates λ, obtained through analysis and simulation. In the numerical evaluation of (37.4), the integral is discretized in steps of 0.1 and converted into a set of linear equations. Further, instead of evaluating all the entries of the coefficient matrix, we compute the entries for one row and use the same for the other rows for a given threshold region. This significantly reduces the computation time and maintains the accuracy of the results as observed from the close match between the analytical and simulation results. It can be observed that the recharge due to inherent idle periods can increase the number of customers served to about three times the battery's nominal capacity of 100 charge units, particularly at low arrival rates (e.g., 350 packets served at $\lambda = 0.3$).

Battery Power Management in Portable Devices

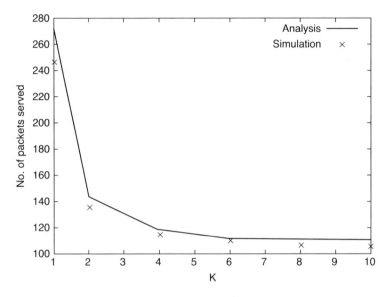

Figure 37.6. Number of Packets Served versus k in the Nonexhaustive Service System with Vacation at High Arrival Rates

In [40], it has been observed that the system with exhaustive service and vacations in Section 37.3.1.1 does not provide much gain. The nonexhaustive system with vacation in Section 37.3.1.2, however, can give significant battery life gain, but at the expense of increased delay performance due to the intentional vacations.

Figure 37.6 shows the number of packets served versus K in the nonexhaustive system with vacations at high arrival rates, obtained through the analysis in Section 37.3.1.2 as well as simulations, for $\delta = 0.5$. Here it is assumed that there are always packets in the queue to be served (approximating high arrival rates). It is observed that this approach gives significant gains in the number of packets served at high arrival rates, particularly for small values of K (for example, K = 1, 2). We have also observed that the approximate analysis is reasonably accurate even for moderate values of λ (in the range > 0.4) for those values of K (= 1, 2) that provide significant gains in terms of the number of packets served. Thus, the nonexhaustive service system with vacations can be applied on the traffic of delay-tolerant applications (e-mail, file transfers, etc.) to increase battery life in wireless mobile devices. In [40], a delay constrained algorithm that puts a limit on the maximum delay has also been presented.

Next, we consider the performance of the dual battery scheme. The following system parameters are considered:

- N = 100 charge units for each battery
- Number of thresholds P = 3

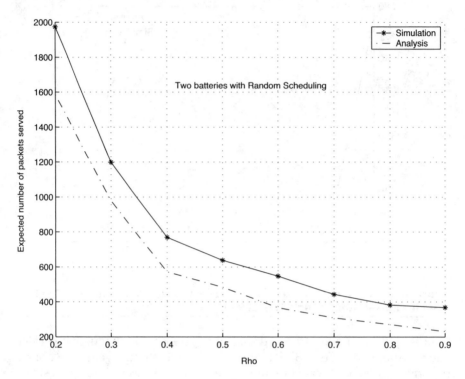

Figure 37.7. Expected Number of Packets Served versus ρ in the Dual Battery Scheme (p = 0.5; μ = 1)

- Threshold values $\theta_1 = 75$, $\theta_2 = 50$, $\theta_3 = 25$
- Recharge slope values $r_1 = 0.4$, $r_2 = 0.3$, $r_3 = 0.2$, $r_4 = 0.1$
- Service time parameter $\mu = 1$ and $\delta = 0.5$.

Figure 37.7 shows the expected number of packets served, C, as a function ρ when the battery scheduling probability p = 0.5, obtained through the analysis. The results obtained through simulations (simulating the same analytical system model but without the approximations made in the analysis) are also plotted. It is observed that the dual battery scheme offers an increased number of packets served (more than 2N, which will be the number of packets served if the batteries continuously serve packets without any idle or vacation periods), particularly at low arrival rates where the recharge due to idle periods can be more. It is further observed that the analytical results reasonably match with the simulation results. The difference between the performances predicted by analysis and simulations is mainly due to the approximation we made in deriving the busy period distribution of the batteries. The possible pulsed discharge during a busy period was approximated as a continuous discharge. Because of this, the

Battery Power Management in Portable Devices

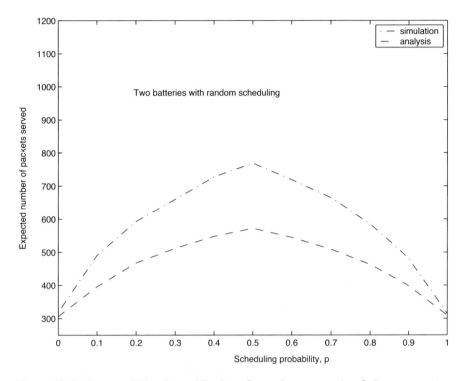

Figure 37.8. Expected Number of Packets Served versus p (ρ = 0.4)

charge recovery is not fully accounted for and this is the reason why the analysis always underestimates the performance compared to the simulations, as observed in Figure 37.7. Also, the delay performance of the dual battery scheme will be the same as that of the single battery scheme without vacation (i.e., same as the delay in a simple M/M/1 queue [40]).

Figure 37.8 shows the expected number of packets as a function of the battery scheduling probability, p, for a given ρ value of 0.4. It is noted that (as expected in random scheduling) the value of p that maximizes the expected number of packets is 0.5. The match between analysis and simulation is very close for p = 0 and 1. This is because when p = 0 or 1, the system essentially behaves like a single battery scheme (i.e., one battery will continue to serve until it expires and only then will the second battery start serving). Because of this, only one battery serves during a given busy period. Hence, the approximation made to obtain the busy period distribution is not required when p = 0 or 1. Figure 37.8 also illustrates that the dual battery scheme (0 < p < 1) serves increased number of packets compared to an equivalent single battery scheme (p = 0 or 1).

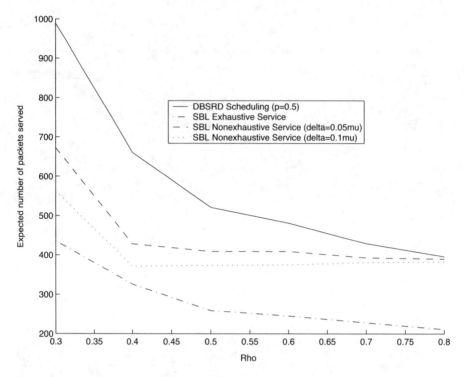

Figure 37.9. Expected number of packets versus ρ for the Dual Battery Scheme (DBS-RD scheme with ρ = 0.5; SBL-ES scheme with μ^{-1} = 0.02 sec. and δ = 0.05 μ, and 0.1 μ)

37.4.1 Lithium Ion Battery Simulation Results

We also studied the performance of the proposed dual battery scheme using the battery simulation program developed by the Chemical Engineering Department, University of California — Berkeley [36]. We evaluated the performance for the dual battery scheme with random scheduling (DBS-RD scheme) as follows. First, we implement the packet arrival process, queueing, and battery scheduling algorithm in a separate program. We run this program to obtain traces of the busy and idle periods of the two batteries. These busy and idle period traces are then given as inputs to the Berkeley lithium ion battery simulation program, which incorporates the actual (nonlinear) discharge/recharge characteristics of the battery. We run this battery simulation program until both batteries fall below their cutoff voltages. Statistics are collected during these simulation runs to obtain the expected number of packets served and the mean packet delay.

Following a similar procedure, for comparison purposes, we evaluated the performance of two single-battery-like (SBL) schemes, in which the first battery will continue to serve packets until it drops below the cutoff voltage and only then will the second battery start serving. We consider a

Battery Power Management in Portable Devices

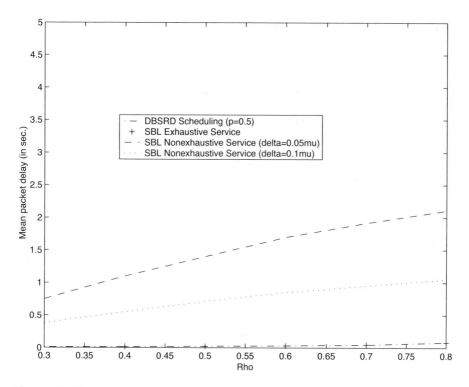

Figure 37.10. Mean Delays versus ρ for the Dual Battery Scheme (DBS-RD scheme with p = 0.5, SBL-ES scheme with μ⁻¹ = 0.02 sec., and δ = 0.05 μ and 0.1 μ)

SBL scheme with exhaustive service (i.e., no intentional vacation) and another SBL scheme with nonexhaustive service (i.e., allow intentional vacations). In nonexhaustive service scheme, the battery takes a exponentially distributed vacation time (with parameter δ) after continuously serving L packets. We compare the performance of DBS-RD scheme, SBL exhaustive service (SBL-ES) scheme, and SBL Nonexhaustive Service (SBL-NS) scheme. Because the total theoretical capacity is taken to be same for the SBL and the DBS-RD schemes, the performance difference between these schemes arise mainly due to the way in which the batteries are discharged in each scheme.

Figure 37.9 and Figure 37.10 show the performance comparison between the DBS-RD, SBL-ES, and SBL-NS schemes. The following observations can be made from Figure 37.9 and Figure 37.10. The DBS-RD scheme performs better than SBL-ES scheme (no intentional vacation) in terms of expected number of packets served. Their delay performances, however, are almost the same. The number of packets served is increased in SBL-NS scheme (compared to SBL-ES scheme) by allowing intentional vacations once every L packets served. Even with intentional vacations with L = 4, the SBL-NS scheme performs poorer than the DBS-RD scheme in terms of number of packets served. Further, because of intentional vacations, the delay performance of the SBL-NS scheme is much poorer than the DBS-RD scheme. Thus, the proposed dual battery architecture can provide both increased number of packets served as well as lesser mean delay compared to single battery schemes with intentional vacations.

37.5 An Optimal Scheduling Problem

In the above sections, a scheduling (of battery) policy was chosen *a priori*. Now, we formulate it as an optimal control problem and find an optimal policy. The notation is changed to facilitate the problem formulation in this section. We consider a slotted system, each slot being of unit length. The system has M batteries. At the beginning of each slot a decision is made regarding the battery to use in that slot and the number of bits to be transmitted. The decision is based on the queue length, capacity, and the charge of each battery at the beginning of that slot. We use the following notation:

q_n = queue length at time n
$c_n(i)$ = the leftover capacity in battery i at time n
c_n = $(c_n(1),\cdots,c_n(M))^T$
$c(i)$ = total capacity of battery i
$n_n(i)$ = the charge in battery i at time n
n_n = $(n_n(1),\cdots,n_n(M))^T$
a_n = number of new packets arriving in slot i
(q_n, c_n, n_n) = state of the system at the beginning of slot n

In slot n, we take action (B_n, s_n) where:

B_n = the battery chosen to use in slot n
s_n = Number of packets/bits served in the nth slot, $0 \leq s_n \leq q_n$

We will also use the following notation:

$\bar{n}_n$ = amount of charge consumed in the nth slot
e_i = $(0,0,\cdots,0,1,0,\cdots,0)^T$
                     ↑
               ith position
$p_n(i)$ = amount of recharge in slot n at battery i if that battery is not used in slot n
p_n = $(p_n(i),\cdots,p_n(n))^T$

The distribution of r.v. $\bar{n}_n$ depends on s_n and other environmental conditions. Also, the distribution of $p_n(i)$ depends on $(n_n(i), c_n(i))$. The system evolves as:

$$q_{n+1} = q_n - s_n + a_n$$

$$n_{n+1}(i) = n_n(i) + (1 - e_{Bn})^T p_n - e^T_{Bn} \bar{n}_n$$

Under our assumptions (q_n, c_n, n_n) is a Markov chain. If $n_n(i) = 0$ then the ith battery is dead and cannot be recharged again. Define τ to be the (stopping) time when all batteries are dead.

The cost function to be minimized should try to keep the mean delays small and at the same time maximize the number of packets/bits transmitted in the life time of the battery. We consider:

$$\max E\left[\sum_{k=1}^{\tau} s_k - \beta \sum_{k=1}^{\tau} q_k\right]$$

where β is an appropriate positive constant that determines the weight to be given to the queueing delay. This is a problem in Markov decision theory.

When $\beta = 0$, this problem has been considered in [45] (under somewhat more specific assumptions). It is pointed out that the usual algorithms of value or policy iteration are computationally complex. Exploiting the special structure of the problem a simple algorithm is obtained that is linear in the number of states. They have also explored the simple policy of using the battery with the maximum charge at the beginning of the slot. It has been found to be close to optimal.

Notes

1. The term battery level is used to denote the amount of charge present in the battery.
2. Actually this is an approximation (in fact, an upper bound) because the last cycle may not be complete.

References

1. J. Lorch and A.J. Smith, Software strategies for portable computer energy management, *IEEE Personal Commun.*, vol. 5, no. 3, pp. 60–73, June 1998.
2. Energy management in personal communications and mobile computing, *IEEE Personal Commun.*, [special issue], June 1998.
3. J. Pouwelse, K. Langendoen, and H. Sips, Dynamic voltage scaling on a low-power microprocessor, *Proc. MobiCom 2001*, July 2001.
4. A. Sinha and A. Chandrakasan, Energy efficient real-time scheduling, *Proc. Int'l. Conf. on Computer Aided Design (ICCAD)*, November 2001.
5. F. Douglis, P. Krishnan, and B. Marsh, Thwarting the power-hungry-disk, *Proc. 1994 Winter USENIX Conf.*, pp. 293–306, January 1994.

6. E. Shih, S-H. Cho, N. Ickes, R. Min, A. Sinha, A. Wang, and A. Chandrakasan, Physical layer driven algorithm and protocol design for energy efficient wireless sensor networks, *Proc. MobiCom,* July 2001.
7. N. Bambos and J.M. Rulnick, Mobile power management for wireless communication networks, *Wireless Networks,* Vol 3., pp. 3–14, 1997.
8. T.S. Rappaport, *Wireless Communications: Principles & Practice,* Upper Saddle River, Prentice Hall, 1996.
9. Bluetooth SIG, Bluetooth Baseband Specifications Version 1.0B. http://www.bluetooth.com.
10. A. Rangarajan, V. Sharma, and S.K. Singh, Information theoretic and communication-theoretic power allocation for fading channels, *IEEE Symposium on Information Theory (ISIT),* 2003.
11. S. Singh and C.S. Raghavendra, PAMAS: Power aware multi-access protocol with signaling for ad-hoc networks, *ACM Computer Communications Review,* pp. 5–26, July 1998.
12. K.M. Sivalingam, J.-C. Chen, P. Agrawal, and M. Srivastava, Design and analysis of low-power access protocols for wireless and mobile ATM networks, *ACM/Baltzer Wireless Networks,* vol. 6, no. 1, pp. 73–87, 2000.
13. A. Chockalingam and M. Zorzi, Energy efficiency of media access protocols for mobile data networks, *IEEE Trans. on Commun.,* vol. 46, no. 11, pp. 1418–1421, November 1998.
14. P. Lettieri, C. Schurgers, and M.B. Srivastava, Adaptive link layer strategies for energy efficient wireless networking, *Wireless Networks,* vol. 5, no. 5, pp. 339–355, 1999.
15. M. Goyal, A. Kumar, and V. Sharma, Optimal power allocation for multiple access fading channels with minimum rate guarantees, *Proc. IEEE ICPWC 2002,* New Delhi, 2002.
16. M. Zorzi and R.R. Rao, Error control and energy consumption in communications for nomadic computing, *IEEE Trans. on Computers,* vol. 46, no. 3, pp. 279–289, 1997.
17. P.M. Soni and A. Chockalingam, Analysis of link layer backoff schemes on point-to-point Markov fading channels, *IEEE Trans. Communication,* vol. 51, no. 1, pp. 29–32, January 2003.
18. W. Heinzelman, A. Chandrakasan, and H. Balakrishnan, Energy efficient routing protocols for wireless microsensor networks, *Proc. 33rd Hawaii Int'l. Conf. on System Sciences (HICSS'00),* January 2000.
19. M. Woo, S. Singh, and C.S. Raghvendra, Power aware routing in mobile ad hoc networks, *Proc. MobiCom'98,* pp. 181–190, 1998.
20. E.M. Royer and C.-K. Toh, A review of current routing protocols for ad-hoc mobile wireless networks, *IEEE Personal Commun.,* pp. 46–55, April 1999.
21. J-H. Chang and L. Tassiulas, Energy conserving routing in wireless ad-hoc networks, *Proc. InfoCom'2000,* 2000.
22. B.J. Prabhu and A. Chockalingam, A routing protocol and energy efficiency techniques in Bluetooth scatternets, *Proc. IEEE ICC'2002,* April–May 2002.
23. J.E. Wieselthier, G.D. Nguyen, and A. Ephremides, On the construction of energy-efficient broadcast and multicast trees in wireless networks, *Proc. InfoCom'2000,* 2000.
24. B. Chen, K. Jamieson, H. Balakrishnan, and R. Morris, Span: An energy-efficient coordination algorithm for topology maintenance in ad-hoc wireless networks, *Proc. MobiCom'2001,* July 2001.
25. R. Wattenhofer, L. Li, V. Bahl, and Y.M. Wang, Distributed topology control for power efficient operation in multihop wireless ad-hoc networks, *Proc. InfoCom'2001,* April 2001.
26. M. Zorzi and R. Rao, Energy efficiency of TCP in a local wireless environment, *ACM/Baltzer Mobile Networks and Applications.* vol. 6, no. 3, pp. 265–276, June 2001.
27. V. Tsaoussidis, H. Badr, X. Ge, and K. Pentikousis, Energy/throughput tradeoffs of TCP error control strategies, *Proc. IEEE Symp. on Computers and Commun. (ISCC 2000),* July 2000.

28. A. Chockalingam and M. Zorzi, Wireless TCP performance with link layer FEC/ARQ, *Proc. IEEE ICC'99*, June 1999.
29. V. Tsaoussidis and H. Badr, TCP-Probing: Towards an error control scheme with energy and throughput performance gains, *The 8th IEEE Conf. on Network Protocols (ICNP'2000)*, November 2000.
30. P. Agrawal, J.-C. Chen, S. Kishore, P. Ramanathan, and K.M. Slivalingam, Battery power sensitive video precessing in wireless networks, *Proc. PIMRC'98*, September 1998.
31. T.F. Fuller, M. Doyle, and J. Newman, Relaxation phenomenon in lithium-ion insertion cells, *J Electrochem. Soc.*, vol. 141, no. 4, pp. 982–990, April 1994.
32. H.D. Linden, *Handbook of Batteries*, 2nd ed., New York: McGraw-Hill, 1995.
33. R.M. Lafollette and D. Bennion, Design fundamentals of high power density, pulsed discharge, lead-acid batteries. II: Modeling, *J Electrochem. Soc.*, vol. 137, no. 12, pp. 3701–3707, December 1990.
34. R.M. Lafollette, Design and performance of high specific power, pulsed discharge, bipolar lead acid batteries, *10th Annual Battery Conf. on Applications and Advances*, pp. 43–47, January 1995.
35. B. Nelson, R. Rinehart, and S. Varley, Ultrafast pulsed discharge and recharge capabilities of thin-metal film battery technology, *11th IEEE Int'l. Pulsed Power Conf.*, pp. 636–641, June 1997.
36. J.S. Newman, FORTRAN programs for simulation of electrochemical systems. http://www.cchem.berkeley.edu/jsngrp/.
37. C.F. Chiasserini and R.R. Rao, Pulsed battery discharge in communication devices, *Proc. MobiCom'99*, August 1999.
38. C.F. Chiasserini and R.R. Rao, A traffic control scheme to optimize the battery pulsed discharge, *Proc. MilCom'99*, November 1999.
39. C.F. Chiasserini and R.R. Rao, Energy efficient battery management, *Proc. InfoCom 2000*, March 2000.
40. B.J. Prabhu, A. Chockalingam, and V. Sharma, Performance analysis of battery power management schemes in wireless mobile devices, *Proc. IEEE WCNC 2002*, March 2002.
41. P.T. Dahake, Performance analysis of multibattery architecture for wireless mobile devices, ME Thesis, Department of ECE, Indian Institute of Science, January 2003.
42. D. Bertsekas and R. Gallager, *Data Networks*, Upper Saddle River, NJ: Prentice Hall, 1987.
43. L. Kleinrock, *Queueing Systems, vol. 1: Theory*, New York: John Wiley & Sons, 1975.
44. G.E. Roberts and H. Kaufman, *Table of Laplace Transforms*, Philadelphia: W.B. Saunders Company, 1966.
45. S. Sarkar and M. Adamnov, A framework for optimal battery management for wireless nodes, *IEEE J Select. Areas Commun.*, vol. 21, pp. 179–188, 2003.

Section IX
Security and Privacy Aspects

Chapter 38
Challenges in Wireless Security: A Case Study of 802.11

Nikita Borisov

38.1 Introduction

The IEEE® 802.11 standard [16] was designed to allow wireless transmissions of data in local area networks. The standard specified the data link and physical layers to be used in such networks. When it was first published, wireless networks were rare. However, recent years have seen drastic increases in popularity of wireless technologies in general and in the deployment of 802.11-based networks in particular. Today, 802.11, with various updates to the physical-layer specification, is the most popular communication mechanism for wireless networks.

Throughout its lifetime, 802.11 has been at the forefront of the proliferation of wireless communication. This position makes its history instructive regarding wireless communication issues, including wireless security. The 802.11 committee has always been aware of the unique threats to wireless communications and has introduced measures to defend from them. However, both the design of these measures and the way in which they were deployed have been fraught with security flaws and pitfalls. This chapter discusses the history of 802.11 security, detailing various mechanisms used, the problems therein, and some proposed solutions. It is intended to provide insight to future researchers and practitioners working in the field of wireless communications.

38.1.1 Overview of 802.11

The 802.11 standard concerns itself with communication between stations (STAs) and access points (APs, sometimes also called base stations). An STA is any device wishing to communicate using 802.11 and an AP is used to manage connecting to the wireless network and forwarding data. APs

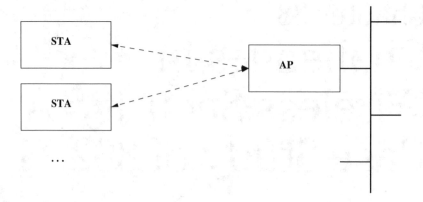

Figure 38.1. 802.11 Communications Model

are usually connected to other, non-802.11 networks; the most common scenario is an 802.3 connection to an Internet Protocol (IP) router (which provides further connections to an organizational intranet and the Internet). Figure 38.1 shows a common scenario.

The AP and STAs exchange frames of data, governed by the Media Access Control (MAC) layer and physical (PHY) layer specifications. 802.11 defines two types of frames — management and data. The management frames are used to establish an association between an STA and an AP. After an STA is associated, it can send and receive data frames; the AP will forward the data onto other STAs or onto the external network. 802.11 also supports ad hoc mode, where STAs communicate directly with each other without association. Most 802.11 networks are managed by APs; some of the following discussion regarding, for example, association, applies only to managed networks.

38.1.2 History

The first 802.11 standard was published in 1997 [15]. At the time, the designers realized that protecting the users' privacy was going to be a big concern, seeing as data was being broadcast over the radio frequency (RF) spectrum, available to anyone within range. Unlike static installations, one could no longer trust the data link layer to be free of attacks. The response from the 802.11 committee was to design the Wired Equivalent Privacy (WEP) protocol. The protocol aimed to make link layer attacks infeasible through the use of encryption.

The design of WEP was influenced, among other factors, by export regulations, which restricted the key size used by WEP to only 40 bits. The export restrictions were eventually relaxed; at the same time, the practical

Challenges in Wireless Security: A Case Study of 802.11

security of 40-bit cryptography was becoming more and more questionable [9]. The demand for better security caused the 802.11 committee to form a task group to update the standard for greater key lengths (a popular extension implemented in industry).

In the course of the proceedings, it was noted that there were serious flaws in the design of WEP [23] and that these problems would not be corrected simply by increasing the key size. Around the same time, the increasing deployment of 802.11 prompted the research community to investigate its security. Researchers at University of California — Berkeley and University of Maryland — College Park independently discovered many of the problems in WEP and developed several new attacks [3, 5]. These new findings prompted a more serious redesign of WEP.

Another popular extension requiring standardization has to do with authentication and key management. The original standard left these areas largely unaddressed, prompting various solutions from 802.11 vendors. The technologies eventually converged on using EAP [4] and RADIUS [19]; their use within the 802 protocol suite was standardized in 802.1X [17]. An important goal of the task group was to integrate 802.1X into the new security protocols.

During the redesign process, a new attack on the underlying encryption algorithm used in WEP was developed [11] and implemented [21]. This was likely the most serious of the attacks because it could be so easily automated; it uncovered another flaw that was to be addressed in the redesign.

Currently, the task group has proposed two standards updates:

1. Temporal Key Integrity Protocol (TKIP) [7] is a revision of the WEP protocol intended to be used in the short term. It addresses the most serious issues of WEP, reusing some of the building blocks of the original design, so that it can run on current hardware with only a firmware update.
2. Counter Mode with Cipher Block Chaining Message Authentication Code Protocol (CCMP) [7] is a longer term solution. It is a more substantial redesign, using the latest encryption standards and novel authenticated modes.

Both CCMP and TKIP rely on 802.1X for key management.

The new standards offer hope of much improved security. However, as of this writing, TKIP and CCMP have not been fully ratified. Nearly all 802.11 networks are using the original, flawed version of WEP and it is fair to expect that many of the networks will remain insecure for years into the future.

38.2 Wireless Security Threats

Wireless networking brings with it new security threats. Many organizations choose to simplify security management by assuming that all threats originate outside the internal network. Their security policy is therefore implemented by monitoring and controlling network traffic from and to the Internet by means of firewalls and intrusion detection systems. Although some security mechanisms are implemented internally (authentication for file servers, e-mail, etc.), they typically receive less administrative attention and are less robust than the external protections. The main mechanism for protecting internal network links is physical security.

Physical security is insufficient to protect wireless links, as anyone within radio range of the building can receive and broadcast transmissions. The typical range of 802.11 transmissions can be several hundred feet indoors and over 1000 feet outdoors. This makes it easy to carry out an attack from a nearby parking lot. Furthermore, using a directional antenna with direct line-of-sight, it is possible to extend the range to several miles (figures exceeding ten miles have been reported).

The possibility of such attacks is far from theoretical in the case of 802.11. In the recent years, the practice of war-driving through urban areas looking for vulnerable APs has gained much popularity. In part this is due to 802.11's success: wide-scale deployment has made 802.11 equipment plentiful and inexpensive. A cheap wireless card attached to a laptop is sufficient to initiate an attack. Good amplifying antennae are within reach of more serious attackers; others employ common-knowledge tricks to extend wireless range using household items.

The general principle, which applies to all types of wireless networks, is that the same devices that are used for legitimate access can be also used for attack. This is especially true for devices with reprogrammable firmware, which are common today. Even if such devices were to enforce restrictions to prevent people from using them maliciously, a firmware update is sufficient to remove them. Reverse engineering the firmware may be a costly task, but once it is finished, the information can quickly spread and the marginal cost of modifying each device is small.

When the physical link cannot be trusted, other steps must be taken to protect the security of the data. First of all, some scheme to preserve confidentiality must be used, as it is trivial to listen in on communications. It is perhaps less clear that integrity and authenticity protection are also necessary. Violating integrity requires active attacks, where the attacker transmits instead of just receiving. These attacks are harder to mount than passive ones, but they are certainly within reach of many and can potentially be more damaging than passive ones. Also, many attacks use the lack of integrity protection to attack confidentiality; it is quickly becoming common wisdom in the cryptography community that effective confidentiality protection without corresponding integrity safeguards is infeasible.

An important aspect of authentication is to prove to the user of the network the true identity of the network itself. In wired contexts, a user can be reasonably certain that the wires being used connect to the correct network; however, wireless clients can be mislead to use different, fake networks. An attacker able to lure a user onto his own network has a great opportunity to control the user experience and set up further security compromises.

38.3 Encryption

Confidentiality is typically achieved by means of encryption. An important task for any protocol designer is to choose an encryption algorithm. National Institute of Standards and Technology (NIST), with help from the National Security Agency (NSA), have attempted to simplify this decision by creating the Data Encryption Standard (DES) [12], and more recently, the Advanced Encryption Standard (AES) [13]. However, other nonstandard algorithms are frequently used for various reasons, ranging from the (usually erroneous) belief that they will provide greater security to specialized protocol or hardware requirements.

The 802.11 committee decided to use the RC4 algorithm [20]. There are several reasons why they might have done so. RC4 is a simple algorithm and is therefore easy to implement in both hardware and software. A software implementation of RC4 is several times faster than DES [6]. It is a proprietary algorithm, but details of its operation have been publicly known for some time and it has been subjected to a fair amount of analysis in the open literature. RC4 also simplifies compliance with export regulations: at the time that 802.11 was first drafted, a common restriction was to disallow export of devices capable of using encryption keys longer than 40 bits. RC4 uses a variable key length and a 40-bit version of RC4 was therefore used in 802.11. DES uses a 56-bit key and as such was not permitted for export (though there are modifications to DES to reduce the key size).

Another difference between DES and RC4 is that RC4 is a stream cipher. What this means is instead of operating directly on data to be encrypted, RC4 generates a continuous pseudorandom keystream. This keystream is then combined with the data to be encrypted (plaintext) using the exclusive-OR function to produce the ciphertext (encrypted data). On the other side, the receiver generates the identical pseudorandom keystream and uses it to recover the plaintext (original data). This operation is shown in Figure 38.2.

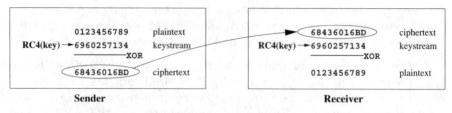

Figure 38.2. Stream Cipher Operation

38.3.1 Keystream Reuse

To use stream ciphers, it is important to make sure that the sender and receiver are synchronized with respect to the portion of the keystream being used. This makes stream ciphers a popular choice for session-based applications, where the session guarantees can be used to ensure proper synchronization. It is also important to never reuse the same portion of the keystream; otherwise, if one were to exclusive-OR the two ciphertexts using that keystream, the keystream would cancel out, leaving the exclusive-OR of the two plaintexts. The individual plaintexts can then be recovered by statistical analysis if there is some knowledge about the nature of the underlying data. To avoid this problem, session-based applications will usually generate a new key to be used for each session, resulting in new, previously unused keystream.

The 802.11 protocol does not make session guarantees that would enable sender and receiver to synchronize their position in the keystream. Instead, each frame uses the beginning portion of the keystream. To avoid keystream reuse, WEP introduced initialization vectors (IVs): a 24-bit number, the IV, is concatenated with the secret key to obtain the encryption key used as input to RC4 to generate the keystream:

$$keystream = RC4(shared\ key\ ||\ IV)$$

In other words, instead of each frame being encrypted with a single shared key, it is encrypted with one of 2^{24} keys. The value of the initialization vector is transmitted in the clear so that the recipient may be able to generate the same keystream.

The standard recommends, though does not require, that IVs be changed with every frame. In practice, nearly all implementations increment the IV with each frame. Because secret keys are typically long-lived, the space of 16 million IVs is quickly exhausted and IV collisions, where the same IV is used to encrypt multiple frames, will occur. An endpoint that sends a lot of traffic can go through 2^{24} frames in less than a day; a busy installation where every endpoint uses the same key (a common scenario) will exhaust the space quicker still. An IV collision means the same keystream is used for two different frames, providing an easy path to plaintext recovery. Moreover, once the space is exhausted, every new frame will manifest an IV collision. Hence, after intercepting several hours' to several days' worth of traffic, an attacker will be able to decrypt a large fraction of all traffic.

38.3.2 RC4 Weaknesses

If the last vulnerability arose from a misuse of RC4, the next one is due to a weakness in the RC4 algorithm itself. RC4 is a fairly new cipher; the details of its implementation were being kept a trade secret by RSA Security until they were leaked in 1994. Early on, researchers noticed that there were detectable correlations exhibited in the first few bytes of the keystream. For some time now, the standard recommendation from RSA Security has been to discard the first 256 bytes of the keystream; however, the 802.11 standard does not implement this measure. (Note that in 802.11 such a precaution would have a significant performance impact, because the keystream is reinitialized at every frame.)

A more recent analysis of RC4 revealed a more detailed and serious vulnerability [11]. It identified a class of weak keys, which would cause the first few bytes of the keystream to have significant correlations with the key material. The weak keys can be used to mount a related-key attack applicable to the way that RC4 is used in WEP. Namely, the attack exploits the fact that part of the key is secret and static, while another part (the IV) is known and changes with each frame. The attack uses correlations to recover the secret component of the key.

To demonstrate the feasibility of this attack, an implementation was built by researchers at AT&T labs [21]. They showed that the attack can be successful after 5 to 6 million packets and its running time can be significantly reduced if assumptions about the secret key (such as that it is an ASCII [American Standard Code for Information Interchange] text string) are made. This attack is quite devastating because it can be fully automated and it successfully recovers the secret key, fully compromising the security of WEP. All that is necessary is knowledge of the first few bytes of the plaintext, which almost always consist of a well-known IP encapsulation header. Tools implementing this attack are readily available [22]; this is by far the most serious problem facing 802.11 networks today.

38.3.3 New Standards

The short-term solution adopted by the 802.11 committee (TKIP) addresses these problems by increasing the size of the IV (renamed packet sequence number) to 48-bits and using a mixing function to derive the per-packet key from the secret key, sequence number, and the transmitter MAC address. The mixing function is a small cipher, used to foil related key attacks by removing easily observable correlations between the sequence numbers and the per-packet key. The larger size of sequence numbers prevents rollovers that would result in collisions. Furthermore, TKIP mandates that a fresh secret key be used for every session (provided by 802.1X), ensuring that no collisions occur between sessions. Incorporating the MAC address into the per-packet key guards against collisions between different endpoints using the same secret key.

The long-term solution (CCMP) uses AES for encryption [13]. AES is a block cipher, which is more appropriate for encrypting packet-based data. Furthermore, the standardization efforts resulted in extensive analysis of the cipher. Although AES is quite recent, it is already one of the most well-studied encryption algorithms, second only to DES. This is not a guarantee against the potential of flaws discovered in the future, but it is the strongest argument for using a cipher available to us today. In retrospect, had the 802.11 designers chosen a standard algorithm in their initial design, the problems described in this section would have been avoided.

38.4 Integrity Protection

As evidenced in the name "Wired Equivalent Privacy," integrity protection was not a high priority in the 802.11 security design. Nevertheless, the WEP protocol format includes an integrity check value (ICV). The ICV is computed by taking the CRC-32 checksum of the plaintext data and then encrypting it using the RC4 keystream. Although the cyclic redundancy check (CRC) function is good for detecting random errors, it turns out not to be secure against a malicious attacker, especially when it is used in conjunction with a stream cipher.

The main problem is that CRC-32 is a public function; in other words, anyone can compute the CRC-32 of a sequence of bytes. This allows an easy attack on the integrity of a frame where the attacker knows the original plaintext. The attacker may substitute a new value for the plaintext, compute its CRC, and then use the malleability of the stream cipher to modify the original packet:

$$\textit{frame} = (P\,||\,CRC(P)) \oplus \textit{keystream}$$

$$\textit{frame}' = (P'\,||\,CRC(P')) \oplus (P\,||\,CRC(P)) \oplus \textit{frame}$$

$$= (P'\,||\,CRC(P')) \oplus (P\,||\,CRC(P)) \oplus (P\,||\,CRC(P)) \oplus \textit{keystream}$$

$$= (P'\,||\,CRC(P')) \oplus \textit{keystream}$$

Where $\oplus$ is the exclusive-OR operation, P and P' are the original and the modified plaintexts, and *frame* and *frame'* are the original and modified frames, respectively. Upon decrypting *frame'*, the recipient will see a correct encoding of P.

Challenges in Wireless Security: A Case Study of 802.11

In fact, it is not necessary for the attacker to know the entire contents of the frame to perform an integrity attack. CRC is a linear function with respect to the exclusive-OR operation. This mathematical property means that it is possible to compute the change to the CRC of a message resulting from changing a portion of the message, without knowledge of the rest of the message or the original CRC. An attacker can therefore modify the known part of the plaintext, compute the CRC change, and modify the affected parts of the frame, leaving the rest of the frame undisturbed.

38.4.1 Integrity-Based Attacks

The lack of effective integrity protection enables a host of new attacks on the WEP protocol. One such attack involves message redirection. Because messages sent over an 802.11 network are typically destined for an IP router, an attacker could potentially change the destination IP address in the message, causing it to be directed to another location. Some tricks are necessary to correctly update the (encrypted) IP header checksum, as it is not linear with respect to the exclusive-OR function. If the wireless network is eventually connected to the Internet, this attack may be used to learn the contents of a message by redirecting it to an IP address controlled by the attacker.

Another possible attack is a reaction attack. This attack can also be used to decrypt a message; it relies on the behavior of the Transmission Control Protocol (TCP) checksum. The TCP checksum is not linear with respect to exclusive-OR; however, it does have the following mathematical property: if we take a message and flip bit i and bit i + 16, the TCP checksum will be undisturbed if and only if the values of the two bits are different. The insecurity of the CRC function allows an attacker to flip the two bits and adjust the ICV appropriately. Then the attacker watches to see whether a TCP acknowledgment is sent (acknowledgments are easy to identify by the size of the frame). If it is, this means that the TCP checksum was valid and hence the bits at the two positions were different; otherwise, the bits must have been the same. By repeated applications of this reaction attack, the entire contents of the message can be deduced starting with 16 known bit values.

Another reaction attack uses the ICV itself to recover plaintext [2]. It requires knowledge of a length of keystream sufficient to encrypt a simple message that would result in an easily detectable response (such as a Dynamic Host Configuration Protocol (DHCP) discover message or an Internet Control Message Protocol (ICMP) echo request). The attack proceeds to inductively learn more of the keystream by sending longer and longer versions of the message, incrementing the length by one byte each time. It computes the CRC of the slightly larger message and encrypts it using the known portion of the keystream and a guess for the extra unknown byte. As before, it sends a message and watches for a response. If one is received, the guess for the keystream byte was correct; otherwise, a new value is tried. In this way, n bytes of the keystream can be recovered after about 128 n tries on average.

As these attacks demonstrate, even if protecting message integrity is not a concern in itself, failing to ensure integrity protection opens the door to many attacks on the privacy of the message. There is a growing sense in the cryptography research community that simple encryption without integrity protection is meaningless.

38.4.2 Replay Attacks

A goal related to integrity protection is to prevent replay attacks, consisting of the retransmission of a previously sent message and getting it accepted as a new one. The danger posed by replays depends on the protocol; TCP [18], for example, ignores repeated receipt of a message, but other, nonidempotent protocols may carry out some action twice. In some cases, it may be possible to replay an entire dialog between a client and a server. WEP does not offer any means of replay protection. Although the standard encourages the use of unique initialization vectors in each frame, it does not specify how the IVs should be chosen. As such, a compatible implementation cannot make assumptions about what order IVs will be chosen or how frequently they will be reused and must therefore accept messages with repeated IVs. Note that the IV space is small enough that even a strict IV sequencing discipline would not entirely eliminate the possibility of replays.

One consequence of the lack of replay protection is that attacks on message integrity are much easier to carry out. If a system prevents replays, a message modified by the attacker must be received before the original message. In practice, this means that the attacker is only able to attack those messages that are received by the attacker but not by the intended recipient. (An amplified or directional antenna or complicated jamming techniques may be helpful to increase the number of such messages.) If a system further insists that messages be received in the order they are sent, the attack must take place before any subsequent messages sent by the true sender are received.

However, without replay protection, such measures are unnecessary. Modified versions of messages sent (and successfully received) some time in the past will still be accepted. In the case of 802.11, this allows greater time for any potential analysis to recover the original plaintext of the message, which is necessary for attack. It is also possible to modify a message in multiple ways and get all versions accepted. The reaction attacks described above rely on this property. As well, recovering the entire plaintext of a single message allows the attacker to inject an unlimited number of arbitrary messages of the same (or shorter) length into the network.

38.4.3 New Protocols

Both TKIP and CCMP employ strong cryptographic message integrity protection. TKIP uses a new message integrity code called Michael [10] to replace the ICV of WEP and CCMP uses the CCM encryption mode [24], which guarantees both confidentiality and integrity. The protocols also successfully avoid replay attacks, by enforcing an ordering discipline on frame sequence numbers. Because the underlying protocol layers may only discard frames, but not reorder them, it is a simple matter of only accepting frames with sequence numbers higher than the last one. The sequence number is protected by the integrity algorithms, such that it is impossible to change the sequence number of a frame without invalidating the integrity check.

38.5 Authentication and Access Control

One of the desired goals of 802.11 is to control access to the network. As such, it is necessary to authenticate the identities of the users of the network. Effective access control also requires integrity protection, as otherwise it may be possible for someone to hijack an authenticated session and use it to gain access to the network.

The 802.11 defined two authentication methods:

1. Open system authentication — access to the network is unrestricted
2. Shared key authentication — authenticates the users by having them prove knowledge of the WEP secret key

In shared key authentication, the AP sends a sequence of 128 challenge bytes in an unencrypted frame. The STA being authenticated must produce a WEP-encrypted frame containing the same bytes before it can successfully associate with the AP.

Unfortunately, this method is insecure: the weak integrity protection in WEP allows an attacker to produce a valid response to the challenge without knowing the WEP key. The attacker needs to intercept a single authentication challenge–response exchange. Then both the plaintext and ciphertext of the response frame will be know to the attacker, as the plaintext is contained in the challenge frame. As seen in Section 38.4, this allows the attacker to modify the plaintext contents of the frame. To respond to future challenges from the AP, the attacker replaces the challenge bytes in the intercepted response with those in the new challenge. The lack of replay detection allows this response frame to be used repeatedly. Moreover, this

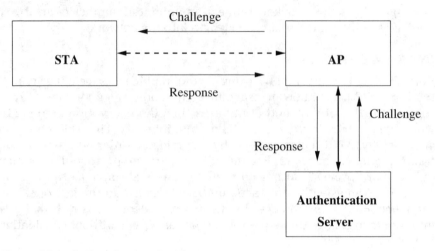

Figure 38.3. Authentication Architecture

frame can also be used to inject any new 128-byte messages into the network, as per Section 38.4.2.

38.5.1 Authentication Extensions

In addition to being insecure, the shared key authentication method was difficult to integrate with a good key management scheme. Typically, the WEP secret key would serve as a network password, known to all users of the network, and therefore hard to keep secret or change. A much more desirable scheme was to authenticate users with individual passwords.

Many vendor-specific extensions were developed to support this model. The authentication would usually be performed not by the AP, but by another authentication server; this way, a central point of management could be used for maintaining user authentication data. The AP would be responsible for forwarding authentication challenges from the server to the STA and forwarding responses back, as shown in Figure 38.3. Eventually, the vendors standardized on using Extensible Authentication Protocol (EAP) [4] and Remote Authentication Dial-In User Service (RADIUS) [19] to encapsulate these communications. This framework allowed any EAP-based authentication method to be used with the same AP and STA hardware, offering organizations the choice of using different authentication servers and the flexibility to change authentication methods without upgrading 802.11 hardware. The 802.1X standard [17] formally defines this authentication architecture.

Several other types of authentication extensions are in common use. One of them performs authentication at a higher layer in the network. For example, Internet kiosks limit IP access until the user logs in through a Web site. This, of course, still leaves the possibility of hijacking an authenticated session. Usually, it is not even necessary to attack the weaknesses in WEP privacy and integrity protection as WEP is not commonly used in these situations.

Another common authentication method is based on restricting access based on MAC address of the 802.11 device. This method simplifies management of authentication because it requires no user involvement. However, this scheme is easily defeated: the MAC address is not authenticated and is broadcast unencrypted with each message. It is therefore easy for an attacker to learn an authorized MAC address and then adjust his own to match. Several common 802.11 devices in fact enable both of these tasks through well-supported interfaces.

38.5.2 Mutual Authentication

In addition to the network authenticating its users, the users must also be able to determine the authenticity of the network they are using. This is especially important in the 802.11 setting, where there is no *a priori* way to tell which network a client is using. Most equipment will select APs based on the system identifiers (SSID) and signal strength. It is therefore trivial for an attacker to set up a fake AP that clients will connect to.

A fake AP can be a starting point for several types of attack. It can install itself as a man-in-the-middle, forwarding messages to and from the real AP. Being in the middle, it is able to delay, alter, delete, and inject new messages at ease. This ability simplifies attacks on integrity and confidentiality of the message.

More serious is the case where the secret key is derived through an exchange with the unauthenticated AP. In this case, the fake AP can cause the STA to use a known key, bypassing the privacy protections. The fake AP can mimic the appearance of a proper network by forwarding all data onto the Internet; a more determined attacker may choose to provide fake versions of the services the client is accessing.

The EAP focuses mainly on authenticating the client. The protocol is structured in terms of challenges sent from the authentication server to the client and responses from the client to those challenges. Some methods, such as EAP-TLS (TLS — Transport Level Security) [1] provide mutual authentication (though in EAP-TLS the authentication of the network is optional); others, such as EAP-MD5 [4], do not.

38.5.3 New Protocols

Both TKIP and CCMP mandate the use of 802.1X to handle authentication, allowing a wide variety of possible authentication methods. Because TKIP and CCMP both provide strong integrity protection, a secure authentication method will provide effective access control. However, many existing methods have weaknesses, such as lack of mutual authentication; care must be exercised in picking one with an appropriate level of security.

38.6 Key Management

The WEP protocol relies upon a shared secret key to encrypt messages. However, key management procedures to set up this shared key were left unspecified, leaving vendors and network administrators to come up with their own solutions. A common one was to rely on a single shared key and distribute it manually to all users. This method had the advantage of requiring no additional software or hardware infrastructure, because virtually all 802.11 equipment allows the option to set the WEP key manually.

Because this key had to be distributed to many people and many systems, the chances of compromise were high. Further, these keys would by necessity be long-lived, as manual key updates in even a medium-sized organization would be expensive and time-consuming. WEP did allow the option of more than one WEP key being in use on the network at the same time, which could be used to establish a transition period for key changes.

In addition, the network key chosen was often insecure. Most software tools provided a way to specify the key as an ASCII string; it was not uncommon to use a five letter English word as a password. The space of such words is quite small and easily susceptible to dictionary attacks. Other tools used a means to hash down arbitrary-length pass-phrases into a key. Unfortunately, some of the more common ones had flaws reducing the effective key length and increasing the feasibility of a brute-force search [14].

Once again, demand for better solutions was met by vendor extensions. Many were integrated with the authentication extensions, where the authentication method would also generate a session key upon success. Because these keys are short-lived and per user, the window of opportunity for the attacks described in the previous sections was greatly reduced, though not entirely eliminated. Another simple extension was to use public key methods such as the Diffie–Hellman key exchange [8] to establish temporary session keys. Once again, there was the advantage over long-lived keys; however, in the absence of authentication there is the possibility of a man-in-the-middle attack. The details of most of the key management extensions are proprietary and thus it is difficult to assess their security.

38.6.1 New Protocols

Both TKIP and CCMP obtain fresh session keys using the 802.1X standard, improving on the static keys of the original 802.11 standard. However, 802.1X does not specify how key management should be done, but rather provides an architecture for interfacing with it, so there is no guarantee that even with these new improvements to the standard that the key

Challenges in Wireless Security: A Case Study of 802.11

management problem with be addressed adequately. Hopefully, future deployment of TKIP and CCMP will be accompanied with the use of well-analyzed secure key management and authentication schemes.

38.7 Conclusion

The security measures contained in the 802.11 standard have many serious flaws. The WEP protocol achieves none of its stated goals. To learn from its history, we must understand why it failed. The previous sections detailed the technical mistakes, yet there are some general observations to be made as well.

A big problem affecting the design of 802.11 is that of priorities. For example, integrity protection was not originally considered important; it was not even stated as a goal in the standard. Correspondingly, the mechanisms used for integrity protection are the weakest from the standpoint of cryptography. In some sense, the entire security suite was designed to thwart only casual eavesdropping [16]. To be sure, some of the attacks described are fairly involved. However, effective automation can make even the most complex attacks accessible to everyone. Today, widely available tools requiring minimal knowledge and effort on the part of the user are able to bypass the security mechanisms of WEP.

Regardless of priorities, trade-offs must always be made between security and performance, extensibility, and other conflicting goals. However, one must be careful when attempting to strike a balance between these goals. Even a slight compromise can change a protocol secure against all but the most determined and resourceful opponents to one that can be easily bypassed and offers little real security. It also must be kept in mind that future technology developments during the lifetime of a protocol can offer improvements in areas such as performance, but as far as security is concerned, they usually bring new avenues of attack and faster and more accessible implementations. It is therefore important to be more conservative in the security design.

It is clear that WEP could have benefited from greater cryptographic expertise and outside review; many of its security flaws are well-known in the cryptography literature. However, design of secure protocols is a fundamentally difficult task and even protocols that have received much more attention than WEP have been later found to contain serious vulnerabilities. So it is even more important to reduce the scope of the problem by relying on existing standards and established practices whenever possible and to reuse components of past designs that have been successful.

If the original 802.11 standard provides an illustration of what not to do when designing a secure protocol, the new proposed revisions are good examples of how to properly address the security issues in wireless networks. Hopefully, future protocol designers will learn from this experience and employ secure designs from the outset. As has been seen in the years since the security flaws in WEP were discovered, correcting protocol flaws after widespread deployment is a lengthy process.

References

1. B. Aboba and D. Simon. PPP EAP TLS authentication protocol. Internet Request for Comments RFC 2716, Internet Engineering Task Force, October 1999.
2. W.A. Arbaugh. An inductive chosen plaintext attack against WEP/WEP2. IEEE Document 802.11-01/230, May 2001.
3. W.A. Arbaugh, N. Shankar, Y.C.J. Wan, and K. Zhang. Your 802.11 wireless network has no clothes. *IEEE Wireless Communications,* December 2002.
4. L. Blunk and J. Vollbrecht. PPP extensible authentication protocol (EAP). Internet Request for Comments RFC 2284, Internet Engineering Task Force, March 1998.
5. N. Borisov, I. Goldberg, and D. Wagner. Intercepting mobile communications: The insecurity of 802.11. In *7th Annual International Conference on Mobile Computing and Networking,* July 2001.
6. A. Bosselaers. Fast implementations on the Pentium . http://www.esat.kuleuven.ac.be/~bosselae/fast.html.
7. N. Cam-Winget, R. Housley, D. Wagner, and J. Walker. Security flaws in 802.11 data link protocols. *Communications of the ACM,* vol. 46, no. 5, May 2003.
8. W. Diffie and M. Hellman. New directions in cryptography. In *IEEE Transactions on Information Theory,* pp. 74–84, June 1977.
9. D. Doligez. SSL challenge virtual press conference. http://pauillac.inria.fr/~doligez/ssl/press-conf.html, 1995.
10. N. Ferguson. Michael: An improved MIC for 802.11 WEP. IEEE Document 802.11-02/020r0, January 2002.
11. S. Fluhrer, I. Mantin, and A. Shamir. Weaknesses in the key schedule algorithm of RC4. In *4th Annual Workshop on Selected Areas of Cryptography,* 2001.
12. National Bureau of Standards. Data encryption standard. Federal Information Processing Standards Publication 46, January 1977.
13. National Institute of Standards and Technology. Announcing the advanced encryption standard (AES). Federal Information Processing Standards Publication 197, November 2001.
14. T. Newsham. Cracking WEP keys. Presented at the 2001 BlackHat Conference.
15. LAN MAN Standards Committee of the IEEE Computer Society. Wireless LAN medium access control (MAC) and physical layer (PHY) specifications. IEEE Standard 802.11, 1997 edition, 1997.
16. LAN MAN Standards Committee of the IEEE Computer Society. Wireless LAN medium access control (MAC) and physical layer (PHY) specifications. IEEE Standard 802.11, 1999 edition, 1999.
17. LAN MAN Standards Committee of the IEEE Computer Society. Port-based network access control. IEEE Standard 802.1X, 2001.
18. J. Postel. Transmission control protocol. Internet Request for Comments RFC 793, Internet Engineering Task Force, September 1981.
19. C. Rigney, S. Willens, A. Rubens, and W. Simpson. Remote authentication dial in user service (RADIUS). Internet Request for Comments RFC 2138, Internet Engineering Task Force, June 2000.
20. R.L. Rivest. *The RC4 Encryption Algorithm*. RSA Data Security, Inc., March 1992. (Proprietary).
21. A. Stubblefield, J. Ioannidis, and A. Rubin. Using the Fluhrer, Mantin, and Shamir attack to break WEP. In *2002 Network and Distributed Systems Security Symposium,* 2002.

22. The Shmoo Group. Airsnort homepage. http://airsnort.shmoo.com.
23. J.R. Walker. Unsafe at any key size: an analysis of the WEP encapsulation. IEEE Document 802.11-00/362, October 2000.
24. D. Whiting, R. Housely, and N. Ferguson. AES encryption and authentication using CTR mode and CBC-MAC. IEEE Document 802.11-02/001r2, May 2002.

Chapter 39
Security for Mobile Agents: Issues and Challenges

Paolo Bellavista, Antonio Corradi, Corrado Federici, Rebecca Montanari, and Daniela Tibaldi

Abstract

Mobile agent (MA) technology raises significant security concerns and requires a thorough security framework with a wide range of strategies and mechanisms for the protection of both agent platform and MAs against possibly malicious reciprocal behavior. The security infrastructure should have the ability to flexibly and dynamically offer different solutions to achieve different qualities of security depending on application requirements. The chapter presents the security threats that typically arise in MA applications and describes the currently available proposed countermeasures to protect both nodes and MAs. In addition, the chapter surveys the state-of-the-art research activities about integrated security supports in MA systems and identifies open research issues and ongoing research work.

39.1 Security: a Missing Link for MAs' Acceptance

The convergence of the Internet with wireless communications has raised new challenges in the support of user and terminal mobility, in facing heterogeneity, and in adapting to the dynamic changes in the network infrastructure [1]. The new scenario seems a suitable application area for computing paradigms that exploit the notion of code mobility, defined as the capability to dynamically change the binding between software components and their location of execution [2]. As mobile networks gain widespread acceptance and ubiquitous environments start to emerge, the ability to change the locations where applications can execute becomes an increasingly important requirement. For example, we can think to heterogeneous and resource-limited portable devices that can benefit from the possibility to download on-demand device-specific software components and discard them when no longer needed.

Along this line, the MA paradigm with its properties of autonomy, asynchronicity, and local resource exploitation particularly suits the peculiarities of the new Internet scenario. The typical proposed applications for MAs include information retrieval, network management, electronic commerce, and service provisioning support in telecommunication systems [3, 4]. Recently, additional proposals are starting to explore the adoption of MAs for implementing novel middleware solutions for mobile computing [1]. These solutions advocate the adoption of MAs at service provision time to act as middleware proxies over the fixed network on behalf of users or devices. These MA-based middleware proxies can autonomously carry on operations, even in case of temporary device disconnection and can migrate dynamically, either to follow device movements or to operate locally to the needed resources.

However, the widespread acceptance and adoption of the MA technology is currently delayed by several complex security problems still to be completely solved. MAs have fostered even more the traditional security issues to the limit. Compared to the client–server model, the MA paradigm offers greater opportunities for performing various attacks because MA systems provide a distributed computing infrastructure on which applications belonging to different (usually untrusted) users can execute concurrently [5, 6]. Additionally, the execution sites hosting MAs may be managed by different authorities with different and possibly conflicting objectives and may communicate across untrusted communication infrastructures, such as the Internet.

The MA technology introduces another dimension of complexity. Unlike other kinds of mobile code, such as applets that are pulled once and in one hop to remote systems, MAs may move through a series of systems with different levels of trust and potential distrust (multiple-hop) [7]. MA principals and owners of the execution environment are usually different: this raises new threats specific to the MA technology. Because hosting nodes have complete control over MA execution, in principle they can do anything to the agent, to both its code and its state.

These various and different issues have led MA security to receive increasing attention in recent years from both a theoretical and a practical point of view. Without an appropriate security level for agents, MA applications could only execute in trusted environments, and could not be deployed in the Internet scenario. Various security techniques and initial frameworks have been developed mainly in the area of site protection from potentially malicious agents [8]. The complementary issue of protecting agents from malicious execution sites represents a new and challenging research area that calls for new security models and frameworks, which up to now have been investigated only by few researchers [8–11].

39.2 Security Requirements

The term mobile agent contains two distinct concepts — mobility and agency — and refers to a self-contained and identifiable software component that can move across the network (hence mobile) and act on behalf of users (hence agent) [12]. According to this definition, in principle, we can consider a MA as the equivalent of a well-trained English butler who knows user needs, likes, and habits and moves in the open Internet landscape to securely carry out application-specific goals on its user's behalf [13]. Such a well-trained butler would be equipped with security credentials, such as stamped passports, that make it identifiable, permit interaction with other agents and with hosting environments, and allow the signing of agreements and contracts [14].

However, this vision has various security implications that still need to be solved to transform our desiderata into reality. To exemplify the security requirements and issues raised by the MA technology, let us consider the case of a shopping MA that has to find the most convenient offer for a flight ticket. Suppose that Bob with his palmtop is willing to access a flight-ticket booking service (FBS), while he is driving on the way back from work, to urgently search for and book the cheapest Rome-to-London flight ticket. FBS provisioning relies on middleware proxies to accommodate portable devices with strict resource constraints [1]. In particular, Bob accesses FBS via a lightweight client running on his access device. Before starting a FBS provisioning session, the client requires Bob to authenticate. After a successful authentication, a middleware mobile proxy, called in the following Alfred (the well-trained English butler), is instantiated to represent Bob over the fixed network and to support Bob's shopping operations.

A trusting relationship should be established between Bob and Alfred: Bob should be confident that Alfred will only do what he asks without abusing or misusing its autonomy capabilities. It should not be possible for an air travel agency to persuade Alfred to purchase a flight from London to Dublin. Neither should Alfred be able to make the commitment to purchase the London-to-Dublin flight. If a trusting relationship exists, Bob may also delegate Alfred to sign the flight purchase contract on his behalf when one convenient offer is found.

Suppose now that Alfred generates a shopping MA and delegates it the flight searching and booking operations. The shopping agent could migrate among the various air travel agencies nodes to locally operate on needed resources, whereas Alfred could closely follow Bob's movements. Once it has completed its tasks, the shopping agent returns to Alfred the result of the computation. The shopping agent should be granted the same rights and be submitted to the same restrictions as Alfred.

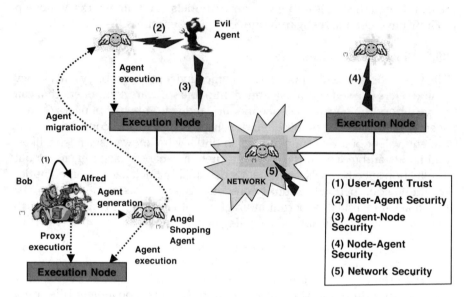

Figure 39.1. Security Threats in Mobile Agent Systems

In this scenario, several security issues arise and several attacks are possible, as Figure 39.1 shows. Because MA execution is at the mercy of hosting nodes, how is it possible to ensure that these nodes do not manipulate the results of the shopping agent computation and do not illegally access the MA's private information? If no countermeasures are taken, for instance, an intermediate air travel agency could intentionally alter or cancel the flight ticket offers previously collected by the MA in order to gain a bargain advantage. If the shopping agent carries Bob's private key, the air travel agency could also steal the key to forge a purchase transaction commitment.

The hosting execution environment should also need protection from the shopping agent. A malicious shopping agent may try to gain privileged or unrestricted access to information belonging to the current execution environment or may try to misuse services offered by the execution environment to attack the hosting node itself or to use the node as the basis for probing and exploiting the vulnerabilities of other systems [6]. For example, unregulated access to the file system may permit agents to install a virus or a Trojan horse, whereas unrestricted access to node resources may allow the shopping agent to produce denial of service attacks or to jeopardize operativity of hosting sites by exhausting memory, storage system, and central processing unit (CPU) cycles.

Recent research efforts have mainly focused on the development of ad hoc security techniques that can only partially circumvent the aforementioned security threats. For example, no viable solution exists that can convince Bob to give his signing key to Alfred (and hence to the shopping agent) and to trust air travel agencies to use it properly.

We still also lack a general comprehensive security framework that can provide solutions for addressing, in a uniform and coherent manner, all the security issues arising in MA applications. That framework should define how to express security requirements, how to administer them, how to distribute them to relevant entities, and how to enforce them. The desired behavior of both agent platforms and agents should be defined at a high level of abstraction, separately from agent — system code to facilitate the dynamic adaptation of security controls to the evolution of both the execution environment and the application requirements.

Differently from the traditional approach to MA security, security concerns should inform every phase of MA system development from requirement engineering to design, implementation, testing, and deployment [15]. The ultimate challenge is, therefore, unifying security with system engineering. Just as MA system engineers analyze and select system features to answer to functional requirements, security engineers must develop applicable security models and deploy those security measures that can make available different qualities of security service depending on specific security requirements of applications and on the most suitable trade-off between security and efficiency.

39.3 Security Countermeasures

This section is intended to outline the key features and to point out the limitations of the most common protection techniques adopted to protect both agent platforms and agents against reciprocal malicious behavior.

39.3.1 User–Agent Trust

Any well-trained English butler should be delegated with proper rights to carry out his tasks. Delegation is a means by which users can trust agents to make decisions on their behalf.

Several trust-management systems based on the notion of delegation have been proposed in recent years in decentralized systems to allow one entity to give some of its authority to other entities (e.g., the Simple Public Key Infrastructure [SPKI] [16] and Attribute Certificates [ACs] [17]).These systems have recognized fruitfulness even for supporting delegation in MA systems and are starting to be integrated in several MA system projects [18, 19].

SPKI exploits public-key-based certificates for supporting both authentication and authorization [16]. SPKI certificates securely bind a user and his public key with a set of rights and specify whether the user is authorized to delegate his rights. SPKI supports a simple notion of delegation that makes it suitable only for small networks. The main weakness is that it does not allow the user to specify any other constraints on delegation. For instance, it is not possible to impose limitations on the entities to whom the user can delegate his rights or specify which rights the user can delegate.

Another emerging technique for delegation is based on the exploitation of ACs that securely bind users with their rights, but do not contain the user public key [17]. ACs support a flexible delegation model because it is possible to specify not only whether the AC owner is authorized to delegate some of his privileges, but also the conditions that must hold for the delegation to take place. For example, the AC issuer can impose some restrictions on the entities to which the AC owner can delegate his privileges, on the delegation chain length, and also on the subset of privileges that the AC owner can delegate.

39.3.2 Protecting Agent Platforms

The problem of protecting agent platforms from malicious agent behavior requires the host to perform various security checks both when an agent arrives and while it is executing. Before executing an agent, the hosting node should guard against malicious agent logic, which is defined as a set of instructions that cause a site security policy to be violated. If the agent code proves to be secure, the host should authenticate the incoming agent and should mediate agent operations on needed resources by means of access control checks.

39.3.2.1 Secure Agent Code. Research on provably secure code has been undertaken for several years and targets verifying that a piece of code is secure before it begins execution. Some research activities in the field of safe programming languages can enhance the development of code safety by proposing solutions based on strong typing, restricted memory-reference manipulations, and runtime-supported memory allocation and deallocation [8]. Another formal technique that can be used to develop provably secure code is the proof carrying code that forces the agent code producer to formally prove that the mobile code has the safety properties required by the hosting agent platform. The proof of the code correct behavior is transmitted to the hosting node that can validate the received code [20]. The major problems with proof carrying code techniques are the need of a standard formalism for establishing safety policies and the significant performance overhead sometimes unacceptable to both transmit and verify proofs.

A more pragmatic approach, adopted in most MA systems, is to trust a specific piece of agent code because one trusts the developer or supplier of the agent code. This technique requires ways to verify through cryptographic means that a particular piece of agent code originates from the trusted party.

Another approach based on path history logs can be exploited to allow hosting platforms to decide whether to execute an incoming agent [21]. The underlying idea is to maintain an authenticable record of the prior platforms visited by an agent, so that a newly visited platform can determine whether to process the agent and what resource constraints to apply. Computing a path history requires each agent platform to add a signed entry to the agent path, indicating its identity and the identity of the next platform to be visited and to supply the complete path history to the next platform.

Another technique for detecting malicious agent logic uses a state appraisal function that becomes part of the agent code and guarantees that the agent state has not been tampered with by malicious entities [22]. The success of this technique relies on the possibility to predict alterations to the agent state and to prepare countermeasures in the form of appraisal functions. The state appraisal function is produced by the agent author and it is signed together with the rest of the agent. The visited platform uses this function to verify that the agent is in a correct state and to determine what privileges to grant to the agent.

39.3.2.2 Agent Authentication. Authentication mechanisms are required for associating agents with responsible principals, where principals represent the subjects that request operations, such as an individual, a corporation, a service provider, and a network administrator. For authenticating incoming agents, agent principals can be associated with personal public/private keys and can be forced to digitally sign agents to ensure the correct identification of their responsible party. The public-key-based authentication process safely verifies the correspondence between principal identities and keys. Most authentication solutions based on public key cryptography delegate key lifecycle management to public key infrastructures (PKIs) [23]. Authentication can also ascertain the paternity of agents by associating them with their responsible role. A role-based model facilitates the administration and management of a large number of principals, by simplifying the dynamic handling of principals and permissions.

39.3.2.3 Agent Authorization. Research in the authorization field has focused on two main areas. Some researchers attempt to realize access control mechanisms, whereas others are concerned about finding languages for expressing access control requirements.

Access control mechanisms can enforce the control of agent behavior at runtime and can limit access to resources. The early sandboxing technique is a typical example [24]. However, the rigidity of the Sandbox Model makes it inadequate for complex agent-based applications. The JDK™ (Java Development Kit) 1.2 security architecture evolves from the sandbox by introducing fine-grained, extensible access control structures for a wide range of applications and clearly separates the enforcement mechanism from the security policy specification [25]. However, the JDK 1.2 provides support for traditional access control lists only, so more sophisticated access control requires further extensions to this architecture. Another proposal for controlling the execution of MAs written in the Tool Command Language (Tcl) scripting language is the Safe-Tcl security framework, which uses at least two interpreters — one regular for trusted code and a safe interpreter for untrusted code [8]. When untrusted code executing in the safe interpreter executes a command requiring access to a system resource, the trusted interpreter evaluates whether access should be granted or denied.

Several MA systems adopt simple access control lists (ACLs) to implement access control in MA applications. However, ACLs exhibit limitations in enforcing all the complex access controls necessary in MA applications. The complexity of access control decisions in MA-based applications derives from both static and dynamic considerations. On the one hand, it is mandatory to consider static attributes, such as the identity of the source code implementer, the host from where the code was loaded, and the identity or role of the principal on behalf of whom the mobile agent is executing. On the other hand, it is necessary to take into account also dynamic attributes relating to the current context in which the MA operates. An agent may be granted different permissions depending on the current time, the current application state, or the state of the resources that the code is accessing. In most current solutions based on ACLs, complex access control constraints are often directly hard coded into the applications, thus imposing reconfiguration, rebuilding, or even rewriting of significant parts of the application at any policy change.

Several research approaches are emerging that propose language-based solutions to separate policy from access control implementation [26–29]. From the research field in mobile code technologies, entirely procedural languages have been developed to restrict MAs' operations depending on their historical behavior in addition to common discriminators like the MA source location or the identity or role of its user [26]. Other languages combine procedural and declarative rules. They describe both the minimal set of capabilities the hosting node must grant to enable an incoming MA to perform its task, as well as the trust conditions to be evaluated to determine its trustworthiness [27]. In addition, more expressive logic-based declarative languages not specifically proposed for MA systems could be considered. They can be extended to cater for mobility and employed in MA systems to provide access control decisions that can take into account temporal and application-dependent dynamic aspects [28–30].

39.3.3 Protecting Agents

The main issues to be comprehensively addressed to protect agents against malicious hosts are agent execution, integrity, and secrecy. Protecting agent execution requires the system to ensure that agents are not hijacked to untrusted destinations that may present agents with a false environment, thus causing them to execute incorrectly; that accepts do not commit to unwilling actions; and do not suffer of premature termination or starvation due to unfair administrator's policies that fail to provide necessary system resources (such as access to files, communication channels, or CPU time). Protecting agent integrity requires the identification of agent tampering, either of its code or of its state, by malicious execution hosts. Providing agent secrecy requires hiding the agent code and state parts from the site responsible for its execution.

There is no universal and general solution to the problem of agent protection. Little can be done to ensure correct agent execution. A possible security practice could be to develop an agent with the necessary support services for its execution embedded into its code. For example, agents may have the crypto code to perform sensitive operations or to resolve names. This would make agents independent from the services offered by hosts, but it would simply reduce the possibility of hosts to attack agents and would also significantly increase the transmission costs because of the bigger agent bytecode size.

Only a few partial approaches have been proposed to either prevent or detect attacks against agent integrity and secrecy. Prevention techniques are aimed at turning attacks into computationally unfeasible or useless tasks and detection techniques are aimed at allowing agent-responsible users to verify whether some attacks have been carried out against their agents once they return back.

With regard to the prevention of attacks to agent integrity and secrecy, some proposals try to overcome the problem by not permitting MAs to move to untrusted hosts. Others face the problem by taking organizational measures [31]. Only trusted parties are permitted to set up an agent platform and to host agent execution, but this is currently too restrictive a requirement.

Some mainly rely on special tamper-proof hardware that avoids unauthorized modification of agent code or state by executing the agent in a physically sealed environment. Even the administrators of the agent execution system do not have access to this execution environment [32]. However, the cost of tamper-proof hardware makes this approach impractical for a large number of users. Tamper-proof hardware can only be deployed in closed corporate systems or secure military networks [11]. In addition, this approach clashes with openness and limits the scalability of applications as it forces agents to migrate only to nodes that are equipped with tamper-proof hardware.

Others try to address the problem by adopting algorithms that obfuscate agent code and data to make it difficult for hostile environments to analyze code and data structures at least for the time the agent executes in a host. State variable splitting and recomposition, hardly predictable program flow by using runtime data values dependencies are examples of obfuscation operations [33]. This approach has, however, two main limitations: the difficulty to identify in advance an agent lifetime that minimizes the possibility to successfully inspect agent data and code and the need for a global time clock to check expiration time.

The prevention of attacks to agent secrecy seems extremely difficult to guarantee. A few researchers address the issue by using the concept of computing with encrypted functions [34]. The key idea behind that approach is to have the agent platform execute agent code embodying an encrypted function without being able to discern the original unencrypted function. However, this technique is generally valid because it applies only to certain classes of polynomials and rational functions. In addition, it still remains to be seen whether agent platform administrators will be willing to run agents executing arbitrary encrypted functions that cannot be interpreted.

Other approaches control the possibility of an agent to reveal private information [35]. Using the environmental key generation method, agent private information can be encrypted and only disclosed when some predefined environmental condition has been met.

Detection techniques can be exploited to discern if any tampering with agent code or state has occurred. Among the proposals that adopt a detection strategy some present cryptographic approaches to encapsulate the results of agent computation at each visited platform for subsequent verification with the aim of preserving the forward integrity of results collected by the agents during its roaming. Some of these solutions employ centralized solutions that rely on Trusted Third Parties (TTP) to encapsulate partial results. The TTP can be used to track the agent execution at each intermediate step by recording partial results [36]. The central role of a TTP may cause some inefficiencies that can be overcome by more distributed protocols [37–39]. Another method, called execution tracing, for detecting unauthorized modifications of agent code or state relies on nonmodifiable traces of agent computation created at each intermediate execution host [40]. Each single trace is a sequence formed by an identifier of executed instructions along with values of internal state variables involved. A signed hash of the trace is sent from node to node upon conclusion along with final agent execution state until the agent returns back to its sender node. If the agent owner suspects that some host behaves maliciously, it starts a check procedure by asking each involved host for its copy of the trace, by locally executing the returned agent and by comparing the computation results with the received traces. The major drawbacks of the approach are not only the management of the trace logs, but also their size and the intensive message exchanges between the sender nodes and all visited nodes at any integrity check.

Finally, some solutions partially achieve integrity by replicating agents and by exploiting replication to compare the results obtained by following different paths [37, 41]. These approaches ensure that destroying or tampering with one or more agents does not compromise the correctness of agent computation because even in the case of attacks, a still adequate number of agents is expected to return with meaningful results.

39.4 Overview of Security Solutions in MA-Based Systems

This section is devoted to examining the different directions of security solutions provided by the most diffused MA systems. Despite differences and peculiarities in supported security models, it is possible to recognize some common features. Because a large number of MA systems are based on Java™ programming language, most MA security solutions rely on the Java Security Model and have been developed on top of the Java security architecture. In particular, earlier MA systems rely on the use of the Sandbox Model and of the Java Security Manager to limit the scope of the agent while executing [24], whereas more recent MA systems propose flexible solutions based on the more evolved Java2 security [25].

Most MA systems mainly focus on security solutions for the protection of agent platforms, whereas the issue of protecting agents from possibly malicious hosting execution sites is often neglected and rarely addressed.

With regard to agent platform protection, the most common approaches rely on cryptographic mechanisms that verify the identity of the agent owner, assign access restrictions to the agent based on the owner identity or role, and execute agents in a secure environment that can enforce these restrictions.

In the Concordia™ system, the agent platform protection is achieved through agent authentication and resource access control [42]. Any Concordia agent has a unique identity associated with the identity of the user that has launched it. Resource control is based on the Java 1.1 Security Model and relies on simple access control lists that allow or deny access to resources only on the basis of agent identities.

Voyager® system implements a proprietary security manager to restrict agent operations with the drawback of modifying the security manager implementation if any security policy changes [43]. In Grasshopper, customizable access control lists rule agent access to resources on the basis of the identity of the agent and of the group that it belongs to [44].

The AGLETS® system provides an AGLET Security Manager to implement own security policies [31]. The behavior of the Security Manager cannot be changed directly, but via a graphical user interface (GUI) tool or directly editing policy files. In the AGLET Security Model, agents can access resources depending on their associated principles (i.e., entities that can be authenticated). Examples of defined principles include the agent, the agent developer, and the agent owner.

Dartmouth Agents [D'Agents] authenticate the owner of an agent, assign resource limits to the agent based on this identity, and enforce those limits [45]. Resources are divided into indirect resources that are accessed through another agent and built-in resources that are directly accessible through language primitives. The agent that controls an indirect resource enforces its own access restrictions. For built-in resources there are some absolute access policies that are enforced by the server.

Ajanta protects hosting resources through an ad hoc security manager that uses identity-based access control lists to grant or deny agent access [46]. The Ajanta Security Manager is used only for mediating access to system-level resources. For all application-defined resources, Ajanta uses a proxy-based mechanism where a proxy intercepts agent requests and denies or grants access based on its own security policy and on the agent's credentials (i.e., tamper-proof information including the agent owner).

The MARISM-A (an Architecture for Mobile Agents with Recursive Itinerary and Secure Migration) platform provides security mechanisms to protect migration, confidentiality, and integrity of agents [47]. MARISM-A is based on Java and is similar to other existing agent platforms, like Grasshopper, but its main different feature is its simple extensibility. In fact, its core provided features, like confidentiality, and authentication for agent communication, can be extended with some more complex security mechanisms. For example, an extension of MARISM-A Core Security Model exploits SPKI features to realize a role-based access control model.

The Secure and Open Mobile Agents (SOMA) support to agent platform protection is built on top of the Java 1.2 Security Model that allows both identity-based and role-based access control [48]. The use of the Ponder language permits the specification of flexible and fine-grained authorization policies, separately from agent code and system access control mechanisms, that can control agent access to resources on the basis of composite factors, both static (identity or role of agent owner/creator) and dynamic (time, application state, resource state). In addition, the integration of Ponder within SOMA allows the system not only to define agent rights, but also agent duties.

Similar security approaches to the SOMA system are proposed in the NOMADS environment [49]. NOMADS is an example of MA system that provides dynamic access and resource control and a policy-based approach to host security. Safe execution of agents is based on the ability of NOMADS to control the resources accessed and consumed by agents. The resource control mechanism allows control over the rate and quantity of resources used by agents. These resource control mechanisms, built on a custom Java Virtual Machine (JVM™) called Aroma [50], complement Java's access control mechanisms and help in making the NOMADS system secure against malicious agents.

With regard to agent protection against malicious hosting nodes, Concordia, Voyager, and Grasshopper systems do not currently support integrated solutions. AGLETS address the issue by relying on an organizational approach that prevents agents to move toward untrusted execution environments. In contrast, the Ajanta system provides a wide range of integrated techniques to at least detect any illegal tampering to agent code and state. Similar approaches are provided in the SOMA platform [39].

39.5 Open Issues and Directions of Work in Secure MA Systems

This section examines some of the most relevant open issues that still need to be faced before the MA technology is ready for commercial applications and overviews the primary current directions of work in MA systems.

With regard to host protection, an open research issue relates to the development of effective countermeasures against denial of service. A hostile host may create clones of authorized agents and use them to attack other nodes. The detection of cloning is still a difficult problem in MA environments. Only a few proposals tackle this issue, but solutions are expensive in terms of computational overhead and limit the autonomy and efficiency of the agent execution by relying on interactive protocols [51]. To reduce the risks of denial of service attacks resulting from a poorly programmed or malicious agent overuse of critical resources, authorization should be associated with monitoring techniques to exclude excessive resource consumption and with auditing mechanisms to record all MA activities. Few solutions are starting to investigate tools and mechanisms for combined resource and access control management [49, 52]. However, there are several technological limitations to overcome. Most approaches rely on standard Java security mechanisms that can only either grant or deny access to a particular service [53]. Changes to the Java Virtual Machine are necessary to minimize the impact of denial of service attacks and to provide meaningful accounting, at the expense of portability, however.

In addition, the concept of secure cascaded delegation should be regarded as an ongoing research issue in need of further investigation. Cascaded delegation occurs when one agent (initiator) may authorize other agents (delegates) to perform some tasks with its rights. Secure delegation implies the ability to reconstruct the complete delegation chain and to verify the proper authorization of agents that claim to act on behalf of someone else. However, several practical issues may arise when considering open MA environments. Due to the unsolved agent protection attacks from malicious nodes, the delegation process can occur only in trusted nodes. In addition, in several application scenarios for both legal reasons and accountability, auditing mechanisms should be available to keep track of the initiator of delegation and of the delegated permissions in a nonrepudiable manner.

Trust issues remain also a major reason for hampering a final acceptance of MAs [54]. For MAs to undertake more sophisticated missions and to operate effectively without the responsible user supervision, users will have to trust their agents to maintain conformance to the desired objectives. The user trust into agents cannot be simply reduced to delegating agents specific rights to MAs (see Section 39.3.1); users need to rely on mechanisms to constantly monitor and rule agent behavior within the bounds of desired constraints.

Another crucial security requirement when deploying MAs in pervasive computing environments is the dynamic establishment of trust relationships between MAs and visited nodes. The dynamic determination of execution nodes requires dynamic establishment and enforcement of access control and the integration of access control with probing of resource availability so as to influence agent migration itineraries [55].

Among the various described open issues, the ones that are starting to most attract research interest are in the area of agent trustworthiness and dynamic security policy negotiation. If users and administrators can be assured that agent behavior conforms to desired constraints and objectives, everyone can stop worrying and start loving agents [14]. In addition, the need for dynamic policy negotiation is increasingly emerging in those application scenarios where a primary requirement is the dynamic formation of temporary coalitions. The following sections provide and overview of some of the emerging proposals in these two areas.

39.5.1 *Agents and Trust*

The key to implementing a trustworthy MA system is to devise an infrastructure mechanism that controls agent autonomy and dynamically adjusts governing strategies to deal with changing priorities and contexts [14]. However, this is not an easy task.

A heading research direction growing considerably in these last years is the integration of MA platforms with policy-based management solutions [56]. Policies that constrain the behavior of system components are becoming an increasingly popular approach to dynamic application adjustability in both academia and industry. Bradshaw et al. [57] points out all the benefits of policy-based approaches, including reusability; efficiency; extensibility; context-sensitivity; verifiability; support for both simple and sophisticated components; protection from poorly designed, buggy, or malicious components; and reasoning about component behavior.

In particular, policy-based network management has been the subject of extensive research over the last decade [56]. Policies are often applied to automate network administration tasks, such as configuration, security, recovery, and quality of service.

In the field of MA systems, policies can be a powerful means to model agent behavior in terms of rights and duties. Explicit policies governing human–agent interaction, based on careful observation of work practice and with an understanding of current social science research, can help ensure that effective and natural coordination, appropriate levels and modalities of feedback, and adequate predictability and responsiveness to human control are maintained [57].

The development of policy-driven MA systems requires the consideration of some general requirements. A basic requirement is the choice of an expressive language for policy specification. In modern interorganizational MA environments, languages should include the ability to group policies that apply to agents to simplify the management of large-scale complex agent systems. As a final consideration, the implementation of policy specifications should not introduce prohibitive performance costs that would limit the deployment of the MA technology.

There is already some research work integrating policy-based management with MA systems [58, 59]. This works focuses on the use of policies for controlling agent-related security aspects. Policies specify complex and flexible access controls for agent-to-resource and agent-to-agent interactions and express which actions agents must perform in response to events. An important advantage of a policy-based approach to MA system security is the possibility to specify, represent, and manipulate policy information independently from the components in charge of policy interpretation and cleanly separated from agent–system code. Modern interorganizational environments require sophisticated security policies that are difficult to implement in current MA platforms where policies are often directly hard coded into applications, thus making difficult the reconfiguration required at any policy change. Separation of concerns enables policies to be dynamically changed and modifies the behavior of MAs by simply loading or unloading relevant policies without the need of reimplementing agent code from scratch.

39.5.2 *Dynamic Configuration of Access Control*

Dynamic migration implies that a MA may not know in advance which nodes are going to host its execution throughout its lifetime. Because target nodes and MAs are typically owned by different authorities, their interests can conflict with respect to resource access and utilization. Thus, there should be a mechanism for *a priori* acquisition of resource availability and for the dynamic establishment of access control between agents and nodes. The knowledge of the availability of resources in a particular site may show several benefits. MAs can exploit the visibility of available resources to adjust their behavior accordingly and to reduce the risk of undesired task failure during execution. Resource availability visibility can also help MAs decide whether it is more profitable to stay in a locality than to move and explore a new computing environment. When entering a new locality, a MA should be provided with the mechanisms needed to express its desired resource visibility and to dynamically negotiate access control policies to obtain access to needed resources. Policy negotiation allows MAs to enlarge the set of accessible resources with no risk of illegitimate resource usage for the hosting nodes.

Some research activities are starting to emerge to provide controlled resource visibility and to support policy negotiation [55, 60]. The COSMOS (a COntext-aware Security middleware for MObile agent Systems) framework dynamically determines the MA contexts and effectively rules the access to them, by taking into account different types of metadata (user profiles and system/user-level authorization policies), expressed at a high level of abstraction and cleanly separated from the service logic [60]. As a distinctive feature, COSMOS provides MAs entering a new locality with a controlled visibility of the directly accessible physical/logical resources and of the other MAs locally executing (active context views). Active context views contain resources that MAs are willing to access and that the COSMOS access control function has qualified as accessible.

COSMOS addresses also the privacy issues that arise when exploiting MAs for building context-aware services [1]. In fact, context awareness requires computing devices to gather, collect, and propagate up to the service level both user- and environment-specific information to permit more informed service management strategies. However, the visibility of user-specific information, such as user location, could be exploited to infer user tasks or preferences, thus to violate user privacy. COSMOS protects user privacy by enabling users to specify which personal context information they are willing to make public. User specifications, carried in the MA state part, guide and automate COSMOS access controls to user personal context information.

With regard to dynamic establishment of access control, a negotiation model integrated in the FarGo system is described in [55]. The model makes resource providers and consumers reach an agreement on the allowed access and on the amount of resource utilization. The goal of negotiation is to identify a set of permissions acceptable to both resource providers and consumers; the result of the negotiation is a contract. Because the negotiation is closely tied to the application level rather than depending on the infrastructure level, it is the application to choose to invoke the negotiation and to decide which negotiation strategy to apply.

39.6 Conclusions

Mobile code-based programming models have recently gained wide prominence for their appealing features in terms of flexibility, extensibility, and efficiency. In particular, MAs have attracted a great research interest and are emerging in mobility-enabled scenarios as an enabling technology for the design, implementation, and deployment of both advanced Internet services and middleware solutions. However, the MA technology poses severe security risks and calls for novel security mechanisms and frameworks.

Great research efforts have been devoted to develop countermeasures for the security threats arising in MA systems. Although a great number of mechanisms currently exist, further improvements are still necessary through either the incremental refinements of available protection mechanisms to reduce processing and storage overhead or the combination of complementary mechanisms to form a more effective protection scheme. In addition, no single solution to all the problems of agent execution, integrity, and secrecy seems to exist, unless tamper-proof hardware is introduced, which is likely to be overhead-prone. Currently no manufacturer produces these devices at low costs even though the recent Trusted Computing Platform Alliance (TCPA)-Palladium (recently renamed Trusted Computing Group (TCG)– NGSCB [Next Generation Secure Computer Base]) represents a significant step toward this goal. Even if tamper-proof hardware devices would gain wider diffusion, we would still need to certify that the manufacturers do not introduce malicious back doors.

The development of thorough security frameworks is another research direction that requires more investigation, experimentation, and experience. The state-of-the-art MA systems point out that there is very little support for fully integrated security frameworks that can provide the required degree of flexibility. Most current security frameworks lack a clear separation between policies and security mechanisms and provide monolithic security solutions where applications cannot choose their suitable trade-off between security, scalability, and performance. In addition, very few frameworks provide the required support to protect agents against malicious hosts.

We can conclude that albeit security is crucial for wider diffusion of the MA paradigm, it is still in a somewhat immature state. Issues related to trust, delegation, and security policy negotiation are nontrivial to solve and only pioneer solution attempts are starting to emerge. In addition, intense standardization efforts should be directed to promote wider acceptance of security supports and to facilitate their use, thus leveraging the adoption of the MA technology in fielded systems.

Acknowledgments

Work supported by the Italian Ministero dell'Istruzione, dell'Università e della Ricerca (MIUR) in the framework of the FIRB WEB-MINDS Project "Wide-Scale Broadband Middleware for Network Distributed Services" and by the National Research Council (CNR) in the framework of the Strategic IS-MANET Project "Middleware Support for Mobile *Ad Hoc* Networks and their Application." We also thank Niranjan Suri for valuable comments and suggestions.

References

1. P. Bellavista, D. Bottazzi, A. Corradi, R. Montanari, and S. Vecchi, Mobile Agent Middlewares for Context-Aware Applications, Section III, *Handbook of Mobile Computing*, Boca Raton, FL: CRC Press, 2003.
2. A. Fuggetta, G.P. Picco, and G. Vigna, Understanding Code Mobility, *IEEE Transactions on Software Engineering*, vol. 24, no. 5, 1998.
3. M. Baldi and G.P. Picco, Evaluating the Tradeoffs of Mobile Code Design Paradigms in Network Management Applications, *20th International Conference on Software Engineering (ICSE'98)*, Los Alamitos, CA, 1998.
4. R.H. Glitho and T. Magedanz, Eds., Special Issue on Applicability of Mobile Agents to Telecommunications, *IEEE Network*, vol. 16, no. 3, 2002.
5. M.S. Greenberg, J.C. Byington, T. Holding, and D.G. Harper, Mobile Agents and Security, *IEEE Communications Magazine*, vol. 36, no. 7, 1998.
6. W. Jansen, Countermeasures for Mobile Agent Security, *Computer Communications*, vol. 23, no. 17, 2000.
7. G. Knoll, N. Suri, and J.M. Bardshaw, Path-Based Security for Mobile Agents, *Electronic Notes in Theoretical Computer Science*, vol. 58, no. 2, 2002.
8. G. Vigna, Mobile Agents and Security, *Lecture Notes in Computer Science*, vol. 1419, New York: Springer-Verlag, 1998.
9. R. Oppliger, Security Issues Related to Mobile Code and Agent-Based Systems, *Computer Communications*, vol. 22, no. 12, 1999.
10. N. Borselius, Mobile Agent Security, *Electronics and Communication Engineering Journal*, vol. 14, no. 5, 2002.
11. J. Zachary, Protecting Mobile Code in the Wild, *IEEE Internet Computing*, vol. 7, no. 2, 2003.
12. V.A. Pham and A. Karmouch, Mobile Software Agents: an Overview, *IEEE Communications Magazine*, vol. 36, no. 7, 1998.
13. N. Negroponte, Agents: from Direct Manipulation to Delegation, *Software Agents*, Menlo Park, CA: AAAI Press, 1997.
14. J. Bradshaw, G. Cabri, and R. Montanari, Taking Back Cyberspace, *IEEE Computer*, vol. 36, no. 7, 2003.
15. P.T. Devanbu and S. Stubblebine, Software Engineering for Security: a Roadmap, *International Conference on Software Engineering (ICSE'00)*, Limerick, Ireland, 2000.
16. J. Howell and D. Kotz, A Formal Semantics for SPKI, *6th European Symposium on Research in Computer Security (ESORICS 2000)*, *Lecture Notes in Computer Science*, vol. 1895, New York: Springer-Verlag, 2000.
17. J. Linn and M. Nystrom, Attribute Certification: an Enabling Technology for Delegation and Role-Based Controls in Distributed Environments, *4th ACM Workshop on Role-Based Access Control (RBAC'99)*, Fairfax, VA, 1999.

Security for Mobile Agents: Issues and Challenges

18. W.A. Jansen, A Privilege Management Scheme for Mobile Agent Systems, *5th International Conference on Autonomous Agents*, Montreal, Canada, May 2001.
19. S. Mudumbai, A. Essiari, and W. Johnston, Anchor Toolkit — a Secure Mobile Agent System, *3rd International Symposium on Mobile Agents (ASA/MA'99)*, Palm Springs, CA, October 1999.
20. G. Necula, Proof Carrying Code, *24th ACM Symposium on Principle of Programming Languages*, Paris, 1997.
21. D. Chess, B. Grosof, C. Harrison, D. Levine, C. Parris, and G. Tsudik, Itinerant Agents for Mobile Computing, *IEEE Personal Communications*, vol. 2, no. 5, 1995.
22. W. Farmer, J. Guttman, and V. Swarup, Security for Mobile Agents: Authentication and State Appraisal, *4th European Symposium on Research in Computer Security*, Rome, 1996.
23. W. Ford and M. Baum, *Secure Electronic Commerce*, Upper Saddle River, NJ: Prentice Hall, 1997.
24. L. Gong, Java Security: Present and Near Future, *IEEE Micro*, vol. 17, no. 3, 1997.
25. L. Gong, *Inside Java 2 Platform Security*, Boston: Addison Wesley, 1999.
26. G. Edjlali, A. Acharya, and V. Chaudhary, History-Based Access Control for Mobile Code, *5th ACM Conference on Computer and Communications Security*, San Francisco, 1998.
27. M. Blaze, J. Feigenbaum, J. Ioannidis, and A.D. Keromytis, The Role of Trust Management in Distributed Systems Security, *Secure Internet Programming: Issues in Distributes and Mobile Object Systems*, New York: Springer-Verlag, 1999.
28. S. Jajodia, P. Samarati, and V.S. Subrahmanian, A Logical Language for Expressing Authorizations, *IEEE Symposium on Security and Privacy*, 1997.
29. V. Varadharajan, C. Crall, and J. Pato, Authorization in Enterprise-Wide Distributed System: a Practical Design and Application, *14th Annual Computer Security Applications Conference*, 1998.
30. E. Lupu, M. Sloman, N. Dulay, and N. Damianou, Ponder: Realising Enterprise Viewpoint Concepts, *4th International Enterprise Distributed Object Computing Conference*, Makuhari, Japan, 2000.
31. D. Lange and M. Oshima, *Programming and Deploying Java Mobile Agents with Aglets*, Boston: Addison Wesley, 1998.
32. U.G. Wilhelm, S. Staamann, and L. Butty, Introducing Trusted Third Parties to the Mobile Agent Paradigm. In J. Vitek and C. Jensen, Eds., *Secure Internet Programming: Security Issues for Mobile and Distributed Objects*, New York: Springer-Verlag, 1999.
33. F. Hohl, Time Limited Blackbox Security: Protecting Mobile Agents from Malicious Hosts, *Mobile Agents and Security, Lecture Notes in Computer Science*, vol. 1419, New York: Springer-Verlag, 1998.
34. T. Sander and C. Tschudin, Protecting Mobile Agents Against Malicious Hosts, *Mobile Agents and Security, Lecture Notes in Computer Science,* vol. 1419, New York: Springer-Verlag, 1998.
35. J. Riordan and B. Schneier, Environmental Key Generation toward Clueless Agents, *Mobile Agents and Security, Lecture Notes in Computer Science,* vol. 1419, New York: Springer-Verlag, 1998.
36. A. Corradi, M. Creminini, and C. Stefanelli, Security Models and Abstractions in a Mobile Agent Environment, *IEEE Workshop on Collaboration in Presence of Mobility*, Stanford, CA, 1998.
37. B. Yee, A Sanctuary for Mobile Agents, *DARPA Workshop on Foundations for Secure Mobile Code*, Monterey, CA, 1997.
38. G. Karjoth, N. Asokan, and C. Gülcü, Protecting the Computation Results of Free-Roaming Agents, *2nd International Workshop on Mobile Agents*, Stuttgart, Germany, 1998.
39. A. Corradi, M. Cremonini, R. Montanari, and C. Stefanelli, Mobile Agents Integrity for Electronic Commerce Applications, *Information Systems*, vol. 24, no. 6, 1999.
40. G. Vigna, Cryptographic Traces for Mobile Agents, *Mobile Agents and Security, Lecture Notes in Computer Science*, vol. 1419, New York: Springer-Verlag, 1998.

41. B. Shneider, Toward Fault Tolerant and Secure Agentry, *11th International Workshop on Distributed Algorithms, Lecture Notes in Computer Science,* vol. 1320, New York: Springer-Verlag, 1997.
42. D. Wong, N. Paciorek, T. Walsh, J. DiCelie, M. Young, and B. Peet, Concordia: an Infrastructure for Collaborating Mobile Agents, *1st International Workshop on Mobile Agents, Lecture Notes in Computer Science,* vol. 1219, New York: Springer-Verlag, 1997.
43. ObjectSpace — Voyager, http://www.recursionsw.com.
44. IKV++ — GrassHopper, http://www.ikv.de/products/grasshopper/.
45. R. Gray, G. Cybenko, D. Kotz, and D. Rus, D'Agents: Security in a Multiple-Language, Mobile-Agent System, *Mobile Agent Security, Lecture Notes in Computer Science,* vol. 1419, New York: Springer-Verlag, 1998.
46. A. Tripathi, Mobile Agent Programming in Ajanta, *19th IEEE International Conference on Distributed Computing Systems Workshop (ICDCS'99),* Austin, TX, 1999.
47. S. Robles, J. Mir, J. Ametller, and J. Borrell, Implementation of Secure Architectures for Mobile Agents in MARISM-A, *4th International Workshop on Mobile Agents for Telecommunication Applications (MATA'02), Lecture Notes in Computer Science,* vol. 2521, New York: Springer-Verlag, 2002.
48. R. Montanari, C. Stefanelli, and N. Dulay, Flexible Security Policies for Mobile Agents Systems, *Microprocessors and Microsystems,* vol. 25, no. 2, 2001.
49. N. Suri, J.M. Bradshaw, M.R. Breedy, P.T. Groth, G.A. Hill, R. Jeffers, T.S. Mitrovich, B.R. Pouliot, and D.S. Smith, NOMADS: Toward a Strong and Safe Mobile Agent System, *4th International Conference on Autonomous Agents,* Barcelona, 2000.
50. N. Suri, J.M. Bradshaw, M.R. Breedy, K.M. Ford, P.T. Groth, G.A. Hill, and R. Saavedra, State Capture and Resource Control for Java: the Design and Implementation of the Aroma Virtual Machine, http://nomads.coginst.uwf.edu/.
51. J. Baek, D. Lee, and R.S. Ramakrishna, A Design of Protocol for Detecting an Agent Clone in Mobile Agent Systems and its Correctness Proof, *8th Annual ACM Symposium on Principles of Distributed Computing,* Atlanta, 1999.
52. P. Bellavista, A. Corradi, and C. Stefanelli, Java for On-Line Distributed Monitoring of Heterogeneous Systems and Services, *The Computer Journal,* vol. 45, no. 6, 2002.
53. J.M. Bradshaw, M. Greaves, H. Holmback, T. Karygiannis, W. Jansen, B.G. Silverman, N. Suri, and A. Wong, Agents for the Masses, *IEEE Intelligent Systems,* vol. 14, no. 2, 1999.
54. A. Patrick, Building Trustworthy Software Agents, *IEEE Internet Computing,* vol. 6, no. 6, 2002.
55. Y. Gidron, I. Ben-Shaul, O. Holder, and Y. Aridor, Dynamic Configuration of Access Control for Mobile Components in FarGo, *Concurrency and Computation: Practice and Experience,* vol. 13, no. 1, 2001.
56. S. Wright, R. Chadha, and G. Lapiotis, Special Issue on Policy Based Networking, *IEEE Network,* vol. 16, no. 2, 2002.
57. J.M. Bradshaw, P. Beautement, L. Bunch, and S.V. Drakunov et al., Making Agents Acceptable to People, *Handbook of Intelligent Information Technology,* Amsterdam: IOS Press, 2003 (in press).
58. A. Corradi, N. Dulay, R. Montanari, and C. Stefanelli, Policy-Driven Management of Mobile Agent Systems, *International Workshop on Policies for Distributed Systems and Networks — Policy 2001, Lecture Notes in Computer Science,* vol. 1995, New York: Springer-Verlag, 2001.
59. J.M. Bradshaw, N. Suri, A.J. Canas, R. Davis, K. Ford, R. Hoffman, R. Jeffers, and T. Reichherzer, Terraforming Cyberspace, *Computer,* vol. 34, no. 7, 2001.
60. P. Bellavista, R. Montanari, and D. Tibaldi, COSMOS: a Context-Centric Access Control Middleware for Mobile Environments, *5th International Worshop on Mobile Agents for Telecommunication Applications (MATA'03), Lecture Notes in Computer Science,* vol. 2881, New York: Springer-Verlag, 2003.

Chapter 40
Security, Trust, and Privacy in Mobile Computing Environments

Lalana Kagal, Jim Parker, Harry Chen, Anupam Joshi, and Tim Finin

40.1 Introduction

With the advent of mobile and pervasive computing, new models of distributed communication and computation are being introduced, leading to systems that are open in that they do not preidentify a set of known participants and dynamic in that the participants change regularly and not just due to occasional failures. It is interesting to note that this evolution is occurring at many levels — communication, infrastructure, and application. At the communication level, for example, mobile ad hoc networks treat nodes as autonomous routers, requiring new techniques to protect against malicious or faulty nodes that subvert or blackhole packets [5]. Similarly, as applications become more sophisticated and intelligent, they require greater degrees of decision making and autonomy so that they can interact spontaneously with other peers that happen to be in their vicinity. The long-range vision is described as societies of intelligent, autonomous agents that are goal-directed and adaptive. But even today, we find the new levels of autonomy emerging in infrastructures like the grid computing, Web services, and pervasive computing. These systems must exchange information about services offered and sought and their associated security and privacy policies, negotiate for information sharing, and monitor for and report on suspicious or anomalous behavior.

Securing these open dynamic environments presents several new challenges. As a concrete instance, consider providing a secure and privacy enhancing pervasive computing environment in spaces such as an office, hospital, school, or subway stop. The space will be filled with devices and agents offering and seeking services. As people move, agents on their personal devices detect, and are detected by, the pervasive infrastructure. The new devices must discover the services of interest from the infrastructure and other devices in the vicinity, negotiate for access, control information exchange, and monitor for suspicious events to be reported to the community. Shared knowledge models (ontologies) and norms of behavior (policies) will undergird the society of communication and cooperating applications, agents, and devices. Addressing this grand challenge will require contributions not just from diverse areas within computer science, but also from other disciplines such as policy, law, and various social sciences.

Without appropriate security and privacy mechanisms, these exciting new ideas will be hobbled and the applications they enable will not be deployed or be found socially acceptable. For example, the Defense Advanced Research Project Agency (DARPA) LifeLog program was recently forced to eliminate many of the more exciting possibilities from its scope because good privacy mechanisms were not available. Notice also the split in the computing community (United States Association for Computing Machinery ([USACM] versus ACM Special Interest Group on Knowledge Discovery in Data and Data Mining [SIGKDD]) on the issue of data mining and the Tactical Information Assistant (TIA) program. We must develop new models for security and privacy that work in such highly distributed, open, and dynamic systems and that will find immediate applications in grid computing, semantic Web, and pervasive computing. We identify three topics where new challenges are emerging — trust-based security, computational policies, and knowledge sharing.

Security and privacy based on authentication is not enough in open systems where principals may be able to provide authentication, but are otherwise unknown to the system and hence not authorizable for specific actions. Traditional role-based approaches also fare poorly. Such environments are common on the Web and in envisioned pervasive computing environments. A solution is to make security and privacy decisions based on attributes related to trust for which a principal can provide evidence, such as proof of key attributes, a signed statement from a trusted source delegating a permission, or undertaking an obligation in return for access. Human societies use trust and reputation to make decisions about requests for service where a right to that service is not preestablished and social networks are an important way of transferring trust and reputation [3]. Such societies have overlapping systems of behavioral norms, constraints, and rules. We are overconstrained, so we cannot always satisfy all of them, but deviating too much or too often has its consequences — loss of reputation, penalty clauses, imposition of sanctions, etc. These mechanisms need to be understood and computational analogues developed for

Security, Trust, and Privacy in Mobile Computing Environments

computational agents to better support information sharing and control in human societies.

One possible approach is to use policies — explicit representations of constraints and rules that govern or inform an agent's or system's behavior. Policies can define permissions, obligations, norms, and preferences for an agent's actions and interactions with other agents and programs [2, 4]. Explicit policies, especially those expressed in high-level declarative languages, can be used as the basis for electronic contracts and provide a sublanguage useful for the negotiation for agreements and commitments. We believe that explicit policies for security, trust, and privacy are promising areas for research. We describe in this chapter our efforts to develop meaningful machine interpretable policies for security and privacy.

One focus is to design policy languages that are simultaneously expressive enough, intuitive and understandable by humans, and writable by nonprogrammers. Eventually, we would like to be able to reason at a high level over the policies, answering hypothetical questions about the limitations and vulnerabilities in the security and privacy systems they model. (Will this policy allow X to happen and if so under what circumstances?)

Another focus is to actually create policies for privacy in pervasive computing environments. Finally, because trust at the application layer is best built upon trusted, secure lower protocol layers, we close the chapter with an approach that uses understanding of routing policies for wireless intrusion detection. Pervasive computing is built using mobile wireless nodes and intrusion detection becomes less traditional and more decentralized. The relevance is to develop, nurture, and build trust in an environment with processes and nodes that detect and react to malicious entities in order to protect the integrity of trust relationships.

We start this chapter by describing how policies can be used to secure pervasive environments. We provide an overview of our language and illustrate its usefulness through an example. We then proceed to privacy and start by describing several privacy mechanisms. We discuss our notions of privacy management and compare it to several existing schemes. We conclude this chapter with a discussion of intrusion detection in mobile ad hoc networks. In each of these focus areas, we bring out relevant work from the community, as well as our own solution efforts.

40.2 Policies and Their Role in Security in Pervasive Computing Systems

40.2.1 Introduction

Policies influence the behavior of entities within their domain by providing sets of rules that users have to conform to. Policy-based security is often used in systems that consist of a large number of users and entities and where configurations like access rights, behaviors, responsibilities, etc., change often. We believe that policy-based security is the most optimal for pervasive environments as it allows the system to modify how heterogeneous entities act without modifying their internal mechanism.

Policies generally require application-specific information to reason over, forcing researchers to create policy languages that are bound to the domains for which they were developed. This prevents policy languages from being flexible or being usable across domains. To enable entities in pervasive computing systems, which consist of different domains and systems, to understand and interpret policies correctly, we propose that they be represented in a semantic language like Resource Description Framework Schema (RDFS) [8], DARPA Agent Markup Language plus Ontology Inference Layer (DAML+OIL) [9], or Web Ontology Language (OWL) [10]. We believe that using a semantic language allows different systems to share a model of policies, roles, and other attributes [11, 12].

Describing comprehensive policies for pervasive computing systems may not always be feasible given the large number of entities and the fact that not all entities can be predetermined. We believe that policies should be as simple as possible and control should be decentralized; that is, authorization should be possible by more than just a few key entities.

Due to the large number of entities in the environment, it may not be possible to identify them accurately or even predetermine the users of each service. Therefore, we suggest developing policies based on attributes of users' services and their context.

We have developed a policy specification language, Rei, which promotes security in pervasive environments. Rei, which is grounded in RDFS, is based on deontic concept [13–15] and includes constructs for rights, prohibitions, obligations, and dispensations. (A dispensation can be thought of as a waived obligation.) Rei provides decentralized control through the use of speech acts. These are commands that can be used by authorized entities to dynamically modify policies. Rei supports individual policies as well as group- and role-based policies in a uniform manner by allowing domain dependent representations for roles or groups to be included in the conditions of the policy rules.

In the following sections, we describe Rei, compare it with other related security research, and discuss our future work.

40.2.2 Related Work

In this section, we compare our work with related research in distributed security systems. Extensible Access Control Markup Language (XACML) [16] is a language in Extensible Markup Language (XML) for expressing access policies. This work is similar to ours in that it allows control over actions and supports resolution of conflicts. However, it does not provide support for speech acts making decentralized control difficult, adding domain specific information is time-consuming, it does not allow types of actions to be defined, its conflict resolution is not across policies, and as it is based in XML, it does not benefit from the interoperability and extensibility provided by an ontology language.

Role Based Access Control (RBAC) [17–20] is one of the better-known methods for access control, in which relations are established between users' roles and permissions' roles. It is difficult, however, to apply the RBAC Model for systems in which it is not possible to assign roles to all users in advance. In addition, it is typically not possible to change access rights of a particular entity without modifying the roles. Simple RBAC is rather restrictive, as every time access rights of an agent change, its role has to change as well. Our policy language allows access rights (and the other deontic constructs) to be associated with different credentials and properties of entities and not roles alone. More sophisticated RBAC Models include work on delegation between roles [21] and assigning roles to users that are outside the system [22, 23]. This research considers delegating the entire set of permissions associated with a set of roles of the delegator (the entity doing the delegation) to the delegatee (the entity receiving the delegation). Our work allows the policy to dictate what subset of permissions a delegator is able to delegate, to whom, and under what conditions. In fact, our policy language is able to represent most RBAC Models. Herzberg [22] uses properties of certificates to map users outside the domain-to-domain specific roles. This is very similar to our work; however, instead of mapping properties of certificates to roles, we map properties of the user to access rights directly. Though this may seem like a drawback, we have found that this provides greater control in specifying access rights to foreign users as certain types of rights can be given, like the read rights on a certain resource, the right to print to all the color printers on the fifth floor, and the right to delete all the files belonging to your colleague.

PolicyMaker, one of the first distributed trust management schemes proposed, allows credentials, policies, and trust relationships to be represented in terms of filters in a programming language that can be safely interpreted by the environment [39]. The basic function of PolicyMaker is to be able to answer queries about trust. It does not verify certificates or signatures, but allows the functionality to be handled by an external program, enabling the application to use any kind of authentication scheme. KeyNote is the next generation of PolicyMaker and uses credentials to directly authorize actions [24] [40]. KeyNote, however, provides a specific language (C-like expressions) for representing assertions (policies and credentials) and includes cryptographic signature verification. Though they are both elegant solutions, PolicyMaker and KeyNote work best in certificate-based systems and are not easily extensible. Delegation is controlled by a delegation depth and simple conditions on delegation. On the other hand, in Rei the permission to delegate is a separate permission and includes constraints not only on who can delegate, but also whom they can delegate to. As Rei is capable of reasoning over domain knowledge in ontology languages, it is possible to develop constraints over attributes of requesters, actions, and the environment at different levels of abstraction. This is not possible in either PolicyMaker or KeyNote as they use a programming language for describing assertions. Rei is also capable of modeling both authorization and obligation policies, whereas KeyNote and PolicyMaker can only be used for authorization.

Ponder is a declarative object-oriented language for policy specification in distributed systems [41]. Ponder defines two kinds of policies: basic and composite. Basic policies such as authorization and obligation policies are rules governing behavior options and are described as a set of subjects and a set of targets. On the other hand, composite policies such as role policies allow the policies relating to organizational units to be grouped. Ponder provides several tools for managing Ponder policies including graphical tools for browsing/updating policies, tools for syntactic and semantic analysis of policy specifications, and tools for transforming Ponder language specifications directly into XML or Java code that can be interpreted at runtime. Ponder also includes a conflict detection tool to detect overlaps and conflicts between policies [42]. Though Ponder provides the specification of interfaces for enforcement agents that can enforce authorization and obligation policies, it does not provide any implementation for them.

Ponder is used to describe policies at a lower level than Rei and is more tightly coupled to the environment. It requires that the environment be fairly static because the specific names of the entities to be controlled and their interfaces need to be predetermined. It also does not use ontologies and thus lacks the ability to handle subjects, actions, and targets at different levels of abstraction. After specifying a policy, it is compiled by the Ponder compiler into a Java class and then represented at runtime by a Java object. Due to this, runtime changes to policy are not possible unlike Rei.

40.2.3 Approach

Our policy language allows policies to be specified over actions in terms of rights, prohibitions, obligations, and dispensations [4]. We believe that most policies can be expressed as what an entity can or cannot do and what it should or should not do in terms of actions, services, conversations, etc., making our language capable of describing a large variety of policies ranging from security policies to conversation and management policies. The policy language has some domain independent ontologies like concepts of permissions, obligations, etc., but can reason over specific domain ontologies that are used by entities in the system as well.

Though the actual execution of services or actions is outside the system, the policy language includes a representation of them that allows more contextual information to be captured and allows for greater understanding of the action and its parameters. The action ontology includes properties for preconditions, effects, target resources, etc. This allows policies to be defined over different aspects of an action and not just its identity. For example, it is possible to say that no students of the Computer Science department can perform any action that causes the temperature in the lab to go higher than 75-degrees Fahrenheit. Composite actions can be composed using action operators of `sequence`, `nondeterministic choice`, `once`, and `iteration`.

The language includes two constructs for specifying meta-policies that are used to resolve conflicts: setting the modality preference (negative over positive or vice versa) for a set of entities or stating the priority between rules and policies [26]. For example, it is possible to say that in case of conflict, the federal policy always overrides the state policy.

Another important aspect of the language is that it models speech acts like `delegation`, `revocation`, `request`, and `cancel` that allow policies to be less exhaustive and allow for decentralized security control. Delegations are important to pervasive environments because services (mobile or otherwise) may not be able to project who will use them or preestablish all the desirable requirements of the entities that should use them. Delegations allow access rights of an entity to be propagated to a set of trusted entities, without explicitly changing its policy or requirements. The other speech acts modeled are `revocation`, which nullifies an existing `right` (whether policy-based or delegation-based), `request`, by which an entity can request another entity for a right or to perform an action on its behalf, and `cancel`, with which an entity can cancel any previously made request. Table 40.1 contains an example Rei policy in Notation3, which is a concise representation of RDF.

40.2.4 Discussion

As computationally enabled devices become more prevalent, as environments get more intelligent, and as semiautomated entities become more useful, the need for more automated security in these environments becomes more important. We believe that policy management is the most effective security mechanism in these environments. In this section, we described the specifications of our policy language, Rei, which we have designed and developed for dynamic distributed environments like pervasive systems and the semantic Web.

Table 40.1. Rei Policy in Notation3

```
All graduate students have the right to delegate the right to
print to the LaitHPPrinter in the LAIT lab to undergraduate
students.
@prefix rdfs:  <http://www.w3.org/2000/01/rdf-schema#>.
@prefix rei:   <http://www.csee.umbc.edu/~lkagal1/rei-simplified#>.
@prefix univ:  <http://www.csee.umbc.edu/~lkagal1/univ#>.
@prefix time:  <http://daml.umbc.edu/ontologies/time#>.
@prefix : <#>.
:x a rei:Variable.
:y a rei:Variable.
:R a rei:Right;
  rei:agent rei:x;
  rei:constraint [a rdfs:statement;
    rdfs:subject y;
    rdfs:predicate rdfs:type;
    rdfs:object univ:Undergraduate];
  rei:action [a rei:Delegate;
    rei:actor x;
    rei:receiver y;
    rei:content [ a univ:Printing;
      rei:actor y;
      rei:target univ:LaitHPPrinter;
      rei:location univ:Lait ]].
:cseepolicy a rei:Policy.
:cseepolicy rei:grants [a rei:granting;
  rei:to x;
  rei:deontic R;
  rei:requirement [a rdfs:statement;
    rdfs:subject v1;
    rdfs:predicate rdfs:type;
    rdfs:object univ:Graduate]].
```

Some of the advantages of our work include:

- Semantic language — our language reasons over policies and ontologies in RDFS that unambiguous interpretation
- Distributed framework — our framework includes speech act specifications that can be used to modify policies dynamically namely: delegation, revocation, request, and cancel.
- Mechanisms for conflict resolution — Rei provides constructs for conflict detection and resolution

We are currently in the process of moving our policy language to a more efficient version of Prolog, XSB. Though we have designed and developed a similar policy-based security frameworks for pervasive environments [27], we have not completely switched to Rei and are using an older version of the language. Another issue involves conflicts. Conflicts are detected and resolved at runtime; however, we believe detecting policy conflicts statically also has several practical uses. We are investigating a feasible solution for predetermining conflicts in Rei policies.

40.3 Toward Privacy Protection in Pervasive Computing Environments

40.3.1 Introduction

In pervasive computing environments, hidden sensors constantly monitor the situational conditions of the users and share their information with other computing systems to provide services to the users. Because the dissemination of this information is often invisible to the users, it creates great concerns for privacy. We describe a new architecture called the Context Broker Architecture (CoBrA) that aims to protect user privacy in a pervasive context-aware environment [31].

Context is any information that can be used to characterize the situation of a person or a computing entity [38]. We believe an understanding of context should include information that describes location, system capabilities, services offered and sought, the activities and tasks in which people and computing entities are engaged, and their situational roles, beliefs, and intentions [36].

Privacy is about control (e.g., who has what information, what information can be shared with whom under what circumstances) [29]. Although research in privacy and security is similar in a way that they both deal with the access to information, they differ from each other in a number of ways. Privacy is about information self-determination: the ability to decide what information about you goes where. Security offers the ability to be confident that those decisions are respected.

CoBrA aims to address the following two issues in user privacy:

1. How to represent privacy policies that users can use to control the dissemination of their contextual information.
2. How to develop a meta-processing mechanism that can prevent the leaking of private information through inference (e.g., it is easy to find out the home address of a person once you know that person's home phone number).

40.3.2 Previous Work

User privacy research in the pervasive computing can be broadly categorized into two groups. The first group concentrates on defining design principles for guiding the implementation of the privacy systems and the second group focuses on the implementations of the privacy protection mechanisms.

40.3.3 Design Principles for Building Privacy Systems

Design principles for privacy systems are the guidelines set for the developers to implement and validate their privacy systems. The following are three key design principles that have been proposed [28, 29, 37].

40.3.3.1 Notice and Consent. This principle states that the systems should attempt to inform users of when and what information about them is being captured and to whom the information is being made available (i.e., notice) and the designers should attempt to empower users to stipulate what information they project and who can hold of it (i.e., consent). Central to these two principles is the explicit involvements of the users in the flow of the information process (i.e., information can only be shared if explicit permissions are granted by the users and users are notified whenever their personal information is captured by the systems). Although the explicit involvements of the users can help them to control and to have a better understanding of how their personal information is used, sometimes requiring users to be actively involved can be distractive. Furthermore, it is not always feasible to require users to give consents to each and every type of information that the systems will use. For example, when information is used by hundreds and thousands of devices in the environment, even the user with the most patience will be annoyed if that user is required to give consents to each one of the devices before any service is provided to him.

40.3.3.2 Proximity and Locality. The principle of proximity and locality proposes that instead of announcing each and every use of the user information, systems should be developed with certain rules for using and sharing user information according to their context. As users enter a matched context, their privacy rules are automatically applied. This principle overcomes the problem of requiring users to give consents to the use of individual contextual information.

40.3.3.3 Anonymity and Pseudonymity. This principle states that the designers should attempt to facilitate the disassociation of the user identity (ID) and information that is associated with the users. Anonymity can be defined as "the state of being not identifiable within a set of subjects" [29]. The larger the set of subjects is, the stronger is the anonymity. Pseudonymity is an alternative that allows for a more fine grained control of anonymity in a system. With pseudonymity, users can be assigned a certain ID; therefore, same user can be repeatedly identified until he changes to a different ID. Using the same pseudonym more than once allows the holder to personalize a service or establish a reputation while always offering the possibility to be disassociated with the role whenever that user wishes. In contrast to the principles of notice and consent, this principle enables user privacy protection without requiring users to give explicit consents to individual pieces of contextual information.

40.3.4 Implementations of the Privacy Systems

To motivate and demonstrate user privacy protections in pervasive computing environments, a number of privacy systems have been prototyped.

Based on our literature survey, we can categorize these prototype systems [29, 30, 35] according to the design principles that the individual system follows.

40.3.4.1 Notice and Consent. The Privacy Aware System (pawS) described by Langheinrich [29] and the Conceptual Privacy Model described by Lederer et al. [35] are two systems that exploit the principle of notice and consent. In pawS, upon entering a pervasive computing environment, the user is immediately notified about the privacy policies of difference services, which describe the type of information that the services are about to collect. In the system, each user has a personal agent that manages that his privacy preferences. As the user agent receives the privacy policies from the environment, it compares the policies with the user's privacy preferences. According to the rules that the user has defined, the user agent grants or denies the permission for the services to use a particular type of personal information.

Lederer et al. developed a similar infrastructure for privacy protection. In this system, all pervasive computing services are required to notify the users about the information that is to be used. They must request consents from the users before using the associated information. Additionally, in their system, users have the option to log the notifications that are announced by the services for subsequent review.

40.3.4.2 Proximity and Locality. Implementations that follow the principle of proximity and locality usually exploit the locality information of the users for enforcing access restrictions to their personal information. A key advantage of this approach is that the use of locality information can relieve the users from specifying a large set of privacy policies for each type of information that the system uses.

In the system developed by Lederer et al., users can define different privacy preferences (or privacy policies) for different faces. Faces are metaphoric terms defined by the system for describing situations at different localities (e.g., secure shopper, cocktail party, hanging out with friends). A key difference between this privacy system and pawS is in the use of locality information. After a user enters a pervasive computing environment that matches one of the predefined faces, the privacy preferences of the associated face are automatically uploaded to the surrounding system.

40.3.4.3 Anonymity and Pseudonymity. In a pervasive computing environment, users constantly interact with the surrounding services and devices. It is inevitable that often the users need to reveal their personal information (e.g., their locations, identities) to the systems to receive personalized services. Privacy systems that build on the principle of anonymity and pseudonymity attempt to protect the privacy of the users by hiding the true identities of the users from the computing systems.

Mist is a communication infrastructure developed by Al-Muhtadi et al. [30] that attempts to preserve user location privacy in a pervasive computing environment. Central to this infrastructure is the use of anonymous communication. In a Mist network, a collection of communication nodes called portals are linked to each other in a hierarchal structure. Each node in the hierarchy represents an abstraction of a physical location in the environment (e.g., a room, a building, a department). In this structure, the leaf nodes represent rooms and the nonleaf nodes represent buildings and campuses. The task of the leaf nodes is to detect the location of the users and route this location information to the nonleaf nodes to be shared by the location-aware services. Location-aware services register callbacks with the nonleaf nodes in the network to be notified about the location of a user. As the nonleaf nodes receive user location information from the leaf nodes, the sources of the information are made anonymous. Because nonleaf nodes do not have knowledge about the detailed location information of the users, location privacy is effectively protected from location-aware services.

40.3.5 Context Broker Architecture

CoBrA is an agent-based architecture for supporting context-aware computing in intelligent spaces. Intelligent spaces are physical spaces (e.g., living rooms, vehicles, corporate offices, and meeting rooms) populated with intelligent systems that provide pervasive computing services to users [31]. By context, we mean an understanding of a location, its environmental attributes (e.g., noise level, light intensity, temperature, and motion) and the people, devices, objects, and software agents it contains.

Central to our architecture is the presence of an intelligent context broker (or broker for short). In a smart space, a context broker has the following responsibilities:

- Provide a centralized model of context that can be shared by all devices, services, and agents in the space.
- Acquire contextual information from sources that are unreachable by the resource-limited devices.
- Reason about contextual information that cannot be directly acquired from the sensors (e.g., intentions, roles, temporal and spatial relations).
- Detect and resolve inconsistent knowledge that is stored in the shared model of context.
- Protect user privacy by enforcing policies that the users have defined to control the sharing and the use of their contextual information.

Figure 40.1 shows a high-level design of the broker and its relationship with agents in an intelligent space.

Security, Trust, and Privacy in Mobile Computing Environments

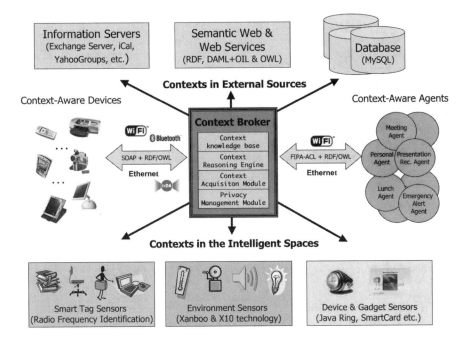

Figure 40.1. High-Level Design of the Broker and Its Relationship with Agents in an Intelligent Space

To protect user privacy in smart spaces, the design of CoBrA follows the principle of proximity and locality, exploiting the locality information of the users for enforcing access restrictions to their personal information. For different types of smart space (e.g., intelligent meeting room, smart vehicle), CoBrA defines different specialized access control models for protecting the privacy of the users. An access control model consists of a set of inference rules that the broker uses to decide the permission for revealing a user's contextual information.

As different users in the same smart space may desire different levels of privacy protection, CoBrA allows users to modify the default access control model of the smart space by providing their own privacy policies. A privacy policy (or policy) is a set of declarative rules that a user defines to restrict the access to that user's personal information. For example, upon entering a smart space, the user authenticates that user's identity and informs the context broker of that user's privacy policy. The broker then reasons about the policy to determine the access control rules that are imposed by the policy. If these rules differ from the rules in the default access control model, the broker will create a personalized access control model of the user. This model will be used, instead of the default model, to guide the broker's reasoning in deciding the appropriate permissions to reveal the user's contextual information.

40.3.6 Privacy Policy Language

In CoBrA, the representation of the privacy policy extends the Rei policy language. Rei is a policy language that defines a set of ontology concepts for modeling rights, prohibitions, obligations, and dispensations in the domain of security. The key advantage of using Rei to develop a new privacy policy language is in its built-in support for the modeling of security objects (i.e., rights, prohibitions, obligations, and dispensations) and its ability to interoperate with the emerging semantic Web languages.

For example, to define a projector device that has the right to access the location context of the user Bob only if the device is located in the same room as Bob, using the has construct, the following rule can be defined:

```
Has(projector,right(accessContext(bob,location),
colocated(bob,projector))).
```

For example, to prohibit a location tracking service from accessing the location context of the user Bob when the service is not authenticated by the context broker, using the has construct, the following rule can be defined:

```
Has(locTracker,prohibition(accessContext(bob,location),
not(authBy(locTracker,broker)))).
```

Protecting the privacy of a user sometimes means hiding the details of certain information from the computing entities in the environment. The Rei language provides users the necessary constructs to define rules to grant or deny the access to their contextual information. To give users a fine-grained control on how the broker can share their contextual information, we introduce additional language constructs to allow granularity parameters to be specified for the contextual information. For example, a student Alice, who is a participant in a meeting, does not want other people to know the specific room that she is in, but she does want others to know that she is on campus. A policy can be defined as the following:

```
Has(broker,right(shareContext(alice,location))),
granularity(location,raduis(1,mile)).
```

This rule states that the broker has the right to share Alice's location information, but for only information that describes a physical location that has the geographic radius larger than 1 mile. In other words, if some service asks the broker if Alice is on the campus, the broker will reply, "yes," and if some service asks the broker if Alice is located in a particular building on the campus, the broker will reply, "Unknown."

Security, Trust, and Privacy in Mobile Computing Environments

40.3.7 Meta-Reasoning with Policies

In a pervasive computing environment, the ubiquitous access to vast amounts of information creates a new problem for user privacy. Although users can define policies to control the dissemination of their situational information, such policies cannot always guard against the possibility of others to deduce the private information of the user through different means of knowledge acquisition (e.g., inference, data mining). In the design of CoBrA, we attempt to address the problem of inference, which is to develop mechanisms to prevent the leaking of information that could potentially be used to deduce a user's private information. Some typical examples of the problem of inference are the following:

- If someone knows the home phone number of a user, it is possible to acquire the mailing address of the user by looking up the white page service on the Internet.
- If someone knows the e-mail address of a user (e.g., someone@host.mil or someone@host.gov), based on the domain name part of the e-mail address, it is possible to infer that the user probably works for one of the U.S. government agencies.

To address the problem of inference, our context broker implementation will allow special meta-reasoning rules to be configured for different types of smart spaces [31]. These meta–reasoning rules help the broker to decide the permissions for revealing the types of contextual information that is not explicitly constrained by the privacy policies. For example, if the policy of a user specifies that no location information should be shared with any nonmeeting-related services, the broker will attempt to keep secret the user's daily schedule because from the daily schedule it is possible to determine the whereabouts of the user; if the policy of a user specifies that his home address should be revealed to others in a meeting, the broker will attempt to keep secret his home phone number. The following are examples of the meta–reasoning rules expressed in Prolog:

```
mayKnow(X,location(Y)):- know(X,dailySchedule(Y)).
mayKnow(X,homeAdd(Y)):- know(X,phoneNum(Y)).
```

40.3.8 Discussion

With the emergence of pervasive computing, user privacy is a critical research issue that must be addressed to bring about users' trust in using the systems. Although the research in both privacy and security deal with the control of information accesses, they are fundamentally different from each other in the purpose of controlling information accesses and the approaches that are used to achieve their distinctive objectives.

We have reviewed a number of design principles (i.e., the principles of notice and consent, the principles of proximity and locality, and principles of anonymity and pseudonymity) for guiding the development of privacy systems in pervasive computing environments. In the design of the CoBrA, we have adopted the principle of proximity and locality, which is to exploit the use of locality information to relieve users from specifying a large set of privacy policies for each type of information that the system uses.

In CoBrA, the use of security policy language (i.e., Rei) and meta-reasoning distinguishes itself from the previous privacy systems, such as pawS and the system developed by Lederer et al. In the future, we plan to demonstrate the feasibility of our architecture by prototyping intelligent meeting room systems called EasyMeeting that will exploit the context broker to provide user privacy protections.

40.4 Intrusion Detection in Mobile Ad Hoc Networks

40.4.1 Introduction

Mobile computing presents challenges differing from those of fixed computing. Disconnected operations, the lack of fixed infrastructure, authentication, security, and trust are but a few of the issues with which mobile computing nodes have to cope. Operating in the mobile environment poses an adequate challenge, but unfortunately these challenges become all the more critical with the threat of an unsecure and untrusted environment. We live in a world where convenience often leads to abuses and advantages are taken by those who prey on the trust of others. Detection of these activities is necessary to ensure the safety of information transiting mobile networks. For pervasive computing to be successful, users need to feel safe and trust that they will not fall victim through their mobile computing devices.

40.4.2 Environments and Devices

Mobile computing environments can be broken into two types — infrastructure and ad hoc. An infrastructure environment is described as having a fixed access point through which all nodes participating in the network must communicate. The access point forwards all traffic to other nodes whether internal or external to the network. Access points can also provide a variety of services from access control and authentication to address assignment through Dynamic Host Control Protocol (DHCP).

Ad hoc networks have no centralized access points. Instead, the ad hoc network is a series of point-to-point connection channels that exist between a node and all other network nodes within communication range. Communication channels can either be infrared (IR) or radio frequency (RF). Although IR can be used for short line of site communication, RF is used for more physically widespread networks. Each node has the ability to route or forward traffic from (or to) its immediate neighbors. Routing algorithms must be in place for the nodes to successfully route network traffic throughout the network. With no centralized point of access into the

Security, Trust, and Privacy in Mobile Computing Environments

network, each node must have a way to ensure that trust exists between itself and any other node that can influence the network.

Ad hoc network nodes can be divided into three broad categories — portable, mobile, and sensor. Although each device has the ability to connect and dynamically configure itself into a network, the main differences come in resource constraints. Portable devices, for example, usually have fixed power and no real constraint on resources, such as processor or memory. A good example would be laptop computers that tend to be brought into an environment and switched on. Mobile devices can be described as those that are in motion most of the time. These devices include PDAs, phones, and other devices that would typically be carried by a person in their daily routine. Mobile devices usually have battery power, but can be recharged. These devices are the most prevalent today and there is general growth in processor power and memory size, but due to smaller, lighter batteries, power constraints still exist. Sensor devices are by far the most restrictive in resource availability. These devices usually are deployed into some environment and have a battery that cannot be recharged. Sensors are usually limited in their processor power and memory size due to packaging.

40.4.3 Intrusion Detection

Intrusion detection can be divided into three categories — host, network, and distributed. Host intrusion detection software runs on a device to monitor activity on the network stack. The software analyzes activity and looks for signs of malicious activity aimed toward the device. Network intrusion devices run independent of the network nodes and passively monitor network activity looking for events that will trigger an intrusion alarm. Distributed network intrusion is the best of both worlds by sharing detection information among hosts. This information could be shared and independently analyzed by all hosts or could be shared to a centralized point. The environment and capabilities of the network nodes drive the decision whether to use host, network, or distributed network intrusion.

Additionally, intrusion detection algorithms are divided into two types — pattern matching and anomaly detection. In pattern matching algorithms, the detector monitors network activity and looks for patterns of network events that indicate ongoing intrusion activity. These are a lot like virus scanners as they rely on searching for known patterns that were previously associated with network attacks. Anomaly detectors work to find events that are outside of the normal activity profile for a particular network or network device. In this case, the system compares current rates of activity (i.e., Transmission Control Protocol synchronize [TCP SYN] packets) with some baseline threshold looking for abnormal levels.

Wired networks can use any combination of host, network, or distributed platforms with pattern matching or anomaly detection. For wired networks, an independent network intrusion detection device can be placed in position to see all network traffic. This allows the device to passively detect network attacks from a centralized viewing point for the entire network. This technique can also be used for the mobile infrastructure environment where each mobile node must connect into a centralized access point. The detector uses a wireless network interface to monitor communication channels and can be placed in proximity of the wireless access point to enable monitoring of all network activity. This technique, however, cannot be used in the ad hoc network case, because the network can grow and contract over a geographical region where there may be no centralized point from which to monitor all network activity due to wireless propagation constraints.

40.4.4   Ad Hoc Network ID

Intrusion detection in wireless ad hoc networks calls for solutions that involve host-based monitoring of activity. This activity may be shared as in the distributed case, but a centralized sharing point may not be available. Ad hoc networks by their nature need to be more open than traditional wired networks. These networks exist for the convenience of their users, but must provide an environment of trust and protection or risk being shunned by an untrusting public. Due to the transient nature of the ad hoc networks, it will be important for detectors to recognize malicious activity and isolate offending nodes quickly. It also stands to reason that attacks against mobile ad hoc nodes will need to be quick and aggressive because target mobile nodes may stay in communication range with a malicious node for only a short period of time.

Intrusion detection is usually combined together with some form of reaction to detected intruders. The action can be anywhere from recognizing and logging malicious activity to terminating all network connection to or from the offending node. Proper reaction to intruders is just as important as detecting intruders. Malicious nodes that go unchallenged can continue their activity and potentially cause great harm to the network. On the other hand, some research shows the benefits of forgiveness and allowing nodes that have been excluded from the network due to malicious activity, to reenter the network after showing some period of reformed activity.

As described earlier, mobile devices can range from having relatively few resource constraints to being extremely resource limited. Many resource constrained nodes (i.e., sensor devices) will gradually restrict activity as power supplies begin to decrease below set thresholds. These so called selfish nodes may stop forwarding traffic for others although they will still need to send and receive network packets in order to operate. The ad hoc network intrusion detector should be able to distinguish between selfish nodes and malicious nodes [5]. Selfish nodes should not be punished and

Security, Trust, and Privacy in Mobile Computing Environments

some distinguishing feature needs to be considered to separate them from being identified as being malicious toward the network. In particular, we can view this problem in a policy sense: the routing algorithm implicitly imposes a policy governing the behavior of the nodes. What the IDS looks for is violations of this policy. This policy driven approach allows us to avoid the problems that a pattern recognition driven centralized approach will have in the ad hoc environment. Further, it should be noted that resource limited devices may not have the capabilities to collect, store, and analyze network traffic to distinguish network attacks with computationally or memory intensive algorithms. Lightweight algorithms will need to be developed for and used on resource limited devices.

40.4.5 Research

As stated earlier, intrusion detection in ad hoc networks involve host-based techniques. If one uses the Open Systems Interconnection (OSI) network stack as a point of reference, the most fundamental intrusion detection techniques exist in the data link and network layers. Essentially, detection involves each ad hoc node promiscuously monitoring network traffic within range and attempting to detect some form of malicious activity. If we assume a symmetric communication channel between node A and node B, then we are assured that both nodes can monitor network traffic generated from the other, barring collisions with traffic from other network nodes. The following scenario has been proposed [32]: Node A is sending traffic to node C through node B. Node A saves certain parameters of each packet sent to node C through node B. Node A then monitors node B's transmissions and compares each packet with the table of parameters to make sure that each packet sent to node C is forwarded with no modifications. If a modification is detected, then node A counts this as malicious activity. If a packet is not detected as having been forwarded within some time-out period, then this also is noted as malicious activity. If we take this idea and expand the responsibility of the nodes, then each node will attempt to detect malicious activity on packet traffic occurring within communications range [5]. Figure 40.2 has traffic going from node A to node E via node B and node D. Note that node A can monitor traffic transmitted from node B and node B can monitor traffic transmitted from node D. Node C can monitor traffic transmitted from node B and node D.

Promiscuously snooping network traffic means that each node must constantly be aware of its immediate neighbors. Traffic being sent to a destination outside of a node's communication range cannot possibly be tracked for correct handling. Additionally, the type of network routing algorithm used can determine how effectively snooping can be implemented.

There are many routing methods used in mobile ad hoc networks. It is rather ironic that one of the methods, Dynamic Source Routing, although

MOBILE COMPUTING HANDBOOK

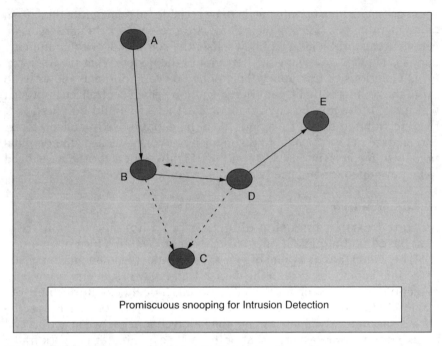

Figure 40.2. Promiscuous Snooping for Intrusion Detection

considered rather unsecure, actually provides good protection against malicious activity. Strict source routing allows other nodes to monitor the packet as it moves through the network. As the packet moves from source to destination, the source route is strictly adhered to and any deviation or modification can easily be detected by neighbor promiscuous nodes. The modifying node can then be identified and properly restricted, along with warnings being sent back to the source node for a route adjustment. It should also be noted that intrusion detection in promiscuous nodes needs to recognize and keep track of routing messages that include route adjustments. Nodes that could be misidentified as having misrouted packets in route messages showing link outages are not considered.

One of the problems with intrusion detection is having false positives (falsely accusing good nodes) and false negatives (failing to detect real intruders); the goal is to implement algorithms that decrease both statistics. Because there is no central authority in ad hoc networks, there is research into distributed voting schemes that involves some threshold of detection by multiple nodes before a bad node is admonished. One example is to use a clustering algorithm to organize the ad hoc network into groups with clusterheads. Figure 40.3 shows an ad hoc network partitioned into clusters. Any node within a cluster can report malicious activity

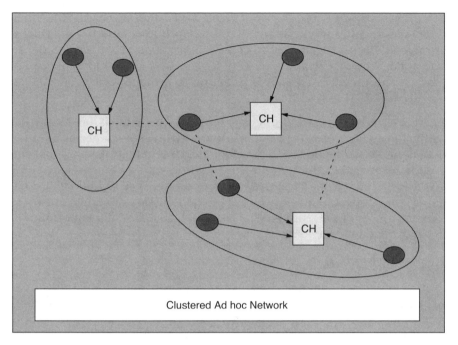

Figure 40.3. Clustered Ad Hoc Network

directly to its clusterhead. The clusterhead will request information from other clusterheads in the network to determine if other nodes have also detected malicious activity centering on the accused node. This takes the form of a vote where each clusterhead will be invited to vote positive if there have been reported incidents of malicious activity, negative if the accused node is within two hops of the clusterhead but no malicious activity has been reported, or neutral if the accused node is not within a two-hop radius of the clusterhead. The votes are arranged to have accusations made by first hand accounts of malicious activity. A quorum must vote and the majority vote determines if the node should be considered malicious. If there is a positive vote, then information about the node is flooded throughout the network and the accused node is ostracized. There is additional research into forgiveness and reentry into the network for nodes committing violations [5].

There is quite a bit of research into combining the detection of malicious activity in the network stack with monitoring in the application layer. This technique attempts to detect malicious activity that would look normal to the data link and network layer, but may be attempting to gain unauthorized entry into the device through an application or service vulnerability. There have been a couple of proposed frameworks for building host intrusion detection software. One framework focuses on using information from multiple layers within the node to get an in-depth look at network attacks. Intrusion detection agents act to collect information in and around the ad hoc node. When an attack is detected, agents on neighboring nodes are advised and they immediately take action to protect devices and the ad hoc network [33].

Another area of research focuses more on the establishment of trust within the ad hoc network [5]. This framework involves gathering information about network nodes and developing a trust rating system. Trust can then be propagated throughout the network, with greater weight being given for advice from an already trusted node. Distributed trust is an important concept and has been researched outside of ad hoc networks for some time now. These concepts seem to have particular importance in the area of ad hoc networks.

40.4.6 Multiple Malicious Nodes

Detection of individual malicious nodes is a problem, but multiple nodes working together present a much tougher scenario. The detection of malicious nodes working together requires detection information from multiple layers be shared between devices. A combination of pattern matching and anomaly detection can be used. There is ongoing research in this area and as the sophistication of network intruders increases, this area will become important in the future.

40.4.7 Directional Antennas and Power Control

One of the main concerns in RF wireless ad hoc networks is power consumption and waste during transmission. The RF hardware components tend to be a large source of power drain in the ad hoc node. Fine tuning the transmit power to be just large enough to reach the intended target is the first step in reducing waste. The second, and probably more important step, is to develop and steer the transmission beam into a long, narrow ellipse directed at the target with minimal side lobes [34]. Figure 40.4 shows a normal transmission range of node D encompassing node B, node C, and node E. The figure also shows that using directional antenna techniques, the broadcast beam can be narrowed down and aimed to form a narrow ellipse effectively eliminating the ability of node B and node C to monitor traffic destined for node E. Using these techniques, RF power consumption can be finely tuned for precise transmission to an intended target. There is real interest in this ongoing research. The fundamental approach to intrusion detection in wireless networks relies on each node promiscuously monitoring network traffic within its range. Although directional antenna technology will improve power consumption, it may radically limit the ability of nodes to promiscuously monitor network traffic.

Security, Trust, and Privacy in Mobile Computing Environments

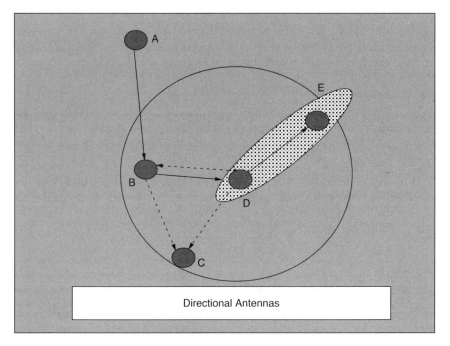

Figure 40.4. Directional Antennas

40.4.8 Discussion

Intrusion detection in mobile ad hoc networks is a problem filled with balancing intrusion detection method complexity against the constraints of the mobile ad hoc node. With still relatively small bandwidth, care should be taken in developing robust intrusion detection algorithms that do not overwhelm the network with a lot of traffic data. Mobile ad hoc networks provide a significant challenge for intrusion detection techniques. Although traditional host techniques for detection of attack against applications and services can work in the wireless world, there is an additional need to work at the network layers for a robust intrusion detector. Mobile ad hoc wireless networks are dynamic and can contain nodes with a wide range of capabilities and constraints. It will be important that a mobile ad hoc intrusion detection system be able to self-organize and provide adequate protection for the network as a whole. Techniques such as promiscuous snooping of network traffic combined with upper layer detection information can be combined in effective ways. Intrusion detection agents can be used to pull the information together and communicate with other agents within the network to recognize and work against intruders. Techniques like cluster voting can work to reduce the amount of false positive accusations within a network while also reducing the ability of a single malicious node to influence a network into denial of service against some target node. As the number of malicious nodes increases, the problem becomes tougher: as those malicious nodes start to cooperate and work together, the problem becomes even more complex.

40.5 Conclusion

In this chapter, we have introduced the problems that arise when security and privacy are sought to be introduced in open, dynamic systems. We show how these problems are not solved by authentication alone and how policy driven approaches for security and privacy may be more appropriate at all levels of the system from networking to applications. We then described our work in creating policy languages to support security, the use of such languages in enforcing privacy, and finally the use of a policy abstraction to detect intrusions.

References

1. Denker, G., Kagal, L., Finin, T., Paoucci, M., and Sycara, K., Security for DAML Web Serviced: Annotation and Matchmaking, in *Proceedings of 2nd International Semantic Web Conference*, October 2003. http://umbc.edu/~finin/papers/iswc03a.pdf.
2. Kagal, L., Finin, T., and Joshi, A., A Policy Based Approach to Security for the Semantic Web, in *Proceedings of 2nd International Semantic Web Conference*, October 2003. http://umbc.edu/~finin/papers/iswc03b.pdf.
3. Ding, L. and Finin, T., Trust Based Knowledge Outsourcing for Semantic Web Agents, in *Proceedings of 2003 IEEE/WIC International Conference on Web Intelligence*, Halifax, Canada, October 2003. http://umbc.edu/~finin/papers/wi03.pdf.
4. Kagal, L., Finin, T., and Joshi, A., A Policy Language for Pervasive Systems, in *Proceedings of 4th IEEE International Workshop on Policies for Distributed Systems and Networks*, Lake Como, Italy, June 2003. http://umbc.edu/~finin/papers/policy03.pdf.
5. Buchegger, S. and Le Boudec, J., Nodes Bearing Grudges: Toward Routing Security, Fairness, and Robustness in Mobile Ad Hoc Networks, in *Proceedings of 10th Euromicro Workshop on Parallel, Distributed, and Network-Based Processing*, 2002
6. Undercoffer, J., Pinkston, J., Joshi, A., and Finin, T., A Target-Centric Ontology for Intrusion Detection, *Knowledge Engineering Review*, 2004 (in press).
7. Undercoffer, J., Joshi, A., and Pinkston, J., Modeling Computer Attacks: an Ontology for Intrusion Detection, in *Proceedings of 6th International Symposium on Recent Advances in Intrusion Detection, Lecture Notes in Computer Science*, vol. 2516, New York: Springer-Verlag, September 2003.
8. Brickley, D. and Guha, R.V., RDF Vocabulary Description Language 1.0: RDF Schema, W3C Working Draft, April 30, 2002. http://www.w3.org/TR/rdf-schema.
9. Horrocks, I. et al., DAML+OIL Language Specifications. http://www.daml.org.
10. Dean, M. et al., Web Ontology Language (OWL) Reference Version 1.0, 2002. http://www.w3.org/TR/2002/WD-owl-ref-20021112/.
11. Bradshaw, J.M. et al., Representation and Reasoning for DAML-Based Policy and Domain Services in KAoS and NOMADS, in *Proceedings of the Autonomous Agents and Multi-Agent Systems Conference (AAMAS 2003)*, Melbourne, Australia, 2003.
12. Kagal, L., Finin, T., and Joshi, A., Developing Secure Agent Systems Using Delegation Based Trust Management, in *Proceedings of Security of Mobile MultiAgent Systems (SEMAS 02) Held at Autonomous Agents and MultiAgent Systems (AAMAS 02)*, 2002.

13. Mally, E., *The Basic Laws of Ought: Elements of the Logic of Willing*, 1926.
14. Meyer, J. and Weiringa, R., Deontic Logic: a Concise Overview, *Deontic Logic in Computer Science,* pp. 3–16, New York: John Wiley & Sons, 1993.
15. von Wright, G.H., A Note on Deontic Logic and Derived Obligation, *Mind*, 1956.
16. Godik, S. and Moses, T., OASIS eXtensible Access Control Markup Language (XACML), OASIS Committee Specification cs-xacml-specification-1.0, November 2002.
17. Ferraiolo, D. and Kuhn, R., Role-Based Access Controls, in *Proceedings of 15th {NIST}–{NCSC} National Computer Security Conference,* 1992.
18. Guiri, L., A New Model for Role-Based Access Control, in *Proceedings of 11th Annual Computer Security Application Conference*, pp. 249–255, New Orleans, December 11–15, 1995.
19. Sandhu, R.S., Role-Based Access Control, in M. Zerkowitz, Ed., *Advances in Computers*, vol. 48, San Diego, CA: Academic Press, 1998.
20. Yialelis, N., Lupu, E., and Sloman, M., Role-Based Security for Distributed Object Systems, in *Proceedings of 5th IEEE Workshops on Enabling Technologies: Infrastructure for Collaborative Enterprises (WET ICE '96)*, 1996.
21. Barka, E. and Sandhu, R., Framework for Role-Based Delegation Models, *Annual Computer Security Applications Conference,* 2000.
22. Herzberg, A. et al., Access Control Meets Public Key Infrastructure: Or Assigning Roles to Strangers, in *Proceedings of 2000 IEEE Symposium on Security and Privacy,* Oakland, CA, May 2000.
23. Hildmann, T. and Barholdt, J., Managing Trust between Collaborating Companies Using Outsourced Role Based Access Control, in *Proceedings of 4th ACM Workshop on Role-Based Access Control*, Fairfax, VA, 1999.
24. Blaze, M. et al., The KeyNote Trust Management System Version, *Internet RFC 2704,* September 1999.
25. Keromytis, A., Ioannidis, S., Greenwald, M., and Smith, J., The STRONGMAN Architecture, in *Proceedings of 3rd DARPA Information Survivability Conference and Exposition (DISCEX III)*, Washington, D.C., April 22–24, 2003.
26. Moffett, J. and Sloman, M., Policy Conflict Analysis in Distributed Systems Management, *Journal of Organizational Computing,* 1993.
27. Undercoffer, J., Perich, F., Cedilnik, A., Kagal, L., Joshi, A., and Finin, T., A Secure Infrastructure for Service Discovery and Management in Pervasive Computing, *The Journal of Special Issues on Mobility of Systems, Users, Data, and Computing,* 2003.
28. Campbell, R. et al., Toward Security and Privacy for Pervasive Computing, in *Proceedings of International Symposium on Software Security.*
29. Abowd, G.D., Brumitt, B., and Shafer, S., Privacy by Design — Principles of Privacy-Aware Ubiquitous Systems, in *Proceedings of UbiComp 2001*. http://www.inf.ethz.ch/vs/publ/papers/privacy-principles.pdf.
30. Al-Muhtadi, J. et al., Routing through the Mist: Privacy Preserving Communication in Ubiquitous Computing Environments, in *Proceedings of the International Conference of Distributed Computing Systems (ICDCS 2002).*
31. Chen, H., Finin, T., and Joshi, A., An Ontology for Context-Aware Pervasive Computing Environments, *Knowledge Engineering Review,* [Special Issue on Ontologies for Distributed Systems], 2003.
32. Marti, S. et al., Mitigating Routing Misbehavior in Mobile Ad Hoc Networks, in *Proceedings of MobiCom 2000*, pp. 255–265, Boston, 2000.
33. Zhang, Y. and Lee, W., Intrusion Detection in Wireless Ad-Hoc Networks, in *Proceedings of the 6th Annual International Conference on Mobile Computing and Networking*, pp. 275–283, Boston, 2000.
34. Yi, S., Pei, Y., and Kalyanaraman, S., On the Capacity Improvement of Ad Hoc Wireless Networks Using Directional Antennas, in *Proceedings of the 4th ACM International Symposium on Mobile Ad Hoc Networking and Computing,* pp. 98–107, Annapolis, MD, 2003.

35. Lederer, S., Dey, A.K., and Mankoff, J., Everyday Privacy in Ubiquitous Computing Environments. Presented at the *UbiComp 2002 Workshop on Socially-Informed Design of Privacy-Enhancing Solutions in Ubiquitous Computing,* September 29, 2002.
36. Chen, H., Finin, T., and Joshi, A., Semantic Web in a Pervasive Context-Aware Architecture, in *Proceedings of Artificial Intelligence in Mobile System 2003, UbiComp 2003,* Seattle, WA, October 12, 2003.
37. Bellotti, V. and Sellen, A., Design for Privacy in Ubiquitous Computing Environments, in *Proceedings of 3rd European Conference on Computer Supported Cooperative Work, (ECSCW 93),* pp. 77–92, 1993.
38. Salber, D., Dey, A.K., and Abowd, G.D., The Context Toolkit: Aiding the Development of Context-Enabled Applications, in *Proceedings of CHI'99,* pp. 434–441.
39. Blaze, M., Feigenbaum, J., and Lacy, J., Decentralized Trust Management, in *Proceedings of IEEE Conference on Privacy and Security,* Oakland, CA, pp. 164–173, May 6, 1996.
40. Keromytis, A., Ioannidis, S., Greenwald, M., and Smith, J., The Strongman Architecture, *Third DARPA Information Survivability Conference and Exposition (DISCEX III),* Washington, D.C., April 22-24, 2003.
41. Damianou, N., Dulay, N., Lupu, E., and Sloman, M., The Ponder Policy Specification Language, *The Policy Workshop 2001,* Bristol, U.K., Jan. 2001.
42. Lupu, E.C. and Sloman, M., Conflicts in Policy-Based Distributed Systems Management, *IEEE Transactions on Software Engineering,* vol. 25, no. 6, pp. 852–869, Nov./Dec. 1999.

Index

Note: Italized pages refer to illustrations.

A

ABR, *see* Associativity Based Routing (ABR)
Absolute validity interval (AVI), 365–366, 368–373, 377, 383
Abstract User Interface Markup Language (AUIML), 60
Abukmail studies, 731–743
Acceptable power to send (APTS), 558
Acceptance, mobile agent security, 941–942
Access control, 933–936, 955–956
Access control lists (ACLs), 948
Access cost, energy-aware web caching, 795
Access points (APs), *see also* Base stations (BSs)
    IEEE 802.11, 923–924
    routing and mobility management, 638
    transmitter power control, 806
Acer, 734
ACID properties, 842–843, 846
ACK, *see* Acknowledgement (ACK)
Acknowledgement (ACK)
    ad hoc network security, 494
    energy-aware web caching, 783–784, 801
    energy efficient wireless networks, 695
    MANET localized broadcasting, 683
    medium access control, *546,* 547, 549–552, *551, 553*
    message delivery, 245
    routing and mobility management, 644
    transmitter power control, 809
ACLs, *see* Access control lists (ACLs)
ACM Special Interest Group on Knowledge Discovery in Data and Data Mining (SIGKDD), 962
ACPI, *see* Advanced Configuration and Power Interface (ACPI)
Acquaviva studies, 747–776
Acquisition of context, 300–301
ACs, *see* Attribute Certificates (ACs)

Action contexts, 300
Active Badges, 310
Active Bat location systems, 301
Active Location Predicate, 268
Active Path Set (APS), 476–477
Active state, wearable computers, 15
Activity-based model, mobility tracking, 172, *172*
Activity-based Travel Demand Modeling Approach, 289
Adaptation Agents for Nomadic Users (Monads), 322, 327, 329
Adaptation commands and policies, 80–81
Adaptec, 734
Adapters, ISAM, 80
Adaptive algorithmic power optimization, *see also* Optimization
    application-assisted technique, 766–769, *767–769*
    basics, 747–749, 755–756, 776, *776*
    buffer insertion, 763–764
    clock setting, 760–763, *762–763*
    collaborative power management, 764–776
    communication energy optimization, 769–773, *770–772*
    computational energy optimization, 773–776, *774–775*
    decoding, 760–764
    encoding, 756–760
    energy, 749–755, *754*
    motion estimation, 757
    MPEG-4, 752–753, 759–790
    multimedia workload, 765–766, *766*
    operating systems, 765–769
    optimization, 749–755
    scalability, 749–755, *754*
    server-assisted technique, 769–776
    source coding, 750–753, *751,* 757–759
    stand-alone power management, 756–764
    transmission, 757–759

987

MOBILE COMPUTING HANDBOOK

voltage setting, 760–763, *762–763*
wavelets, 750–752, *751*
Adaptive clustering algorithm, 619
Adaptive Data Replication (ADR) algorithm, 210
Adaptive instantiation of software, 81
Adaptive Multi-Rate (AMR) codec, 125
Adaptive pervasive execution, 81
Adaptive prefetch approach, 412
Adaptive Routing using Clusters (ARC), 616, 620–621
Adaptive threshold scheme, 185
Adaptive value-based prefetch (AVP) scheme, 405, 415–416
Addressing, multimedia messaging service, 124–125
Ad hoc caching, *783,* 789–794, *790, 793–794, 801,* 801–802
Ad hoc networks
    cache management, 356–358
    energy awareness, web caching, 789–794, *799–800,* 799–801
    routing and mobility management, 638–639, *639,* 657–658
    security, trust, and privacy, 976–984
Ad hoc network security
    anticipated unknown attacks, 500
    basics, 483–486
    detection, 493–494
    distance vector routing, 490–491
    evaluation, 500
    future directions, 499–500
    IEEE 802.11, 486–487
    intrusion detection system, 495–496
    key management, 496–499
    link-layer security, 486–487
    link state routing, 491–492
    localized trust, 499
    MAC vulnerabilities, 486–487
    message authentication primitives, 488–489
    network layer security, 487–496
    proactive approach, 489–493
    protected packet forwarding, 493–495
    reaction, 494–495
    reactive approach, 493–495
    security in depth, 499–500
    source routing, 489–490
    trusted third party, 497–498
    trust management, 496–499
    unknown attacks, 500
    web-of-trust, 498–499
    WEP vulnerabilities, 487
Ad Hoc On-Demand Distance Vector (AODV) protocol

ad hoc network security, 486, 490, 492–493, 495
energy-aware web caching, 785
MANET security, 468
quality of service, 583–584, 592–593
routing and mobility management, 647, 650–651, 656
scalable clustered network architecture, 614, 617
self-policing MANETs, 437, 441
Ad hoc routing and protocols
    energy awareness, web caching, 785–788, *786*
    energy-efficient wireless networks, 700–703
    management, 641–656
Ad hoc traffic indication map (ATIM), 696
Ad hoc wireless networks, *see also* Mobile ad hoc networks (MANETs)
    attributed task graphs, 503–539
    energy-efficient utilization, 689–704
    localized broadcast, neighbor designation, 663–685
    medium access control mechanisms, 543–564
    quality of service routing, 569–606
    routing and mobility management, 637–658
    scalable clustered network architecture, 611–633
    security, 457–480, 483–500
    self-policing, 435–454
ADR, *see* Adaptive Data Replication (ADR) algorithm
ADUs, *see* Application data units (ADUs)
Advanced Configuration and Power Interface (ACPI)
    power awareness, 732, 734
    power management, 720, 736
Advanced Encryption Standard (AES), 927, 930
Advanced Mobile Application Support Environment (AMASE), 323–325, 327
Advanced Mobile Phone Service (AMPS), 247–248
Advanced power management (APM), 732, 734, 736
Advanced RISC machine (ARM), 734, 742
AES, *see* Advanced Encryption Standard (AES)
Agarwal studies, 896
Agents, *see* Mobile agents, security
AGLETS system, 951–953
Aho studies, 739
Ahson studies, 121–145
AIM, *see* America Online Instant Messenger (AIM)
Ajanta
    locating mobile objects, 198, 218, 222, *223*
    mobile agent security, 952–953

Index

Algorithmic power optimization, multimedia workload
  application-assisted technique, 766–769, *767–769*
  basics, 747–749, 755–756, 776, *776*
  buffer insertion, 763–764
  clock setting, 760–763, *762–763*
  collaborative power management, 764–776
  communication energy optimization, 769–773, *770–772*
  computational energy optimization, 773–776, *774–775*
  decoding, 760–764
  encoding, 756–760
  energy, 749–755, *754*
  motion estimation, 757
  MPEG-4, 752–753, 759–790
  multimedia workload, 765–766, *766*
  operating systems, 765–769
  optimization, 749–755
  scalability, 749–755, *754*
  server-assisted technique, 769–776
  source coding, 750–753, *751*, 757–759
  stand-alone power management, 756–764
  transmission, 757–759
  voltage setting, 760–763, *762–763*
  wavelets, 750–752, *751*
Algorithms
  localized broadcast, neighbor designation, 669–672
  message delivery, 229
  scalable clustered network architecture, 617–622, 627–629, *629*
ALICE, 99
Al-Karaki studies, 569–607
Al-Muhtadi studies, 972
ALOHA carrier sensing, 545
Alpha, 741
Always-update strategy, 175
Alzoubi studies, 678
AMASE, *see* Advanced Mobile Application Support Environment (AMASE)
Amdahl's Law, 724, 727–728
American Standard Code ofr Information Interchange (ASCII), 929, 936
America Online Instant Messenger (AIM), 303, 305, 307–308
Amnesic terminals (AT) approach, 362
  ache management, 352–353
AMPS, *see* Advanced Mobile Phone Service (AMPS)
AMR, *see* Adaptive Multi-Rate (AMR) codec
Analytic mobility, PCS networks
  basics, 877–878

cellular systems, *878,* 878–879
cost, 886–887
distributions, 886
location update, 881–886
Markov walk models, *879,* 879–881, *881*
paging, 886–887
performance evaluation, *881,* 887–891, *888–889*
recursive Markov analysis, 883–885
system models, 878–881
tracking, location, 882
two renewal processes, 882–883, *883*
updates, location, 882
Andrew File system, 406
Angular interference reduction, 560–563, *561*
Announcements, 231, 233–234
Anonymity, 970–972
Anticipated unknown attacks, 500
Antientropy, 862
Anycasting, dynamic task-based, 505
AODV, *see* Ad Hoc On-Demand Distance Vector (AODV) protocol
AP, *see* Access points (APs)
Apache Struts, 63–64
API, *see* Application Processing/Programming Interface (API)
APM, *see* Advanced power management (APM)
Application-assisted technique, 766–769, *767–769*
Application data units (ADUs), 516, 532, 536
Application Processing/Programming Interface (API)
  adaptive algorithmic power optimization, 770–771
  context-awareness, 303
  grid and cluster computing, 97, 111
  ISAM, 86
  pervasive application development, 57, 59
  power awareness, 740
Applications
  environments, pervasive application development, 58–59
  ISAM, 92
  solutions, power awareness, 740–741
APS, *see* Active Path Set (APS)
APTS, *see* Acceptable power to send (APTS)
ARAN, *see* Authenticated Routing for Ad hoc Networks (ARAN)
ARC, *see* Adaptive Routing using Clusters (ARC)
Architecture
  ISAM, *75–76,* 75–77
  locating mobile objects, 201–215
  mobile transaction models, 852–853
  multimedia messaging service, *122,* 122–126
  power awareness, 733–736

989

MOBILE COMPUTING HANDBOOK

Architecture-based wireless networks, 346, 347–356, 348
Architecture for Mobile Agents with Recursive Itinerary and Secure Migration (MARISM-A), 952
ARC protocol, 625, 625–626
Aref, Samet and, studies, 258–259
Aref and Samet studies, 259
Ariadne, 442, 489
ARM, see Advanced RISC machine (ARM)
Aroma, 953
ARTour Web Express, 796
AS algorithm, see Asynchronous stateful (AS) algorithm
ASCII, see American Standard Code ofr Information Interchange (ASCII)
ASP.NET Mobile Controls, 60
ASR, see Automatic speech recognition (ASR) engines
Assembly tuples, 41
Association rules, 394–396
Associativity Based Routing (ABR), 647, 651
Asynchronous stateful (AS) algorithm, 349–350
Asynchronous Transfer Mode (ATM), 249, 696
AT, see Amnesic terminals (AT) approach
ATIM, see Ad hoc traffic indication map (ATIM)
ATM, see Asynchronous Transfer Mode (ATM); Asynchronous transfer mode (ATM)
Atomicity, 846, see also ACID properties
Atomic transactions, 850
Attacks, 438–439, see also Security
AT&T Labs, 310, 929
Attribute Certificates (ACs), 945–946
Attributed task graphs
 ADU delay distributions, 536–537, 536–537
 algorithms and protocols, 516–528
 basics, 503–506, 537–539
 data flow tuple representation model, 511
 dilation, 528–531, 530
 disconnections, 527–528
 distributed algorithm, 522–528, 523, 529
 effective throughput, 529, 532–533, 533
 embedding, 514, 515, 531, 531–532
 greedy algorithm, 519–522, 520–521
 handling device mobility, 525–526, 525–527
 nonpreassigned tasks, 510
 performance evaluation, 516, 528–537, 529
 polynomial-time embedding algorithm, 517–519, 518, 520–521
 preassigned tasks, 510
 preliminaries, 507–508
 problem formulation, 517
 reinstantiation, 534, 535
 task execution, 507–513

 task graph instantiation, 516–528
 tasks and task graphs, 507–516, 509
 taxonomy of tasks, 510
Augustin studies, 73–93
AUIML, see Abstract User Interface Markup Language (AUIML)
Authenticated Routing for Ad hoc Networks (ARAN), 488, 492–493
Authentication
 ad hoc network security, 496
 IEEE 802.11, 933–936
 MANET security, 459
 mobile agent security, 947
Authorization
 MANET security, 460
 mobile agent security, 947–948
 self-policing MANETs, 437
Automatic contextual reconfiguration, 302
Automatic home agent discovery, 220
Automatic repeat request (ARQ), 692, 895
Automatic speech recognition (ASR) engines, 298
Availability, ad hoc network security, 484
Availability adaptation, 108–110
Average Effective Throughput (AvgEffT), 516, 527, 532
AvgEffT, see Average Effective Throughput (AvgEffT)
AVI, see Absolute validity interval (AVI)
AVP, see Adaptive value-based prefetch (AVP) scheme
Awerbuch studies, 492, 495

B

Backbone-based message delivery, 243–246, 244, 246
Backoff algorithm, MAC, 544, 546, 549, 550, 551, 554, 556
Bagrodia studies, 95–118
Baker, Bansal and, studies, 445
Baker, Marti, Giuli, Lai and, studies, 444
Balakrishnan studies, 839–874
Bambos studies, 805–835
Banavar studies, 53–69
Bandwidth efficiency, cache management, 355–356
Bansal and Baker studies, 445
Bao and Garcia-Luna-Aceves studies, 558
Barbara and Imielinski studies, 362, 407
Barbosa studies, 73–93
Base stations (BSs), see also Access points (APs)
 cache management, 339, 347
 connectivity, message delivery, 249
 IEEE 802.11, 923

Index

routing and mobility management, 637
transmitter power control, 806
Basu studies, 503–539
Bat System, 310
Batteries
    grid and cluster computing, 100, 102
    life target, power-aware cache management, 415–416
    power awareness, 733–734
    power management, mobile computers, 710–712, *711,* 725–727, *726–727*
    self-policing MANETs, 437
Battery power management, portable devices
    analysis, 900–909, *901*
    basics, 893–899
    discharge/recharge model, 899–900, *900*
    exhaustive service, 904
    extensions, 903–909
    generalizations, 903–909
    lithium ion simulation results, *914–915,* 914–916
    multi-battery system, *905,* 905–909
    nonexhaustive service, 904
    performance results, *910–913,* 910–916
    relaxation phenomenon, 896–899, *897*
    scheduling problem, 916–917
    system model, 899–900
    vacations, 904
Bat Ultrasonic Location System, 310
Bayou
    basics, 862–863
    communication cost and scalability, 872
    comparisons, *865, 867*
    consistency and concurrency, 869
    infrastructure requirements and compatibility, 870
    Lime, 26
Becker studies, 275–293
Bellas studies, 736
Bellavista studies, 315–332, 941–957
Bellman-Ford algorithm, 642
Bellosa studies, 738
Benini studies, 747–776
Beowulf systems, 98, 110, 112
Berkeley Internet Name Domain (BIND), 156, 160
Bessel function, 903
Best quality (BQ) path, 588
BGP, *see* Border Gateway Protocol (BGP)
Bharghavan, Sinha, Sivakumar and, studies, 667
Bharghavan studies, 554
BICONN-AUGMENT, 704
Binary exponential backoff scheme, 549
BIND, *see* Berkeley Internet Name Domain (BIND)

Binder Manager, 326
BIP, *see* Broadcast Incremental Power (BIP)
Bit-sequenced (BS) scheme
    cache invalidation, 362, 372–373, 377, 383
    energy efficient cache invalidation, 429, *429–430,* 429–431
    power-aware cache management, 407–408
BLiMST, *see* Broadcast Link-based MST (BLiMST)
BLU, *see* Broadcast Least-Unicast-cost (BLU)
Bluetooth
    battery power management, 895
    grid and cluster computing, 99, 102, 117
    MANET security, 463
    mobile agent middlewares, 317
    modeling distributed applications, 504
    power-aware cache management, 403
    wearable computers, 15
Bordercast Resolution Protocol (BRP), 631
Border Gateway Protocol (BGP), 245
Borisov studies, 923–938
Borriello, Hightower and, studies, 300
Bottazzi studies, 315–332
BQ path, *see* Best quality (BQ) path
Bradshaw studies, 954
BRANs, *see* Broadband radio access networks (BRANs)
Broadband radio access networks (BRANs), 638
Broadcast Incremental Power (BIP), 703
Broadcasting time stamp scheme, 407
Broadcast Least-Unicast-cost (BLU), 703
Broadcast Link-based MST (BLiMST), 703
Broadcast search, message delivery, 232–240
BRP, *see* Bordercast Resolution Protocol (BRP)
BSs, *see* Base stations (BSs)
BS scheme, *see* Bit-sequenced (BS) scheme
Buchegger studies, 435–454
Buffers
    adaptive algorithm power optimization, 763–764
    transmitter power control, 811–816, 820
Bulk operations, Lime, 28
Busy tone, medium access control, 544, 546, 557–558, *559*
Buttyán and Hubaux studies, 440

C

CA, *see* Certification Authority (CA)
Cache Consistency Model, 405–406
Cache hit rate, 376
Cache invalidation schemes
    AVI, 366–373
    basics, 361–362, 383, 387

991

MOBILE COMPUTING HANDBOOK

client algorithm, 371–373
database size, 376–383, *379–380*
evaluation, performance, 376–383, *377*
existing schemes, 362–364
explicit invalidation, 372–373, *374*
IAVI, 368–369
implicit invalidation, 371–372
invalidation report generation, 369–370
metrics, performance, 375–376
performance study, 373–383
server algorithm, 369–373, *370*
system model, 373, *375*
temporal data model, 364–366
update rate, *377, 381–382, 384–386*
validity period, 366–368, *367*
Cache management, *see also* Power-aware cache management
  ad hoc networks, 356–358
  architecture-based wireless networks, *346, 347*–356, *348*
  bandwidth efficiency, 355–356
  basics, 337–342, *338–339*, 358
  consistency strategies, 346–347
  cooperative caching, 356–358
  energy awareness, web caching, 795–796
  energy efficiency, 355–356
  handling disconnections, 355
  invalidation-based strategy, 348–355
  problems, 356–358
  stateful approaches, 348–350, *350–352*
  stateless approaches, 350–355, *354–355*
  state of the art, 342–346, *344*
  taxonomy of schemes, *346*, 346–347
  TTL-based strategy, 355–356
Cache time, DNS location management, 157–158, *158*
Caching, *see also* Web caching
  eager caching, 207
  energy efficient invalidation, 421–432
  hoarding, 389–400
  invalidation, 361–387, 421–432
  lazy caching, 207
  level caching, 208
  locating mobile objects, *206,* 206–209, *209*
  management, 337–358
  mobile transaction models, 856–857
  power-aware management, 403–418
  simple caching, 208
Cali studies, 552, 556
Call arrival pattern, *174,* 174–175
Call to mobility ratio, 205
Candidate set construction, 396
Cao studies, 403–418
Capability/application requirements match, 5

Capacitance load reduction, 735
Capra studies, 311
Cardei studies, 689–705
Care-Free Sleep theorem, 698
Care-of-address (CoA), 219–220, 657
Carnegie Mellon's Wearable Computers project, 3, *see also* Wearable computing
Carrier Sense Multiple Access (CSMA) protocol, 654
Carrier Sense Multiple Access with Collision Avoidance (CSMA/CA) protocol, 544, 695
Carrier sense threshold (CSThresh), 549–550
Case studies, *see also* IEEE 802.11
  locating mobile objects, 218–223
  power management, mobile computers, 721–727, *722*
Castelluccia, Montenegro and, studies, 445
Cavalheiro studies, 73–93
CBRP, *see* Cluster-Based Routing Protocol (CBRP)
CC-EMBED, 517
CCMP, *see* Counter Mode with Cipher Block Chaining Message Authentication Code Protocol (CCMP)
CCN, *see* Current Computation Node (CCN)
CCS, *see* Common Channel Signaling (CCS) network; Credit clearance service (CCS)
CDF, *see* Cumulative density function (CDF)
CDMA, *see* Code division multiple access (CDMA)
CDS, *see* Connected dominating set (CDS)
CEDAR, *see* Core extraction distributed ad hoc routing (CEDAR)
Cell history, 169–170
Cellular systems, *see also* Mobility tracking
  distance-based location update, *178*
  graph model, *167*
  grouping of location areas, *181*
  hexagonal configuration, *169*
  location accuracy matrix, *190*
  mesh configuration, *168*
  movement-based location update, *177*
  network plan, *164*
  one-dimensional network, *168*
  oscillating user location, *179*
  overlapped location areas, *180*
  paging, *188*
  PCS networks, analytic mobility models, *878,* 878–879
  reporting cells, *183–184*
  ring paging, *188*
  routing and mobility management, 637
  3 by 3 network, *170*
  transitional diagram, *171*

Index

two-dimensional network, *168–169*
unbounded nonreporting cells, *184*
Cellular telephones, 403, 731, 893
Centralized control, 579
Central processing unit (CPU)
    adaptive algorithmic power optimization, 761–763
    energy efficient cache invalidation, 423
    power management, 716, 718–719
    speed-setting, dynamic, 722–725
Centrino processor, 96
Certification Authority (CA)
    ad hoc network security, 497–498
    MANET security, 464–465, 468
Certification reply (CREP), 498
Certification request (CREQ), 498
CGI, *see* Common Gateway Interface (CGI)
CGSR, *see* Cluster Gateway Switch Routing (CGSR)
Chandy-Lambert algorithm, 235, 238, 247
Channel listening, 784–785
Chan studies, 361–387
Chen, H., studies, 961–984
Chen, T.W., studies, 595
Chen, Y., studies, 584
Chen, Y.F., studies, 297–312
Chen and Yang, Zhong, studies, 441
Chiasserini and Rao studies, 898
Chockalingham studies, 893–917
CHTML, *see* Compact HTML
Chung studies, 773
Classifications
    localized broadcast, neighbor designation, 665–672
    location-dependent queries, 270–271, *271*
    scalable clustered network architecture, 613–617
Clear-to-send (CTS) packets
    ad hoc network security, 486
    energy-aware web caching, 782–784, 801
    energy efficient wireless networks, 695
    medium access control, 547, 554
Client algorithm, cache invalidation schemes, 371–373
Client failures, 338
Client process, DNS location management, 152, 154–155
Clock setting, 760–763, *762–763*
Cluster architectures and protocols, *615*, 615–617, *618*
Cluster-based algorithm, 677–681, *679–681*
Cluster-Based Routing Protocol (CBRP), 616, 620–621, 623, *625*, 625–626

Clustered data model
    basics, 851–852
    communication cost and scalability, 871
    comparisons, *865–866*
    consistency and concurrency, 864
    infrastructure requirements and compatibility, 869
Cluster Gateway Switch Routing (CGSR)
    routing and mobility management, 642, 645–646, *646*, 658
    scalable clustered network architecture, 616, 620, 623, *623*, 623–624
Cluster member table, 623
Cluster topology configuration, 116
CMOS circuitry components, 734–735
CoA, *see* Care-of-address (CoA)
CoBrA, *see* Context Broker Architecture (CoBrA)
CoD, *see* Code on Demand (CoD)
Coda and Coda file system
    cache management, 343–344
    hoarding, 391–393, 399–400
    Lime, 26
    mobile transaction models, 860
    power-aware cache management, 404, 406
Code, mobile agent security, 946–947
Code base, 45
Code deployment, 117
Code division multiple access (CDMA)
    battery power management, 895
    quality of service, 578, 584, 586, 592, 595
    transmitter power control, 806, 818
Code generation, 739
Code mobility, 318
Code on Demand (CoD), 45, 319
Cohen studies, 53–69
Coherency, energy-aware web caching, 795–796
Collaboration, mechanics, 19
Collaborative power management, 764–776
Collaborative Reputation (CORE), 450, *see also* Reputation systems
Collisions
    energy awareness, web caching, 782–784, *783*
    medium access control, *546*, 547, 549–552, *551, 553*, 554, 556
Columbia University, 8
Commerical databases, 869–871
Common Channel Signaling (CCS) network, 202
Common Gateway Interface (CGI), 154–155, 158–161
Common Object Request Broker Architecture (CORBA)
    context-awareness, 305
    grid and cluster computing, 99
    mobile agent middlewares, 321–322, 325

993

MOBILE COMPUTING HANDBOOK

Communication
   energy awareness, web caching, 789–791, *790*
   energy optimization, 769–773, *770–772*
   grid and cluster computing, 112–113, *114*
   mobile transaction models, 845, 871–872
   power awareness, 736–738
   quality of service, 577
   security, trust, and privacy, 976
CompactFlash card, SanDisk, 100
Compact HTML, 60
Compaq iPAQ laptops, 112
Compaq Itsy, 721–722, 725
Compensatable transaction, 850
Compilation, power awareness, 738–740
Composer, 67
Composite Capabilities/Preferences Profile (CC/PP), 312, 322
Compound predicate (CP), 264
Comprehensive protection, 484
Compression-based scheme, 185–186
Compromise of devices, 484
Computational energy optimization, 773–776, *774–775*
Computation grid, *see* Grid and cluster computing
Computing contest, 300
Conceptual Privacy Model, 971
Concordia system, 951, 953
Concurrency, mobile transaction models, 844, 856–857, 864–869
CONFIDANT, *see* Cooperation Of Nodes, Fairness In Dynamic Ad-hoc NeTworks (CONFIDANT)
Confidentiality, 437, 460, 496, *see also* Security
Conflict resolution, 968
Congestion control, 549, *550*
CONNECT, 704
Connected/disconnected mode, 863
Connected dominating set (CDS), *674*
   localized broadcast, neighbor designation, *674, 676–677, 677*
   MANET localized broadcasting, 666–670
Connectivity, 57, 619
Consent, 970–971
Consistency, 346–347, 864–869, *see also* ACID properties
Contention rule, 921
Content of invalidation report, 423–425
Context-awareness
   actions, 302
   adaptation, 75, *79*
   basics, 315
   issues, 330–332
   lessons learned, 330–332
   middlewares, 326–330, *327*

mobile agents, 318–322, *319*
mobile computing and context awareness, 316–318
pervasive application development, 56, 66
research directions, 326–330, *327*
self-policing MANETs, 445
supports, 322–326, *324*
triggered actions, 302
wearable computing, 5–6, *7,* 8, 12–17, *14, 16–17,* 21
Context-awareness, mobile computing
   acquisition of context, 300–301
   basics, 297–299, 301–305, 312
   context, 299–300
   contextware, 306–310
   future directions, 311–312
   IETF OPES group, 303
   iMobile, 303–305, *304, 306,* 307–308, *309*
   location determination/service platform integration, 306, *307*
   location sensing techniques, 301
   middleware, 306–310
   motivation, 297–299
   Parlay, 302–303
   privacy, 308–310
   research initiatives and projects, 310–311
   sensor data acquisition, 300–301
   technology independent interfaces, 302–305
COntext-aware Security middleware for MObile agent Systems (COSMOS), 956
Context Broker Architecture (CoBrA), 969, 972–976, *973*
Context sources, 58
Context-triggered actions, 302
Contextual command applications, 302
Contextware, 306–310
CoolAgents, 328
Cooltown project, 328
Cooperation Of Nodes, Fairness In Dynamic Ad-hoc NeTworks (CONFIDANT), 444, 447–452
Cooperative caching, 356–358
Coopnet, 357
CORBA, *see* Common Object Request Broker Architecture (CORBA)
CORE, *see* Collaborative reputation (CORE)
Core Extraction Distributed Ad hoc Routing (CEDAR)
   quality of service, 597–598
   routing and mobility management, 655
   scalable clustered network architecture, 616, 628–630, *630*
CoreLime, 49
Cornell University, 67

Index

Corradi studies, 315–332, 941–957
COSMOS, *see* COntext-aware Security middleware for MObile agent Systems (COSMOS)
Cost, PCS networks, 886–887
Cost-based page replacement algorithm, 344
Cotransactions, *see* Reporting and cotransactions model
Cougar, 67
Counter-based scheme, localized broadcast, 666
Countermeasures, mobile agent security, 945–951
Counter Mode with Cipher Block Chaining Message Authentication Code Protocol (CCMP), 925, 930, 933, 935–937
Covered backbone, message delivery, 245
CP, *see* Compound predicate (CP)
C-130 project, 10
CPU, *see* Central processing unit (CPU)
CRC, *see* Cyclic redundancy codes (CRC)
Credit clearance service (CCS), 441
CREP, *see* Certification reply (CREP)
CREQ, *see* Certification request (CREQ)
Cricket location support system, 301, 310
Crusoe processor, 732, 734–735
Cryptography
    ad hoc network security, 496–497
    MANET security, 462, 465–467, 469–470, 475, 479–480
    self-policing MANETs, 441–442
CSMA, *see* Carrier Sense Multiple Access (CSMA) protocol
CSMA/CA protocol, Carrier Sense Multiple Access with Collision Avoidance (CSMA/CA) protocol
CSThresh, *see* Carrier sense threshold (CSThresh)
CTS, *see* Clear-to-send (CTS) packets
Cumulative ADU delay distributions, 535
Cumulative density function (CDF), 902–903, 907
Cumulative probability distributions (CPFs), 536–537
Currency of location information, 200
Current Computation Node (CCN), 699
Cyclic redundancy codes (CRC), 355, 930–931

D

DAA, *see* Data access agent (DAA)
DAG, *see* Directed acyclic graph (DAG)
D'Agents, 323, 328, 952
Dai, Wu and, studies, 669
Dai and Wu studies, 669
DAML+OIL, *see* Defense Advanced Research Project Agency (DARPA)
DARPA, *see* Defense Advanced Research Project Agency (DARPA)
Dartmouth Agents, 323, 952
Dartmouth University, 68
da Silva studies, 73–93
Das studies, 667
Data
    consistency, 844
    downloading, 831–833
    forwarding, 473–477
Data access agent (DAA), 857
Data and queries, 256–258
Databases for Moving Objects (DOMINO), 258
Database size, 376–383, *379–380*
Data-centric naming, 66
Data Encryption Standard (DES), 927
Data mining, 392–396
DBLA, *see* Distance-based location area (DBLA) strategy
DBS-RD, *see* Dual battery scheme with random scheduling (DBS-RD)
DBTMA, *see* Dual Busy Tone Multiple Access (DBTMA)
DCB, *see* Double-covered broadcast (DCB) algorithm
DCF, *see* Distributed coordination function (DCF)
DCF interframe spacing (DIFS), 547
DCT, *see* Discrete cosine transform (DCT)
DCTS, *see* Directional CTS (DCTS)
DDCH, *see* Distributed Database Coverage Heuristic (DDCH)
Deafness, 560
Decoding, 760–764
Deep prefetching, 835, *see also* Prefetching
Defense Advanced Research Project Agency (DARPA)
    Agent Markup Language plus Ontology Inference Layer (DAML+OIL), 964
    LifeLog program, 962
Delay constraint, 165
Delivery report, *126,* 129–132, *133–136*
Dell laptops, 112
Denial of Service (DoS)
    ad hoc network security, 486–487, 493, 497
    MANET security, 459, 462
Depth 1 look-ahead, 833–835
DES, *see* Data Encryption Standard (DES)
Descriptive naming, 66
Design-time adaptation, 61–62
Destination Sequenced Clustered Routing (DSCR), 616, 620, 623, *624,* 624–625
Destination-Sequenced Distance Vector (DSDV)
    ad hoc network security, 486, 488–490
    battery power management, 896

995

energy-aware web caching, 785
quality of service, 583, 592, 596
routing and mobility management, *639*, 642–644, *643*, 646, 658
scalable clustered network architecture, 614
self-policing MANETs, 442
Detection
ad hoc network security, 493–494
self-policing MANETs, 442–445, 452–453
Deterministic algorithms, 666–669
Device-independent views, 60–63, 69
Devices
compromise/theft, 484
heterogeneity, 104
platform heterogeneity, 56–58
security, trust, and privacy, 976–977
Dey studies, 300
DHCP, *see* Dynamic Host Configuration Protocol (DHCP)
Dict being, 83–86, 89, 91
Diengagement, 32
Difference Cache Consistency Model, 418
Diffie-Hellman algorithm, 463, 936
Diffusing computations, *241*, 241–243
Digital Equipment Corporation, 17
Digital Ink, 4
Digital pulse code modulation/discrete cosine transform (DPCM/DCT), 757
Digital signal processor (DSP), 754
Dijkstra algorithm, 290, 645
Dijkstra-Scholten algorithm, 241–242
Directed acyclic graph (DAG), 739
Directional antennas
medium access control, 544, 560–563, *561*
security, trust, and privacy, 982, *983*
Directional CTS (DCTS), 560
Directional hidden terminals, 560
Directional history, 170, *170–171*
Directional MAC (DMAC), 560
Directional reception capability, 563
Directional RTS (DRTS), 560
Directory-based schemes, 358
Direct sequence spread spectrum (DSSS), 554, 799
Discharge/recharge model, 899–900, *900*
Disconnectable application, 56
Disconnections, 355, 389–390, 398, *see also* Hoarding
Disconnect/reconnect, 92
Discovery procedure, 471–472
Discrete cosine transform (DCT), 750
Discrete Time Markov Chain (DTMC), 693
Discrete wavelet transform (DWT), 750
Display update, Lime, *42*, 43–44

Distance-based location area (DBLA) strategy, *167*, 177–178, *178*
Distance-based scheme, localized broadcast, 666
Distance Routing Effect Algorithm for Mobility (DREAM), 655
Distance vector
ad hoc network security, 490–491
routing and mobility management, 641
scalable clustered network architecture, 614
Distributed applications modeling
ADU delay distributions, 536–537, *536–537*
algorithms and protocols, 516–528
basics, 503–506, 537–539
data flow tuple representation model, 511
dilation, 528–531, *530*
disconnections, 527–528
distributed algorithm, 522–528, *523, 529*
effective throughput, *529*, 532–533, *533*
embedding, 514, *515, 531*, 531–532
greedy algorithm, 519–522, *520–521*
handling device mobility, *525–526*, 525–527
nonpreassigned tasks, 510
performance evaluation, 516, 528–537, *529*
polynomial-time embedding algorithm, 517–519, *518, 520–521*
preassigned tasks, 510
preliminaries, 507–508
problem formulation, 517
reinstantiation, *534*, 535
task execution, 507–513
task graph instantiation, 516–528
tasks and task graphs, 507–516, *509*
taxonomy of tasks, 510
Distributed computing, message delivery, 228
Distributed coordination function (DCF)
energy efficient wireless networks, 695
medium access control, 543–544, 546–552
Distributed coordination function interframe spacing (DIFS), 547
Distributed Database Coverage Heuristic (DDCH), 617
Distributed environment, hoarding, 398–399
Distributed framework, security, 968
Distributions, PCS networks, 886
DMAC, *see* Directional MAC (DMAC)
DNS, *see* Domain Name System (DNS)
Domain Name System (DNS)
basics, 149–151, 160–161
cache time, 157–158, *158*
client process, 154–155
DNS server, 152–154
experiments, *156*, 156–160
nomadic host location management, 151–155, *152*

Index

scalability analysis, 158–160, *159*
security, 155
server process, 154
time-to-live, 157–158, *158*
web browser, 154–155
web server, 154
Dominant pruning, *674–675,* 675
Dominating set (DS), 666
DOMINO, *see* Databases for Moving Objects (DOMINO)
DoS, *see* Denial of Service (DoS)
Double-covered broadcast (DCB) algorithm, 684–685
Douceur studies, 445
Downloading, 398, 831–833
Doze mode, 863
DPCM/DCT, *see* Digital pulse code modulation/discrete cosine transform (DPCM/DCT)
DRCI, *see* Dual-Report Cache Invalidation (DRCI) scheme
DREAM, *see* Distance Routing Effect Algorithm for Mobility (DREAM)
DRTS, *see* Directional RTS (DRTS)
DRTS/DCTS-based DMAC, *561,* 562–563
DRTS/oCTS-based DMAC, 562
Druid, 60
DS, *see* Dominating set (DS)
DSCR, *see* Destination Sequenced Clustered Routing (DSCR)
DSDV, *see* Destination-Sequenced Distance Vector (DSDV)
DSP, *see* Digital signal processor (DSP)
DSR, *see* Dynamic Source Routing (DSR)
DSRT, *see* Dynamic Soft Real-Time (DSRT) processor scheduler
DSSS, Direct sequence spread spectrum (DSSS)
DTMC, *see* Discrete Time Markov Chain (DTMC)
Du, D.Z. studies, 689–705
Du, Y. studies, 337–358
Dual battery scheme with random scheduling (DBS-RD), 914–915
Dual Busy Tone Multiple Access (DBTMA), 556, 558
Dual-Report Cache Invalidation (DRCI) scheme
  cache invalidation, 363
  cache management, 353–354
  energy efficient cache invalidation, 427–428
Dunham, Ren and, studies, 270
Dunham studies, 255–272, 839–874
Durability, 846, *see also* ACID properties
DVS, *see* Dynamic processor voltage scaling (DVS)

DWT, *see* Discrete wavelet transform (DWT)
Dynamic, unpredictable availability, 104
Dynamic case, mobility tracking, 181–182, 184
Dynamic CPU speed-setting, 722–725, *723–725*
Dynamic Host Configuration Protocol (DHCP)
  DNS location management, 150–151
  IEEE 802.11, 931
  MANET security, 478
  quality of service, 579
  security, trust, and privacy, 976
Dynamic processor voltage scaling (DVS), 749, 775
Dynamic programming formulation, 827–829, *828*
Dynamic Soft Real-Time (DSRT) processor scheduler, 766
Dynamic Source Routing (DSR)
  ad hoc network security, 486, 488–489, 492, 495
  battery power management, 896
  energy-aware web caching, 785, 787
  modeling distributed applications, 532
  quality of service, 583, 598–599
  routing and mobility management, 647–650, 656
  scalable clustered network architecture, 614, 617
  security, trust, and privacy, 979–980
  self-policing MANETs, 437, 441, 444
Dynamic task-based anycasting, 505
Dynamic topology, quality of service, 576–577
Dynamic tuning, MAC, 556, *557*

E

Eager caching, 207
EAP, *see* Extensible Authentication Protocol (EAP)
EC-MAC, *see* Energy Conserving MAC (EC-MAC)
Economic model, grid and cluster computing, 117
Economic systems, self-policing MANETs, 446
EEHF, *see* Environmental Energy Harvesting Framework (EEHF)
Efficiency, ad hoc network security, 484
Efficient data downloading, 831–833
Egenhofer studies, 259
EIA/TIA, *see* Electronics Industry Association/Telecommunication Industry Association's (EIA/TIA) Standard 41 (IS-41)
EIFS, *see* Extended interframe spacing (EIFS)
E911 initiative, 299

997

Electronics Industry Association/Telecommunication Industry Association's (EIA/TIA) Standard 41 (IS-41), 201, 657
El-Rewini studies, 149–161
Embedding, 505
EMERALD, *see* Experimental Machine Example-based Reasoning and Learning Disciple (EMERALD)
EMN, *see* Enterprise Messaging Network (EMN)
EMS, *see* Enhanced Messaging Service (EMS)
Enago, 324–325
Encoding, adaptive algorithm power optimization, 756–760
Encryption, IEEE 802.11, 927–930, *928*
End-host reaction, 494
Energy, 749–755, *754*
Energy-aware processors, 734
Energy aware web caching
  ad hoc caching, 791–794, *793–794*
  ad hoc cooperative caching, *783,* 789–794, *790, 793–794, 801,* 801–802
  ad hoc networks, 789–794, *799–800,* 799–801
  ad hoc routing protocols, 785–788, *786*
  basics, 779–781, 802
  cache management, 795–796
  channel listening, 784–785
  collisions, 782–784, *783*
  communication, 789–791, *790*
  evaluation, 798–802
  local caching, 795–798
  MAC layer, 781–785
  power-aware communication, 781–789
  prefetching, 796–798, *797*
  routing layer, 785–788
  transport layer, 788–789
Energy Conserving MAC (EC-MAC), 696, 783–784
Energy efficiency, 355–356
Energy efficient cache invalidation
  basics, 421–423, 431–432
  bit sequences with bit count, *429–430,* 429–431
  content of invalidation report, 423–425
  invalidation mechanisms, 425–426
  invalidation schemes, 426
  selective dual-report invalidation scheme, 427–428, *428*
  selective invalidation schemes, 426–431
  taxonomy of strategies, 423–426
  update log structure, 426
Energy-efficient wireless networks
  ad hoc routing, 700–703
  basics, 689–692, 705

  energy harvesting, 693–695
  multicast tree construction, 702–703
  network topology, 703–704
  node activity, 695–700
  power-aware broadcast, 702–703
  power-aware link layer adaptation, 692–693
  power-aware medium access control, 695–697
  protocols, 701–702
  radio activity, 695–700
  scheduling activity, 695–700
  WSN, 697–698
Energy harvesting, 693–695
Energy management solutions, 736
Energy-Quality curves (E-Q), 749, 753–755
Engagement, Lime, 32
Enhanced CONFIDANT, 447–449
Enhanced Messaging Service (EMS), 121
Enterprise Messaging Network (EMN), 305
Environmental Energy Harvesting Framework (EEHF), 694
Environments, security, 976–977
E-Q, *see* Energy-Quality curves (E-Q)
Erlang distribution, 907
Ethernet
  grid and cluster computing, 98, 105
  Lime, 39
  message delivery, 249
ETSI, *see* European Telecommunications Standards Institute (ETSI)
European Telecommunications Standards Institute (ETSI), 638
Evaluation
  ad hoc network security, 500
  energy awareness, web caching, 798–802
  wearable computing, 17–21
Evaluation, cache invalidation schemes, 376–383, *377*
Example systems, wearable computing, *7–8,* 7–17, *10*
Execution aspects, ISAM, 91–93, *93*
Execution controller, ISAM, 82
Execution Environment for High Distributed Applications (EXEHDA), 80–83
EXEHDA, *see* Execution Environment for High Distributed Applications (EXEHDA)
Exhaustive service, battery power management, 904
Expanding ring paging, 187–188, *188*
Experimental Machine Example-based Reasoning and Learning Disciple (EMERALD), 215
Experiments
  DNS location management, *156,* 156–160
  grid and cluster computing, 111–116
Expire, DNS location management, 153

Index

Explicit invalidation, 372–373, *374*
Exposed terminal problem, 557–558, *559*
Extended interframe spacing (EIFS), *546,* 547, 549–552, *551, 553,* 560
Extensible Access Control Markup Language (XACML), 964
Extensible Authentication Protocol (EAP), 934–935
Extensible Markup Language (XML)
   ISAM, 91–92
   Lime, 49
   mobile agent middlewares, 322
   security, trust, and privacy, 964–966
   simulation models, 290
Extensions
   battery power management, 903–909
   IEEE 802.11, 934–935
   message delivery, 239–240, 243–246, *244, 246*

F

False invalid period (FIP), 366, 368–369, 371
False invalid rate, 376
False valid period (FVP), 366, 368–371
False valid rate, 375–376
Familiar Linux, 112
FAR, *see* Furthest away replacement (FAR)
FarGo system, 956
Fast Ethernet, 100, 112
Fast Track infrastructure, 99
FCFS, *see* First come first service (FCFS) policy
FDMA, *see* Frequency division multiple access (FDMA)
FEC, *see* Forward error correction (FEC)
Federal Communications Commission (FCC), 299
Federated tuple space, 32–33, *35*
Federici studies, 941–957
Ferscha studies, 300
Ficus, 404
Field service engineers (FSEs), 11–12, 17–18
FIFO, *see* First in, first out (FIFO) channels
Finin studies, 961–984
Finite impulse response (FIR), 754–755
FIP, *see* False invalid period (FIP)
FIPA, *see* Foundation for Intelligent Physical Agents (FIPA)
FIR, *see* Finite impulse response (FIR)
First class transactions, 861
First come first service (FCFS) policy, 373
First in, first out (FIFO) channels, 232, 236, *247,* 247–249
Fisheye State Routing (FSR), 597, 614, 644
Fixed-assignment channel access, 545
Flat approach, hoarding, 394–395

Flat routing protocols, 614–615, *624,* 624–625
Flinn and Satyanarayanan studies, 721
Flooded packet identifier (FPI), 655–656
Flooding method, 655–656
Flow Oriented Routing Protocol (FORP), 656
Fluhrer-Mantin-Shamir attack, 487
Fluid-flow model, 173
Folding@home, 95
Formats, multimedia messaging service, 125–126
FORP, *see* Flow Oriented Routing Protocol (FORP)
Forward bypass pointer, 208
Forward error correction (FEC), 692, 895–896
Forwarding, multimedia messaging service, 143–145, *144–145*
Forwarding pointers, 211–215, *212, 214–215*
Forward node selection process, 672–673
Foundation for Intelligent Physical Agents (FIPA), 322, 325
Four-way handshake, 615
FPI, *see* Flooded packet identifier (FPI)
FQHN, *see* Fully Qualified Host Name (FQHN)
Fragmentable objects, 857
Frainer studies, 73–93
Free neighbor, 674
Frequency division multiple access (FDMA), 578, 806, 818
Frequency scaling, 735
FSR, *see* Fisheye State Routing (FSR)
Fully connected/disconnected mode, 863
Fully Qualified Host Name (FQHN), 149–150
Furthest away replacement (FAR), 345
Future directions
   ad hoc network security, 499–500
   context-aware mobile computing, 311–312
   grid and cluster computing, 116–118
   hoarding, 400
   mobile agent security, 953–956
   multimedia messaging service, 145
   power-aware cache management, 416–418, *417*
FVP, *see* False valid period (FVP)

G

GAF, *see* Geographic Adaptive Fidelity (GAF)
Gamma function, 908
Garcia-Luna-Aceves, Bao and, studies, 558
Garcia-Luna-Aceves, Wang and, studies, 562
Gartner Dataquest, 101
Gauss-Markov Model, 171–172
GCM, *see* Global Communication Manager (GCM)

GCORE, *see* Grouping with cold update set retention (GCORE)
GCS, *see* Global Concurrency Submanager (GCS)
General Magic, 318
Generic mobility models, 283–285, *285–286*
GeoCast, *see* Geographic Addressing and Routing (GeoCast)
Geocoding, 58
Geographic Adaptive Fidelity (GAF), 616, 627, 700
Geographic Addressing and Routing (GeoCast), 655
Geographical contexts, 300
Geographic Information Systems (GIS), 258
GeoPlex peer service, 722
Georgia Tech Context Toolkit, 328
Georgia Tech wearable computer, 7
Geyer studies, 73–93
GIF, *see* Graphics Interchange Format (GIF)
Gigabyte Ethernet, 100
GIM, *see* Global Interface Manager (GIM)
GIR, *see* Group invalidation report (GIR)
GIS, *see* Geographic Information Systems (GIS)
Gitzenis studies, 805–835
Giuli, Lai and Baker, Marti, studies, 444
Global broadcasting, 667
Global certification order, 861–862
Global Communication Manager (GCM), 852
Global Concurrency Submanager (GCS), 852
Global Interface Manager (GIM), 853
Global Mobile Information Systems Simulation Library, 290
Global mobile transaction manager (GMTM), 870
Global Object-Based Environment (GLOBE), 198, 220–222, *223*
Global Positioning System (GPS)
  context-awareness, 301
  energy-aware web caching, 787
  MANET localized broadcasting, 666
  mobile agent middlewares, 327–328
  quality of service, 598–600, 603
  routing and mobility management, 655
  scalable clustered network architecture, 612
  simulation models, 277
Global reaction, ad hoc network security, 494
Global Recovery Manager (GRM), 853
Global Schedule Submanager (GSS), 852
Global serializability, 861
Global State Routing (GSR), 644
Global System for Communication-Mobile Application Part (GSM-MAP), 657
Global System for Mobile (GSM)
  battery power management, 895
  context-awareness, 308

locating mobile objects, 201
mobile agent middlewares, 317
multimedia messaging service, 121
Global Transaction Manager (GTM), 852
Global Transaction Manager Participants (GTMPs), 852
Global Virtual Data Structures (GVDS), 49
Globus, 107, 110
GloMoSim simulator, 118, 290
GLS, *see* Grid Location Service (GLS)
GMTM, *see* Global mobile transaction manager (GMTM)
Gnutella, 99
Gomez studies, 558
3GPP, *see* Third Generation Partnership Project (3GPP)
GPS, *see* Global Positioning System (GPS)
GPSR, *see* Greedy Perimeter Stateless Routing (GPSR)
Graph-based Mobility Model, 282, 285
Graphics Interchange Format (GIF), 125
Grasshopper, 324–325, 953
Gravity model, mobility tracking, 173
Gray Code, 735, 738
GreedyEmbed, 522
Greedy Perimeter Stateless Routing (GPSR), 655
Greenberg, Gutwin and, studies, 19
Grid and cluster computing
  availability adaptation, 108–110
  background, 98–99
  basics, 95–98, *97*
  communication overhead, 112–113, *114*
  experiments and analysis, 111–116
  future directions, 116–118
  interlocutor, 107–108
  job management, 108–110
  key challenges, 104–105
  LEECH, 104–111, *106*
  major component, 106–107
  minion, 108
  minor component, 106–107
  motivation, 99–104, *101–102*
  RSA decryption, 113–116, *115*
Grid and cluster software overhead, 104
Grid Location Service (GLS), 616, 628–629, *631*, 632
GRM, *see* Global Recovery Manager (GRM)
Grossglauser and Tse studies, 612
Grouping with cold update set retention (GCORE), 362
Group Invalidation Report (GIR), 353–354, 363, 427
GSM, *see* Global System for Mobile (GSM)

Index

GSM-MAP, *see* Global System for Communication-Mobile Application Part (GSM-MAP)
GSR, *see* Global State Routing (GSR)
GSS, *see* Global Schedule Submanager (GSS)
GTM, *see* Global Transaction Manager (GTM)
GTMPs, *see* Global Transaction Manager Participants (GTMPs)
GUI being, 83–89, *86–89*
Guidelines, power management, 727–729
GUIDE project, 287
Gupta, S., studies, 337–358
Gupta, V., studies, 487
Guting studies, 258
Gutwin and Greenberg studies, 19
GVDS, *see* Global Virtual Data Structures (GVDS)

H

Haas, Papadimitratos and, studies, 442
Haas, Zhou and, studies, 497
Haas studies, 457–480
Hand-held computers, 403
Handshake, four-way, 615
Hardware
　power awareness, 733–736
　power management, 716–719, *717, 719*
Harvest, 357
Hash-based schemes, 357
Havinga, Smit and, studies, 735
Helal, Loy and, studies, 737, 740
Helal studies, 731–743, 839–874
Herzberg studies, 965
Hetergeneous nodes and networks, QoS, 577–578, 606
Hewlett-Packard Cooltown project, 328
Hewlett-Packard interface bus (HPIB), 741
Hidden terminals, 545, 547, 560
Hierarchical protocols, 202–204, *203–204*, 646–647
Hierarchical Routing (HR), 642, 646–647, 658
Hierarchical State Routing (HSR)
　quality of service, 597
　scalable clustered network architecture, 616, 623, *623*, 623–624
High Performance Radio Local Area Network (HIPERLAN), 582, 638, 645
Hightower and Borriello studies, 300
HIPERLAN, *see* High Performance Radio Local Area Network (HIPERLAN)
Hiper studies, 211
Historical developments, IEEE 802.11, 924–925
Hitachi, 734
Hit rate, 376

Hit ratio, 341
HLC, *see* Home Location Cache (HLC)
HLR, *see* Home Location Register (HLR)
HMAC, 486, 488–490
Hoarding
　association rule-based techniques, 394, 396
　basics, 389–390
　cache management, 344
　candidate set construction
　Coda, 391–392
　comparison of techniques, 399
　data mining techniques, 392–396
　distributed environment, 398–399
　future directions, 400
　partitioning, 394–395
　program trees, 396–398, *397*
　SEER, 392–394, *393*
Hodes studies, 538
Holoparadigm, 83, 86
Home address, 219
Home Location Cache (HLC), 347–348
Home Location Register (HLR)
　locating mobile objects, 201, 204, 206, 210–213
　routing and mobility management, 657
Host-independent models, *64*, 64–66, 69
Host-level tuple space, 32
HPIB, *see* Hewlett-Packard interface bus (HPIB)
HSR, *see* Hierarchical State Routing (HSR)
HTML, *see* Hypertext Markup Language (HTML)
HTTP, *see* Hypertext Transfer Protocol (HTTP)
Hu, Johnson and Perrig studies, 442
Hu, Perrig and Johnson studies, 442
Huang, L., studies, 95–118
Hubaux, Buttyán and, studies, 440
Hu studies, 408
Hybrid networks, web caching
　ad hoc caching, 791–794, *793–794*
　ad hoc cooperative caching, *783*, 789–794, *790, 793–794, 801*, 801–802
　ad hoc networks, 789–794, *799–800*, 799–801
　ad hoc routing protocols, 785–788, *786*
　basics, 779–781, 802
　cache management, 795–796
　channel listening, 784–785
　collisions, 782–784, *783*
　communication, 789–791, *790*
　evaluation, 798–802
　local caching, 795–798
　MAC layer, 781–785
　power-aware communication, 781–789
　prefetching, 796–798, *797*
　routing layer, 785–788
　transport layer, 788–789

1001

Hybrid protocols, 652–653
Hybrid strategies, 186
Hypertext Markup Language (HTML), 59–60, 317, 325
Hypertext Transfer Protocol (HTTP)
    cache management, 342, 356
    context-awareness, 303, 305, 308
    DNS location management, 155
    multimedia message service, 123
    multimedia messaging service, 126

I

IARP, *see* Intra-Zone Routing Protocol (IARP)
IAVI, *see* Invalidation by absolute validity interval (IAVI)
IBM Research, 68
ICMP, *see* Internet Control Message Protocol (ICMP)
ICP, *see* Internet Caching Protocol (ICP)
ICV, *see* Integrity Check Value (ICV)
IDA, *see* Information Dispersal Algorithm (IDA)
IDEALINK, 19–21, *20*
Identifier-based algorithm, 618
Identity, self-policing MANETs, 453–454
Idleness and idle state, 15, 761
IEEE 802.11
    access control, 933–936
    ad hoc network security, 485
    authentication, 933–936
    basics, 923–925, *924*, 937
    encryption, 927–930, *928*
    energy-aware web caching, 782, 784–785, 789
    energy efficient wireless networks, 690, 695–696
    extensions, authentication, 934–935
    grid and cluster computing, 102, 117
    historical developments, 924–925
    integrity-based attacks, 931–932
    integrity protection, 930–933
    key management, 936–937
    keystream reuse, 928–929
    Lime, 39
    MAC vulnerabilities, 486–487
    MAC vulnerabilities, ad hoc network security, 486–487
    medium access control, 543–552, 554, 563
    modeling distributed applications, 504
    mutual authentication, 935
    new protocols, 933, 935–937
    new standards, 929–930
    power-aware cache management, 403
    RC4 weakness, 929
    replay attacks, 932
    routing and mobility management, 638, 654
    WEP vulnerabilities, 487
    wireless security threats, 926–927
IEEE 802.11i/WPA, 485, 487
IERP, *see* Inter-Zone Routing Protocol (IERP)
IETF, *see* Internet Engineering Task Force (IETF)
IETF OPES group, 303
IFS, *see* Interframe spacing (IFS)
Image tuples, 41
Imielinski, Barbara and, studies, 362, 407
iMobile, 303–305, *304, 306,* 307–308, *309*
iMobile Enterprise Edition (iMobile EE), 305
Implementations, 39–44, 970–972
Implicit invalidation, 371–372
Independent channel, 812–815, *814,* 820–823, *823–824*
Information access model, 285–288
Information Dispersal Algorithm (IDA), 476
Information dissemination, 450
Information infrastructures, 627–632
Infraestrutura de Suporte às Aplicações Móveis, see ISAM
Infrastructure, self-policing MANETs, 437
Infrastructured wireless LAN, 543
Infrastructureless wireless LAN, 543
Infrastructure requirements and compatibility, 844, 869–871
Input/output modalities, 4–5
Inquiry, location, 165
INRT, *see* Intermediate Node Reply Token (INRT)
INS, *see* Intentional Naming System (INS); International Naming System (INS)
Integrated Mobility Model, 285
Integrity, *see also* Security
    ad hoc network security, 496
    attacks, 931–932
    IEEE 802.11, 930–933
    self-policing MANETs, 437
Integrity Check Value (ICV), 930–931
Intelligent paging, 188–190, *190*
Intelligent power drainage, 733–734
Intentional naming, 66
Intentional Naming System (INS), 310
Interaction modalities, pervasive application development, 57
Interfaces
    evaluation methodology, 5
    multimedia messaging service, *122–124,* 123–124
    multimodal, 57
    technology independent, 302–305
Interface tuple space (ITS), 29–33
Interference reduction, 558–563, *559, 561*
Interframe spacing (IFS), 547, *548*

Index

Interlocutor, 107–108
Intermediate Node Reply Token (INRT), 473
International Naming Resolvers, 538
International Naming System (INS), 538
Internet Caching Protocol (ICP), 342
Internet Control Message Protocol (ICMP), 583–584, 931
Internet Engineering Task Force (IETF)
    DNS location management, 149
    locating mobile objects, 218
    mobile transaction models, 842
    routing and mobility management, 638, 641–642, 657
Internet-MANET interaction, 605
Internet Protocol (IP)
    context-awareness, 303
    DNS location management, 149
    energy-aware web caching, 781
    MANET security, 478
    mobile transaction models, 840
    routing and mobility management, 655
Internet Service Provider (ISP), 157
InterNIC, 159
Inter-Zone Routing Protocol (IERP), 631
Intra-Zone Routing Protocol (IARP), 630, 652–653
Intrusion detection, 495–496, 976–984
Intrusion detection systems (IDSs), 495
Invalidation, *see also* Energy Efficient cache invalidation
    cache management, 348–355
    mechanisms, 425–426
    power-aware cache management, 405–410
    report size, 375
Invalidation by absolute validity interval (IAVI), 368–369, 373, 376–377, 383, 387
Invalidation report, 369–370, 403, 423–425
Invalidation schemes, 426, *see also* Cache invalidation schemes
Invisible computing, 53
Involvement level of MSCs, 250
In-zone communication, 791
I/O management, 737
IP, *see* Internet Protocol (IP)
IR-based cache invalidation model, 405–407, 418
IS-41, *see* Electronics Industry Association/Telecommunication Industry Association's (EIA/TIA) Standard 41 (IS-41)
ISAM
    adaptation commands and policies, 80–81
    adapters, 80
    application model, 74–75
    architecture, *75–76,* 75–77

    basics, 73–74, 93
    context, 78–80, *79*
    execution aspects, 91–93, *93*
    EXEHDA, 80–83
    GUI being, 86–89, *86–89*
    ISAMadapt, 77–81, *78*
    ISAMadapt Development Environment software, 78
    ISAMadaptEngine, 77, 80
    ISAMadapt IDE, 91, *91*
    ISAMcontextServer, 79
    ISAM Pervasive Environment, 76
    Print being, 89–90
    Spell being, 89, *90*
    WalkEd, *83–85,* 83–93
Isolation, 846, *see also* ACID properties
Isolation-only transactions
    basics, 860–862
    comparisons, *865, 867*
    consistency and concurrency, 869
    infrastructure requirements and compatibility, 870
ISP, *see* Internet Service Provider (ISP)
Issarny studies, 779–802
ISS processes, 887, 890–891
Issues
    middleware, context-aware applications, 330–332
    mobile agent security, 953–956
    mobile transaction models, 842–845, 872–873
    routing and mobility management, 656–657
    transmitter power control, 805–809
    wearable computing, 4–6
Itsy, 721–722, 725

J

Jana studies, 297–312
Java
    ISAM, 86
    Lime, 27
    mobile agent middlewares, 318
    pervasive application development, 57
    power awareness, 740
    security, trust, and privacy, 966
Java Agent Development (JADE), 323–324
Java Development Kit (JDK), 947
Java Native Interface, 328
Java/RMI, 45
Java2 security, 951
Java Security Model, 951
JavaSpaces, 36, 47, 49
Java Virtual Machine (JVM), 320, 326, 328, 953
JDK, *see* Java Development Kit (JDK)

1003

Jigsaw assembly game example, 39–44, *42*
Jing studies, 362
Jini network technology, 45–46, 325, 537
Job management, 108–110
Joey transactions, 858
Johnson, Hu, Perrig and, studies, 442
Johnson and Perrig, Hu, studies, 442
Joint gateway, 620
Joint Photographic Experts Group (JPEG), 125, 750
Joshi studies, 961–984
JPEG, *see* Joint Photographic Experts Group (JPEG)
J2SE, 91
JVM, *see* Java Virtual Machine (JVM)

K

Kagal studies, 961–984
Kalubandi studies, 543–564
Kamal studies, 569–607
Kangaroo model
 basics, 857–858
 communication cost and scalability, 872
 comparisons, *865, 867*
 consistency and concurrency, 864, 868
 infrastructure requirements and compatibility, 870
Kastidou studies, 197–224
Kazaa, 99
KDC, *see* Key distribution center (KDC)
Ke studies, 503–539
Key challenges, 104–105
Key distribution center (KDC), 497
Key management, 496–499, 936–937
KeyNote, 965–966
Keystream reuse, 928–929
K-hop zone-based algorithm, *674,* 681–682, *683*
Khurana studies, 149–161, 554
Kim, Lim and, studies, 675
Kim studies, 543–564
Klaim model, 48
Kong studies, 483–500
Ko studies, 562
Kravets, Yi, Naldburg and, studies, 442
Kravets and Krishnan studies, 737
Kremer studies, 739
Krishnan, Kravets and, studies, 737
Kubach studies, 275–293
Kuenning studies, 392
Kumar studies, 765

L

Lai and Baker, Marti, Giuli, studies, 444
LAM, *see* Location Accuracy Matrix (LAM)
Lamparter, Plaggemeier and Westhoff studies, 438
Lam studies, 361–387
Landmark Ad Hoc Routing (LANMAR)
 routing and mobility management, 647
 scalable clustered network architecture, 616, 623, *624,* 624–625
Land Warrior, 8
Langheinrich studies, 971
LANMAR, *see* Landmark Ad Hoc Routing (LANMAR)
LANs, *see* Wireless local area networks (LANs)
Laplace Transform, 903
Laptop computers
 grid and cluster computing, 100–103, 112
 modeling distributed applications, 507
 power awareness, 731
 power management, 723
 security, trust, and privacy, 977
LAQ, *see* Location-aware query (LAQ)
LAR, *see* Location-Aided Routing (LAR)
Last phase, hoarding, 398
Layer integration, quality of service, 605
Lazy caching, 207
LCA for routing backbone, 617–628
LC path, *see* Least-cost (LC) path
LDAP, *see* Lightweight Directory Access Protocol (LDAP)
LDQs, *see* Location-dependent queries (LDQs)
LEAP, *see* Lightweight Extensible Agent Platform (LEAP)
Least Cluster Change (LCC), 620–621, 645
Least-cost (LC) path, 588
Least recently used (LRU) location
 cache management, 337
 hoarding, 391, 397
 locating mobile objects, 208
Le Boudec studies, 435–454
Lederer studies, 971, 976
Lee, Xu and, studies, 257
Lee, Zhang and, studies, 440, 495
LEECH, *see* Leveraging Every Existing Computer out tHere (LEECH)
Lee studies, 257, 543–564, 611–633
Lessons learned, middleware, 330–332
Lettieri and Srivastava studies, 895
Level caching, 208
Leveraging Every Existing Computer out tHere (LEECH)
 architecture overview, 105
 availability adaptation, 108–110

Index

background, 98
basics, 96–98
communication overhead, 112–113, *114*
experiments and analysis, 111–116
future directions, 116–118
grid and cluster computing, 96, 98
grid/cluster system, 105–106, *106*
interlocutor, 107–108
job management, 108–110
key challenges, 104–105
major component, 106–107
minion, 108
minor component, 106–107
motivation, 103
programming model, 110–111
RSA decryption, 113–116, *115*
LeZi-update scheme, 185
LGeo-based protocols, 616, 631–632
LGPL, *see* Library General Public License (LGPL)
Li, Wu and, studies, 669
Liao studies, 591
Library General Public License (LGPL), 50
LifeLog program, *see* Defense Advanced Research Project Agency (DARPA)
Lightweight Directory Access Protocol (LDAP), 323, 325
Lightweight Extensible Agent Platform (LEAP), 323–324
LILT, *see* Local Information Link-State Topology (LILT)
Lim and Kim studies, 675
Limbo, 47–48
Lime, *see* Linda in a Mobile Environment (LIME)
Lin and Liu studies, 593
Linda in a Mobile Environment (LIME)
  application example, 39–44
  availability, 50
  basics, 25–29, 47–50
  design, 40–44
  display update, *42,* 43–44
  implementation, 39–44
  Linda, 28
  middleware functionality, 44–47
  model setting, 29
  reactions, 36–37, *38*
  reconciling different forms of mobility, 32–34, *34*
  requirements, 40
  scope of operations restriction, 34–36, *35*
  service provision, 46–47
  system configuration access, 38–39
  transient sharing, *30,* 31–32, 44–46
  tuple space, *30,* 30–31, 40–41
  user actions, 41–43, *42*

Line of defense, clear, 483
Link cluster architecture (LCA), 611–613, 616
Link-layer security, 486–487
Link quality, 578
Link state
  ad hoc network security, 491–492
  message delivery, 250
  routing and mobility management, 641–642, 644–645
  scalable clustered network architecture, 614, 631
Link state update (LSU), 491–492
Lin studies, 586
LINT, *see* Local Information No Topology (LINT)
Linux operating system, 39, 98, 722
Linux Wristwatch, 721
Li studies, 552, 612
Lithium ion batteries, *see* Batteries
Little studies, 503–539
Liu, Lin and, studies, 593
Liu studies, 839–874
Live expert help desk, *10,* 10–11, 18
LLog-based protocol, 629–631
LNG algorithms, *619,* 621–622, *622,* 626–627, *627*
Load power, 725–727, *726–727*
Local algorithms, 669–672, *671*
Local anchoring scheme, 212
Local area networks (LANs), 105
Local broadcasting, 669
Local caching, 795–798
Local Information Link-State Topology (LILT), 704
Local Information No Topology (LINT), 704
Locality, 970–971
Localized broadcast, neighbor designation
  basics, 663–665, *664–665,* 686
  CDS-based algorithm, *674,* 676–677, *677*
  classification, 665–672
  cluster-based algorithm, 677–681, *679–681*
  counter-based scheme, 666
  deterministic algorithms, 666–669
  distance-based scheme, 666
  dominant pruning, *674–675,* 675
  forward node selection process, 672–673
  global broadcasting, 667
  K-hop zone-based algorithm, *674,* 681–682, *683*
  local algorithms, 669–672, *671*
  local broadcasting, 669
  location-based scheme, 666
  multi-point relays, 673–674, *673–674*
  neighbor-designating broadcast algorithms, 672–677
  partial dominant pruning, *674,* 675–676, *676*

1005

probabilistic algorithms, 665–666
quasi-global broadcasting, 667–668
quasi-local broadcasting, 668–669
reliable broadcast algorithm, *674*, 683–685, *685*
total dominant pruning, *674*, 675–676, *676*
Localized detection, 493
Localized trust, 497, 499
Local serializability, 861
Location Accuracy Matrix (LAM), 189–190
Location-Aided Routing (LAR), 655
Location area, mobility tracking, 178–182
Location-Aware Query (LAQ), 266–268
Location-based schemes and protocols, 655, 666
Location-based services
   context-aware mobile computing, 297–312
   middleware for context-aware applications, 315–332
   query processing, 255–272
   simulation models and tool, 275–293
Location Binding, 268–269
Location-dependent access, 287–288
Location-dependent queries (LDQs)
   basics, 255–256, 271–272
   data and queries, 256–258
   location-aware queries, 266–268
   location-dependent queries, 268–270
   location relatedness, 260–271
   moving object database, 258, 270
   query classification, 270–271, *271*
   query model, 260–271, *261*, *263*
   spatial database management, 258–260
   translation steps, 271–272
Location-Dependent Query, 268
Location determination/service platform integration, 306, *307*
Location inquiry strategies, 186–190
Location management
   Domain Name System, 149–161
   locating mobile objects, 197–224, *217–219*
   message delivery, 227–251
   mobility tracking, 163–190, *165*
Location Query, 266
Location relatedness, 260–271
Location sensing techniques, 301
Location-sensitive transaction operations, 873
Location simulation models, 277–278
Location Trace Aided Routing (LOTAR), 655
Location update, PCS networks, 881–886
Logical mobility, Lime, 26
Login/logout, ISAM, 91–92
LOTAR, *see* Location Trace Aided Routing (LOTAR)
Lou, Wu and, studies, 665

Lou and Wu studies, 675, 682, 684
Lou studies, 663–686
Low-power design and measurement, 741–743
Loy and Helal studies, 737, 740
LRU, *see* Least recently used (LRU) location
LSG algorithms and protocols, 618–621, *619*, 622–626
LSU, *see* Link state update (LSU)
Lu, Peng and, studies, 675–676
Lu, S., studies, 483–500
Lu, Y.H., studies, 763
Luo studies, 483–500

M

MA-based middlewares, 326–330, *327*
MA-based supports, 322–326, *324*
MAC, *see* Medium access control (MAC); Message Authentication Code (MAC)
MACA, *see* Multiple Access with Collision Avoidance (MACA)
MAC protocol data unit (MPDU), 554
Mahgoub studies, 149–161
Main solution tracks, 440–447
Maintenance complexity, 58
Major component, LEECH, 106–107
Management cost, location, 166–167
Management of cache, *see* Cache management
MANETs, *see* Mobile ad hoc networks (MANETs)
MANET security
   basics, 457–459, 478–480
   data forwarding, 473–477
   discovery procedure, 471–472
   goals, 459–460
   Neighbor Lookup Protocol, 469–471
   priority-based query handling, 472
   route maintenance procedure, 472–473
   routing, secure, 467–473
   secure message transmission protocol, *475*, 475–477, *477*
   secure routing protocol extension, 473
   threats and challenges, 460–462
   trust management, 463–467, *465–466*
MARISM-A, *see* Architecture for Mobile Agents with Recursive Itinerary and Secure Migration (MARISM-A)
Marker arrival rule, 235
Markov chain, 810–811, 882, 884
Markovian Channel Model, 812
Markovian Movement Model, 169–170, 178
Markov Mobility Model, 178, 877
Markov Walk Model, 877–881, *879*, *881*, 884, 887, 891
MARS, 48–49

Index

Marti, Guili, Lai and Baker studies, 444
Martin studies, 709–729
Marti studies, 493
MAs, *see* Mobile agents (MAs)
MASIF, *see* Mobile Agent System Interoperability Facility (MASIF)
Massachusetts Institute of Technology, 310
Massively parallel processor (MPP) systems, 98
Master/apprentice (live expert) help desk, 6, 8, *10*, 10–11, 18
Master eligibility rule, 621
Matched capability/application requirements, 5
Max-Min D-Cluter Formation (MMDF), 617, 628
MBONE, *see* Multicast backbone (MBONE)
MC, *see* Mobile clients (MC)
MCB, *see* Mobile control block (MCB)
MCDS, *see* Minimum connected dominating set (MCDS)
MDAT, *see* Multi-Device Authoring Tool (MDAT)
M-Destination-Sequenced Distance Vector (M-DSDV), 643–644
MDS, *see* Minimum dominating set (MDS)
MDSTPM, *see* Multi-database transaction processing manager (MDSTPM)
Mean response time, 375
Mechanics of collaboration, 19
Medium access control (MAC)
   acknowledgement packets, *546,* 547, 549–552, *551, 553*
   adaptive algorithmic power optimization, 754
   angular interference reduction, 560–563, *561*
   backoff algorithm, 549, *550,* 554, 556
   basics, 543–545, 563–564
   battery power management, 894
   busy tone, 557–558, *559*
   clear-to-send packets, 547, 554
   collisions, *546,* 547, 549–552, *551, 553,* 554, 556
   congestion control, 549, *550*
   directional antenna, 560–563, *561*
   distributed coordination function, 546–552
   DRTS/DCTS-based DMAC, *561,* 562–563
   DRTS/oCTS-based DMAC, 562
   dynamic tuning, 556, *557*
   energy awareness, web caching, 781–785
   energy-aware web caching, 781, 787
   energy efficient wireless networks, 691
   exposed terminal problem, 557–558, *559*
   extended interframe spacing, *546,* 549–552, *551, 553*
   hidden terminal problem, 547
   IEEE 802.11, 546–552, 924, 935
   interference reduction, 558–563, *559, 561*
   interframe spacing, 547, *548*
   multi-hop RTS MAC, 563
   node treatment, *555,* 556
   oRTS/oCTS-based DMAC, *561, 562*
   protocols, 545–552
   quality of service, 571, 578
   radial interference reduction, 558–560, *559*
   random access MAC, 545–546, *546*
   receiver-oriented multiple access, 563
   request-to-send packets and mechanism, 547, 554
   spatial channel utilization, 557–563, *564*
   temporal channel utilization, 552–563
   transmission power control, 558–560, *559*
   transmission schedule, 563
   transmitter power control, 807
Memory
   bottleneck, power management, 722–725, *723–725*
   power awareness, 737, 739
Memoryless movement model, 168
MEMS, *see* Micro Electro Mechanical Systems (MEMS)
Message arrival rule, 235
Message Authentication Code (MAC), 471
Message authentication primitives, 488–489
Message delivery
   algorithm development, 229
   announcement delivery, 233–234
   backbone-based type, 243–246, *244, 246*
   base station connectivity, 249
   basics, 227–232, 251
   broadcast search, 232–240
   diffusing computations, *241,* 241–243
   distributed computing, 228
   extensions, 239–240, 243–246, *244, 246*
   FIFO channels, *247,* 247–249
   involvement level of MSCs, 250
   mobile agents, 240
   mobile computing and environment, 228, 231, *231*
   mobile unit tracking, *241,* 241–243
   model and problem definitions, 231–232
   motivation, 233
   multicast, 240
   multiple announcement deliveries, 239
   multiple RBSs per MCS, 249
   rapidly moving mobile units, 239–240
   reality check, 246–251
   reliable delivery on links, 249–250
   route delivery, 240
   snapshot algorithm, 233–238, *235–237*
   storage requirements, 250–251
   tracking for delivery, 241–246

1007

Message Passing Interface (MPI), 97, 105, 107, 109–111, 113
Messages, *see* Multimedia messaging service (MMS)
Message transmission list (MTL), 644
Meta-reasoning, 975
Metrics, cache invalidation schemes, 375–376
MHTTP, *see* Mobile HTTP (MHTTP)
Michael, 933
Michiardi and Molva studies, 444
Microdrive, 100, 716–717, 728
Micro Electro Mechanical Systems (MEMS), 311
Microsoft, 60
Middleware, *see also* specific type of middleware
  context-aware mobile computing, 306–310
  Lime, 44–47
  query processing, 272
Middleware, context-aware applications
  basics, 315
  issues, 330–332
  lessons learned, 330–332
  MA-based middlewares, 326–330, *327*
  MA-based supports, 322–326, *324*
  mobile agents, 318–322, *319*
  mobile computing and context awareness, 316–318
  research directions, 326–330, *327*
MIDI, *see* Musical Instrument Digital Interface (MIDI)
Migration, mobile transaction models, 852
MILD, *see* Multiplicative Increase and Linear Decrease (MILD)
MIME, *see* Multipurpose Internet Mail Extensions (MIME)
Minimum connected dominating set (MCDS), 666, 668
Minimum dominating set (MDS), 627
Minimum set covering (MSC), 627
Minion, 108
Minor component, LEECH, 106–107
Mist, 972
MJPEG, *see* Motion JPEG (MJPEG)
MM1, *see also* Multimedia messaging service (MMS)
  delivery report, *133*
  delivery report request, *135*
  delivery report status codes, *136*
  immediate and deferred message retrieval, *141*
  message forward, *144*
  message forward request, *144*
  message forward response, *145*
  notification, *139*
  notification request, *140*
  notification response, *140*
notification response/status codes, *141*
read-reply report, *136*
read-reply report request, *137*
read-reply report status codes, *137*
retrieval acknowledgment, *143*
retrieval response, *142*
retrieval response errors, *143*
submit request, *127*
submit response, *128*
submit response errors, *130*
MM4, *see also* Multimedia messaging service (MMS)
  delivery report request, *134*
  delivery report response, *135*
  forward response errors, *133*
  message forward report, *131*
  message forward request, *132*
  message forward response, *132*
  read-reply report request, *138*
  read-reply report response, *139*
MMAC, *see* Multi-hop RTS MAC (MMAC)
MMDF, *see* Max-Min D-Cluter Formation (MMDF)
MMS, *see* Multimedia messaging service (MMS)
MobiDE Workshop, 256
Mobile ad hoc networks (MANETs), *see also* Ad hoc wireless networks
  cache management, 341–342, 357
  desired protocol characteristics, 640
  Internet interactions, 605
  Lime, 26
  localized broadcasting, 663–686
  medium access control mechanisms, 543–564
  modeling distributed applications, 503–539
  quality of service, 569–606
  routing and mobility management, 640
  scalable clustered network architecture, 611–633
  securing and security, 457–480, 483–500, 605
  self-policing, 435–454
Mobile agents, security
  acceptance, 941–942
  access control, 955–956
  agent platforms, 946–948
  agents, 954–955
  authentication, 947
  authorization, 947–948
  basics, 941, 957
  code, 946–947
  countermeasures, 945–951
  future directions, 953–956
  open issues, 953–956
  protecting agents, 948–951
  requirements, 943–945, *944*

Index

solutions overview, 951–953
trust, 954–955
user-agent trust, 945–946
Mobile agents (MAs)
  locating mobile objects, 222–223
  message delivery, 240
  middleware, 240, 318–322, *319*
Mobile Agents with Recursive Itinerary and Secure Migration, Architecure (MARISM-A), 952
Mobile Agent System Interoperability Facility (MASIF), 322, 325
Mobile Applications Support Infrastructure, *see* ISAM
Mobile clients (MC), 422
Mobile Communication and Computing Architecture (MoCCA)
  basics, *7*
  issues, *4*
  prestored procedures, 17
  prestored text and graphics, 11
  team collaboration, 8, 13
Mobile computing
  context awareness, 316–318
  grid and cluster computing, 95–118
  ISAM, 73–93
  Lime, 25–50
  message delivery, 228, 231, *231*
  multimedia messaging service, 121–145
  pervasive application development, 53–69, 73–93
  wearable computing, 3–22
Mobile control block (MCB), 855
Mobile hosts, transaction models, 841
Mobile HTTP (MHTTP), 356
Mobile Internet Protocol (Mobile IP)
  DNS location management, 149–150
  grid and cluster computing, 99
  locating mobile objects, 198, 218–220
  mobile transaction models, 840
Mobile IP, *see* Mobile Internet Protocol (Mobile IP)
Mobile object location
  Ajanta, 222
  architectures, 201–215
  basics, 197–200, *199*, 223–224
  caching, *206*, 206–209, *209*
  call to mobility ratio, 205
  case studies, 218–223
  forwarding pointers, 211–215, *212*, *214–215*
  Globe, 220–222
  hierarchical scheme, 202–204, *203–204*
  location management techniques, 215–218, *217–219*
  mobile agents systems, 222–223
  mobile IP case study, 218–220
  partitions, 205–206
  replication, *209–211*
  taxonomy, 215–218, *217–219*
  two-tier scheme, 201–202
  Voyager, 223, *223*
Mobile Office Workstations using GSM Links (MOWGLI), 355–356
Mobile Pentium III processor, 734
Mobile station integrated services digital network (MSISDN), 124–125
Mobile support centers (MSCs), 231, 233, 235, 238–240, 246–248, 250
Mobile support stations (MSSs), 422, 841
Mobile switching stations (MSSs), 339, 347, 637
Mobile Transmission Control Protocol (MTCP), 356
Mobile unit tracking, message delivery, *241*, 241–243
Mobility
  models, quality of service, 605
  self-policing MANETs, 437
  simulation models, 279–285
  trace, 172–173
Mobility tracking, *see also* Cellular systems
  activity-based model, 172, *172*
  adaptive threshold scheme, 185
  always-update strategy, 175
  basics, 163–164, *164*, 190
  call arrival pattern, *174*, 174–175
  cell history, 169–170
  compression-based scheme, 185–186
  delay constraint, 165
  directional history, 170, *170–171*
  distance-based, *167*, 177–178, *178*
  dynamic case, 181–182, 184
  expanding ring paging, 187–188, *188*
  fluid-flow model, 173
  Gauss-Markov model, 171
  gravity model, 173
  hybrid strategies, 186
  inquiry, location, 165
  intelligent paging, 188–190, *190*
  location area, 178–182
  location inquiry strategies, 186–190
  location management, 165, *165*
  management cost, location, 166–167
  Markovian model, 169–170
  memoryless movement model, 168
  mobility trace, 172–173
  movement-based strategy, 176–177, *177*
  network topology, 167–168, *167–169*
  never-update strategy, 175–176
  paging area, 187

1009

pattern, 168–173
Poisson model, 174
profile-based scheme, 185
random walk movement model, 168
reporting center, 182–184
shortest distance model, 171
simultaneous networkwide search, 186–187
static case, 179, *179–181,* 182–184, *183–184*
time-based strategy, 176
update, location, 165, 175–186
MOCA, 498, 537
MoCCa, *see* Mobile Communication and Computing Architecture (MoCCA)
Models, *see also* Distributed applications modeling; Simulation models and tool; Survey, mobile transaction models
  activity-based model, 172
  battery power management, 899–900
  cache invalidation schemes, 364–366, 373, *375*
  discharge/recharge, 899–900, *900*
  fluid-flow, 173
  generic mobility models, 283–285, *285–286*
  gravity model, 173
  grid and cluster computing, 117
  host-independent models, *64,* 64–66, 69
  information access model, 285–288
  IR-based cache invalidation model, 405–407, 418
  ISAM, 74–75
  Lime, 29
  location simulation models, 277–278
  memoryless movement model, 168
  message delivery, 231–232
  migration, 852
  mobile hosts, 841
  mobile transaction models, 845–849
  mobility, 279–285
  PCS networks, analytic mobility models, 878–881
  QoS, 605
  query, 260–271
  random mobility, 280–282, *280–282*
  reference model, 841–842, *842*
  relocation, 844–845
  shortest distance, 171
  spatial information, 278–279, *279*
  temporal data, 364–366
  time-based, 858–859, 868, 870
  transmitter power control, 825–827
  usage, *285,* 290–293, *291–292*
  user interface, 4
  user mobility, 288–293
Model-view-controller (MVC) structure, *see* Pervasive application development

MOD queries, *see* Moving object database (MOD) queries
Modulation schemes, 807, 819
Moh studies, 611–633
Molva, Michiardi and, studies, 444
Monads, *see* Adaptation Agents for Nomadic Users (Monads)
Monitor, 453
Montanari studies, 315–332, 941–957
Montenegro and Castelluccia studies, 445
Moore's Law, 96, 101
Morris studies, 612
MOST, *see* Moving Object Spatio-Temporal (MOST) Data Model
Motion estimation, 757
Motion JPEG (MJPEG), 329
Motion Picture Experts Group (MPEG)
  adaptive algorithmic power optimization, 765, 771–772, 775
  adaptive algorithm power optimization, 752–753, 759–790
  mobile agent middlewares, 329
  multimedia messaging service, 125
  power awareness, 740
Motion state thresholds, *17*
Motivation
  context-aware mobile computing, 297–299
  grid and cluster computing, 99–104, *101–102*
  message delivery, 233
  power awareness, 731–732
Motorola, 734
Movement-based strategy, 176–177, *177*
Movement Dynamics Model, 284–285, 288, 290
Moving object database (MOD) queries, 258, 270
Moving Object Spatio-Temporal (MOST) Data Model, 258
MOWGLI, *see* Mobile Office Workstations using GSM Links (MOWGLI)
MPATH, *see* Multipath routing algorithm (MPATH)
MPDU, *see* MAC protocol data unit (MPDU)
MPEG, *see* Motion Picture Experts Group (MPEG)
MPI, *see* Message Passing Interface (MPI)
MPICH/MPICH-G2, 110, 112–114, 116
MPP, *see* Massively parallel processor (MPP) systems
MPRs, *see* Multi-point relays (MPRs)
MSC, *see* Minimum set covering (MSC)
MSCs, *see* Mobile support centers (MSCs)
MSS, *see* Mobile support stations (MSSs)
MSSs, *see* Mobile switching stations (MSSs)
MTCP, *see* Mobile Transmission Control Protocol (MTCP)

Index

MTL, *see* Message transmission list (MTL)
Multi-battery system, *905*, 905–909
Multicast backbone (MBONE), 230
Multicast-based schemes, 358
Multicasting protocol, 655–656
Multicast message delivery, 240
Multicast tree construction, 702–703
Multi-class traffic, 604
Multi-database transaction processing manager (MDSTPM)
   basics, 852–853
   communication cost and scalability, 871
   comparisons, *865–866*
   consistency and concurrency, 868
   infrastructure requirements and compatibility, 870
Multi-device application, 55
Multi-Device Authoring Tool (MDAT), 62
Multi-hop RTS MAC (MMAC), 563
Multi-hop wireless network participation, 105
Multimedia applications, *see* Adaptive algorithmic power optimization
Multimedia messaging service (MMS), *see also* MM1; MM4
   addressing, 124–125
   architecture, *122*, 122–126
   basics, 121–122
   delivery report, *126*, 129–132, *133–136*
   forwarding, 143–145, *144–145*
   future directions, 145
   3GPP documents, 146
   interfaces, *122–124*, 123–124
   messages, 126
   notification, 134–137, *139–141*
   read-reply reports, *126*, 133–134, *136–139*
   retrieval, 138–143, *141–143*
   submission, *126–128*, 126–129, *130*
   supported formats, 125–126
   technical specifications, 125
   transfer, 129, *131–133*
   WAP forum documents, 146
Multimedia workload, adaptive algorithmic power optimization
   application-assisted technique, 766–769, *767–769*
   basics, 747–749, 755–756, 776, *776*
   buffer insertion, 763–764
   clock setting, 760–763, *762–763*
   collaborative power management, 764–776
   communication energy optimization, 769–773, *770–772*
   computational energy optimization, 773–776, *774–775*
   decoding, 760–764
   encoding, 756–760
   energy, 749–755, *754*
   motion estimation, 757
   MPEG-4, 752–753, 759–790
   multimedia workload, 765–766, *766*
   operating systems, 765–769
   optimization, 749–755
   scalability, 749–755, *754*
   server-assisted technique, 769–776
   source coding, 750–753, *751*, 757–759
   stand-alone power management, 756–764
   transmission, 757–759
   voltage setting, 760–763, *762–763*
   wavelets, 750–752, *751*
Multimodal application, 55
Multimodal interfaces, 57
Multimode transmission links, 819–824
Multipath protocol, 656
Multipath routing algorithm (MPATH), 656
Multiple Access with Collision Avoidance (MACA), 696–697
Multiple announcement deliveries, 239
Multiple malicious nodes, 982
Multiple object spaces, 82
Multiple RBSs per MCS, 249
Multiplicative Increase and Linear Decrease (MILD), 554, 556
Multi-point relays (MPRs)
   localized broadcast, neighbor designation, 673–674, *673–674*
   MANET localized broadcasting, 670
   quality of service, 582–583
   routing and mobility management, 645
Multipurpose Internet Mail Extensions (MIME), 126, 305
Multi-transmitter/multi-receiver case, 817–819
Murphy studies, 25–50, 227–251
Musical Instrument Digital Interface (MIDI), 125
Mutual authentication, IEEE 802.11, 935
MVC structure, *see* Pervasive application development
Myrinet, 98

N

NACK, *see* Negative acknowledgement (NACK)
Nahrsted, Yuan and, studies, 766
Nahrstedt, Shah and, studies, 587
Naldburg and Kravets, Yi, studies, 442
NAP, *see* Neighbor Addition Protocol (NAP)
Napster, 99
Napuri studies, 562
National Institute of Standards and Technology (NIST), 927

1011

MOBILE COMPUTING HANDBOOK

National Laboratory for Applied Network Research (NLANR), 342
National Security Agency (NSA), 927
Navigator 1, 728
Navigator 2, 7, 7
Near Term Digital Radio (NTDR), 616
Negative acknowledgement (NACK), 683, 809
Neighbor Addition Protocol (NAP), 704
Neighbor designation, localized broadcast
  basics, 663–665, *664–665,* 686
  CDS-based algorithm, *674,* 676–677, *677*
  classification, 665–672
  cluster-based algorithm, 677–681, *679–681*
  counter-based scheme, 666
  deterministic algorithms, 666–669
  distance-based scheme, 666
  dominant pruning, *674–675,* 675
  forward node selection process, 672–673
  global broadcasting, 667
  K-hop zone-based algorithm, *674,* 681–682, *683*
  local algorithms, 669–672, *671*
  local broadcasting, 669
  location-based scheme, 666
  multi-point relays, 673–674, *673–674*
  neighbor-designating broadcast algorithms, 672–677
  partial dominant pruning, *674,* 675–676, *676*
  probabilistic algorithms, 665–666
  quasi-global broadcasting, 667–668
  quasi-local broadcasting, 668–669
  reliable broadcast algorithm, *674,* 683–685, *685*
  total dominant pruning, *674,* 675–676, *676*
Neighbor Lookup Protocol (NLP), 469–471, 491–492
Neighbor prefetching, 833–835
Neighbor Reduction Protocol (NRP), 704
Neighbor Watch, 453
NetMan, 8
Network Address Translation, 150
Network dynamics, 484
Network failures, 338, 575
Network File System (NFS), 406
Networking-oriented simulators, 118
Network layer security, 487–496
Network partitioning, 606
Network security, *see* Ad hoc network security
Network Simulator 2 (NS-2), 290
Networks of workstations (NOW), 98
Network topology, 167–168, *167–169,* 703–704
Network transparency, 872
Never-update strategy, 175–176
New Tahoe TCP, 896

New Transaction Management System
  basics, 863–864
  communication cost and scalability, 872
  comparisons, *865, 867*
  consistency and concurrency, 869
  infrastructure requirements and compatibility, 871
NFS, *see* Network File System (NFS)
NiagaraCQ, 68
Ning and Sun studies, 438
NIST, *see* National Institute of Standards and Technology (NIST)
Ni studies, 664
NLANR, *see* National Laboratory for Applied Network Research (NLANR)
NLP, *see* Neighbor Lookup Protocol (NLP)
NLR-Query, *see* Non-Location-Related Query (NLR-Query)
Nodes
  cooperation, 606
  energy-efficient wireless networks, 695–700
  MANET security, 471–472
  medium access control, *555, 556*
  mobile agent security, 953
  multiple malicious, 982
  position identification, 605
  promiscuous mode, 474
  quality of service, 605–606
  security, trust, and privacy, 982
  self-policing MANETs, 436–440
  treatment, 555
Nomadic host location management, 151–155, *152*
NOMADS, 952–953
Noncompensating transactions, 849
Nondeterministic lifetime, 843
Nonexhaustive service, 904
Non-Location-Related Query (NLR-Query), 265
Nonrepudiation, 437, 460, 496, *see also* Security
Nonvital components, 848
Normal state, wearable computers, 15
Notebook computers, *see* Laptop computers
Notice, security, 970–971
Notification of messages, 134–137, *139–141*
NOW, *see* Networks of workstations (NOW)
NRP, *see* Neighbor Reduction Protocol (NRP)
NS-2, *see* Network Simulator 2 (NS-2)
NSA, *see* National Security Agency (NSA)
NS simulator, 118
NTDR, *see* Near Term Digital Radio (NTDR)

O

Object Invalidation report (OIR), 353–354, 363, 427–428

Index

Object Management Group (OMG), 322
Observation-based Cooperation Enforcement in Ad hoc Networks (OCEAN), 445, 450–451
Obstacle Mobility Model, 282–283
OCEAN, *see* Observation-based Cooperation Enforcement in Ad hoc Networks (OCEAN)
OCTS, *see* Omnidirectional CTS (oCTS)
OIR, *see* Object invalidation report (OIR)
OLSR, *see* Optimized Link State Routing (OLSR)
OMG, *see* Object Management Group (OMG)
Omnidirectional CTS (oCTS), 560
Omnidirectional RTS (oRTS), 560
On-demand protocols
  routing and mobility management, 647–652
  scalable clustered network architecture, 614, *625*, 625–627, *627*
Online look-ahead heuristics, 830–835
Open GIS Consortium, 261
Open Nested Transactions Models, 845–846, 851, 854
Open Pluggable Edge Services (OPES), 303
Open Shortest Path First (OSPF), 250, 491, 600
Open Systems Interconnection (OSI), 872, 979
Operating systems
  adaptive algorithm power optimization, 765–769
  power awareness, 736–738
  power management, 716
Operational systems, QoS, 605
Operators, 68
OPES, *see* Open Pluggable Edge Services (OPES)
OPNET simulator, 118
Optimization, 411–416, 749–755, *see also* Adaptive algorithmic power optimization
Optimized Link State Routing (OLSR)
  energy-aware web caching, 785
  MANET localized broadcasting, 673
  quality of service, 582–583
  routing and mobility management, 645
Options evaluation, power management, 728–729
Organizational contexts, 300
ORTS, *see* Omnidirectional RTS (oRTS)
ORTS/oCTS-based DMAC, *561*, 562
OSI, *see* Open Systems Interconnection (OSI)
OS/2 operating system, 396
OSPF, *see* Open Shortest Path First (OSPF)
OWL, *see* Web Ontology Language (OWL)

P

Packet arrivals, 815–816
Packet forwarding, 809–824
Packet prioritization, 605

Packet Purse Model, 440–441
Packet Trade Model, 440
Paging
  mobility tracking, 187–190, *188, 190*
  PCS networks, analytic mobility models, 886–887
palmOne, 99
Palm OS operating system, 101, 743
PAMAS, *see* Power-Aware Multi-Access Protocol with Signaling (PAMAS)
PAN, *see* Personal area network (PAN)
PANDA, *see* Positional attribute-based next-hop determination approach (PANDA)
Papadimitratos and Haas studies, 442
Papadimitratos studies, 457–480
PAR, *see* Prefetch-access-ratio (PAR)
Parker studies, 961–984
Parlay, 302–303
Partial dominant pruning, *674*, 675–676, *676*
Partially disconnect mode, 863
Partitioning
  hoarding, 394–395
  locating mobile objects, 205–206
  network, 606
Pascoe studies, 300
Path Rater, 444, 451, 494
Pattern, 168–173, 739
Paul and Westhoff studies, 445
pawS, *see* Privacy Aware System (pawS)
Payment systems, 440–441, 446
PCA, *see* Personal Client Architecture (PCA) processors
PCF, *see* Point coordination function (PCF)
PCM, *see* Power Control MAC (PCM)
PCMA, *see* Power Controlled Multiple Access (PCMA)
PCS networks, analytic mobility models
  basics, 877–878
  cellular systems, *878*, 878–879
  cost, 886–887
  distributions, 886
  location update, 881–886
  Markov walk models, *879*, 879–881, *881*
  paging, 886–887
  performance evaluation, *881*, 887–891, *888–889*
  recursive Markov analysis, 883–885
  system models, 878–881
  tracking, location, 882
  two renewal processes, 882–883, *883*
  updates, location, 882
PCSP, *see* Precomputation-based Selective Probing (PCSP)
PDA, *see* Personal digital assistant (PDA)

1013

PDF, *see* Probability density function (PDF)
PDFs, *see* Portable document format (PDF) files
Peer-to-peer (P2P) programs
    cache management, 342–343, 356–358
    energy-aware web caching, 791
    grid and cluster computing, 99
PeerWare, 49–50
Pendulum Instruction Set Architecture (PISA), 741
Peng and Lu studies, 675–676
Perfect server (PS) scheme, 373, 376
Performance
    analytic mobility, PCS networks, 877–891
    battery power management, 893–917, *910–913*, 910–916
    cache invalidation schemes, 373–383
    grid and cluster computing, 117
    mobile transaction models, 873
    PCS networks, analytic mobility models, *881,* 887–891, *888–889*
    routing and mobility management, 656–657
    survey, 839–873
PerkIP, 840, 842
Per object profile approach, 210
Perrig, Hu, Johnson and, studies, 442
Perrig and Johnson, Hu, studies, 442
Persistent messages, 231
Personal area network (PAN), 805
Personal Client Architecture (PCA) processors, 732
Personal digital assistant (PDA)
    battery power management, 893
    cache invalidation, 361
    context-awareness, 308
    energy-aware web caching, 781
    grid and cluster computing, 95–96, 98–101, 112, 116
    ISAM, 91
    mobile agent middlewares, 317
    modeling distributed applications, 507
    power-aware cache management, 403
    power awareness, 731
    power management, 723
Personal-Java software, 39, 91
Pervasive application development
    application environments, 58–59
    basics, 53–56, 68–69, 73–93
    connectivity, 57
    context-aware applications, 66
    design-time adaptation, 61–62
    development, 56–68
    device-independent views, 60–63, 69
    device platform heterogeneity, 56–58
    host-independent models, *64,* 64–66, 69
    interaction modalities, 57
    maintenance complexity, 58
    platform capabilities, 57
    platform-independent controllers, 63–64, 69
    presentation transcoding, 59–60
    runtime adaptation, 60–61
    source-independent context data, 66–69
    user interface, 57
    visual tools, 62–63
Pervasive computing environments, 963–976
PGP, *see* Pretty Good Privacy (PGP)
Phan studies, 95–118
Physical context, context-awareness, 300
Physical mobility, Lime, 26
Picco studies, 25–50
PIFS, *see* Point coordination function interframe spacing (PIFS)
PIMA, *see* Platform-Independent Model for Applications (PIMA)
Piquer, Tanter and, studies, 329
PISA, *see* Pendulum Instruction Set Architecture (PISA)
Pitoura studies, 197–224, 256
PIX, 344
Plaggemeier and Westhoff, Lamparter, studies, 438
Platform capabilities, 57
Platform-independent controllers, 63–64, 69
Platform-Independent Model for Applications (PIMA), 538
Playstation 2, 96, 98, 112, 116
PLU, *see* Probabilistic location update (PLU) scheme
PocketPC, 99
Point coordination function interframe spacing (PIFS), 547
Point coordination function (PCF), 696
Poisson Call Arrival Model, 178
Poisson model, 174
Policies, security, 963–968, 975
Policy language, 966–967
PolicyMaker, 965–966
Politecnico di Milano, 49
Ponder, 966
Popularity, energy-aware web caching, 795
Portable devices, battery power management
    analysis, 900–909, *901*
    basics, 893–899
    discharge/recharge model, 899–900, *900*
    exhaustive service, 904
    extensions, 903–909
    generalizations, 903–909
    lithium ion simulation results, *914–915,* 914–916
    multi-battery system, *905,* 905–909
    nonexhaustive service, 904

Index

performance results, *910–913*, 910–916
relaxation phenomenon, 896–899, *897*
scheduling problem, 916–917
system model, 899–900
vacations, 904
Portable document format (PDF) files, 509
Positional attribute-based next-hop determination approach (PANDA), 599
Power analyzer, 742
Power and energy relationship, 709
Power-aware cache management, *see also* Cache management
   adaptive prefetch approach, 412
   adaptive value-based prefetch scheme, 415–416
   basics, 403–405, 416–418, *417*
   battery life target, 415–416
   bit sequences scheme, 407–408
   broadcasting time stamp scheme, 407
   cache consistency model, 406
   future directions, 416–418, *417*
   invalidation, 405–410
   IR-based invalidation model, 406–407
   optimization, 411–416
   power level adaptation, 416
   prefetching, 409–410
   scheme basics, 411–412
   UIR-based invalidation model, 408, *408*
   value-based prefetch scheme, 413–414, *414*
Power-aware communication, 781–789
Power-Aware Multi-Access Protocol with Signaling (PAMAS)
   battery power management, 895
   energy-aware web caching, 783–785
   energy efficient wireless networks, 696
Power awareness
   broadcasting, 702–703
   link layer adaptation, 692–693
   medium access control, 695–697
   protocols, routing and mobility management, 654–655
Power awareness, management
   application level solutions, 740–741
   architectural design, 736
   architecture techniques, 733–736
   basics, 731–733, 743
   batteries, 733–734
   capacitance load reduction, 735
   CMOS circuitry components, 734–735
   code generation, 739
   communication, 736–738
   compilation, 738–740
   energy-aware processors, 734
   energy management solutions, 736
   frequency scaling, 735
   hardward techniques, 733–736
   intelligent power drainage, 733–734
   I/O management, 737
   low-power design and measurement, 741–743
   memory management, 737
   memory operands reduction, 739
   motivation, 731–732
   operating systems, 736–738
   pattern matching, 739
   power analyzer, 742
   power consumption equation, 734
   PowerScope, 741–742
   remote task execution, 739–740
   reordering instruments, 738–739
   scheduling, 738
   SimpleScalar, 742–743
   smart batteries, 733–734
   software applications, 738–741
   taxonomy, research and industry solutions, 732–733
   voltage scaling, 735
   Wattch project, 742–743
Power consumption, 105, 734
Power control, 982, *983*
Power Controlled Multiple Access (PCMA)
   medium access control, 556, 558
   transmitter power control, 810, 816–817, 819
Power Control MAC (PCM), 560
Power decision, 829–830
Power level adaptation, 416
Power limitations, QoS, 577
Power management
   adaptive algorithmic power optimization, 747–776
   Amdahl's Law, 727–728
   awareness and management techniques, 731–743
   basics, 709, 729
   batteries, 710–712, *711*, 725–727, *726–727*
   case study, 721–727, *722*
   dynamic CPU speed-setting, 722–725, *723–725*
   energy-aware web caching, 779–802
   guidelines, 727–729
   hardware states, 716–719, *717*, *719*
   hybrid networks, 779–802
   load power, 725–727, *726–727*
   memory bottleneck, 722–725, *723–725*
   multimedia workload, 747–776
   options evaluation, 728–729
   power and energy relationship, 709
   power supplies, *712*, 712–715, *714–715*

1015

software, 719–721
transmitter power control, 805–835
Power optimization, adaptive algorithms
    application-assisted technique, 766–769, 767–769
    basics, 747–749, 755–756, 776, 776
    buffer insertion, 763–764
    clock setting, 760–763, 762–763
    collaborative power management, 764–776
    communication energy optimization, 769–773, 770–772
    computational energy optimization, 773–776, 774–775
    decoding, 760–764
    encoding, 756–760
    energy, 749–755, 754
    motion estimation, 757
    MPEG-4, 752–753, 759–790
    multimedia workload, 765–766, 766
    operating systems, 765–769
    optimization, 749–755
    scalability, 749–755, 754
    server-assisted technique, 769–776
    source coding, 750–753, 751, 757–759
    stand-alone power management, 756–764
    transmission, 757–759
    voltage setting, 760–763, 762–763
    wavelets, 750–752, 751
PowerPoint software, 121
PowerScope, 721, 740–743
Powersmart, 734
Power supplies, 712, 712–715, 714–715
P2P, *see* Peer-to-peer (P2P) programs
PPRM, *see* Processor/power resource management (PPRM)
Precision of location information, 200
Precomputation-based Selective Probing (PCSP), 588–589
Preemption, QoS, 604
Prefetch-access-ratio (PAR), 337, 412
Prefetching
    cache management, 337, 344
    deep prefetching, 835
    energy aware web caching, 796–798, 797
    ISAM, 81
    power-aware cache management, 405, 409–410, 413–414, 414
    transmitter power control, 824–835
Presentation transcoding, 59–60
Prestored and static capabilities
    basics, 5, 9–10, 11
    evaluation, 17, 19
    example systems, 7
Pretty Good Privacy (PGP), 498

Prewrite
    basics, 855
    communication cost and scalability, 871
    comparisons, 865–866
    consistency and concurrency, 868
Primitives for Object Scheduling (PRIMOS), 82
PRIMOS, *see* Primitives for Object Scheduling (PRIMOS)
Print being, 83–86, 89–90
Priority-based query handling, 472
Privacy, 308–310, 974, *see also* Security, trust, and privacy
Privacy Aware System (pawS), 971
Proactive approach, 489–493
Proactive protocols, 623, 623–624
Proactive synthetic assistant, 6, 7, 8–9, 12–17, 14, 16–17
Probabilistic algorithms, 665–666
Probabilistic location update (PLU) scheme, 184
Probability density function (PDF), 904, 906–909
Procedures, prestored, *see* Prestored and static capabilities
Processor/power resource management (PPRM), 766
Profile-based scheme, mobility tracking, 185
Programmable tuple spaces, 48
Programming language support, 873
Program trees, hoarding, 396–398, 397
Prolog, XSB, 968
Promera, 4
Pro-Motion
    basics, 853–855
    communication cost and scalability, 871
    comparisons, 865–866
    consistency and concurrency, 868
    infrastructure requirements and compatibility, 870
Protected packet forwarding, 493–495
Protecting agents, 948–951
Protocols
    authentication and access control, 935–936
    comparisons, routing and mobility management, 643, 648, 653, 654
    conventional, routing and mobility management, 640–641
    energy-efficient wireless networks, 701–702
    integrity protection, 933
    key management, 936–937
    self-policing MANETs, 437
Proximate selection, 302
Proximity, security, 970–971
Proxying, 852
Pseudonymity, 970–972
PSNR, 748, 752, 757

Index

PSP strategy, 887
PS scheme, *see* Perfect server (PS) scheme
PSTN, *see* Public switched telephone networks (PSTN)
Public Key Infrastructure (PKI), *see also* Simple Public Key Infrastructure (SPKI)
   ad hoc network security, 498
   MANET security, 463, 465, 474
   mobile agent middleware, 321
   mobile agent security, 947
Public switched telephone networks (PSTN), 637
PulsON Time Modulated Ultra Wideband technology, 301
Purdue University, 49
Pure ALOHA, 545

Q

Qayyum studies, 673
QoS, *see* Quality of service (QoS)
Quality of service (QoS)
   adaptive algorithmic power optimization, 747–750
   awareness, routing and mobility management, 654–655
   battery power management, 895
   location management, 165
   MANET security, 468
   mobile agent middlewares, 329
   modeling distributed applications, 512
   routing and mobility management, 638
   transmitter power control, 806, 808
Quality of service (QoS), MANETs
   bandwidth based calculation based scheme, 592–595, *593–594*
   basics, 569–576, *570, 573*, 606–607
   challenges, 576–579
   current trends, 579–604, *580*
   flat networks, 581–595
   future directions, 604–606
   hierarchical protocols, 595–598, *596, 598*
   metrics, 575–576
   position-based protocol, 598–600, *601*
   power-aware routing, 600–604
   predictive protocols, 587–589
   proactive protocols, 582–583
   reactive protocols, 583–587, *585–586*
   routing support challenges, 576–579
   ticket-based probing scheme, 589–592, *590*
Quasi-global broadcasting, 667–668
Quasi-local broadcasting, 668–669
Quasi-Local Model, 668
Query classification, 270–271, *271*

Query processing, location-dependent
   basics, 255–256, 271–272
   data and queries, 256–258
   location-aware queries, 266–268
   location-dependent queries, 268–270
   location relatedness, 260–271
   moving object database, 258, 270
   query classification, 270–271, *271*
   query model, 260–271, *261, 263*
   spatial database management, 258–260
   translation steps, 271–272

R

Radial interference reduction, 558–560, *559*
Radio activity, 695–700
Radio base stations (RBSs), 231, 246, 249
Radio Frequency Identification (RFID), 328
RADIUS, *see* Remote Authentication Dial-In User Service (RADIUS)
Raghavan and Snoeren studies, 441
Random access MAC, 545–546, *546*
Random access method, 545
Random Direction Model, 281
Randomized clustering algorithm, 619
Random mobility models, 280–282, *280–282*
Random Walk Model
   analytic mobility models, 877, 880
   mobility tracking, 168–169, 172, 178
   simulation models, 280
Random Waypoint Mobility Model, 289–290
Random Waypoint Model, 281, 283, 285
Rao, Chiasserini and, studies, 898
Rao, Zorzi and, studies, 896
Rapidly moving mobile units, 239–240
RAT, *see* Reverse Address Translation (RAT)
RBAC, *see* Role Based Access Control (RBAC)
RBSs, *see* Radio base stations (RBSs)
RC4 algorithm, 927–929
R-Destination-Sequenced Distance Vector (R-DSDV), 643
RDF, *see* Resource Description Framework (RDF)
RDFS, *see* Resource Description Framework Schema (RDFS)
Reactions, 36–37, *38*, 493–495
Read-reply reports, *126*, 133–134, *136–139*
Reality check, 246–251
Real studies, 73–93
Real-time operating system (RTOS) kernel, 765
Real-Time Transport Protocol/Real-Time Streaming Protocol (RTP/RTSP), 303
Receiver-Oriented Multiple Access (ROMA), 563
Receive threshold (RXThresh), 549–550
Reconciliation, Lime, 32–34, *34*

1017

MOBILE COMPUTING HANDBOOK

Recursive Markov analysis, 883–885
Redemption, self-policing MANETs, 451–452
Red Hat Linux, 98, 112
Reduced instruction set computing (RISC) architecture, 721
Reed Solomon coding, 693
Reference model, mobile transactions, 841–842, *842*
Refresh, DNS location management, 153
Rei, 964, 966, 968, 974, 976
Relaxation phenomenon, 896–899, *897*
Reliable broadcast algorithm, *674*, 683–685, *685*
Reliable delivery on links, 249–250
Relocation, mobile transaction models, 844–845
Remote Authentication Dial-In User Service (RADIUS), 934
Remote Evaluation (REV), 319–320
Remote task execution, 739–740
Ren and Dunham studies, 270
Reno TCP, 896
Reorderable objects, 857
Reordering instruments, 738–739
Replay attacks, 932
Replicated job assignments, 109–110
Replication, *209–211*
Reporting and cotransactions model
 basics, 850–851
 communincation cost and scalability, 871
 comparisons, *865–866*
 consistency and concurrency, 864
 infrastructure requirements and compatibility, 869
Reporting center, 182–184
Reporting transaction, 850
Reputation systems, 442–454, *see also* Collaborative Reputation (CORE)
Request power to send (RPTS), 558
Requests for comments (RFCs), 218
Request-to-send (RTS) packets and mechanism
 ad hoc network security, 486
 energy-aware web caching, 782–783, 801
 energy efficient wireless networks, 695
 medium access control, 547, 554
RERR, *see* Route error (RERR) packet
Research
 context-aware mobile computing, 310–311
 middleware, context-aware applications, 326–330, *327*
 security, trust, and privacy, 979–982, *980–981*
Resource constraints, 483
Resource Description Framework (RDF), 329
Resource Description Framework Schema (RDFS), 964
Resources, 438, 577

Response and response systems, 442–445, 451
Responsive channel, 816–817
Restrictions, *see* Scope of operations restriction
Retrieval of messages, 138–143, *141–143*
Retry, DNS location management, 153
REV, *see* Remote Evaluation (REV)
Reverse Address Translation (RAT), 150
Reverse bypass pointer, 208
Revocation rule, 921
RFCs, *see* Requests for comments (RFCs)
RFID, *see* Radio Frequency Identification (RFID)
RISC, *see* Reduced instruction set computing (RISC) architecture
Robustness, 484
Robust Security Network/AES Counter Mode with CBC-MAC protocol (RSN/AES-CCMP), 485, 487
Role Based Access Control (RBAC), 965
ROMA, *see* Receiver-Oriented Multiple Access (ROMA)
Roman studies, 25–50, 227–251
Rome system, 68
Round-trip, regenerated artifact, 62
Route delivery, 240
Route discovery, 614, 617, 647
Route disruption, 438
Route error (RERR) packet, 583, 602
Route invasion, 438
Route maintenance, 472–473, 614, 647
Route reply (RREP) packet
 ad hoc network security, 490–492
 quality of service, 583, 586, 593
Route request (RREQ) packet
 ad hoc network security, 490–492
 quality of service, 583, 586, 593–595, 602
 routing and mobility management, 649–650
Routing, secure, 467–473
Routing and mobility management
 ad hoc networks, 638–639, *639*, 657–658
 ad hoc on-demand distance vector routing, 650–651
 ad hoc protocols, 641–656
 associativity based routing, 651
 basics, 637, 658
 challenges, 656–657
 clusterhead gateway switch routing, 645–646, *646*
 comparison of protocols, *643*, *648*, 653, *654*
 conventional protocols, 640–641
 destination-sequenced distance vector, *639*, 642–644, *643*
 dynamic source routing, 647–650
 flooding method, 655–656
 hierarchical protocols, 646–647

Index

hybrid protocols, 652–653
issues and challenges, 656–657
link state protocol, 644–645
location-based protocol, 655
MANETs, desired protocol characteristics, 640
multicasting method, 655–656
multipath protocol, 656
on-demand protocols, 647–652
performance issues and challenges, 656–657
power-aware protocols, 654–655
QoS-aware routing, 654–655
table-driven protocols, 642–647
temporally ordered algorithm, 653
wireless protocol, 644
zone protocol, 652–653
Routing fidelity, 627
Routing layer, 785–788
RPTS, *see* Request power to send (RPTS)
RREP, *see* Route reply (RREP) packet
RREQ, *see* Route request (RREQ) packet
RSA decryption, 113–116, *115*
RSN/AES-CCMP, *see* Robust Security Network/AES Counter Mode with CBC-MAC protocol (RSN/AES-CCMP)
RTOS, *see* Real-time operating system (RTOS) kernel
RTP/RTSP, *see* Real-Time Transport Protocol/Real-Time Streaming Protocol (RTP/RTSP)
RTS, *see* Request-to-send (RTS) packets and mechanism
Rudenko studies, 739
Ruiz studies, 95–118
Runtime adaptation, 60–61
RXThresh, *see* Receive threshold (RXThresh)

S

SA, *see* Security Association (SA)
Saadawi, Xu and, studies, 552
SACCS, *see* Scalable asynchronous cache consistency scheme (SACCS)
SafeTCL, 318
Saga, 848–849
Sailhan studies, 779–802
Salutation, 325
Samaras studies, 197–224
Samet, Aref and, studies, 259
Samet and Aref studies, 258–259
Samsung, 734
Sandbox Model, 947, 951
SanDisk CompactFlash card, 100
Sankar studies, 637–658
Satyanarayanan, Flinn and, studies, 721

Save/restore session, 92
Saygin studies, 389–400
SBL, *see* Single-battery-like (SBL) schemes
SBS, *see* Smart battery system (SBS)
Scalability
   adaptive algorithm power optimization, 749–755, *754*
   ad hoc network security, 484
   DNS location management, 158–160, *159*
   grid and cluster computing, 117–118
   LEECH, 104
   mobile transaction models, 845, 871–872
   pervasive application development, 58
Scalable asynchronous cache consistency scheme (SACCS), 349–350
Scalable clustered network architecture, 616
   algorithms, 617–622, 627–629, *629*
   ARC protocol, *625*, 625–626
   basics, 611–613, 632–633, *633–634*
   CBRP protocol, *625*, 625–626
   CEDAR protocol, 630, *630*
   CGRS protocol, *623*, 623–624
   classification of protocols, 613–617
   cluster architectures and protocols, *615*, 615–617
   cluster architectures protocols, 617, *618*
   DSCR protocol, *624*, 624–625
   flat routing protocols, 614–615, *624*, 624–625
   GLS protocol, *631*, 632
   HSR protocol, *623*, 623–624
   information infrastructures, 627–632
   LANMAR protocol, *624*, 624–625
   LCA for routing backbone, 617–628
   LGeo-based protocols, 631–632
   LLog-based protocol, 629–631
   LNG algorithms, *619*, 621–622, *622*, 626–627, *627*
   LSG algorithms and protocols, 618–621, *619*, 622–626
   on-demand protocols, *625*, 625–627, *627*
   proactive protocols, *623*, 623–624
   ZHLS protocol, *631*, 631–632
   zone routing protocol, *630*, 630–631
Scarce resources, 577
Scheduling, 695–700, 738, 916–917
Schilit and Theimer studies, 300
Schilit studies, 299
Schneier studies, 442
Scope of operations restriction, 34–36, *35*
SDCI, *see* Selective Dual-Report Cache Invalidation (SDCI) scheme
SEAD, *see* Secure Efficient Distance (SEAD)
Second class transactions, 861
Second-hand information, weighting, 452

1019

Secure Ad Hoc On-Demand Distance Vector (Secure-AODV) protocol, 468–469, 488
Secure and Open Mobile Agents (SOMA)
　mobile agent middlewares, 322–324, 326, 328–329
　mobile agent security, 952–953
Secure-AODV, *see* Secure Ad Hoc On-Demand Distance Vector (Secure-AODV) protocol
Secure Efficient Distance (SEAD), 442, 488
Secure Link State Routing (SLSP), 491–492
Secure Message Transmission Protocol (SMTP), 473, *475*, 475–477, *477*
Secure routing, self-policing MANETs, 441–442
Secure Routing Protocol (SRP)
　ad hoc network security, 487
　MANET security, 467, 469
　security, mobile ad hoc networks, 473
　self-policing MANETs, 442
Secure Socket Layer (SSL), 117, 155
Security, *see also* Security, trust, and privacy; specific type of security
　DNS location management, 155
　grid and cluster computing, 117
Security, ad hoc networks
　anticipated unknown attacks, 500
　basics, 483–486
　detection, 493–494
　distance vector routing, 490–491
　evaluation, 500
　future directions, 499–500
　intrusion detection system, 495–496
　key management, 496–499
　link-layer security, 486–487
　link state routing, 491–492
　localized trust, 499
　802.11 MAC vulnerabilities, 486–487
　message authentication primitives, 488–489
　network layer security, 487–496
　proactive approach, 489–493
　protected packet forwarding, 493–495
　reaction, 494–495
　reactive approach, 493–495
　security in depth, 499–500
　source routing, 489–490
　trusted third party, 497–498
　trust management, 496–499
　unknown attacks, 500
　web-of-trust, 498–499
　802.11 WEP vulnerabilities, 487
Security, IEEE 802.11
　access control, 933–936
　authentication, 933–936
　basics, 923–925, *924*, 937
　encryption, 927–930, *928*

extensions, authentication, 934–935
historical developments, 924–925
integrity-based attacks, 931–932
integrity protection, 930–933
key management, 936–937
keystream reuse, 928–929
mutual authentication, 935
new protocols, 933, 935–937
new standards, 929–930
RC4 weakness, 929
replay attacks, 932
wireless security threats, 926–927
Security, MANETs
　basics, 457–459, 478–480
　data forwarding, 473–477
　discovery procedure, 471–472
　goals, 459–460
　Neighbor Lookup Protocol, 469–471
　priority-based query handling, 472
　route maintenance procedure, 472–473
　routing, secure, 467–473
　secure message transmission protocol, *475*, 475–477, *477*
　secure routing protocol extension, 473
　threats and challenges, 460–462
　trust management, 463–467, *465–466*
Security, mobile agents
　acceptance, 941–942
　access control, 955–956
　agent platforms, 946–948
　agents, 954–955
　authentication, 947
　authorization, 947–948
　basics, 941, 957
　code, 946–947
　countermeasures, 945–951
　future directions, 953–956
　open issues, 953–956
　protecting agents, 948–951
　requirements, 943–945, *944*
　solutions overview, 951–953
　trust, 954–955
　user-agent trust, 945–946
Security, trust, and privacy
　ad hoc networks, 976–984
　anonymity, 970–972
　approach, 966–967, *968*
　basics, 961–963, 967–968, 975–976, 983–984
　consent, 970–971
　context broker architecture, 972–973, *973*
　design principles, 969–970
　devices, 976–977
　directional antennas, 982, *983*
　environments, 976–977

Index

implementations, 970–972
intrusion detection, 976–984
locality, 970–971
meta-reasoning, 975
multiple malicious nodes, 982
notice, 970–971
pervasive computing environments, 963–976
policies, 963–968, 975
power control, 982, *983*
privacy policy language, 974
proximity, 970–971
pseudonymity, 970–972
related work, 964–966
research, 979–982, *980–981*
Security Association (SA), 469
Security-aware ad hoc routing, 442
Security-aware routing, 442
Security in depth, ad hoc network security, 499–500
SEDF, *see* Slacked Earliest Deadline First (SEDF)
SEER, 392–394, *393,* 399
Selective Cache Invalidation scheme, 426
Selective Dual-Report Cache Invalidation (SDCI) scheme, 427–428, *428,* 431
Selective invalidation schemes, 426–431
Self-policing ad hoc networks
  Ariadne, 442
  attacks, 438–439
  basics, 435–436, 445–447, 454
  CONFIDANT, 444, 450–452
  context-aware interface, 445
  CORE, 444–445, 450–453
  cryptography, 441–442
  detection, 452–453
  detection systems, 442–445
  enhanced CONFIDANT, 447–449
  identity, 453–454
  information dissemination, 450
  main solution tracks, 440–447
  mobile ad hoc networks, 447–454
  Monitor, 453
  Neighbor Watch, 453
  node misbehavior, 436–440
  OCEAN, 445, 451–452
  Path Rater, 444, 451
  payment systems, 440–441
  redemption, 451–452
  reputation systems, 442–445, 447–454
  response, 451
  response systems, 442–445
  second-hand information, weighting, 452
  secure efficient distance, 442
  secure routing, 441–442
  security-aware routing, 442
  spurious ratings, 449–450
  time, weighting, 451–452
  traffic diversion, 438–439
  type of information, 450–451
  Watchdog, 444, 446, 451, 453
  weighting, 451–452
Semantic language, 968
Semantics-based mobile transaction models
  basics, 856–857
  communication cost and scalability, 871–872
  comparisons, *865–866*
  consistency and concurrency, 868
  infrastructure requirements and compatibility, 870
SenSay, 7, 12
Sensing technologies, context-awareness, 300–301
Sensor data acquisition, 300–301
Sensor-MAC (SMAC), 697, 782, 784–785
Sentient Computing, 311
Serializability, 861
Serial number, 153
Server algorithm, 369–373, *370*
Server-assisted technique, 769–776
Server failures, 338
Server process, 152, 154
Service discovery, 104
Service discovery agent (SDA), 108
Service Location Protocol (SLP), 325, 537
Service provision, Lime, 46–47
Session guarantees, 862–863
Seti@home, 95
Seydim studies, 255–272
Shah and Nahrstedt studies, 587
Sharma studies, 893–917
Sharp Zaurus 6500 PDA, 91
Shih studies, 895
Shortest distance model, 171
Short interframe spacing (SIFS), 547
Short Message Service (SMS)
  context-awareness, 303
  mobile agent middlewares, 317
  multimedia messaging service, 121
  wearable computers, 15
Siewiorek studies, 3–22, 709–729
SIFS, *see* Short interframe spacing (SIFS)
SIG, *see* Signature (SIG) approach
SIGKDD, *see* Special Interest Group on Knowledge Discovery in Data and Data Mining (SIGKDD)
Signature (SIG) approach, 352–353, 362
Simple caching, 208
Simple Message Transfer Protocol (SMTP), 129, 133

1021

MOBILE COMPUTING HANDBOOK

Simple Network Management Protocol (SNMP), 328, 579
Simple Object Access Protocol (SOAP), 332
Simple predicate (SP), 262, 264
Simple Public Key Infrastructure (SPKI), 945, *see also* Public Key Infrastructure (PKI)
SimpleScalar, 742–743, 774
Simulation models and tool, *see also* Models
  advanced mobility models, 282–283, *283*
  basics, 275–276
  existing mobility models, 279–283
  generic mobility models, 283–285, *285–286*
  information access model, 285–288
  location-dependent access, 287–288
  location models, 277–278
  mobility, 279–285
  random mobility models, 280–282, *280–282*
  software architecture, *288–289*, 288–290
  spatial information models, 278–279, *279*
  spatial model, 276–279
  usage, *285*, 290–293, *291–292*
  user mobility modeling, 288–293
  Zipf distribution, 286–287
Simultaneous networkwide search, 186–187
Single-battery-like (SBL) schemes, 914–915
Single-mode transmission links, 824–835
Single-mode wireless links, 809–819
Sinha, Sivakumar and Bharghavan studies, 667
SIP, *see* System-in-package (SIP)
Sivakumar and Bharghavan, Sinha, studies, 667
Slacked Earliest Deadline First (SEDF), 894
Sleep mode, 626
Slotted ALOHA, 545
SLP, *see* Service Location Protocol (SLP)
SLSP, *see* Secure Link State Routing (SLSP)
SMAC, *see* Sensor-MAC (SMAC)
Smailagic studies, 3–22, 709–729
Smart batteries, 733–734
Smart battery system (SBS), 733
Smart Dust, 311
SMIL, *see* Synchronized Multimedia Integration Language (SMIL)
Smit and Havinga studies, 735
SMS, *see* Short Message Service (SMS)
SMTP, *see* Secure message transmission protocol (SMTP); Simple Message Transfer Protocol (SMTP)
SMT protocol, *see* Secure Message Transmission Protocol (SMTP)
Snapshot algorithm, 233–238, *235–237*
SNMP, *see* Simple Network Management Protocol (SNMP)
Snoeren, Raghavan and, studies, 441
SOA, *see* Start of Authority (SOA)

SOAP, *see* Simple Object Access Protocol (SOAP)
SOC, *see* System-on-chip (SOC)
Software, *see also* specific type of software
  pervasive application development, 57
  power awareness, 738–741
  power management, mobile computers, 719–721
  simulation models, *288–289,* 288–290
Solaris operating system, 112
Solar middleware, 328
Solar system, 68
Solutions overview, 951–953
SOMA, *see* Secure and Open Mobile Agents (SOMA)
Sony devices, Palm OS operating system, 101
Soroker studies, 53–69
Source coding, 750–753, *751,* 757–759
Source-independent context data, 66–69
Source routing, 489–490
Sox audio player, 723, 725–726
SP, *see* Simple predicate (SP)
SPAN algorithm, 621–622, 626–627
Spanning-tree CDS (STCDS), 667–668
Sparse Topology and Energy Management (STEM), 700
Spatial channel utilization, 557–563, *564*
Spatial database management, 258–260
Spatial information models, 278–279, *279*
Spatial Model, 276–279, 284–285, 288–289
Special Interest Group on Knowledge Discovery in Data and Data Mining (SIGKDD), 962
Speed-setting, 761
SpeedStep, 734
Spell being, 83–84, 86, 89, *90,* 91
Sphinx speech recognition, 722
SPKI, *see* Simple Public Key Infrastructure (SPKI)
Split transactions, 846–848
Spurious ratings, 449–450
Squirrel, 357–358
Srivastava, Lettieri and, studies, 895
SSL, *see* Secure Socket Layer (SSL)
SSP, *see* Storage Service Provider (SSP) chains
ST200, 773
Stand-alone power management, 756–764
Stanford University, 68
Starter being code, 92
Start of Authority (SOA), 153, 160–161
Stateful approaches, 348–350, *350–352*
State information maintenance, 578
Stateless approaches, 350–355, *354–355*
State of the art, 342–346, *344*
Static capabilities, *see* Prestored and static capabilities

Index

Static case, mobility tracking
  location area, 179, *179–181*
  reporting center, 182–184, *183–184*
STCDS, *see* Spanning-tree CDS (STCDS)
STEM, *see* Sparse Topology and Energy Management (STEM)
Stepanov studies, 275–293
Storage requirements, message delivery, 250–251
Storage Service Provider (SSP) chains, 215
Strict transactions, 851–852
StrongArm
  adaptive algorithmic power optimization, 755, 764, 773
  power management, 718, 721
Structural properties, transmitter power control, 820–823, *823–824,* 829–830
Structured Query Language-like models, 67–68
Structure Query Language, 873
Submission, multimedia messaging service, *126–128,* 126–129, *130*
Subrata studies, 163–190
Sun, Ning and, studies, 438
Sun Blade 100 workstation, 107, 112
Survey, mobile transaction models
  antientropy, 862
  applicable models, 845–849
  approaches, 849–864
  architecture, 852–853
  basics, 839–845, 873–874
  Bayou, 862–863, *865, 867,* 869–870, 872
  caching, 856–857
  characteristics, 842–844
  clustered data model, 851–852, 864, *865–866,* 869, 871
  Coda file system, 860
  commerical databases, 869–871
  communication cost, 871–872
  comparisons, 864–872, *865–867*
  compatibility, 869–871
  concurrency, 856–857, 864–869
  consistency, 864–869
  cotransactions, 850–851, 864, *865–866,* 869, 871
  defined, 842–844
  fragmentable objects, 857
  infrastructure requirements and compatibility, 869–871
  isolation-only transactions, 860–862, *865, 867,* 869–870
  issues, 842–845, 872–873
  kangaroo model, 857–858, *865, 867,* 868, 870, 872
  migration, 852
  multi-database transaction processing manager, 852–853, *865–866,* 868, 870–871
  New Transaction Management System, 863–864, *865, 867,* 869, 871–872
  noncompensating transactions, 849
  open nested transactions, 845–846
  Prewrite, 855, *865,* 866, 868, 871
  Pro-Motion, 853–855, *865–866,* 868, 870–871
  proxying, 852
  reference model, 841–842, *842*
  reorderable objects, 857
  reporting and cotransactions model, 850–851, 864, *865–866,* 869, 871
  requirements, 869–871
  saga, 848–849
  scalability, 871–872
  semantics-based, 856–857, *865–866,* 868, 870–872
  session guarantees, 862–863
  split transactions, 846–848
  strict transactions, 851–852
  time-based model, 858–859, 868, 870
  two-tier replication, 859–860, 862, *865, 867,* 869–870, 872
  weak consistency, 862
  weak transactions, 851–852
Sybil attack, 453–454
Synchronized Multimedia Integration Language (SMIL), 126
System configuration access, Lime, 38–39
System-in-package (SIP), 740
System models
  battery power management, 899–900
  PCS networks, analytic mobility models, 878–881
  transmitter power control, 825–827
System-on-chip (SOC), 740

T

Table-driven protocols, 614, 642–647
Tactical Information Assistant-Prototype (TIA-P), 7, 8
Tactical Information Assistant (TIA) program, 962
Tahoe TCP, 896
Tails, message delivery, 243
Tait studies, 396
Tan studies, 363, 421–432
Tanter and Piquer studies, 329
Tapestry, 67
Task-based anycasting, 505
Task graph abstraction
  ADU delay distributions, 536–537, *536–537*
  algorithms and protocols, 516–528

1023

MOBILE COMPUTING HANDBOOK

basics, 503–506, 537–539
data flow tuple representation model, 511
dilation, 528–531, *530*
disconnections, 527–528
distributed algorithm, 522–528, *523, 529*
effective throughput, *529,* 532–533, *533*
embedded time, *531,* 531–532
embedding task graphs, 514, *515*
greedy algorithm, 519–522, *520–521*
handling device mobility, *525–526,* 525–527
nonpreassigned tasks, 510
performance evaluation, 528–537, *529*
performance evaluation metrics, 516
polynomial-time embedding algorithm, 517–519, *518, 520–521*
preassigned tasks, 510
preliminaries, 507–508
problem formulation, 517
reinstantiation, *534,* 535
task execution, 507–513
task graph instantiation, 516–528
tasks and task graphs, 507–516, *509*
taxonomy of tasks, 510
Taxonomies
cache management, *346,* 346–347
energy efficient cache invalidation, 423–426
locating mobile objects, 215–218, *217–219*
power awareness, 732–733
TBRPF, *see* Topology Broadcast Based on Reverse-Path Forwarding (TBRPF)
TCG-NGSCB, *see* Trusted Computing Group (TCG)-NGSCB (Next Generation Secure Computer Base)
TCL, *see* Tool Command Language (TCL)
TCP, *see* Transmission Control Protocol (TCP); Transport Control Protocol (TCP)
TCPA-Palladium, *see* Trusted Computing Platform Alliance (TCPA)-Palladium
TCP SYN, *see* Transmission Control Protocol synchronize (TCP SYN) packets
TDMA, *see* Time division multiple access (TDMA)
Team collaboration
basics, 6
evaluation, 19–21, *20*
example systems, 8, *10,* 11–12, *13*
Technological contexts, 300
Technology independent interfaces, 302–305
Telescript technology, 318
Temporal channel utilization, 552–563
Temporal data model, 364–366
Temporal Key Integrity Protocol (TKIP), 925, 929–930, 933, 935–937

Temporally Ordered Routing Algorithm (TORA)
energy-aware web caching, 785
routing and mobility management, 652–653, 656, 658
TESLA, *see* Timed Efficient Stream Loss-Toleranct Authenticaiton (TESLA)
Text to Speech (TTS), 298
Theft of devices, 484
Theimer, Schilit and, studies, 300
Thick-client application, 56, 65
Thin-client application, 56, 64–65
Third generation (3G) technology, 102, 303, 325
Third Generation Partnership Project (3GPP), 121, 123, 125, 146
Third party trust, *see* Trusted Third Party (TTP)
Threats, mobile ad hoc networks, 460–462
3G technology, *see* Third generation (3G) technology
3GPP, *see* Third Generation Partnership Project (3GPP)
TIA, *see* Tactical Information Assistant (TIA) program
Tian studies, 275–293
TIA-P, *see* Tactical Information Assistant-Prototype (TIA-P)
Tibaldi studies, 941–957
TIM, *see* Traffic indication map (TIM)
Time, weighting, 451–452
Time-based model, 858–859, 868, 870
Time-based strategy, 176
Time contexts, 300
Timed Efficient Stream Loss-Tolerant Authentication (TESLA), 442, 489
Time division multiple access (TDMA)
context-awareness, 308
quality of service, 578, 581, 584, 586, 592–593, 595–596
transmitter power control, 806, 818
Time stamp (TS) approach
cache invalidation, 362, 372–373, 376–377, 383
cache management, 352, 354
energy efficient cache invalidation, 426
power-aware cache management, 407
Time-to-live (TTL)
cache consistency strategy, 355–356
cache management, 345–346
DNS location management, 152–154, 157–158, *158,* 160–161
energy-aware web caching, 793, 795
modeling distributed applications, 513
Tiny OS, 311
Tiwari studies, 721
TKIP, *see* Temporal Key Integrity Protocol (TKIP)

Index

TLS, *see* Transport Layer Security (TLS)
Tool Command Language (TCL), 318
Topology, dynamic, 576–577
Topology Broadcast Based on Reverse-Path
 Forwarding (TBRPF), 645
TORA, *see* Temporally Ordered Routing
 Algorithm (TORA)
Total dominant pruning, *674,* 675–676, *676*
Touring machine, 8
Tracking, location, 882
Tracking for delivery, 241–246
Traffic diversion, 438–439
Traffic indication map (TIM), 696
Transaction relocation, 873
Transcoding, 60
Transfer of messages, 129, *131–133*
Transient data corruption/loss, 575
Transient sharing, Lime, *30,* 31–32, 44–46
Translation steps, 271–272
Transmeta processors, 732, 734–735
Transmission, adaptive algorithm power
 optimization, 757–759
Transmission Control Protocol/Internet Protocol
 (TCP/IP), 356
Transmission Control Protocol (TCP)
 ad hoc network security, 487
 battery power management, 896
 IEEE 802.11, 931–932
 locating mobile objects, 220
 medium access control, 551
 modeling distributed applications, 516, 532,
  539
Transmission Control Protocol synchronize (TCP
 SYN) packets, 977
Transmission power control, 544, 558–560, *559*
Transmission range, 549
Transmission schedule, 563
Transmitter power control
 basics, 805
 buffer, 811–816, 820
 data downloading, 831–833
 deep prefetching, 835
 depth 1 look-ahead, 833–835
 dynamic programming formulation, 827–829,
  828
 efficient data downloading, 831–833
 independent channel, 812–815, *814,* 820–823,
  823–824
 issues, 805–809
 multimode transmission links, 819–824
 multi-transmitter/multi-receiver case, 817–819
 neighbor prefetching, 833–835
 online look-ahead heuristics, 830–835
 packet arrivals, 815–816

 packet forwarding, 809–824
 PCMA algorithms, 816–817
 power decision, 829–830
 prefetching, 824–835
 responsive channel, 816–817
 single-mode transmission links, 824–835
 single-mode wireless links, 809–819
 structural properties, 820–823, *823–824,*
  829–830
 system model, 825–827
Transport Control Protocol (TCP), 788–789
Transport layer, 788–789
Transport Layer Security (TLS), 117
Travelstar, 717
TreeEmbed, 522
TreeTG, 535
Treo smartphone, 15
Triggers, 68
Trust, 954–955, *see also* Security, trust, and
 privacy; Trusted Third Party (TTP)
Trusted Computing Group (TCG)-NGSCB (Next
 Generation Secure Computer Base), 957
Trusted Computing Platform Alliance (TCPA)-
 Palladium, 957
Trusted Third Party (TTP), 497–498, 950, *see also*
 Trust
Trust management
 ad hoc network security, 496–499
 security, mobile ad hoc networks, 463–467,
  465–466
TS, *see* Time stamp (TS) approach
Tse, Grossglauser and, studies, 612
TSpaces, 36, 47, 49
TTL, *see* Time-to-live (TTL)
TTS, *see* Text to Speech (TTS)
TuCSoN, 48–49
Tuples and tuple spaces, 28, *30,* 30–31, 40–41
Two-location algorithm (TLA), 180–181
Two renewal processes, 882–883, *883*
Two-tier replication
 basics, 859–860
 Bayou, 862
 communication cost and scalability, 872
 comparisons, *865, 867*
 consistency and concurrency, 869
 infrastructure requirements and compatibility,
  870
Two-tier scheme, 201–202

U

Ubiquitous and robust access control (URSA)
 framework, 494

1025

UDDI, *see* Universal Description, Discovery and Integration (UDDI)
UIML, *see* User Interface Markup Language (UIML)
UIR, *see* Updated invalidation report (UIR) method
Ultra Wideband, 102
UMA, *see* Uniform memory access (UMA)
Uniform memory access (UMA), 98
Uniform Resource Name (URN) scheme, 222
Uninterruptible state, 15
United States-American Standard Code for Information Exchange (US-ASCII), 125
United States Association for Computing Machinery (USACM), 962
Universal Description, Discovery and Integration (UDDI), 332
Universal resource locator (URL), 308
Universal Transverse Mercator (UTM) grid, 277
Universal tuple space, 48
University College of London, 49
University of Bologna, 49
University of California, 311, 698, 897, 914, 925
University of Colorado, 742
University of Illinois at Urbana-Champaign, 98
University of Maryland, 925
University of Michigan, 742
UNIX, 344, 860
UNIX NOW, 98
Unknown attacks, 500
Update, location, 165, 175–186
Updated invalidation report (UIR) method
  cache invalidation, 363, 387
  power-aware cache management, 405, 408, *408*, 416, 418
Update log structure, 426
Update rate, 377, *381–382, 384–386*
Updates, location, 882
Update time stamp (UTS), 365–366
URN, *see* Uniform Resource Name (URN) scheme
URSA, *see* Ubiquitous and robust access control (URSA) framework
USACM, *see* United States Association for Computing Machinery (USACM)
Usage, simulation models, *285,* 290–293, *291–292*
User actions, Lime, 41–43, *42*
User-agent trust, 945–946
User-based partitioning approach, 394–395
User context, 300
User interface, pervasive application development, 57
User Interface Markup Language (UIML), 60
User interface models, 4
User mobility modeling, 288–293

Users with finite memory, 827
User Trip Model, 284–285, 288–290
User Virtual Environment (UVE), 89, 92
Utilicom Longranger, 704
UTM, *see* Universal Transverse Mercator (UTM) grid
UTS, *see* Update time stamp (UTS)
UVE, *see* User Virtual Environment (UVE)

V

Vacations, 904
Validity period, 366–368, *367*
Value-based prefetch (VP), 405, 413–414, *414*
Varghese studies, 227–251
Variable-size data items, 826
VC, *see* Virtual circuit (VC)
Vecchi studies, 315–332
Venus, 391
VID, *see* Virtual class ID (VID)
Video object planes (VOPs), 753
Virtual circuit (VC), 586
Virtual class ID (VID), 111
Visitor Location Registers (VLRs), 201, 206, 211–213
Visual tools, pervasive application development, 62–63
VLRs, *see* Visitor Location Registers (VLRs)
Voltage scaling, 735
Voltage setting, 760–763, *762–763*
VOPs, *see* Video object planes (VOPs)
Voyager
  locating mobile objects, 198, 218, 223, *223*
  mobile agent security, 951, 953
VP, *see* Value-based prefetch (VP)
Vulnerable period, 545
Vulnerable region, 546
VuMan 3, 4, 7, *7*

W

WalkEd (Walking Editor), 83, *83–85,* 83–93
Wang and Garcia-Luna-Aceves studies, 562
WANs, *see* Wireless wide area networks (WANs)
WAP, *see* Wireless Access Protocol (WAP); Wireless Application Protocol (WAP)
Ward studies, 300
Warren studies, 709–729
Washall-Floyd algorithm, 519
WASP, *see* Web Architecture for Service Platforms (WASP)
Watchdog
  ad hoc network security, 493–495
  self-policing MANETs, 444, 446, 451, 453

Index

Wattch project, 742–743
WaveLAN personal computer, 690, 799
Wavelets, 750–752, *751*
WBMP, *see* Wireless bitmap (WBMP)
W3C P3P, *see* World Wide Web Consortium's Platform for Privacy Preferences (W3C's P3P)
Weak consistency, 862
Weak transactions, 851–852
Wearable computing
  basics, 3–4, 21–22
  context-aware collaboration, 6, *7,* 12–17, *14, 16–17,* 21
  evaluation, 17–21
  example systems, *7–8, 7–17, 10*
  issues, 4–6
  master/apprentice (live expert) help desk, 5, 8, *10,* 10–11, 18
  proactive synthetic assistant, *7,* 8–9, 12–17, *14, 16–17*
  procedures, prestored and static capabilities, 5, 7, 9–10, *11,* 17, *19*
  team collaboration, 5, 8, *10,* 11–12, *13,* 19–21, *20*
Web Architecture for Service Platforms (WASP), 303
Web browser, DNS location management, 154–155
Web caching
  ad hoc caching, 791–794, *793–794*
  ad hoc cooperative caching, *783,* 789–794, *790, 793–794, 801,* 801–802
  ad hoc networks, 789–794, *799–800,* 799–801
  ad hoc routing protocols, 785–788, *786*
  basics, 779–781, 802
  cache management, 795–796
  channel listening, 784–785
  collisions, 782–784, *783*
  communication, 789–791, *790*
  evaluation, 798–802
  local caching, 795–798
  MAC layer, 781–785
  power-aware communication, 781–789
  prefetching, 796–798, *797*
  routing layer, 785–788
  transport layer, 788–789
WebExpress, 355
Web-of-trust, 497–499
Web Ontology Language (OWL), 964
Web server, DNS location management, 154
Web Services Description Language (WSDL), 332
Website META Language (WML), 60
Weighted fair queueing (WFQ), 600
Weighting, self-policing MANETs, 451–452
Weinmiller studies, 554, 556
Weiser studies, 53, 299
WEP, *see* Wired Equivalent Privacy (WEP) protocol
Westhoff, Lamparter, Plaggemeier and, studies, 438
Westhoff, Paul and, studies, 445
WFQ, *see* Weighted fair queueing (WFQ)
WGS 84, *see* World Geodetic System 1984 (WGS 84)
What You See Is What You Get (WYSIWYG), 62–63
Wide area network (WAN), 638
Wideband Code Division Multiple Access (W-CDMA)/CDMA2000, 96
Windows operating system, Lime, 39
Wired Equivalent Privacy (WEP) protocol
  ad hoc network security, 485, 487
  IEEE 802.11, 924–925, 928, 930, 932, 934–938
Wireless Access Protocol (WAP)
  context-awareness, 303, 308
  forum documents, 146
  multimedia messaging service, 121, 123, 128
Wireless Application Protocol (WAP), 317
Wireless bitmap (WBMP), 125
Wireless fidelity (Wi-Fi), 317, 504
Wireless Local Area Networks (LANs)
  adaptive algorithmic power optimization, 769–773
  energy-aware web caching, 779
  grid and cluster computing, 96
  hoarding, 398
  medium access control, 543
  routing and mobility management, 637
  transmitter power control, 805
Wireless Markup Language (WML), 325
Wireless Markup Language/X Hypertext Markup Language (WML/XHTML), 126
Wireless network interface (WNI), 404
Wireless protocol, routing and mobility management, 644
Wireless Routing Protocol (WRP), 642, 658
Wireless security threats, 926–927
Wireless sensor networks (WSN), 691, 697–698
Wireless Session Protocol (WSP), 124, 126
Wireless Transport Layer Security (WTLS), 117
Wireless wide area networks (WANs), 398
WML, *see* Website META Language (WML); Wireless Markup Language (WML)
WML/XHTML, *see* Wireless Markup Language/X Hypertext Markup Language (WML/XHTML)
Woo studies, 896
Working set method, 211

World Geodetic System 1984 (WGS 84), 277
World Wide Web Consortium's Platform for
   Privacy Preferences (W3C's P3P), 303,
   309, 312
Wormhole attacks, 489
Wristwatch, 721
WRP, *see* Wireless Routing Protocol (WRP)
WSDL, *see* Web Services Description Language
   (WSDL)
WSN, *see* Wireless sensor networks (WSN)
WSP, *see* Wireless Session Protocol (WSP)
WTLS, *see* Wireless Transport Layer Security
   (WTLS)
Wu, C.H., studies, 877–891
Wu, Dai and, studies, 669
Wu, J., studies, 663–686
Wu, K., studies, 362
Wu, Lou and, studies, 675, 682, 684
Wu and Dai studies, 669
Wu and Li studies, 669
Wu and Lou studies, 665
Wu studies, 674
WYSIWYG, *see* What You See Is What You Get
   (WYSIWYG)

X

X10, 305
XACML, *see* Extensible Access Control Markup
   Language (XACML)
Xerox Palo Alto Research Center, 67
XForms, 60
XML, *see* Extensible Markup Language (XML)
XQuery-like language, 68
Xscale architecture, 734
XScale processor, 96
Xu and Lee studies, 257
Xu and Saadawi studies, 552
Xu studies, 364

Y

Yamin studies, 73–93
Yang, Zhong, Chen and, studies, 441
Yang studies, 483–500
Ye studies, 483–500
Yi, Naldburg and Kravets studies, 442
Yuan and Nahrsted studies, 766
Yuen studies, 361–387
Yu studies, 543–564, 611–633

Z

Zerfos studies, 483–500
Zero-Configuration, 478
Zhang and Lee studies, 440, 495
Zhang studies, 483–500
ZHLS, *see* Zone-based Hierarchical Link State
   (ZHLS)
Zhong, Chen and Yang studies, 441
Zhou and Haas studies, 497
ZigBee, 504
Zipf distribution, 286–287
Ziv and Lempel compression algorithm, 185
Zomaya studies, 163–190
Zone-based Hierarchical Link State (ZHLS), 616,
   628, *631*, 631–632
Zone protocol
   routing and mobility management, 652–653
   scalable clustered network architecture, *630*,
      630–631
Zone Routing Protocol (ZRP)
   energy-aware web caching, 786, 789–790, 798
   routing and mobility management, 652–653,
      656, 658
   scalable clustered network architecture, 616,
      629
Zorzi and Rao studies, 896
ZRP, *see* Zone Routing Protocol (ZRP)